MW00845885

Handbook of Neurophotonics

Series in Cellular and Clinical Imaging

The purpose of the series is to promote education and new research using cellular and clinical imaging techniques across a broad spectrum of disciplines. The series emphasizes practical aspects, with each volume focusing on a particular theme that may cross various imaging modalities. Each title covers basic to advanced imaging techniques as well as detailed discussion dealing with interpretations of the studies. The series provides cohesive, complete, and state-of-the-art cross-modality overviews of the most important and timely areas within cellular and clinical imaging.

Super-Resolution Imaging in Biomedicine
edited by Alberto Diaspro, Marc A. M. J. van Zandvoort

Imaging in Photodynamic Therapy
edited by Michael R. Hamblin, Yingying Huang

Optical Probes in Biology
edited by Jin Zhang, Sohum Mehta, Carsten Schultz

Coherent Raman Scattering Microscopy
edited by Ji-Xin Cheng, Xiaoliang Sunney Xie

Natural Biomarkers for Cellular Metabolism
Biology, Techniques, and Applications
edited by Vladimir V. Ghukasyan, Ahmed A. Heikal

Handbook of Neurophotonics
edited by Francesco S. Pavone, Shy Shoham

For more information about this series, please visit:
[https://www.crcpress.com/Series-in-Cellular-and-Clinical-Imaging/book-series/CRCSERCELCLI]

Handbook of Neurophotonics

Edited by

Francesco S. Pavone and Shy Shoham

CRC Press is an imprint of the
Taylor & Francis Group, an **informa** business

First edition published 2020
by CRC Press
6000 Broken Sound Parkway NW, Suite 300, Boca Raton, FL 33487-2742

and by CRC Press
2 Park Square, Milton Park, Abingdon, Oxon, OX14 4RN

Library of Congress Cataloging-in-Publication Data

Names: Pavone, Francesco S., editor. | Shoham, Shy, editor.
Title: Handbook of neurophotonics / edited by Francesco S. Pavone, Shy Shoham.
Other titles: Series in cellular and clinical imaging.
Description: First edition. | Boca Raton, FL : CRC Press, 2020. | Series: Series in cellular and clinical imaging | Includes bibliographical references and index.
Identifiers: LCCN 2019056020 (print) | LCCN 2019056021 (ebook) | ISBN 9781498718752 (hardback) | ISBN 9780429194702 (ebook)
Subjects: MESH: Optical Imaging--methods | Microscopy--methods | Brain Mapping--methods | Neuropathology--methods | Optogenetics
Classification: LCC RC78.7.D53 (print) | LCC RC78.7.D53 (ebook) | NLM WN 195 | DDC 616.07/54--dc23
LC record available at https://lccn.loc.gov/2019056020
LC ebook record available at https://lccn.loc.gov/2019056021

ISBN: 978-1-4987-1875-2 (hbk)
ISBN: 978-0-429-19470-2 (ebk)

Typeset in Minion
by Deanta Global Publishing Services, Chennai, India

Contents

Preface

THIS BOOK IS INTENDED as a state-of-the-art overview for the rapidly emerging field of Neurophotonics, which lies at the intersection of Neuroscience, optical imaging, and manipulation technologies. While the field's roots trace back at least half a century, the past decade has been transformative for it, as dramatic technological advances have enabled researchers to measure, control, and track molecular events in live neuronal cells, and increasingly in intact, fully functioning nervous systems. This monumental change is greatly improving our understanding of brain function and dysfunction, and is beginning to have a profound impact on our ability to manage and treat brain-related diseases. As a result, neurophotonic techniques such as optogenetics and optical brain-machine-interfaces are already having a broad cultural impact, even among non-scientists. Across the globe, recent large-scale scientific projects including the US BRAIN initiative and the European Human Brain Project are investing heavily in the development of neurophotonic technologies, and optical manufacturers and suppliers are increasingly interested and invested in this technology segment.

In this book, we've brought together some of the top researchers and practitioners in the field to provide a measured and holistic view, emphasizing what's been done, lessons learned, and a realistic outlook on future possibilities including the clinical applications of advanced optical technologies for imaging and manipulating brain structure. Existing resources on brain optical imaging provide only a limited, often dated view with a relatively narrow technology focus. Given the recent and continued growth in this field, we saw the pressing need for a coherent source of information, both for newcomers as an introduction, and for experts as an update and reference. In order to achieve this objective, we've aimed to address both fundamental and more advanced aspects. The intended audience is intentionally broad, to include the wide range of non-experts who need a rapid single source of information as an update on the field, as well as those who are already working in the area and are interested in detailed review of specific application areas. Among those we believe will benefit are active researchers in biomedical science, imaging, and optical engineering, as well as students in biomedical imaging and microscopy.

We've divided chapters into three main sections. The first part covers functional and structural neuroimaging methods such as multiphoton, light sheet, and light field microscopies together with tissue preparation protocols, superresolution, and optoacoustic imaging. The next section addresses neurophotonic control and perturbation, including optogenetics, laser axotomy, and single synapse stimulation. The third section turns to human and clinical neurophotonics from optical coherence tomography to high resolution diffuse optical tomography, acousto optic monitoring, and highlights in the most promising areas such as vision restoration, cochlear prostheses, and neurophotonic biopsy.

All the authors are to be commended for their work in presenting a cutting-edge overview of their topic. Throughout the book, we've endeavored to highlight not just the technology, but again, to show the reader how these novel methods are becoming critical to research breakthroughs and advances in clinical practice.

Editors

Francesco S. Pavone is full professor at the University of Florence in the Department of Physics and at the European Laboratory for Non-Linear Spectroscopy (LENS), and group leader at the Biophotonics Laboratories. He obtained a PhD in optics in 1993 and spent two years as postdoctoral fellow at the *Ecole Normale Superieure* with the group of Claude Cohen Tannoudjy (Nobel Prize, 1997). His research group is involved in developing new microscopy techniques for high-resolution and high-sensitivity imaging, and laser manipulation purposes. These techniques have been applied in single-molecule biophysics, single-cell imaging, and optical manipulation. He is also engaged in tissue imaging research, for which nonlinear optical techniques have been applied to skin and neural tissue imaging. He is the author of more than 100 peer-reviewed journal articles and has delivered more than 60 invited talks. He coordinates various European projects and has organized international congresses. He is director of the international PhD program at LENS. He is on the editorial board of the journal *Neurophotonics* and is a principal investigator for the Human Brain Project, an EU Flagship Initiative.

Shy Shoham is an associate professor at the Department of Biomedical Engineering in the Technion – Israel Institute of Technology. He was born in Rehovot, Israel, and holds a BSc degree in Physics from Tel Aviv University and a PhD in Bioengineering from the University of Utah. After completing his PhD, he was a Lewis Thomas Postdoctoral Fellow at the department of Molecular Biology, Princeton University. In 2005 he joined the Technion's faculty of Biomedical Engineering, where he established the Neural Interface Engineering Laboratory. His lab focuses on the development of implant-less retinal prostheses aimed at restoring vision loss from outer-retinal degenerative diseases, and on developing advanced technologies for acoustic neuromodulation, microscopic neuro-imaging and for bioengineering brain-like tissues. He is a recipient of a starting grant from the European Research Council, and of the Daniel Shiran and Juludan awards for engineering advances in biomedicine, and is a member of the editorial boards of *Journal of Neural Engineering*, and of *Translational Vision Science & Technology*.

Contributors

Taner Akkin
Faculty of Engineering
Holon Institute of Technology
Holon, Israel

Anna Letizia Allegra Mascaro
Neuroscience Institute, National Research Council
Pisa, Italy
European Laboratory for Non-linear Spectroscopy University of Florence
Florence, Italy

Eleonora Aronica
Academic Medical Center
Department of (Neuro)Pathology and Swammerdam Institute for Life Sciences
Center for Neuroscience, University of Amsterdam
Amsterdam, Netherlands

Jean C. Augustinack
Faculty of Engineering
Holon Institute of Technology
Holon, Israel

Michal Balberg
Faculty of Engineering
Holon Institute of Technology
Holon, Israel

Adi Schejter Bar-Noam
Faculty of Biomedical Engineering
Technion – Israel Institute of Technology
Haifa, Israel

Shai Berlin
The Ruth and Bruce Rappaport Faculty of Medicine
Technion-Israel Institute of Technology
Haifa, Israel
Helen Wills Neuroscience Institute
University of California
Berkeley, California

Klaus Becker
Chair for Bioelectronics, FKE
Technical University Vienna
Vienna, Austria
Section of Bioelectronics, Center for Brain Research
Medical University Vienna
Vienna, Austria

David Boas
Boston University
College of Engineering
Boston, Massachusetts

Raghav K. Chhetri
HHMI Janelia Research Campus
Ashburn, Virginia

Daniel Choquet
Interdisciplinary Institute for Neuroscience
University of Bordeaux
Centre National de la Recherche Scientifique (CNRS)
Bordeaux Imaging Center, UMR 3420, US 4 University of Bordeaux
CNRS, INSERM
Bordeaux, France

Hans-Ulrich Dodt
Chair for Bioelectronics, FKE
Technical University Vienna
Vienna, Austria
Section of Bioelectronics, Center for Brain Research
Medical University Vienna
Vienna, Austria

Daniela Duc
ARC Centre of Excellence for Electromaterials Science
Faculty of Science, Engineering and Technology, Swinburne University of Technology
Melbourne, Australia

Hod Dana
Department of Neurosciences
Lerner Research Institute, Cleveland Clinic Foundation
Cleveland, Ohio

Adam T. Eggebrecht
Mallinckrodt Institute of Radiology
Washington University School of Medicine
St Louis, Missouri

Bruce Fischl
Faculty of Engineering
Holon Institute of Technology
Holon, Israel

Shaun Gietman
ARC Centre of Excellence for Electromaterials Science
Faculty of Science, Engineering and Technology, Swinburne University of Technology
Melbourne, Australia

Marie Louise Groot
LaserLab Amsterdam
VU University, De Boelelaan
Amsterdam, Netherlands
Neuroscience Campus Amsterdam, VU University
Amsterdam, Netherlands

Anne-Sophie Hafner
Interdisciplinary Institute for Neuroscience
University of Bordeaux
Centre National de la Recherche Scientifique (CNRS)
Bordeaux, France

Christian Hahn
Chair for Bioelectronics, FKE
Technical University Vienna
Vienna, Austria
Section of Bioelectronics, Center for Brain Research
Medical University Vienna
Vienna, Austria

Brad A. Hartl
LaserLab Amsterdam
VU University, Amsterdam, Netherlands
Neuroscience Campus Amsterdam, VU University
Amsterdam, Netherlands

Leore R. Heim
Department of Physiology and Pharmacology
Sackler Faculty of Medicine
Tel Aviv University
Tel Aviv, Israel

Maximillian Hoffmann
NeuroCure Cluster of Excellence, Charité & Humboldt University
Berlin, Germany

Roarke Horstmeyer
Bioimaging and Neurophotonics Lab
NeuroCure Cluster of Excellence
Charité & Humboldt University
Berlin, Germany

Sander Idema
Dept. of Neurosurgery, VU University Medical Center
Amsterdam, Netherlands
Amsterdam Brain Tumor Center, VU University Medical Center
Amsterdam, Netherlands

Ehud Y. Isacoff
Department of Molecular and Cell Biology
University of California
Berkeley, California
Physical Bioscience Division
Lawrence Berkeley National Laboratory
Berkeley, California

Nina Jährling
Chair for Bioelectronics, FKE
Technical University Vienna
Vienna, Austria
Section of Bioelectronics, Center for Brain Research
Medical University Vienna
Vienna, Austria

Benjamin Judkewitz
Bioimaging and Neurophotonics Lab
NeuroCure Cluster of Excellence, Charité & Humboldt University
Berlin, Germany

Philipp J. Keller
HHMI Janelia Research Campus
Ashburn, Virginia

Oded Klavir
Department of Psychobiology
University of Haifa
Haifa, Israel

Ender Konukoglu
Faculty of Engineering
Holon Institute of Technology
Holon, Israel

Thomas Knöpfel
Department of Medicine
Imperial College London
London, UK

Mathias Mahn
Department of Neurobiology
Weizmann Institute of Science
Rehovot, Israel
Department of Psychobiology
University of Haifa
Haifa, Israel

Tobias Nöbauer
Laboratory of Neurotechnology and Biophysics
The Rockefeller University
New York, New York

Nikolay Kuzmin
LaserLab Amsterdam
VU University, Amsterdam, Netherlands
Neuroscience Campus Amsterdam, VU University
Amsterdam, Netherlands

Laura Marcu
LaserLab Amsterdam
VU University,
Amsterdam, Netherlands
Neuroscience Campus Amsterdam, VU University
Amsterdam, Netherlands

Simon E. Moulton
ARC Centre of Excellence for Electromaterials Science
Faculty of Science, Engineering and Technology, Swinburne University of Technology
Melbourne, Australia

Shir Paluch
Department of Ophthalmology, Tech4Health and Neuroscience Institutes
NYU Langone Health, New York, New York

Arvind P. Pathak
Russell H. Morgan Department of Radiology and Radiological Science
Departments of Biomedical and Electrical Engineering
Sidney Kimmel Comprehensive Cancer Center
The Johns Hopkins University School of Medicine
Baltimore, Maryland

Chiara Paviolo
Univ. Bordeaux, LP2N - Institut d'Optique & CNRS
Talence (Bordeaux), France

Revital Pery-Shechter
Ornim Medical Ltd.
Kfar Saba, Israel

Francesco Saverio Pavone
European Laboratory for Non-linear Spectroscopy University of Florence
Florence, Italy
Department of Physics and Astronomy
University of Florence
Florence, Italy

and

National Institute of Optics,
National Research Council
Pisa, Italy

Marko Pende
Chair for Bioelectronics, FKE
Technical University Vienna
Vienna, Austria
Section of Bioelectronics, Center for Brain Research
Medical University Vienna
Vienna, Austria

Daniel Razansky
Institute for Biological and Medical Imaging
Technical University of Munich and Helmholtz Center Munich
Munich, Germany

Claus-Peter Richter
Department of Otolaryngology
Feinberg School of Medicine at Northwestern University
Departments of Biomedical Engineering and Communication Sciences and Disorders at Northwestern University,
Evanston, Illinois

Shani Rosen
Department of Ophthalmology, Tech4Health and Neuroscience Institutes
NYU Langone Health, New York, New York

Haowen Ruan
Electrical Engineering Department
California Institute of Technology
Pasadena, California

Shamira Sridharan
LaserLab Amsterdam
VU University, Amsterdam, Netherlands
Neuroscience Campus Amsterdam, VU University
Amsterdam, Netherlands

Inna Sabdyusheva-Litschauer
Chair for Bioelectronics, FKE
Technical University Vienna
Vienna, Austria
Section of Bioelectronics, Center for Brain Research
Medical University Vienna
Vienna, Austria

Saiedeh Saghafi
Chair for Bioelectronics, FKE
Technical University Vienna
Vienna, Austria
Section of Bioelectronics, Center for Brain Research
Medical University Vienna
Vienna, Austria

Janaka Senarathna
Russell H. Morgan Department of Radiology and Radiological Science
The Johns Hopkins University School of Medicine
Baltimore, Maryland

Shy Shoham
Department of Ophthalmology, Tech4Health and Neuroscience Institutes
NYU Langone Health, New York, New York

Ludovico Silvestri
European Laboratory for Non-linear Spectroscopy (LENS)
Sesto Fiorentino, Italy
National Institute of Optics, National Research Council (CNR-INO)
Sesto Fiorentino, Italy

Chenchen Song
Laboratory for Neuronal Circuit Dynamics
Imperial College London, London, UK

Eran Stark
Department of Physiology and Pharmacology
Sackler Faculty of Medicine
Tel Aviv University
Tel Aviv, Israel
Sagol School of Neuroscience
Tel Aviv University
Tel Aviv, Israel

Paul R. Stoddart
ARC Training Centre in Biodevices
Faculty of Science, Engineering and Technology
Swinburne University of Technology
Hawthorn, Australia

Naofumi Suematsu
Department of Otolaryngology
Feinberg School of Medicine at Northwestern University
Chicago, Illinois

Nitish V. Thakor
Department of Biomedical Engineering
The Johns Hopkins University School of Medicine
Baltimore, Maryland

Betty M. Tyler
Department of Neurosurgery
Johns Hopkins University School of Medicine
Baltimore, Maryland

Xiaodong Tan
Department of Otolaryngology
Feinberg School of Medicine at Northwestern University
Chicago, Illinois

Changhuei Yan
Electrical Engineering Department
California Institute of Technology
Pasadena, California

Alipasha Vaziri
Laboratory of Neurotechnology and Biophysics
The Rockefeller University
New York, New York
Research Institute of Molecular Pathology
Vienna, Austria
The Kavli Neural Systems Institute
The Rockefeller University
New York, New York

Martina Wanis
Chair for Bioelectronics, FKE
Technical University Vienna
Vienna, Austria
Section of Bioelectronics, Center for Brain Research
Medical University Vienna
Vienna, Austria

Hui Wang
Faculty of Engineering
Holon Institute of Technology
Holon, Israel

Pieter Wesseling
Department of Pathology
VU University Medical Center, Amsterdam
Princess Maxima Center for Pediatric Oncology
Amsterdam, Netherlands

Muriah D. Wheelock
Department of Psychiatry
Washington University School of Medicine
St Louis, Missouri

Philip C. de Witt Hamer
Department of Neurosurgery
VU University Medical Center
HV Amsterdam, Netherlands
Amsterdam Brain Tumor Center, VU University Medical Center
Amsterdam, Netherlands

Nan Xia
Department of Otolaryngology
Feinberg School of Medicine at Northwestern University
Chicago, Illinois

Yingyue Xu
Department of Otolaryngology
Feinberg School of Medicine at Northwestern University
Chicago, Illinois

Ofer Yizhar
Department of Neurobiology
Weizmann Institute of Science
Rehovot, Israel

Hang Yu
Department of Biomedical Engineering
Columbia University
New York, New York

I

Function and Structural Neurophotonic Imaging

CHAPTER 1

Miniaturized Optical Neuroimaging Systems

Hang Yu, Janaka Senarathna, Betty M. Tyler, Nitish V. Thakor, and Arvind P. Pathak

CONTENTS

1.1 INTRODUCTION

For over a century, optical microscopy has been a major tool for neuroscience research. Historically, optical microscopy enabled the earliest neuroscientists to elucidate the 'neuron doctrine' from static histological brain samples (Llinás, 2003). Now, optical microscopy has evolved to capture the dynamics of neurons in action with real-time imaging from a variety of mammalian and non-mammalian organisms. Unlike functional Magnetic Resonance Imaging (fMRI) and Positron Emission Tomography (PET), optical imaging techniques allow us to monitor brain function with unprecedented spatio-temporal details. In conjunction with the ever-expanding repertoire of contrast agents that label neural activity (Lin and Schnitzer, 2016), such as genetically encoded calcium indicators (GECI) (Tian et al., 2009), fluorescent voltage sensors (Gong et al., 2015) or neurochemically activated glutamate sensors (Marvin et al., 2013), optical imaging tools are able to monitor neural and neural-related activities directly with high specificity and sensitivity.

Optical imaging is very flexible in terms of spatio-temporal resolution and choice of contrast agents, permitting neuroscientists to probe neural systems from multiple perspectives. For example, two-photon laser scanning microscopy (2PM) permits deep imaging of the brain at sub-micron resolutions, but with a limited field of view (FoV, e.g. 0.5×0.5 mm^2). In contrast, wide-field single photon microscopy can be optimized to image neural activity at the mesoscale (~ 100 um) (Ma et al., 2016; Xiao et al., 2017) while covering a larger FoV (e.g. 10×10 mm^2). Besides monitoring neural activity, optical neuroimaging can also track neural-related events. For example, laser speckle contrast imaging (LSCI) and optical intrinsic signal imaging (OISI) can record the changes in cerebral blood flow and oxygen saturation that accompany neural firing (Dunn et al., 2003; Jones et al., 2008).

Historically, most optical neuroimaging experiments were conducted while the animal was anesthetized and immobilized within a stereotaxic restraint. This limited the type of experiments one could perform, as well as prevented the correlation of brain activity with animal behavior. Additionally, many studies have shown that anesthetics have a broad effect on brain function, including neural excitability, vascular response, and functional connectivity, which can confound the interpretation of optical imaging data (Bonhomme et al., 2011; Masamoto and Kanno, 2012). Studies of awake, behaviorally engaged animals have become the new mainstay of current neuroscience research. This technical advance has been made possible by two complementary approaches: (i) utilizing optical brain imaging in head-restrained, awake animals (Dombeck et al., 2007), and (ii) utilizing miniaturized head-mounted microscopes to image the brains of unrestrained animals (Yu et al., 2015). In the head-restrained approach, a head-immobilization system is retrofitted to a conventional benchtop light microscopy system. The animal is trained to become accustomed to the head-fixation rig. During the experiment, the animal is usually placed on a treadmill or trackball, and its responses to tasks such as visual stimulation (Keck et al., 2013), whisker stimulation (Sofroniew et al., 2014), or locomotion (Dombeck et al., 2007) are recorded. The animal is usually imaged for up to an hour or longer depending on the animal's tolerance of the head-restraint. In most miniature neuroimaging systems, a miniaturized head-mounted microscope (see Section 1.2 for more details) is attached to the cranium of the animal. This arrangement permits the animal to move about and interact freely with its environment during the imaging session. Without the need for head-fixation, the animal can be imaged over extended durations. Moreover, the miniaturized imaging approach permits a wider range of animal behaviors to be studied, especially those that are not feasible when using a head-restraint. Examples of such behaviors include motor tasks that require the motion of head and neck, social interactions such as mating or agonistic behaviors, or exploring a new environment.

Miniaturized optical imaging devices have additional advantages over the traditional benchtop systems. For example, when combined with chronic cranial window preparations, the same neural environment (i.e. neural circuits, glial and vascular network) can be monitored over days, weeks, or even months (Holtmaat et al., 2009). This is an invaluable tool for observing the anatomical/functional changes of the brain during perturbations, such as learning a new task or when characterizing the progression of brain-related diseases (e.g. stroke, cancer, Alzheimer's disease), during which conventional head-fixed imaging might prove challenging due to advanced disease burden. Additionally, miniaturized neuroimaging systems are compatible with a wide spectrum of behavioral assays used in neuroscience research. Most current miniaturized imaging systems are standalone devices built from off-the-shelf components (except for *miniaturized two-photon imaging systems* – see Section 1.3.3), they are low cost, and they have the potential for mass production. These factors lower the barrier for adopting miniaturized optical neuroimaging systems and encourage their use as an entry-level device for laboratories that are new to *in vivo* imaging.

To conclude, miniaturized optical imaging systems serve as an efficacious platform for imaging in awake, freely moving animals. These systems have yielded novel insights into our understanding of the relationship between behavior and neural activity. The rest of this chapter will focus on discussing approach (ii), i.e. miniaturized optical neuroimaging systems, in detail.

1.2 DESIGN CRITERIA FOR MINIATURIZED OPTICAL NEUROIMAGING DEVICES

To date, the miniaturization of a variety of optical neuroimaging systems has been achieved, including single-photon fluorescence microscopy (Flusberg et al., 2005), two-photon fluorescence microscopy (Piyawattanametha et al., 2009), light-sheet microscopy (Engelbrecht et al., 2010), laser speckle contrast imaging (Miao et al., 2011; Senarathna et al., 2012), optical intrinsic signal imaging (Liu et al., 2013), diffuse optical imaging (Holzer et al., 2006; Sutin et al., 2012) and photoacoustic tomography (Tang et al., 2015). Also, certain designs of microendoscopes (Flusberg et al., 2005) have been employed for imaging of awake animals. Finally, the widespread availability of genetically modified (i.e. transgenic) mice, preclinical disease models, and optical contrast agents/probes, have resulted in most awake animal imaging experiments being conducted with rodents.

1.2.1 Fiber Optic-Adapted Systems versus Discrete Systems

Miniaturized optical neuroimaging systems are either fiber optic-adapted or discrete systems. Regardless of their design, a miniaturized optical neuroimaging setup consists of three major parts: (i) a headpiece, (ii) wired/optic fiber connector, and (iii) external hardware components. A generalized setup illustrating these three major components is shown in Figure 1.1.

Since the 1990s, researchers have been developing fiber optic-based imaging systems to perform imaging in awake, freely moving animals (Rector and Harper, 1991). In a fiber optic-adapted system, an optical-fiber bundle is employed to transmit the illumination and detected light signals (either fluorescence or reflectance) to and from a conventional bench-top microscopy system. In early versions of this approach, the optical fibers were in direct contact with the brain tissue without the use of any lens (Ferezou et al., 2006). As a result, the detection efficiency and spatial resolution for this setup were low by today's standards.

FIGURE 1.1 A generalized setup of a miniaturized optical neuroimaging system. The system consists of three major components: (i) a headpiece, (ii) wired/optical fiber connector, and (iii) external hardware components.

Nevertheless, these pioneering studies provided useful data and experience in imaging of awake, freely moving animals. Moreover, the idea of using fiber optics for miniaturized neuroimaging devices was implemented and extended to other imaging modalities (Engelbrecht et al., 2008; Liu et al., 2013). In more recent versions, the headpiece includes a miniaturized objective lens for optical coupling, relaying the light to and from the optical fiber. This arrangement increases the numerical aperture, which enables cellular-resolution imaging in awake, freely moving animals.

The development of small "footprint" illumination sources (e.g. LEDs, laser diodes) and compact light detectors enabled an alternative approach for awake animal imaging, i.e. a discrete imaging system that does not rely on conventional benchtop microscope components. In such a discrete imaging system, most of the microscope components, including a light source, lens, and detectors, are all miniaturized and housed within the headpiece. However, the discrete systems reported to date still need an external wire to connect to external hardware components, which are either too heavy or difficult to miniaturize. These external components often include a power supply unit and an image acquisition controller, as well as digital storage media for the acquired images. Novel focusing mechanisms using motorized stages or manual focusing with high-precision threads has also been implemented. Finally, the design of the microscope head-mount needs to account for proper optical coupling and stable mechanical mating of the miniaturized microscope to the animal's skull.

1.3 APPLICATIONS OF MINIATURIZED OPTICAL NEUROIMAGING DEVICES

1.3.1 Miniaturized Single-Photon Fluorescence Microscopes

Owing to its relatively straightforward optical set-up, single-photon fluorescence microscopy is the most common modality to be miniaturized by different research laboratories for awake animal neuroimaging. Early miniaturized designs utilized an optical-fiber bundle to relay illumination and emission light from a regular fluorescence benchtop system. A pilot study by Ferezou et al. (2006) employed voltage sensitive dye imaging to study the response to whisker stimulation in freely moving mice. A custom-made multi-fiber optic array was placed in direct contact with the whisker barrel cortex to relay excitation/emission light to and from a standard benchtop imaging system (Figure 1.2a–b). Using this system, the authors demonstrated that whisker stimulation evoked sensory responses that depended strongly on the animal's behavior (e.g. quiet vs. actively whisking). This type of experiment could not have been performed on anesthetized animals with conventional benchtop fluorescence systems.

Ghosh et al. (Ghosh et al., 2011) reported an elegant design of a discrete, single-photon fluorescence microscope. Weighing only 1.9 g, the headpiece could be easily carried by an adult mouse. A blue LED served as the excitation light source. The blue light was collected by a drum lens, passed through an excitation filter and finally deflected by a dichroic mirror. The emitted fluorescence was first collected by a GRIN (gradient refractive index) lens, then passed through the dichroic mirror and the emission filter, and then focused onto a complementary metal–oxide–semiconductor (CMOS) detector by an achromatic lens (shown in Figure 1.2c). The system supported a maximum field of view of 600 × 800 μm, with a magnification ratio of 5× and a lateral resolution of 2.5 μm. The system allowed data acquisition at 36 frames per second (fps) for the full field of view or a 100-fps data rate for a cropped region of 300 × 300 pixels. This microscope was first employed to study hemodynamics in freely moving mice. The researchers focused on the lobules V and VI of the cerebellar vermis, an area that is thought to be responsible for coordinating the forelimbs and hind limbs. Blood plasma was labeled with fluorescein-dextran administered via a tail vein injection. With post-processing, they could track the speed of red blood cells and changes in capillary diameter with a temporal resolution of 2 s; a representative image of the vasculature is shown in Figure 1.2d. The authors categorized mouse behavior into three different states: rest, walking, and running. Statistical analysis showed that walking and running states exhibited significant increases in blood flow and capillary diameter compared to the rest state. They also used this microscope to study the calcium dynamics of Purkinje neurons in the cerebellar vermis. Ca^{2+} signals from 200 Purkinje neurons across nine cerebellar microzones were monitored in unencumbered mice. The authors showed that individual microzones exhibited large-scale, synchronized Ca^{2+} activity during locomotion. Inspired by this design, a research group at UCLA has started the Miniscope project (www.miniscope.org) to make these miniature fluorescence microscopes accessible to the broader neuroscience community.

1.3.2 Miniaturized Intrinsic Contrast Microscopes

Typically, native fluorescence is absent from most biological tissues. Therefore, a fluorescent marker or tag must be provided exogenously (e.g. viral transfection, or genetically modified mouse lines). Cerebral blood flow is tightly regulated by neural function through a cascade of signaling mechanisms. This phenomenon is termed neurovascular coupling or functional hyperemia. Several camera-based optical imaging modalities,

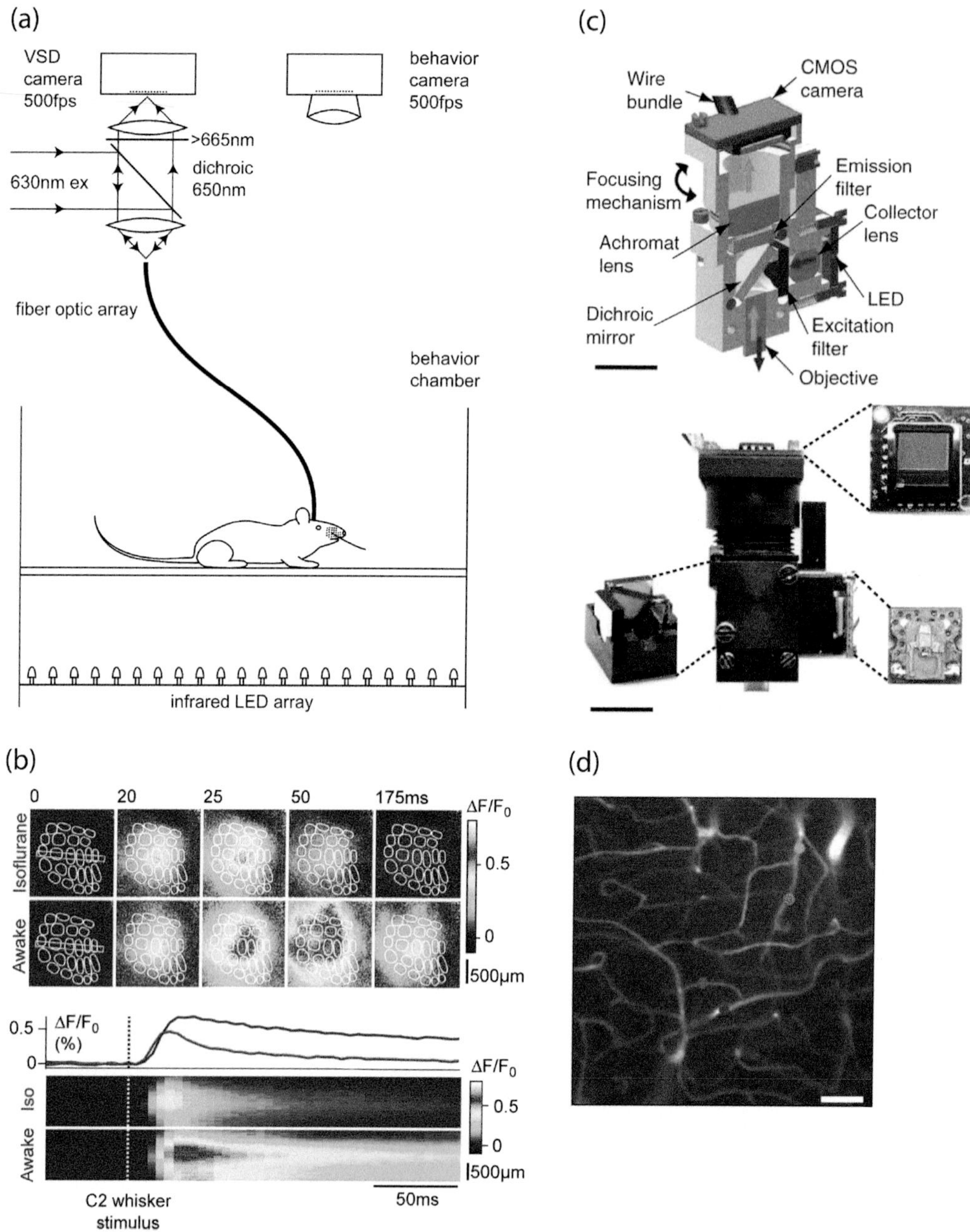

FIGURE 1.2 **(a)** Schematic of the awake imaging rig with fiber optic array reported by Ferezou et al. (2006). **(b)** Cortical responses to C2 whisker perturbation captured by voltage sensitive dye between isoflurane-anesthetized and awake state (averaged data from 30 trials). Adapted from Ferezou et al., (2006) with permission from Elsevier. **(c)** Schematic of the discrete single-photon microscope reported by Ghosh et al., (2011). **(d)** Vasculature captured by the miniaturized microscope in (c) with fluorescein-dextran labeling. The image shown is the standard deviation from 10-second acquisition to emphasize vasculature. (Scale bar = 50 μm). (Adapted from Ghosh et al. (2011) with permission from Springer Nature.)

such as optical intrinsic signal imaging (OISI) and laser speckle contrast imaging (LSCI) can monitor *in vivo* cerebral hemodynamics with intrinsic contrast (Dunn et al., 2003). OISI is based on the change in hemoglobin absorption during functional activation (Malonek and Grinvald, 1996). Hemoglobin (Hb) can exist as either deoxy-hemoglobin (HbR, Hb not bound to oxygen) or oxy-hemoglobin (HbO, Hb bound to oxygen). HbR and HbO have their own distinct wavelength-dependent absorption spectra. By acquiring images at specific wavelengths, one can calculate the change in oxy- and deoxy-hemoglobin concentrations, as well as the total change in hemoglobin during neural activation (Hillman, 2007). The local change in deoxy-hemoglobin is also the basis of the blood oxygenation level-dependent (BOLD) contrast mechanism underlying fMRI studies (Ogawa et al., 1990). Laser speckle contrast imaging relies on the interactions of laser light with red blood cells to infer cerebral blood flow velocity. Moving particles (e.g. red blood cells) cause a time-dependent blurring of the laser-induced interference or speckle pattern. The degree of spatial blurring can be quantified and one can derive the speed of the moving particles (Bandyopadhyay et al., 2005). See reviews by Senarathna et al. and Kazmi et al. for a comprehensive overview of LSCI (Senarathna et al., 2013; Kazmi et al., 2015).

Therefore, the use of OISI/LSCI has become widespread for imaging hemodynamics for functional brain mapping. Several groups have reported the development of miniaturized microscopes with OISI and/or LSCI (Miao et al., 2011; Senarathna et al., 2012; Liu et al., 2013; Sigal et al., 2016; Miao et al., 2017). Liu et al. recently reported an elegant dual-mode miniaturized microscope that combined LSCI and OISI (Liu et al., 2013). The system schematic is shown in Figure 1.3a. The headpiece weighs 1.5 g. A 660 nm

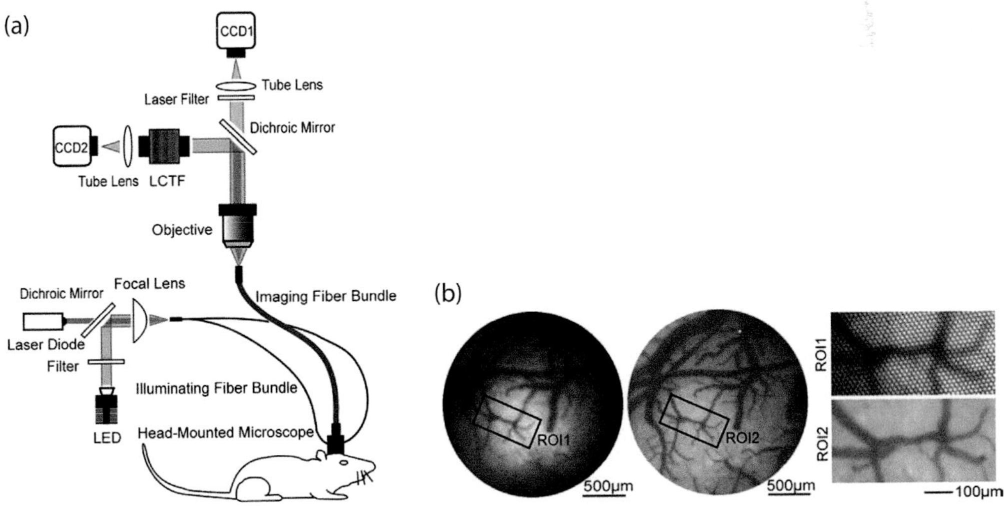

FIGURE 1.3 **(a)** Schematic of the dual-modal miniaturized microscope reported by Liu et al. (2013). **(b)** Comparison between the miniaturized microscope in (a) and a benchtop microscope. Left, an image acquired by the miniature system. Middle, the same field of view acquired by a benchtop microscope. Right, two zoomed-in images acquired by both systems. (Adapted from Liu et al. (2013) with permission from the Optical Society of America.)

laser diode and white light emitting diode (LED) were combined via a dichroic mirror for LSCI and OISI imaging, respectively. The mixed illumination was transmitted to the cortex via a multi-mode optical-fiber bundle. The cortex was imaged with a gradient-index (GRIN) objective lens, and the detected signals were transmitted to a benchtop detector system via another optical-fiber bundle. The mixed signals from two imaging modalities were separated by another dichroic mirror. Two camera detectors (CCD1 and CCD2) were employed for each imaging modality respectively, a 660 nm laser filter was used for the LSCI camera (CCD1), and liquid crystal tunable filter was used to sequentially select 550 nm and 625 nm for dual-color OISI imaging (CCD2). Figure 1.3b depicts a comparison between the head-mounted microscope and a similarly equipped benchtop microscope. The same field of view from part of the exposed cortex was imaged with both systems. As shown by the magnified images on the right panel, the head-mounted microscope was able to resolve most of the micro-vessels. However, the spatial resolution of the miniaturized microscope was limited by the spacing between imaging fibers. The authors also employed this head-mounted microscope to study KCl-induced cortical spread depression (CSD) between anesthetized and awake states in rats. This dual-modality imaging platform could capture CSD-related changes in cerebral blood flow and hemodynamics at high temporal resolution. The authors concluded that the CSD duration was significantly higher under anesthesia than in the awake state (reflected by prolonged cerebral blood flow (CBF), oxyhemoglobin (HbO), and total hemoglobin (HbT) changes under isoflurane anesthesia).

1.3.3 Miniaturized Two-Photon Fluorescence Microscopes

Two-photon laser scanning microscopy (2PM) has been widely adopted in neuroscience research (see Chapter 3 for more details). The reduced scattering from biological tissues at near-infrared wavelengths enables two-photon microscopy to image deeper into the brain (Theer and Denk, 2006). The quadratic dependence of fluorescence on excitation intensity confines the excitation profile to a diffraction-limited spot, which provides superior optical sectioning and better signal to background ratio than single-photon excitation (Helmchen and Denk, 2005). The miniaturization of two-photon microscopy for imaging in awake animals has been reported in a variety of design configurations (Helmchen et al., 2001; Göbel et al., 2004b; Engelbrecht et al., 2008; Piyawattanametha et al., 2009; Sawinski et al., 2009). Nevertheless, there are challenges of using two-photon microscopy for functional neuroimaging. For example, the field of view for 2PM is usually limited to several hundred microns. The point scanning requirement of 2PM for image formation limits the frame rate, which makes it more susceptible to motion artifact during behavioral imaging experiments (Helmchen et al., 2001). This section will only cover several key design considerations for miniaturizing 2PM. Readers can refer to an excellent review by Helmchen et al. (2013) for a detailed protocol describing the design and construction of miniaturized 2PM.

Typically, a commercial tunable Ti:Sapphire femtosecond laser system is used for two-photon excitation. The need for high power and ultrashort laser pulses (~100 fs) makes fiber optic-based systems the only viable approach for miniaturizing the two-photon microscope. However, there are issues with delivering the ultrashort laser pulses with regular step-index single-mode glass optic fibers. Within such optic fibers, the laser pulse broadens due to dispersion and nonlinear effects, limiting the efficiency of two-photon excitation. The positive group velocity dispersion can be compensated by prechirping the laser pulse before coupling to the optic fiber, with double passing the laser beam through a pair of diffraction gratings (Göbel et al., 2004b), or two prism arrays (Sawinski et al., 2009). Several other approaches that do not require prechirping compensation have also been reported, such as using a hollow-core photonic crystal fiber for Ti:Sapphire laser pulse delivery (Göbel et al., 2004a), and implementing a double-clad fiber and 1.55 μm wavelength laser for excitation (Murari et al., 2011). Recent advances in compact ultrafast laser sources (Taira et al., 2007; Huang et al., 2016; Niederriter et al., 2017) provide promising alternative light sources for miniaturized two-photon systems.

Another key component of a miniaturized two-photon microscope is the scan engine. For most of the reported miniaturized 2PM systems, the scanner is incorporated into the headpiece. One approach for implementing a compact scan engine is to move the freestanding tip of the illumination fiber with piezo actuators, in which case the fiber tip is placed in optical conjugate with the focal plane of the objective lens. The scanning pattern at the focal plane would be a magnified version of the fiber tip displacements. Several scanning patterns, such as the Lissajous scan (Helmchen et al., 2001), or the spiral scan (Lee et al., 2010), can be achieved through this approach. A "piezolever fiber scanner" reported by (Sawinski and Denk, 2007) can implement random access scanning and video rate frame scanning. The excitation laser beam can also be modulated by a microelectromechanical (MEMS) scanner, which is widely used in microendoscope applications (Tran et al., 2004; Shin et al., 2007; Jung et al., 2008). Miniaturized two-photon microscopes using a MEMS scanner were reported by Piyawattanametha et al. and Zong et al. (Piyawattanametha et al., 2009; Zong et al., 2017). Göbel et al. used the same coherent fiber bundle for excitation and detection, alleviating the need for a compact scanning device (Göbel et al., 2004b). In this set-up, a regular galvanometric scan mirror pair from a benchtop 2PM system generated the scanning pattern that was delivered to the fiber bundle. Nevertheless, the spatial resolution of this system was limited by the spacing between the individual optical fibers.

An elegant example of utilizing a miniaturized two-photon microscope for imaging awake, freely moving animals was demonstrated by Sawinski et al. (2009). As shown in Figure 1.4a–c, the head-mounted unit weighed 5.5 g, which could be carried by an adult or even an adolescent rat. A custom-designed 0.9 numerical aperture (NA) objective lens provided better fluorescence excitation and detection than a regular GRIN lens pair. A large core multi-mode plastic fiber was used to deliver the fluorescent signals to a benchtop detector. The compact fiber scanner in the headpiece achieved a 10.9 Hz frame rate with 64 × 128 pixels scanning. This miniaturized microscope could resolve individual soma from layer 2/3 of the visual cortex with a penetration depth of more than 200 μm. For their experiment, the primary visual cortex was stained with two fluorescent dyes, a calcium indicator (Oregon green BAPTA-1, OGB), and sulforhodamine 101 (SR101) for labeling astrocytes. The experiment was conducted on a semicircular track on which the animal could move freely, and visual stimuli were delivered by three monitors placed at the ends

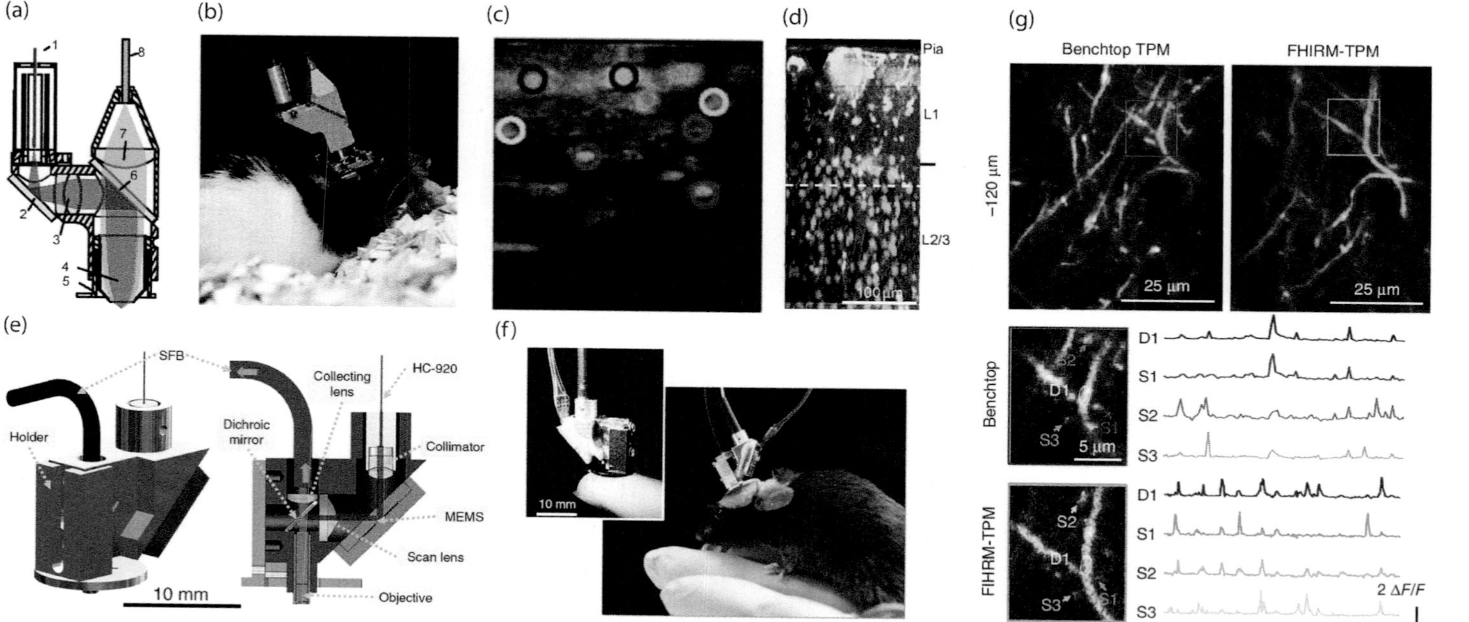

FIGURE 1.4 **(a)** Schematic of the miniaturized two-photon microscope reported by Sawinski et al. (2009); 1. single-mode excitation fiber; 2 folding mirror; 3, tube lens; 4, objective; 5, focusing flange; 6, beam splitter; 7, collimation lens; and 8, multi-mode collection fiber. **(b)** Photo of a rat with the head-mounted microscope in (a). **(c)** Representative image (10-frame average) from layer 2/3 of a rat visual cortex acquired by the miniaturized two-photon microscope in (a). The neurons (green channel) were labeled with Oregon green BAPTA-1, OGB1 and astrocytes (orange channel) were labeled with sulforhodamine 101. **(d)** Side projection of an image stack acquired by a conventional benchtop two-photon microscope. The labeling for neurons and astrocytes is the same as in (c). The dashed line approximates the image plane of the miniaturized two-photon microscope. (Adapted from Sawinski et al. (2009) with permission from Proceedings of the National Academy of Sciences.) **(e)** Schematic of the miniaturized two-photon microscope reported by Zong et al., (2017) **(f)** Photos of the miniaturized microscope in (e) on a fingertip and mounted on a mouse. **(g)** A comparison between the miniaturized and benchtop microscope. Top, Ca^{2+} signals from dendrites and spines of a mouse expressing GCaMP6f (average of 240 frames). Bottom, left panel shows two zoomed-in images of one dendritic shaft (D1) and three spines (S1, S2, and S3). Right panel shows their Ca^{2+} dynamics over 100 seconds. The images were acquired by 512 × 512 pixels scanning at 8Hz within an 80 μm field of view. (Adapted from Zong et al. (2017) with permission from Springer Nature.)

and apex of the track. Four infrared LEDs placed on the headpiece allowed the animal's position and head orientation to be monitored with a synchronized infrared camera. The authors observed robust Ca^{2+} transients from the labeled neurons, comparable to results from benchtop two-photon microscopy in both anesthetized and awake animals (Stosiek et al., 2003; Greenberg et al., 2008). A significant increase in Ca^{2+} transients was observed when the animal swept its gaze across different monitors with some neurons showing preferential firing to a certain monitor location.

Another recent miniature two-photon microscope (Zong et al., 2017) realized fast high-resolution imaging of freely moving mice. Weighing 2.15 g, the miniaturized microscope can perform 256 × 256 pixels scanning at 40 Hz over a 130 μm field of view (shown in Figure 1.4.e–f). A hollow-core photonic crystal fiber (HC-920) was custom-made to transmit 920 nm fs laser pulses with minimal nonlinear pulse broadening. By inserting a glass tube that compensated the negative dispersion from HC-920, the system could deliver 920 nm, ~100 fs ultrashort laser pulses to the sample. A custom-designed MEMS scanner was incorporated into the system for video-rate imaging, which also supported line and random access scanning. By using a high NA miniature compound objective lens with a matching collimator and scan lens, the system achieved a lateral resolution of 0.64 μm and an axial resolution of 3.35 μm. Activity within individual dendritic spines could be resolved using this miniaturized microscope. Minimal motion artifact was observed during experiments involving free movement of the animals such as tail suspension, stepping down from a stage, or interacting with its sibling. Figure 1.4.g presents a comparison between the miniature two-photon microscope (FHIRM-2PM) with a benchtop 2PM. The images illustrate the Ca^{2+} signals from dendrites and spines from a mouse expressing GCaMP6f. The authors were able to image the same mouse for approximately four hours across different behavioral paradigms.

1.4 CHALLENGES OF MINIATURIZED NEUROIMAGING SYSTEMS

1.4.1 Miniaturized Microscope Weight Constraints

A major design criterion of miniaturized neuroimaging systems is the additional weight it imposes on the animal. For certain types of fiber optics-based systems, the weight of the headpiece can be minimized to ~1 g (Flusberg et al., 2008), since only a compact objective lens and optical fiber need to be incorporated into the headpiece. For a discrete imaging system, most of the microscope components are housed within the headpiece. Therefore, it is critical that the weight of the miniaturized imaging system is taken into consideration during design. A similar argument can be made for the size of the headpiece. These concerns with the weight and size of the headpiece might be partially resolved by using a weight-bearing system (e.g. a commutator or spring mass system (Tang et al., 2015)), or switching to a larger species of animal (e.g. rat rather than mouse). Nevertheless, care must be taken to ensure that normal animal behaviors are not encumbered by the presence of the headpiece. A training period for the animal to habituate to the headpiece may also be necessary prior to experimentation and image acquisition.

1.4.2 Motion Artifacts

Susceptibility to motion artifact is another issue with imaging in awake, freely moving animals. Besides motion from respiration and heartbeats that are common with benchtop imaging, miniaturized imaging systems are susceptible to additional motion caused by an animal's body movement or the relative displacement between the animal and the headpiece. Imaging modalities that require beam scanning (such as miniaturized 2PM) are more sensitive to motion artifact since the position of the scanning pattern is used to reconstruct the image. Off-line image registration and motion correction algorithms (Guizar-Sicairos et al., 2008; Greenberg and Kerr, 2009; Miao et al., 2010) can remove the effects of gross motion from the image plane. Researchers have also developed motion correction methods applicable to a single frame image (Dombeck et al., 2007). However, images exhibiting displacements too severe to be corrected, or deviations from the image plane should be excluded from further analysis.

1.4.3 Interpretation of Animal Behaviors

To study the correlation between animal behavior and the corresponding neural activity, it is imperative to establish a comprehensive animal monitoring system. Owing to the rapid neural dynamics during behavioral experiments, the monitoring system needs to be synchronized with the miniaturized imaging system. Depending on the task that the animal performs, different types of sensors can be utilized for behavioral monitoring. Typically, an infrared video camera is used to document the animal's activities during the experiment. In conjunction with infrared LEDs placed on the headpiece (Sawinski et al., 2009; Ghosh et al., 2011), the orientation and position of the head can be recorded in two dimensions. The running speed of the animal can be monitored using accelerometers or a treadmill (Dombeck et al., 2007). Advanced systems (Ou-Yang et al., 2011) are able to track the animal's locomotion and pose in three dimensions. Experiments with miniaturized microscope systems involve image acquisition for much longer than is typical of anesthetized and head-restrained experiments. Therefore, the resulting large dataset from these experiments calls for high-throughput, automatic algorithms that quantify the animal's behaviors with high precision (Matsumoto et al., 2013; Hong et al., 2015; Wiltschko et al., 2015).

1.4.4 Fiber Optic-Adapted versus Discrete Systems

There is no well-defined *a priori* rationale for choosing either miniaturization approach because each has its pros and cons. Most of the imaging modalities reported so far, such as single-photon fluorescence microscopy, laser speckle contrast, and optical intrinsic signal imaging, can be achieved using either method. Examples include a fiber optics-based single-photon fluorescence system (Flusberg et al., 2008), a discrete one-photon fluorescence system (Ghosh et al., 2011), a fiber optics-based laser speckle contrast imaging/optical intrinsic signal imaging system (Liu et al., 2013), and a discrete laser speckle contrast imaging system (Senarathna et al., 2012).

Miniaturized fiber optic systems permit the use of high-end light sources and/or detectors designed for conventional benchtop imaging systems. For example, ultrafast (femtosecond) laser pulses can be delivered through optical fiber for two-photon excitation

(Göbel et al., 2004a). For a fiber optics-based system, the illumination light source can be easily switched between different imaging modalities. Similarly, the detected light signals can be easily split to different channels for multi-color or multi-modality imaging. Theoretically, a miniaturized fiber optic system will have a lighter headpiece (see Section 1.4.1 on *Miniaturized Microscope Weight Constraints*). Miniaturization might be easier for fiber optic-based systems than for discrete systems, since only a part of the microscope needs to be miniaturized. Nevertheless, the optical fiber is much stiffer than most electrical wire connectors used in discrete systems. This can affect the animal's mobility and certain types of behaviors. The efficiency of light collection through fiber coupling is low, which may also affect the image quality. The efficiency of emission light collection also depends on the bending of the fiber, which can confound the interpretation of some experimental results.

Since discrete systems do not rely on bulky benchtop microscope components, they can be easily transported, which gives researchers the flexibility to perform their experiments in a wider range of environments (e.g. within home cage systems or laminar flow hoods). Discrete systems are usually built from off-the-shelf components, which implies that they are relatively low cost and mass producible. Discrete systems are more amenable to experiments (such as social interactions) that involve imaging multiple animals simultaneously. Studies have also shown that discrete systems are more robust and less susceptible to motion artifacts than fiber optics-based system (Ghosh et al., 2011). However, since a higher degree of miniaturization is needed for discrete systems, their development may require more design iterations than those required when building a miniaturized fiber optic-based microscope. Finally, imaging parameters, such as magnification and field of view are not easily adjustable in discrete systems.

1.5 FUTURE DIRECTIONS

Over the past few decades, we have seen extensive development of prototype miniaturized optical neuroimaging systems. These miniaturized systems often achieve imaging performance comparable to benchtop microscopy systems. Additionally, the compact, portable features of these head-mounted microscopes have opened a wide range of behavioral imaging experiments that were unfeasible with benchtop systems. However, the field of miniaturized neuroimaging is still relatively young, and we expect further developments for this exciting new area of neuroimaging. Some of these developments are discussed in the ensuing sections.

1.5.1 Optogenetic Control of Neural Circuits

Another advantage of miniaturized microscope development is the ability to integrate it with the extant thriving optogenetic toolbox (Fenno et al., 2011). Through the development of red-shifted calcium indicators (Inoue et al., 2015) and optogenetic actuators (Prakash et al., 2012; Lin et al., 2013), researchers can now probe and manipulate neural activity deep within the rodent brain. More importantly, these red-shifted calcium indicators and optogenetic actuators are spectrally separated. This permits us to image and manipulate the neural activity simultaneously, with minimum cross-talk. Similar to these benchtop studies (Rickgauer et al., 2014; Packer et al., 2015), miniaturized microscopes have the potential to achieve all-optical recording and manipulation of the neural circuits in awake, freely moving animals (Szabo et al., 2014).

1.5.2 Novel Imaging Applications

Future applications of miniaturized microscopes need not be restricted to probing the relationship between behaviors and neural dynamics, but can also be applied to the investigation of brain diseases in preclinical models. These microscopes can monitor the onset of seizures, track neuroplasticity and rehabilitation after stroke, and record neural degradation in Alzheimer's disease. Such microscopes could also be used to image drug delivery, track the efficacy of stem-cell therapy, and longitudinally image the progression and response to treatment of brain tumors or brain metastases in preclinical models. Finally, the use of such microscopes doesn't need to be limited to imaging the brains of freely moving animals. For example, one can image the cellular activities of the spinal cord using the miniaturized microscope approach (Sekiguchi et al., 2016).

1.5.3 Towards a True Tetherless Imaging System?

All the miniaturized microscope systems reported so far require either optical fiber or electrical wire connections to external components. Although miniaturized microscopes have broadened the kinds of behavioral assays that can be performed during optical neuroimaging, the tether still limits the scope of experiments to a predefined behavioral arena and constrains the imaging duration of each experimental session. Several research groups, including ours, have reported prototypes (Senarathna et al., 2012; Miao et al., 2017) that relocate the external microscope/electronic components to a backpack that can be worn by the animal during behavioral experiments. This approach overcomes some of the disadvantages of using a tether. With advances in wireless power transmission and communication, we anticipate that a true tetherless, miniaturized optical imaging system can be built in the near future. Progress in related fields, such as compact wireless-powered LEDs for optogenetic control (Wentz et al., 2011; Montgomery et al., 2015) may be modified to act as the light source for tetherless imaging systems. A high-speed video rate CMOS camera with ultra-low power consumption (14μW) (Zhang et al., 2016) could be a strong candidate for the light detector in tetherless imaging systems.

Over the past decade, we have seen an exponential growth in the development and application of miniaturized optical neuroimaging systems in awake, freely moving animals. Further development of this relatively new field will benefit from advances in novel benchtop microscopy systems, as well as the miniaturization of light sources and light detectors. We believe that these compact imaging systems will eventually become an indispensable tool in modern neuroscience research.

REFERENCES

Bandyopadhyay R, Gittings A, Suh S, Dixon P, Durian D (2005) Speckle-visibility spectroscopy: a tool to study time-varying dynamics. *Review of Scientific Instruments* 76:093110.

Bonhomme V, Boveroux P, Hans P, Brichant JF, Vanhaudenhuyse A, Boly M, Laureys S (2011) Influence of anesthesia on cerebral blood flow, cerebral metabolic rate, and brain functional connectivity. *Current Opinion in Anaesthesiology* 24:474–479.

Dombeck DA, Khabbaz AN, Collman F, Adelman TL, Tank DW (2007) Imaging large-scale neural activity with cellular resolution in awake, mobile mice. *Neuron* 56:43–57.

Dunn AK, Devor A, Bolay H, Andermann ML, Moskowitz MA, Dale AM, Boas DA (2003) Simultaneous imaging of total cerebral hemoglobin concentration, oxygenation, and blood flow during functional activation. *Optics Letters* 28:28–30.

Engelbrecht CJ, Johnston RS, Seibel EJ, Helmchen F (2008) Ultra-compact fiber-optic two-photon microscope for functional fluorescence imaging in vivo. *Optics Express* 16:5556–5564.

Engelbrecht CJ, Voigt F, Helmchen F (2010) Miniaturized selective plane illumination microscopy for high-contrast in vivo fluorescence imaging. *Optics Letters* 35:1413–1415.

Fenno L, Yizhar O, Deisseroth K (2011) The development and application of optogenetics. *Neuroscience* 34:389.

Ferezou I, Bolea S, Petersen CC (2006) Visualizing the cortical representation of whisker touch: voltage-sensitive dye imaging in freely moving mice. *Neuron* 50:617–629.

Flusberg BA, Jung JC, Cocker ED, Anderson EP, Schnitzer MJ (2005) In vivo brain imaging using a portable 3.9 gram two-photon fluorescence microendoscope. *Optics Letters* 30:2272–2274.

Flusberg BA, Nimmerjahn A, Cocker ED, Mukamel EA, Barretto RP, Ko TH, Burns LD, Jung JC, Schnitzer MJ (2008) High-speed, miniaturized fluorescence microscopy in freely moving mice. *Nature Methods* 5:935–938.

Ghosh KK, Burns LD, Cocker ED, Nimmerjahn A, Ziv Y, El Gamal A, Schnitzer MJ (2011) Miniaturized integration of a fluorescence microscope. *Nature Methods* 8:871–878.

Göbel W, Nimmerjahn A, Helmchen F (2004a) Distortion-free delivery of nanojoule femtosecond pulses from a Ti:sapphire laser through a hollow-core photonic crystal fiber. *Optics Letters* 29:1285–1287.

Göbel W, Kerr JN, Nimmerjahn A, Helmchen F (2004b) Miniaturized two-photon microscope based on a flexible coherent fiber bundle and a gradient-index lens objective. *Optics Letters* 29:2521–2523.

Gong Y, Huang C, Li JZ, Grewe BF, Zhang Y, Eismann S, Schnitzer MJ (2015) High-speed recording of neural spikes in awake mice and flies with a fluorescent voltage sensor. *Science* 350:1361–1366.

Greenberg DS, Houweling AR, Kerr JN (2008) Population imaging of ongoing neuronal activity in the visual cortex of awake rats. *Nature Neuroscience* 11:749–751.

Greenberg DS, Kerr JN (2009) Automated correction of fast motion artifacts for two-photon imaging of awake animals. *Journal of Neuroscience Methods* 176:1–15.

Guizar-Sicairos M, Thurman ST, Fienup JR (2008) Efficient subpixel image registration algorithms. *Optics Letters* 33:156–158.

Helmchen F, Denk W (2005) Deep tissue two-photon microscopy. *Nature Methods* 2:932–940.

Helmchen F, Fee MS, Tank DW, Denk W (2001) A miniature head-mounted two-photon microscope: high-resolution brain imaging in freely moving animals. *Neuron* 31:903–912.

Helmchen F, Denk W, Kerr JN (2013) Miniaturization of two-photon microscopy for imaging in freely moving animals. *Cold Spring Harbor Protocols, 2013*(10), pdb-top078147.

Hillman EM (2007) Optical brain imaging in vivo: techniques and applications from animal to man. *Journal of Biomedical Optics* 12:051402.

Holtmaat A, Bonhoeffer T, Chow DK, Chuckowree J, De Paola V, Hofer SB, Hübener M, Keck T, Knott G, Lee W-CA (2009) Long-term, high-resolution imaging in the mouse neocortex through a chronic cranial window. *Nature Protocols* 4:1128–1144.

Holzer M, Schmitz C, Pei Y, Graber H, Abdul R-A, Barry J, Muller R, Barbour R (2006) 4D functional imaging in the freely moving rat. In: *28th Annual International Conference of the IEEE Engineering in Medicine and Biology Society, 2006 (EMBS'06)*, pp. 29–32. IEEE.

Hong W, Kennedy A, Burgos-Artizzu XP, Zelikowsky M, Navonne SG, Perona P, Anderson DJ (2015) Automated measurement of mouse social behaviors using depth sensing, video tracking, and machine learning. *Proceedings of the National Academy of Sciences of the United States of America* 112:E5351–E5360.

Huang L, Mills AK, Zhao Y, Jones DJ, Tang S (2016) Miniature fiber-optic multiphoton microscopy system using frequency-doubled femtosecond Er-doped fiber laser. *Biomedical Optics Express* 7:1948–1956.

Inoue M, Takeuchi A, Horigane S, Ohkura M, Gengyo-Ando K, Fujii H, Kamijo S, Takemoto-Kimura S, Kano M, Nakai J (2015) Rational design of a high-affinity, fast, red calcium indicator R-CaMP2. *Nature Methods* 12:64–70.

Jones PB, Shin HK, Boas DA, Hyman BT, Moskowitz MA, Ayata C, Dunn AK (2008) Simultaneous multispectral reflectance imaging and laser speckle flowmetry of cerebral blood flow and oxygen metabolism in focal cerebral ischemia. *Journal of Biomedical Optics* 13:044007.

Jung W, Tang S, McCormic DT, Xie T, Ahn Y-C, Su J, Tomov IV, Krasieva TB, Tromberg BJ, Chen Z (2008) Miniaturized probe based on a microelectromechanical system mirror for multiphoton microscopy. *Optics Letters* 33:1324–1326.

Kazmi SS, Richards LM, Schrandt CJ, Davis MA, Dunn AK (2015) Expanding applications, accuracy, and interpretation of laser speckle contrast imaging of cerebral blood flow. *Journal of Cerebral Blood Flow and Metabolism* 35:1076–1084.

Keck T, Keller GB, Jacobsen RI, Eysel UT, Bonhoeffer T, Hübener M (2013) Synaptic scaling and homeostatic plasticity in the mouse visual cortex in vivo. *Neuron* 80:327–334.

Lee CM, Engelbrecht CJ, Soper TD, Helmchen F, Seibel EJ (2010) Scanning fiber endoscopy with highly flexible, 1-mm catheterscopes for wide-field, full-color imaging. *Journal of Biophotonics* 3:385.

Lin JY, Knutsen PM, Muller A, Kleinfeld D, Tsien RY (2013) ReaChR: a red-shifted variant of channelrhodopsin enables deep transcranial optogenetic excitation. *Nature Neuroscience* 16:1499–1508.

Lin MZ, Schnitzer MJ (2016) Genetically encoded indicators of neuronal activity. *Nature Neuroscience* 19:1142–1153.

Liu R, Huang Q, Li B, Yin C, Jiang C, Wang J, Lu J, Luo Q, Li P (2013) Extendable, miniaturized multi-modal optical imaging system: cortical hemodynamic observation in freely moving animals. *Optics Express* 21:1911–1924.

Llinás RR (2003) The contribution of Santiago Ramon y Cajal to functional neuroscience. *Nature Reviews Neuroscience* 4:77–80.

Ma Y, Shaik MA, Kim SH, Kozberg MG, Thibodeaux DN, Zhao HT, Yu H, Hillman EM (2016) Wide-field optical mapping of neural activity and brain haemodynamics: considerations and novel approaches. *Philosophical Transactions of the Royal Society B* 371:20150360.

Malonek D, Grinvald A (1996) Interactions between electrical activity and cortical microcirculation revealed by imaging spectroscopy: implications for functional brain mapping. *Science* 272:551.

Marvin JS, Borghuis BG, Tian L, Cichon J, Harnett MT, Akerboom J, Gordus A, Renninger SL, Chen T-W, Bargmann CI (2013) An optimized fluorescent probe for visualizing glutamate neurotransmission. *Nature Methods* 10:162–170.

Masamoto K, Kanno I (2012) Anesthesia and the quantitative evaluation of neurovascular coupling. *Journal of Cerebral Blood Flow and Metabolism* 32:1233–1247.

Matsumoto J, Urakawa S, Takamura Y, Malcher-Lopes R, Hori E, Tomaz C, Ono T, Nishijo H (2013) A 3D-video-based computerized analysis of social and sexual interactions in rats. *PloS one* 8:e78460.

Miao P, Lu H, Liu Q, Li Y, Tong S (2011) Laser speckle contrast imaging of cerebral blood flow in freely moving animals. *Journal of Biomedical Optics* 16:090502.

Miao P, Rege A, Li N, Thakor NV, Tong S (2010) High resolution cerebral blood flow imaging by registered laser speckle contrast analysis. *IEEE Transactions on Biomedical Engineering* 57:1152–1157.

Miao P, Zhang L, Li M, Zhang Y, Feng S, Wang Q, Thakor NV (2017) Chronic wide-field imaging of brain hemodynamics in behaving animals. *Biomedical Optics Express* 8:436–445.

Montgomery KL, Yeh AJ, Ho JS, Tsao V, Iyer SM, Grosenick L, Ferenczi EA, Tanabe Y, Deisseroth K, Delp SL (2015) Wirelessly powered, fully internal optogenetics for brain, spinal and peripheral circuits in mice. *Nature Methods* 12: 969–974.

Murari K, Zhang Y, Li S, Chen Y, Li M-J, Li X (2011) Compensation-free, all-fiber-optic, two-photon endomicroscopy at 1.55 μm. *Optics Letters* 36:1299–1301.

Niederriter RD, Ozbay BN, Futia GL, Gibson EA, Gopinath JT (2017) Compact diode laser source for multiphoton biological imaging. *Biomedical Optics Express* 8:315–322.

Ogawa S, Lee T-M, Kay AR, Tank DW (1990) Brain magnetic resonance imaging with contrast dependent on blood oxygenation. *Proceedings of the National Academy of Sciences of the United States of America* 87:9868–9872.

Ou-Yang T-H, Tsai M-L, Yen C-T, Lin T-T (2011) An infrared range camera-based approach for three-dimensional locomotion tracking and pose reconstruction in a rodent. *Journal of Neuroscience Methods* 201:116–123.

Packer AM, Russell LE, Dalgleish HW, Häusser M (2015) Simultaneous all-optical manipulation and recording of neural circuit activity with cellular resolution in vivo. *Nature Methods* 12:140–146.

Piyawattanametha W, Cocker ED, Burns LD, Barretto RPJ, Jung JC, Ra H, Solgaard O, Schnitzer MJ (2009) In vivo brain imaging using a portable 2.9 g two-photon microscope based on a microelectromechanical systems scanning mirror. *Optics Letters* 34:2309–2311.

Prakash R, Yizhar O, Grewe B, Ramakrishnan C, Wang N, Goshen I, Packer AM, Peterka DS, Yuste R, Schnitzer MJ (2012) Two-photon optogenetic toolbox for fast inhibition, excitation and bistable modulation. *Nature Methods* 9:1171–1179.

Rector D, Harper R (1991) Imaging of hippocampal neural activity in freely behaving animals. *Behavioural Brain Research* 42:143–149.

Rickgauer JP, Deisseroth K, Tank DW (2014) Simultaneous cellular-resolution optical perturbation and imaging of place cell firing fields. *Nature Neuroscience* 17:1816–1824.

Sawinski J, Denk W (2007) Miniature random-access fiber scanner for in vivo multiphoton imaging. *Journal of Applied Physics* 102:034701.

Sawinski J, Wallace DJ, Greenberg DS, Grossmann S, Denk W, Kerr JN (2009) Visually evoked activity in cortical cells imaged in freely moving animals. *Proceedings of the National Academy of Sciences of the United States of America* 106:19557–19562.

Sekiguchi KJ, Shekhtmeyster P, Merten K, Arena A, Cook D, Hoffman E, Ngo A, Nimmerjahn A (2016) Imaging large-scale cellular activity in spinal cord of freely behaving mice. *Nature Communications* 7 (11450).

Senarathna J, Murari K, Etienne-Cummings R, Thakor NV (2012) A miniaturized platform for laser speckle contrast imaging. *IEEE Transactions on Biomedical Circuits and Systems* 6:437–445.

Senarathna J, Rege A, Li N, Thakor N (2013) Laser speckle contrast imaging: theory, instrumentation and applications. *IEEE Reviews in Biomedical Engineering* 6: 99–110.

Shin H-J, Pierce MC, Lee D, Ra H, Solgaard O, Richards-Kortum R (2007) Fiber-optic confocal microscope using a MEMS scanner and miniature objective lens. *Optics Express* 15:9113–9122.

Sigal I, Koletar MM, Ringuette D, Gad R, Jeffrey M, Carlen PL, Stefanovic B, Levi O (2016) Imaging brain activity during seizures in freely behaving rats using a miniature multi-modal imaging system. *Biomedical Optics Express* 7:3596–3609.

Sofroniew NJ, Cohen JD, Lee AK, Svoboda K (2014) Natural whisker-guided behavior by head-fixed mice in tactile virtual reality. *The Journal of Neuroscience* 34:9537–9550.

Stosiek C, Garaschuk O, Holthoff K, Konnerth A (2003) In vivo two-photon calcium imaging of neuronal networks. *Proceedings of the National Academy of Sciences of the United States of America* 100:7319–7324.

Sutin J, Wu W, Ruvinskaya L, Franceschini MA (2012) Methods for simultaneous optical and electrical measurement of neurovascular coupling in awake rats. In: *Biomedical Optics*, Miami (Florida, USA), *p BSu3A. 90: Optical Society of America.*

Szabo V, Ventalon C, De Sars V, Bradley J, Emiliani V (2014) Spatially selective holographic photoactivation and functional fluorescence imaging in freely behaving mice with a fiberscope. *Neuron* 84:1157–1169.

Taira K, Hashimoto T, Yokoyama H (2007) Two-photon fluorescence imaging with a pulse source based on a 980-nm gain-switched laser diode. *Optics Express* 15:2454–2458.

Tang J, Xi L, Zhou J, Huang H, Zhang T, Carney PR, Jiang H (2015) Noninvasive high-speed photoacoustic tomography of cerebral hemodynamics in awake-moving rats. *Journal of Cerebral Blood Flow and Metabolism* 8: 1224 –1232.

Theer P, Denk W (2006) On the fundamental imaging-depth limit in two-photon microscopy. *JOSA A* 23:3139–3149.

Tian L, Hires SA, Mao T, Huber D, Chiappe ME, Chalasani SH, Petreanu L, Akerboom J, McKinney SA, Schreiter ER (2009) Imaging neural activity in worms, flies and mice with improved GCaMP calcium indicators. *Nature Methods* 6:875–881.

Tran PH, Mukai DS, Brenner M, Chen Z (2004) In vivo endoscopic optical coherence tomography by use of a rotational microelectromechanical system probe. *Optics Letters* 29:1236–1238.

Wentz CT, Bernstein JG, Monahan P, Guerra A, Rodriguez A, Boyden ES (2011) A wirelessly powered and controlled device for optical neural control of freely-behaving animals. *Journal of Neural Engineering* 8:046021.

Wiltschko AB, Johnson MJ, Iurilli G, Peterson RE, Katon JM, Pashkovski SL, Abraira VE, Adams RP, Datta SR (2015) Mapping sub-second structure in mouse behavior. *Neuron* 88:1121–1135.

Xiao D, Vanni MP, Mitelut CC, Chan AW, LeDue JM, Xie Y, Chen ACN, Swindale NV, Murphy TH (2017) Mapping cortical mesoscopic networks of single spiking cortical or sub-cortical neurons. *eLife* 6:e19976.

Yu H, Senarathna J, Tyler BM, Thakor NV, Pathak AP (2015) Miniaturized optical neuroimaging in unrestrained animals. *NeuroImage* 113:397–406.

Zhang J, Xiong T, Tran T, Chin S, Etienne-Cummings R (2016) Compact all-CMOS spatiotemporal compressive sensing video camera with pixel-wise coded exposure. *Optics Express* 24:9013–9024.

Zong W, Wu R, Li M, Hu Y, Li Y, Li J, Rong H, Wu H, Xu Y, Lu Y (2017) Fast high-resolution miniature two-photon microscopy for brain imaging in freely behaving mice. *Nature Methods* 24: 713–719.

CHAPTER 2

Functional Imaging with Light-Sheet Microscopy

Raghav K. Chhetri and Philipp J. Keller

CONTENTS

2.1 INTRODUCTION

Brain imaging techniques are essential for advancing our understanding of the basic principles underlying neural circuit function, connectivity across neuronal populations and entire brain regions, the functional development of the early nervous system, and the mechanisms of neural circuits underlying behavior. The central tenet of imaging a nervous system in action is the spatiotemporal scale at which neuronal activity occurs, thereby setting it apart as a unique challenge rarely encountered in other bio-imaging scenarios. Additionally, imaging the activity from a population of neurons in the intact brain of a behaving animal makes this task even more challenging, which has motivated researchers to engage a variety of advanced techniques to push through this boundary. Among optical methods, light-sheet microscopy is well-positioned to tackle the numerous challenges encountered in functional imaging of the nervous system, as it provides a unique combination of strengths: light-sheet microscopy offers high spatial resolution, high imaging speed, good physical coverage of partially opaque specimens, and low energy load on the specimen for long-term imaging under physiological conditions. In this chapter, we discuss recent efforts in recording neuronal activity using light-sheet microscopy and the opportunities it has carved in deepening our understanding of the inner workings of the brain.

2.2 PRINCIPLES OF LIGHT-SHEET MICROSCOPY

The conceptual foundation of light-sheet microscopy dates back to 1902 when Siedentopf and Zsigmondy developed a microscope, which they termed the "ultramicroscope", to study the scattering of visible light from sub-wavelength colloidal particles (Siedentopf and Zsigmondy, 1902). Unlike a traditional light microscope, the design of the ultramicroscope utilized an illumination beam orthogonal to the detection lens. This simple technical concept is the foundation of all modern light-sheet microscopy techniques, which rely on illuminating a thin section of a fluorescently labeled sample and collecting the emitted photons from the entire illuminated section along a detection axis positioned orthogonally to the illumination axis. This concept of separate illumination and detection, in combination with recent technical advancements in light sources, opto-mechanical components, fluorescent labels, detectors, and computing frameworks, has positioned light-sheet microscopy as a powerful technique for fluorescence imaging in the life sciences.

Various high-resolution optical techniques exist that can capture the full 3D extent of a physically sectioned sample (Osten and Margrie, 2013). However, methods that rely on physical sectioning obviously are not suitable for *in vivo* imaging. Thus, optical sectioning methods that leave the live specimen intact become necessary. Optical sectioning can be performed on images acquired from a standard epi-fluorescent microscope using post-processing algorithms to separate in-focus regions from out-of-focus regions. Confocal microscopes also perform optical sectioning by rejecting out-of-focus light using a confocal pinhole in the detection path. However, these methods always expose the entire depth of the sample to fluorescence excitation light which leads to photobleaching and phototoxicity and can irreversibly damage the specimen when imaged over an extended period of time. Additionally, the rejection of out-of-focus light is not reliably achieved in highly scattering samples and is evident as a drop off in image contrast at larger depths. Light-sheet microscopy tackles these shortcomings of conventional techniques by implementing an efficient optical sectioning strategy that only illuminates a thin section of the specimen. The entire illuminated section is then rapidly imaged with a camera, which offers data acquisition rates that are several orders of magnitude higher than those achievable with point-scanning techniques. Volumetric imaging is achieved by sequentially illuminating different depth sections, and since signal photons are only emitted from a thin section of the sample at any time during this process, light-sheet microscopy also does not suffer as severely from background in highly scattering samples. Light-sheet microscopy thus outperforms conventional fluorescence microscopy techniques in its ability to rapidly image

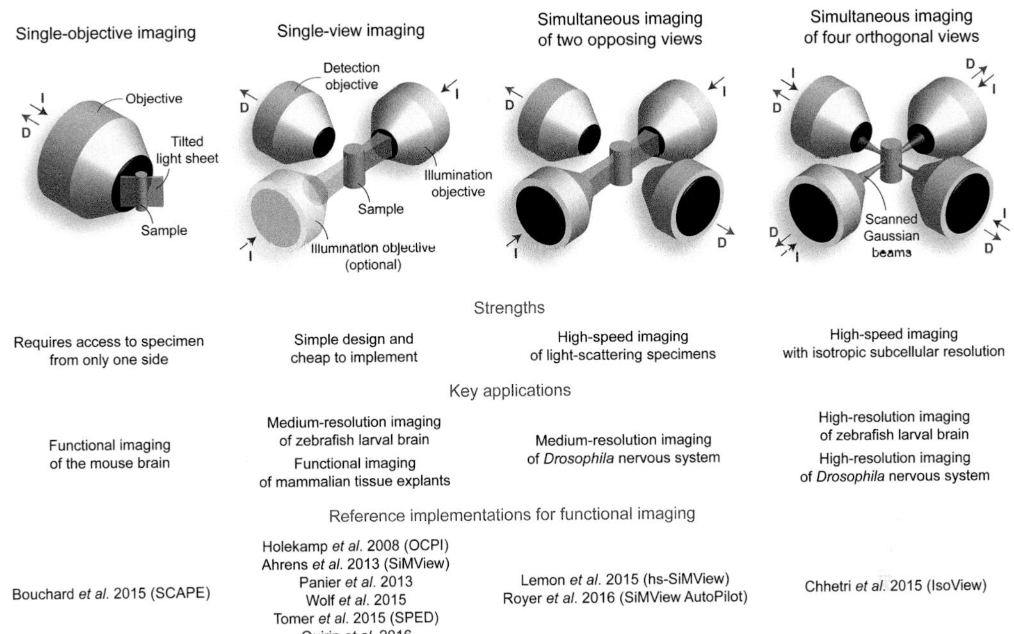

FIGURE 2.1 Main classes of light-sheet microscopy techniques for functional imaging. Overview of the main classes of light-sheet microscopy techniques designed to address the respective challenges associated with functional imaging in various biological model systems. This overview includes a brief summary of the key strengths and reference implementations presented to date for each class.

live specimens over an extended period of time with minimal photobleaching and phototoxicity while achieving excellent resolution and larger penetration depths.

There are many variations of light-sheet microscopy methods, and yet the general principle remains the same, as illustrated in Figure 2.1. The key approach in all light-sheet microscopes is the orthogonal illumination of a thin volume section with a "sheet" of light (Voie et al., 1993; Fuchs et al., 2002; Huisken et al., 2004) or with a rapidly scanning "pencil-beam" (Keller et al., 2008). Fluorescent light emitted by this thin section is then captured by a wide-field detection system to form an image. To acquire a 3D volumetric dataset, either the sample is translated stepwise across a stationary light-sheet or the light-sheet is scanned step-by-step across a stationary sample to illuminate different volume sections (Ahrens et al., 2013). Iterating this 3D acquisition procedure then results in a temporal recording of the imaged volume.

2.2.1 Live-Imaging of Multi-Cellular Organisms

Recent advances in light-sheet microscopy have made it possible to image complex multi-cellular organisms in their entirety while they undergo rapid morphological changes during their development (Keller et al., 2008; Tomer et al., 2012). The key to capturing such dynamics at the sub-cellular level across the entire biological system is achieving a favorable combination of the following factors: (1) high spatial resolution;

(2) high temporal resolution; (3) good physical coverage of the specimen; (4) long-term imaging capability; and (5) low phototoxicity and photodamage. High spatial resolution ensures that neighboring cells and sub-cellular features are distinctly resolved in a multi-cellular organism. High temporal resolution is crucial for capturing fast dynamic processes such as cell shape changes and cell migration. Good physical coverage of the specimen is needed for a system-level analysis of whole-tissue morphogenesis and development of entire embryos. Long-term imaging capability is crucial to attaining uninterrupted tracking of the dynamic processes across the entire developmental timescale. Lastly, the interrogation of the biological specimen needs to be minimally invasive and should not disrupt the system under study, and as such, photobleaching and phototoxicity need to be kept to a minimum to avoid perturbing normal growth and development of the specimen. The number of photons emitted from the illuminated section of the specimen is the common resource all five of these parameters are competing for.

The low energy load in light-sheet microscopy experiments enables optimal utilization of the photon budget and affords whole-animal imaging with high spatial resolution and high temporal resolution at the same time. Experimental demonstrations to this end include high-resolution imaging of small multi-cellular organisms such as *Caenorhabditis elegans* (roundworm) embryos (Wu et al., 2013; Chen et al., 2014), invertebrates such as *Drosophila melanogaster* (fruit fly) embryos/larvae (Tomer et al., 2012; Chhetri et al., 2015), and vertebrates such as *Danio rerio* (zebrafish) embryos/larvae (Keller et al., 2008; Chhetri et al., 2015). These samples encompass a wide range of physical sizes; the dimensions of *Caenorhabditis elegans* embryos are $50 \times 30 \times 30\ \mu m^3$, *Drosophila melanogaster*'s embryonic dimensions are $500 \times 200 \times 200\ \mu m^3$, and *Danio rerio* embryonic dimensions are $700 \times 700 \times 700\ \mu m^3$. Owing to their different sizes, shapes, and levels of transparency, each of these animal models presents a unique imaging challenge and can be optimally tackled with a unique imaging solution. For instance, *C. elegans* embryos are small and transparent, and thus are ideal for imaging with short, thin light-sheets (a few μm) from one or at maximum two (orthogonal) views, as has been demonstrated using lattice light-sheet microscopy (Chen et al., 2014) and dual-view inverted selective plane illumination microscopy (diSPIM) (Wu et al., 2013).

Imaging large specimens such as zebrafish embryos that comprise a large, light-scattering central yolk cell and partially opaque specimens such as *Drosophila* embryos presents additional challenges with limited depth penetration that cannot be resolved using lattice light-sheet microscopy and diSPIM. Additionally, the use of short, thin light-sheets may not be ideal in these large specimens since temporal resolution must be sacrificed to acquire the large number of images that are needed to cover the entire specimen volume. Thus, imaging these large vertebrate and higher invertebrate embryos at a high temporal resolution necessitates using longer, slightly thicker light-sheets which consequently reduces axial resolution across the imaging volume. However, this reduction in axial resolution can be recovered by combining the usage of thicker light-sheets with orthogonal multi-view imaging (Swoger et al., 2007), which facilitates the reconstruction of a 3D image of the specimen with near-isotropic resolution. The acquisition of at least four orthogonal views (comprising two sets of pairwise opposing views of the sample) additionally improves physical coverage of large, non-transparent specimens.

2.2.2 Imaging at High Spatiotemporal Resolution

First-generation laser light-sheet fluorescence microscopes acquired volumetric data by translating the sample sequentially across the stationary light-sheet and detection focal plane. These setups utilized motorized stages that can translate as well as rotate the mounted specimen, and were used quite successfully for long-term developmental imaging with relaxed temporal sampling requirements. For imaging processes that demand higher spatiotemporal resolution, the physical scanning of the specimen is often slow and thus inadequate. Increasing the scanning speed of the (typically rather soft) specimen introduces motion artifacts into the acquired images, but more importantly, also has the potential to perturb the physical state of the specimen, particularly in long-term imaging applications. In order to achieve a higher temporal resolution, instead of translating the specimen for depth-sectioning, both the detection focus and the light-sheet can be synchronously translated across a stationary specimen. This scheme typically utilizes a combination of high-speed piezoelectric scanners to translate the detection objectives and galvanometric mirrors to synchronously position the light-sheet coplanar with the detection focal plane (Ahrens et al., 2013). A physical rotation of the specimen can then facilitate the acquisition of multi-view image data and subsequent reconstruction of images with isotropic resolution, as was demonstrated by imaging *Drosophila* and zebrafish embryos (Swoger et al., 2007). However, due to the inherent delay in rotating the specimen in multiple orientations, some of the faster cellular dynamics can potentially still be missed or temporally under-sampled, thereby making it difficult to accurately register information from multiple views during post-processing. The limitation of inadequate temporal resolution in multi-view imaging of large specimens at a high spatial resolution was recently addressed using isotropic multi-view light-sheet microscopy (IsoView) (Chhetri et al., 2015). IsoView addresses the issue of attaining sub-cellular, isotropic resolution while retaining high temporal resolution in light-sheet imaging of large, non-transparent specimens (up to 800 × 800 × 800 μm^3 in size). IsoView utilizes an orthogonal arrangement of four shared illumination and detection objectives (Figure 2.1). As such, the lateral and axial dimensions, along which resolution are high and low, respectively, are permuted for orthogonal imaging arms. Thus, the registration of the image content from the four views and subsequent multi-view deconvolution results in near-isotropic volumetric data, with a system resolution of 400–450 nm in all spatial dimensions. IsoView enables fast-volumetric imaging via three unique modes of imaging: (1) sequential imaging of orthogonal views; (2) simultaneous imaging of orthogonal views using non-overlapping emission spectra; and (3) simultaneous four-view imaging by spatially offsetting the orthogonal light-sheet scans in the vertical direction and matching the active row of pixels in the respective sCMOS detectors operated in confocal line-scanning mode. To demonstrate the utility of IsoView for fast developmental imaging, simultaneous two-color imaging was performed in a gastrulating *Drosophila* embryo at a volumetric rate of 0.25 Hz (75 planes/volume for each view, acquiring eight views in total), which offered not only high enough temporal resolution to capture key events during gastrulation but also isotropic, μm-level resolution for reliably distinguishing neighboring cells and morphological features across the entire embryo.

2.3 FUNCTIONAL IMAGING OF THE NERVOUS SYSTEM USING LIGHT-SHEET MICROSCOPY

Recently engineered genetically encoded calcium indicators (GECIs), such as GCaMPs, RCaMPs, etc., that change fluorescence levels depending on calcium concentration allow neuronal activity to be measured optically (Looger and Griesbeck, 2012). Unlike whole-cell patch clamp electrophysiology, which constitutes a direct measurement of neuronal activity, calcium imaging is an indirect measure from which neuronal spiking activity can be extracted. The basic working principle in calcium imaging is the influx of calcium during an action potential of a neuron followed by a quick restoration of the resting potential as calcium is returned to the extracellular space and the compartments in the endoplasmic reticulum. This brief surge of calcium, which is reported by GECIs, thus represents an indirect activity measure for a firing neuron. Although whole-cell patch clamping is still considered the gold standard in measuring neuronal activity, it is often not ideally suited to recapitulate how different parts of a neuron interact or how assemblies of neurons communicate. Recording neuronal activity using optical methods overcomes these shortcomings and offers several advantages. First, imaging neuronal activity is less invasive than inserting electrodes directly onto a cell, which can irreversibly damage the cell. Second, optical imaging facilitates gathering data from a much larger population of neurons, thereby enabling correlative analysis of activities across different neuronal sub-populations and brain regions. Lastly, optical methods also provide anatomical information and enable pairing of neuronal activity with cell identity when neurons expressing calcium sensors are co-labeled for cell types. Although various optical techniques (Wilt et al., 2009) exist for recording from multiple neurons, the number of simultaneously recorded neurons in the brain of the animal is typically small. This constraint primarily arises from limitations in the speed at which volumetric data can be collected across the brain of an animal.

Recording the activity from a population of neurons demands much higher acquisition speeds than the majority of dynamical processes encountered in developmental imaging. Various high-speed optical imaging modalities, such as two-photon point-scanning microscopy, two-photon random access microscopy, and light-field microscopy among others, aim to meet the demands of high spatiotemporal resolution in neuronal imaging, and each offer distinct advantages and utility. Two-photon microscopy, which confines fluorescence excitation to the illumination focus, excels at imaging inside scattering tissues and captures images with high spatial resolution at depths inaccessible to conventional single-photon techniques. However, two-photon point scanning microscopy requires the excitation spot to be sampled sequentially across the entire imaging volume, which limits the overall volumetric acquisition speed. In sparsely labeled samples, the speed bottleneck of two-photon imaging is overcome by two-photon random access microscopy, which samples a limited number of spots distributed across a large sample volume, thereby increasing the temporal rate at which the regions of interest are sampled (Grewe et al., 2010; Sofroniew et al., 2016). Light field microscopy, in which the spatial and angular distributions of the emitted fluorescent light are captured simultaneously using a microlens array positioned conjugate to the image plane, acquires volumetric information from the entire sample volume onto a single camera chip and thus offers exceptionally high volumetric acquisition speed (Levoy et al., 2006; Prevedel et al., 2014). However, the compression of the entire light-field onto a single two-dimensional sensor reduces the spatial resolution across the acquired volume, which often constrains the interrogation of neuronal activities to a local group of neurons instead of single neurons. Compared to these complementary high-speed neuroimaging techniques, light-sheet microscopy offers a unique approach for large-scale *in vivo* recording of neural activity in transparent and small non-transparent samples at high spatiotemporal resolution and over an extended period of time.

Below we discuss various recent implementations of light-sheet methods in imaging neuronal activity in zebrafish, *Drosophila*, and mammalian brains.

2.3.1 Light-Sheet Functional Imaging in Zebrafish

The larval zebrafish is an excellent model organism for studies in developmental biology and neuroscience. Owing to its compact size and transparency, zebrafish naturally lend themselves to optical imaging and have thus been used for numerous imaging-based studies in developmental biology and neuroscience. The suitability of zebrafish for light-sheet imaging combined with the availability of GECIs in larval zebrafish have made it possible to image activity in large population of neurons across the zebrafish brain. Panier et al. and Ahrens et al. first demonstrated the utility of light-sheet microscopy for functional imaging in larval zebrafish, and captured single-neuron level functional activity from thousands of neurons in the zebrafish brain (Ahrens et al., 2013; Panier et al., 2013) (Figure 2.2a–f). Panier et al. used GCaMP3 to label the neurons in the brain of 5–9 dpf larval zebrafish, and performed light-sheet imaging by paralyzing the zebrafish and embedding them in 1.8% agarose. In this study, the authors demonstrated functional recording from about 30% of the neurons across the brain by sequentially sampling five image planes (with a separation of 8 μm between planes) at a volumetric rate of 4 Hz. A functional recording from a single z-plane at a repetition rate of 10–20 Hz was also demonstrated for a duration of 30–60 minutes. Although one-third of the entire brain was imaged in this study, volumetric imaging was performed by rapidly moving the sample (followed by a short pause after each movement to allow the specimen to come to rest) and thus constituted a limiting factor in attaining a finer depth sectioning across the entire brain while maintaining the same high volumetric imaging rate. Ahrens et al. instead employed piezo-based volumetric imaging, which allows keeping the specimen stationary and moving the light-sheet and detection objective in synchrony. In this latter study, the authors recorded activity from the entire volume of the brain of a GCaMP5G labeled larval zebrafish at 0.8 Hz (41 z-planes/volume with the planes 5 μm apart), covering more than 80% of all neurons at single-cell resolution. These imaging experiments captured for the first time single-neuron level functional activity across almost the entire brain volume of 800 × 600 × 200 μm^3 in an intact, live vertebrate. Thereby, this study demonstrated a key advantage of light-sheet microscopy in interrogating neuronal activity patterns in disparate brain regions and subsequently allowing characterization of correlations across the entire brain. These computational analyses

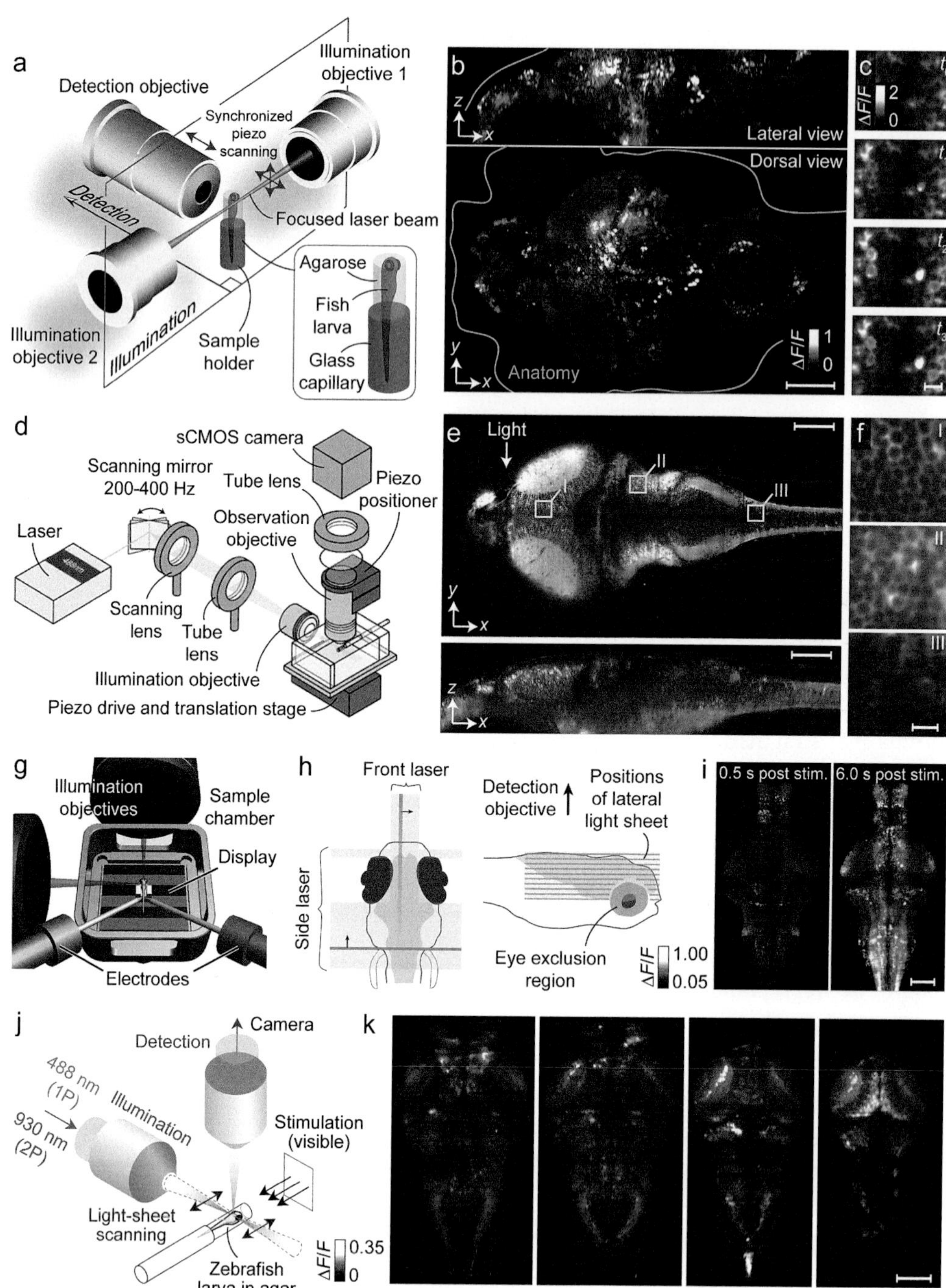

FIGURE 2.2 Light-sheet microscopy techniques for zebrafish whole-brain functional imaging. Light-sheet microscopy techniques for single-view functional imaging of the zebrafish larval brain by Ahrens et al. (2013) (a–c), Panier et al. (2013) (d–f), Vladimirov et al. (2014) (g–i), and Wolf et al. (2015) (j, k). (a) Whole-brain, neuron-level functional imaging in larval zebrafish using laser-scanning light-sheet microscopy (Keller et al., 2008). The zebrafish is embedded in agarose gel and positioned in front of the detection lens. The 4 μm-thick light-sheet is generated by fast vertical scanning of a laser beam focused inside the fish by one or both illumination objectives

FIGURE 2.2 (CONTINUED)
oriented orthogonal to the detection objective. Fluorescence is recorded with a fast sCMOS camera. Volumetric imaging is performed by scanning the light-sheet across the sample and moving the detection objective so that the light-sheet always coincides with the focal plane (Ahrens et al., 2013). (b) Whole-brain, neuron-level activity, reported by a genetically encoded calcium indicator in a week-old Tg(elavl3:GCaMP5G) fish. Panels show lateral and dorsal projections of changes in fluorescence intensity ($\Delta F/F$) at one point in time. Whole-brain volumes were recorded in intervals of 1.39 s. (c) Activity at the single-neuron level in a sub-region of a single slice from the volume visualized in (b). Activity is shown superimposed on the anatomy. Slices represent activity recorded at consecutive intervals of 1.39 s. (d) Light-sheet microscope design implemented by Panier et al. (2013) for zebrafish whole-brain functional imaging. (e) Dorsal and lateral views of the brain-volume reconstructed from a complete 3D stack. (f) Enlarged views of the three different sub-regions highlighted by green rectangles in (e). (g) Light-sheet microscope design by Vladimirov et al. (2014) for whole-brain functional light-sheet imaging of fictively behaving zebrafish. A larval zebrafish receives visual input from a display underneath, and intended motor output is recorded electrically from the tail. To avoid exposing the eyes to direct laser light, two laser beams are utilized, one scanning from the side but skipping over the eyes, and the second scanning the tissue between the eyes from the front. The detection objective, located above the chamber, is not shown. (h) Schematic of the laser-scanning strategy in the microscope implementation by Vladimirov et al. (2014). (i) Whole-brain functional imaging during behavior in a Tg(elavl3:GCaMP6s) fish. The panels show dorsal maximum-intensity projections of $\Delta F/F$ over the entire volume during stimulus presentation (averaged over 24 trials), superimposed on the anatomical map (gray). Cells and neuropil distributed across the brain show activity during the optomotor response. (j) Light-sheet microscope implementation by Wolf et al. (2015) for two-photon light-sheet functional imaging of visually evoked neural activity. (k) Nine images were acquired at 1 Hz across a 5-day-old larva brain (90 ms exposure per image), of which four planes are shown here. On the sections shown, the average neural response ($\Delta F/F$), measured in the first second following a 3,600 μW cm^{-2} flash, is color coded (120 flashes, 10 s intervals between flashes). (Panels and captions (a–c) were adapted with permission from Ahrens et al. (2013), (d–f) from Panier et al. (2013), (g–i) from Vladimirov et al. (2014) and (j, k) from Wolf et al. (2015).) Scale bars, 100 μm (b, e, i, k), 10 μm (c, f).

revealed two functionally defined neuronal circuits, termed the hindbrain oscillator and the hindbrain spinal circuit.

For studies relying on visual stimulus to generate behavior, the experimental setups in Panier et al. (2013) and Ahrens et al. (2013) leave the retinas of the zebrafish exposed to the excitation light, which can disrupt behavior and neural processing. To circumvent this problem, Vladimirov et al. presented a scheme that avoids scanning the laser beam across the eyes of the zebrafish and thus prevents direct stimulation of the photoreceptors by the imaging laser (Vladimirov et al., 2014) (Figure 2.2g–i). The authors were then able to present a visual stimulus via a projection screen in the larval zebrafish's field of view and simultaneously monitor the cellular-level neural activity during visual-motor behavior using the expression of cytoplasmic GCaMP6s. Near-complete coverage of the entire brain in the fictively behaving larval zebrafish was obtained at a high volumetric rate of 3 Hz (40 z-planes/volume with the planes separated by 5 μm) in this study. The authors successfully captured forward optomotor responses initiated by the movement of the grating pattern projected onto the screen, and also tested the adaptation of the zebrafish's motor output in response to the strength of the visual signal. An alternative light-sheet imaging approach for avoiding the activation of the retinas and other photosensitive cells was presented by Wolf et al. (2015) (Figure 2.2j,k). This approach relies on two-photon excitation, using a light-sheet at a wavelength of 930 nm in order to significantly reduce the response of the fish's visual system to light-sheet illumination. Visual stimulation was presented in the form of a series of blue flashes of increasing intensity, and the flash-evoked neuronal activity in the brain of a zebrafish larva expressing GCaMP5G was recorded at a volumetric rate of 1 Hz (9 z-planes/volume with the planes 8 μm apart). Additionally, the authors presented a direct comparison of flash-evoked response when imaging is carried out using single-photon versus two-photon excitation. In order to obtain similar contrast and signal-to-noise ratios, average laser powers on the order of several 100 mW and several 100 μW were needed for two-photon excitation and one-photon excitation, respectively. A 4 Hz single-plane recording showed that the flashes evoked stronger responses in various regions of the brain during two-photon imaging compared to single-photon imaging, which establishes two-photon light-sheet microscopy as a useful alternative for functional imaging assays that require minimizing the possible impact of the laser illumination beam on neural processing.

Alternatives to high-speed piezo scanning for volumetric imaging have also been explored for light-sheet microscopy (Tomer et al., 2015; Quirin et al., 2016) (Figure 2.3). These approaches involve extending the depth of field of the detection optics (Zalevsky and Ben-Yaish, 2007; Mouroulis, 2008) and attaining depth-sectioning by simply scanning the light-sheet across the sample. These approaches thus leave the local vicinity of the specimen motion-free, as the detection objective remains stationary, and also offer a way to image from a select few arbitrary regions within the imaging volume at a high repetition rate (Quirin et al., 2016). Similar to piezo-based high-speed light-sheet microscopes, the speed in these approaches are also currently limited by the frame rate of the detector, the pixel dwell time required to collect images at a reasonable SNR, and most importantly, the photon budget and light tolerance of the specimen. One such approach applied to neuronal imaging in larval zebrafish is SPED (Spherical-aberration-assisted Extended Depth-of-field light-sheet microscopy) (Tomer et al., 2015). In this method, the depth of field is extended by inserting a thick block of optical material with an altered refractive index ($n = 1.454$) between the detection objective and the sample, which in effect introduces a large spherical aberration and elongates the detection point-spread-function (PSF) (Figure 2.3c,d). In this manner, as the light-sheet is scanned across the sample, an image volume is captured with the stationary detection objective. At 4× magnification, the authors demonstrated 12 Hz imaging (40 z-planes/volume with 5 μm separation between adjacent planes) of the brain and 6.23 Hz imaging (39 z-planes/volume with 5 μm separation between adjacent planes) of the central nervous system (CNS) including the spinal cord in GCaMP6s-expressing zebrafish larva. The authors also demonstrated 4.14 Hz imaging (39 z-planes/volume with 5 μm separation between adjacent planes) of the brain and spinal cord at a slightly higher magnification of 10×. It should be noted that, unlike other light-sheet microscopy techniques, SPED critically relies on detection objectives with low numerical apertures (NA of 0.25–0.28), since the z-range of the

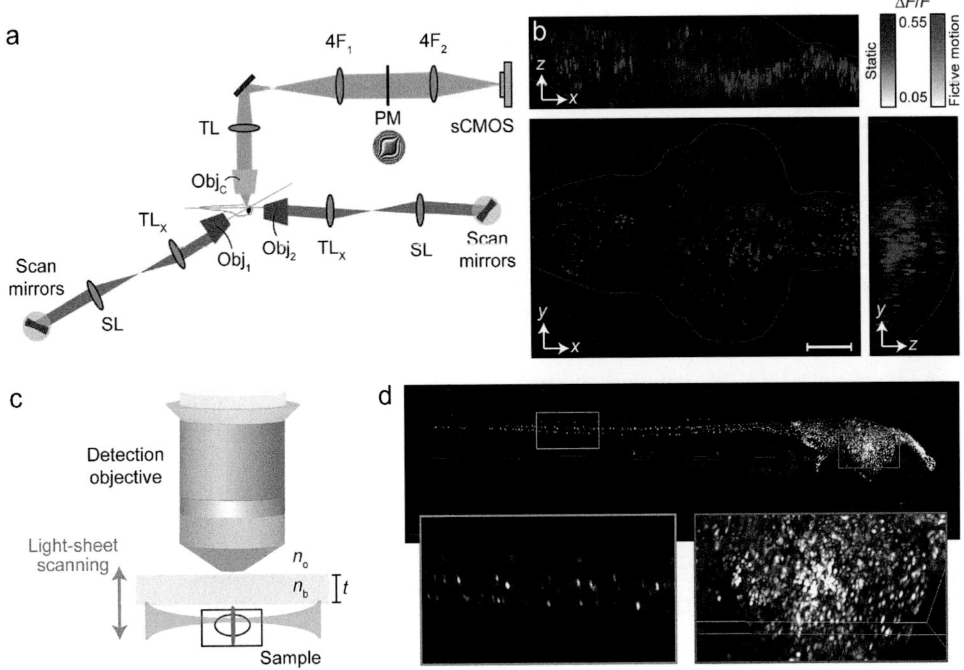

FIGURE 2.3 Zebrafish functional imaging with axially elongated detection point-spread functions. Light-sheet microscopy techniques with axially elongated detection point-spread functions by Quirin et al. (2016) (a, b) and Tomer et al. (2015) (c, d). (a) Schematic of the excitation and imaging path in the technique by Quirin et al. (2016) using cubic phase pupil encoding. (b) Demonstration of functional imaging capabilities after data restoration. The images show maximum-intensity projections conveying the spatial specificity of neurons responding to visual motion. (c) Schematic of the imaging path in the SPED technique by Tomer et al. (2015). A block of higher (or lower) refractive index (n_b) material is placed between the objective and the sample to induce spherical aberrations that elongate the PSF. The thickness of the block is denoted as t. (d) Volume renderings of 10 dpf Tg(elavl3:H2B-GCaMP6s) zebrafish larvae imaged with a 4×/0.28NA objective. Cyan and magenta boxes provide magnified views. Image volumes of ten consecutive time points were collapsed into one volume by taking the maximum values voxel-wise across the recording duration. The bounding box size is 0.75 mm × 2.99 mm × 0.48 mm. (Panels and captions (a, b) were adapted with permission from Quirin et al. (2016), (c, d) from Tomer et al. (2015).) Scale bar 100 μm (b).

SPED imaging volume depends fundamentally on the free working distance and numerical aperture of the objective. Since the photon efficiency of the imaging system is proportional to the square of the numerical aperture, SPED imaging thus taxes the photon budget, which in turn limits spatiotemporal resolution and increases phototoxicity and photodamage. The microscope's lateral resolution, which is proportional to the numerical aperture of the detection objective, is degraded for the same reason. Extended depth-of-field imaging furthermore reduces axial resolution (from around 2 μm in conventional functional imaging experiments to around 5 μm in SPED), since axial resolution is only determined by the illumination PSF in SPED. Another extended depth-of-field approach applied to neuronal imaging in larval zebrafish involves the use of a cubic phase mask

(Dowski and Cathey, 1995) in the detection path of a light-sheet microscope (Quirin et al., 2016) (Figure 2.3a,b). In this method, the wavefront is encoded with a spatially dependent phase retardation, and a high-resolution volumetric data set is recovered via deconvolution using the PSFs that result from system modulation with the cubic phase mask. Unlike SPED, in which the induced spherical aberrations become non-linear for high-NA objectives, the extension of the depth-of-field using a phase mask does not pose any limitation on the NA of the detection objective. As such, high-NA objectives (e.g. NA of 0.8 used in Quirin et al.) can be utilized, offering a higher photon efficiency and also collecting higher spatial frequencies from the sample that are otherwise missed in SPED (NA of 0.25–0.28 used in Tomer et al., 2015). Using the cubic phase mask aided light-sheet microscope, Quirin et al. (2016) imaged the brains of zebrafish larvae expressing GCaMP6s or GCaMP6f at a volumetric rate of 1.5 Hz (40 z-planes/volume with the planes 5 μm apart) in the presence of periodic visual stimuli. Owing to the ability to access arbitrary planes quickly, the authors were also able to image from three axial planes across an axial range of 160 μm at a repetition rate of 33 Hz.

Although the transparency of zebrafish is a great match for light-sheet microscopy, the relatively large size of these specimens often makes it challenging to maintain high resolution and high contrast particularly at larger imaging depths. Maintaining these factors at a high volumetric imaging rate further complicates this task. Recently, we presented an effective strategy for high-speed, isotropic multi-view imaging with the development of IsoView microscopy (Figure 2.4a). Using IsoView, four-view image data of the whole brain of a larval zebrafish expressing GCaMP6s was acquired at a volumetric rate of 1 Hz (67 planes/volume for each view, with 6 μm separation between adjacent planes), which captured large-scale brain activity and reliably resolved neighboring cells even in the deep regions of the brain. Figure 2.4b shows high-resolution images acquired using IsoView contrasted against images acquired using conventional light-sheet methods. A closer look at the deep regions outlined in Figure 2.4b shows that IsoView microscopy reliably distinguishes neighboring cells even in such deep regions (Figure 2.4c) and thus faithfully captures the activity traces of single neurons that frequently suffer from signal cross-talk between neighbors in conventional light-sheet imaging.

To obtain high-resolution images of living specimens at all spatial locations and over long periods of time (hours to days) adds another layer of difficulty, since the optical properties of complex multi-cellular organisms are not only heterogeneous across the specimen volume, but also change significantly over time. Consequently, the spatial overlap of light-sheets and detection focal planes, which is crucial for recording high-resolution images in light-sheet microscopy, suffers and thus spatial resolution and contrast in the images are degraded. Recovering and maintaining perfect coplanarity of light-sheets and the detection focal planes requires an imaging framework that automatically adapts to these changing imaging conditions. Such a framework for spatiotemporally adaptive imaging has recently been developed and released as the open-source *AutoPilot* project (Royer et al., 2016). This framework consists of (1) a multi-view light-sheet microscope capable of digitally adjusting

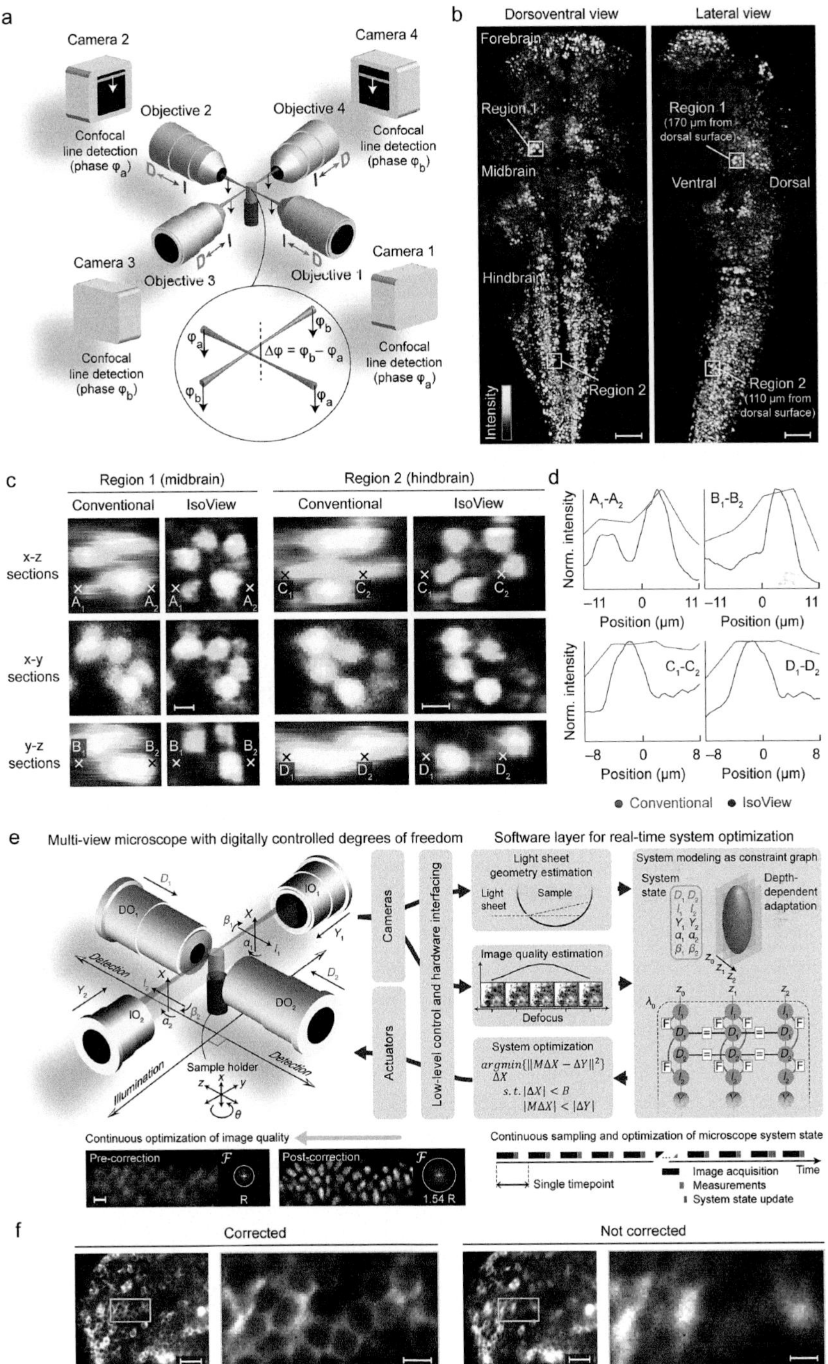

FIGURE 2.4 Light-sheet microscopy techniques for high-resolution zebrafish whole-brain functional imaging. Light-sheet microscopy techniques by Chhetri et al. (2015) (a–d) and Royer et al. (2016) (e, f) that enable whole-brain functional imaging with high temporal resolution and high spatial resolution. (a) Isotropic Multiview (IsoView) light-sheet microscopy by Chhetri et al. (2015). The IsoView microscope consists of four orthogonal arms for simultaneous light-sheet illumination

FIGURE 2.4 (CONTINUED)
and fluorescence detection. The specimen is located at the center of this arrangement. Volumetric imaging consists of sweeping light-sheets across the sample and translating detection planes with objective piezo positioners. In the primary mode of IsoView microscope operation, illumination and detection are performed at the same time in all arms, and cross-talk is avoided via spatially matched beam scanning and confocal detection using a phase offset in orthogonal arms (spatial separation). (b) IsoView whole-brain functional imaging in larval zebrafish. Dorsoventral and lateral projections of IsoView functional imaging data of the brain of a 3-day-old larval zebrafish expressing nuclear-localized GCaMP6 throughout its nervous system. (c) Side-by-side comparison of conventional image data (acquired using single-view functional imaging, see Figure 2.2) and multi-view deconvolved IsoView image data for x–y, x–z, and y–z slices from the two deep-tissue image regions highlighted in (b). Optical path lengths inside the brain relative to lateral and dorsal surfaces were 140 μm and 170 μm for region 1 and 80 μm and 110 μm for region 2. (d) Side-by-side comparison of intensity profiles in conventional image data and multi-view deconvolved IsoView image data for the four linear segments (A1–A2, B1–B2, C1–C2, and D1–D2) indicated in (c). Note that FWHM size measurement results do not directly quantify local resolution, as the results have not been corrected for the finite physical size of the analyzed structures (cell somas typically have a diameter of at least 3 μm). (e) Spatiotemporally adaptive light-sheet microscopy by Royer et al. (2016). Overview of the fully automated light-sheet microscopy framework for spatiotemporally adaptive imaging, which consists of (1) a multi-view light-sheet microscope with ten digitally adjustable degrees-of-freedom that control 3D offsets and 3D angles between light-sheets and detection focal planes, and (2) a real-time software layer that autonomously monitors image quality throughout the imaging volume and automatically and continuously adjusts these degrees of freedom to optimize spatial resolution and image quality across the sample in space and time. (f) Spatiotemporally adaptive whole-brain functional imaging in larval zebrafish. (b) Side-by-side comparison of image quality and spatial resolution in adaptively corrected and uncorrected image data of a representative midbrain region after 11 hours of whole-brain functional imaging in a 4-day-old Tg(elav3:GCaMP6f) zebrafish larva. Enlarged views of the image regions marked by orange boxes are shown next to each overview image. Note that non-adaptive imaging fails to resolve individual cell identities, whereas adaptive imaging recovers and maintains single-cell resolution. (Panels and captions (a–d) were adapted with permission from Chhetri et al. (2015), (e, f) from Royer et al. (2016).) Scale bars, 50 μm (b), 5 μm (c, e; f, enlarged views), 20 μm (f, overview images).

the positions of detection planes as well as the positions and angles of light-sheets in three dimensions; and (2) a software control layer that monitors and continuously optimizes image quality across the specimen volume in real-time (Figure 2.4e). Using this spatiotemporally adaptive light-sheet microscope, whole-brain functional imaging in zebrafish larva expressing nuclear-localized GCaMP6f across the nervous system has been demonstrated at a volumetric rate of 3 Hz (41 planes/volume with 5 μm separation between adjacent planes) over a period of 24 hours. Spatial resolution and contrast in the adaptively corrected images are substantially improved compared to the uncorrected image data (Figure 2.4f): the *AutoPilot* framework maintains high image quality even across a sample volume as large as the zebrafish larval brain and recovers cellular and sub-cellular structures in many anatomical regions that are not resolved by conventional, non-adaptive light-sheet microscopy.

2.3.2 Light-Sheet Functional Imaging in *Drosophila*

The fruit fly *Drosophila melanogaster* is a popular model organism in neuroscience due to its small size, genetic tractability, the availability of a powerful arsenal of genetic tools, its remarkable repertoire of complex behaviors, and its importance as a model system for many neurodegenerative diseases. Light-sheet microscopy has recently expanded the utility of fluorescence imaging in *Drosophila* embryos by parallelizing the acquisition of multi-view image data in long-term live imaging experiments of *Drosophila* developmental dynamics. Unlike zebrafish, *Drosophila* embryos and larvae are partially opaque and exhibit higher cell densities in many parts of their central nervous system, which presents a challenge for volumetric imaging at high resolution despite their relatively smaller size. Building upon SiMView light-sheet microscopy (Tomer et al., 2012), Lemon et al. designed a high-speed microscopy framework for multi-view functional imaging (hs-SiMView) with one-photon and two-photon excitation, and successfully demonstrated functional imaging of neural activity across entire nervous system explants of third-instar *Drosophila* larvae (approximately $500 \times 200 \times 200\ \mu m^3$ in size), utilizing two opposing sCMOS cameras for streaming multi-view volumetric image data at a sustained data rate of 1 GB/s (Lemon et al., 2015). This method involved simultaneously imaging from two opposing views to overcome the limited physical coverage achieved with single-view imaging in a highly light-scattering sample such as the CNS of *Drosophila* (Figure 2.5a). To attain rapid optical sectioning, the light-sheets were swept using galvanometer scanners and the detection focal planes were synchronously moved across a stationary specimen using high-speed piezo positioners. Using one-photon excitation, the authors demonstrated 5 Hz (37 z-planes/camera with 4–6 μm separation between adjacent planes) volumetric imaging for a 1-hour period in *Drosophila* third instar larval CNS explants expressing GCaMP6s. Using two-photon excitation (940 nm) volumetric imaging was demonstrated at a volume rate of 2 Hz for a 1-hour period. The two-photon assay thus complements the faster 5 Hz single-photon recording capabilities (488 nm) with an imaging mode that sacrifices temporal resolution but improves depth penetration and signal-to-background ratio. This work constitutes a first demonstration of functional imaging of neuronal activity at near cellular resolution throughout the entire CNS of a higher invertebrate (Figure 2.5c,d). A further improvement in resolution and isotropy was achieved with the development of IsoView microscopy (Chhetri et al., 2015). Using IsoView, we recently performed whole-animal functional imaging of late-stage *Drosophila* embryos expressing GCaMP6s throughout the nervous system at a volumetric rate of 2 Hz (40 planes/volume with 5.2 μm separation between planes; 4 views/volume). In these experiments, we recorded whole-nervous-system activity patterns associated with various motor behaviors such as forward crawling, backward crawling, and turning. The high, isotropic resolution provided by IsoView was evident throughout the nervous system, as axon bundles, neighboring cell somas, and in some instances even individual axons were distinctly resolved, even during fast specimen movements. Figure 2.5e shows dorsal- and lateral-view projections of an embryo imaged with IsoView. A closer look at a deep region in the ventral nerve cord (Figure 2.5f) demonstrates the high-resolution of IsoView image data in all three dimensions and shows that neighboring soma with a diameter of 2–3 μm are resolved as separate structures. IsoView also

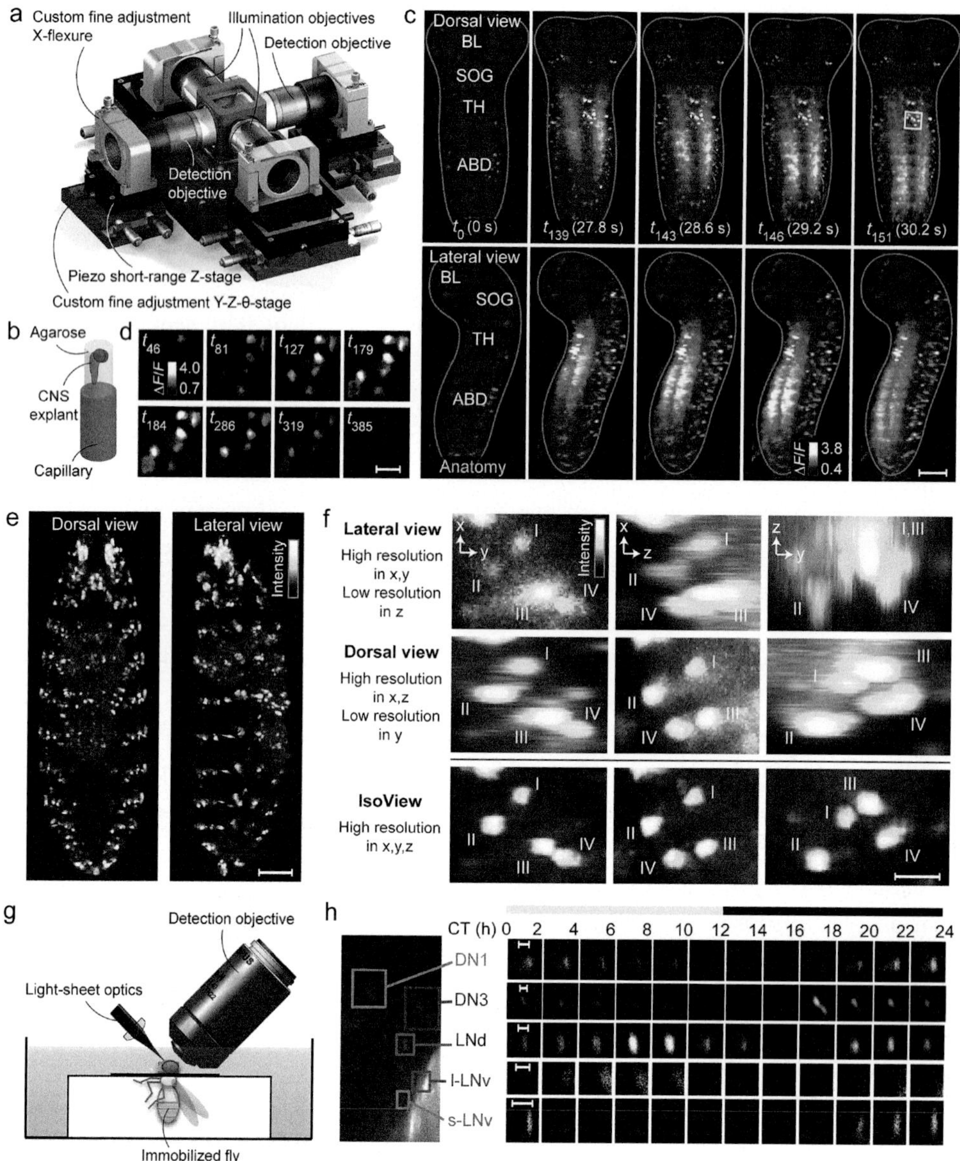

FIGURE 2.5 Light-sheet microscopy techniques for *Drosophila* whole-CNS functional imaging. Light-sheet microscopy techniques for functional imaging of the *Drosophila* embryonic, larval and adult nervous system by Lemon et al. (2015) (a–d), Chhetri et al. (2015) (e, f) and Liang et al. (2016) (g, h). (a) hs-SiMView light-sheet microscope design by Lemon et al. (2015) for whole-CNS functional imaging in strongly light-scattering *Drosophila* CNS explants. The illustration shows the hs-SiMView microscope core for functional imaging, including the central specimen chamber, two illumination objectives for bi-directional fluorescence excitation with scanned laser light-sheets, and two opposing detection objectives mounted on high-speed piezo stages. The 3D volumes covered by the two piezo-operated detection objectives are matched with a precision of a few micrometers using custom Y-Z-theta fine adjustment stages and objective X-flexures. (b) For optimal optical access, the CNS of a *Drosophila* third instar larva is extracted by surgery and embedded in a soft, transparent agarose cylinder supported by a glass capillary for mounting in the hs-SiMView

FIGURE 2.5 (CONTINUED)
light-sheet microscope. The CNS explant is then transferred to the microscope's specimen chamber filled with physiological saline. (c) Whole-CNS functional imaging at 5Hz of a *Drosophila* third instar larval CNS expressing 57C10-GAL4,UAS-GCaMP6s, using hs-SiMView light-sheet microscopy. Imaging was performed with one-photon excitation at 488 nm, maintaining a constant imaging speed of 370 frames per second (491 MB per second) for a period of 1 h. Image panels show maximum-intensity projections of $\Delta F/F$ (color look-up-table) and CNS anatomy (grey, gamma-corrected GCaMP6s baseline fluorescence) from dorsal (top) and lateral (bottom) views, for five time points during a backward locomotor sequence. Outline indicates CNS boundary. (d) Image sequence showing changes in $\Delta F/F$ for cell bodies in the ROI indicated by a white rectangle in panel (c). This example sequence demonstrates slow changes in $\Delta F/F$ across a bout of locomotor waves. Images are median filtered. ABD, abdomen; BL, brain lobes; SOG, suboesophageal ganglion; TH, thorax. (e) IsoView whole-animal functional imaging in embryonic *Drosophila* by Chhetri et al. (2015). (a) Dorsal- and lateral-view maximum-intensity projections of multi-view deconvolved IsoView image data of a stage 17 *Drosophila* embryo expressing GCaMP6s throughout the nervous system. The underlying four-view image data were recorded in 800 ms. A false-color look-up table is used for better visibility of high-dynamic-range images. (f) Improving resolution and isotropy by IsoView functional imaging. Side-by-side comparison of raw anisotropic image data (conventional single-view image data, top and middle rows) and multi-view deconvolved IsoView image data (bottom row) for an optical section of the ventral nerve cord of the specimen shown in (e), using the same false-color look-up table. Roman numerals identify locations of somas with high GCaMP6s fluorescence. (g) Illustration of light-sheet method for long-term *in vivo* imaging of the *Drosophila* adult brain. The head is immersed in saline while the body remains in an air-filled enclosure. (h) Ca^{2+} activity patterns in circadian pacemaker neurons *in vivo*. Left: A representative image of tim>GCaMP6s signals showing the locations of five identifiable pacemaker groups. Right: Representative images showing 24-hour Ca^{2+} activity patterns of five identifiable groups. (Panels and captions (a–d) were adapted with permission from Lemon et al. (2015), (e, f) from Chhetri et al. (2015) and (g, h) from Liang et al. 2016.) Scale bars, 50 μm (c, e), 10 μm (d, f), 20 μm (h).

offers long-term functional imaging capability, which was demonstrated by performing functional imaging of an entire *Drosophila* embryo over developmental timescales: development and functional maturation of the nervous system were captured from the onset of neuronal cell differentiation up to the first-instar larval stage at a volumetric rate of 2 Hz (35 planes/volume with 6.8 μm separation between planes; 4 views/volume) over a 9-hour period, at the end of which the fully formed larva crawled out of the imaging volume.

Functional *in vivo* imaging in adult *Drosophila* was also recently demonstrated using objective-coupled planar illumination (OCPI) microscopy (Liang et al., 2016) (Figure 2.5g). OCPI is a light-sheet microscopy method in which a miniaturized illumination arm is coupled directly to the detection objective, and the light-sheet is generated using a beam passing through a single-mode optical fiber, a light collimator, and a cylindrical lens (Holekamp et al., 2008). In this method, as the piezo-positioner moves the detection objective for depth-sectioning, the illumination sheet moves along with it as a result of the physical coupling, and thus offers a geometrically compact alternative to three-dimensional imaging without physically moving the sample or scanning the light-sheet. Additionally, to minimize optical path lengths in the sample, both the illumination and detection axes are tilted by approximately 45° with respect to a horizontally mounted sample. To perform OCPI imaging of the brains of living adult flies, the authors made cranial holes and monitored GCaMP6s fluorescence in five of the eight major pacemaker neurons over a duration of 24 hours. Using volumetric stacks obtained in 10-minute intervals for 24 hours, the authors were able to identify a systematic change in calcium dynamics in pacemaker neurons as a function of the time of day (Figure 2.5h).

2.3.3 Light-Sheet Functional Imaging in Mammalian Brains

A majority of functional imaging with light-sheet microscopy in the mammalian brain has thus far been performed on *ex vivo* tissues. Using OCPI (Figure 2.6a), Holekamp et al. demonstrated the utility of light-sheet microscopy in imaging regions of mammalian brains by capturing fast calcium dynamics from a population of neurons up to a depth of 150 μm from the surface of an excised tissue (Holekamp et al., 2008). This work not only represents the first application of light-sheet microscopy to live imaging of mammalian neural tissues but it also marks the first use of light-sheet microscopy for functional imaging in general. In this study, intact vomeronasal epithelium from male mice, labeled with the calcium-sensitive fluorescent dye Oregon green BAPTA-1, was excised and imaged using OCPI microscopy to study the pheromone-sensing neurons of the mouse vomeronasal organ (VNO) in response to chemical stimulation (Figure 2.6b). By scanning the plane of illumination in tandem with the detection focal plane at a volumetric rate of 0.167 Hz (40–50 z-planes/stack with 5 μm separation between the planes; each stack acquired in two seconds), the authors were able to simultaneously record the response of hundreds of VNO neurons to chemical stimulation for up to a few hours. Holekamp et al. (2008) furthermore monitored calcium dynamics in 88 cells within a single field of 700 × 100 μm^2 at a repetition rate of 200 Hz, capturing small changes (~0.2%) in fluorescence due to spontaneous activity of the neurons. Building upon this original demonstration, the investigators further visualized neuronal activity via the expression of GCaMP2 across the entire depth of the vomeronasal epithelium at a volumetric rate of 0.2 Hz (40 z-planes/stack with 5 μm separation between the planes; each stack acquired in two seconds), typically over a duration of ~1 hour (Turaga and Holy, 2012). In this case, the time-lapse image data of the 700 × 175 × 200 μm^3 tissue volume captured spontaneous and chemical-stimulus-driven activities from thousands of sensory neurons. In a recent study, even larger image volumes (713 × 712 × 400 μm^3) of the mouse accessory olfactory bulb (AOB) were acquired using OCPI at a volumetric rate of 0.2 Hz (50 z-planes/stack with 8 μm separation between the planes) (Hammen et al., 2014). These image stacks covered the entire anterior AOB and a third of the posterior AOB. To probe the connectivity of the VNO to the AOB, the authors delivered chemical stimuli to the VNO and recorded the subsequent response in the densely packed glomerular layer of the AOB using GCaMP2 for a duration of up to 1 hour and 40 minutes. To analyze the roles of neuronal cell types present in mice in relation to gender, hormones, and sensory experience, a large-scale recording from approximately 10,000 neurons expressing GCaMP2 or GCaMP3 in the VNO was recently demonstrated using OCPI (Xu et al., 2016). In this study, the authors imaged two intact VNO regions (710 × 125 × 282 μm^3), constituting about one-seventh of the entire VNO tissue volume,

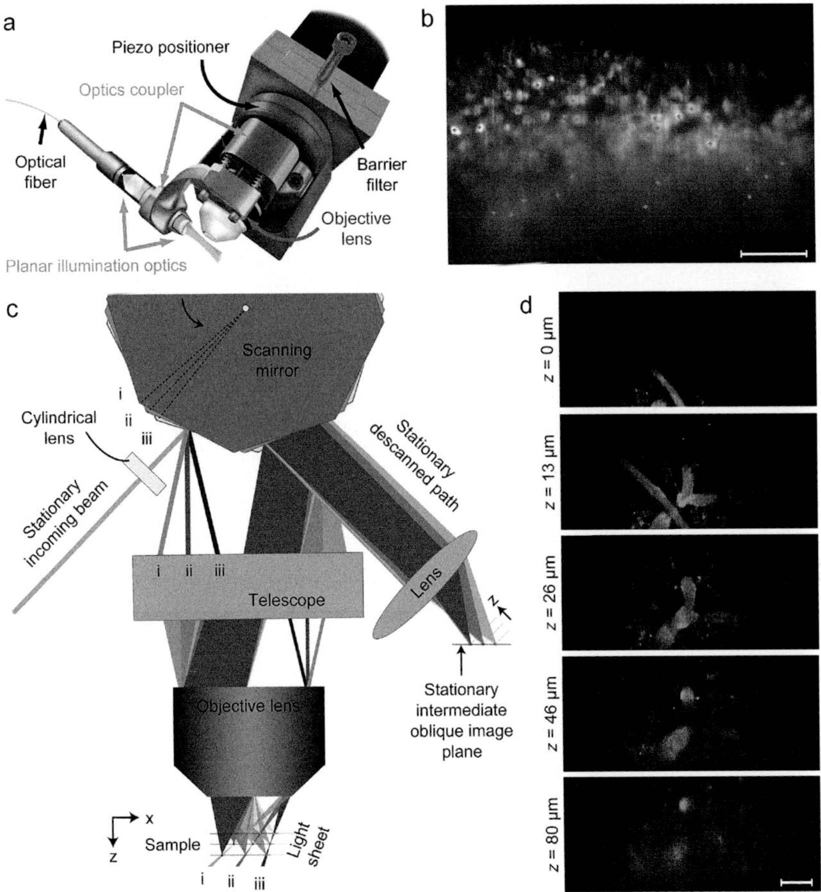

FIGURE 2.6 Light-sheet microscopy techniques for functional imaging of mammalian neural tissues. Light-sheet microscopy techniques by Holekamp et al. (2008) (a, b) and Bouchard et al. (2015) (c, d) for functional imaging of mammalian neural tissues. (a) Schematic of an Objective-Coupled Planar Illumination (OCPI) light-sheet microscope by Holekamp et al. (2008). Laser light for fluorescence excitation is provided by an optical fiber and is shaped into a light-sheet ~3–5 μm thick using two lenses. The light-sheet is coplanar with the focal plane of the detection objective. By coupling the illumination optics to the objective lens, the alignment of the light-sheet with the detection focal plane is maintained while scanning a sample volume using the piezoelectric positioner attached to the detection objective. Objective lens and illumination optics are designed for water immersion applications. (b) Probing of responses to chemical stimuli of single VNO neurons by high-speed 3D calcium imaging. The image shows a single optical section of an intact VN epithelium labeled with Oregon Green BAPTA-1. Purple dots indicate the positions of single neurons. (c) Illustration of SCAPE's scanning–descanning geometry by Bouchard et al. (2015), which sweeps an oblique light-sheet back and forth across the sample while the descanned detection plane remains stationary. The only moving component is the slowly oscillating polygonal scanning mirror. (d) SCAPE imaging in an awake, behaving mouse with intravascular Texas red dextran (red) and GCaMP6f in superficial dendrites from layer 5 neurons (green). The images show individual x′–y′ planes extracted from a single 350 × 800 × 105 μm (x′–y′–z′) SCAPE volume acquired in 0.1 s (each plane is an average of five sequential time points). Panels and captions (a, b) were adapted with permission from Holekamp et al. (2008) and (c, d) from Bouchard et al. (2015). Scale bars, 50 μm (b), 100 μm (d).

for a total of 26 VNO preparations, and identified 17 physiological types of vomeronasal neurons. These studies constitute an illustration of fast, comprehensive, *ex vivo* neuronal recordings of anatomically inaccessible regions of a mammalian brain, such as the VNO and the AOS, over an extended duration using light-sheet microscopy.

Light-sheet microscopy has the potential to be a mainstay for high-speed imaging in laboratories studying organotypic slices of mammalian brains, especially considering the extensive resources to build low-cost systems afforded by the openSPIM project (Pitrone et al., 2013). Recently, a compact inverted light-sheet microscope, based on the openSPIM design, was presented and successfully applied to the imaging of calcium dynamics in brain slices of rat pups (Yang et al., 2016). In this study, glutamate-uncaging-evoked calcium transients were captured in two-dimensional time-lapse images acquired at a frame rate of 200 Hz. Additionally, action-potential-evoked calcium influx along an axon in an organotypic rat hippocampal slice virally infected by GCaMP6s was also recorded as a two-dimensional time-lapse at a frame rate of 30 Hz. Contrasted with conventional methods for recording calcium transients in which a line-scan is performed over a predefined region, this method enabled calcium transients to be recorded across the entire two-dimensional area, thereby capturing events that would have otherwise been missed.

Unlike in zebrafish and *Drosophila*, the geometry of light-sheet microscopes with orthogonal illumination and detection objectives limits access to an intact mammalian brain for *in vivo* imaging. Thus, methods that can generate a tilted light-sheet and image fluorescence with a single objective are required to probe the neurons in a live mammalian brain. One such solution was recently introduced as a light-sheet design termed SCAPE (Swept, Confocally-Aligned Planar Excitation) microscopy (Bouchard et al., 2015). Unlike most other light-sheet microscopes, SCAPE utilizes a single objective for oblique light-sheet illumination as well as detection from the light-sheet plane (Figure 2.6c). Thus, a single, stationary objective configuration of SCAPE makes sample positioning and alignment as simple as with conventional upright or inverted microscopes. SCAPE microscopy implements a unique scanning and de-scanning scheme, such that a scanning polygonal mirror located behind the objective sweeps the light-sheet across the sample and an adjacent facet of the same polygonal mirror de-scans in the imaging path, thereby capturing images of the optically sectioned illumination planes without any physical translation of the objective or the sample. Using SCAPE, the authors imaged superficial cortical layers ($350 \times 800 \times 105\ \mu m^3$) in head-fixed, awake mice expressing GCaMP6f in layer 5 pyramidal neurons at a volumetric rate of 10 Hz (100 angular sampling steps in 350 μm), and demonstrated spontaneous calcium transients in superficial dendrites (Figure 2.6d). Additionally, a large volume ($600 \times 650 \times 134\ \mu m^3$) of the superficial cortex in a head-fixed, awake mouse expressing GCaMP5G in layer 5 pyramidal neurons was also imaged for a duration of 180 seconds at a volumetric rate of 10 Hz (240 angular sampling steps in 600 μm) using 2×2 camera binning. Upon subsequent analysis of the calcium onset and decay dynamics, the authors were able to reliably differentiate dendrites in the imaged volume based on their unique firing dynamics. These image data of the brain of awake mice illustrate the utility of SCAPE microscopy for rapid volumetric imaging in the superficial layers of the cortex. In the present implementation of SCAPE microscopy, the authors show that spatial resolution is worse than for two-photon microscopy, even in superficial regions of the mouse cortex, but they also demonstrate that SCAPE offers sufficient spatial resolution for meaningful analyses of neuronal dynamics at substantially higher sampling rates.

2.4 PERFORMANCE AND DESIGN CHOICES

Whole-brain functional imaging experiments stand out among other types of light-sheet microscopy experiments in particular with respect to their unique performance requirements. Calcium imaging of entire zebrafish larval brains, *Drosophila* larvae or mammalian neural tissues requires scanning relatively large volumes (typically ranging from 500 × 200 × 200 μm³ to 800 × 500 × 400 μm³ or more) at high volumetric imaging rates of at least 1 Hz, ensuring proper temporal sampling of the calcium indicator. At the same time, it is desirable – if not crucial – to achieve high spatial resolution in order to ensure that the functional activity of individual neurons can be reliably distinguished from the activity of their respective neighbors. When executing complex experimental workflows involving, e.g. behavior assays with multiple measurement conditions, long-term imaging capability may furthermore be essential for the successful acquisition of a meaningful dataset. Achieving this combination of high temporal resolution and high spatial resolution across a large imaging volume over extended periods of time poses technical challenges and requires careful consideration of the most appropriate microscope design choice in each experiment. Further difficulties can arise when performing behavioral assays that require avoiding exposure of the specimen to excitation laser light at visible wavelengths or when working with specimens that do not offer optical access from two orthogonal directions. In this section we will discuss the main performance criteria outlined above, the capabilities and limitations of different approaches, and the general performance implications of using certain types of light-sheet microscope designs. A performance comparison of various light-sheet microscopes discussed in this chapter is summarized in Table 2.1.

2.4.1 Temporal Resolution

The primary speed bottleneck in state-of-the-art light-sheet microscopes is camera performance. Irrespective of the volume imaging strategy used in a particular light-sheet microscope implementation, limitations in camera frame rate are the first obstacle that is encountered in the imaging workflow when striving to maximize data rates. Two-photon light-sheet microcopy currently represents the only exception to this rule; in this latter case, the primary limitation is the signal rate, as will be discussed further below.

For example, high-speed, high-resolution functional imaging of a zebrafish brain volume of 830 × 500 × 400 μm³ with IsoView light-sheet microscopy is limited to a volume rate of 2 Hz, since the corresponding data rate (340 million voxels per second) marks the performance limit of the microscope's Orca Flash 4.0 sCMOS cameras: considering an overhead of at least 1 ms per frame to accommodate the finite exposure time, the camera operates at a duty cycle of >90% in this scenario (Chhetri et al., 2015). Of course, it is possible to increase this volume rate beyond 2 Hz by sacrificing spatial resolution. This decision in turn allows using a smaller image frame size and thus a correspondingly higher frame rate. An example to this end is zebrafish whole-brain functional imaging with SPED

TABLE 2.1 Comparison of Functional Imaging Experiments Performed with Different Light-Sheet Microscopy Methods

	Imaging Method		hs-SiMView		IsoView		SPED	2P LSFM	SCAPE	OCPI
	Reference		Lemon et al., 2015		Chhetri et al., 2015		Tomer et al., 2015	Wolf et al., 2015	Bouchard et al., 2015	Holekamp et al., 2008
	Experiment		Whole-CNS Functional Imaging		Whole-Animal Functional Imaging	Whole-Brain Functional Imaging	Whole-Brain Functional Imaging	Functional Imaging of Brain Regions	Functional Imaging of Brain Regions	Functional Imaging of Brain Regions
	Model System		*Drosophila* Third-Instar Larval CNS Explant		*Drosophila* First-Instar Larva	Larval Zebrafish	Larval Zebrafish	Larval Zebrafish	Mouse	Mouse VNO Explant
	Imaging Geometry		Two Illumination Arms Two Detection Arms		Four Illumination Arms Four Detection Arms		One Illumination Arm One Detection Arm	One Illumination Arm One Detection Arm	Single Objective	One Illumination Arm One Detection Arm
	Fluorescence Excitation		One- Photon	Two-Photon	One-Photon		One-Photon	Two-Photon	One-Photon	One-Photon
Detection Optics	Detection Objective Manufacturer		Nikon		Special Optics		Olympus	Olympus	Olympus	Olympus
	Detection Magnification		16×		16×		4×	20×	20×	20×
	Detection Numerical Aperture		0.8		0.714		0.28	1.0	0.5	0.5
	Relative Photon Efficiency[1]		0.64 (×2)		0.51 (×2)		0.08	1	0.25	0.25
Resolution	Sampling-Limited System Resolution[2]	X [μm]	0.8		0.8		3.25	0.8	5.0	2.6
		Y [μm]	0.8		0.8		3.25	0.8	6.5	2.6
		Z [μm]	8.0		0.8		10.0	16.0	6.6	10.0
	Relative 3D Resolution[3]		10		1		206	20	419	132
	3D Resolution Anisotropy[4]		9.0		0		2.1	19.0	0.3	2.8
Volume	Image Size [pixels]		1,216 × 572	1472 × 700	1,504 × 704	2,043 × 1,231	554 × 246	2,048 × 2,048	240 × 200	1004 × 1002
	Volume Size	X [μm]	494	598	610	830	900	800	600	1,300
		Y [μm]	232	284	290	500	400	800	650	1,300
		Z [μm]	180	220	210	400	200	72	134	200
Speed	Acquisition Speed [10^6 voxels s^{-1}]		257	185	296	674	65	38	19	8
	Acquisition Speed[5] [volumes s^{-1}]		2 × 5	2 × 2	4 × 2	4 × 1	12	1	10	0.2

[1] The photon collection efficiency is proportional to the square of the numerical aperture (NA) and is normalized to the value obtained with the highest-NA (1.0) objective in this table. Note that absolute photon efficiency is affected by additional factors and depends on experiment-specific settings.

[2] Sampling-limited resolution considers not only the size of the point-spread-function but also the three-dimensional size of voxels in the image data. According to the Nyquist sampling theorem, effective resolution in each dimension cannot be better than two times the voxel size along the respective dimension. The long axial step size between image planes in hs-SiMView, SPED, 2P LSFM, SCAPE, and OCPI experiments is a key factor limiting effective axial resolution for these methods.

[3] Three-dimensional (3D) resolution is defined as the product of resolution values in x-, y-, and z-dimensions and is normalized to the value obtained for the highest-resolution technique (IsoView).

[4] Resolution anisotropy is defined as the ratio of sampling-limited resolution values along those dimensions exhibiting the lowest and highest spatial resolution, respectively, minus one. Thus, if sampling-limited resolution is identical in all dimensions, resolution anisotropy is zero.

[5] For microscopes that acquire multiple views of the full specimen volume simultaneously (hs-SiMView, IsoView), acquisition speed is listed as [number of views] × [volume acquisition rate].

microscopy, covering a $900 \times 400 \times 200\ \mu m^3$ volume at 12 Hz (Tomer et al., 2015). The 2.6-fold increase in volume rate in this example (note that the IsoView imaging volume is 2.3-fold larger) is not related to the different volume scanning approach used in SPED, but is rather the result of a reduction in spatial sampling and spatial resolution: the SPED experiment is performed at 4× magnification and acquires images of the brain with a frame size that is 18.5-fold smaller than the respective frame size in IsoView (which uses the same camera model but employs a higher, 16× magnification). If we also consider that raw axial resolution is 2-fold lower in SPED, it follows that the 3D size of voxels in the raw image data is 32-fold larger in SPED than in IsoView. Despite the higher volume rate, the data throughput in the SPED experiment is in fact only 65 million voxels per second, i.e. this SPED experiment provides 5-fold lower data throughput than, e.g. a single detection arm of the IsoView microscope.

It should be noted that there are some common misconceptions regarding the sources of limitations in volumetric imaging rates in light-sheet microscopy. In particular, the notion that imaging rates are currently not only limited by camera performance but also by piezo performance when using piezo-based volume scanning is incorrect. Even for the heaviest, highest-performing detection objectives (such as custom objective designs that combine high numerical apertures with large working distances, weighing in typically at 300–500 g), volume rates of 20 Hz are in principle feasible for a 250 μm travel range and 10 Hz for a 750 μm travel range. As outlined above, when acquiring well-sampled, high-resolution volumetric image data piezo performance thus surpasses the data throughput capabilities of existing sCMOS camera technology approximately 10-fold. Moreover, when using low-weight, low-NA objectives (such as those used in SPED), existing piezo technology offers even higher volume rates, ranging up to several hundred Hz for 200–900 μm travel ranges (e.g. Piezosystem Jena piezo models nanoX 200/400 and nanoSX 400/800).

Importantly, although alternative volume imaging approaches that avoid piezo scanning can be advantageous when it is necessary to minimize motion in the vicinity of the sample, these approaches also inherently suffer from some disadvantages. Possible alternatives to piezo scanning include the use of electric tunable lenses for remote focusing (Fahrbach et al., 2013) and swept oblique light-sheet imaging (Bouchard et al., 2015), as well as extended depth of field imaging mediated by spherical aberrations (Tomer et al., 2015) or pupil encoding with cubic phase masks (Quirin et al., 2016). Microscope designs using electric tunable lenses suffer from the fact that optical quality of tunable lenses is lower than that of conventional lenses and image quality is additionally reduced when acquiring images at a significant distance from the native focal plane of the detection objective. This latter issue is also encountered in microscope designs utilizing swept oblique light-sheet imaging and pupil encoding with cubic phase masks. Lateral and axial resolution in SCAPE are further decreased by an intrinsic 2-fold reduction in detection NA. Extended depth of field imaging mediated by spherical aberrations or pupil encoding with cubic phase masks also reduces axial resolution because it eliminates the contribution of the detection objective to the axial component of the microscope's PSF. Axial resolution is thus reduced 2–3-fold in SPED and in light-sheet imaging with cubic phase masks when compared to piezo-based volumetric imaging. SPED furthermore suffers from a reduction in lateral resolution and photon-efficiency as it is intrinsically limited to the use of low-magnification objectives with low numerical aperture. A typical SPED whole-brain functional imaging experiment with a 4×/0.28NA objective reduces sampling-limited lateral resolution 5-fold and photon-collection efficiency 13-fold compared to a conventional piezo-based whole-brain functional imaging experiment with a high-quality 20×/1.0NA objective (Ahrens et al., 2013). Finally, it should be noted that with the advent of more advanced camera technology the bottleneck in future generations of light-sheet microscopy will shift from maximum camera speed to the maximum achievable fluorescence signal rate under physiological conditions. As volumetric imaging rates increase, correspondingly higher laser power densities are needed to maintain the same signal-to-noise ratio under otherwise identical imaging conditions (i.e. assuming that spatial resolution and spatial sampling are maintained as well). The speed of image acquisition will thus ultimately be limited by the saturation of fluorophores and the maximum light dose the biological specimen tolerates.

2.4.2 Spatial Resolution

Performance with respect to spatial resolution varies substantially over the current spectrum of light-sheet microscopy techniques for functional imaging. Of course, it is generally desirable to achieve high resolution in any study that requires faithful measurement of neuronal activity at the single-cell level. Achieving this goal, however, is complicated by the need to image a relatively large field-of-view in typical whole-brain functional imaging studies. At the core of this issue is the fact that resolution is highly anisotropic in conventional light-sheet microscopy for large field-of-view imaging, i.e. axial resolution is often 5- to 10-fold worse than lateral resolution (Ahrens et al., 2013; Panier et al., 2013; Lemon et al., 2015; Tomer et al., 2015; Wolf et al., 2015; Quirin et al., 2016). There are currently only a handful of techniques that overcome this particular limitation and offer high lateral as well as high axial resolution. Unfortunately, most of these techniques rely on principles that increase spatial resolution at the expense of temporal resolution, which makes them intrinsically unsuited to large-volume functional imaging. For example, Bessel beam and lattice light-sheet microscopy (Planchon et al., 2011; Chen et al., 2014) offer excellent spatial resolution on the order of 300 nm; however, these techniques achieve high resolution by constructing very thin light-sheets over a short field-of-view (typically with a length of around 100 μm). Covering a zebrafish larval brain with a lateral cross-section of 800 × 600 μm^2 would thus require extensive tiling, since it is physically not possible to create such thin sheets of light over larger field-of-views with a reasonable illumination power density. Even if a different, yet-to-be-identified physical approach would enable the construction of light-sheets with such properties in the future, such microscopes would still intrinsically be too slow for most functional imaging applications: imaging a 200 μm deep brain volume with a 300 nm thick light-sheet would require acquiring in the order of 1,000 images to achieve sufficiently high z-sampling, i.e. ~10-fold more images than are needed for conventional illumination strategies using thicker light-sheets. In order to overcome these limitations and provide high spatial resolution and high temporal resolution at the same time, the multi-view imaging techniques diSPIM and IsoView have been developed (Wu et al., 2013; Chhetri et al., 2015). As discussed earlier, IsoView microscopy offers an imaging mode that is particularly well-suited to functional imaging applications: this mode relies on the use of thick light-sheets to maximize temporal resolution while performing simultaneous acquisition of four orthogonal views to maximize spatial resolution. This spatial-resolution enhancing mechanism does not sacrifice temporal resolution and is also not limited to small volumes.

Importantly, the increase in spatial resolution offered by techniques such as IsoView and diSPIM is critical for cellular resolution imaging of the whole brain of a zebrafish larva or an entire *Drosophila* larva. Conventional light-sheet microscopy utilizing piezo-based volumetric imaging already struggles with cellular resolution imaging in deep regions of the brain and thus suffers from a significant reduction in single-cell coverage (Ahrens et al., 2013). This situation is even worse with extended depth of field imaging techniques such as SPED which degrade not only lateral resolution but also suffer from a more than 2-fold decrease in axial resolution (Tomer et al., 2015). For an imaging technique with anisotropic PSF, this latter dimension is the most critical factor influencing the ability to perform cellular resolution imaging. A careful analysis of imaging speed and spatial resolution requirements is thus essential to deciding on the optimal imaging technique for a given experiment.

When imaging large, multi-cellular organisms, spatial resolution can be further improved by using complementary optimization strategies such as adaptive imaging and adaptive optics. These imaging strategies adapt the microscope's degrees of freedom to the optical properties of the specimen as a function of space and time, in order to compensate for aberrations and improve the overall geometry between light-sheet and detection focal plane. Adaptive light-sheet microscopy suitable for functional imaging already exists (Royer et al., 2016), using design principles that minimize the number of measurements needed to map out the specimen's optical properties and thus maximize the microscope bandwidth used for functional imaging.

2.4.3 Mode of Fluorescence Excitation and Imaging Geometry

Complementary to these performance considerations in the domain of temporal and spatial resolution, certain types of functional imaging experiments introduce additional technical requirements and constraints. For example, behavioral assays using visual stimulation may require the use or exclusion of certain wavelengths for whole-brain functional imaging, so as not to disrupt the behavior under investigation. This type of requirement can be effectively addressed, e.g. by using two-photon excitation. However, one needs to consider that signal rates in two-photon light-sheet microscopy are lower than in one-photon light-sheet microscopy, which impacts signal-to-noise ratio and imaging speed. A side-by-side comparison in *Drosophila* larval CNS explants with hs-SiMView light-sheet microscopy showed that a volume rate of 2 Hz marks the upper limit of two-photon functional imaging in this specimen when striving for image data with a signal-to-noise ratio suitable for automated, quantitative computational analyses (Lemon et al., 2015). By contrast, one-photon functional imaging still offered high image quality at volume rates of 5 Hz and was ultimately limited by camera speed rather than the signal rate. As discussed above, two-photon light-sheet imaging has also been very successfully demonstrated for the zebrafish larval brain (Wolf et al., 2015). Since the field-of-view is even larger in this scenario (which in turn reduces signal rate when using two-photon excitation at a given illumination power density), imaging speed was limited to 1 Hz for a subset of the total brain volume, enabling the acquisition of ten images per second at a 100 ms exposure time per frame (which is contrasted by the ~1 ms exposure time per frame usually used for whole-brain functional imaging with one-photon excitation). When using two-photon excitation it is furthermore important to avoid tissue regions exhibiting pigmentation, as high light absorption at infrared wavelengths in these regions typically results in severe photodamage of pigment-producing cells.

Finally, we note that the decision on the optimal imaging assay also depends on the level of optical access offered by the specimen. Although multi-objective light-sheet microscopes generally offer the highest image quality and highest spatial resolution (for the reasons discussed above), these optical geometries necessitate access to the specimen from at least two orthogonal directions. While this does not represent a practical constraint for most model systems, such as *C. elegans*, *D. melanogaster*, and zebrafish, it is a prohibitive limitation for *in vivo* imaging of the mouse brain. SCAPE microscopy, which facilitates illumination and detection via a single objective, offers an excellent solution for functional imaging in this model system (Bouchard et al., 2015).

2.5 PROCESSING AND ANALYSIS OF LIGHT-SHEET FUNCTIONAL IMAGING DATA

Light-sheet whole-brain or whole-CNS functional image datasets are typically relatively large and encode detailed information about complex dynamic processes across the nervous system. The sheer size and complexity of these images thus introduce substantial challenges for post-acquisition data management and data analysis. In this section, we will discuss the key computational steps involved in the analysis of functional image data acquired with state-of-the-art light-sheet microscopy (Figure 2.7).

A typical one-hour whole-brain functional recording at 4 Hz in larval zebrafish using SiMView light-sheet microscopy (Tomer et al., 2012; Ahrens et al., 2013; Lemon et al., 2015) comprises 14,400 image stacks with a size of 246 MB each (2048 × 1536 × 41 voxels, corresponding to a volume of 800 × 600 × 200 μm^3) or 3.4 TB in total. Typical data volumes are similar for other model systems and other light-sheet microscope implementations: a 4.5-hour whole-CNS functional recording at 2 Hz in embryonic/larval *Drosophila* using IsoView microscopy (Chhetri et al., 2015) consists of 32,400 time points comprising four image stacks each with a size of 84 MB per stack (four orthogonal views, 1728 × 728 × 35 voxels, corresponding to a volume of 700 × 300 × 170 μm^3) or 10.4 TB in total. In both examples, calcium dynamics are recorded simultaneously for a fairly large number of neurons – more than 10,000 in early *Drosophila* first instar larvae and approximately 100,000 across the zebrafish larval brain.

The first key requirement for datasets of this size is an appropriate data handling and storage solution, in particular if such experiments are performed on a daily basis. An effective hardware and software solution is not only important for long-term data storage but also for computational efficiency in all subsequent image processing and visualization steps. At the hardware level, image data are typically streamed locally to solid state disks combined in RAID-0 arrays. At the end of each microscopy experiment, the data is then transferred to file servers with relatively inexpensive high-capacity hard disk drives combined in RAID-6 arrays (Ahrens et al., 2013; Chhetri et al., 2015; Lemon et al., 2015). High-capacity file servers capable of storing 500–1,000 TB of raw image data with this level of redundancy can be purchased at a cost of $80 per TB at the time of writing of this chapter. However, it is usually not practical to store data in an uncompressed file format – this approach would dramatically increase storage costs, reduce data throughput during image processing, and introduce overhead during data inspection. Thus, it is highly recommended to use a file format that combines high compression ratios, a lossless compression scheme with high read and write speeds and a data structure that offers access to spatial or temporal sub-regions of the 4D image data with little to no overhead (Amat et al., 2015). This latter strategy of partitioning image data in small 4D data chunks is extremely valuable both for interactive data visualization and for high-throughput image processing: data visualization rarely involves viewing all of the data at the same time across space and time, and efficient image processing demands that computations are parallelized across the entire data set to the maximum extent possible, which in turn requires distributing chunks of image data to different processing cores. A powerful solution to this problem, which combines all features in a single, open-source file format termed KLB, has been introduced by Amat et al. (2015) (Figure 2.7, **Step 1**). KLB offers the same lossless compression ratio as the 3D JPEG2000 format, but improves read and write speeds more than 3-fold. In contrast to JPEG2000, KLB can also take full advantage of modern multi-core processors with large numbers of processing cores and uses a block-based scheme that subdivides large volumetric image data as sets in small data chunks (with freely adjustable spatial and/or temporal dimensions) to form powerful synergies with data visualization and high-throughput image processing tasks.

Before moving on to the main image processing workflow that facilitates the extraction and analysis of the functional signals encoded in the time-lapse image data, additional computational steps are often needed that vary with the type of light-sheet microscopy used for data acquisition. For example, multi-view imaging is a powerful concept (Swoger et al., 2007) that can be harnessed for increasing physical coverage of partially opaque specimens, such as *Drosophila* embryos and larvae (Tomer et al., 2012; Chhetri et al., 2015; Lemon et al., 2015), and for improving spatial resolution (Wu et al., 2013; Chhetri et al., 2015). However, acquiring multi-view image data in turn demands efficient approaches to multi-view image registration, fusion, and/or deconvolution to take full advantage of the information encoded in complementary views. Several powerful solutions to this end exist in the form of open-source software packages, including methods for interest-point-based image registration (Preibisch et al., 2010), content-based image registration (Chhetri et al., 2015), and high-throughput 3D multi-view deconvolution that takes full advantage of modern graphics processing units (GPU) and multi-core central processing units (CPU) (Chhetri et al., 2015) (Figure 2.7, **Step 1**). The latter software is capable of multi-view deconvolution of 10 TB data sets within about two days on an image processing workstation equipped with a high-end CUDA-enabled graphics card in the price range of $500–$1,000.

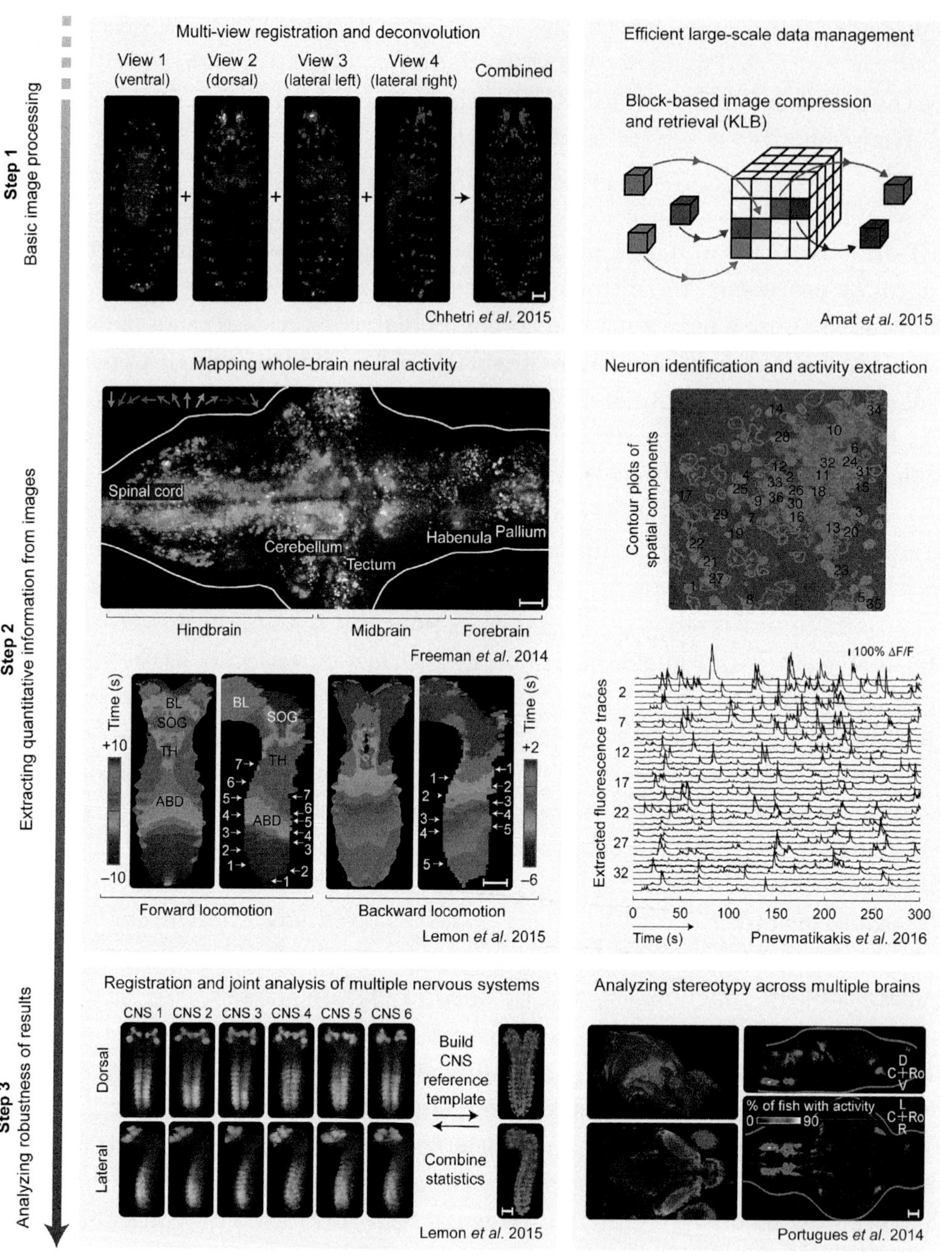

FIGURE 2.7 Processing and analysis of light-sheet functional image data. Overview of key computational steps in the processing and analysis of large-scale light-sheet functional image data, categorized as "basic image processing" (step 1, top), "extracting quantitative information from images" (step 2, middle) and "analyzing robustness of results" (step 3, bottom). From top to bottom, this overview figure includes methodological examples from the studies listed below. Multi-view registration and deconvolution (Chhetri et al., 2015): Images show maximum-intensity projections of single- and multi-view deconvolved IsoView image data of a stage 17 *Drosophila* embryo expressing GCaMP6s throughout the nervous system. Computations were performed using a high-throughput GPU-based implementation of the Lucy–Richardson algorithm, which achieves a data throughput of 1 Gigavoxel per minute (including I/O) on Tesla K40 GPUs when executing 20 iterations of the algorithm. Efficient large-scale

FIGURE 2.7 (CONTINUED)
data management (Amat et al., 2015): The Keller Lab Block (KLB) lossless image compression format combines high compression ratios, fast read/write speeds, and a flexible block architecture that enables efficient access to arbitrary regions of interest. Inspired by Parallel BZip2, a common Linux compression module, images are partitioned in 5D blocks and all blocks are compressed in parallel using BZip2. Both reading and writing operations are parallelized and scale linearly with the number of cores in the CPU. The KLB source code is accompanied by a simple API for interfacing the open-source C++ code with various platforms, as well as an interface file for the SWIG tool, which can be used to autogenerate wrapper code for various languages, including Java, C#, Python, Perl, and R. Mapping whole-brain neural activity (Freeman et al., 2014): analysis of direction tuning across the brain of larval zebrafish presented with moving whole-field visual stimuli in a setup that combines light-sheet imaging with visual stimulation and behavior (Vladimirov et al., 2014). The direction tuning maps are derived by fitting every voxel (with ~100 voxels per neuron) with a tuning-curve model that separately describes the temporal response profile and the tuning to direction. Color indicates preferred direction; saturation, tuning width (i.e. circular variance); brightness; response strength. White means responsive, but without unidirectional tuning. Image shows a dorsal maximum intensity projection through 39 planes covering 195 μm. Mapping whole-brain neural activity (Lemon et al., 2015): Mapping whole-CNS activity timing for forward and backward locomotor waves (dorsoventral and lateral slices are shown to the left and right, respectively). To create these maps, the timing of activity was evaluated across all detected wave events in one specimen (forward: $n = 30$, backward: $n = 70$). Intuitively, these maps show, for each part for the CNS, the time during locomotor wave windows when activity increases. Arrows mark relative progression of locomotor waves on dorsal/ventral sides of the VNC (ascending numbers). Forward and backward wave window sizes were defined as [−10 s, 10 s] and [−6 s, 2 s] (centered on waves in ventral nerve cord) to ensure wave propagation was captured throughout the ventral nerve cord and overlap of events was avoided. Neuron identification and activity extraction (Pnevmatikakis et al., 2016): demonstration of a method for simultaneously identifying the locations of neurons, demixing spatially overlapping components, and denoising and deconvolving the spiking activity from the slow dynamics of the calcium indicator. The method was applied to mouse *in vivo* GCaMP6s data. The top panel shows contour plots of inferred spatial components superimposed on the correlation image of the raw data. The components are sorted in decreasing order based on the maximum temporal value and their size. Contour plots of the first 200 identified components are shown, and the first 36 components are numbered. The bottom panel shows extracted ΔF/F fluorescence traces for the first 36 components. Registration and joint analysis of multiple nervous systems (Lemon et al., 2015): spatial registration of the nervous systems of six *Drosophila* larvae. All CNS explants used for functional imaging expressed GCaMP panneuronally (57C10-GAL4,UAS-GCaMP6s, shown in green) and tdTomato in anatomically defined regions within the larval CNS (58B03-LexA,LexOP-tdTomato, shown in magenta; expression in mushroom bodies and neuropil regions). To register nervous systems, a CNS template (shown to the right) was constructed from GCaMP reference image stacks representing each of the six independent time-lapse experiments. Image data from each experiment was then registered to the CNS reference template using non-linear spatial registration techniques. Analyzing stereotypy across multiple brains (Portugues et al., 2014): three-dimensional volume registration across larval zebrafish brains. Left: an individual brain (green) is morphed onto a reference brain (magenta) by performing an affine transformation, followed by non-rigid alignment. Right: spatial maps encoding the percentage of fish ($n = 13$) imaged that show activity at each voxel (after registration) for all voxels within the brain, depicted as maximum projections from two orthogonal views. (All figures and captions were adapted with permission from the respective sources listed in the bottom right corner of each panel.) Scale bars, 30 μm (step 1, left), 40 μm (step 2, top left), 50 μm (step 2, bottom left; step 3).

At the core of a typical analysis pipeline for functional image data are the processing steps that follow the basic data handling and processing tasks described above. These steps typically involve the use of computational approaches to functional image segmentation, regression, dimensionality reduction, clustering, and other approaches to mapping of activity across the brain or nervous system (Keller and Ahrens, 2015) (Figure 2.7, **Step 2**). Since light-sheet based functional imaging has not yet been widely adopted, it is not uncommon that new custom computational tools are developed and used in each new study involving light-sheet microscopy data. We note, however, that several general computational toolkits are already available. In particular, Freeman et al. (2014) and Pnevmatikakis et al. (2016) present methodologies and algorithms to this end that likely form a good starting point for data exploration in many investigations. Moreover, fast and simple software tools for mapping activity at the voxel level across entire nervous systems in response to external stimuli or robust internal events were presented by Lemon et al. (2015). This toolkit enabled a systematic, unbiased search for neurons across the *Drosophila* nervous system with distinct and robust activity signatures during different types of motor programs.

Following the in-depth analysis of a given whole-brain or whole-CNS functional imaging dataset, it is often desirable to evaluate the robustness and significance of the respective findings across multiple specimens. This final step typically involves the registration of image data or neural activity maps derived from multiple brains or nervous systems in 3D (Figure 2.7, **Step 3**). Such computations are usually complicated by the fact that overall geometry and local neuronal morphologies differ significantly across multiple specimens, which in turn demands non-linear registration methods to maximize spatial correspondence of individual neurons or (more realistically) local anatomical regions comprising small populations of neurons across all datasets. Typical processing strategies to this end first construct reference brains or entire reference nervous systems and then map individual datasets onto these reference scaffolds, e.g. by taking advantage of non-linear registration algorithms of the open-source framework Advanced Normalization Tools (https://github.com/stnava/ANTs). Example workflows that follow this concept were presented and applied by Lemon et al. (2015) for the *Drosophila* larval CNS and by Portugues et al. (2014) for the zebrafish larval brain.

2.6 OUTLOOK

The recent advancements in light-sheet imaging discussed in this chapter have opened a window into neuronal imaging in multi-cellular organisms with high spatiotemporal resolution. Light-sheet microscopy for functional neuronal imaging is still in its early days, and we expect further improvements of this technique as some of the still outstanding challenges are addressed.

One obvious set of future advancements in light-sheet imaging will be the mitigation of light scattering and optical aberrations, which manifest themselves as a degradation in image quality at larger imaging depths and impact more severely in non-transparent tissues. Adaptive imaging techniques (Ji et al., 2010; Wang et al., 2015) and automated framework capable of optimizing the image quality across the specimen volume in real time (Royer et al., 2016) can address some of these issues and thus further complement the imaging capabilities of light-sheet methods. Light-sheet microscopy is also poised to benefit greatly from recently developed clearing methods (Höckendorf et al., 2014; Marx, 2016), which largely mitigate light scattering in *ex vivo* tissues, and present a unique opportunity to image large tissue volumes. When combined with light-sheet microscopy these approaches offer a way to obtain images of the entire cleared volume at high spatial resolution. It would thus in principle already be feasible with existing methodology to first map activity in a neural tissue or an entire brain/CNS at cellular resolution, then fix and chemically clear the specimen, and finally acquire an anatomical map of the entire tissue with spatial resolution of 100 nm or better, in particular when light-sheet microscopy is paired with the recently developed Expansion Microscopy (ExM) methods (Chen et al., 2015; Chozinski et al., 2016). Combining these high-resolution anatomical and high-speed functional measurements enabled by light-sheet microscopy presents a great opportunity to interpret and understand functional activity patterns in a structural/anatomical context.

Light-sheet microscopy also stands to benefit directly from ongoing advancements in complementary optical tools for neuronal studies, such as the development of red-shifted calcium sensors (Dana et al., 2016), voltage sensors (Peterka et al., 2011), and the expanding optogenetics toolkit (Fenno et al., 2011). As red-shifted calcium sensors make it feasible to image from deeper regions of the brain, high-speed light-sheet microscopes would record neuronal activities at a high spatiotemporal resolution from regions that are typically inaccessible to GFP-based sensors. Additionally, a high-speed, two-color recording from two distinct population of cells would immediately be possible, particularly with light-sheet microscopes that allow simultaneous multi-color imaging, such as the IsoView microscope. The development of voltage sensors with high signal-to-noise ratio, brightness, and photostability would enable a way to record the membrane potential directly, and when imaged with high-speed light-sheet microscopes, electrical activity could be systematically measured across a relatively large population of neurons at the same time. Successful large-volume voltage imaging is expected to demand substantial improvements in the acquisition speed of light-sheet microscopy. Efforts to this end will require further conceptual advances in light-sheet microscopy but will also benefit from the ongoing advancement of sCMOS camera technology and related detector technology. Another avenue of potential future improvements is the integration of high resolution, high-speed volumetric imaging of the nervous system using light-sheet microscopy and the delivery of precise and targeted optogenetics manipulation. An implementation of an online data analysis platform for instructing perturbations, paired with light-sheet microscopy for volumetric imaging at high spatiotemporal resolution would then enable functional connectivity mapping as well as online hypothesis testing in a behaving animal. With the ongoing rapid advancements in light-sheet microscopy and the development of complementary optical tools, we will undoubtedly see a steadily expanding role of light-sheet microscopy in elucidating the function of the nervous system.

REFERENCES

Ahrens MB, Orger MB, Robson DN, Li JM, Keller PJ (2013) Whole-brain functional imaging at cellular resolution using light-sheet microscopy. *Nature Methods* 10:413–420.

Amat F, Höckendorf B, Wan Y, Lemon WC, McDole K, Keller PJ (2015) Efficient processing and analysis of large-scale light-sheet microscopy data. *Nature Protocols* 10:1679–1696.

Bouchard MB, Voleti V, Mendes CS, Lacefield C, Grueber WB, Mann RS, Bruno RM, Hillman EMC (2015) Swept confocally-aligned planar excitation (SCAPE) microscopy for high-speed volumetric imaging of behaving organisms. *Nature Photonics* 9:113–119.

Chen B-C, Legant WR, Wang K, Shao L, Milkie DE, Davidson MW, Janetopoulos C, Wu XS, Hammer JA, Liu Z, English BP (2014) Lattice light-sheet microscopy: imaging molecules to embryos at high spatiotemporal resolution. *Science* 346:1257998.

Chen F, Tillberg PW, Boyden ES (2015) Expansion microscopy. *Science* 347:543–548.

Chhetri RK, Amat F, Wan Y, Höckendorf B, Lemon WC, Keller PJ (2015) Whole-animal functional and developmental imaging with isotropic spatial resolution. *Nature Methods* 12:1171–1178.

Chozinski TJ, Halpern AR, Okawa H, Kim H-J, Tremel GJ, Wong ROL, Vaughan JC (2016) Expansion microscopy with conventional antibodies and fluorescent proteins. *Nature Methods* 13:1–7.

Dana H, Mohar B, Sun Y, Narayan S, Gordus A, Hasseman JP, Tsegaye G, Holt GT, Hu A, Walpita D, Patel R, Macklin JJ, Bargmann CI, Ahrens MB, Schreiter ER, Jayaraman V, Looger LL, Svoboda K, Kim DS (2016) Sensitive red protein calcium indicators for imaging neural activity. *eLife* 5:e12727.

Dowski ER, Cathey WT (1995) Extended depth of field through wave-front coding. *Applied Optics* 34:1859–1866.

Fahrbach FO, Voigt FF, Schmid B, Helmchen F, Huisken J (2013) Rapid 3D light-sheet microscopy with a tunable lens. *Optics Express* 21:21010.

Fenno L, Yizhar O, Deisseroth K (2011) The development and application of optogenetics. *Annual Review of Neuroscience* 34:389–412.

Freeman J, Vladimirov N, Kawashima T, Mu Y, Sofroniew NJ, Bennett DV, Rosen J, Yang C-T, Looger LL, Ahrens MB (2014) Mapping brain activity at scale with cluster computing. *Nature Methods* 11:941–950.

Fuchs E, Jaffe J, Long R, Azam F (2002) Thin laser light sheet microscope for microbial oceanography. *Optics Express* 10:145–154.

Grewe B, Langer D, Kasper H, Kampa BM, Helmchen F (2010) High-speed in vivo calcium imaging reveals neuronal network activity with near-millisecond precision. *Nature Methods* 7:399–405.

Hammen GF, Turaga D, Holy TE, Meeks JP (2014) Functional organization of glomerular maps in the mouse accessory olfactory bulb. *Nature Neuroscience* 17:953–961.

Höckendorf B, Lavis LD, Keller PJ (2014) Making biology transparent. *Nature Biotechnology* 32:1104–1105.

Holekamp TF, Turaga D, Holy TE (2008) Fast three-dimensional fluorescence imaging of activity in neural populations by objective-coupled planar illumination microscopy. *Neuron* 57:661–672.

Huisken J, Swoger J, Del Bene F, Wittbrodt J, Stelzer EHK (2004) Optical sectioning deep inside live embryos by selective plane illumination microscopy. *Science* 305:1007–1009.

Ji N, Milkie DE, Betzig E (2010) Adaptive optics via pupil segmentation for high-resolution imaging in biological tissues. *Nature Methods* 7:141–147.

Keller PJ, Ahrens MB (2015) Visualizing whole-brain activity and development at the single-cell level using light-sheet microscopy. *Neuron* 85:462–483.

Keller PJ, Schmidt AD, Wittbrodt J, Stelzer EHK (2008) Reconstruction of zebrafish early embryonic development by scanned light sheet microscopy. *Science* 322:1065–1069.

Lemon WC, Pulver SR, Höckendorf B, McDole K, Branson K, Freeman J, Keller PJ (2015) Whole-central nervous system functional imaging in larval Drosophila. *Nature Communications* 6:7924.

Levoy M, Ng R, Adams A, Footer M, Horowitz M (2006) Light field microscopy. *ACM Transactions on Graphics* 25:924.

Liang X, Holy TE, Taghert PH (2016) Synchronous Drosophila circadian pacemakers display nonsynchronous Ca2+ rhythms in vivo. *Science* 351:976–981.

Looger LL, Griesbeck O (2012) Genetically encoded neural activity indicators. *Current Opinion in Neurobiology* 22:18–23.
Marx V (2016) Optimizing probes to image cleared tissue. *Nature Methods* 13:205–209.
Mouroulis P (2008) Depth of field extension with spherical optics. *Optics Express* 16:12995–13004.
Osten P, Margrie TW (2013) Mapping brain circuitry with a light microscope. *Nature Methods* 10:515–523.
Panier T, Romano SA, Olive R, Pietri T, Sumbre G, Candelier R, Debrégeas G (2013) Fast functional imaging of multiple brain regions in intact zebrafish larvae using selective plane illumination microscopy. *Frontiers in Neural Circuits* 7:65.
Peterka DS, Takahashi H, Yuste R (2011) Imaging voltage in neurons. *Neuron* 69:9–21.
Pitrone PG, Schindelin J, Stuyvenberg L, Preibisch S, Weber M, Eliceiri KW, Huisken J, Tomancak P (2013) OpenSPIM: an open-access light-sheet microscopy platform. *Nature Methods* 10:598–599.
Planchon TA, Gao L, Milkie DE, Davidson MW, Galbraith JA, Galbraith CG, Betzig E (2011) Rapid three-dimensional isotropic imaging of living cells using Bessel beam plane illumination. *Nature Methods* 8:417–423.
Pnevmatikakis EA, Soudry D, Gao Y, Machado TA, Merel J, Pfau D, Reardon T, Mu Y, Lacefield C, Yang W, Ahrens M, Bruno R, Jessell TM, Peterka DS, Yuste R, Paninski L (2016) Simultaneous denoising, deconvolution, and demixing of calcium imaging data. *Neuron* 89: 285–299.
Portugues R, Feierstein CE, Engert F, Orger MB (2014) Whole-brain activity maps reveal stereotyped, distributed networks for visuomotor behavior. *Neuron* 81:1328–1343.
Preibisch S, Saalfeld S, Schindelin J, Tomancak P (2010) Software for bead-based registration of selective plane illumination microscopy data. *Nature Methods* 7:418–419.
Prevedel R, Yoon Y-G, Hoffmann M, Pak N, Wetzstein G, Kato S, Schrödel T, Raskar R, Zimmer M, Boyden ES, Vaziri A (2014) Simultaneous whole-animal 3D imaging of neuronal activity using light-field microscopy. *Nature Methods* 11:727–730.
Quirin S, Vladimirov N, Yang C-T, Peterka DS, Yuste R, Ahrens M (2016) Calcium imaging of neural circuits with extended depth-of-field light-sheet microscopy. *Optics Letters* 41:855.
Royer LA, Lemon WC, Chhetri RK, Wan Y, Coleman M, Myers E, Keller PJ (2016) Real-time adaptive light-sheet microscopy recovers high resolution in large living organisms. *Nature Biotechnology* 34: 1267–1278.
Siedentopf H, Zsigmondy R (1902) Uber Sichtbarmachung und Größenbestimmung ultramikoskopischer Teilchen, mit besonderer Anwendung auf Goldrubingläser. *Annals of Physics* 315:1–39.
Sofroniew NJ, Flickinger D, King J, Svoboda K (2016) A large field of view two-photon mesoscope with subcellular resolution for in vivo imaging. *eLife* 5:1–20.
Swoger J, Verveer P, Greger K, Huisken J, Stelzer EHK (2007) Multi-view image fusion improves resolution in three-dimensional microscopy. *Optics Express* 15:8029–8042.
Tomer R, Khairy K, Amat F, Keller PJ (2012) Quantitative high-speed imaging of entire developing embryos with simultaneous multiview light-sheet microscopy. *Nature Methods* 9:755–763.
Tomer R, Lovett-barron M, Kauvar I, Broxton M, Deisseroth K (2015) SPED light sheet microscopy : fast mapping of biological system structure and function. *Cell* 163:1796–1806.
Turaga D, Holy TE (2012) Organization of vomeronasal sensory coding revealed by fast volumetric calcium imaging. *The Journal of Neuroscience* 32:1612–1621.
Vladimirov N, Mu Y, Kawashima T, Bennett DV, Yang C-T, Looger LL, Keller PJ, Freeman J, Ahrens MB (2014) Light-sheet functional imaging in fictively behaving zebrafish. *Nature Methods* 11:883–884.
Voie AH, Bruns DH, Spelman FA (1993) Orthogonal-plane fluorescence optical sectioning: three-dimensional imaging of macroscopic biological specimens. *Journal of Microscopy* 170:229–236.
Wang K, Sun W, Richie CT, Harvey BK, Betzig E, Ji N (2015) Direct wavefront sensing for high-resolution in vivo imaging in scattering tissue. *Nature Communications* 6:7276.

Wilt BA, Burns LD, Wei Ho ET, Ghosh KK, Mukamel EA, Schnitzer MJ (2009) Advances in light microscopy for neuroscience. *Annual Review of Neuroscience* 32:435–506.

Wolf S, Supatto W, Debrégeas G, Mahou P, Kruglik SG, Sintes J-M, Beaurepaire E, Candelier R (2015) Whole-brain functional imaging with two-photon light-sheet microscopy. *Nature Methods* 12:379–380.

Wu Y, Wawrzusin P, Senseney J, Fischer RS, Christensen R, Santella A, York AG, Winter PW, Waterman CM, Bao Z, Colón-Ramos DA, McAuliffe M, Shroff H (2013) Spatially isotropic four-dimensional imaging with dual-view plane illumination microscopy. *Nature Biotechnology* 31:1032–1038.

Xu PS, Lee D, Holy TE (2016) Experience-dependent plasticity drives individual and sex differences in pheromone-sensing neurons. *Neuron* 91:878–892.

Yang Z, Haslehurst P, Scott S, Emptage N, Dholakia K (2016) A compact light-sheet microscope for the study of the mammalian central nervous system. *Scientific Reports* 6:26317.

Zalevsky Z, Ben-Yaish S (2007) Extended depth of focus imaging with birefringent plate. *Optics Express* 15:7202–7210.

CHAPTER 3

Two-Photon Microscopy in the Mammalian Brain

Hod Dana and Shy Shoham

CONTENTS

3.1 FUNCTIONAL TWO-PHOTON LASER-SCANNING MICROSCOPY

Two-photon fluorescence microscopy, first introduced in 1990 (Denk et al., 1990), dramatically improved the ability to see inside strongly scattering biological tissues. It has therefore emerged as a highly suitable modality for neural imaging and for noninvasive fluorescence microscopy in living animals (Zipfel et al., 2003). Development of membrane-permeable calcium sensitive fluorescent indicators and their *in vivo* demonstration for multi-cell loading in 2003 (Stosiek et al., 2003), combined with two-photon microscopy techniques, enabled the investigation of neural populations *in vivo* with the ability to monitor each neuron separately (Ohki et al., 2005; Garaschuk et al., 2006). The fluorescence of these indicators depends on Ca^{2+} concentration, and since the initiation of an action potential in neurons is associated with a fast inward Ca^{2+} flux, followed by slower return to baseline (Smetters et al., 1999), calcium concentration inside the neuron follows the action potentials pattern. In addition, temporal deconvolution and other spike inference methods can be applied to the data, allowing for the separation of temporally adjacent action potentials and therefore the improvement of temporal resolution (Yaksi and Friedrich 2006; Deneux et al., 2016; Pnevmatikakis et al., 2016).

Two-photon laser-scanning microscopy (TPLSM) was the first implementation of two-photon fluorescence microscopy (Denk et al., 1990) and is today's standard two-photon imaging method deep inside scattering biological tissue (Helmchen and Denk, 2005; Sofroniew et al., 2016). In Laser Scanning Microscopy (LSM), a laser beam is usually focused on a diffraction-limited focal spot with sub-micron diameter, and then raster scanned across a two- or three-dimensional specimen. In linear (single-photon) fluorescence microscopy, a dye molecule emits fluorescence after being excited by the absorption of a *single* photon with an appropriate wavelength. In contrast, in two-photon fluorescence microscopy, the same dye molecule is excited by the near-simultaneous (within approximately 0.5 femtosecond (Helmchen and Denk, 2005)) absorption of two photons, each carrying half of the required energy (i.e. having twice the wavelength of one-photon excitation) (Denk et al., 1990). Therefore, TPLSM is a non-linear LSM method, which depends on the square of the excitation intensity. The use of TPLSM requires an ultrafast laser as a light source. This laser emits powerful short pulses (typically, tens of nanojoules during ~100 femtosecond pulse) that are essential to increase the probability of almost simultaneous two photon absorption in a dye molecule.

Several physical properties contribute to two-photon microscopy's advantages as a biological imaging modality. First, the sample's fluorescence is proportional to the square of the light intensity, and therefore essentially only occurs near the focal volume of the excitation beam, allowing for the accessing of only a small volume within the sample being observed. In addition, the non-linear nature of the signal strength improves the data's Signal to Noise Ratio (SNR), allows for efficient use of the light source's power compared to one-photon imaging methods (and therefore less bleaching and thermal damaging), and enables image reconstruction using a relatively simple procedure. Third, the central wavelength of the light source used is typically in the near infra-red (NIR), which is in the "optical window" for biological samples; at these wavelengths (usually 600–1,100 nm, but see also Horton et al., 2013; Ouzounov et al., 2017), absorbance in living tissue is weak and the photons are scattered less than at shorter wavelengths. Another advantage is the large wavelength difference between the excitation light (NIR usually) and the emitted fluorescence (usually in the visible range), enabling the use of dichroic mirrors or filters to separate them. Finally, because of the small size of the focal spot and the necessary near-simultaneous absorption of two photons, the majority of the fluorescence signal is contributed by absorption of ballistic (non-scattered) photons in the focal volume, where photon density is maximal. This has two important consequences that contribute to depth penetration: (1) scattered photons have a negligible contribution to the fluorescence signal, and (2) all of the emitted fluorescence signal can be collected (without a pinhole) and attributed to the focal spot. The effect of scattering is mainly limited to exponential attenuation of the signal power according to Beer-Lambert law: $I = I_0 \exp(-z/l_s)$ and $F = F_0 \exp(-2 \cdot z/l_s)$, where I and F are the ballistic photons' intensity and fluorescence signal respectively, I_0 and F_0 are their value on the scattering medium surface, z is the depth inside the tissue, and l_s is the mean free path (MFP). Figure 3.1 illustrates of one- and two-photon excited volumes.

The enhanced SNR and reduced tissue absorbance and scattering in two photon fluorescence microscopy allowed the maximal penetration depth into a brain tissue to increase from less than 100 µm up to 500 µm (Svoboda et al., 1997; Oheim et al., 2001). This new maximal imaging depth is determined by the exponential attenuation of excitation energy (due to scattering), and the maximal intensity available from a femtosecond laser light-source. One way to circumvent this limitation is to use amplified femtosecond lasers. These provide more powerful pulses at a lower repetition rate and similar average power. Indeed, the use of amplified pulses (~2µJ instead of ~10nJ), increases the maximum penetration depth to 1,400 µm *in vitro* and 1,000 µm *in vivo* (Theer et al., 2003; Theer and Denk 2006). Interestingly, the maximal depth penetration of amplified pulses is no longer determined by the maximal light source intensity (it was achieved with 30% of the maximal amplified laser power), but rather by an undesirable out-of-focus (OOF) fluorescence signal. An in-depth study of the OOF phenomena (Theer and Denk 2006) showed that it is inherent in the physics of the system. Since one must exponentially increase the intensity of the light source to overcome the scattering-related signal attenuation (see Section 3.3) and penetrate deeper into the tissue, when the intensity reaches a certain level, it will be high enough to create significant fluorescence outside the focal volume. Simulation showed that most of the OOF signal was generated close to the stained tissue surface. This OOF signal masks the fluorescence from the focal volume.

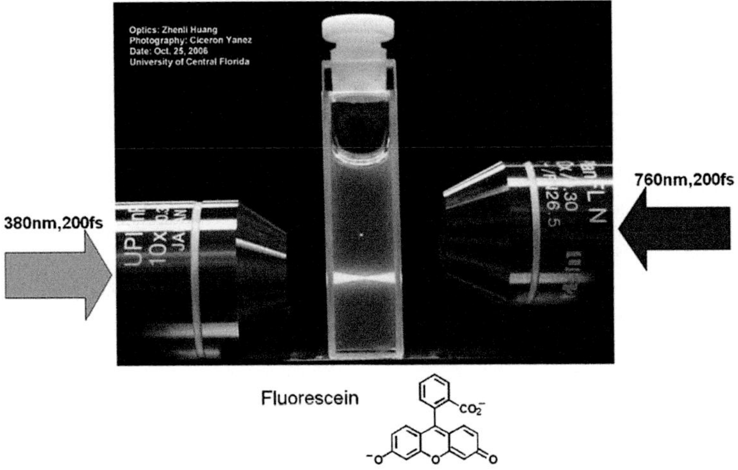

FIGURE 3.1 Single-photon excitation vs two-photon excitation. The fluorescence generated by two-photon excitation is restricted to a small volume around the focal point of the objective. In contrast, single-photon excitation is generated along the laser path. (Image by Z-L Huang, U. Central Florida.)

Another challenge associated with functional two-photon fluorescence microscopy is achieving high-speed three-dimensional (3D) imaging (Kim. et al., 2008). Two-photon imaging is inherently relatively slow because volumes are scanned voxel by voxel. The development of a high-speed two-photon microscopy method can offer a powerful tool for a variety of biomedical investigations. In many cases, high speeds of acquisition are not crucial, and indeed several important studies of cortical neural responses were published using sequential planar acquisition at 1 Hz frame rates (Ohki et al., 2005; Ohki et al., 2006; Kara and Boyd 2009). However, the ability to monitor calcium changes that are associated with single action potentials ("spiking activity") requires an order of magnitude improvement in temporal resolution. It would also be extremely useful to image a full volume of a neural network instead of a plane. Research in the area of high-speed two-photon microscopy focuses on two directions: understanding the fundamental limitations of high-speed deep tissue two-photon imaging (Kim et al., 2008) and the practical implementation of high-speed two-photon microscopes (Gobel et al., 2007; Kim et al., 2008; Field et al., 2010; Grewe et al., 2010; Cheng et al., 2011; Katona et al., 2012; Prevedel et al., 2016; Yang et al., 2016; Hillman et al., 2018). The first demonstration of real-time volumetric imaging (using a spiral scan pattern) was published in 2007 (Gobel et al., 2007). Since then, two other types of rapid 3D imaging systems were introduced, based on spatiotemporal multiplexing of the laser beam (Field et al., 2010; Chen et al., 2011; Yang et al., 2016), and on rapid acousto-optic deflectors (AODs) which enables to sample only pre-defined regions of interest in the volume (Katona et al., 2012; Cotton et al., 2013). Both the spiral scanning and the AOD-based systems are based on speeding up the *serial* scanning process of a single diffraction-limited spot, while spatiotemporal multiplexing is based on an almost *parallel* data acquisition process of different points with no, or with a very short (~3 nsec), temporal delay between them. The advantages and disadvantages of serial and parallel acquisition will be discussed in the following sections.

3.2 RAPID SERIAL LATERAL SCANNING FOR FUNCTIONAL TPLSM

The serial nature of TPLSM dictates a relatively slow acquisition rate. Standard galvanometric scanning systems acquire typically 0.1–4 frames/sec (Kim. et al., 2008), which limits the ability of TPLSM imaging systems to monitor dynamic biological processes, such as neuronal activity patterns. In this section we will review state of the art methods for enhancing serial data acquisition rate in two-photon microscopy, based on both mechanical and inertia-free scanning techniques.

3.2.1 Rapid Mechanical Scanning

Serial data acquisition refers to acquiring the volumetric data voxel by voxel and is easily implemented in two-photon microscopy, since when an ultrafast laser beam is focused on a diffraction limited spot, the non-linear nature of two-photon excitation guarantees that the excitation occurs only in a small volume around the focal point of the beam. Standard TPLSM setups consist of an ultrafast laser, an attenuation unit which controls the light intensity level, a scanning unit, and a detection unit which measures the fluorescence signal intensity from the focal spot. Enhancing the performance of such a system lies in speeding up the scanning process.

Most common scanning units contain a set of two galvanometric mirrors, a fast mirror and a slow mirror, which receive a control signal and deflect the incoming light at a specific angle. Three lenses (scan, tube, and objective lenses) transform this angle to a linear translation in the focal plane of the objective (see detailed description in Tsai et al., 2002). The fast mirror scans a line, and the slow mirror moves this line laterally to scan a plane. Galvanometric mirrors offer flexible imaging capabilities, with different scanning speed and amplitude, and are very convenient to control. However, their maximal speed is limited to ~1ms per line in raster scanning mode, which may be sufficient for a single plane imaging with frame rate of ~4 frames/sec (assuming 256 × 256 pixels per image), but is too slow for 3D functional imaging. The galvanometric mirrors may be replaced with faster scanning mirrors, such as polygonal mirrors and resonant mirrors.

Polygonal mirrors are shaped as multi-faceted polygons and are rapidly rotated in a constant speed to scan a single line within 33–120 μs. Scanning in the orthogonal axis may be performed with galvanometric mirror, and imaging rate in such a system was enhanced up to 200 frames/sec. Polygonal mirror drawbacks are that scanning range cannot be modified and depends on the polygon geometry, and that there is a "dead scanning time" when the laser beam is moving from one polygonal facet to the next one and laser beam intensity must be attenuated. Figure 3.2a shows a schematic design of a polygonal mirrors-based rapid microscope (Kim et al., 1999).

Resonant mirrors (Figure 3.2b) are galvanometric mirrors that scan the sample at their resonant frequency, usually 4–12 KHz, which enables scanning a single line in up to 40 μs (using bi-directional scanning). Combined with a galvanometric mirror in the slow axis, enhanced frame rates of 130 frames/sec were demonstrated (Fan et al., 1999), and current commercial systems can acquire data from a small field of view at hundreds of frames/sec. The main drawback of scanning using resonant mirrors is that scanning profile is sinusoidal, which means that scanning speed is not constant. Therefore, laser intensity is constantly modulated during scanning in order to compensate for the difference in pixel dwell time and to maintain the same SNR in the entire field of view. Another drawback is that scanning frequency is fixed and only its amplitude may be changed. Both polygonal and resonant scanners require a dedicated control system for mapping the exact scanner position to the sampled point in the sample, and usually suffer from lower image quality compared to galvanometric-mirror scanning systems (Kim et al., 2008). To summarize, both polygonal and resonant mirrors offer a significant improvement in scanning a single plane (without axial movement). However, they require constructing more sophisticated control and image formation systems, and reduce the microscope's flexibility either by forcing a constant field of view size, or by forcing a constant line scanning rate.

A different mechanical scanning approach for enhancing TPLSM data acquisition rates is based on changing the commonly used 2D raster scanning to a more time-efficient custom line scanning (Gobel et al., 2007; Lillis et al., 2008; Rothschild et al., 2010;

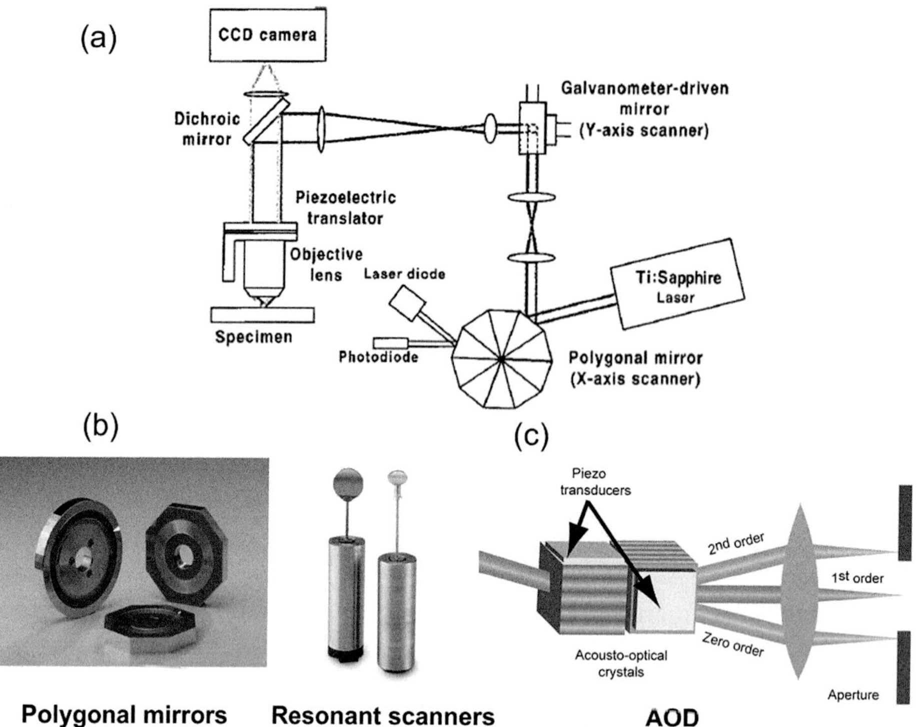

FIGURE 3.2 Rapid serial scanning for functional TPLSM. (a) Rapid polygonal mirror based TPLSM microscope design. The polygonal mirror scans the fast axis and a galvanometric mirror scans the slow axis in a raster scanning of up to 100 frames/sec. (Image is taken from Kim et al. (1999).) (b) Commercially available rapid serial scanners: polygonal mirrors, resonant scanners, and AOD.

Garion et al., 2014). This approach is based on designing a custom line trajectory (optionally in 3D) that will pass through as many cells of interest as possible. Such scanning methods reduce the "dead time" in which cells aren't being scanned, but are sensitive to motion artifacts, since cells' movement may result in scanning an empty space instead of a cell. An implementation of this approach to scan 3D volumes is rapid scanning of 3D spiral shapes (Gobel et al., 2007), where a galvanometric mirror pair and a piezoelectric device, which axially moved the objective lens, scan a custom line in 3D volume. Spiral line with varying diameter was shown to efficiently cover the sampled volume (see Figure3.3b), and to allow volumetric scanning rate of up to 10 volumes/sec for volume dimensions of 200 μm × 200 μm × 200 μm, which contains several hundreds of cells. The achieved sample rate is sufficient for detection of a single action potential induced change in fluorescence signal of a calcium indicator (Kerr et al., 2005). However, the main drawback of that system is that the imaged volume is sparsely sampled, for example, dividing the volume into 15 μm cubes showed that only 50% of these cubes were sampled (Gobel et al., 2007). Sparse sampling makes the volumetric measurements sensitive to the value of a single sampled point; moreover, in the case of *in vivo* imaging where sample movements are expected, imaging results may be strongly influenced by movement artifacts.

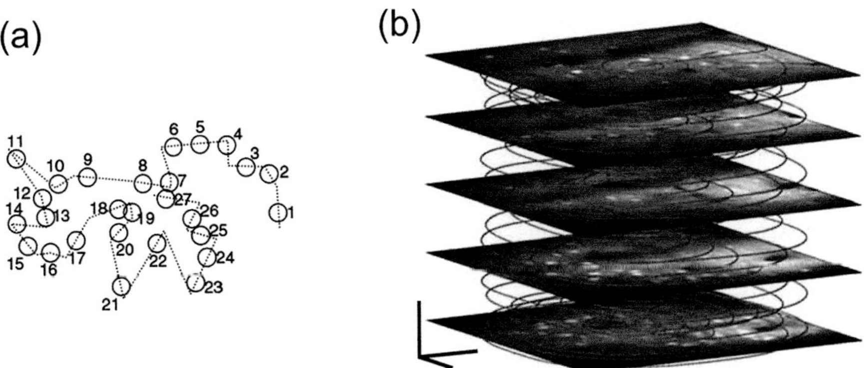

FIGURE 3.3 Time efficient line scanning in two and three dimensions. (a) Custom-line scans predefined neural cells in area A1 stained with Fluo-4 Ca^{+2} indicator, with temporal resolution of 300Hz. (Image is taken from Rothschild et al. (2010).) (b) Efficient spiral line scanning of a volume with temporal resolution of 10 volumes/sec. (Image is taken from Gobel et al. (2007).)

3.2.2 Rapid Inertia-Free Scanning

A different solution for rapid 3D sampling is based on non-mechanical movement of the laser beam by AODs. AOD devices, shown in Figure 3.1b, are based on an acoustic wave that propagates in a transparent medium, changing its refractive index periodically, and creating a phase grating that diffracts the beam toward a known angle (depending on the grating period). AODs main advantage is their fast response time – approximately 10 µs are required to jump between any two points in the volume (Grewe et al., 2010), without limitation on the distance between them. AODs may be used as an alternative to the fast galvanometric mirror in a standard scanning setup, but an effective implementation of their unique capabilities is a hopping imaging system, also known as random-access system. The imaging procedure is as follows: an initial volume scan is performed, then regions of interest are chosen and marked, and imaging is performed by rapid hopping between these regions. The imaging cycle time is the sum of the acquisition time per single point and the transition time between points, multiplied by the number of sampled points. AODs were used for various rapid imaging applications for many years (Milton et al., 1983; Goldstein et al., 1990; Duemani Reddy et al., 2008; Chen et al., 2011); their main drawbacks were a significant addition of both spatial and temporal dispersion for an ultrafast laser pulse, and low diffraction efficiency. A complicated setup was required in order to compensate for the dispersion, and the low efficiency prevented application of these systems to *in vivo* experiments. Improvements in commercial lasers' performance, together with development of simple and effective dispersion compensation solutions (Zeng et al., 2006; Katona et al., 2012), enabled the application of such imaging systems for *in vivo* imaging of 2D (Grewe et al., 2010) and 3D scenes (Katona et al., 2012; Cotton et al., 2013). The rapid response time of AOD-based systems also enables basic correction of movement artifacts during an imaging session, by sampling a chosen selected area of the field of view for movement detection, and calibrating the signals being sent to the AOD scanners according to its exact location in each measured time point. To summarize, AOD-based imaging systems today offer the maximal serial data acquisition rate, by minimizing the transition time from one point to the other and overcoming the mechanical limitations of standard scanning methods. AODs have been combined with a scanning galvo mirror for rapid raster scan planar imaging at rates of 1,000 frames/sec in custom-built microscopes (Chen et al., 2011) and also in commercially available systems. However, the robustness of these systems for movement artifacts hasn't been broadly tested (see more in Section 3.5.2), and they are still fundamentally limited by the inherent tradeoff between the number of sampled points and the imaging rate. Such fundamental limits may only be overcome by parallel data acquisition.

3.3 BEAM PROPAGATION INSIDE A SCATTERING MEDIUM

The scattering properties of biological tissues limit the use of linear imaging methods, such as widefield or confocal microscopy, for efficiently imaging deep inside brain tissue. Nonlinear optical methods, such as TPLSM, are more suitable for this task. For TPLSM, light scattering generally results in exponential decay of the fluorescent signal, but without significant reduction in resolution (but see Theer and Denk, 2006 for the fundamental limitation that scattering imposes on deep-tissue TPLSM). However, when parallel non-linear methods for recording brain activity are considered, then the scattering properties of the tissue may impose several additional limitations on the parallelization level, imaging depth, and fluorescence signal detection. Therefore, in this section, we will elaborate on light-tissue interactions inside a scattering medium before describing parallel non-linear acquisition methods in Section 3.4.

Propagation characteristics of a beam in a scattering medium is a subject of interest for many fields of physics and engineering, such as astrophysics, meteorology, oceanic optics, light detection and ranging (LIDAR), and optical imaging (van de Hulst and Kattawar, 1994; McLean et al., 1998). Light propagation in a scattering-absorbing medium is governed by the Radiative Transfer Equation (RTE), a linear integro-differential equation, given by:

$$\frac{\partial}{\partial t}\psi(\hat{s},\vec{r},t) = -\hat{s}\cdot\nabla\psi(\hat{s},\vec{r},t) - (\sigma_s + \sigma_a)\psi(\hat{s},\vec{r},t) + \int P\left(\hat{s}'\cdot\hat{s}\right)\psi\left(\hat{s}',\vec{r},t\right)d\hat{s}'$$

$\psi(\hat{s},\vec{r},t)$ is the radiant power per unit solid angle, per unit area perpendicular to the direction of propagation, $\hat{s}$ is the beam direction, $\vec{r}$ is the beam location, and t is the time. σ_s and σ_a are the scattering and absorption cross-sections respectively (in units of area) and $P\left(\hat{s}'\cdot\hat{s}\right)$ is the tissue phase function, that gives the probability of a beam that comes in a direction $\hat{s}'$ to be scattered to a direction $\hat{s}$. This probability depends on the angle between the two directions. The left-hand side term represents temporal changes in local beam intensity, the first right-hand side term represents beam divergence, the second one represents beam attenuation due to scattering and absorption, and the third term considers beams from different directions that were scattered into $\hat{s}$ direction.

Since the RTE has no general analytical solution (Sergeeva et al., 2006), approximated solutions, under some assumptions, were developed. The best-known approximation is the diffusion approximation (Wang and Wu, 2007), which holds for thick samples and gives analytical results that agree with experimental data. However, our interest in this chapter is in a different regime of light propagation, where the entering beam remains nearly collimated and fluorescent signal is related to ballistic photons only.

A photon that experiences a scattering event changes its initial direction of propagation. We use two angles to describe the photon direction: polar angle, θ, and azimuthal angle, φ. By assuming a small homogeneous volume around the scattering position, we treat φ as uniformly distributed; the θ (which is the angle with the former propagation direction) distribution function is called the phase function, and its evaluation is of crucial importance for an exact beam propagation analysis. A frequently used phase function for biological tissues optics is the Henyey-Greenstein (HG) phase function (Henyey and Greenstein, 1941), which exhibits a good agreement with experimental data together with mathematical simplicity. The phase function is used to calculate the anisotropic factor g, which is the average cosine of θ. For most biological tissues g≥0.9, indicating a strong forward-scattering behavior (Cheong et al., 1990). The HG phase function uses g as a parameter and therefore, it can be matched to the experimentally measured g of a tissue. The HG function is given by:

$$p(\cos\theta) = \frac{1-g^2}{2\left(1+g^2-2g\cos\theta\right)^{3/2}}, \quad \theta \in [0,\pi]$$

Since biological tissues, including brain tissue, are forward-scattering, which means a small mean change in beam orientation as a result of a scattering event, then the RTE is usually approximated under the assumption that each scatter results in a small change in θ angle (small angles approximation) (van de Hulst and Kattawar, 1994; Borgers and Larsen, 1995; McLean et al., 1998; Prinja and Pomraning, 2001; Kim and Keller, 2003; Sergeeva et al., 2006). Even under this assumption, there is no analytical solution for the propagation of a pulsed pencil beam that hits a homogeneous tissue, and numerical methods must be introduced.

Another approach introduces assumptions for the tissue impulse response (i.e. the propagation characteristics of a pulsed pencil beam that hits the tissue surface, which may serve as a mathematical basis for solving more complex cases, also known as Green's function) by means of statistical analysis. The beam statistical moments may be calculated analytically up to the second order in each variable, without approximations (Lutomirski et al., 1995). From these moments, a probability function may be estimated (McLean et al., 1998) and serve as an approximated Green's function solution to the beam propagation problem. This approach was used to evaluate the effect of light scattering on the maximal imaging depth of TPLSM systems (Theer and Denk, 2006) and showed good agreement with experimental measurements.

Monte Carlo simulations offer an alternative approach to solving the beam propagation problem, by numerically simulating a large number of photons that propagate through a tissue with known parameters (tissue dimensions, absorption and scattering coefficients, and phase function) and experience absorption and scattering. These simulations are able to model non-homogeneous media, which are a more suitable model for realistic tissues, and do not assume small angle scattering. This detail may be of great importance, since the small angle scattering assumption may result in a significant error when HG phase function is assumed (Pomraning, 1992). Monte-Carlo simulations are considered to be the gold standard in modeling light-tissue interactions and usually showed better accuracy than other methods. There are several free-access Monte Carlo simulation codes (Wang et al., 1995; Boas et al., 2002; Fang and Boas 2009; Yu et al., 2018) that may be used to investigate pulse propagation in three-dimensional complex tissues. The main disadvantage of Monte Carlo methods is that they are the most time-consuming, even though the use of a graphics processing unit (GPU) can shorten the processing time (Yu et al., 2018). These methods also suffer from the drawback of most numerical methods, in that they provide no analytical insight into the process they analyze.

3.4 PARALLEL ACQUISITION FOR RAPID FUNCTIONAL TWO-PHOTON MICROSCOPY

Parallelization of the data acquisition process, in both the illumination and detection parts of the microscope, will enable significantly increased imaging speeds, generally up to the level of parallelization. Suggested parallelization approaches vary between illumination of several diffraction-limited spots, a line, and an entire plane simultaneously. The non-linear characteristics of two-photon fluorescence, as well as basic optical principles and interactions with scattering biological media, impose several limitations on each suggested technique that will be introduced below. Two-photon absorption depends on the squared intensity of the laser beam, and if a laser illuminates a spot with intensity I and excite a fluorescence F, then splitting this beam to N identical spots will result with I/N intensity and F/N^2 fluorescence evolved from each spot. This signal power degradation can be compensated by increasing laser power, up to its maximal power or to the safety limit (Podgorski and Ranganathan, 2016), but eventually laser power will limit either the level of parallelization or the imaging depth inside a scattering media. Therefore, using lasers with higher excitation efficiency is typically of crucial importance for any parallel data acquisition method.

The first attempt to enhance the two-photon microscopy acquisition rate by parallelization involved line illumination, a concept that was borrowed from confocal microscopy (Wilson, 1990; Brakenhoff et al., 1996). Illuminating a line seems very attractive, since with a line-illumination system, a plane may be illuminated in the same time that it takes to illuminate a single line in a standard TPLSM system (~1 msec). However, significant deterioration of the axial sectioning, as well as the required increase in laser power prevented the use of such systems in biological imaging. In addition, such system required use of a CCD to detect the fluorescence (the majority of parallel methods do), which makes the imaging sensitive to tissue scattering of the fluorescence signal as it propagates from inside the tissue to the detector.

3.4.1 Multifocal Multiphoton Microscopy

A different parallelization approach for enhancing data acquisition without significant loss of axial sectioning is multifocal multiphoton microscopy (MMM) (Bewersdorf et al., 1998;

Buist et al., 1998), where a laser beam is divided by a microlens array to several beamlets, and each beamlet is focused on a different spot in the sample space and separately detected by a detector. Design considerations play a significant role in this case. For example, the spatial arrangement of focal points affects the axial sectioning properties of the system, where proximity of 10 μm results in interference with the beamlets and reduces performance (Buist et al., 1998). Moreover, a smart arrangement of the foci distribution and scanning method, e.g. a helical pattern on a rotating disk, may improve scanning speed or system efficiency. If the foci are distant enough, they may be detected by a multi-anode photomultiplier tube (Kim et al., 2007), which makes the imaging system more robust to tissue scattering, comparable to standard TPLSM (as long as scattering doesn't mix one focus' fluorescence with another focus' detector). Finally, the distribution of light between different beamlets isn't uniform, because of the Gaussian profile of the laser beam, which results in a non-uniform image. Custom optical solutions for the aforementioned problems exist, for example, time multiplexing of the multiple foci (Andresen et al., 2001) may eliminate interference between adjacent foci, and beam splitters or diffractive optical elements may be used for achieving a more uniform illumination (Fittinghoff et al., 2000; Nielsen et al., 2001; Sacconi et al., 2003), at the cost of increased complexity. In summary, MMM systems were demonstrated to achieve imaging rates of up to 600 frames/sec by combining MMM with fast resonant scanners (Bahlmann et al., 2007), perform well in scattering media, and maintain a relatively good resolution, comparable to TPLSM. The flexibility of defining the exact number of foci may be used to maximize the laser power usage for a known imaging depth inside a tissue. However, its applicability to scanning a volume, and not just a 2D plane, is not straightforward. Moreover, the necessary design consideration and the complexity of the optical setup prevent wide use of such system and limit their use to small community of optics experts. A possible simple alternative for creating multiple focal spots (or any other 2D pattern) is to use holographic shaping of the laser beam, for example, by using a spatial light modulator (SLM) (Nikolenko et al., 2008). Phase SLMs can digitally modulate the phase front of the illumination beam – a lens is then used to perform a spatial Fourier transform and a mathematical algorithm is used to calculate the necessary phase mask for a desired illumination pattern (Golan et al., 2009). Holographic projection systems can also be used to multiplex the laser beam and focus each spot on a different location in 3D by adding the right phase coding to individual spots (Quirin et al., 2013). Detection of these multiplexed signals imposes another challenge, which is hard to address with existing systems. For a relatively simple case, where the excitation is limited to a small number of spots, and the sample is sparsely labeled, a single PMT detector can detect the projection of all the imaged spots simultaneously (Yang et al., 2016) (Figure 3.4).

3.4.2 Temporal Focusing Multiphoton Microscopy

Wide-field illumination offers a potential solution for fast volume imaging, but its major drawback is that it lacks any optical sectioning. In 2005, Oron et al. (2005) and Zhu et al. (2005) offered an approach that combines widefield illumination with optical sectioning. Their concept is to illuminate a relatively large plane (typical diameter of ~100 μm was demonstrated), but to manipulate the laser pulse's temporal width, so that it reaches its

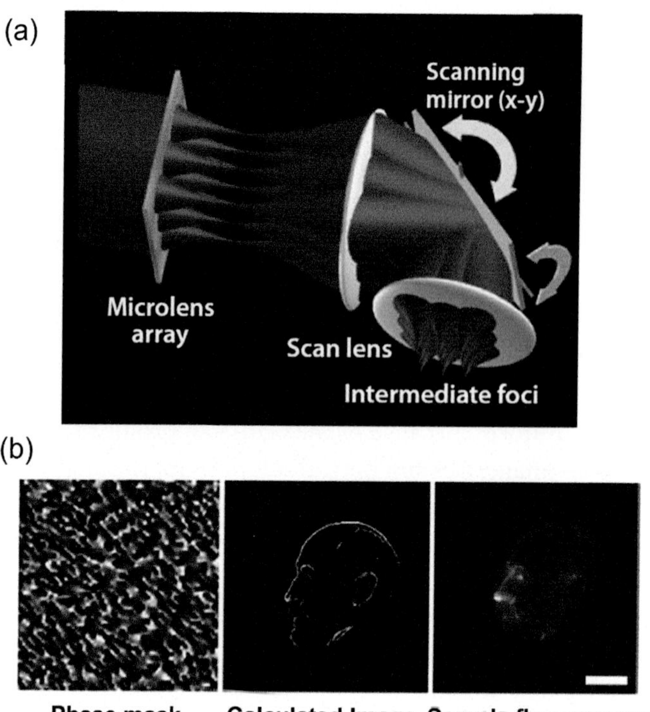

FIGURE 3.4 Parallel excitation for rapid functional imaging. (a) MMM illustration. A microlens array creates multiple foci that are scanned using scanning mirrors and a scan lens. (Image is taken from Wilt et al. (2009).) (b) demonstration of SLM holographic microscope to illuminate flexible 2D patterns. (Image is taken from ref. Nikolenko et al. (2008).)

shortest duration at the objective's focal plane. By axial scanning of the objective lens along the optical axis, an image of the entire volume of interest can be acquired. In TPLSM the pulse duration remains constant as it propagates, but the beam's spatial dimensions change (i.e. the beam is spatially focused), while in this approach the beam's spatial dimensions remain relatively unchanged but the temporal duration changes, is was termed temporal focusing (TF).

For linear optical methods the pulse's temporal duration has no significant effect on the fluorescence signal detected from a sample. However, two-photon absorption is a non-linear process and the emitted fluorescence signal power is inversely proportional to the pulse duration. This may be intuitively understood by thinking about the pulse as a group of photons that travel together at the same direction in space. Between the photons there is an average time difference, which is proportional to the pulse duration. When we stretch the pulse, we expand this time difference (since the pulse is stretched in time, it is also stretched in space) and therefore the probability for a near-simultaneous absorption of two photons is reduced. The reduced probability is translated to reduced fluorescence. A more rigorous treatment requires calculation of $\langle I(t)\rangle$ and $\langle I^2(t)\rangle$, when $\langle\bullet\rangle$ represents temporal averaging, and results in an inversely proportional linear relation between pulse duration and fluorescent emission (Zipfel et al., 2003). Therefore, the temporal focusing approach decreases the fluorescence in any plane except the focal plane, thus achieving optical sectioning capabilities.

The basic concept of TF is to use the optical system to image a scattering object (e.g. a diffraction grating or diffuser plate) onto the system's focal plane. Since each point at the scatterer's surface scatters the light at various angles, a point near the scatterer (and its image point near the focal plane) receives photons from various locations on the scatterer. Therefore, at that point there are path differences between the arriving photons, and the pulse duration is temporally stretched due to this geometrical dispersion. By Fermat's principle, all the optical paths from the scattering surface to the system's focal plane are equal. Therefore, the pulse duration at that plane remains equal to the original pulse duration and shorter than in any other near plane (see Figure 3.5b). A mathematical model describing the TF phenomenon calculated this temporal stretching, showing good agreement with experimental measurements (Durst et al., 2006). Theoretical and experimental results are shown in Figure 3.5.

Temporally focused illumination can be divided into three categories: Widefield Illumination TEmporal Focusing (WITEF), Line Illumination TEmporal Focusing (LITEF), and PATtern illumination TEmporal Focusing (PATEF). WITEF is based on spreading each laser pulse to illuminate the entire field of view, and requires the simplest optical setup: a diffraction grating, a tube lens, and an objective lens in a 4f configuration (Oron et al., 2005). The optical sectioning effect is purely due to the change in laser pulse duration. LITEF may be implemented by either replacing the tube lens with a cylindrical lens (Zhu et al., 2005), or by adding a cylindrical lens, which focuses the laser beam to a line on the diffraction grating (Tal et al., 2005). In this configuration the sectioning capabilities, as well as the fluorescence signal power, were reported to be improved compared to WITEF (Tal et al., 2005; York et al., 2011; Dana et al., 2013). Further modeling and measurements showed how the addition of spatial light focusing increases the probability of near-simultaneous two-photon absorption and therefore the optical sectioning of LITEF systems. Moreover, the enhanced LITEF optical sectioning was robust to light-scattering effects down to a depth of one millimeter inside a tissue phantom (Dana et al., 2013). These characteristics make LITEF more suitable than WITEF for rapid volumetric imaging. A LITEF-based system was used for such an application inside a lightly-scattering engineered tissue, recording activity from 1,500 cells with 200 frames/sec and a 10Hz volumetric acquisition rate (Dana et al., 2014). However, the fluorescence was detected with an EMCCD camera, which makes this method sensitive to scattering effects of the emitted fluorescent signal and less suitable for imaging deep inside the brain. Another implementation of a TF-based system for rapid imaging used a temporally focused, cell-sized focal spot for efficient excitation of somata, while recording the signal with a PMT detector (Prevedel et al., 2016). This approach applies TF in a TPLSM-like configuration of exciting a single spot (or a very small plane). This approach works well for recording inside a scattering medium, but without any parallelization of the acquisition process, and therefore no significant increase in acquisition rates compared to TPLSM systems.

Combining TF optical setup with computer generated holography approaches (Papagiakoumou et al., 2008, Therrien et al., 2011) allows illuminating flexible

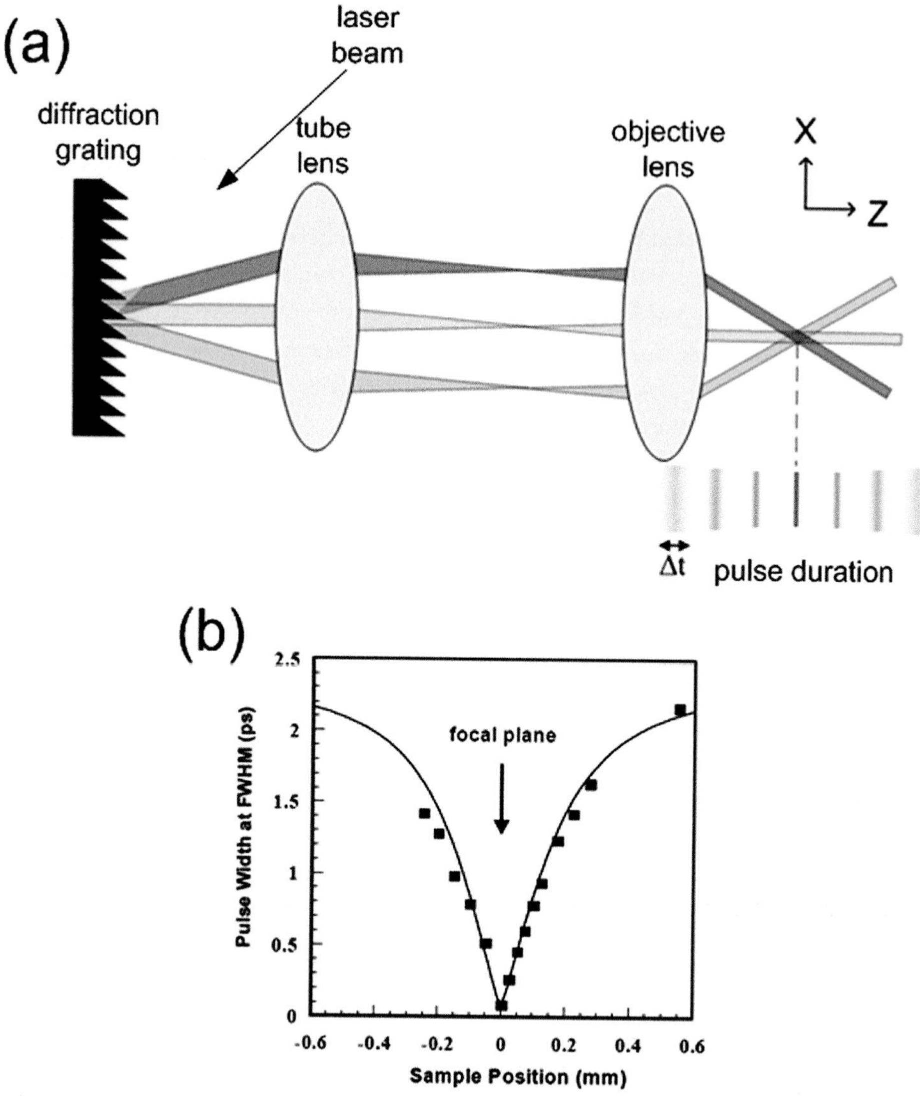

FIGURE 3.5 Basics of temporal focusing microscopy. (a) Light propagation scheme of wide-field TF. A laser beam impinges upon a diffraction grating and each spectral component is diffracted towards a different angle. Two lenses, usually a tube lens and an objective lens, in a 4f configuration image the grating surface onto the objective's focal plane. The result of the change in spectral separation during pulse propagation results in variable temporal duration, as illustrated in the figure. (b) Autocorrelator measurements of laser pulse duration as function of distance from the focal plane. (Taken from ref. Zhu et al., 2005.)

two-dimensional patterns on the sample plane. In this case the pattern is determined by shaping the illuminated laser beam wavefront, usually using an SLM, and using a lens to Fourier transform the phase pattern to an intensity pattern on the diffraction grating's surface. Interestingly, even though computer generated holography is capable of generating 3D patterns, the use of TF limits its capabilities to a single plane. This limitation can be bypassed by adding an additional SLM into the optical setup, projecting the temporally focused pattern in 3D (Hernandez et al., 2016; Pégard et al., 2017). Therefore, PATEF may also be seen as a generalization of holographic shaping of the laser beam with enhanced axial sectioning due to the TF effect. PATEF offers an unprecedented capability of tailoring a suitable illumination pattern to any application, but with the cost of increased complexity and reduced system's efficiency due to losses in the SLM devices, which makes it less suitable for large-scale imaging applications, but very attractive for single-cell two-photon stimulation of light-sensitive ion channels (Andrasfalvy et al., 2010; Papagiakoumou et al., 2013; Rickgauer et al., 2014; Hernandez et al., 2016; Pégard et al., 2017). The unique combination of large-scale optically sectioned illumination is also useful for several other optical applications. Therefore, TF-based systems were also shown to be useful in applications such as super-resolution microscopy (Vaziri et al., 2008), micro-machining (Vitek et al., 2010), lithographic microfabrication (Kim and So 2010; Vitek et al., 2010), and fluorescence lifetime imaging (Choi et al., 2012).

3.4.3 Light Sheet Microscopy

Selective Plane Illumination Microscopy (SPIM) is an alternative parallelization approach for simultaneously illuminating a plane. Originally, the plane was illuminated from the side, and imaging from above the sample, perpendicular to the illuminated plane (Huisken et al., 2004). This method was further generalized to implement two-photon SPIM with either Gaussian or Bessel beams (Planchon et al., 2011; Truong et al., 2011). SPIM offers rapid acquisition combined with optically sectioned images, similar to temporal focusing. However technical issues limit the applicability of light-sheet microscopy to mammalian *in vivo* experiments, because the illumination and detection must be handled from two orthogonal directions. In order to adjust SPIM for *in vivo* imaging, a miniaturized SPIM probe, consisting of an objective lens and a prism, was designed (Engelbrecht et al., 2010). However, it requires pushing a relatively large-scale fiber and prism into the brain, causing damage to the tissue.

A different implementation of planar illumination approach, called swept confocally aligned planar excitation (SCAPE), has overcome this fundamental limitation of SPIM by imaging an oblique plane inside the sample through the objective lens, and using the same objective to image the illuminated plane onto a CCD detector (Bouchard et al., 2015), reaching imaging rates of 10–20 volumes/sec for volumes of 600 μm × 650 μm × 134 μm. The current published SCAPE system is based on linear fluorescence, but attempts to build a two-photon SCAPE system were reported (Yu et al., 2017) and a similar system with two-photon excitation capabilities was presented (Kumar et al., 2018).

3.5 RAPID AXIAL SCANNING FOR VOLUMETRIC FUNCTIONAL IMAGING

3.5.1 Rapid Mechanical Scanning of the Objective Lens

Implementation of fast axial (z-axis) scanning is more complicated than that of rapid lateral (xy plane) scanning, since objective lenses are designed to work in a certain plane (the focal plane) and a naive attempt to use an objective lens to image or illuminate a different plane will cause significant image aberrations (Botcherby et al., 2007). Therefore, aberration-free rapid axial scanning would have to physically move the objective or to find a way to correct these aberrations. Rapid mechanical objective movement for axial scanning based on a piezo-electric device (Gobel et al., 2007), was demonstrated to scan a depth of approximately 200 μm at 10 Hz, which is often a sufficient rate for detecting single spikes with commonly used fluorescent calcium sensors (Chen et al., 2013; Dana et al., 2016). Therefore, mechanical scanning of the objective lens offers a simple and accessible solution for rapid axial scanning. However, this solution has some severe drawbacks. First, the objective lens is located close to the sample and moving it rapidly could vibrate the sample. Moreover, there are situations, like neural imaging of freely moving animal, where the objective lens is implanted into an animal's head, and such movement is not feasible (Sawinski et al., 2009; Ghosh et al., 2011; Ziv et al., 2013). Second, a typical objective lens is relatively heavy, and new objective models for large field of view imaging are usually even bigger and heavier (Chen et al., 2016; Sofroniew et al., 2016; Stirman et al., 2016). Therefore, the axial mechanical movement would be limited by the objective's inertia. Third, when a photo-stimulation system is built into the microscope (Andrasfalvy et al., 2010; Papagiakoumou et al., 2010; Anselmi et al., 2011), the objective is used for both excitation and detection, and scanning it causes a typically unwanted coupling between the detection and excitation subsystems.

3.5.2 Remote Scanning of the Focal Spot

Remote scanning is defined as an axial shift of the imaging system focal plane without moving the objective or the sample. Because of the challenges associated with mechanical scanning of the objective lens, remote scanning solutions can be very useful for rapid imaging systems and multi-modal microscopy systems. Current remote scanning techniques are mainly implemented by two AODs (Duemani Reddy et al. 2008; Katona et al., 2012), an SLM (Nikolenko et al., 2008; Dal Maschio et al., 2011; Yang et al., 2016), by multiplexing the laser beam and focusing each beamlet on a different plane (Stirman et al., 2016), or by adding a tunable lens to the optical path (Grewe et al., 2011; Kong et al., 2015). Alternatively, remote scanning may be implemented by adding an additional imaging arm, conjugated to the original imaging system, and moving a mirror that scans the focal spot (Hoover et al., 2011; Botcherby et al., 2012; Sofroniew et al., 2016).

3.5.2.1 Remote Scanning by Lens Tuning

Electrically tunable lenses (ETLs) are filled with a liquid polymer and have a flexible membrane surface. By changing the pressure on the membrane surface, its curvature is changed and the lens focal length is tuned. Similarly, ultrasound lenses are tunable lenses, which their properties are modulated by an ultrasonic wave, generated by small piezoelectric devices. These tunable lenses are usually used in combination with a second standard lens (also called an offset lens) in a telescope configuration. Tuning the focal length of the tunable lens will result in divergence or convergence of the beam, and will axially scan the focal point when focused by an objective lens. A scanning range of 700 μm was demonstrated as well as rapid imaging of two layers separated by 40 μm *in vivo* (Grewe et al., 2011). ETL-based remote scanning was used to acquire 3D structure of the mouse retina *in vivo* with two-photon functional recording of retinal ganglion cells activity (Bar-Noam Schejter et al., 2016). Currently, ETL performance is too slow for multi-plane scanning (10–15 ms minimal time required for axial plane shift (Grewe et al., 2011; Nakai et al., 2015)), and they also have a side-effect of changing the field of view diameter for different focal shifts, due to the different illumination beam divergence. Ultrasonic lenses are much faster and can scan a depth of over 100 μm within less than a microsecond (Kong et al., 2015). However, they require more sophisticated control systems to monitor the focal spot's exact position, and more complex integration with the laser source and scanning mirrors. To date, tunable lenses are not used extensively for functional imaging due to these limitations.

Inertia-free scanning of the objective focal spot may also be implemented by an AOD pair (Duemani Reddy et al. 2008; Katona et al., 2012). Each AOD adds a linear chirp in an opposite direction to the beam's phase-front and together they focus it on a different plane, equivalent to adding an "adaptive lens" into the optical path (Kaplan et al., 2001). A complete 3D AOD-based scanning unit is built from four AOD scanners, and establishes a random-access imaging microscope, where each point in the imaged volume may be accessed independently of the previously imaged point location. This type of microscope requires a sophisticated optical setup, which includes spatial and angular dispersion compensation units (Grewe et al., 2010; Katona et al., 2012; Cotton et al., 2013). 3D manipulation of the focal spot (including axial scanning) by controlling the laser beam's phase-front may also be implemented holographically, using a phase SLM to add a controlled quadratic phase to the beam's phase-front (Nikolenko et al., 2008; Dal Maschio et al., 2011; Yang et al., 2016). In both of these methods the spatial resolution is deteriorated for large shifts from the objective focal plane because of aberrations that will be discussed in the following section. Random-access systems are considered very efficient, since sampling may be limited to points of interest only, and not to empty space. However, once the imaging system spatial sampling density is low, such as in these random-access systems, motions artifacts become a major concern. In order to correct for such artifacts, it was suggested to combine dense sampling of reference planes within the sample (Cotton et al., 2013), and use them to correct for tissue movements. The reliability of random-access system under realistic conditions of tissue movements is hard to estimate and yet to be broadly tested.

Temporal multiplexing of the laser beam enables the establishment of a hybrid method of TPLSM-MMM which is not as fast as MMM, but performs better in scattering media, requires a simpler setup and may be easily expanded to image 3D volume (Field et al., 2010; Cheng et al., 2011; Stirman et al., 2016). In this method, a laser beam is split into several identical beamlets, then each beamlet is directed to a different optical path which creates a temporal delay of several nanoseconds in their arrival time to the sample. The beamlets may be focused on different regions in the same plane, or be focused on different planes by changing each beamlet divergence or convergence with a weak lens.

Scanning is performed using a single scanning unit for all the beamlets, which may be chosen from the scanning devices that were discussed earlier. The temporal multiplexing of the beamlets enables detection of their fluorescence with a single rapid detector without interference by the adjacent beams, which makes this technique as robust as TPLSM to scattering effects. The parallelization level, which equals the number of beamlets, is limited by several factors such as the technical complexity of merging all of them to a single optical axis, the fluorescence signal decay time, which is typically ~3 nsec, and the laser source repetition rate, which is typically 80 MHz. Therefore, for commonly used laser sources and fluorescence sensors, the beam can be multiplexed for maximal number of four beamlets.

3.5.2.2 Remote Scanning by a Moving Mirror

Conventional attempts to focus the laser beam on a different plane than the focal plane, for example, by adding a quadratic phase to the laser beam that was mentioned in the previous section, results in resolution deterioration due to gradual aberrations, which limits the scanning range. These aberrations become more significant when high NA objectives are used. This is because a parabolic phase-front is added to the illumination beam for shifting the focal plane, but an addition of a parabolic profile phase will shift the focal plane only under the paraxial approximation (Botcherby et al., 2007), which is justified only for low NA lenses. Remote scanning using a moving mirror and a second objective lens is an elegant way of performing aberration-free high-NA imaging in virtually any desired plane. This method was shown to be useful for both two-photon excitation (Hoover et al., 2011; Botcherby et al., 2012; Sofroniew et al., 2016) and the detection of tilted planes by a CCD (Anselmi et al., 2011). The basic concept is that a collimated excitation beam passes through a secondary objective lens and a mirror. Naturally the beam will focus to the focal plane of the objective. In this technique, the mirror is located at distance d from the focal plane, the light is reflected from the mirror back to the objective, and its wavefront after passing through the objective lens is identical to a wavefront of a point source located at distance 2d from the focal plane. Two lenses are used to image the secondary objective back aperture to the main objective (the one that is located in front of the sample) back aperture. These two objectives are chosen to be as similar as possible, and this similarity causes the light to focus to a distance 2d from the main objective focal plane to a diffraction-limited spot. Similarly, this concept may also be used in the opposite direction for image detection. In this case a planar image is located at distance d from the focal plane and direct imaging of it will result in a low-quality image. By imaging the main objective back aperture to a secondary objective back aperture, this image will be focused to a distance d from the focal plane, and placing a mirror at distance d/2 from the focal plane will reflect the light back as it is located from the focal plane of the objective, and it may be imaged to a CCD without aberrations (Anselmi et al., 2011). Since in this case the objective's movement is replaced by movement of a lightweight mirror, significant improvement of the axial scanning speed is possible, up to KHz rates (Botcherby et al., 2012) (Figure 3.6).

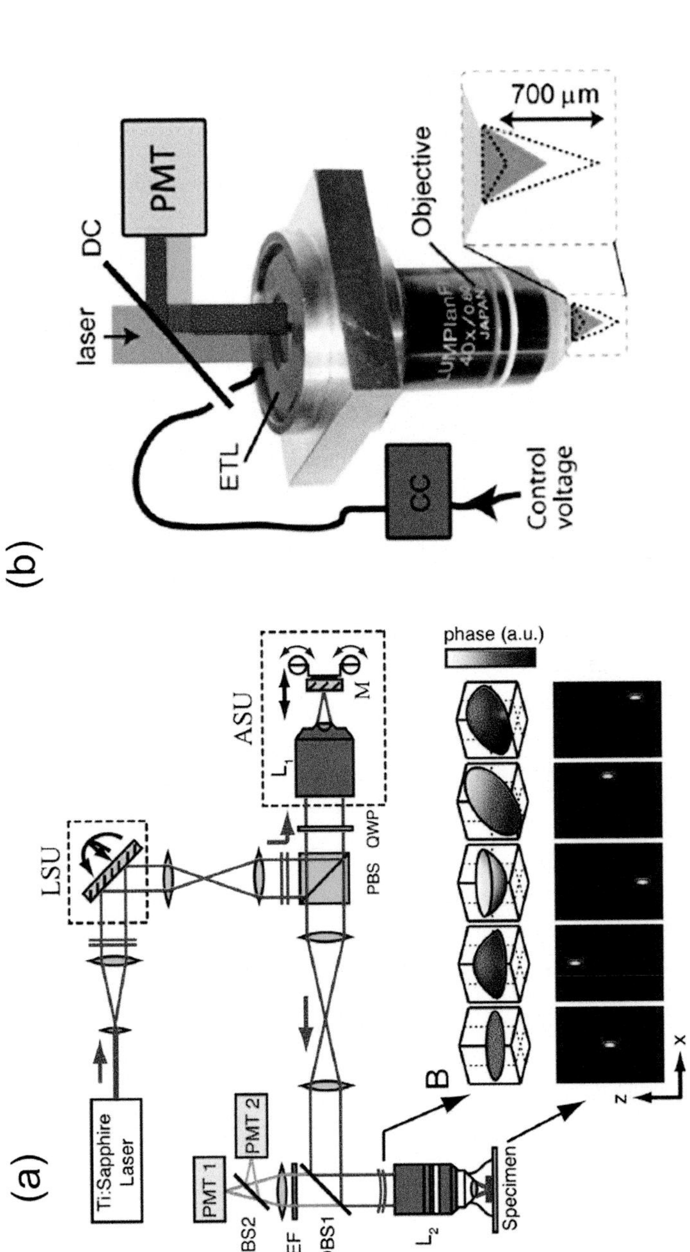

FIGURE 3.6 Remote scanning of the focal spot. (a) Remote scanning by a second objective and a moving mirror. The lateral scanning unit (LSU) consists of standard galvanometric mirrors, the axial scanning unit (ASU) with objective L_1 and moving mirror M performs axial scanning of the focal spot of objective L_2. Insets show the beam phase front at the entrance pupil of L_2 and movement of the focal spot. PSB – polarizing beam splitter, QWP – quarter wave plate, EF – emission filter. Image is taken from ref. Botcherby et al. (2012). (b) Remote scanning by electrically tunable lens. The ETL and an offset lens are mounted before the objective lens, the ETL curvature is controlled by an external voltage, and changes the incoming laser beam divergence. Scanning range of 700 μm was demonstrated using this method. (Image is taken from Grewe et al. (2011).)

REFERENCES

Andrasfalvy, B., B. Zemelman, J. Tang and A. Vaziri (2010). "Two-photon single-cell optogenetic control of neuronal activity by sculpted light." *PNAS* **107**: 11981–11986.

Andresen, V., A. Egner and S. W. Hell (2001). "Time-multiplexed multifocal multiphoton microscope." *Optics Letters* **26**(2): 75–77.

Anselmi, F., C. Ventalon, A. Bègue, D. Ogden and V. Emiliani (2011). "Three-dimensional imaging and photostimulation by remote-focusing and holographic light patterning." *PNAS* **108**(49): 19504–19509.

Bahlmann, K., P. T. So, M. Kirber, R. Reich, B. Kosicki, W. McGonagle and K. Bellve (2007). "Multifocal multiphoton microscopy (MMM) at a frame rate beyond 600 Hz." *Optics Express* **15**(17): 10991–10998.

Bar-Noam Schejter, A., N. Farah and S. Shoham (2016). "Correction-free remotely scanned two-photon in vivo mouse retinal imaging." *Light: Science and Applications* **5**: e16007.

Bewersdorf, J., R. Pick and S. W. Hell (1998). "Multifocal multiphoton microscopy." *Optics Letters* **23**(9): 655–657.

Boas, D., J. Culver, J. Stott and A. Dunn (2002). "Three dimensional Monte Carlo code for photon migration through complex heterogeneous media including the adult human head." *Optics Express* **10**(3): 159–170.

Borgers, C. and E. W. Larsen (1995). *The Fermi Pencil Beam Approximation*. International Conference on Mathematics and Computation, Reactor Physics, and Environmental Analyses, Portland, OR.

Botcherby, E. J., R. Juskaitis, M. J. Booth and T. Wilson (2007). "Aberration-free optical refocusing in high numerical aperture microscopy." *Optics Letters* **32**(14): 2007–2009.

Botcherby, E. J., C. W. Smith, M. M. Kohl, D. Débarre, M. J. Booth, R. Juškaitis, O. Paulsen and T. Wilson (2012). "Aberration-free three-dimensional multiphoton imaging of neuronal activity at kHz rates." *PNAS* **109**(8): 2919–2924.

Bouchard, M. B., V. Voleti, C. S. Mendes, C. Lacefield, W. B. Grueber, R. S. Mann, R. M. Bruno and E. M. Hillman (2015). "Swept confocally-aligned planar excitation (SCAPE) microscopy for high-speed volumetric imaging of behaving organisms." *Nature Photonics* **9**(2): 113–119.

Brakenhoff, G., J. Squier, T. Norris, A. C. Bliton, M. Wade and B. Athey (1996). "Real-time two-photon confocal microscopy using a femtosecond, amplified Ti:sapphire system." *Journal of Microscopy* **181**(3): 253–259.

Buist, A., M. Muller, J. Squier and G. Brakenhoff (1998). "Real time two-photon absorption microscopy using multi point excitation." *Journal of Microscopy* **192**(2): 217.

Chen, J. L., F. F. Voigt, M. Javadzadeh, R. Krueppel and F. Helmchen (2016). "Long-range population dynamics of anatomically defined neocortical networks." *eLife* **5**: e14679.

Chen, T.-W., T. J. Wardill, Y. Sun, S. R. Pulver, S. L. Renninger, A. Baohan, E. R. Schreiter, R. A. Kerr, M. B. Orger and V. Jayaraman (2013). "Ultrasensitive fluorescent proteins for imaging neuronal activity." *Nature* **499**(7458): 295–300.

Chen, X., U. Leischner, N. L. Rochefort, I. Nelken and A. Konnerth (2011). "Functional mapping of single spines in cortical neurons in vivo." *Nature* **475**(7357): 501–505.

Cheng, A., J. T. Goncalves, P. Golshani, K. Arisaka and C. Portera-Cailliau (2011). "Simultaneous two-photon calcium imaging at different depths with spatiotemporal multiplexing." *Nature Methods* **8**(2): 139–142.

Cheong, W.-F., S. A. Prahl and A. J. Welch (1990). "A review of the optical properties of biological tissues." *IEEE Journal of Quantum Electronics* **26**(12): 2166–2185.

Choi, H., D. S. Tzeranis, J. W. Cha, P. Clémenceau, S. J. de Jong, L. K. van Geest, J. H. Moon, I. V. Yannas and P. T. So (2012). "3D-resolved fluorescence and phosphorescence lifetime imaging using temporal focusing wide-field two-photon excitation." *Optics Express* **20**(24): 26219–26235.

Cotton, R. J., E. Froudarakis, P. Storer, P. Saggau and A. S. Tolias (2013). "Three-dimensional mapping of microcircuit correlation structure." *Frontiers in Neural Circuits* **7**: 151.

Dal Maschio, M., A. M. De Stasi, F. Benfenati and T. Fellin (2011). "Three-dimensional in vivo scanning microscopy with inertia-free focus control." *Optics Letters* **36**(17): 3503–3505.

Dana, H., N. Kruger, A. Ellman and S. Shoham (2013). "Line temporal focusing characteristics in transparent and scattering media." *Optics Express* **21**: 5677–5687.

Dana, H., A. Marom, S. Paluch, R. Dvorkin, I. Brosh and S. Shoham (2014). "Hybrid multiphoton volumetric functional imaging of large-scale bioengineered neuronal networks." *Nature Communications* **5**: ncomms4997.

Dana, H., B. Mohar, Y. Sun, S. Narayan, A. Gordus, J. P. Hasseman, G. Tsegaye, G. T. Holt, A. Hu and D. Walpita (2016). "Sensitive red protein calcium indicators for imaging neural activity." *eLife* **5**: e12727.

Deneux, T., A. Kaszas, G. Szalay, G. Katona, T. Lakner, A. Grinvald, B. Rózsa and I. Vanzetta (2016). "Accurate spike estimation from noisy calcium signals for ultrafast three-dimensional imaging of large neuronal populations in vivo." *Nature Communications* **7**: 12190.

Denk, W., J. H. Strickler and W. W. Webb (1990). "Two-photon laser scanning fluorescence microscopy." *Science* **248**(4951): 73.

Duemani Reddy, G., K. Kelleher, R. Fink and P. Saggau (2008). "Three-dimensional random access multiphoton microscopy for functional imaging of neuronal activity." *Nature Neuroscience* **11**(6): 713–720.

Durst, M. E., G. Zhu and C. Xu (2006). "Simultaneous spatial and temporal focusing for axial scanning." *Optics Express* **14**(25): 12243–12254.

Engelbrecht, C. J., F. Voigt and F. Helmchen (2010). "Miniaturized selective plane illumination microscopy for high-contrast in vivo fluorescence imaging." *Optics Letters* **35**(9): 1413–1415.

Fan, G., H. Fujisaki, A. Miyawaki, R. K. Tsay, R. Y. Tsien and M. H. Ellisman (1999). "Video-rate scanning two-photon excitation fluorescence microscopy and ratio imaging with cameleons." *Biophysical Journal* **76**(5): 2412–2420.

Fang, Q. and D. A. Boas (2009). "Monte Carlo simulation of photon migration in 3D turbid media accelerated by graphics processing units." *Optics Express* **17**(22): 20178–20190.

Field, J. J., K. E. Sheetz, E. V. Chandler, E. E. Hoover, M. D. Young, S. Y. Ding, A. W. Sylvester, D. Kleinfeld and J. A. Squier (2010). "Differential multiphoton laser scanning microscopy." *IEEE Journal of Selected Topics in Quantum Electronics* **18**(1): 14–28.

Fittinghoff, D., P. Wiseman and J. Squier (2000). "Widefield multiphoton and temporally decorrelated multifocal multiphoton microscopy." *Optics Express* **7**(8): 273–279.

Garaschuk, O., R. I. Milos and A. Konnerth (2006). "Targeted bulk-loading of fluorescent indicators for two-photon brain imaging in vivo." *Nature Protocols* **1**(1): 380–386.

Garion, L., U. Dubin, Y. Rubin, M. Khateb, Y. Schiller, R. Azouz and J. Schiller (2014). "Texture coarseness responsive neurons and their mapping in layer 2–3 of the rat barrel cortex in vivo." *eLife* **3**: e03405.

Ghosh, K. K., L. D. Burns, E. D. Cocker, A. Nimmerjahn, Y. Ziv, A. El Gamal and M. J. Schnitzer (2011). "Miniaturized integration of a fluorescence microscope." *Nature Methods* **8**(10): 871–878.

Gobel, W., B. M. Kampa and F. Helmchen (2007). "Imaging cellular network dynamics in three dimensions using fast 3D laser scanning." *Nature Methods* **4**(1): 73–79.

Golan, L., I. Reutsky, N. Farah and S. Shoham (2009). "Design and characteristics of holographic neural photo-stimulation systems." *Journal of Neural Engineering* **6**: 066004.

Goldstein, S. R., T. Hubin, S. Rosenthal and C. Washburn (1990). "A confocal video-rate laser-beam scanning reflected-light microscope with no moving parts." *Journal of Microscopy* **157**(1): 29–38.

Grewe, B. F., D. Langer, H. Kasper, B. M. Kampa and F. Helmchen (2010). "High-speed in vivo calcium imaging reveals neuronal network activity with near-millisecond precision." *Nature Methods* **7**(5): 399–405.

Grewe, B. F., F. F. Voigt, M. Van't Hoff and F. Helmchen (2011). "Fast two-layer two-photon imaging of neuronal cell populations using an electrically tunable lens." *Biomedical Optics Express* **2**(7): 2035–2046.

Helmchen, F. and W. Denk (2005). "Deep tissue two-photon microscopy." *Nature Methods* **2**(12): 932–940.

Henyey, L. G. and J. L. Greenstein (1941). "Diffuse radiation in the galaxy." *Astrophysics Journal* **93**: 70–83.

Hernandez, O., E. Papagiakoumou, D. Tanese, K. Fidelin, C. Wyart and V. Emiliani (2016). "Three-dimensional spatiotemporal focusing of holographic patterns." *Nature Communications* **7**: 11928.

Hillman, E. M., V. Voleti, K. Patel, W. Li, H. Yu, C. Perez-Campos, S. E. Benezra, R. M. Bruno and P. T. Galwaduge (2018). "High-speed 3D imaging of cellular activity in the brain using axially-extended beams and light sheets." *Current Opinion in Neurobiology* **50**: 190–200.

Hoover, E. E., M. D. Young, E. V. Chandler, A. Luo, J. J. Field, K. E. Sheetz, A. W. Sylvester and J. A. Squier (2011). "Remote focusing for programmable multi-layer differential multiphoton microscopy." *Biomedical Optics Express* **2**(1): 113–122.

Horton, N. G., K. Wang, D. Kobat, C. G. Clark, F. W. Wise, C. B. Schaffer and C. Xu (2013). "In vivo three-photon microscopy of subcortical structures within an intact mouse brain." *Nature Photonics* **7**(3): 205–209.

Huisken, J., J. Swoger, F. Del Bene, J. Wittbrodt and E. H. K. Stelzer (2004). "Optical sectioning deep inside live embryos by selective plane illumination microscopy." *Science* **305**(5686): 1007–1009.

Kaplan, A., N. Friedman and N. Davidson (2001). "Acousto-optic lens with very fast focus scanning." *Optics Letters* **26**(14): 1078–1080.

Kara, P. and J. D. Boyd (2009). "A micro-architecture for binocular disparity and ocular dominance in visual cortex." *Nature* **458**(7238): 627–631.

Katona, G., G. Szalay, P. Maák, A. Kaszás, M. Veress, D. Hillier, B. Chiovini, E. S. Vizi, B. Roska and B. Rózsa (2012). "Fast two-photon in vivo imaging with three-dimensional random-access scanning in large tissue volumes." *Nature Methods* **9**(2): 201–208.

Kerr, J. N. D., D. Greenberg and F. Helmchen (2005). "Imaging input and output of neocortical networks in vivo." *PNAS* **102**(39): 14063–14068.

Kim, A. D. and J. B. Keller (2003). "Light propagation in biological tissue." *Journal of the Optical Society of America A* **20**(1): 92–98.

Kim, D. and P. T. C. So (2010). "High-throughput three-dimensional lithographic microfabrication." *Optics Letters* **35**(10): 1602–1604.

Kim, K. H., C. Buehler, K. Bahlmann, T. Ragan, W. C. A. Lee, E. Nedivi, E. L. Heffer, S. Fantini and P. T. C. So (2007). "Multifocal multiphoton microscopy based on multianode photomultiplier tubes." *Optics Express* **15**(18): 11658–11678.

Kim, K. H., C. Buehler and P. T. C. So (1999). "High-speed, two-photon scanning microscope." *Applied Optics* **38**(28): 6004–6009.

Kim, K. H., B. K., T. Ragan, D. Kim and P. T. C. So (2008). "High speed imaging using multiphoton excitation microscopy." In *Handbook of Biomedical Nonlinear Optical Microscopy*. Eds. B. R. Masters and P. T. C. So. Oxford University Press.

Kong, L., J. Tang, J. P. Little, Y. Yu, T. Lämmermann, C. P. Lin, R. N. Germain and M. Cui (2015). "Continuous volumetric imaging via an optical phase-locked ultrasound lens." *Nature Methods* **12**(8): 759.

Kumar, M., S. Kishore, J. Nasenbeny, D. L. McLean and Y. Kozorovitskiy (2018). "Integrated one- and two-photon scanned oblique plane illumination (SOPi) microscopy for rapid volumetric imaging." *Optics Express* **26**(10): 13027–13041.

Lillis, K. P., A. Eng, J. A. White and J. Mertz (2008). "Two-photon imaging of spatially extended neuronal network dynamics with high temporal resolution." *Journal of Neuroscience Methods* **172**(2): 178–184.

Lutomirski, R. F., A. P. Ciervo and G. J. Hall (1995). "Moments of multiple scattering." *Applied Optics* **34**(30): 7125.

McLean, J. W., J. D. Freeman and R. E. Walker (1998). "Beam spread function with time dispersion." *Applied Optics* **37**(21): 4701–4711.

Milton, G., C. Ireland and J. Ley (1983). "Electro-optic and acousto-optic scanning and deflection." *Optical Engineering* **3**.

Nakai, Y., M. Ozeki, T. Hiraiwa, R. Tanimoto, A. Funahashi, N. Hiroi, A. Taniguchi, S. Nonaka, V. Boilot and R. Shrestha (2015). "High-speed microscopy with an electrically tunable lens to image the dynamics of in vivo molecular complexes." *Review of Scientific Instruments* **86**(1): 013707.

Nielsen, T., M. Fricke, D. Hellweg and P. Andresen (2001). "High efficiency beam splitter for multifocal multiphoton microscopy." *Journal of Microscopy* **201**(3): 368–376.

Nikolenko, V., B. O. Watson, R. Araya, A. Woodruff, D. S. Peterka and R. Yuste (2008). "SLM microscopy: scanless two-photon imaging and photostimulation with spatial light modulators." *Frontiers in Neural Circuits* **2**: 5.

Oheim, M., E. Beaurepaire, E. Chaigneau, J. Mertz and S. Charpak (2001). "Two-photon microscopy in brain tissue: parameters influencing the imaging depth." *Journal of Neuroscience Methods* **111**(1): 29–37.

Ohki, K., S. Chung, Y. H. Ch'ng, P. Kara and R. C. Reid (2005). "Functional imaging with cellular resolution reveals precise micro-architecture in visual cortex." *Nature* **433**(7026): 597–603.

Ohki, K., S. Chung, P. Kara, M. Hübener, T. Bonhoeffer and R. C. Reid (2006). "Highly ordered arrangement of single neurons in orientation pinwheels." *Nature* **442**(7105): 925–928.

Oron, D., E. Tal and Y. Silberberg (2005). "Scanningless depth-resolved microscopy." *Optics Express* **13**: 1468–1476.

Ouzounov, D. G., T. Wang, M. Wang, D. D. Feng, N. G. Horton, J. C. Cruz-Hernández, Y.-T. Cheng, J. Reimer, A. S. Tolias, N. Nishimura and C. Xu (2017). "In vivo three-photon imaging of activity of GCaMP6-labeled neurons deep in intact mouse brain." *Nature Methods* **14**: 388.

Papagiakoumou, E., F. Anselmi, A. Begue, V. de Sars, J. Gluckstad, E. Y. Isacoff and V. Emiliani (2010). "Scanless two-photon excitation of channelrhodopsin-2." *Nature Methods* **7**(10): 848–854.

Papagiakoumou, E., A. Bègue, B. Leshem, O. Schwartz, B. M. Stell, J. Bradley, D. Oron and V. Emiliani (2013). "Functional patterned multiphoton excitation deep inside scattering tissue." *Nature Photonics* **7**(4): 274–278.

Papagiakoumou, E., V. de Sars, D. Oron and V. Emiliani (2008). "Patterned two-photon illumination by spatiotemporal shaping of ultrashort pulses." *Optics Express* **16**: 22039–22047.

Pégard, N. C., A. R. Mardinly, I. A. Oldenburg, S. Sridharan, L. Waller and H. Adesnik (2017). "Three-dimensional scanless holographic optogenetics with temporal focusing (3D-SHOT)." *Nature Communications* **8**(1): 1228.

Planchon, T. A., L. Gao, D. E. Milkie, M. W. Davidson, J. A. Galbraith, C. G. Galbraith and E. Betzig (2011). "Rapid three-dimensional isotropic imaging of living cells using Bessel beam plane illumination." *Nature Methods* **8**(5): 417–423.

Pnevmatikakis, E. A., D. Soudry, Y. Gao, T. A. Machado, J. Merel, D. Pfau, T. Reardon, Y. Mu, C. Lacefield and W. Yang (2016). "Simultaneous denoising, deconvolution, and demixing of calcium imaging data." *Neuron* **89**(2): 285–299.

Podgorski, K. and G. Ranganathan (2016). "Brain heating induced by near-infrared lasers during multiphoton microscopy." *Journal of Neurophysiology* **116**(3): 1012–1023.

Pomraning, G. C. (1992). "The Fokker-Planck operator as an asymptotic limit." *Mathematical Models and Methods in Applied Science* **2**(1): 21–36.

Prevedel, R., A. J. Verhoef, A. J. Pernía-Andrade, S. Weisenburger, B. S. Huang, T. Nöbauer, A. Fernández, J. E. Delcour, P. Golshani and A. Baltuska (2016). "Fast volumetric calcium imaging across multiple cortical layers using sculpted light." *Nature Methods* **13**(12): 1021–1028.

Prinja, A. K. and G. C. Pomraning (2001). "A generalized Fokker-Planck model for transport of collimated beams." *Nuclear Science and Engineering* **137**(3): 227–235.

Quirin, S., D. S. Peterka and R. Yuste (2013). "Instantaneous three-dimensional sensing using spatial light modulator illumination with extended depth of field imaging." *Optics Express* **21**(13): 16007–16021.

Rickgauer, J. P., K. Deisseroth and D. W. Tank (2014). "Simultaneous cellular-resolution optical perturbation and imaging of place cell firing fields." *Nature Neuroscience* **17**(12): 1816–1824.

Rothschild, G., I. Nelken and A. Mizrahi (2010). "Functional organization and population dynamics in the mouse primary auditory cortex." *Nature Neuroscience* **13**(3): 353–360.

Sacconi, L., E. Froner, R. Antolini, M. Taghizadeh, A. Choudhury and F. Pavone (2003). "Multiphoton multifocal microscopy exploiting a diffractive optical element." *Optics Letters* **28**(20): 1918–1920.

Sawinski, J., D. J. Wallace, D. S. Greenberg, S. Grossmann, W. Denk and J. N. D. Kerr (2009). "Visually evoked activity in cortical cells imaged in freely moving animals." *PNAS* **106**(46): 19557–19562.

Sergeeva, E. A., K. M. Yu and A. V.Priezzhev (2006). "Propagation of a femtosecond pulse in a scattering medium: theoretical analysis and numerical simulation." *Quantum Electronics* **36**(11): 1023.

Smetters, D., A. Majewska and R. Yuste (1999). "Detecting action potentials in neuronal populations with calcium imaging." *Methods* **18**(2): 215–221.

Sofroniew, N. J., D. Flickinger, J. King and K. Svoboda (2016). "A large field of view two-photon mesoscope with subcellular resolution for in vivo imaging." *eLife* **5**: e14472.

Stirman, J. N., I. T. Smith, M. W. Kudenov and S. L. Smith (2016). "Wide field-of-view, multi-region, two-photon imaging of neuronal activity in the mammalian brain." *Nature Biotechnology* **34**: 857.

Stosiek, C., O. Garaschuk, K. Holthoff and A. Konnerth (2003). "In vivo two-photon calcium imaging of neuronal networks." *PNAS* **100**(12): 7319.

Svoboda, K., W. Denk, D. Kleinfeld and D. W. Tank (1997). "In vivo dendritic calcium dynamics in neocortical pyramidal neurons." *Nature* **385**(6612): 161–165.

Tal, E., D. Oron and Y. Silberberg (2005). "Improved depth resolution in video-rate line-scanning multiphoton microscopy using temporal focusing." *Optics Letters* **30**(13): 1686–1688.

Theer, P. and W. Denk (2006). "On the fundamental imaging-depth limit in two-photon microscopy." *Journal of the Optical Society of America A* **23**(12): 3139–3149.

Theer, P., M. T. Hasan and W. Denk (2003). "Two-photon imaging to a depth of 1000 um in living brains by use of a Ti:Al2O3 regenerative amplifier." *Optics Letters* **28**(12): 1022–1024.

Therrien, O., B. Aubé, S. Pagès, P. D. Koninck and D. Côté (2011). "Wide-field multiphoton imaging of cellular dynamics in thick tissue by temporal focusing and patterned illumination." *Biomedical Optics Express* **2**(3): 696–704.

Truong, T. V., W. Supatto, D. S. Koos, J. M. Choi and S. E. Fraser (2011). "Deep and fast live imaging with two-photon scanned light-sheet microscopy." *Nature Methods* **8**(9): 757–760.

Tsai, P., N. Nishimura, E. Yoder, E. Dolnick, G. White and D. Kleinfeld (2002). "Principles, design, and construction of a two photon laser scanning microscope for in vitro and in vivo brain imaging." *In vivo Optical Imaging of Brain Function*: 113–171.

van de Hulst, H. C. and G. W. Kattawar (1994). "Exact spread function for a pulsed collimated beam in a medium with small-angle scattering." *Applied Optics* **33**(24): 5820–5829.

Vaziri, A., J. Tang, H. Shroff and C. Shank (2008). "Multilayer three-dimensional super resolution imaging of thick biological samples." *PNAS* **105**(51): 20221–20226.

Vitek, D. N., D. E. Adams, A. Johnson, P. S. Tsai, S. Backus, C. G. Durfee, D. Kleinfeld and J. A. Squier (2010). "Temporally focused femtosecond laser pulses for low numerical aperture micromachining through optically transparent materials." *Optics Express* **18**(17): 18086–18094.

Vitek, D. N., E. Block, Y. Bellouard, D. E. Adams, S. Backus, D. Kleinfeld, C. G. Durfee and J. A. Squier (2010). "Spatio-temporally focused femtosecond laser pulses for nonreciprocal writing in optically transparent materials." *Optics Express* **18**(24): 24673–24678.

Wang, L., S. L. Jacques and L. Zheng (1995). "MCML--Monte Carlo modeling of light transport in multi-layered tissues." *Computer Methods and Programs in Biomedicine* **47**(2): 131–146.

Wang, L. V. and H.-I. Wu (2007). "Radiative transfer equation and diffusion theory." In *Biomedical Optics Principles and Imaging*. John Wiley & Sons, Inc.

Wilson, T. (1990). *Confocal Microscopy*.

Wilt, B. A., L. D. Burns, E. T. W. Ho, K. K. Ghosh, E. A. Mukamel and M. J. Schnitzer (2009). "Advances in light microscopy for neuroscience." *Annual Review of Neuroscience* **32**: 435.

Yaksi, E. and R. W. Friedrich (2006). "Reconstruction of firing rate changes across neuronal populations by temporally deconvolved Ca2+ imaging." *Nature Methods* **3**(5): 377–383.

Yang, W., J.-E. K. Miller, L. Carrillo-Reid, E. Pnevmatikakis, L. Paninski, R. Yuste and D. S. Peterka (2016). "Simultaneous multi-plane imaging of neural circuits." *Neuron* **89**(2): 269–284.

York, A. G., A. Ghitani, A. Vaziri, M. W. Davidson and H. Shroff (2011). "Confined activation and subdiffractive localization enables whole-cell PALM with genetically expressed probes." *Nature Methods* **8**(4): 327.

Yu, H., P. T. galwaduge, V. Voleti, K. Patel, W. Li, M. A. Shaik and E. M. Hillman (2017). *Two-Photon Swept Confocally Aligned Planar Excitation Microscopy (2P-SCAPE)*. Optics in the Life Sciences Congress, San Diego, CA, Optical Society of America.

Yu, L., F. Nina-Paravecino, D. Kaeli and Q. Fang (2018). "Scalable and massively parallel Monte Carlo photon transport simulations for heterogeneous computing platforms." *Journal of Biomedical Optics* **23**(1): 010504.

Zeng, S., X. Lv, C. Zhan, W. R. Chen, W. Xiong, S. L. Jacques and Q. Luo (2006). "Simultaneous compensation for spatial and temporal dispersion of acousto-optical deflectors for two-dimensional scanning with a single prism." *Optics Letters* **31**(8): 1091–1093.

Zhu, G., J. van Howe, M. Durst, W. Zipfel and C. Xu (2005). "Simultaneous spatial and temporal focusing of femtosecond pulses." *Optics Express* **13**: 2153–2159.

Zipfel, W. R., R. M. Williams and W. W. Webb (2003). "Nonlinear magic: multiphoton microscopy in the biosciences." *Nature Biotechnology* **21**(11): 1369–1377.

Ziv, Y., L. D. Burns, E. D. Cocker, E. O. Hamel, K. K. Ghosh, L. J. Kitch, A. El Gamal and M. J. Schnitzer (2013). "Long-term dynamics of CA1 hippocampal place codes." *Nature Neuroscience* **16**(3): 264.

CHAPTER 4

Light Field Microscopy for *In Vivo* Ca^{2+} Imaging

Tobias Nöbauer and Alipasha Vaziri

CONTENTS

4.1 INTRODUCTION

Light Field Microscopy (LFM) is an emerging technique for neuroimaging with a high potential for dissemination in the neuroscience community. This computational imaging method offers outstandingly high-volume acquisition rates (up to 100 Hz), and fields-of-view large enough for whole-brain imaging of neuronal activity – as reported by genetically encoded calcium indicators – in small model organisms, such as larval zebrafish and *C. elegans*. LFM stands out from competing imaging methods due to its simplicity and versatility. The technique is uniquely scalable to larger fields-of-view, since acquisition of an entire volume is performed in a single shot, i.e. without the need for time-consuming scanning of an excitation beam or detection optics. In recent developments, LFM has been shown to be capable of imaging neural activity deep in scattering tissue such as the mammalian neocortex. Moreover, the approach was recently shown to allow for imaging whole-brain activity in freely swimming fish by LFM-enabled closed-loop sample positioning combined with LFM imaging at 70 Hz.

In what follows, we start by outlining the basic principles and performance trade-offs involved in Light Field Microscopy on a general level (Section 4.2). On a practical level, we provide an overview of optomechanical hardware implementations and alignment strategies (Section 4.3), before discussing the specific properties and requirements involved in *in vivo* imaging of Ca^{2+} activity and surveying results obtained in *C. elegans*, zebrafish larvae, and head-restrained awake mice (Section 4.4). We give an overview of signal extraction algorithms suitable for *in vivo* imaging and the computational cost involved with established and emerging LFM reconstruction strategies.

The general idea of recording and manipulating light fields is far-reaching and continues to inspire a wealth of new developments and variations. We conclude this chapter by discussing a number of recent proposals aimed at further increasing the acuity and scope of this elegant and powerful approach (Section 4.5).

4.2 PRINCIPLES OF LFM

4.2.1 Basic Principles

Before discussing light-field imaging, we would like to review a few relevant principles of regular, classical wide-field microscopes and, with certain differences, those of photographic cameras. These devices in essence record a two-dimensional array of light intensity values $I(x, y)$ – or simply, an image. The lenses in between the object and the image sensor are arranged such that light rays originating from each point in a given plane within the object (the focal plane) intersect at (i.e. are focused on) a single point in the plane where the image sensor is placed. Rays originating from planes other than the focal plane are not focused onto the sensor: they do not intersect the sensor plane at a single point, but across a certain area. As this area becomes larger than the pixel size of the sensor, the corresponding point in the object appears blurred in the image, or even becomes entirely blended into a low-contrast – and often noisy – background.

Even in this simple ray optics picture, it is obvious that merely recording a 2D image (i.e. the intersection of the sensor plane with a set of light rays) does not capture all the information that would be available. This is because each ray has a direction of propagation, which, if recorded, would allow tracing the ray backwards into object space. By capturing the direction of propagation of at least two rays for each source point in object space, the 3D position of these source points could be inferred. This had already been realized by Leonardo da Vinci during his deliberations on the central perspective; the technology for doing so practically was, however, not yet available at that time.

To fully define a ray in 3D, one possibility is to specify its point of intersection with two planes rather than just one (or equivalently, one intersection point plus two angles). The simplest device that can achieve this is a pinhole camera: all the rays that reach a point on the image plane must have passed through the pinhole, which acts as the second intersection point and fully defines the rays. However, not enough information is collected by a single pinhole camera to infer information on the axial position of a source. To understand this, let us assume for clarity that the pinhole is infinitesimally small. In this case, the acceptance range for incoming rays is reduced to a single ray from each source point. To triangulate the axial position of a source, however, the direction of propagation of two different rays emanating from the same source would be required. Since in an ideal pinhole camera, only one ray per source point is captured, no depth information is recorded at all; such a camera has an infinite depth of field (DOF) and there is no image blur for certain axial source positions. Unlike a camera with a focusing lens, an ideal pinhole camera does not even allow for axial optical sectioning by focusing one specific plane and blurring all others.

To recover sufficient angular information for depth resolution, at least two pinhole perspectives would be required. This is in principle possible by exposing a single camera at several different positions, or by using several cameras in parallel. However, since pinhole cameras gather only a small fraction of the light that reaches the pinhole plane, this approach is too inefficient for practical purposes.

In contrast to a pinhole camera, a photographic camera with a focusing lens collects rays from a range of angles for each source point and thus is capable of capturing more light than a pinhole camera. The angular directions of the rays, however, do not get recorded, but rather averaged over, which is just what leads to blur outside of the focal plane. In a typical photographic camera, an adjustable iris diaphragm is placed in the optical path such that it allows adjustment of the range of incident ray angles being captured. Such an element, known as the aperture stop, allows a choice between a quasi-pinhole mode (aperture stop closed, small range of ray angles captured, deep DOF), and a wide-open mode (iris opened, large range of ray angles captured, shallow DOF).

The concept of fully capturing the direction and position of all light rays propagating in a volume is known as plenoptic imaging, and the quantity containing all the captured information is the plenoptic function. In its most general form, the plenoptic function of time-independent ray optics is $L(x,y,z,\vartheta,\varphi)$, which gives ray radiance* as a function of ray intercept (x,y,z) and ray propagation angle (ϑ,φ). Fully measuring the values of this 5D function for an arbitrary light field can be exceedingly complex. However, an important

* In the SI system, the unit of radiance is Watts per steradian and square meter.

simplification can be made if a scene contains no occlusions, or if we are only interested in finding its convex hull (i.e. its surface): without occlusions, the radiance of a ray along its direction of propagation does not change. Thus, after having recorded the plenoptic function at one axial position z_1, its value at all other axial positions z_i can be calculated. In more abstract terms, the plenoptic function reduces from a 5D to a 4D structure. The practical meaning of this simplification becomes more obvious when considered the other way around: a given ray is fully defined – or, equivalently, its plenoptic function fully determined – if its intersection coordinates (s,t) and (u,v) with only two planes of different axial positions $z_1$and z_2 are known (Figure 4.1a). In the absence of occlusions, a plenoptic function containing several rays can thus be measured by recording the intersection

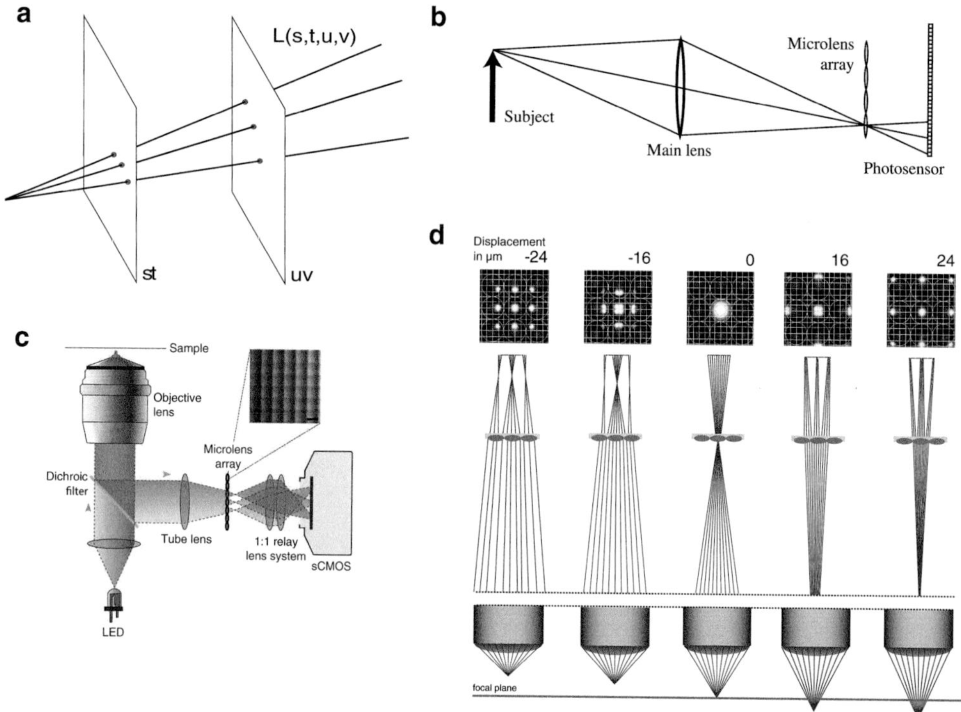

FIGURE 4.1 Principles of light field imaging. (a) Illustration of the 4D light field function L(s,t,u,v): in the case of no absorption and no occlusions, all rays emitted from an object into one half-space can be parametrized by their intersection coordinates (s,t) and (u,v) with two planes. (From: Hoffmann, M., MSc. thesis, University of Vienna, 2014, unpublished.) (b) Schematic of a light-field camera, consisting of a main lens, microlens array and a sensor. The main lens forms an image of the object onto the microlens array. The microlens array encodes the axial position of the object into a characteristic pattern detected by the sensor. (From: Ng et al. (2005).) (c) Schematic of a Light Field Microscope with epi-fluorescence illumination. (From: Prevedel et al. (2014).) (d) Schematic ray diagrams and simulated point-spread functions illustrating image formation in a light field microscope for a point source at five different axial displacements from the front focal plane of the microscope objective. White circles in point-spread function plots indicate projected microlens positions. (From: Hoffmann, M., MSc. thesis, University of Vienna, 2014, unpublished.)

coordinates of each such ray at two different planes, resulting in a 4D quantity $L(s,t,u,v)$, which we refer to as the light field in what follows. The question remains how this measurement of the light field can be achieved practically in an efficient way.

The first concept of a light-field imager was put forward by Lippmann in 1908 (Lippmann, 1908). He proposed to expose a photographic emulsion film through an array of lenses, behind each of which one distinct perspective onto the scene would be recorded. To re-display the 3D scene, light would be shined through the back of the transparent, developed film, re-creating the light field that had been recorded. In 1992, Adelson and Wang realized a light-field camera using a microlens array in front of a CCD array sensor (1992), an approach which was further simplified and expanded by Ng et al. in 2005 (Figure 4.1b); in their implementation, a main lens forms an image of a scene onto a microlens array, in the back focal plane of which the sensor array is placed. In this classical light-field camera configuration, the (s,t) and (u,v) planes are chosen in a very specific and efficient way: first, the microlenses segment the image into macropixels, which we index by (s,t). Each of the microlenses forms an image of the main lens aperture on the sensor, as seen from its particular position. Let us index the sensor pixels that are illuminated by light passing through a particular microlens by indices (u,v), relative to the microlens origin. The (u,v) -indexed pixels are then illuminated by light passing through a particular sub-aperture of the main lens – or, put differently, the (u,v) indices can be interpreted as the position on the main lens that a ray passes through.

In a light field microscope (LFM; see Figure 4.1c for a schematic of an LFM in epi-fluorescence configuration), the situation is different from the case of a light-field camera in two relevant ways:

First, a microscope is object-space telecentric: a stop in a plane conjugate to the back focal plane of the objective – usually physically located inside the microscope objective – restricts the angular spread of accepted rays to a well-defined range that is independent of lateral position in the field of view (FOV), and symmetric about the optical axis. Thereby a microscope produces orthographic views of the scene, and not perspective views, as in a standard, non-telecentric photographic camera. If the microscope uses a so-called infinity-corrected objective and forms the image using a tube lens (as is the case in practically all modern standard microscopes), it is also image-space telecentric. In effect, the sensor pixels with the same position relative to the nearest microlens (that is, the same (u,v) indices) all capture the same range of ray angles emanating from object space, irrespective of the microlens coordinate (s,t), which is not the case in the light-field camera configuration discussed above.

This telecentric layout is the standard light field microscope configuration that we will discuss in what follows, particularly when discussing experimental aspects in Section 4.3. In contrast to the non-telecentric case, this design has the advantage that the images behind each microlens are all of equal size, and concentric with the respective microlenses, which simplifies modelling of the optical system as well as alignment, and tends to lead to smaller optical aberrations (Levoy et al., 2006; Levoy, 2006).

The second important difference between a camera and a microscope in general is that in a camera, due to the large demagnification, diffraction effects can be neglected and a ray-optical description is sufficient. In a microscope, however, as we will discuss in Section 4.2.3 below, diffraction cannot be ignored, and the point-spread-function (PSF) has to be calculated taking diffraction effects into account. Figure 4.1d illustrates the typical resulting LFM PSF patterns observed for five different axial positions of a point source in an LFM.

4.2.2 Performance and Design Trade-Offs

Introducing a microlens array (MLA) into the native image plane provides angular and, via triangulation, axial resolution at the expense of spatial (lateral) resolution: in first approximation, the **lateral resolution** in object space drops to the size of a lenslet d_{ML} (we assume square lenslets), divided by the magnification M of the microscope:

$$\Delta x \approx \frac{d_{ML}}{M}$$

Spatial frequencies higher than $\sim 1/\Delta x$ are aliased into the digitized data, but may partly be recovered by deconvolution (Broxton et al., 2013), which may lead to a resolution increase beyond the estimates given in this section. We discuss deconvolution in the following section.

It is noteworthy that in a well-designed LFM, the total information bandwidth of the optical system, loosely defined as the number of resolvable spots per field-of-view area, remains the same as in a non-LFM, wide-field configuration. Crucially, however, the information budget is invested in a way such that angular (axial) resolution is obtained, at the expense of lateral resolution.

To avoid wasting neither information transmitted by the objective nor available sensor pixels, commonly the F-numbers of the MLA and the output F-number of the microscope are matched:

$$F_{MLA} = F_{out}$$

$$F_{MLA} = \frac{f_{ML}}{d_{ML}}$$

$$F_{out} = \frac{M}{2\text{NA}}$$

where f_{ML} is the microlens focal length, d_{ML} the microlens diameter, M the microscope magnification, and NA the numerical aperture of the microscope objective. To maximize light throughput, it is advantageous to place the microlenses on a quadratic grid of pitch d_{ML}, so that adjacent microlenses exactly touch each other.

The **axial resolution** of an LFM can be estimated from depth-of-field considerations as (Levoy et al., 2006)

$$\Delta z \approx \frac{(2+N)\lambda n}{2\text{NA}^2}$$

where λ is the wavelength, n the refractive index, and N the number of diffraction-limited spots per microlens width. Let us repeat here that this estimate does not take into account the effect of deconvolution, as discussed in the following section.

In standard implementations in the context of neuronal imaging, N is usually in the range of 10 to 15, due to restrictions on available camera pixel sizes in conjunction with resolution requirements. Using the Sparrow criterion for resolvable spot size R_{obj}, the diameter d_{ML} of one lenslet is

$$d_{ML} = NR_{obj} = N\frac{0.47\lambda M}{\text{NA}}.$$

Levoy et al. (2006) estimate the **axial range** (i.e. the axial size of the sample volume that can be captured in one exposure) as

$$\text{Range}_z \approx \frac{\left(2+N^2\right)\lambda n}{2\text{NA}^2}$$

The choice of objective NA is the most important parameter for determining axial resolution and range: increasing the NA improves (decreases) the resolution, but also decreases the axial range. Choosing the microlens pitch larger for a fixed NA increases the ratio of the number of resolvable axial planes to axial range at the expense of lateral resolution.

4.2.3 Recovering Volumetric Data

To quickly asses the 3D contents of an LFM raw frame, it is useful to compute the set of focal slices that would be observed when stepping the focal plane of a non-LFM microscope through the axial range. This can be done using a simple shift-and-add algorithm termed "refocusing" in the LFM literature (Ng et al, 2005). The depth-of-field (DOF) of the refocused slices can be chosen as small as the DOF that the bare microscope would have; light sources from outside of that DOF are blurred.

For many applications, including Ca^{2+} imaging, it is necessary to reconstruct the 3D distribution of light sources (such as fluorescently labelled neurons) in order to find the positions of the individual cells and to demix their signals from each other and any background. This is not achieved by the refocusing procedure.

An approach for full 3D deconvolution of the LFM raw data with a simulated PSF of the LFM optical system in the absence of scattering was originally proposed by Broxton et al. (2013) and implemented in the context of *in vivo* Ca^{2+} imaging by Prevedel et al. (2014). (MATLAB code implementing PSF simulation and deconvolution is available as Supplementary Software with that article).

Broxton's approach simulates propagation of a spherical wave originating from a set of source positions through an objective that obeys the Abbe sine condition, and a tube lens. The phase shift caused by the MLA is applied, and the resulting wavefront propagated to the sensor plane. In this way, a PSF can be realized that varies as a function of the axial source position. Laterally, the PSF also varies as the source is moved across FOV, but is periodic with the microlens pitch (demagnified into object space). For a given source volume V (written as a $n_v \times 1$ vector of all n_v voxels), the LFM image on the sensor, I (written as a $n_p \times 1$ vector of all n_p pixels on the camera), can be calculated by forming the matrix product between V and an $n_p \times n_v$ matrix P, each row of which is the computed PSF for the corresponding voxel in object space):

$$\mathrm{I} = \mathrm{PV}$$

The periodicity of the PSF can be exploited by re-writing this (large) matrix product into a convolution with a spatially varying, but smaller kernel, making the problem computationally more tractable (see also Section 4.4.5).

The full forward model of LFM imaging is then formed by describing each pixel value as a Poissonian random variable, and allowing for a background b:

$$\mathrm{I} \sim \mathrm{Poisson}\left(\mathrm{PV} + b\right)$$

The inverse problem can be solved using a variety of algorithms. The well-known Richardson–Lucy algorithm (Richardson, 1972; Lucy, 1974) is a standard choice; it is an iterative procedure that corresponds to a scaled gradient descent and is guaranteed to converge to the maximum-likelihood estimate for the source volume V by iterating the following basic update step:

$$\mathrm{V}^{i+1} = \mathrm{V}^{i}\,\frac{\mathrm{P}^{\mathrm{T}}\mathrm{I}}{\mathrm{PV}^{i}}$$

where the upper indices i and $i{+}1$ indicate the iteration number, and the superscript T indicates matrix transpose. The Richardson–Lucy algorithm can be modified to integrate a range of constraints and regularizations, such as non-negativity, sparsity, or total variation.

Deconvolving LFM raw data has been shown to result in superresolution: spatial frequencies higher than the reciprocal values of the lateral and axial resolution estimates for an LFM – as given above – in general are not captured at sufficiently high sampling frequency by the LFM camera. This leads to aliasing, i.e. a misrepresentation of the exceedingly high spatial frequencies in the form of lower-frequency features in the sampled image. However, in an LFM, the sampling density and hence the exact effects of aliasing are not constant with depth; when tracing all possible rays backwards from the sensor pixels into object space, it becomes apparent that the rays do not sample object space uniformly across the axial range. In some z-planes, the rays are evenly spread out laterally, resulting in high resolution (a strong dependence of LFM PSF shape on lateral precision), while in other planes the rays bunch and sample object space more coarsely, leading to an effective lower resolution (Figure 4.2a). Plainly speaking, this has the consequence that several differently sampled views of the scene are contained in a light-field image, each of which contains information on a slightly different range of spatial frequencies. In a manner similar to the recovery of higher-resolution photographic images by computationally combining several

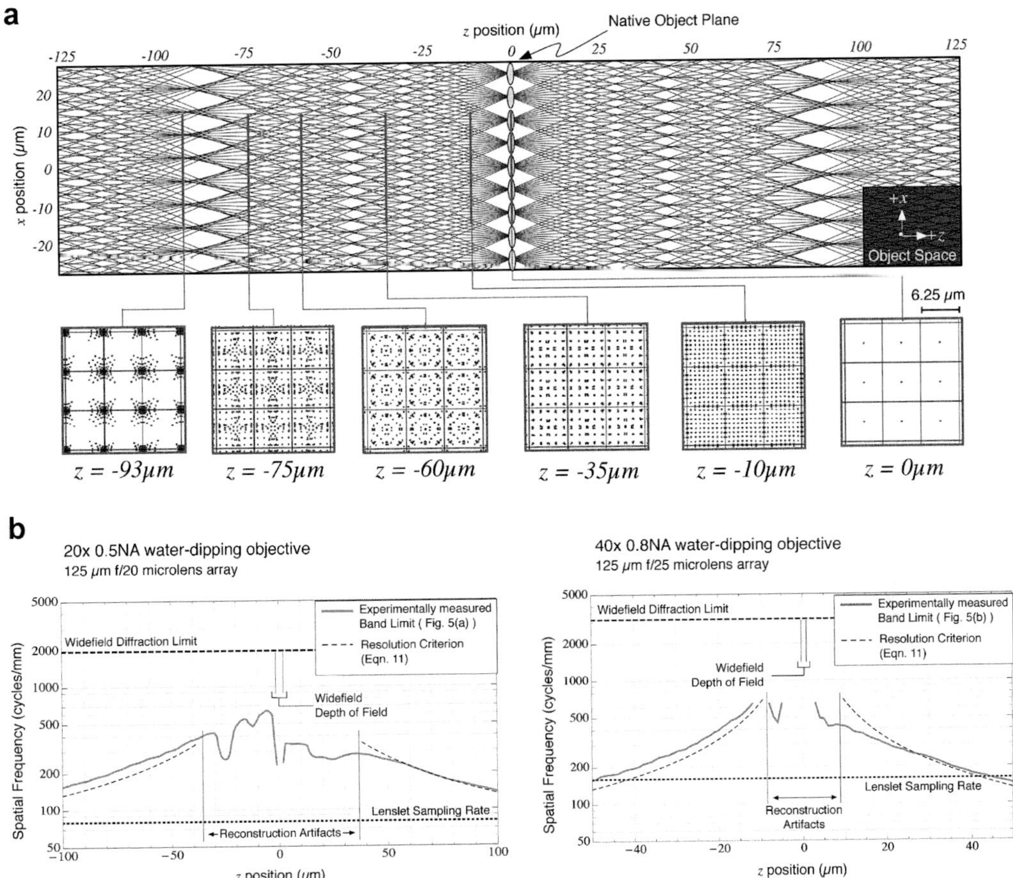

FIGURE 4.2 LFM resolution. (a) Back-projection of the chief rays passing through every pixel on an LFM sensor into the sample volume, illustrating the variation of spatial sampling density with depth. The native object (focal) plane of the microscope at z = 0 is optically conjugate to the microlens array plane in image space (not shown), which is indicated schematically by the de-magnified lenslets at the center of the top panel. Bottom panels show cross-sections through the back-projected rays at the locations indicated by red lines. The partially very dense ray patterns allow for obtaining resolution enhancement by deconvolution. The extremely low ray density at the native object plane gives rise to resolution breakdown and reconstruction artefacts in this region. (From: Broxton et al. (2013).) (b) Experimentally measured LFM band limit versus axial position relative to objective front focal plane, for two different configurations of objective and microlens array: green (orange) line in left (right) panel, respectively. Violet dashed line: Theoretical LFM band limit. Area of resolution breakdown indicated by pairs of vertical lines. Blue dashed line: wide-field diffraction-limited resolution. Black dashed line: pre-deconvolution LFM resolution given by lenslet size demagnified to object space. (From: Broxton et al. (2013).)

lower resolution exposures taken from slightly different viewpoints, as in classical computational superresolution approaches, deconvolution of the LFM raw image with a simulated LFM PSF can partially undo the effects of aliasing and recover the original detail. Depending on depth, this results in a resolution better than the first-order estimates by up to a factor of approximately two in the lateral directions (Broxton et al., 2013), as discussed in more detail in the following.

Due to the aforementioned dependence of effective sampling density on depth, the resolution of LFM (post-deconvolution) is not constant across the volumetric FOV. Lateral resolution generally declines linearly with increasing distance from the native focal plane. In addition, lateral resolution breaks down in the axial region close to the native focal plane, which can lead to artefacts in unconstrained volume reconstructions (Broxton et al., 2013). This is most severely so in the native focal plane, where lateral resolution is effectively undefined. The axial extent of the breakdown region can be estimated as (Broxton et al., 2013)

$$|z| < \frac{d_{ML}{}^2}{2M^2\lambda}$$

where d_{ML} is the microlens pitch, M the microscope magnification, and λ the wavelength.

Outside of this region, the lateral post-deconvolution resolution Δx_{deconv} in dependence of the axial position z is

$$\Delta x_{deconv}(z) \approx \frac{c\lambda M|z|}{d_{ML}}$$

Here, c is a constant that selects a resolution criterion: for example, $c = 1.22$ selects the Rayleigh 2-point criterion, and $c = 0.94$ selects the Sparrow 2-point criterion. The resolution enhancement due to deconvolution is most pronounced at axial positions just outside the breakdown region. With increasing axial distance from the native focal plane, the post-deconvolution resolution falls off to the pre-deconvolution estimate $\Delta x \approx {d_{ML}}/{M}$. The larger the ratio of magnification and microlens pitch, the steeper the fall-off, and the smaller the breakdown region, as shown in Figure 4.2b and discussed in more detail in Broxton et al. (2013). To reduce artefacts, it can be effective to employ wavefront coding techniques (Cohen et al., 2014) in which a phase element, such as a cubic phase plate, is introduced into the beam path to lift the spatial degeneracy of rays near the native focal plane.

4.3 EXPERIMENTAL ASPECTS

4.3.1 Optomechanical Setup

While LFM generally requires comparatively simple optical setups that do not involve any advanced laser systems as an excitation source, some care has to be taken to achieve a suitable stability and adjustability; to this end we give a number of guidelines and considerations in what follows. Due to their widespread use and simpler alignment, we limit our discussion to the case of infinity-corrected microscopes.

Conceptually, since LFM differs from standard wide-field microscopes only by the insertion of a microlens array and a changed camera position, it has proven convenient to use a commercial fluorescence microscope as a platform, and to mount the MLA at the image plane of a camera port. For inverted microscopes, this results in a particularly simple setup, as the microlens array can simply be mounted in a 5-axis kinematic mount (for adjustment of transverse position, tip/tilt angles and rotation about the optical axis) and on a linear translation stage (with a graduated micrometer screw for repeatable adjustment of position along the optical axis), which in turn is attached to the optical table (Figure 4.3a).

To allow for convenient removal or exchange of the MLA during alignment, a precision magnetic base plate can be used. Since MLA focal lengths are usually in the order of tens of microns to millimeters, it may be inconvenient to place the camera sensor directly into the back focal plane of the MLA due to the necessity to remove the filter glass plates that are often permanently attached to the sensor package in front of the active surface, or other mechanical restrictions. Instead, it is possible to use relay optics: depending on FOV, magnification and resolution, standard off-the-shelf optics, such as 1:1 macro objectives or a back-to-back pair of photographic prime lenses, may provide sufficient performance; otherwise, more specialized objectives used in machine vision applications can be explored. Mounting the camera on a micrometer linear stage is useful for repeatable alignment.

When an upright microscope is used as a base (for specimen such as fish or mammals), it is preferable to place a horizontal optical breadboard next to the camera port (as opposed to constructing a vertical assembly), since this allows for convenient and stable placement of components (Figure 4.4b). Since the optical path is strictly linear, an optical rail system placed on thick posts may be sufficient. For exact alignment of the rail parallel to the optical axis, some form of height adjustment (such as adjustable height pillar posts, or height-adjustable feet) is desirable. A detailed alignment procedure for such a setup is given in Figure 4.4c and the legend thereof.

Care must be taken to avoid vignetting – the undesired clipping of rays by optical or mechanical elements – in the fluorescence path. While vignetting in a widefield configuration merely leads to a brightness fall-off with increasing radius in the field-of-view, in LFM it leads to a loss of axial resolution. In a system with vignetting, the opening angle of rays that are transmitted through the system depends on position in the FOV; in LFM, this means that less angular and hence depth information is available for large lateral radii. In inverted microscopes, this is not usually an issue due to sufficiently short optical paths. In upright microscopes, however, especially ones with an adjustable distance between objective and tube lens, vignetting may be encountered more commonly. To diagnose vignetting, it is instructive to evenly illuminate a radiometric slide using Koehler illumination, and measuring brightness fall-off with increasing radial position at the camera port that is to be used for LFM, and with the objective-tube lens distance set to maximum.

If a more compact setup is required, the MLA holder has to be designed such that it is directly integrated into the mounting frame of a camera sensor and the MLA can be positioned at a distance of one focal length of the MLA from the active surface. An example of a three-point adjustable mount can be found in Ng et al. (2005).

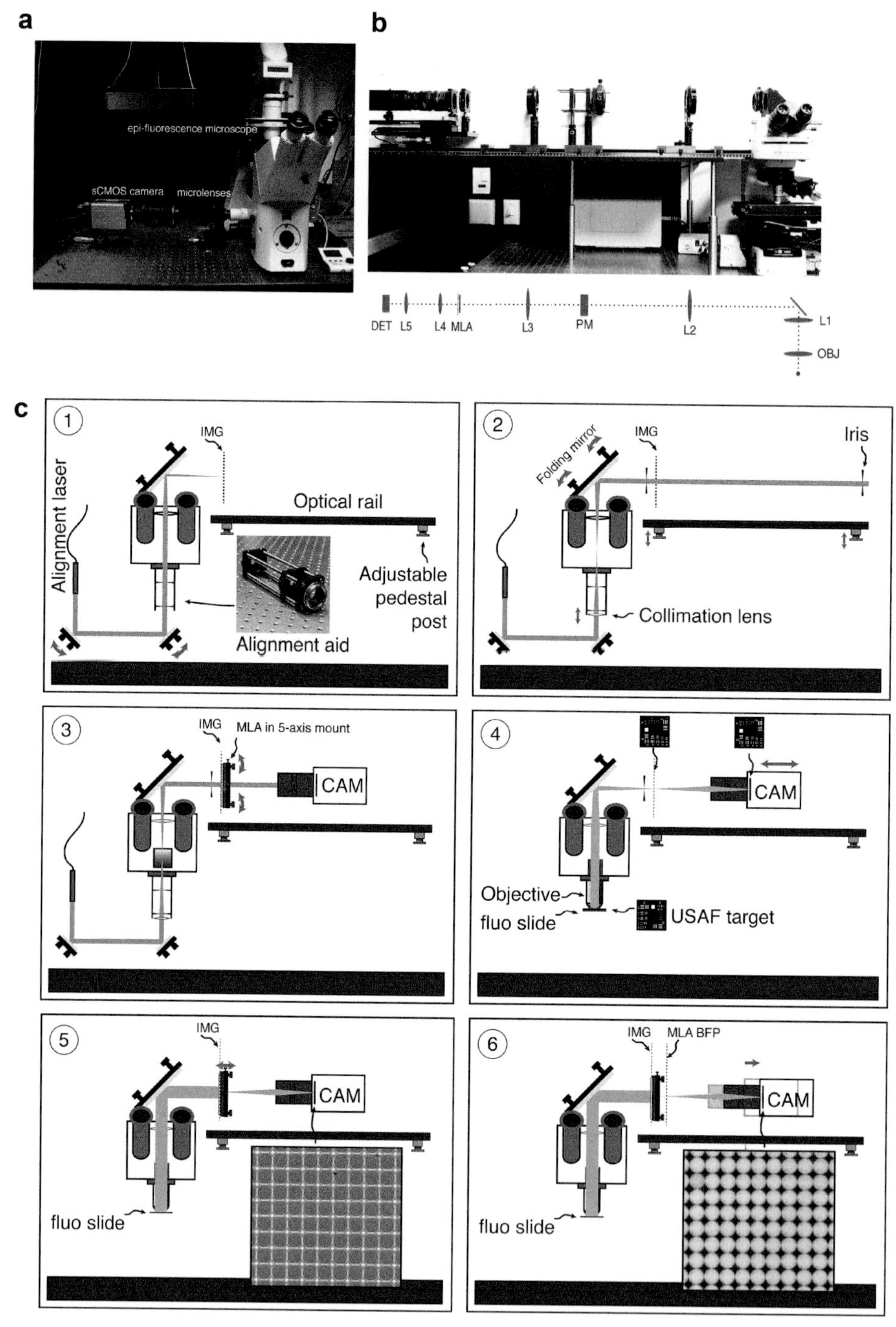

FIGURE 4.3 LFM implementation and alignment. (a) Inverted LFM constructed by placing a microlens array and a camera with relay objective at the side image port of an inverted epi-fluorescence microscope. (b) Upright LFM constructed by mounting optical elements onto an optical rail placed next to the top camera port of an upright epi-fluorescence microscope. MLA: microlens array, L1-L5: lenses, OBJ: objective, DET: detector (camera), PM: phase mask for resolution enhancement by wavefront coding. (From: Cohen et al. (2014).) (c) Illustration of important steps

FIGURE 4.3 (CONTINUED)
in aligning an upright LFM: (1) Screw alignment aid (depicted in inset and described in the main text) into objective thread of microscope. Use pair of mirrors on kinematic mounts to shoot an alignment laser beam exactly through both irises of the alignment tool, so that the laser beam is on the optical axis. (2) Add a lens to the bottom end of the alignment tool and translate up or down until beam emerging from microscope top camera port is collimated. Exchange lens for one with a different focal length if collimation cannot be achieved (distance between tube lens in microscope head and collimation lens has to be equal to sum of their focal lengths for achieving collimation). Mount optical rail on adjustable height pedestal posts to the side of the microscope head. Mount two irises on rail carriers, adjust irises to same height, place at either end of rail, and adjust pedestal post height as well as top folding mirror until laser goes through both irises, such that the optical rail is parallel to the optical axis. (3) Mount MLA in 5-axis mount and onto rail carrier, and camera (CAM) with 1:1 macro objective onto a linear stage (not shown) and rail carrier. Place both on optical rail. Observe back-reflections of laser from MLA and camera objective and adjust back-reflections to coincide with incoming beam, to align MLA orthogonal to optical axis. Since the curved lenslet surfaces should be facing away from the camera for ideal performance, back-reflection from lenslet surfaces will be divergent and hard to see. This can be overcome by carefully pressing a glass slide against the MLA mount or MLA directly for creating a collimated back-reflection. Manually nudge camera until it is orthogonal to optical axis before tightening fixation screws. (4) Remove alignment aid and MLA, screw objective into objective thread, switch to fluorescence illumination and rotate fluorescence filter cube into optical path. Place transmissive negative USAF resolution target (or similar) onto uniformly fluorescing slide and focus via eyepiece. Set camera objective to 1:1 magnification. Translate camera forward or backward until it is focused onto image of USAF target in image plane (labelled IMG) near top camera port of microscope. (5) remove USAF target and focus into fluorescent slide to cause uniform fluorescence to emerge from microscope top port. Place MLA back onto rail. Translate MLA until camera shows the lenslet plane (inset; also observe dust particles on both MLA surfaces). MLA is now in IMG plane, which CAM is imaging. (6) Translate CAM backwards by the focal length of the MLA, so that it images the MLA back focal plane (BFP). If the F-number of microscope output (equal to objective magnification divided by twice the objective NA) and MLA are matched, the CAM will now show exactly touching illuminated circles (inset). This is the basic LFM configuration. See main text for additional fine alignment steps.

4.3.2 Alignment Strategy

A detailed alignment procedure for such a setup can be found in Figure 4.4c and the legend thereof. A number of additional explanations are given in what follows: for alignment of all components, it is practical to first align a weak, collimated laser beam to the optical axis of the bare microscope. To this end, we have found it useful to build a simple alignment tool that can be screwed into the objective holder in place of the objective, and which consists of two irises at a distance of 20 cm or so. Suitable thread adapters or cage systems are readily available. Using two mirrors in kinematic mounts, the laser beam can then be adjusted (tip, tilt, parallel translation) until it passes through the centers of both irises. An additional iris screwed into the camera port can help with increasing the accuracy of alignment to the optical axis by adjusting the laser beam to exit through the center of that iris.

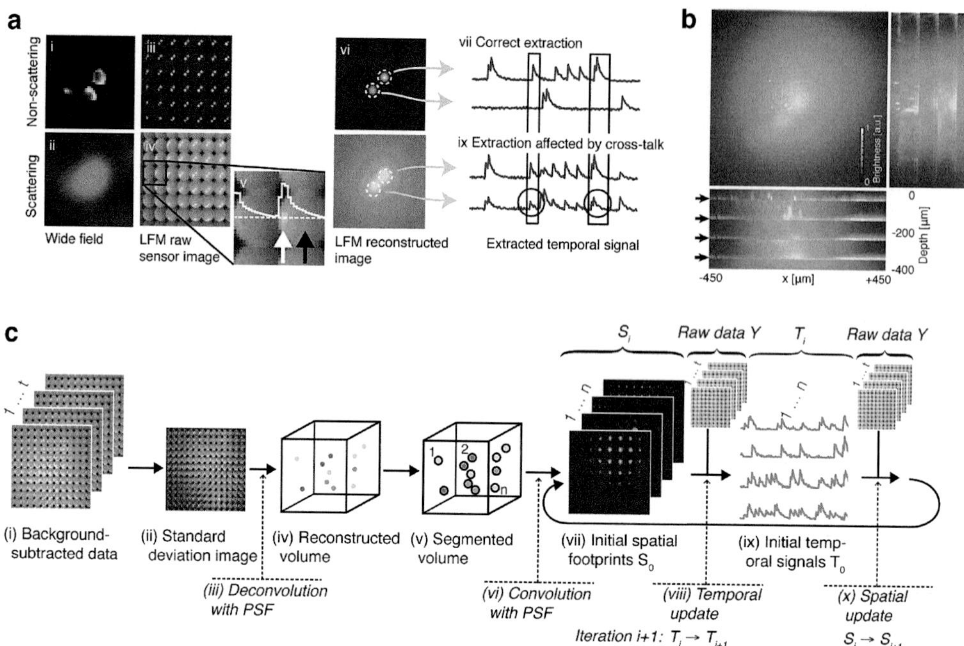

FIGURE 4.4 LFM in scattering medium and SID algorithm (all from: Nöbauer et al., 2017). (a) Scattering effects, illustrated for conventional wide-field and light-field microscopy in comparison. Sub-panels: (i) Underlying ground truth: neurons in the mouse cortex. (ii) Simulated wide-field image of ground truth neurons in the presence of scattering. (iii)–(iv) LFM raw image of ground truth without scattering (iii) and with scattering (iv). (v) Zoom into the indicated region in (iv). Solid white line gives intensity profile along dashed white line. Black arrow highlights brightness gradients originating from directionality information retained in scattered light, while white arrow highlights peak due to ballistic light. (vi) Volumetric reconstruction of LFM image without scattering shown in (iii), projected along axial direction. (vii) Temporal signals extracted from an LFM movie without scattering, from the regions-of-interest indicated as dashed circles in (vi). (viii) Volumetric reconstruction of the LFM image with scattering shown in (iv), projected along axial direction. (ix) Temporal signals extracted from an LFM movie with scattering, from dashed circles in (viii). Black rectangles and red circles highlight regions in which crosstalk from one neuronal signal to the other is observable in the presence of scattering. (b) Stack of five volumetric frames obtained by conventional reconstruction of LFM recordings at increasing depths in the mouse cortex, maximum-intensity projections along three orthogonal directions. The high-intensity planes visible in the side-views (indicated by arrows) are due to reconstruction artefacts in conventional, unconstrained LFM reconstruction. (c) Illustration of key steps in the Seeded Iterative Demixing (SID) algorithm (see text).

Since in most commercial microscopes the tube lens is not easily removable, we recommend adding a (replaceable) focusing lens into the alignment tool, after the irises, with a focal length such that the laser can be focused into the front focal plane of the tube lens. In this configuration, the laser will emerge collimated from the camera port. For exact collimation, an objective translation actuator on the microscope body can be used.

With the laser emerging collimated from the camera port, all other optical elements can be placed orthogonally to the optical axis onto the optical rail by aligning the back-reflections from their surfaces such that they exactly coincide with the incoming beam (as observed on an iris centered onto the beam). For best performance, the plano-convex microlens array should be placed with the curved lenslet surface facing away from the camera. In this orientation, the alignment laser will impinge on the curved surfaces of the lenslets, so the reflection will be diverging and difficult to observe. Pressing a microscope slide against the MLA holder should provide a good-enough approximation for coarse alignment. For aligning the camera chip, care must be taken to align onto the (zeroth order) reflection, not any of the higher diffraction orders created by the pixel grid. Alternatively, a microscope slide can be used as before.

After alignment of all optical elements orthogonal to the optical axis, the rear image plane can be found by temporarily removing the microlens array and focusing the microscope (observing the image in the eyepiece) onto a test slide, such as the USAF resolution target, and translating the camera until it is focused onto the image plane at the camera port. After placing the microlens array back onto its (magnetic) base plate, it can be adjusted to coincide with the image plane by translating it until the lenslets are in focus on the camera image (observing dust particles on the lenslet surface may help; see Figure 4.4c, step 5, for an example image). Finally, the camera is translated backwards by the focal length of the MLA, ideally by means of a graduated micrometer screw actuating a linear stage, which completes the coarse alignment.

For fine alignment, it is helpful to shine collimated light onto the MLA, such that the focal spots formed by the MLA on the camera surface can be observed and used for fine adjustment. This can be done either by expanding the alignment laser to fill the FOV, and again collimating it through the microscope, or by using transmitted Koehler illumination with the aperture stop closed down. A simple image analysis script can help with finding the alignment setting where the size of all focal spots on the camera is minimal and their size distribution narrow.

Finally, place a homogeneously fluorescent slide into the front focal plane of the microscope. If the output F-number of the microscope and the MLA are exactly matched (as discussed above), the LFM camera should now see an array of illuminated filled circles (one behind each microlens) that just touch each other, but do not overlap (Figure 4.4c, step 6). If there is vignetting in the system, full circles will be visible only near the center of the FOV, and the patterns will gradually transform into a sickle-like shape further out. It is advisable to take a snapshot of a fluorescent slide before each imaging session, to be able to conveniently find the microlens position parameters (pitch and offset) needed for reconstruction. To this end, the LFDisplay software available online (LFDisplay, n.d.) is useful; this tool in addition implements the refocusing algorithm described above.

4.4 APPLICATION OF LFM TO CA^{2+} IMAGING

4.4.1 Light-field Imaging of Neural Activity *In Vivo*

In its most general form, the LFM imaging model described in Section 4.2 does not make any assumptions about the object being imaged, the illumination conditions, or the absorption and emission characteristics of the object, other than that it is occlusion-free. In what follows, we introduce a number of more specific considerations that arise when applying LFM to *in vivo* imaging of neural activity as reported by expressing fluorescent proteins known as genetically encoded calcium indicators (GECIs) in neurons.

First, the average **size and spacing of neurons** in various brain tissue types is known, and hence the expected relevant spatial frequency contents of the reconstructed volumes can be estimated. In certain specimen and tissue types, such as parts of the mammalian cortex, it can be assumed that the distribution of neurons in a given volume exhibits a certain degree of sparsity, i.e. that only a fraction of the volume actually contains neurons. Knowledge about such properties can be incorporated into reconstruction algorithms in the form of additional constraints or regularization terms, to obtain solutions with suitable sparsity and smoothness.

Second, **fluorescence is omnidirectional** and incoherent. This assumption is already contained in the forward model described in the previous section, since the PSF simulation given there is based on spherical sources.

Third, in Ca^{2+} imaging, it is of interest to record **time series of neural activity.** The position of neurons within a given volume can generally be considered static (aside from effects of specimen motion; see Section 4.4.7 below). Thus, the challenge in Ca^{2+} imaging poses itself more as an *assignment* problem rather than a *resolution* problem – as when imaging an unknown scene in a single frame. The activity of neurons in a volume may be partially correlated, but in general it is safe to assume that the volume being imaged contains a set of isolated sources that vary with some degree of independence. It therefore is effective to incorporate methods of time series analysis and signal demixing early into any data-processing pipeline, as in the case of the SID algorithm, which is discussed in more detail in the following section.

Fourth, most biological tissues exhibit **scattering** and, to a lesser degree, absorption.* All considerations on LFM imaging so far have assumed ballistic light propagation, i.e. the absence of scattering. In recordings from mostly transparent, **weakly scattering specimen**, such as larval zebrafish or *C. elegans*, or for shallow recordings in the mammalian brain, this assumption may still be justified, and scattering merely leads to an increased and diffuse background and decreased resolution, as well as the appearance of more pronounced artefacts in the resolution breakdown region during deconvolution. In this regime, the purely ballistic models of LFM image formation (Broxton et al., 2013) described above are still applicable, and deconvolution based on these models can be used for obtaining reconstructed volumes with cellular or near-cellular resolution, as discussed in Section 4.4.3 further below.

In many biological applications, however, and most importantly in Ca^{2+} imaging of the **mammalian brain**, scattering has to be taken into account. In the mouse cortex, for example, the scattering length for green light is approximately 50–100 μm (Helmchen and Denk, 2005; Jacques, 2013). As recording depth increases beyond one to two scattering lengths, a naive LFM reconstruction that does not specifically model the effects of scattering results

* Since the absorption length is generally at least an order of magnitude larger than the scattering length in brain tissue, it can be neglected in the applications discussed here.

in increasingly blurred images with reduced signal to background ratio and increased crosstalk between separate neurons (Figure 4.4a), until it finally becomes impossible to differentiate any features in the reconstructions (Figure 4.4b).

In a scattering medium, the ratio of unscattered light decreases exponentially with propagation distance. Fortunately, however, in brain tissue, the distribution of scattering angles is not uniform, but strongly peaked in the forward direction,* so that some information about the initial propagation direction of a light ray is retained even after several scattering events. The distance over which directional information is retained is known as the transport mean free path, which is as much as ten times the scattering length in cortex (Helmchek and Denk, 2005). Beyond the transport mean free path, light propagation increasingly resembles a random walk, and no information about the original direction of propagation is present anymore. However, when imaging at a distance less than the transport mean free path, the angular distribution of scattered light that emerges from the sample surface still has a maximum along the ballistic direction; depending on depth, some remnant ballistic light may also still be detectable. It is in this intermediate regime of scattering where signal extraction algorithms can be designed that to a good degree recover and correctly assign Ca^{2+} activity. One such algorithm, Seeded Iterative Demixing (SID), is discussed in the following section.

4.4.2 Seeded Iterative Demixing

Seeded Iterative Demixing (SID) (Nöbauer, 2017) is a novel computational technique that was recently demonstrated to allow for capturing neuronal dynamics *in vivo* within a volume of 900 × 900 × 260 μm, located as deep as 380 μm in the mouse cortex and hippocampus, and at a very high volume imaging rate of 30 Hz.

SID extends LFM into more strongly scattering tissue such as the mammalian cortex by seeding a machine learning algorithm with remaining unscattered (or very weekly scattered) light and then iteratively demixing the effects of scattering in order to retrieve the neuronal activity signals from deep inside scattering tissue. The principle of SID is illustrated in Figure 4.4c and proceeds as follows:

To extract estimates of the unscattered/weakly scattered LFM images of the individual neurons (neuron footprints) that are contained in the raw video recorded by the LFM camera, it is sufficient to compute a summary statistic, such as the standard deviation, for each pixel of the raw video: pixels that are illuminated by the remaining ballistic light stand out in this summary statistic due to the higher relative activity compared to pixels that receive scattered contributions from several neurons. Alternatively, a non-negative matrix factorization (NMF) of the raw video can be performed, which seeks to re-write the video as a product of a small (compared to the number of pixels) set of spatial components times their respective temporal signals. Given sufficient temporal activity, NMF is effective in separating the superimposed neuron footprints contained in the raw video into separate components, achieving a first demixing and greatly easing the task of fully separating the

* In mouse cortex, the distribution of scattering angles can be modelled by a Henyey-Greenstein distribution with anisotropy factor $g \approx 0.9$ (Helmchek and Denk, 2005).

components in subsequent steps. Prior to the computation of the summary statistic (or the NMF) it is crucial to remove any global background by performing, e.g. a rank-1 (or other low-rank) matrix factorization of the raw video.

Next, SID reconstructs the standard deviation image, or the individual NMF components by Richardson–Lucy deconvolution with the simulated LFM PSF, as described in Section 4.2.3. Depending on the data, it may be helpful to add sparsity and/or smoothness constraints – such as the well-known total variation regularization (Dey et al., 2004) – to the reconstruction problem to achieve more robust and artefact-free results. The reconstructed volumes are then segmented by performing a local maximum search. This yields an initial guess of neuron locations. Each neuron location estimate is used to generate a template of camera pixels that may receive scattered contributions from this neuron. To do so, a 3D Gaussian brightness distribution is placed centered on each individual neuron candidate location and projected forward by convolution with the LFM PSF. The resulting simulated LFM camera image is thresholded, resulting in a template for each neuron.

The core part of the SID algorithm consists in an iterative spatio-temporal demixing procedure that is initialized with the neuron templates found in the previous step, and with random Gaussian time series as temporal components. In an alternating sequence, the spatial component estimates are held fixed while updating the temporal components by gradient descent to approximate the background-subtracted raw data, and vice versa. While both the temporal and spatial sub-problem of this optimization are convex, the combined problem is only bi-convex and not in general guaranteed to converge to the global optimum. However, given a good initial guess as provided by the mostly ballistic contributions found in a previous step, the procedure converges rapidly and reliably to a reasonable solution (Gorski et al., 2007), as can be verified by inspection of the resulting neuron footprints and signals.

The performance of SID can be evaluated by simultaneously acquired two-photon microscopy (2PM) data, which can be regarded the ground truth (Nöbauer et al, 2017). We discuss ground-truth evaluation results for SID in mouse cortex in Section 4.4.4 further below.

4.4.3 Application to (Semi-) Transparent Organisms

The nematode *C. elegans* (Altun et al., 2002) has been a key model organism for neuroscience, due to its comparatively simple nervous system (comprising 302 neurons) yet wide behavioral repertoire (such as mating, thermo- and oxygen-sensing, sleep-like states, and several distinct types of locomotion motifs). Membrane potentials in *C. elegans* are graded (i.e. non-spiking) and propagate at reduced speed compared to action potentials. Hence, the neuronal dynamics of *C. elegans* can be sufficiently characterized at a sampling rate of ~5 Hz. Since *C. elegans* is highly transparent, it lends itself well to Ca^{2+} imaging. Difficulties may arise from the fact that as the animal bends, lensing effects and optical aberrations may occur at its surface, which have to be accounted for in post-processing. Adult worms grow to a diameter of approximately 60 μm, and 1.5 mm in length. Neurons are small (2–3 μm cell body diameter), and partly densely packed, as in the head ganglia.

An LFM design suitable for *C. elegans* Ca^{2+} imaging therefore has to be customized for comparatively high resolution: Prevedel et al. (2014) chose a 40×/0.95 dry objective (on an inverted microscope platform), which resulted in a resolution of 1.36 μm laterally and 2.55 μm axially (characterized by imaging fluorescent beads). The resulting FOV was ~350 × 350 × 30 μm, enough to image the majority of a worm (immobilized in a microfluidic device) while maintaining single-neuron resolution (Figure 4.5a). Prevedel et al. were able to record the activity of neurons in the brain region surrounding the nerve ring and the ventral cord at a volume rate of 5 Hz (Figure 4.5b, c), and to identify a large number of neurons and neuron classes based on location, morphology, and activity patterns. LFM can readily achieve imaging rates much higher than the 5 Hz required for *C. elegans* neural recording;

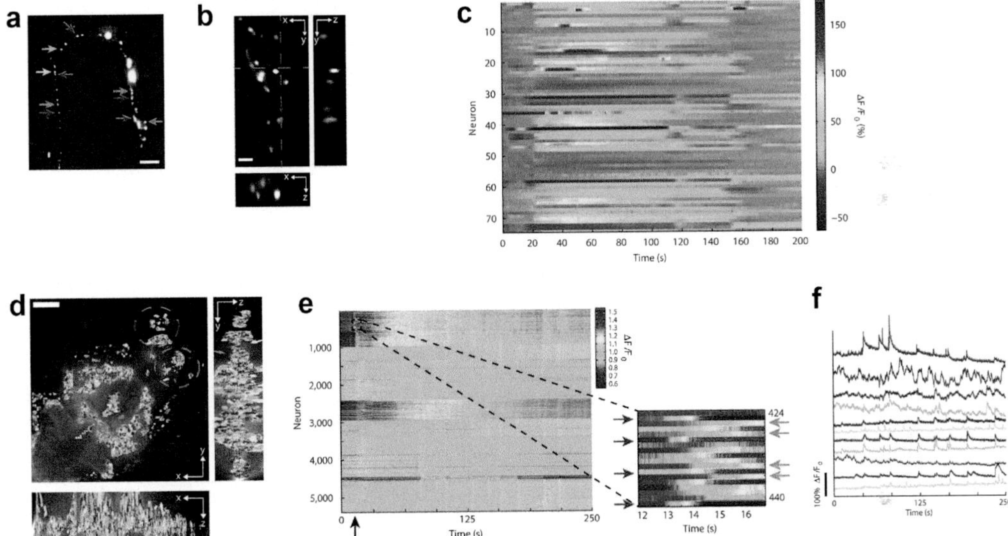

FIGURE 4.5 *In vivo* LFM imaging of Ca^{2+} activity in semi-transparent specimen (all from: Prevedel et al., 2014). (a)–(c): whole-animal Ca^{2+} imaging a of *C. elegans* using light field deconvolution microscopy. (a) Light-field-deconvolved image containing 14 distinct z planes, maximum-intensity projection (MIP). Arrows indicate individual neurons in the head ganglia and ventral cord. Scale bar: 50 μm. (b) Close-up of the *C. elegans* head ganglia region, with the MIP of the xy plane as well as xz and yz cross-sections indicated by the dashed lines. Scale bar: 10 μm. (c) Activity of all 74 neurons identified in an LFM recording of the head ganglia of *C. elegans.* Each row shows a time-series heat map of an individual neuron. Color corresponds to percentage of fluorescence changes normalized to baseline ($\Delta F/F_0$); (d)–(f): Whole-brain Ca^{2+} imaging of larval zebrafish *in vivo.* (d) Light-field-deconvolved volume containing 51 z-planes (maximum-intensity projections along three orthogonal directions). Exposure time: 50 ms per frame, 4 μm z-plane spacing. Neurons labelled with GCaMP5. A total of 5,379 spatial filters (colored patches) were identified by an automatized algorithm based on principal component analysis and independent component analysis, most of which correspond to individual neurons. Scale bar: 100 μm. (e) Ca^{2+} intensity signal ($\Delta F/F_0$) extracted from spatial filters shown in (d). Arrow at ~15 s indicates onset of aversive odor stimulus. Inset on lower right is zoom into area indicated by dashed box: colored arrows highlight neurons with small differences in response onset. (f) Ca^{2+} intensity traces of a manual selection from the cells shown in (e), in olfactory system, midbrain, and hindbrain.

Prevedel et al. exploited this capability to image unrestrained worms at 50 Hz, a frame rate high enough to avoid motion blur and track neurons during locomotion. Combined with a motorized stage for sample re-positioning as well as static fluorescent labelling for neuron tracking, this frame rate was shown to be sufficient for performing whole-brain imaging in freely moving worms. In another recent study, LFM was combined with a microforce sensing probe for studies of *C. elegans* mechano-sensation (Shaw et al., 2016).

Larval zebrafish brains, although transparent, are much larger (approx. 800 × 400 × 300 μm for 5 dpf. larvae) and contain in the order of 100,000 neurons with a typical diameter on the order of 5–7 μm, which generate action potentials. Using a 20×/0.5 water-dipping objective in an upright microscope, a volumetric FOV of ~700 × 700 × 200 μm can be captured and was used to perform LFM at a 20 Hz volume rate, allowing to capture the neuronal dynamics on the whole-brain level and single-cell resolution in larval zebrafish (Prevedel et al., 2014) (Figure 4.5d–f). To achieve this large FOV, a trade-off in LFM design parameters had to be chosen that resulted in a resolution just below single-cell resolution in this organism. However, a signal extraction and demixing approach based on Independent Component Analysis (ICA), as described in Section 4.4.6, was capable of extracting thousands of cellular Ca^{2+} signals from the recordings, revealing detailed dynamics on a 50 ms timescale.

Using the SID signal demixing strategy outlined in the previous section, we were recently able to show improved neuron localization in larval zebrafish by seeding a demixing algorithm with remaining unscattered light. Light scattering in larval zebrafish brain is much weaker than, e.g., in the mammalian cortex, but severe enough to lead to noticeable blurring when imaging structures across the brain. It is therefore beneficial to account for the effects of scattering using a suitable demixing algorithm. By comparing SID-extracted signals to ICA-extracted signals, we were able to find indications that SID avoids well-known pitfalls of the ICA-based approach, such as ghosting and reduced segmentation ability for highly correlated signals. Crucially, SID requires two to three orders of magnitude less computational time than frame-by-frame reconstruction followed by ICA (see Section 4.4.5), enabling high-throughput and scalable zebrafish imaging.

Pégard et al. (2016) demonstrated a computational strategy termed compressive light field microscopy for Ca^{2+} imaging of larval zebrafish, which starts by applying ICA to a time series of raw LFM frames to separate the LFM signatures of active neurons into spatially sparse components. Signatures contained within each such component are then identified by fitting an analytical model for the LFM signature of a point source in a scattering medium (Liu et al., 2015). Finally, the neural activity time series are extracted by solving a non-negative least-squares optimization problem that for each time step finds the most likely brightness of each neuron signature identified in the previous step by comparison with the data. Compressive light field microscopy achieves a computational efficiency roughly comparable to SID. Due to differing recording conditions, it is difficult to accurately compare the neuron identification performance of the two methods. To point out a crucial difference, SID does not rely on a very specific analytical model for the LFM signature of a point source in a scattering medium, but only imposes rather coarse restrictions on the acceptable shape of a candidate signature. It may therefore be suspected that

SID is more tolerant of the optical aberrations and imperfections that inevitably occur in the context of *in vivo* imaging, such as light refraction at tissue boundaries, local variations in scattering parameters, and background. SID verifiably exhibits such robustness when applied to the more strongly scattering conditions of the mammalian cortex and hippocampus, as we were able to show by simultaneous ground-truth recordings, as discussed in Section 4.4.4.

All aforementioned results were achieved in head-fixed specimen; imaging freely behaving zebrafish has been a challenging goal due to the very fast motion bursts, which occur on millisecond timescales and with up to 1 g acceleration (Cong et al., 2017). Recently, a variant of an LFM imaging system (termed XLFM), combined with high-speed, closed-loop 3-axis sample re-positioning was used to demonstrate imaging of whole-brain neural activity during visually evoked prey capture behaviors in freely swimming larval zebrafish (Cong et al., 2017), opening an exciting new perspective for applications of LFM technology. The XLFM design involves an array of tens of millimeter-sized lenses instead of a microlens array with thousands of lenslets. In addition, the lenses are placed in a plane that shows a Fourier transform of the object, rather than an image plane. The specifics of this XLFM optical design are discussed further in Section 4.5.1.

4.4.4 Application to the Mammalian Brain

As discussed above, comparatively much stronger light scattering in the mammalian brain poses a challenge to any imaging method that aims to capture GECI activity from brain regions deeper than one scattering length (50–100 μm in mouse cortex). In naively reconstructed LFM recordings, scattering leads to a range of artefacts and signal crosstalk, as illustrated in Figure 4.4a,b. As outlined in Section 4.4.2 above, the SID algorithm was designed to allow for robust extraction of neuronal signals under these conditions, as is discussed in the following.

Applying SID to Ca^{2+} imaging in cortex of head-fixed, awake mice, we were able to record neuronal activity from volumes of 900 μm in diameter and 260 μm axial extent at a 30 Hz volume rate, and to extract signals from neurons as deep as 380 μm, corresponding to cortical layers I–III and part of layer IV. SID detected over 500 active neurons during a one-minute recording, corresponding to ~10% of labelled neurons (Figure 4.6a–c). By imaging through a window implanted onto the top of the corpus callosum layer, we recorded from 150 neurons in the cell body layer of the hippocampal CA1 region. We used a hybrid two-photon/light field microscopy setup to cross-validate SID versus ground-truth two-photon recordings in a range of planes at increasing depths.

For this purpose, we designed and built a hybrid LFM-2PM, excited GECIs in a single plane at a series of depths in mouse cortex using two-photon scanning excitation, and split fluorescence between an LFM detection arm and a standard 2PM detection arm (based on photomultiplier tubes). This allowed us to perform a rigorous statistical analysis of the data quality and detection statistics, some results of which are illustrated in Figure 4.6d, e: examining true positive neuron detections from SID, we found that the median correlation between SID-extracted and 2PM ground-truth signals decreased only lightly (Figure 4.6d), from 0.92 ± 0.06 at a depth of 100 μm to 0.90 ± 0.06 at 375 μm (median ± s.e.). Furthermore,

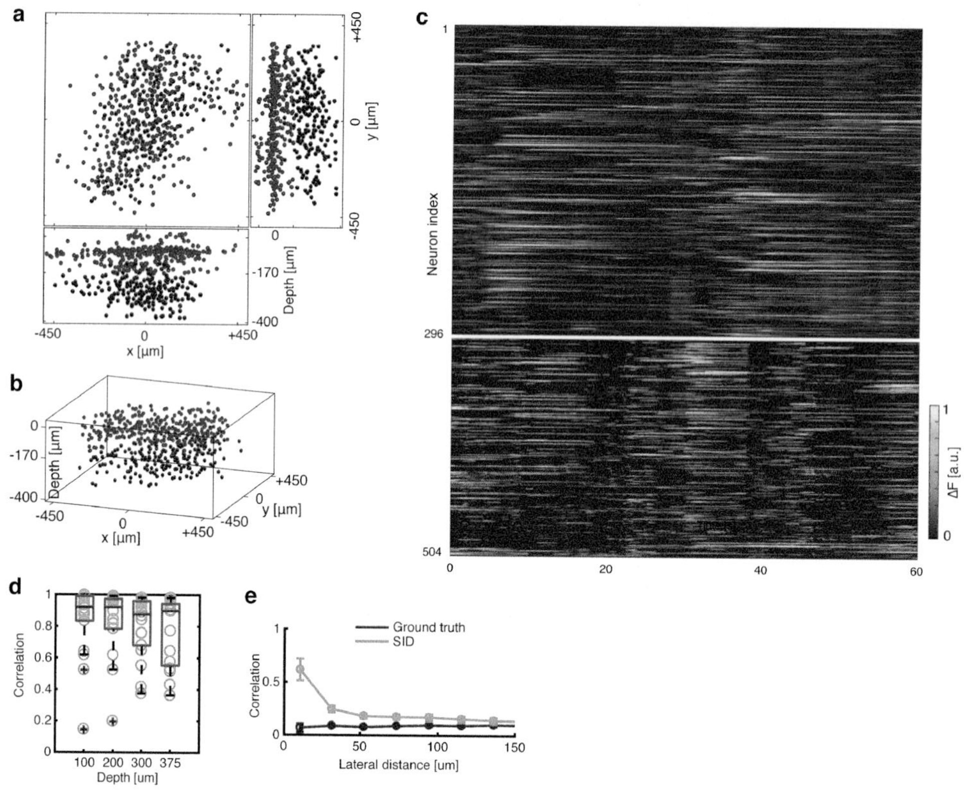

FIGURE 4.6 Video-rate volumetric Ca^{2+} imaging to 380 µm depth in mouse cortex enabled by LFM/SID microscopy (all from: Nöbauer et al., 2017). (a) Neuron locations extracted by Seeded Iterative Demixing (SID) from a 60-second recording at 30 fps in mouse cortex. Neuron locations from two subsequent recordings of different depth ranges are shown together, color indicates recording range: 0–170 µm (blue) and 120–380 µm (red). (b) Isometric perspective onto data shown in (a). (c) Heatmap of neuron activity traces obtained by Seeded Iterative Demixing (SID), corresponding to locations shown in (a)–(b). Upper panel: 296 active, GCaMP6m-expressing neurons found in 0–170 µm depth range. Lower panel: 208 neurons found in 120–380 µm in a subsequent recording. (d) Comparison of SID-extracted signals to ground truth: correlation versus depth between SID signals and their ground truth (2PM-recorded) counterparts, shown for one example recording. Red crosses in (a) are box plot outliers, green circles indicate data points. (e) Pair correlations for uncorrelated (<0.2) neurons extracted by SID and their corresponding ground truth pair correlations (error bars: s.e.m.), versus pair distance, shown for one example recording.

we examined how true correlations between pairs of ground-truth signals get reproduced in the SID data (Figure 4.6e): for pairs of neurons with low correlation in the ground-truth (cross-correlation <0.2), we found an increase of correlations in the corresponding SID-extracted signals (i.e. false positive correlations) for neuron separations smaller than ~20 µm. This provides a limit on the neuron pair distance down to which SID is able to discriminate signals, i.e. to assign neuronal time series correctly in the scattering mouse brain. For neuron pairs with high ground-truth correlations (cross-correlation >0.6), we found no such excess correlation in the corresponding SID neuron pairs, irrespective of neuron distance.

The depth reach of SID is in our current implementation limited by dynamical background fluorescence emanating from outside of the volume modeled in the simulated LFM PSF. More anatomically specific and (soma- or nucleus-) confined labeling and red-shifted GECIs can be expected to greatly ease extraction and further improve SID reach and data quality, along with further improvements in the algorithm. Combined with its very high-volume acquisition rate (only limited by camera frame rate and GECI brightness) and overall versatility and scalability, we expect our approach to get adopted quickly and inspire further innovations in the field of light-field Ca^{2+} imaging.

4.4.5 Computational Cost, Data Management, and Scaling

In the traditional deconvolution-based reconstruction, the computational cost of reconstructing a 3D source volume from an LFM raw frame depends on frame size and the axial and lateral extent of the simulated PSF that is used for deconvolution. In general, complexity rises linearly with the number of image pixels, and quadratically with the axial range of the PSF, as well as the numerical aperture of the objective. Since in conventional **frame-by-frame reconstruction** all frames are treated independently, overall computational cost rises linearly with the number of frames, quickly leading to cluster-scale resource requirements for typical recordings in neuroscience. This scaling can be overcome under certain conditions by approaches such as SID, the computational demands of which are discussed further below.

The core operation that causes high computational cost in an LFM inverse problem solver – such as the Richardson–Lucy algorithm – is the calculation of the camera image that would arise from a given source volume estimate (forward projection) and/or the calculation of a source volume estimate from a given camera frame (backward projection). Both operations can be formulated in a manner reminiscent of convolutions with a spatially varying kernel. This quasi-convolution can be implemented either in a direct approach, i.e. by simply carrying out the necessary multiplications and additions, or based on Fast Fourier Transform (FFT), exploiting the fact that convolution transforms into multiplication in Fourier space. While FFT-based implementations of convolution are known to generally perform better for large input matrices, we found that for the specific quasi-convolution operation required for LFM reconstruction, an implementation using the direct approach (multiplications and additions) can outperform an FFT-based approach. However, achieving this advantage requires careful engineering of memory access patterns and other optimizations. When implemented in a high-level programming language, such as MATLAB or Python, the availability of high-performance FFT library functions will in many cases make an FFT-based strategy more viable. MATLAB implementations of both strategies have been published as a supplement to Prevedel et al. (2014).

Depending on the exact inverse problem solver used, significant computational effort may be required for numerically evaluating gradients of the merit function. We found methods based on adjoint operators, such as Richardson–Lucy, to be more suitable, since the adjoint operation of the forward projection, i.e. the backward projection, can be evaluated efficiently from a quasi-transpose of the LFM PSF matrix.

As a guideline, reconstructing a 900 × 900 px frame recorded with a 20×/0.5 objective (a configuration typical for zebrafish recordings) using a Richardson–Lucy-based solver takes approximately 1 core-hour on an up-do-date server CPU and requires 2 GB of RAM. On a high-end graphics card (GPU) suitable for general purpose computations, the same operation may take approximately 5 minutes.

Due to the high frame rates that LFM allows, it is not rare for a single experiment to result in tens of thousands of frames. Clearly, reconstructing all frames individually requires cluster-scale resources (hundreds of CPU cores or dozens of GPUs) in order to be completed on a timescale of a few hours. Aside from using an academic cluster, it may be advisable to explore commercial cloud-based solutions. Since the individual reconstructions are carried out independently, and since a guaranteed job completion time is not required, the more affordable spot instances can be used to lower costs.

Prior to compression or signal extraction, the reconstructed frames produced by available algorithms can be rather large (tens of megabytes per frame), resulting in terabyte-sized datasets. Storing, transferring, and archiving these can be cumbersome and costly. It is therefore advisable to keep the raw data only (compressed losslessly), and to discard the reconstructed volumes after signal extraction.

In scenarios such as high-speed Ca^{2+} imaging from head-restrained animals, it is evident that much more efficient approaches than naive reconstruction of each individual and full frame can be conceived. Such approaches may exploit the fact that the neuron positions in the imaged volume are largely static – at least up to minor shifts due to motion bursts, which, in turn, may only occur sparsely in time. In addition, in many brain regions and experimental paradigms, much fewer neurons exhibit activity during a recording than would in principle fit into the volume of brain tissue being imaged (sparsity of active neurons in space). If it is acceptable to detect active neurons only, this property can be exploited as well.

The **SID approach** discussed previously implements such a more efficient strategy for recording scenarios where the position of neurons in the field-of-view can be expected to be rather static. SID requires frame-by-frame reconstructions of only a small number of LFM frames, from which it extracts neuron candidate positions. Subsequent processing steps are performed based on the active neurons only (by iteratively updating estimates of neuron image on the camera, and an estimate of neuron temporal signal). This requires forward projections of small regions of sample volume only, namely those that contain active neurons. Thus, SID achieves a substantial reduction in computational cost – up to 1,000-fold for a typical 10,000-frame recording from a cortical volume of 900 × 900 × 260 μm containing thousands of neurons – reducing hardware requirements from cluster-scale to workstation-scale. In addition, SID performs the processing steps of reconstruction and segmentation/signal extraction simultaneously and directly yields extracted signals, without the need of a separate signal extraction stage. Since the standard, frame-by-frame reconstruction method does require such a stage, however, we discuss a number of approaches to that end in the following section.

4.4.6 Signal Extraction

For extracting neuron positions and signals from the frame-by-frame reconstructed volumetric LFM data, a widely used approach is based on the PCA and ICA algorithms (Mukamel et al., 2009). To make this approach viable for the dataset sizes regularly encountered in LFM imaging, it is necessary to select a subset of voxels for further processing, for example, by calculating the standard deviation of each voxel in time (incremental algorithms are available for this task, reducing memory load) and processing only the voxels with the highest standard deviation values. Prior to this, however, it is necessary to remove any global and slow trends in the data – such as the effects of fluorophore bleaching – by fitting a low-order polynomial to the frame means and dividing by the fit value. In addition, background subtraction (by rank-1 matrix factorization) is essential. In addition to that, the entire FOV volume can be tiled into a number of sub-volumes, each of which can be processed individually. Binning time steps during the PCA/ICA steps may also reduce computational load without compromising detection quality. PCA/ICA then factorizes the data into statistically independent components (maximizing a mixture between spatial and temporal independence has proven productive). The resulting spatial filters may contain more than one neuron, so an additional segmentation step is required (either using a local maximum search, or other approaches based on watershed transform or morphological operations). The resulting neuron spatial filters are examined for size and shape criteria. For all accepted filters, the temporal signals are extracted by summing over all pixels in the neuron filter.

By appropriate voxel preselection and tiling, the individual PCA/ICA sub-problems (which are the most resource-intensive steps) can be made small enough to become tractable on a single scientific workstation. Alternatively, sub-jobs may be distributed onto a cluster, for example, by using map-reduce cluster computing frameworks such as Apache Spark, together with the neuroscience analysis library Thunder (Freeman et al., 2014), which implements ICA, among other algorithms.

Constrained non-negative matrix factorization algorithms optimized for Ca^{2+} signal extraction has been shown to exhibit better neuron demixing capabilities (Pnevmatikakis et al., 2016), but similar to PCA/ICA, to our knowledge, no implementations exist that would be capable of processing LFM reconstructed videos without voxel preselection or tiling.

Several approaches exist for inferring firing rates from GECI signals (Pnevmatikakis et al., 2016; Vogelstein et al., 2010; Friedrich et al., 2017) all of which can be combined with the signal extraction algorithms discussed in this section, as well as the SID algorithm.

4.4.7 Susceptibility to Motion

In all Ca^{2+} imaging applications, careful engineering, animal training and acclimation, as well as data post-processing are necessary to minimize and correct for motion artefacts introduced by movement of neurons relative to the optical system. When the animal is fully or partially restrained, it is advisable to construct a maximally rigid specimen stage and head post, and optimize placement and fixation of the cranial window. At the same time, it is beneficial to ensure that forces exerted by the animal are not met with large resistance. Air-suspended spherical and disk-shaped treadmills or elastic linear treadmills have been constructed to this end; a design for a disk-shaped, counter-balanced treadmill for head-fixed mouse imaging was published by Prevedel et al. (2016). Animal training and habituation also contributes greatly to reducing motion. In a suitable head-fixed experiment design, in-plane motion effects can be reduced to fractions of a neuron diameter (i.e. a few microns), and axial motion can often be reduced to a negligible amount.

In comparison to scanning imaging methods – especially random-access scanning – LFM provides increased robustness with respect to motion; this is because it collects unbiased volumetric data at a high frame rate, thus avoiding motion blur. In addition, LFM records instantaneous volumetric snapshots, i.e. there is no delay between acquisition of data points across a volume. Thus, information loss due to motion – such as neurons drifting out of an imaging plane or being missed by a random-access excitation beam – is minimal. Should the set of neurons being imaged shift during a recording, this effect can be corrected for in post-processing by performing volume co-registration using standard software packages on the reconstructed LFM volumes.

The SID algorithm in its current realization is not designed to actively correct motion artefacts. However, motion can be diagnosed and detected by computing simple motion metrics on the LFM raw data, such as the value range of the spatial autocorrelation of the difference from subsequent raw frames (Nöbauer et al., 2017). Frames affected by motion can thus be excluded from SID analysis, so that motion does not affect the quality of signal extraction from the motion-free segments of the recording. Motion glitches shorter than GECI response dynamics can be corrected/interpolated by training a model of GECI response dynamics (Pnevmatikakis et al., 2016) on the motion-free data and using it to extrapolate across motion-affected frames.

4.5 ALTERNATIVE IMPLEMENTATIONS AND FUTURE DIRECTIONS

The standard implementation of an LFM as described in the first sections of this chapter (i.e. a microlens array in an image plane, and a camera in the back focal plane of the microlens array) has proven powerful, but is by no means the only way to sample the 4D light field $L(s,t,u,v)$ in a useful way. In fact, as pointed out by Antipa et al. (2016), even a random diffuser placed in front of the camera leads to a certain (random, but fixed) encoding of both angular and lateral information into the intensity pattern captured by the camera. If this encoding is known (e.g. by exact measurement of the surface of the diffuser as proposed by Antipa et al., or, more naively, by imaging a point source at all possible positions), the inverse problem can in principle be solved and the source volume recovered. However, not all possible encodings may be equally well invertible, and hence this may lead to artefacts or ambiguities in the presence of noise and other disturbances.

In addition to finding an optimal detection strategy, the performance of a (light-field) imaging system may also be enhanced by imprinting additional information onto the scene by means of specifically designed illumination. In the second part of this section, we briefly discuss two quite opposite illumination approaches: random speckle illumination, which uses a series of random illumination patterns to obtain background suppression and increased resolution, and volumetric light field excitation, a method that aims to maximize the invertibility of a light-field imaging dataset by finding the optimal set of illumination configurations.

Prior to that, however, we examine a number of recent proposals and implementations for alternative sampling strategies in light field detection.

4.5.1 XLFM: Segmenting Fourier Space

One alternative approach to sampling the light field is to place the lenslets not into an image plane (i.e. one that is conjugate with the object), as in the light field microscopes discussed at the beginning of this chapter, but in a Fourier plane (i.e. one that is conjugate to the Fourier transform of the object, such as the pupil plane of the objective). Two recent publications (Cong et al., 2017; Zhang et al., 2017) use this arrangement; here we restrict our discussion to the work of Cong et al. (2017), which uses the Fourier-plane LFM configuration (termed XLFM) for Ca^{2+} imaging of freely moving zebrafish. In the XLFM design, a custom array of 27 lenslets (1.3 mm diameter, 26 mm focal length) is placed in a plane conjugate to the objective aperture stop plane (pupil plane), which produces a set of 27 sub-images of the specimen on the camera, each encoding a different slice of the 4D light field (Figure 4.7a). The PSF used for inverting the problem is determined experimentally and is (to first approximation) spatially invariant. The authors report a lateral resolution of 3.4 μm and an axial resolution better than 10 μm over an axial range of 400 μm (peak axial resolution: ~5 μm). A key advantage of this approach is that it avoids the artefacts in the center of the axial range found in the standard LFM design. Depending on the required resolution, the overall effective volumetric FOV can be as large as ⌀800 × 400 μm. In this arrangement, 84% of the light collected by the objective is blocked by the opaque areas in between the lenslets, a loss that is compensated for by using an objective with a numerical aperture much higher than in standard LFM designs. This is possible because in this design, the NA-determined trade-off between axial range and axial resolution (as discussed in Section 4.2.2) is relaxed, leading to greater design flexibility. The computational effort required for reconstructing the 3D source volume is comparable to standard LFM designs, making XLFM an attractive alternative approach to light field microscopy, the applicability of which to scattering samples yet has to be evaluated. XLFM performance decreases with decreasing sparsity of light sources in the sample, which may be a limitation when applied to specimens with very high densities of active neurons.

4.5.2 Diffuser-Encoded Light-Field Imaging

As mentioned above, Antipa et al. (2016) have put forward the interesting proposition to use a random phase plate (diffuser) to encode the 4D light field into a 2D camera image. In a prototype implementation, they show how to determine the exact phase pattern imprinted by the diffuser by recording a focal stack of images of across the diffuser. The recovered phase pattern is used to solve the inverse problem, for which a number of different regularized least-squares formulations are explored. While this approach is not currently designed for, nor has been demonstrated in, the context of *in vivo* Ca^{2+} imaging, it radically highlights the enormous flexibility that exists in designing light-field imagers.

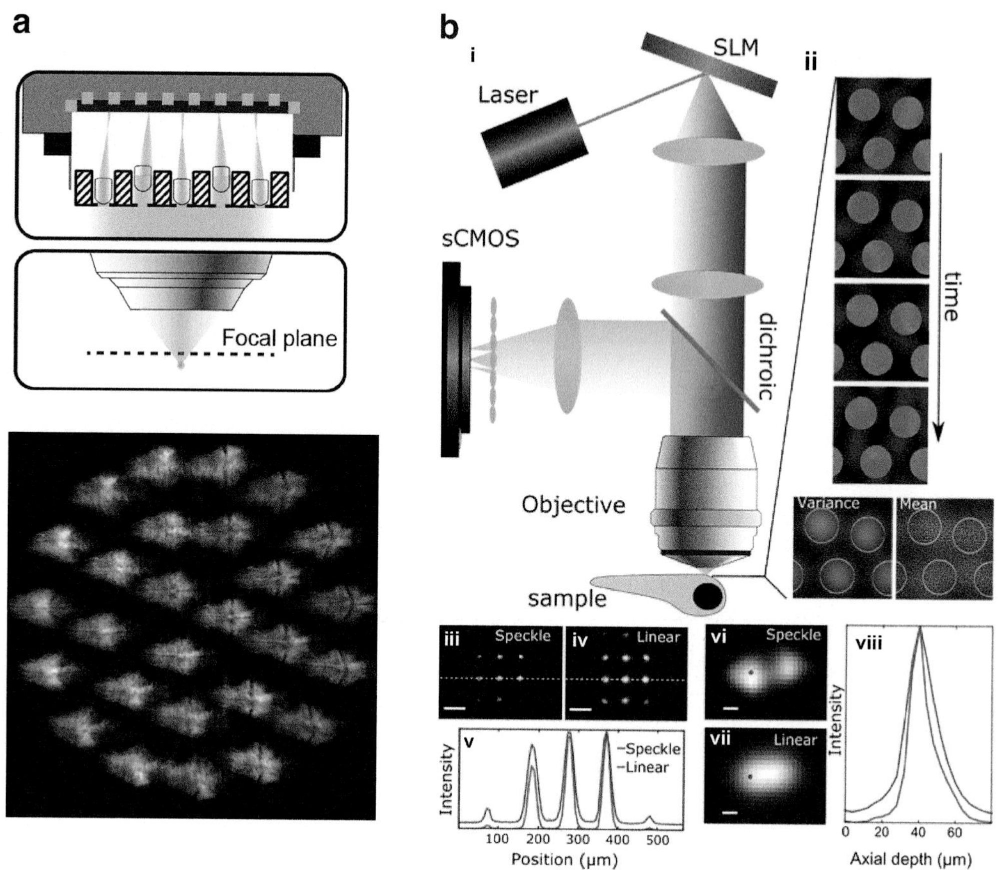

FIGURE 4.7 Alternative implementations: XLFM and SpeckleLFM. (a) XLFM (Cong, 2017). Top: schematic of lenslet positions and image formation in XLFM, for a source below the objective focal plane. Lenslets are positioned in a plane conjugate to the objective pupil plane. Bottom: example of XLFM raw camera frame showing images of zebrafish larva formed by array of lenslets. (b) SpeckleLFM (Taylor, 2018). (i) Experimental schematic. The LFM is constructed with a microlens array placed in the primary image plane of the fluorescence microscope, which then projects onto the chip of a sCMOS camera via a 1:1 image relay (not shown). Controlled speckle illumination is introduced by imaging an SLM with a random phase mask onto the objective back-focal plane. (ii) In SpeckleLFM the fluorescent objects (grey disks) are illuminated with a sequence of independent speckle patterns (blue), and the resulting fluorescence is recorded. The mean and variance of the recorded intensities both form an image of the sample, with the mean converging to the conventional intensity-based image while the variance achieves sharper resolution. (ii, iv) The point spread function (PSF) of a small bead located at z = 30 µm below the focal plane, as recorded from the image variance (iii) and mean (iv). (v) A line profile along the dashed line shows that the speckle PSF (blue) is sharper and includes less background than the corresponding linear PSF (orange). (vi, vii) A zoom-in on two beads that are separated by 4.0 µm, and which are resolved with speckle LFM (vi) but unresolved with linear LFM (vii). (viii) An axial line profile through the points marked in (vi, vii) shows the sharper axial resolution of speckle LFM; speckle LFM and linear LFM respectively achieve an axial full-width half-maximum of 12.5 µm and 19.0 µm. Scale bars 100 µm in (iii, iv), 2 µm in (vi, vii).

4.5.3 Aperture Interference

Kauvar et al. (2017) attempt a more systematic examination of optimal light field acquisition strategies by allowing time-sequential sampling of the aperture plane in an LFM-type setup. While time-sequential sampling reduces attainable frame rate and hence may not be directly applicable to fast *in vivo* imaging, it provides maximal design flexibility for exploratory purposes. The authors identify a number of masks, such as rings, circles, and random patterns, that are scanned across the aperture plane and result in various degrees of invertibility (i.e. uniqueness of the solution to the inverse problem) across a range of signal-to-noise ratios. Some of these configurations outperform standard LFM in terms of spatial resolution and axial field of view, albeit at a significant reduction of achievable frame rate. The theoretical framework employed in this article can serve as an instructive basis for systematic explorations of the space of possible light field implementations.

4.5.4 Light Field Excitation

As demonstrated by Levoy et al. (2009), the idea of light field microscopy can be turned around to create a light field illumination device, i.e. an optical system that is capable of sending beams of light at arbitrary angles of incidence and lateral offset into the sample. This is achieved by illuminating a spatial light modulator, and imaging it into the back focal plane of a microlens array placed in a plane conjugate to the object at the illumination port of a microscope (optionally equipped with light field detection). Thus, illumination configurations such as brightfield, darkfield, headlamp, or oblique incidence can be emulated. Schedl and Bimber (2016) take this idea one step further by devising an algorithm to compute optimal illumination light fields for selectively exciting target objects in a volume and avoiding excitation of other areas. Similar approaches may prove useful in increasing the robustness and selectivity of light-field imaging devices in scattering samples.

4.5.5 Speckle-Illuminated LFM

In a recent article (Taylor et al., 2018), we outline how the scattering properties of tissue can be turned into a resource for enhancing the resolution of an LFM and reducing background and crosstalk (Figure 4.7b): by illuminating zebrafish larvae with random but repeating speckle patterns generated by shining a laser through a diffuser and recording a light field image for each configuration, the variance between sequences of frames can be used to recover the source volume with 1.4-fold increased resolution and strongly suppressed background. The illumination pattern realized within the source volume in this paradigm need not be known; it is merely the varying illumination patterns with high spatial frequency contents that lead to this effect (blind structured illumination).

4.6 CONCLUSION

Understanding the global dynamics of large neuronal populations and the emergence of complex behavior from large-scale integration of information in the nervous system is a fundamental problem in current neuroscience. The advent of genetically encoded fluorescent calcium (Ca^{2+}) indicators (GECIs) has once again made optical imaging a key method in biological research, and a range of methods have been devised to tackle the challenges involved in reading out neuronal activity from increasingly large populations, across brain areas and from deep within scattering brain tissue.

In this chapter, we have tried to outline the elegant principles and enormous potential of Light Field Microscopy as a neuronal imaging approach, the evolution of which has only just begun. The performance and versatility of LFM-inspired techniques have convinced us that the co-development of optical excitation and detection devices with customized machine learning solutions will continue to push the boundaries of neuronal activity readout.

REFERENCES

Adelson, E. H. & Wang, J. Y. A. Single lens stereo with a plenoptic camera. *IEEE Trans. Pattern Anal. Mach. Intell.* **14**, 99–106 (1992).

Altun, Z. F., Herndon, L. A., Wolkow, C. A., Crocker, C., Lints, R. & Hall, D. H. *Wormatlas.* (2002). Available at http://www.wormatlas.org/

Antipa, N., Necula, S., Ng, R. & Waller, L. Single-shot diffuser-encoded light field imaging. In 2016 IEEE Int. Conf. Comput. Photogr (ICCP), pp. 1–11 (2016). doi:10.1109/ICCPHOT.2016.7492880

Broxton, M., Grosenick, L., Yang, S., Cohen, N., Andalman, A., Deisseroth, K. & Levoy, M. Wave optics theory and 3-D deconvolution for the light field microscope. *Opt. Express* **21**, 25418–25439 (2013).

Cohen, N., Yang, S., Andalman, A., Broxton, M., Grosenick, L., Deisseroth, K., Horowitz, M. & Levoy, M. Enhancing the performance of the light field microscope using wavefront coding. *Opt. Express* **22**, 24817–24839 (2014).

Cong, L., Wang, Z., Chai, Y., Hang, W., Shang, C., Yang, W., Bai, L., Du, J., Wang, K. & Wen, Q. Rapid whole brain imaging of neural activity in freely behaving larval zebrafish (Danio rerio). *eLife Sciences* **6**, e28158 (2017).

Dey, N., Blanc-Féraud, L., Zimmer, C., Roux, P., Kam, Z., Olivo-Marin, J.-C. & Zerubia, J. *3D Microscopy Deconvolution using Richardson-Lucy Algorithm with Total Variation Regularization.* 71 (INRIA, 2004). Available at https://hal.inria.fr/inria-00070726/document

Friedrich, J., Zhou, P. & Paninski, L. Fast online deconvolution of calcium imaging data. *PLOS Computational Biology* **13**, e1005423 (2017).

Friedrich, J., Zhou, P. & Paninski, L. Fast active set methods for online deconvolution of calcium imaging data. *ArXiv160900639 Q-Bio Stat* (2016). Available at http://arxiv.org/abs/1609.00639

Gorski, J., Pfeuffer, F. & Klamroth, K. Biconvex sets and optimization with biconvex functions: a survey and extensions. *Math. Methods Oper. Res.* **66**, 373–407 (2007).

Helmchen, F. & Denk, W. Deep tissue two-photon microscopy. *Nat. Methods* **2**, 932–940 (2005).

Jacques, S. L. Optical properties of biological tissues: a review. *Phys. Med. Biol.* **58**, R37 (2013).

Kauvar, I., Chang, J. & Wetzstein, G. Aperture interference and the volumetric resolution of light field fluorescence microscopy. *IEEE ICCP* (2017). Available at http://www.computationalimaging.org/publications/aperture-interference-light-field-fluorescence-microscopy-iccp-2017/

Levoy, M. *Optical Recipes for Light-Field Microscopes.* (Stanford University, 2006). Available at https://graphics.stanford.edu/papers/lfmicroscope/lfmicroscope-optics.pdf

Levoy, M., Ng, R., Adams, A., Footer, M. & Horowitz, M. Light field microscopy. *ACM Trans. Graph.* **25**, 924 (2006).

Levoy, M., Zhang, Z. & Mcdowall, I. Recording and controlling the 4D light field in a microscope using microlens arrays. *J. Microsc.* **235**, 144–162 (2009).

LFDisplay: a real-time system for light field microscopy. Available at http://graphics.stanford.edu/software/LFDisplay/

Lippmann, G. Epreuves reversibles. Photographies integrals. *Comptes Rendus Hebd. Séances Académie Sci.* **146**, 446–451 (1908).

Liu, H.-Y., Jonas, E., Tian, L., Zhong, J., Recht, B. & Waller, L. 3D imaging in volumetric scattering media using phase-space measurements. *Opt. Express* **23**, 14461 (2015).

Lucy, L. B. An iterative technique for the rectification of observed distributions. *Astron. J.* **79**, 745 (1974).

Mukamel, E. A., Nimmerjahn, A. & Schnitzer, M. J. Automated analysis of cellular signals from large-scale calcium imaging data. *Neuron* **63**, 747–760 (2009).

Ng, R., Levoy, M., Bredif, M., Duval, G., Horowitz, M. & Hanrahan, P. *Light Field Photography with a Hand-Held Plenoptic Camera*. (Stanford University, 2005).

Nöbauer, T., Skocek, O., Pernía-Andrade, A. J., Weilguny, L., Martínez Traub, F., Molodtsov, M. I. & Vaziri, A. Video rate volumetric Ca2+ imaging across cortex using seeded iterative demixing (SID) microscopy. *Nat. Methods* **14**, 811–818 (2017).

Pégard, N. C., Liu, H.-Y., Antipa, N., Gerlock, M., Adesnik, H. & Waller, L. Compressive light-field microscopy for 3D neural activity recording. *Optica* **3**, 517 (2016).

Pnevmatikakis, E. A., Soudry, D., Gao, Y., Machado, T. A., Merel, J., Pfau, D., Reardon, T., Mu, Y., Lacefield, C., Yang, W., Ahrens, M., Bruno, R., Jessell, T. M., Peterka, D. S., Yuste, R. & Paninski, L. Simultaneous denoising, deconvolution, and demixing of calcium imaging data. *Neuron* (2016).

Prevedel, R., Verhoef, A. J., Pernía-Andrade, A. J., Weisenburger, S., Huang, B. S., Nöbauer, T., Fernández, A., Delcour, J. E., Golshani, P., Baltuska, A. & Vaziri, A. Fast volumetric calcium imaging across multiple cortical layers using sculpted light. *Nat. Methods* **13**, 1021–1028 (2016).

Prevedel, R., Yoon, Y.-G., Hoffmann, M., Pak, N., Wetzstein, G., Kato, S., Schrödel, T., Raskar, R., Zimmer, M., Boyden, E. S. & Vaziri, A. Simultaneous whole-animal 3D imaging of neuronal activity using light-field microscopy. *Nat. Methods* **11**, 727–730 (2014).

Richardson, W. H. Bayesian-based iterative method of image restoration. *J. Opt. Soc. Am.* **62**, 55 (1972).

Schedl, D. C. & Bimber, O. Volumetric light-field excitation. *Sci. Rep.* **6**, 29193 (2016).

Shaw, M., Elmi, M., Pawar, V. & Srinivasan, M. A. Investigation of mechanosensation in *C. elegans* using light field calcium imaging. *Biomed. Opt. Express* **7**, 2877–2887 (2016).

Taylor, M., Nöbauer, T., Pernia-Andrade, A., Schlumm, F. & Vaziri, A. Brain-wide 3D light-field imaging of neuronal activity with speckle enhanced resolution. *Optica* **5**, 345–353 (2018).

Vogelstein, J. T., Packer, A. M., Machado, T. A., Sippy, T., Babadi, B., Yuste, R. & Paninski, L. Fast non-negative deconvolution for spike train inference from population calcium imaging. *J. Neurophysiol.* **104**, 3691–3704 (2010).

Zhang, M., Geng, Z., Pei, R., Cao, X. & Zhang, Z. Three-dimensional light field microscope based on a lenslet array. *Opt. Commun.* **403**, 133–142 (2017).

CHAPTER 5

Genetically Encoded Activity Indicators

Chenchen Song and Thomas Knöpfel

CONTENTS

5.1 INTRODUCTION

Linking patterns of neuronal electrical activities with behavior is a major goal towards delineating neuronal circuit functions and ultimately understanding the human brain. Microelectrode-based electrophysiology has been, and largely still is, the golden standard for measuring neuronal signals at the level of single neurons while macroscopic electrical measurements such as electroencephalography (EEG) are a traditional tool in basic and clinical research at the systems level. However, both single cell-level and whole brain-level electrophysiology only provide partial insight, as many brain functions emerge from the interactions between groups of cells (neural ensembles) coupled by neural circuits, and therefore need to be understood at the level of neuronal circuits. Mechanistic understanding of neuronal networks underlying behavior and cognitive functions requires experimental approaches to measure neuronal activity patterns with spatiotemporal resolution and coverage beyond the limits of electrode-based methods (Knöpfel et al., 2006; Knöpfel, 2012).

Optical imaging has long been sought to provide the necessary spatiotemporal resolution and coverage for a circuit-centric approach. This circuit-centric concept dates back to the motivation driving the pioneering work of Larry Cohen, Amiram Grinvald, and Roger Tsien, who developed activity-reporting fluorescent dye indicators to monitor neuronal activities in cell cultures, intact brain tissue, and living animals, allowing the simultaneous access to activities of a large number of cells as well as the functions of their connections (Grinvald and Hildesheim, 2004; Baker et al., 2005; Homma et al., 2009). These methodological developments were more recently complemented by genetically encoded activity indicators that enable activity measurements with greater precision and specificity by combining the power of light-based techniques with cell population specificity conferred by genetic manipulations methods (Knöpfel et al., 2006; Knöpfel, 2012).

5.2 FLUORESCENT DYES THAT REPORT NEURONAL ACTIVITIES

A well-established method of optically imaging neuronal activities is through the use of voltage and calcium sensitive low molecular weight (LMW) dyes, often referred to as "organic" dyes. Voltage sensitive dyes (VSDs), when incorporated into plasma membranes, alter their optical properties as a function of the membrane potential. Calcium sensitive dyes optically report changes of cytosolic calcium concentration that is associated with neuronal activity. Both calcium sensitive dyes and VSDs have been instrumental in advancing our understanding of cellular and systems neurophysiology.

5.2.1 Calcium Indicators

Compared to VSDs, calcium sensitive dyes are an indirect approach to monitoring neuronal activity, but are much more extensively used in neurobiology than voltage indicators, thanks to the large optical signal of the calcium sensor. Calcium is a versatile ion responsible for many key functions in cells, but the most pronounced physiological fluctuations of calcium concentration in the cytosol of neuronal cell bodies are associated with action potential (AP) firing. Due to this feature, calcium indicators can be used as surrogate activity reporters, as opposed to VSDs that more directly report membrane potential transients.

Monitoring calcium concentration changes has several limitations. Due to the slower dynamics of intracellular calcium ion transients, in contrast with much faster membrane potential changes, calcium imaging does not resolve high frequency membrane potential fluctuations, or postsynaptic events that do not reach the firing threshold. Subthreshold neuronal signals evoke no (or small) calcium transients (at the level of neuronal somata). Calcium sensors are also limited in reporting narrow APs in GABAergic fast-spiking interneurons that are not associated with large changes in cytosolic calcium transients. Most importantly, calcium imaging is blind to hyperpolarizing activity, which is essential for understanding fast dynamics of cortical and neostriatal function. Calcium imaging therefore paints an incomplete picture of neuronal activity, but due to practical and theoretical limitations of voltage imaging (see Section 5.2.2), calcium imaging is still widely and successfully used as surrogate for neuronal activity.

Excellent low molecular weight (LMW) "organic" fluorescent calcium indicators are readily available for neurophotonic applications (Paredes et al., 2008). These dyes come with fluorescence emission colors covering the visual to near-infrared wavelength spectrum, and have calcium binding constants covering the nanomolar to millimolar range. Although red calcium sensors have recently become available for in vivo imaging, most widely used LMW calcium indicators fluoresce in the green spectral band and have calcium binding constants (Kd values) centered around the dynamic range of the expected calcium transients. For neurons, such candidates include Oregon Green 488 BAPTA and Fluo-4. Since the forward rate constant of calcium binding to LMW calcium indicators is typically diffusion limited, the Kd value limits the backward rate constant. To faithfully monitor fast calcium transients, calcium indicators with higher Kd (trading in signal size) are useful (de Juan-Sanz et al., 2017).

Bulk loading techniques using membrane permeable calcium indicator esters allow intracellular loading of indicators into large populations of cells in vivo (Garaschuk et al., 2006). This approach enables single cell resolution calcium imaging in awake living mice with excellent signal to noise ratio (SNR). A main pitfall of this method is that, in tissues with a mixture of functionally diverse cell classes, the class identity of the imaged cells is unknown. This issue has been addressed by double labeling techniques where staining of the calcium indicator is combined with the use of dyes that appears to preferentially label certain cell types or by performing the experiments in mice where specific cell populations express a genetically encoded fluorescent marker (e.g. green fluorescent protein; GFP) (Kerlin et al., 2010; Nimmerjahn and Helmchen, 2012). This approach, however, is not very convenient and is limited in the possibility of differentiating between the many functionally distinct cell classes found in complex tissues. Indeed, this issue has been the main driving force for the development of genetically encoded calcium indicators (GECIs) that enable genetic targeting of the protein-based calcium activity indicator itself.

5.2.2 Voltage Indicators

Voltage sensitive dyes achieve a direct readout of neuronal electrical activity. Indeed, VSD imaging has allowed measurements of membrane potential transients with high spatial resolution (practically limited only by light scattering) as well as high temporal resolutions (practically limited only by data acquisition rates) in a variety of preparations including molluscs, fish, worms, and mammalians (Grinvald and Hildesheim, 2004; Baker et al., 2005; Homma et al., 2009; Antic et al., 2016). Low molecular weight (LMW) "organic" VSDs that respond quasi-instantaneously (within microseconds) to a change in membrane potential are able to faithfully image fast electrical signaling, including the shape and spread of fast action potentials (Vranesic et al., 1994). LMW VSDs can be used to bulk-stain membranes in tissues or be injected into single cells. Bulk-staining allows voltage imaging from a large number of cells simultaneously and with good spatial coverage but normally not with single cell resolution, while staining individual cells allows imaging of voltage transients in fine dendritic branches and axons with high spatial resolution (Antic et al., 2016). Activities of VSD-stained cells can be imaged by using simple epifluorescence wide-field imaging techniques that are often used at the level of population signals (Grinvald and Hildesheim, 2004; Baker et al., 2005; Homma et al., 2009), but dye-injected labeling of single neurons using VSDs also allows optical recordings that resolve subcellular structures including individual axons/dendrites (Zhou et al., 2007).

VSDs have proven to be very useful both in advancing optical imaging technologies and in furthering our understanding of neuronal network properties. Despite the seminal contribution of VSD imaging in cellular and systems physiology, there are several severe technical limitations inherent to this approach. Extracellularly applied VSDs indiscriminately label all cell types in the stained heterogeneous tissue, including not only different classes of neurons but also glia and vascular cells. This limits the interpretability of VSD population signals and results in a poor SNR, which further complicates optical signal interpretation. Hemodynamic signal had been a severe confounding factor in VSD imaging in the living mammalian brain, where the detected optical response contains a mixture of voltage-related VSD signal and optical reflectance signal from hemoglobin absorbance in the blood flow. However, this issue became less severe with the introduction of "blue" VSDs that have a shifted excitation wavelength for reduced spectral overlap with hemoglobin absorption (Grinvald and Hildesheim, 2004). Additionally, VSDs suffer from other issues including toxicity and phototoxicity, pharmacological side effects, and photobleaching, as well as being inadequate for long-term imaging in vivo due to the invasive dye application method (Knöpfel, 2012; Antic et al., 2016). These practical and conceptual limitations of VSDs motivated the conception and development of genetically encoded voltage indicators (GEVIs). GEVIs (see Section 5.4) directly sense and report transmembrane voltage changes with additional beneficial features including cell type-specificity and reproducible long-term expression of the indicators in vivo. This provides an opportunity for voltage imaging in longitudinal studies of behaviors like learning and memory (Knöpfel et al., 2015).

5.3 GENETICALLY ENCODED CALCIUM INDICATORS (GECIS)

The first GECIs were based on fusion of two color variants of GFP and the Ca^{2+}-binding protein calmodulin (Miyawaki et al., 1997). The two GFP variants were selected such that the emission spectrum of the first overlaps with the absorption spectrum of the second, enabling fluorescence resonance energy transfer (FRET) between these two fluorescent proteins (FP). Ca^{2+} sensitivity of the fluorescence readout is achieved by coupling the conformational change of calmodulin upon calcium binding to a change in distance or orientation between the FRET pair of FPs. While this FRET approach has the advantage of providing a ratiometric signal, a different design principle subsequently became dominant. This second principle is based on the modulation of fluorescence output of circularly permuted variants of GFP (Palmer et al., 2006). The most extensively optimized, and to date most widely used, representatives of this design principle are the GCaMPs (Mao et al., 2008; Chen et al., 2013). The most recent optimization step, with extensive characterization, resulted in the GCaMP6 series, which includes -s (slow kinetics) -m (medium) and -f (fast) variants (Chen et al., 2013). As described above, the larger signal of GCaMP6s comes at the cost of slower kinetics compared to GCaMP6f. More recently, well performing red GECIs based on the circularly permuted FP variant-based design principle have been provided as a tool for improving the depth of in vivo imaging (less scattering) and multicolor functional imaging.

More recently, CaMPARIs, the photoconvertible versions of GCaMP, are generated utilizing a protein Eos that normally fluoresces in the green spectral ranges but permanently shifts to red wavelength band following exposure to ultraviolet (UV) light (Fosque et al., 2015). By fusing Eos with the calcium-sensitive calmodulin domain, CaMPARI creates a permanent snapshot of neurons that are active during the time window of UV exposure, serving as an activity integrator. CaMPARI provides an opportunity to identify and track ex vivo the activity of neurons involved in specific behaviors, where the neurons may be outside the imaging plane or field of view and thus would escape from conventional standard in vivo ("real time") microscopy (Fosque et al., 2015).

5.4 GENETICALLY ENCODED VOLTAGE INDICATORS (GEVIS)

5.4.1 Molecular Structure of GEVIs

In contrast to optogenetic actuators that were engineered by site-directed mutagenesis from naturally occurring light-activated ion channels (e.g. channelrhodopsin2), proteins that exhibit voltage-dependent fluorescence have not been known to exist in nature when the first GEVIs were generated. The basic idea was then to combine protein domains from unrelated host proteins into scaffolds to combine voltage sensing and fluorescence. This led to the voltage sensing domain-based GEVIs, which are based on the molecular fusion of protein components from voltage sensing proteins, such as voltage gated ion channels or the more successful voltage sensitive phosphatases, and fluorescent proteins (Dimitrov et al., 2007; Akemann et al., 2015). More recently, opsin-based GEVIs were discovered (Kralj et al., 2011). These take advantage of a membrane voltage sensitive state in the catalytic photocycle of retinal bound to certain bacteriorhodopsins. GEVIs can be further categorized based on their optical output properties – intensiometric (monochromatic indicators) and ratiometric (differential dual emission) indicators (see Section 5.4.2.1).

5.4.1.1 Voltage Sensing Domain-Based GEVIs

Initial attempts to generate practically useful GEVIs were based on a single fluorescent protein fused to a voltage gated ion channel (Siegel and Isacoff, 1997; Baker et al., 2007). At the same time, the use of a FRET tandem of fluorescent proteins fused to the C-terminal end of an isolated voltage sensing domain of potassium channels was designed and developed (Sakai et al., 2001). This molecular design was based on the concept that movement of voltage sensing domain S4 transmembrane segment is conferred onto a relative movement between the two FPs, thereby changing their FRET efficacy. However, it was only with the use of the voltage sensing domain from *Ciona intestinalis* voltage sensing phosphatases (ciVSP) that general proof of principle was provided for the large set of voltage sensing domain-based GEVIs known to date (Dimitrov et al., 2007). Several modifications from the first prototype of this family, termed VSFP2.3, lead to GEVI scaffolds that resulted in improved performance and practical usefulness. The most fruitful variations include: use of a single FP resulting in monochromatic ("intensiometric") GEVIs (VSFP3s, Arclight) (Lundby et al., 2008; Han et al., 2013); sandwiching the voltage sensing domain between the FP FRET pair (Butterflies) (Akemann et al., 2012; Mishina et al., 2014); and inserting a circularly permuted FP into ciVSP (flicR) (Abdelfattah et al., 2016) or into the ciVSP homologue from *Gallus gallus* (ASAPs) (St-Pierre et al., 2014). Site-directed mutagenesis around key residues (mostly at junctions between protein domains) resulted in further performance optimizations that are best identified by consulting the most recent literature.

5.4.1.2 Opsin-Based GEVIs

Channelrhodopsins are widely used as optogenetic actuators. The photocurrent of these light-activated ion channels and their engineered derivatives is used in optogenetics to actuate membrane potential (Deisseroth and Schnitzer, 2013). The reversal, namely modulation of light output by membrane potential, would not be expected from the principal function of rhodopsins. The finding that Archaerhodopsin 3 (Arch) exhibits a voltage-dependent near-infrared fluorescence therefore came with much surprise to the field (Kralj et al., 2011). The excitement about the fast kinetics of the optical response to membrane voltage changes of Arch was dampened by its low quantum yield (QY 10^{-3}) and the resulting need to use very high illumination densities. This issue has subsequently been addressed by the combination of FPs and opsins with the latter acting as voltage-dependent quenchers (Gong et al., 2014; Zou et al., 2014), and these FRET-opsin GEVIs have been generated with several FP color variants. Moreover, site-directed mutagenesis optimized the voltage dependent steps in the Arch photocycle, yielding better GEVIs that along with specialized microscopes yielded high SNR voltage recordings from single neurons in a brain slice and in vivo (Gong et al., 2015; Werley et al., 2017).

5.4.1.3 Hybrid Voltage Indicators

In addition to GEVIs there are voltage indicator designs that combine a genetically encoded component with a small synthetic molecule, forming a genetically targetable voltage indicator. The first example of this approach is the hybrid voltage sensor (hVOS) where a FP with attached farnesylated and palmitoylated terminal motifs that anchor the protein to the plasma membrane is combined with the non-fluorescent synthetic compound dipicrylamine (DPA) (Chanda et al., 2005). Since DPA is lipophilic but negatively charged, it distributes in the membrane in a voltage-dependent fashion and serves as a voltage sensing FRET acceptor (quencher). Unfortunately, DPA increases the membrane capacitance, so care must be taken to ensure that the concentrations used do not disrupt the native physiological responses (Sjulson & Miesenbock, 2008; Akemann et al., 2009).

Another notable "hybrid" strategy entails activating an organic VSD via an enzyme that is genetically targeted to specific cell populations (Hinner et al., 2006; Miller et al., 2012). This strategy is analogous to the widely-used acetyl ester-modified calcium indicators (that are activated by an endogenous enzyme) and recently has been successfully applied to photo-induced electron transfer (PeT) VSDs (Miller et al., 2012; Liu et al., 2017).

The general concerns regarding these hybrid strategies relate to the practical difficulty of applying the exogenous lipophilic compound to neuronal membranes with good control of membrane concentration and spatial homogeneity in intact tissue.

5.4.2 GEVI Selection for a Particular Application

Selection and application of GEVIs from the large palette of available options requires careful consideration between different indicator properties. Particularly for in vivo experiments, in addition to the SNR-determining characteristics including spectral properties, optical response magnitude, quantum yield, response kinetics, dynamic range, and membrane targeting and localization of the indicator, other factors including expression stability and effects of long-term overexpression are also crucial.

Most GEVI-based voltage imaging experiments in the mammalian brain are performed using mice in vivo or ex vivo. GEVI sequences can be introduced in different ways, such as via viral approaches (e.g. adeno-associated viruses), in utero electroporation, or transgenic approaches (Knöpfel, 2012). Several GEVI transgenic mouse lines are already available that capitalize on the rich and growing toolbox of specific promoter driver lines via the Cre-lox or transactivator systems to restrict indicator expression to specific cell types (Madisen et al., 2015).

5.4.2.1 Choosing between Monochromatic and Ratiometric GEVIs

With respect to instrumentation, GEVIs that emit voltage-dependent fluorescence at only one wavelength are less demanding than GEVIs emitting differential voltage-dependent fluorescence at two wavelengths. The latter require the splitting of fluorescence emission into two detection channels and their synchronized detection by two detectors (or two areas on the same camera when using a dual-view approach). A major drawback of the use of monochromatic GEVIs, however, is the presence of signal components that don't represent membrane voltage but other biological activity dependent parameters such as hemodynamic and other "intrinsic" responses. This caveat is of particular concern in conventional wide-field imaging where population voltage signals are measured as relatively small changes in fluorescence intensity (less than a few percent). Meticulously designed optics and data-processing algorithms may be needed to extract voltage-related signals from the detected optical signals. As an alternative strategy, the use of dual differential emission ("ratiometric") GEVIs with two FPs emitting at different wavelength bands have proven especially beneficial to overcome the issue of signal contamination inherent to monochromatic GEVIs (Akemann et al., 2012; Carandini et al., 2015).

Existing ratiometric GEVIs use FRET as the conceptualized mechanism for the transduction of membrane voltage changes into optical signals. Examples of such ratiometric GEVIs are the VSFP Butterfly family, where the voltage-dependent VSD conformational change results in altered FRET efficiency between the FP pair, and results in anticorrelated fluorescence intensity changes between the FRET donor and acceptor FPs. VSFP Butterfly 1.2 is composed of a ciVSD flanked by a pair of yellow (mCitrine; FRET donor) and red (mKate2; FRET acceptor) FPs (Akemann et al., 2012), and it is capable of consistent membrane targeting and localization, is more effective at detecting subthreshold events, and has enabled in vivo imaging in awake mice (Akemann et al., 2012; Madisen et al., 2015). Further improvement of VSFP Butterfly 1.2 resulted in the chimeric VSFP Butterfly GEVIs, where the transplantation of a portion of the fast Kv3.1 potassium channel into ciVSD S4 transmembrane segment achieved faster kinetic properties (Mishina et al., 2014).

5.4.2.2 Choosing between GEVI Spectral Variants

Fluorescent proteins have been engineered to cover excitation and emission wavelengths virtually over the whole visible spectrum (40–700 nm) with an extension into the near-infrared range (>700 nm; NIR). FP-based GEVIs can also cover (at least in principle) this near-infrared spectral range. Opsin-based GEVIs (without a FP component) have their excitation/emission spectra limited to the red/NIR band but FRET-opsin GEVIs (with an additional FP component) cover a broader spectral range.

Choosing a suitable indicator color therefore depends on the biological question and imaging approach, as well as a compromise between other indicator characteristics such as response kinetics and amplitude. Spectrally shifting new GEVI variants further into the far-red or NIR spectral range has the obvious advantage of deeper tissue penetration with excitation light, reduced tissue scattering, less autofluorescence, and reduced hemodynamic component in the optical signal. Furthermore, NIR GEVIs provide an opportunity for combined optical voltage monitoring with opsin-based optogenetic manipulation, as well as combination with other optical indicators, such as the green calcium indicator GCaMP6, for investigations of neuronal circuit input/output relationships.

5.4.2.3 Kinetic Variants

Many LMW VSDs exploit the molecular Stark effect or photoinduced electron transfer yielding very fast (for Stark effect <1 us) responses to changes in the membrane electric field. The fastest currently known GEVIs exhibit response time constants around 1 ms but many otherwise practically used GEVIs have much slower response time constants (a few ms to tens of ms) (Antic et al., 2016). For faithful monitoring of individual action potentials, a response time constant of 1 ms or less would be needed. However, instrumentation that would enable optical recordings of action potentials from many individual neurons (single cell resolution) in intact tissue is only emerging, and most current voltage imaging experiments aim at recordings of fluctuations in population membrane potentials that are dominated by volume averaged postsynaptic potentials. These voltage signals exhibit a time course that is much slower than that of single action potentials. For these approaches, GEVIs with slower response time constants are the best choice. Thus, despite the thriving efforts in indicator developments to generate faster GEVIs, many natural biological processes (e.g. synaptic potentials, population up-down states) are inherently not fast and many existing GEVIs already have sufficient kinetic sensitivity.

5.5 EQUIPMENT FOR ACTIVITY INDICATOR IMAGING

Optical imaging of neuronal activity requires specialized optoelectronic equipment. Several aspects of specialized microscopes are covered in other chapters of this handbook. This section highlights some issues, without many details, that are relevant when adapting fluorescence microscopes for imaging of activity indicators. The key consideration is that the data acquired with functional imaging have a time-scale. Acquisition of as much information per time bin as possible translates into the requirement to acquire as many signaling photons as possible and at the same time as few non-signaling photons as possible. From this the following guidelines can be derived:

- If necessary, trade in spatial resolution (and even some aberration) for larger photon counts per pixel.
- Use optical filters that maximize excitation and emission intensity and minimize unblocked excitation light in the emission channel.
- Optimize collection of signal photons (large apertures, high detector quantum yield, etc.).
- Resolving small relative changes in fluorescence requires sampling large amounts of photons. In that regime, readout noise of most detectors/cameras is smaller than the shot noise and should therefore not be the dominant decision criterion when selecting a detector/camera.

5.5.1 Wide-Field Epifluorescence Imaging

Functional imaging using calcium and voltage indicators emerged largely from wide-field epifluorescence structural microscopy. For quasi two-dimensional preparations (cultured cells) wide-field epifluorescence is often the best choice, unless subcellular resolution is required. The large emission photon flux (from all points in the field of view in parallel) allows for high temporal resolution in shot noise limited regimes. For this reason, wide-field epifluorescence is also successfully used for voltage imaging in brain slices and in vivo. More recently this classical approach was implemented for large spatial coverage calcium imaging in vivo (Kim et al., 2016). One important limitation of the wide-field epifluorescence method is the lack of means for rejection of scattered photons. The light that is scattered out of the focus decreases spatial resolution and light that is scattered into the focus at the image plane decreases signal size and, more importantly, SNR. The optical configurations discussed in the following sections are designed to address this issue.

5.5.2 Patterned Fluorescence Excitation

Light that is scattered from the surrounding areas into the region of interest (ROI) at the image plane may carry activity signals from neighboring neurons, thus degrading the signal of interest. Even if the scattered photons only represent a constant background, they will add shot noise and reduce the SNR. This issue can be addressed by limiting fluorescence excitation to the ROI. The simplest implementation of this idea is to form a laser beam spot in the object plan ("spot illumination"). A more elegant implementation of this concept is computer-generated holography (CGH) (Foust et al., 2015). In contrast to full-field illumination, patterned excitation minimizes signal degradation arising from non-specific autofluorescence, indicator fluorescence that does not carry the signal of interest, or nearby labelled structures (Foust et al., 2015).

5.5.3 Confocal Laser Scanning Microscopy

Confocal laser scanning microscopy has the advantage of spot illumination while still forming an image. The confocal pinhole rejects additional out of focus light. The main disadvantage of this approach is the inefficient collection of signal photons (scattered photons of interest are largely lost).

5.5.4 Two-Photon Laser Scanning Microscopy

Laser scanning 2P microscopy combines efficient spot illumination with efficient collection of signal photons. Infrared light is less phototoxic since native brain tissue has a relatively low infrared absorption and a low 2P cross section. For these reasons, two-photon line scanning microscopy (2PLSM) is widely used in neuroscience. In combination with intracellular calcium indicators, it allows readout of neuronal action potentials from single cells in highly scattering mammalian brain tissue. As a laser scanning technique, 2PLSM has to compromise between temporal resolution, spatial sampling, and fluorescence excitation. GEVIs have not yet been systematically optimized for 2P imaging. Proof of principle 2P GEVI imaging has been performed using the ratiometric VSFP and the intensiometric ASAPs (Akemann et al., 2013; Chamberland et al., 2017).

5.5.5 Fiber Optic Imaging

Epifluorescence approaches do not allow optical penetration much deeper than 1 mm in mammalian brain tissues, but deeper structures can be imaged by ablating overlying tissue or by inserting optical components into the brain tissue. Emerging implementations of the latter approach are based on GRIN lenses, single core fibers or fiber bundles (Marshall et al., 2016; Miyamoto and Murayama, 2016), and such approaches also allow neuronal activity monitoring in freely moving animals.

5.6 CHOICES DETERMINED BY SCALE AND IMAGING MODALITY

LMW calcium indicators are often preferred for in vitro experiments for practical reasons (indicator loading is relatively simple while gene delivery and use of transgenic mice requires a more costly infrastructure). For in vivo experiments, modern GECIs such as GCaMPs (including red variants) are a better choice under most experimental conditions using either one photon or two photon excitation modalities.

Choosing the most suitable voltage indicator is less straightforward. For subcellular voltage imaging of single cells in combination with patch clamp techniques, LMW VSDs provide an established methodology. For voltage imaging in vivo, GEVIs fundamentally surpassed the possibilities of VSDs. Choosing the most suitable GEVI should be based on the most recent literature because the field is fast-moving, with robust in vivo validation of several "improved" GEVIs still pending. GEVI-encoding genes can be delivered in different ways, including via viral approaches (e.g. adeno-associated viruses, AAV). Several GEVI transgenic mouse lines are also available that utilize the rich and growing toolbox of specific promoter driver lines via the Cre-lox or transactivator systems to restrict indicator expression to specific cell types.

5.6.1 Single Cell-Level Voltage Imaging in Brain Slices

Patch clamp electrophysiology and field potential recordings have been the traditional technique for studying neuronal voltage activity in brain slices. In addition to the high technical difficulty using this approach, subcellular regions such as dendritic spines or axonal boutons remain too small to be easily accessed by standard electrodes. The ease of light access is thus the ideal, and VSDs have previously been used where the dye is injected via the patch electrode to label the neuron for optical imaging. This is technically equally demanding, and has a relatively low success rate. Transgenic mice that express GEVIs in cell populations of interest can serve as a promising alternative approach. However, in dense cell populations such as cortical pyramidal neurons, the overlapping neuropil from different neurons intermingles, thus making optical signal attribution to individual neurons difficult. This issue has been addressed by using titratable gene expression strategies that achieve sparse strong Golgi staining-like indicator. Sparse expression approaches enabled GEVI-based measurements of APs from upper cortical layers of the mouse cortex (Song et al., 2017).

5.6.2 Voltage Imaging In Vivo

Since the discovery of the first VSD more than 50 years ago, voltage imaging approaches have contributed much to our understanding of neurophysiology at the system levels, by mapping the spatiotemporal dynamics of sensory stimulus-evoked and endogenous cortical activity (Grinvald and Hildesheim, 2004). GEVIs improve upon classical VSDs in at least four aspects, by: (i) allowing for non-invasive transcranial imaging in species with thin craniums (such as mice), eliminating the previously compulsory craniotomies for dye staining; (ii) providing reliable recordings from the same neuronal population in a subject over prolonged periods of time for multiple sessions, which is not possible with classical voltage sensitive dyes; (iii) genetically targeting specific cell populations, so the signals originate only from specific neurons of interest in an otherwise diverse population; and (iv) enabling transgenic expression strategies that provide highly reproducible expression of protein indicators in different animals to eliminate between-subject variability.

While the initial proof-of-principle in vivo GEVI experiments used in utero electroporation as a mean to deliver the GEVI gene (Akemann et al., 2010; Akemann et al., 2012), the subsequent generation of transgenic animals with stable and reproducible GEVI expression made this approach applicable in a larger number of laboratories.

A general issue that voltage imaging has to deal with in in vivo applications are the hemodynamic responses that add to the GEVI signal. With GEVIs that emit in the visual spectral range this confound is unavoidable and requires careful data analysis. In addition to the use of ratiometric FRET-based GEVIs that separate voltage response (anticorrelated ratiometric FRET donor/acceptor signals) from the hemodynamic component (correlated donor/acceptor signals), the state-of-the-art approach is to perform combined fluorescence voltage imaging with intrinsic optical signal imaging that measures the optical reflectance from the brain tissue that changes upon modulations of blood volumetric and oxygenation. This can be methodologically achieved via stroboscopic illumination of the cortical window preparation or brain tissue using excitation wavelengths of the GEVI fluorophore (for voltage imaging), a hemoglobin isosbestic point (for total blood volume measurement), and a large differential deoxy- vs oxy-hemoglobin point (for blood oxygenation measurement). Imaging data acquired this way can separate the voltage and hemodynamic response components to not only extract the precise voltage response, but also provide additional insights into neurovascular coupling relationships. Such insights would improve our understanding and interpretation of signals from other common brain imaging methods that use hemodynamic response as the surrogate readout of neuronal activity, including BOLD-fMRI.

5.7 OUTLOOK AND FUTURE DIRECTIONS

LMW calcium indicators and GECI provide the basis of a mature methodology for imaging neuronal activities. However, calcium imaging for network analysis is limited in scope (blindness to activity patterns that are not reflected in easily measurable calcium signals). While green fluorescent GECIs don't leave much room for further improvement (except perhaps of some application-specific refinements), there is still a need for well performing red and near infrared GECIs. More pressing are required improvements of instrumentations for cellular resolution imaging across large brain areas (several mm) and concepts for the analysis of patterns of calcium transients.

Voltage imaging using GEVIs has steadily progressed over the last decade but much space for improvement remains. Needed are GEVIs with reduced photobleaching, increased response bandwidth and optimization for in vivo use. Most useful would be well performing GEVIs that operate in the near infrared spectrum. They would facilitate deep tissue imaging and combination with optogenetic actuators for all-optical electrophysiology. Thus, as to GEVIs, it is probably fair to say that they offer novel methods of measuring membrane activity in neurons, but that further improvement of GEVI variants, instrumentation, and data analysis strategies are necessary to meet their full potential.

ACKNOWLEDGMENTS

Supported by NIH grant MH109091. We are grateful to Dr Srdjan Antic for comments on the manuscript.

REFERENCES

Abdelfattah, A.S., Farhi, S.L., Zhao, Y., Brinks, D., Zou, P., Ruangkittisakul, A., Platisa, J., Pieribone, V.A., Ballanyi, K., Cohen, A.E. & Campbell, R.E. (2016) A bright and fast red fluorescent protein voltage indicator that reports neuronal activity in organotypic brain slices. *The Journal of Neuroscience*, **36**, 2458–2472.

Akemann, W., Lundby, A., Mutoh, H. & Knöpfel, T. (2009) Effect of voltage sensitive fluorescent proteins on neuronal excitability. *Biophysical Journal*, **96**, 3959–3976.

Akemann, W., Mutoh, H., Perron, A., Park, Y.K., Iwamoto, Y. & Knöpfel, T. (2012) Imaging neural circuit dynamics with a voltage-sensitive fluorescent protein. *Journal of Neurophysiology*, **108**, 2323–2337.

Akemann, W., Mutoh, H., Perron, A., Rossier, J. & Knöpfel, T. (2010) Imaging brain electric signals with genetically targeted voltage-sensitive fluorescent proteins. *Nature Methods*, **7**, 643–649.

Akemann, W., Sasaki, M., Mutoh, H., Imamura, T., Honkura, N. & Knöpfel, T. (2013) Two-photon voltage imaging using a genetically encoded voltage indicator. *Scientific Reports*, **3**, 2231.

Akemann, W., Song, C., Mutoh, H. & Knöpfel, T. (2015) Route to genetically targeted optical electrophysiology: development and applications of voltage-sensitive fluorescent proteins. *Neurophotonics*, **2**.

Antic, S.D., Empson, R.M. & Knöpfel, T. (2016) Voltage imaging to understand connections and functions of neuronal circuits. *Journal of Neurophysiology*, **116**, 135–152.

Baker, B.J., Kosmidis, E.K., Vucinic, D., Falk, C.X., Cohen, L.B., Djurisic, M. & Zecevic, D. (2005) Imaging brain activity with voltage- and calcium-sensitive dyes. *Cellular and Molecular Neurobiology*, **25**, 245–282.

Baker, B.J., Lee, H., Pieribone, V.A., Cohen, L.B., Isacoff, E.Y., Knöpfel, T. & Kosmidis, E.K. (2007) Three fluorescent protein voltage sensors exhibit low plasma membrane expression in mammalian cells. *Journal of Neuroscience Methods*, **161**, 32–38.

Carandini, M., Shimaoka, D., Rossi, L.F., Sato, T.K., Benucci, A. & Knöpfel, T. (2015) Imaging the awake visual cortex with a genetically encoded voltage indicator. *Journal of Neuroscience*, **35**, 53–63.

Chamberland, S., Yang, H.H., Pan, M.M., Evans, S.W., Guan, S., Chavarha, M., Yang, Y., Salesse, C., Wu, H., Wu, J.C., Clandinin, T.R., Toth, K., Lin, M.Z. & St-Pierre, F. (2017) Fast two-photon imaging of subcellular voltage dynamics in neuronal tissue with genetically encoded indicators. *eLife*, **6**.

Chanda, B., Blunck, R., Faria, L.C., Schweizer, F.E., Mody, I. & Bezanilla, F. (2005) A hybrid approach to measuring electrical activity in genetically specified neurons. *Nature Neuroscience*, **8**, 1619–1626.

Chen, T.W., Wardill, T.J., Sun, Y., Pulver, S.R., Renninger, S.L., Baohan, A., Schreiter, E.R., Kerr, R.A., Orger, M.B., Jayaraman, V., Looger, L.L., Svoboda, K. & Kim, D.S. (2013) Ultrasensitive fluorescent proteins for imaging neuronal activity. *Nature*, **499**, 295–300.

de Juan-Sanz, J., Holt, G.T., Schreiter, E.R., de Juan, F., Kim, D.S. & Ryan, T.A. (2017) Axonal endoplasmic reticulum Ca(2+) content controls release probability in CNS nerve terminals. *Neuron*, **93**, 867–881 e866.

Deisseroth, K. & Schnitzer, M.J. (2013) Engineering approaches to illuminating brain structure and dynamics. *Neuron*, **80**, 568–577.

Dimitrov, D., He, Y., Mutoh, H., Baker, B.J., Cohen, L., Akemann, W. & Knöpfel, T. (2007) Engineering and characterization of an enhanced fluorescent protein voltage sensor. *PLoS One*, **2**, e440.

Fosque, B.F., Sun, Y., Dana, H., Yang, C.T., Ohyama, T., Tadross, M.R., Patel, R., Zlatic, M., Kim, D.S., Ahrens, M.B., Jayaraman, V., Looger, L.L. & Schreiter, E.R. (2015) Neural circuits. Labeling of active neural circuits in vivo with designed calcium integrators. *Science*, **347**, 755–760.

Foust, A.J., Zampini, V., Tanese, D., Papagiakoumou, E. & Emiliani, V. (2015) Computer-generated holography enhances voltage dye fluorescence discrimination in adjacent neuronal structures. *Neurophotonics*, **2**, 021007.

Garaschuk, O., Milos, R.I. & Konnerth, A. (2006) Targeted bulk-loading of fluorescent indicators for two-photon brain imaging in vivo. *Nature Protocols*, **1**, 380–386.

Gong, Y., Huang, C., Li, J.Z., Grewe, B.F., Zhang, Y., Eismann, S. & Schnitzer, M.J. (2015) High-speed recording of neural spikes in awake mice and flies with a fluorescent voltage sensor. *Science*, **350**, 1361–1366.

Gong, Y., Wagner, M.J., Zhong Li, J. & Schnitzer, M.J. (2014) Imaging neural spiking in brain tissue using FRET-opsin protein voltage sensors. *Nature Communications*, **5**, 3674.

Grinvald, A. & Hildesheim, R. (2004) VSDI: a new era in functional imaging of cortical dynamics. *Nature Reviews Neuroscience*, **5**, 874–885.

Han, Z., Jin, L., Platisa, J., Cohen, L.B., Baker, B.J. & Pieribone, V.A. (2013) Fluorescent protein voltage probes derived from ArcLight that respond to membrane voltage changes with fast kinetics. *PloS one*, **8**, e81295.

Hinner, M.J., Hubener, G. & Fromherz, P. (2006) Genetic targeting of individual cells with a voltage-sensitive dye through enzymatic activation of membrane binding. *Chembiochem: A European Journal of Chemical Biology*, **7**, 495–505.

Homma, R., Baker, B.J., Jin, L., Garaschuk, O., Konnerth, A., Cohen, L.B., Bleau, C.X., Canepari, M., Djurisic, M. & Zecevic, D. (2009) Wide-field and two-photon imaging of brain activity with voltage- and calcium-sensitive dyes. *Methods in Molecular Biology*, **489**, 43–79.

Kerlin, A.M., Andermann, M.L., Berezovskii, V.K. & Reid, R.C. (2010) Broadly tuned response properties of diverse inhibitory neuron subtypes in mouse visual cortex. *Neuron*, **67**, 858–871.

Kim, T.H., Zhang, Y., Lecoq, J., Jung, J.C., Li, J., Zeng, H., Niell, C.M. & Schnitzer, M.J. (2016) Long-term optical access to an estimated one million neurons in the live mouse cortex. *Cell Reports*, **17**, 3385–3394.

Knöpfel, T. (2012) Genetically encoded optical indicators for the analysis of neuronal circuits. *Nature Reviews Neuroscience*, **13**, 687–700.

Knöpfel, T., Diez-Garcia, J. & Akemann, W. (2006) Optical probing of neuronal circuit dynamics: genetically encoded versus classical fluorescent sensors. *Trends in Neurosciences*, **29**, 160–166.

Knöpfel, T., Gallero-Salas, Y. & Song, C. (2015) Genetically encoded voltage indicators for large scale cortical imaging come of age. *Current Opinion in Chemical Biology*, **27**, 75–83.

Kralj, J.M., Douglass, A.D., Hochbaum, D.R., Maclaurin, D. & Cohen, A.E. (2011) Optical recording of action potentials in mammalian neurons using a microbial rhodopsin. *Nature Methods*, **9**, 90–95.

Liu, P., Grenier, V., Hong, W., Muller, V.R. & Miller, E.W. (2017) Fluorogenic targeting of voltage-sensitive dyes to neurons. *Journal of the American Chemical Society*, **139**, 17334–17340.

Lundby, A., Mutoh, H., Dimitrov, D., Akemann, W. & Knöpfel, T. (2008) Engineering of a genetically encodable fluorescent voltage sensor exploiting fast Ci-VSP voltage-sensing movements. *PloS one*, **3**, e2514.

Madisen, L., Garner, A.R., Shimaoka, D., Chuong, A.S., Klapoetke, N.C., Li, L., van der Bourg, A., Niino, Y., Egolf, L., Monetti, C., Gu, H., Mills, M., Cheng, A., Tasic, B., Nguyen, T.N., Sunkin, S.M., Benucci, A., Nagy, A., Miyawaki, A., Helmchen, F., Empson, R.M., Knöpfel, T., Boyden, E.S., Reid, R.C., Carandini, M. & Zeng, H. (2015) Transgenic mice for intersectional targeting of neural sensors and effectors with high specificity and performance. *Neuron*, **85**, 942–958.

Mao, T., O'Connor, D.H., Scheuss, V., Nakai, J. & Svoboda, K. (2008) Characterization and subcellular targeting of GCaMP-type genetically-encoded calcium indicators. *PloS one*, **3**, e1796.

Marshall, J.D., Li, J.Z., Zhang, Y., Gong, Y., St-Pierre, F., Lin, M.Z. & Schnitzer, M.J. (2016) Cell-type-specific optical recording of membrane voltage dynamics in freely moving mice. *Cell*, **167**, 1650–1662 e1615.

Miller, E.W., Lin, J.Y., Frady, E.P., Steinbach, P.A., Kristan, W.B., Jr. & Tsien, R.Y. (2012) Optically monitoring voltage in neurons by photo-induced electron transfer through molecular wires. *Proceedings of the National Academy of Sciences of the United States of America*, **109**, 2114–2119.

Mishina, Y., Mutoh, H., Song, C. & Knöpfel, T. (2014) Exploration of genetically encoded voltage indicators based on a chimeric voltage sensing domain. *Frontiers in Molecular Neuroscience*, **7**, 78.

Miyamoto, D. & Murayama, M. (2016) The fiber-optic imaging and manipulation of neural activity during animal behavior. *Neuroscience Research*, **103**, 1–9.

Miyawaki, A., Llopis, J., Heim, R., McCaffery, J.M., Adams, J.A., Ikura, M. & Tsien, R.Y. (1997) Fluorescent indicators for Ca2+ based on green fluorescent proteins and calmodulin. *Nature*, **388**, 882–887.

Nimmerjahn, A. & Helmchen, F. (2012) In vivo labeling of cortical astrocytes with sulforhodamine 101 (SR101). *Cold Spring Harbor Protocols*, **2012**, 326–334.

Palmer, A.E., Giacomello, M., Kortemme, T., Hires, S.A., Lev-Ram, V., Baker, D. & Tsien, R.Y. (2006) Ca2+ indicators based on computationally redesigned calmodulin-peptide pairs. *Chemistry and Biology*, **13**, 521–530.

Paredes, R.M., Etzler, J.C., Watts, L.T., Zheng, W. & Lechleiter, J.D. (2008) Chemical calcium indicators. *Methods*, **46**, 143–151.

Sakai, R., Repunte-Canonigo, V., Raj, C.D. & Knöpfel, T. (2001) Design and characterization of a DNA-encoded, voltage-sensitive fluorescent protein. *The European Journal of Neuroscience*, **13**, 2314–2318.

Siegel, M.S. & Isacoff, E.Y. (1997) A genetically encoded optical probe of membrane voltage. *Neuron*, **19**, 735–741.

Sjulson, L. & Miesenbock, G. (2008) Rational optimization and imaging in vivo of a genetically encoded optical voltage reporter. *Journal of Neuroscience*, **28**, 5582–5593.

Song, C., Do, Q.B., Antic, S.D. & Knöpfel, T. (2017) Transgenic strategies for sparse but strong expression of genetically encoded voltage and calcium indicators. *International Journal of Molecular Sciences*, **18**.

St-Pierre, F., Marshall, J.D., Yang, Y., Gong, Y., Schnitzer, M.J. & Lin, M.Z. (2014) High-fidelity optical reporting of neuronal electrical activity with an ultrafast fluorescent voltage sensor. *Nature Neuroscience*, **17**, 884–889.

Vranesic, I., Iijima, T., Ichikawa, M., Matsumoto, G. & Knöpfel, T. (1994) Signal transmission in the parallel fiber-Purkinje cell system visualized by high-resolution imaging. *Proceedings of the National Academy of Sciences of the United States of America*, **91**, 13014–13017.

Werley, C.A., Chien, M.P. & Cohen, A.E. (2017) Ultrawidefield microscope for high-speed fluorescence imaging and targeted optogenetic stimulation. *Biomedical Optics Express*, **8**, 5794–5813.

Zhou, W.L., Yan, P., Wuskell, J.P., Loew, L.M. & Antic, S.D. (2007) Intracellular long-wavelength voltage-sensitive dyes for studying the dynamics of action potentials in axons and thin dendrites. *Journal of Neuroscience Methods*, **164**, 225–239.

Zou, P., Zhao, Y., Douglass, A.D., Hochbaum, D.R., Brinks, D., Werley, C.A., Harrison, D.J., Campbell, R.E. & Cohen, A.E. (2014) Bright and fast multicoloured voltage reporters via electrochromic FRET. *Nature Communications*, **5**, 4625.

CHAPTER 6

Functional Optoacoustic Imaging

Daniel Razansky

CONTENTS

6.1 INTRODUCTION

Optical imaging is becoming increasingly attractive for *in vivo* brain imaging due to its inherent sensitivity to blood oxygenation and hemodynamics. In the previous chapters, various optical neuroimaging methods have been described. In the context of microscopic techniques, the main limitation remains the intense photon scattering, which confines the effective imaging depth of even the most advanced wide-field microscopy methods to superficial brain vasculature (Kalchenko et al., 2014), also necessitating removal of the scalp and/or skull. Other advanced techniques, such as two-photon microscopy, can access deeper subcortical structures, such as the hippocampus, by displacing large chunks of overlying structures (Mizrahi et al., 2004). Alternatively, other invasive approaches, e.g. optogenetic fiber probes, are used to visualize and manipulate neurons, in which case the readings of real-time activity are still bounded to small volumes. More generally, volumetric real time observations of fast neural activity with state-of-the-art optical microscopy are limited to depths below ~1 mm and FOV below 1 mm^3 in *scattering* intact brains.

On the other hand, macroscopic optical techniques relying on diffuse light, such as near-infrared spectroscopy (NIRS) and diffuse optical tomography (DOT), make use of the reduced hemoglobin absorption in the 650–950 nm spectral window to visualize optical contrast through several centimeters in highly vascularized mammalian tissues. Promising new developments include the introduction of near-infrared-shifted fluorescent molecules that can be used for *in vivo* labeling of deep tissue functional and molecular processes (Guo et al., 2014). However, in-depth optical observations suffer from severe degradation of the spatial resolution as light transforms from quasi-ballistic into a fully diffusive regime, leading to resolving power in the order of 1 mm for deep brain imaging in rodents (Devor et al., 2012).

The imaging depth limitation of optical microscopy related to light scattering in living tissues can be generally overcome by means of optoacoustic (OA) techniques (Razansky et al., 2012). OA imaging is based upon generation of acoustic waves induced by transient absorption of light energy in tissues, enabling high-resolution imaging with optical contrast at depths of several millimeters to centimeters (Wang and Hu, 2012), orders of magnitude deeper than possible with state-of-the-art microscopy. OA sensing and imaging draws its roots from the discovery of the *photophone* by A. G. Bell and his assistant C. S. Tainter in 1880, which the inventors had predominantly destined for wireless telephony applications (Tainter and Bell, 1880). However, the lack of appropriate laser sources and ultrasound detection technologies has long hindered practical biomedical applications of this fascinating phenomenon. In fact, optoacoustic spectroscopy of biological tissues was first considered in the 1970s (Rosencwaig, 1973) with images from living small animals reported at the turn of the 21st century (Wang et al., 2003). Nowadays, by combining sensitive ultrasound detectors with advanced pulsed laser sources having wide and fast wavelength tunability, multi-spectral optoacoustic tomography (MSOT) was shown to be capable of generating OA responses from tissue that carry spatially resolved spectroscopic information on the underlying absorption contrast (Ntziachristos and Razansky, 2010). Due to its hybrid nature combining optical excitation and acoustic detection, OA reports on the versatile optical absorption contrast, but relative to other optical methods, provides a sort of "super-vision" or, more accurately, "super-hearing" by exploiting the low scattering of ultrasound to break through the barriers imposed by optical diffusion.

In vascularized tissues, highly absorbing hemoglobin manifests excellent OA contrast. It is therefore natural to attain high fidelity OA images of vascular anatomy, dynamic microcirculation, neovascularization, as well as blood oxygenation levels deep within highly diffuse tissues without introduction of contrast agents (Wang et al., 2003; Zhang et al., 2009). Furthermore, many other bio-chromophores exist that have specific spectral signatures allowing them to be distinguished from each other within an integrated absorption signal, including melanin, bilirubin, lipids, and water. Their relative signals contribute additional diverse information about function and/or pathological status of the tissue being examined (Cerussi et al., 2001). Due to versatility and wide availability of optical molecular agents, functional and molecular imaging studies can be done using extrinsic contrast from, e.g. reporter genes or targeted biomarkers that can be genetically expressed or accumulated in a cell- or tissue-specific manner (Razansky et al., 2009). Finally, optoacoustics provides a unique multiscale imaging capacity, allowing bridging the gap between the microscopic and macroscopic realms with the same type of contrast (Deán-Ben et al., 2017).

Despite its great promise, similarly to ultrasonography, optoacoustic imaging possesses several limitations related to imaging through acoustically mismatched areas, such as lungs, skull, and bones. Furthermore, even though OA spatial resolution performance is not directly affected by light scattering, effective penetration is still limited due to light attenuation and signal degradation in deep tissues. Therefore, whole-body imaging with optoacoustics is currently only possible in small animals, whereas clinical applications are limited to relatively superficial, low-attenuating, or otherwise conveniently accessible areas of the body. In the following, the different aspects of OA-mediated functional neuroimaging are covered, including technical essentials, reconstruction strategies and instrumentation, *in vivo* labeling, and dynamic contrast enhancement approaches, as well as the rapidly evolving use of OA in studies on neural activity and brain disorders.

6.2 PRINCIPLES OF OPTOACOUSTIC IMAGING

OA imaging is based upon absorption of light radiation in tissue and conversion of the deposited energy into heat, which in turn results in thermal expansion and mechanical stress propagating in the form of pressure waves (Figure 6.1).

Similar to most optical imaging modalities, light in the visible or near-infrared spectrum (400 nm–1200 nm) is most commonly used for excitation of OA responses due to the relatively weak absorption by water molecules in this spectral region. Of particular importance is the near-infrared spectral window, which allows light penetration of up to several centimeters into optically dense tissues due to reduced absorption by hemoglobin (Ntziachristos et al., 2005). A large variety of bio-chromes are absorbing at the optical wavelengths, which further leads to a high contrast between different tissues with varying chromophore concentration. Figure 6.2 shows absorption spectra of the most dominant intrinsic absorbers in tissue in the visible and near-infrared range (Yao and Wang, 2014).

For efficient generation of OA images, two conditions should be fulfilled. First, the characteristic thermal diffusion length of the medium must remain shorter than the desired spatial resolution of the imaging system during the laser pulse, namely, the so-called thermal confinement regime must be maintained. This corresponds to no leakage of energy out of the effective optical absorption zone. Considering a typical thermal diffusivity of

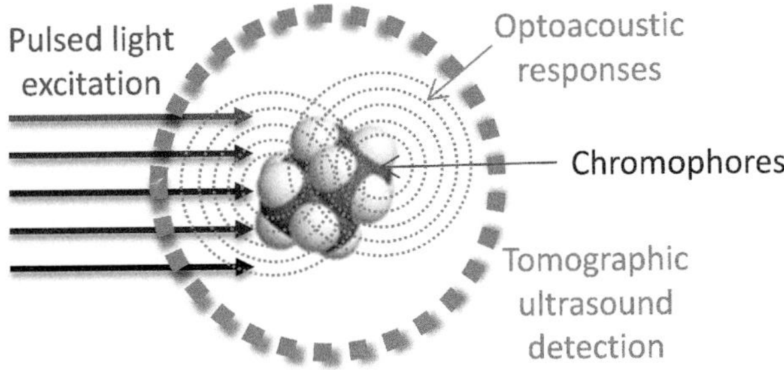

FIGURE 6.1 Basic principle of optoacoustic (OA) imaging. Short pulses of light are absorbed by tissue chromophores, generating pressure (ultrasound) waves propagating through the medium. The latter are recorded by ultrasonic detectors in order to reconstruct the initial distribution of the absorbing molecules.

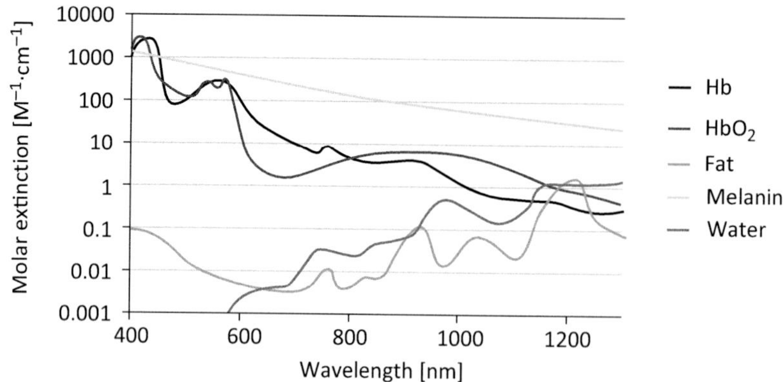

FIGURE 6.2 Absorption spectral profiles of major intrinsic tissue chromophores responsible for generation of optoacoustic responses.

D~0.1 mm²/s in soft biological tissues, this condition can be readily met for typical biological imaging applications that use nanosecond-duration laser pulses (Wang, 2009). In addition, the excitation laser pulse duration should remain shorter than the time required for the generated stress wave to propagate out of the heated region defined by the effective spatial resolution, a condition known as the stress confinement regime. For instance, given a common speed of sound of ~1,500 m/s in soft tissues, the stress confinement condition would be readily fulfilled for laser pulse duration of ~10 ns and typical diffraction-limited spatial resolution of ~100 μm of the imaging system (corresponding to ultrasonic detection bandwidth in the 10 MHz range).

Using pulse durations which satisfy both thermal and stress confinement regimes, the initial local pressure rise p_0 induced by laser energy deposition and subsequent instantaneous heating can be simply expressed via

$$p_0 = \Gamma \cdot H, \tag{6.1}$$

being H the distribution of the absorbed energy density and Γ the dimensionless Grüneisen parameter, which lumps together the thermoelastic properties of the medium, i.e. (Wang, 2009)

$$\Gamma = \beta c^2 / C_p, \tag{6.2}$$

where β is the thermal expansion coefficient, c is the speed of sound, and C_p is the specific (per unit mass) heat capacity at constant pressure.

The initial goal of optoacoustic imaging and tomography is retrieval of the absorbed optical energy density inside the object. However, the generated pressure fields can normally only be measured outside the object, whereas propagation of pressure waves toward the detection point is described by the optoacoustic wave equation (Kruger et al., 1995). The way to obtain optoacoustic images therefore consists in reconstructing distribution of the energy density $H(r)$, i.e. solving or inverting the optoacoustic wave equation given the pressure variations measured around the imaged object.

For biomedical applications, it is the spatial distribution of the optical absorption coefficient $\mu_a(r)$ that is usually of primary interest since it is more directly related to the biochemical composition of the imaged tissue than the absorbed energy density. Both are related via

$$H = \mu_a \cdot \Phi \tag{6.3}$$

where Φ is the light fluence (or photon density) field. Although one may presume that H is a simple product, it in fact depends non-linearly on the absorption coefficient μ_a. This is because the light fluence generally depends on the underlying optical properties, including optical scattering and absorption coefficients. Thus, it is important for image quantification purposes to account for the light fluence, since it may considerably vary as a function of depth while the light attenuation may additionally exhibit a strong dependence on wavelength, an effect know as spectral coloring (Tzoumas et al., 2016).

While reconstructing distribution of the optical absorption coefficient may provide valuable structural (anatomical) information on the underlying tissue contrast, it is the concentrations of different chromophores c_i, not the optical absorption coefficient μ_a, that is mainly of interest from the biological point of view. For instance, in many neuroimaging applications, the hemoglobin-related contrast is of particular interest, providing valuable physiological or functional information on blood oxygenation and cerebral hemodynamics. The relation between concentrations of the different chromophores and the optical absorption coefficient may be expressed via linear superposition, namely,

$$\mu_a(\lambda) = \sum_i \varepsilon_i(\lambda) \cdot c_i + \mu_{BG}(\lambda), \tag{6.4}$$

where $\varepsilon_i(\lambda)$ are the wavelength-dependent molar extinction coefficients of the different intrinsic tissue chromophores or extrinsically administered agents and μ_{BG} is the residual (background) absorption, which may also include noise. To differentiate between contributions of different chromophores, the MSOT imaging method spectroscopically distinguishes between their absorption spectra using illumination at several wavelengths (Razansky, 2012). The process of recovering c_i from multi-spectral measurements is also known as spectral unmixing, which can be combined with calculation of the light fluence or treated as a separate image processing step.

Due to versatility and wide availability of optical molecular agents, sensitive and accurate spectral processing to recover concentration of extrinsic labels may enable highly specific studies of function and molecular targeting. This can be done by resolving spectrally distinct markers that report on neural activity or disease status. Figure 6.3 summarizes the main steps involved in the OA image reconstruction chain.

6.3 INSTRUMENTATION EMBODIMENTS

A typical optoacoustic setup consists of several key components. In the pulsed excitation mode, the tissue is illuminated by laser emitting monochromatic pulses of light with typical duration of some nanoseconds. For deep tissue imaging applications, optical parametric

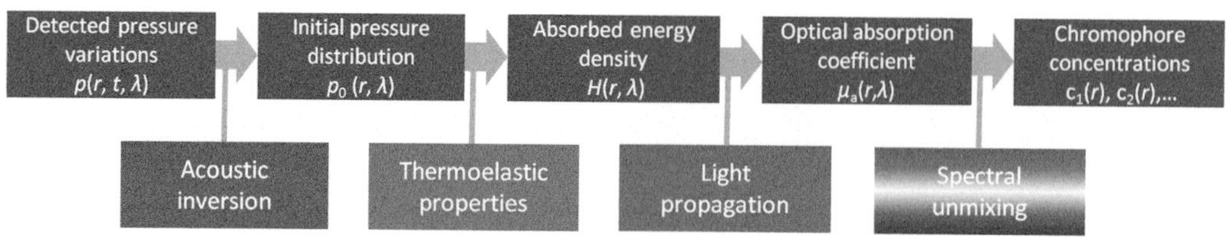

FIGURE 6.3 The various steps involved in the multi-spectral optoacoustic tomography inversion, from the detected pressure variations to the maps of chromophore concentrations.

oscillators are often used to provide wavelength tunability within the spectrum of interest with pulse repetition rate in the order of a few tens of Hertz and per-pulse energies in the millijoule range. In optoacoustic microscopy and other superficial imaging applications, where high per-pulse energies are not required, other types of sources in the microjoule and nanojoule range are considered as well, including high repetition dye lasers (Song et al., 2009), laser diodes (Allen and Beard, 2006), and fiber lasers (Wang et al., 2011).

For tomographic imaging, the optoacoustically generated pressure profiles are captured with detectors surrounding the object. As opposed to pulse-echo ultrasonography (US), the generated OA signals have relatively low amplitude while their frequency content is mainly dependent on the characteristic size of absorbers in the imaged volume. Typically, the signals are of ultrawideband nature containing frequencies from tens of KHz (for large absorbing structures) up to tens of MHz (for small absorbers). Due to the dominating low frequency content of OA waves generated by common biological targets, utilization of physical or synthetic aperture focusing may turn inefficient (Merčep et al., 2015). As a result, OA image formation with linear phased arrays and other types of limited-view tomographic geometries suffers from severe out-of-focus artifacts, impaired contrast, image blurring, and overall lack of quantification abilities (Lutzweiler and Razansky, 2013). Thus, in contrast to pulse-echo US, correct image reconstruction in OA imaging is ideally achieved by an unfocused detection of OA responses from as many tomographic viewing angles as possible around the imaged object.

Acoustic coupling between object and detector is usually ensured by water or coupling gel. For acquisition of spatially resolved data, either a single detector is scanned around/along the object or multiple detectors acquire the data in parallel. The latter further allows for fast data acquisition e.g. rendering images from single laser shots. The signals are then pre-amplified and digitized by a fast data acquisition system. A broad variety of OA set-ups for small animal imaging have been reported, such as optical-resolution microscopy or OR-PAM (Hu et al., 2011), endoscopy (Yang et al., 2012), cross-sectional whole-body tomography (Razansky et al., 2011), as well as real-time volumetric (4D) and 5D imaging systems (Deán-Ben et al., 2014), to name a few examples. Some common implementations are also presented in Figure 6.4.

6.4 IMAGING CEREBRAL HEMODYNAMICS AND LABEL-FREE CONTRAST

As mentioned in the introductory section, the massive prospects and opportunities for imaging the central nervous system are only very partially addressed by the optical

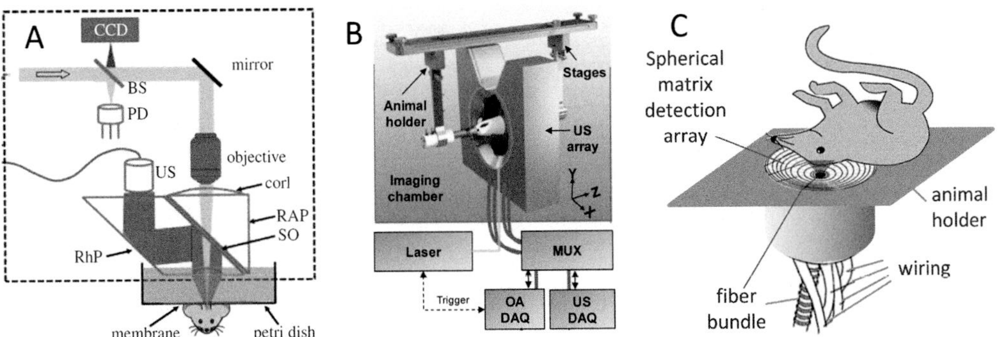

FIGURE 6.4 Examples of typical optoacoustic imaging systems used in functional neuroimaging studies. (a) Optical-resolution photoacoustic microscopy (OR-PAM) scanner used for *in vivo* imaging of mice. Spatial resolution of the system in the lateral direction is determined by the size of the optical focus. Three-dimensional data is acquired by raster scanning of the optical beam along the imaged area. BS, beam splitter; PD, photodiode; CorL, correction lens; RAP, right-angle prism; SO, silicone oil; RhP, rhomboid prism; US, ultrasonic transducer (central frequency: 50 MHz). The CCD is used to view the imaged region. (*Reprinted with permission from* (Hu et al., 2011). © 2011 *Optical Society of America.*) (b) Schematic drawing of a real-time cross-sectional mouse scanner including the animal holder, illumination device, and transducer array. Optical fibers are used to homogeneously illuminate the object for efficient OA signal excitation. The concave array consists of 512 cylindrically focused detection elements covering a 270° angle around the imaged mouse, enabling parallel OA and pulse-echo US data acquisition from single cross-sections in real time. By translating the animal holder, three-dimensional and whole-body data can be acquired. (Reprinted with permission from Merčep et al. (2015). © 2015 – IEEE.) (c) Real-time volumetric imaging system comprising of a spherical matrix ultrasound detection array with the laser illumination provided via a fiber bundle pulled through the center of the array. Rapid acquisition of multi-wavelength data from an entire imaged volume, combined with real-time image reconstruction and spectral unmixing, provides four- and five-dimensional imaging capabilities by rendering volumetric unmixed images of individual chromophores in real time. (Reprinted with permission from Deán-Ben et al. (2014). © 2016 – Macmillan Publishers Ltd.)

neuroimaging methods due to an insufficient spatio-temporal resolution, limited field of view, and/or reliance on indirect observations of neural activity using effects mediated through complex neuro-vascular coupling processes. The vast advances in the development and application of OA imaging techniques now offer a viable alternative for dynamic high-resolution observations of the brain function. The endogenous OA contrast based on spectrally resolved hemoglobin measurements readily provides a remarkable tool for label-free visualization of cerebral hemodynamics under different stimuli (Figure 6.5). OR-PAM enables simultaneous assessment of several key cerebral hemodynamic parameters like oxygenated (HbO) and deoxygenated (HbR) hemoglobin, blood oxygen saturation (SO_2), total hemoglobin (HbT), cerebral blood flow (CBF) and, derived from these parameters, the oxygen extraction function (OEF), and the cerebral metabolic rate of oxygen ($CMRO_2$) (Hu, 2016). The achievable spatial resolutions reach the capillary level and state-of-the-art systems additionally offer temporal resolutions of up to 100 kHz for 1D scans. In one

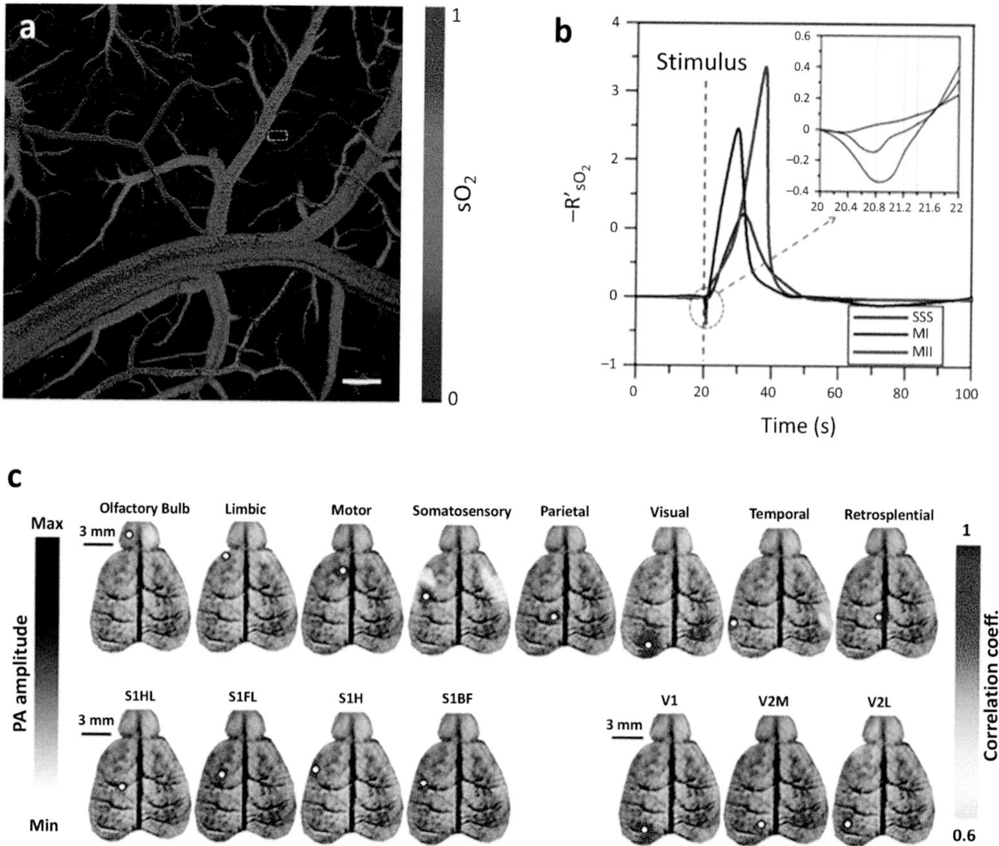

FIGURE 6.5 Optoacoustic imaging of hemodynamic changes in the rodent brain. (a) Representative sO_2 image of the visual cortex acquired by OR-PAM *in vivo* (scalp and skull over the imaged area were removed). Scale bar: 200 μm. (Adapted with permission from Wang et al. (2013). © 2013 National Academy of Sciences.) (b) Oxygen saturation changes of the superior sagittal sinus (SSS) and the contralateral MI and MII arterioles as a function of time obtained from A-line (1D) optoacoustic signals. (Adapted with permission from Liao et al. (2012). © 2012 Sage Publications.) (c) Functional connectivity maps in a live mouse brain acquired with cross-sectional optoacoustic tomography indicating eight main functional regions in the cortex. (Figure adapted with permission from Nasiriavanaki et al. (2014), © 2013 National Academy of Sciences.)

of its earliest embodiments, functional optoacoustic microscopy was able to provide an independent verification of the BOLD-fMRI "initial dip" (Figure 6.5b), demonstrating a decrease in blood oxygenation due to an increase in arteriole HbR in the first few hundreds of milliseconds after an electrical paw stimulus (Liao et al., 2012).

In optical-resolution microscopy, the penetration depth is, however, limited to superficial cortical layers (Yao et al., 2015). In addition, while achieving adequate imaging performance with high-resolution OR-PAM normally necessitates removal of the scalp (and sometimes the skull), tomographic systems, such as those based on ring-shaped transducer arrays, are generally fully non-invasive and have been utilized to image stimulus-evoked (Yao et al., 2015) and resting state (Nasiriavanaki et al., 2013) activity in deep brain

(Figure 6.5c). A 512-element full-ring ultrasonic transducer array with an in-plane resolution of 100 μm was used to image glucose metabolism in the mouse cortex after intravenous injection of a glucose-analogue molecule providing optoacoustic contrast (Yao et al., 2013). A lower resolution 64-element ring transducer array was used to image brain dynamics in awake, freely moving rats (Tang et al., 2015). Alternative real-time cross-sectional optoacoustic imaging systems were configured to acquire coronal sections of the brain of a mouse, which allows visualizing deep-seated structural contrast and oxygenation parameters (Burton et al., 2013).

The recent development of spherically shaped matrix transducer arrays with densely packed elements has taken the tomographic concept a step further, allowing real-time recording of true three-dimensional OA data in real time. One such system with per-pulse wavelength tunability and fast 3D imaging capacity has been shown to be capable of providing spectroscopic information for real-time analysis of hemodynamic changes deep inside the mouse brain (Gottschalk et al., 2015). In this way, changes in blood oxygen saturation, total hemoglobin, blood volume, and oxygenized and deoxygenized forms of hemoglobin in whole mouse brains were monitored with sub-second temporal resolution.

6.5 EXTRINSIC MARKERS FOR FUNCTIONAL OPTOACOUSTIC NEUROIMAGING

Notwithstanding this important progress in advanced hemodynamic imaging, modern neuroimaging work has largely steered towards the development and application of indicators that allow a much more direct monitoring of neuronal activity, most notably calcium- and voltage-sensitive indicators. Genetically encoded calcium indicators (GECIs), and primarily the GCaMP family of indicators have become the workhorse tool for visualizing distributed activity across neuronal populations. In a very recent study, functional optoacoustic neuro-tomography (FONT) has successfully demonstrated imaging of calcium-sensitive protein GCaMP5G (490 nm peak absorption) expressed in a transgenic zebrafish model (Deán-Ben et al., 2016). The study established a very strong correlation between the fluorescence and OA activity-dependent signals of the indicator. It was demonstrated in a generalized seizure-model that calcium neuronal dynamics recorded with optoacoustics could be analyzed deep inside the scattering adult zebrafish brain. Conversely, these dynamics were not observable with concurrent fluorescence imaging, which rendered a blurry appearance lacking the depth information (Figure 6.6).

Unlike small-sized zebrafish brains, effective penetration of visible wavelengths into the mouse brain is limited to the superficial cortical areas due to strong absorption of light by blood. To this end, far-red- and near-infrared-shifted fluorescent proteins have been engineered to provide peak absorption above 680 nm (Shu et al., 2009). In particular, the infrared fluorescent proteins (iRFP) are derived from bacterial phytochromes (Shcherbakova et al., 2015). In contrast to GFP, which only requires oxygen for chromophore maturation, phytochrome-derived proteins incorporate biliverdin as chromophoric substrate. OA detection of iRFP has been demonstrated in living mouse brain using multi-spectral optoacoustic tomography and further validated against fluorescent measurements (Deliolanis et al., 2014). Due to their limited optoacoustic generation efficiency, detection of iRFPs

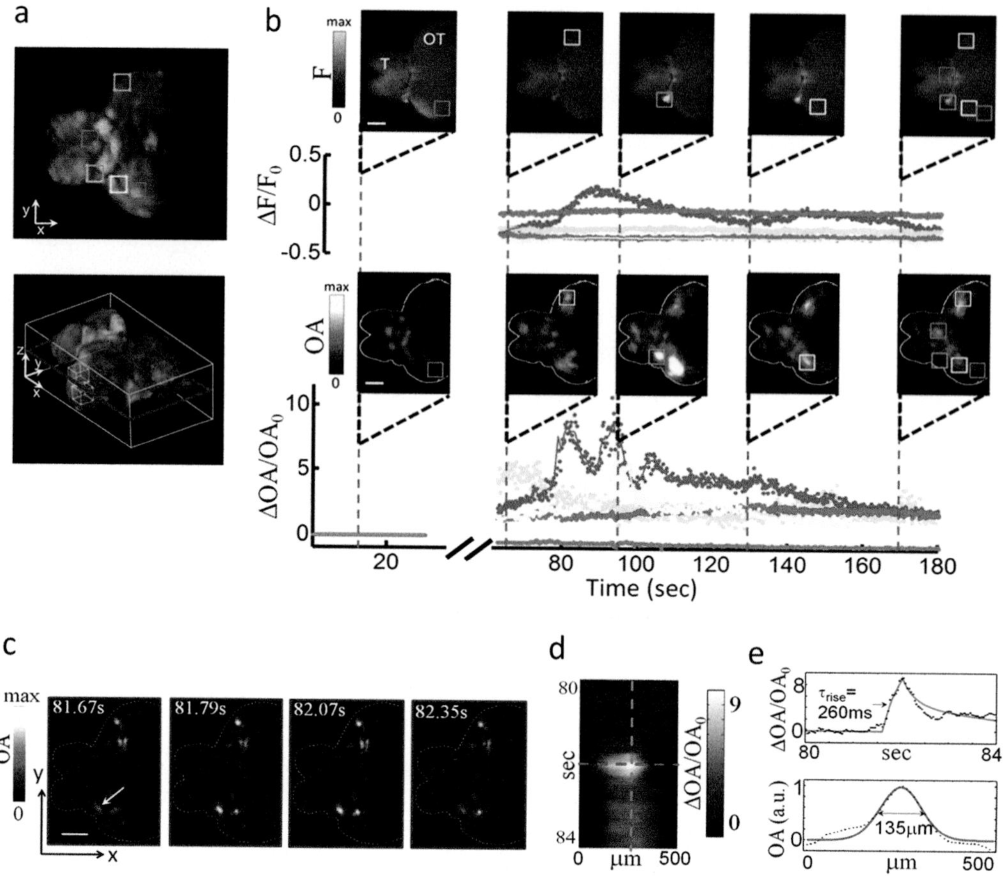

FIGURE 6.6 Functional optoacoustic neuro-tomography (FONT) visualizes neuronal activity using the genetically encoded calcium indicator GCaMP5G. (a) Volumetric optoacoustic images of an adult zebrafish brain. (b) Real-time imaging of calcium activity with epi-fluorescence (top) and optoacoustics (bottom) after injection of the neurostimulant agent pentylenetetrazole (PTZ) into the brain. Temporal traces in the five marked regions of interest are shown. The fluorescence images have a very blurry appearance indicating that the intense light scattering in large brains makes them inaccessible by optical microscopy methods. In contrast, FONT is able to provide high-resolution three-dimensional information regarding real-time neuronal activity in the entire scattering brain. (c) Time-resolved images from a single slice through the 3D data, as indicated in violet in (a). (d) Close-up spatio-temporal resolution analysis of a single line, whose orientation is indicated by an arrow in (c). (e) Temporal and spatial profiles through the image in (d). Scale bars: 500 µm. (Adapted with permission from Deán-Ben et al. (2016). © 2016 – Macmillan Publishers Ltd.)

has only been reported at shallow depths of several millimeters. Specifically designed screening platforms are being used to optimize the performance of existing FPs and chromoproteins (Forbrich et al., 2016), facilitating the engineering of novel genetic reporters for functional optoacoustic imaging and sensing. Since the OA brain imagers employ the widely tunable nanosecond OPO laser technology (Gottshalk et al., 2015), they can be conveniently tuned to work with a large array of functional probes. It is therefore anticipated that the development of far-red and near-infrared calcium- and voltage-sensitive indicators will expedite the use of OA technology into deep tissue functional neuroimaging of mammalian brains.

6.6 OPTOACOUSTIC IMAGING OF BRAIN DISORDERS

Small animals are increasingly used in modeling a large variety of brain disorders, including malignancies and traumatic injuries, seizures, ischemia, stroke, and chronic neurodegenerative conditions, necessitates the development of innovative technologies for their interrogation. To this end, the highly versatile endogenous and exogenous contrast provided by optoacoustics has proven highly instructive for the use in pre-clinical neuroimaging. Changes in hemoglobin gradients and tissue oxygenation are used as the main readout for a range of vascular pathologies and brain injuries. For instance, occlusion of the middle cerebral artery in mice has been used for monitoring focal ischemia and stroke-related hemodynamic changes in the mouse brain with OA imaging (Hu et al., 2011). A similar study of blood flow prior and after blockage of middle cerebral artery showed a buildup of deoxygenated hemoglobin within the affected hemisphere, with the hypoxic area localized next to the stroke region (Kneipp et al., 2014). Monitoring water content of the brain tissue with OA imaging is yet another useful way of observing microcirculatory disorders and has been utilized for investigating cerebral edema (Xu et al., 2011), traumatic brain injury (Yang et al., 2007), and neuro-inflammatory processes (Guevara et al., 2013).

Localized changes in the hemoglobin gradients have been also used as a hallmark in OA studies of brain tumors, with reduced oxygen saturation of the core and enhanced angiogenesis correlating with melanomas grafted in mouse brain (Staley et al., 2010). Expression of integrin $\alpha_V\beta_3$ in human U87 glioblastomas implanted in nude mouse brain was tracked optoacoustically with the hemoglobin saturation and vascularization of neoplastic tissue mapped *in vivo* (Li et al., 2008) (Figure 6.7a–d).

In the field of neurodegenerative diseases, OA microscopy was used to resolve deposits of amyloid plaques in living transgenic mouse model (Hu et al., 2009). In this way, monitoring the amyloid pathology in Alzheimer's disease was performed without invasive skull trepanation or implantation of cranial windows commonly employed in optical neuroimaging studies. Label-free sensing of other endogenous chromophores of the brain, such as neuro-melanin, transferrin, and lipofuscin, hold promise for extending the applications of OA imaging to other degenerative brain diseases affecting key subcortical structures, such as Parkinson's disease, schizophrenia, and depression.

OA imaging also proved highly instructive for studies of pharmacologically induced seizures. In neocortex of adult rats, bicuculline-induced seizures were localized and mapped, with occurrence of paroxysmal activity confirmed using EEG recordings (Zhang et al., 2008). Gottschalk et al. optoacoustically visualized the onset, spread, and termination of 4-Aminopyridine-induced epileptic seizures in a model of focal epilepsy and combined the imaging readings with concurrent electroencephalographic (EEG) recording of the aberrant neural activity (Gottschalk et al., 2017). For the first time, synchronized epileptic foci were visualized in real time throughout the entire mouse brain (Figure 6.7e,f). Longitudinal OA tracking of hemodynamic changes in a generalized epilepsy model was

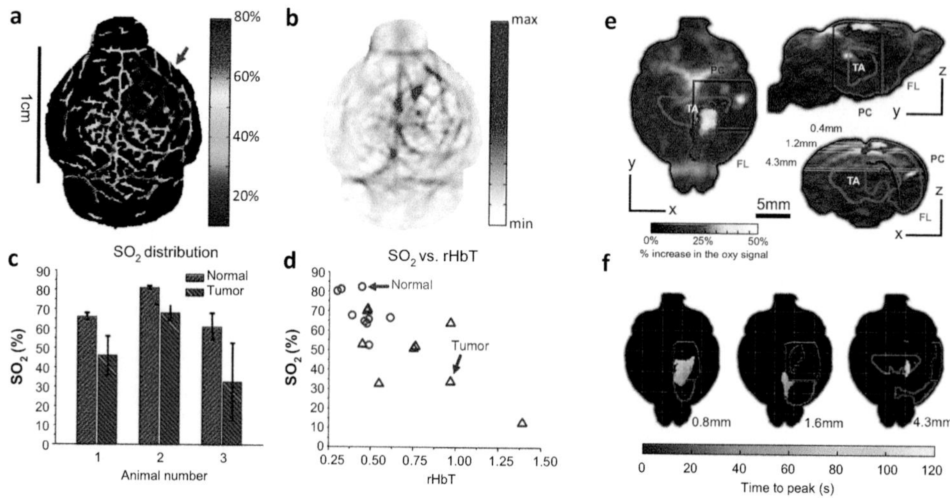

FIGURE 6.7 Optoacoustic imaging of brain disorders. (a) *In vivo* OA image of oxygen saturation (SO_2) in a nude mouse brain implanted with a U87 glioblastoma xenograft. The red arrow indicates the hypoxic region. (b) Total hemoglobin (HbT) image of the same mouse. (c) Comparison of normal and tumor vasculatures in the SO_2 images from triplicate mice. Three normal and three tumor vessels were chosen from each SO_2 image for the calculation. Error bar: one standard deviation. (d) The SO_2 versus the relative HbT in normal and tumor vasculatures from the same triplicate mice as in (b). (Adapted with permission from Li et al. (2008). © 2008 – IEEE.) (e) Intensity plots of the seizure-related hemodynamic activity as identified by correlating the optoacoustic hemodynamic responses to the EEG traces. Maximum intensity projection views along the horizontal, sagittal, and coronal directions show the neuronal activity pattern projected onto the corresponding structural volumetric OA reconstruction. Spatial dimensions of the thalamus (TA, grey), the frontal lobe (FL, green) as well as the parietal-temporal cortex (PC, purple) are outlined. (f) Time to peak analysis of activated areas in the different transverse slices across the brain – the corresponding slice depth is indicated in (e). (Adapted with permission from Gottschalk et al. (2017). © 2016 – SPIE.)

also performed in awake and freely moving rats by directly mounting the detection array onto their head with scalp and skull intact (Tang et al., 2015).

6.7 CHALLENGES AND LIMITATIONS OF FUNCTIONAL OPTOACOUSTIC NEUROIMAGING

From the technical point of view, much like other imaging modalities, optoacoustics comes with its own set of limitations that call for development of advanced solutions. Naturally, imaging depth is restricted due to light attenuation in optically opaque tissues, which also affects the minimal detectable concentrations of intrinsic tissue chromophores and extrinsically administered contrast agents. Some of it can be compensated by increasing the deposited laser energy. However, for *in vivo* applications, especially when keeping the clinical translation path in mind, illumination on the skin surface is limited by the laser exposure safety standards (ANSI, 2000). As a result, the effective imaging depth is usually restricted to regions where the light fluence is sufficiently high to generate detectable pressure variations, typically up to a few centimeters in most soft tissues at the near-infrared wavelengths. To this end, *in vivo* sensitivities in the sub-micromolar range for organic dyes and proteins were reported in whole-body rodent studies (Yao and Wang, 2014). In reality, sensitivity limits are not only affected by molecular weight of the probe employed but also by multiple additional factors, such as the total volume, the spectrum and absorption coefficient of the imaged chromophore, and the noise equivalent pressure (NEP) of the detectors, as well as the level and spectral dependence of background tissue absorption.

When selecting appropriate contrast agents for functional optoacoustic neuroimaging, effects related to their photostability under pulsed nanosecond radiation, such as irreversible bleaching, as well as *in vivo* toxicity have to be carefully considered. The light pulses employed in OA imaging induce photobleaching effects that considerably differ from those known from single- or multi-photon fluorescence imaging (Gottschalk et al., 2015). Many fluorescent proteins are known for their lack of photostability, which can be seen in effects like dark states, blinking, transient absorption, and photobleaching (Nienhaus, 2015). Furthermore, they are known to induce phototoxicity under certain excitation conditions (Liao et al., 2014). While most of these effects are yet unexplored for realistic *in vivo* imaging conditions, OA signal degradation associated to photobleaching effects in fluorescent proteins has already been observed under different imaging settings (Laufer et al., 2013). Photobleaching may even take place for laser energy levels considered safe for human use (Gottschalk et al., 2015). Of particular interest, therefore, are non-fluorescent chromoproteins or enzymatically amplified chromophores, which, in comparison to fluorescent markers, exhibit higher OA conversion efficiency due to the absence of radiative relaxation and ground state depopulation, and also higher photostability (Jiang et al., 2015).

The highly heterogeneous nature of biological tissues may further lead to the appearance of significant image artifacts and compromise imaging performance and quantification. In MSOT imaging, one common challenge is the so-called "spectral coloring". Due to the non-local and non-linear dependence of the light fluence distribution on the optical properties of the object, the optoacoustically extracted spectra of deep-seated tissue chromophores and agents gets corrupted (Tzoumas et al., 2016). Therefore, for improving quantitative determination of chromophore concentrations, one should accurately account for the light distribution in tissue. In addition, hemoglobin in its oxygenated and deoxygenated form is the dominant intrinsic absorber in mammalian tissues in both the visible and most of the near-infrared range of the optical spectrum, which enables the label-free determination of cerebral hemodynamics and blood oxygenation. Yet, when imaging other extrinsic labels of neural activity, the strong blood background may hamper quantified extraction of biomarker concentrations, necessitating sophisticated approaches for efficient separation/unmixing between contributions of the different chromophores.

Finally, advances in research concerning the treatment and progression of neurological diseases call for the development of non-invasive or minimally invasive methods that can acquire longitudinal data with high reproducibility. To this end, various OA imaging implementations have shown capacity for transcranial imaging of the mouse brain (Yao et al., 2013; Kneipp et al., 2014; Gottschalk et al., 2015). Yet, it is generally known that acoustic heterogeneities may introduce artifacts into OA reconstructions, reduce spatial resolution and quantification abilities (Lutzweiler and Razansky, 2013). This is of critical importance when considering non-invasive deep brain imaging through highly mismatched cortical bone with high spatial resolution. Abundance of existing literature on ultrasound transmission through the human skull for diagnostic and therapeutic purposes indicates that it represents a significant barrier for propagation of high-frequency longitudinal wave components (Sun and Hynynen, 1999), which has been confirmed by the recent studies with rodent skulls (Estrada et al., 2016).

6.8 SUMMARY AND CONCLUSIONS

Neuroscience has a fundamental need for more efficient and less intrusive ways to observe brain circuits. In particular, imaging neural activity with high spatio-temporal resolution over the entire intact brain, including deep and normally inaccessible areas, could play a critical role in the attempt to decipher the fundamental operating principles underlying circuit activity and devastating brain disorders. OA imaging is ideally poised to address those needs – it has shown excellent capacity for imaging intrinsic contrast in entire brains of vertebrates and rodents non-invasively; it delivers unmatched temporal resolution in the milliseconds range for true volumetric imaging in real time; it is capable of label-free observations of hemodynamic changes; and it is sensitive to genetic markers of neural activity.

The true multi-scale imaging capabilities of optoacoustics can be better appreciated when comparing its dynamic imaging performance with other neuroimaging modalities (Figure 6.8). While optical microscopy can provide micron-scale spatial resolution (Ntziachristos, 2010) along with proven sensitivity to fast calcium- or voltage-related

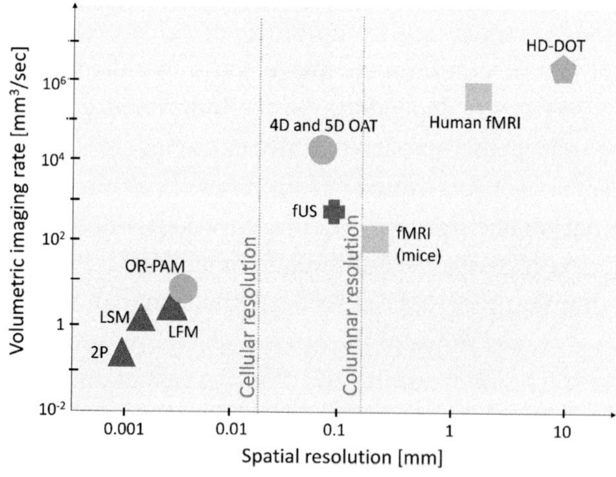

FIGURE 6.8 Comparison of dynamic imaging capabilities of the various state-of-the-art functional neuroimaging modalities used in small animal research and the clinics. Shown are: optical methods based on two photon microscopy (2P) (Kong et al., 2015), light-sheet microscopy (LSM) (Bouchard et al., 2015), and light field microscopy (LFM) (Prevedel et al., 2014); small animal (Bosshard et al., 2015) and human (Moeller et al., 2010) functional magnetic resonance imaging (fMRI); high-density diffuse optical tomography (HD-DOT) (Eggebrecht et al., 2014); functional ultrasound (fUS) (Mace et al., 2011); optical-resolution photoacoustic microscopy (OR-PAM) (Yao et al., 2015); 4D and 5D optoacoustic tomography (4D-5D OAT) (Deán-Ben et al., 2016).

signals, the imaging rate rendered with state-of-the-art volumetric microscopy methods lies in the 1 mm^3/s range, which is insufficient for large-scale recording of brain activity at the whole mouse brain level. Moreover, photon scattering remains the fundamental physical limitation for those methods, and thus the imaged volume cannot be extended beyond several hundreds of microns in the depth direction, further impeding acquisition of dynamic information from large tissue volumes. Deep tissue optical imaging can be alternatively done by means of diffuse optical tomography techniques, which, however, suffer from severely impaired spatial resolution that degrades to about 5–10 mm at centimeter-scale depths (Eggebrecht et al., 2014), depending on the type of imaged tissue. On the opposite edge of the performance scale are functional magnetic resonance imaging (fMRI) (Bosshard et al., 2015) or functional ultrasound (fUS) (Mace et al., 2011). Some of these non-optical imaging methods are excellent in visualizing large tissue volumes, including the entire human brain (Moeller et al., 2010). Yet the contrast is mainly representative of tissue morphology, its mechanical properties, or hemodynamics. Despite significant efforts in the past decade, synthesis of specific contrast agents remains difficult for those modalities, limiting their applicability for direct visualization of fast neural activity.

Owing to its hybrid nature, optoacoustics benefits from both the rich and versatile optical contrast and high (diffraction-limited) spatial resolution associated with the low scattering nature of ultrasonic wave propagation. Much like other optical imaging modalities, OA uses safe nonionizing radiation at the visible and near-infrared wavelengths. OA imaging relies on the absorption of light and is therefore ideally suited for vascular imaging since most of intrinsic contrast is obtained from hemoglobin. This opens a multitude of potential applications in visualization of stimulus-evoked and disease-related vascular changes, neovascularization, blood perfusion, and oxygenation. Its capacity to image spectrally distinctive photo-absorbing agents with high spatio-temporal resolution at depths far beyond the diffusive limit of light opens unprecedented capabilities for functional neuroimaging. In addition, the fundamental ability to optoacoustically track neural calcium dynamics via absorption changes in genetically encoded calcium indicators has been recently demonstrated. Thus, OA imaging provides an excellent platform for multi-scale investigations using the same contrast, from microscopic observations at the single capillary and synapse level to whole-brain imaging in rodents.

Extensive research is underway to address technical challenges associated with the intriguing and highly promising combination of light and sound. Main limitations are currently associated with the lack of reliable and affordable lasers and ultrasound detection technology that can optimally address the unique needs of OA neuroimaging, such as high per-pulse laser energy or repetition rate, ultrawideband detection, high detection sensitivity, and miniaturization. Multiple frontiers are also open in the algorithmic and inverse theory areas, attempting to address challenges related to removal of skull-related image artifacts, image quantification, efficient acquisition and processing of very large data, and multi-spectral data processing. Finally, further advances in activity markers and genetic labeling tools tailored for optimal OA signal transduction are expected to greatly facilitate the application of the OA imaging technology in systems neuroscience and the study of neurological and psychiatric diseases.

REFERENCES

Allen, T.J., P.C. Beard. Pulsed near-infrared laser diode excitation system for biomedical photoacoustic imaging. *Optics Letters*, 31, 3462–3464 (2006).

American National Standards for the Safe Use of Lasers. ANSI Z136.1. American Laser Institute (2000).

Burton, N.C., M. Patel, S. Morscher, W.H.P. Driessen, J. Claussen, N. Beziere, T. Jetzfellner, A. Taruttis, D. Razansky, B. Bednar, V. Ntziachristos. Multispectral opto-acoustic tomography (MSOT) of the brain and glioblastoma characterization. *NeuroImage*, 65, 522–528 (2013).

Cerussi, A.E., A.J. Berger, F. Bevilacqua, N. Shah, D. Jakubowski, J. Butler, R.F. Holcombe, B.J. Tromberg. Sources of absorption and scattering contrast for near-infrared optical mammography. *Academic Radiology*, 8(3), 211–218 (2001).

Deán-Ben, X.L., D. Razansky. Adding fifth dimension to optoacoustic imaging: volumetric time-resolved spectrally-enriched tomography. *Light: Science and Applications*, 3, e137 (2014).

Deán-Ben, X.L., G. Sela, A. Lauri, M. Kneipp, V. Ntziachristos, G.G. Westmeyer, S. Shoham, D. Razansky. Functional optoacoustic neuro-tomography (FONT) for scalable whole-brain monitoring of calcium indicators. *Light: Science and Applications*, 5, e16201 (2016).

Deán-Ben, X.L., T.F. Fehm, S. Gottschalk, S.J. Ford, D. Razanksy. Spiral volumetric optoacoustic tomography visualizes multi-scale dynamics in mice. *Light: Science and Applications*, 6, e16247 (2017).

Deliolanis, N.C., A. Ale, S. Morscher, N.C. Burton, K. Schaefer, K. Radrich, D. Razansky, V. Ntziachristos. Deep-tissue reporter-gene imaging with fluorescence and optoacoustic tomography: a performance overview. *Molecular Imaging and Biology*, 16, 652–660 (2014).

Devor, A., S. Sakadžić, V.J. Srinivasan, M.A. Yaseen, K. Nizar, P.A. Saisan, P. Tian, A.M. Dale, S.A. Vinogradov, M.A. Franceschini, D.A. Boas. Frontiers in optical imaging of cerebral blood flow and metabolism. *Journal of Cerebral Blood Flow and Metabolism*, 32, 1259–1276 (2012).

Duck, F.A. *Physical Properties of Tissues: A Comprehensive Reference Book*. Academic Press (1990).

Gottschalk, S., H. Estrada, O. Degtyaruk, J. Rebling, O. Klymenko, M. Rosemann, D. Razansky. Short and long-term phototoxicity in cells expressing genetic reporters under nanosecond laser exposure. *Biomaterials*, 69, 38–44 (2015).

Gottschalk, S., T.F. Fehm, X.L. Dean-Ben, D. Razansky. Noninvasive real-time visualization of multiple cerebral hemodynamic parameters in whole mouse brains using five-dimensional optoacoustic tomography. *Journal of Cerebral Blood Flow and Metabolism*, 35, 531–535 (2015).

Gottschalk, S., T.F. Fehm, X.L. Deán-Ben, V. Tsytsarev, D. Razansky. Correlation between volumetric oxygenation responses and electrophysiology identifies deep thalamocortical activity during epileptic seizures. *Neurophotonics*, 4, 011007 (2017).

Guevara, E., R. Berti, I. Londono, N. Xie, P. Bellec, F. Lesage, G.A. Lodygensky. Imaging of an inflammatory injury in the newborn rat brain with photoacoustic tomography. *PLoS One*, 8(12), e83045 (2013).

Guo, Z.Q., S. Park, J. Yoon, I. Shin. Recent progress in the development of near-infrared fluorescent probes for bioimaging applications. *Chemical Society Reviews*, 43, 16–29 (2014).

Hu, S., P. Yan, K. Maslov, J.M. Lee, L.V. Wang. Intravital imaging of amyloid plaques in a transgenic mouse model using optical-resolution photoacoustic microscopy. *Optics Letters*, 34(24), 3899–3901 (2009).

Hu, S., E. Gonzales, B. Soetikno, E. Gong, P. Yan, K. Maslov, J.M. Lee, L.V. Wang. Optical-resolution photoacoustic microscopy of ischemic stroke. *Proceedings of SPIE*, 7899 (2011).

Hu, S., K. Maslov, L.V. Wang. Second generation optical-resolution photoacoustic microscopy with improved sensitivity and speed. *Optics Letters*, 36, 1134–1136 (2011).

Hu, S. Listening to the brain with photoacoustics. *IEEE Journal of Selected Topics in Quantum Electronics*, 22, 1–10 (2016).

Jiang, Y.Y., F. Sigmund, J. Reber, X.L. Dean-Ben, S. Glasl, M. Kneipp, H. Estrada, D. Razansky, V. Ntziachristos, G.G. Westmeyer. Violacein as a genetically-controlled, enzymatically amplified and photobleaching-resistant chromophore for optoacoustic bacterial imaging. *Scientific Reports*, 5, 11048 (2015).

Kalchenko, V., D. Israeli, Y. Kuznetsov, A. Harmelin. Transcranial optical vascular imaging (TOVI) of cortical hemodynamics in mouse brain. *Scientific Reports*, 4, 5839 (2014).

Kneipp, M., J. Turner, S. Hambauer, S. Krieg, J. Lehmberg, U. Lindauer, D. Razansky. Functional real-time optoacoustic imaging of middle cerebral artery occlusion in mice. *PLoS ONE*, 9(4), e96118 (2014).

Kruger, R.A., P. Liu, Y.R. Fang, C.R. Appledorn. Photoacoustic ultrasound (PAUS)--reconstruction tomography. *Medical Physics*, 22, 1605–1609 (1995).

Laufer, J., A. Jathoul, M. Pule, P. Beard. *In vitro* characterization of genetically expressed absorbing proteins using photoacoustic spectroscopy. *Biomedical Optics Express*, 4, 2477–2490 (2013).

Li, M.L., J.T. Oh, X.Y. Xie, G. Ku, W. Wang, C. Li, G. Lungu, G. Stoica, L.V. Wang. Simultaneous molecular and hypoxia imaging of brain tumors in vivo using spectroscopic photoacoustic tomography. *Proceedings of the IEEE*, 96(3), 481–489 (2008).

Li, Y., A. Forbrich, J.H. Wu, P. Shao, R.E. Campbell, R. Zemp. Engineering dark chromoprotein reporters for photoacoustic microscopy and FRET imaging. *Scientific Reports*, 6, 22129 (2016).

Liao, L.D., C.T. Lin, Y.Y.I. Shih, T.Q. Duong, H.Y. Lai, P.H. Wang, R. Wu, S. Tsang, J.Y. Chang, M.L. Li, Y.Y. Chen. Transcranial imaging of functional cerebral hemodynamic changes in single blood vessels using in vivo photoacoustic microscopy. *Journal of Cerebral Blood Flow and Metabolism*, 32, 938–951 (2012).

Liao, Z.X., Y.C. Li, H.M. Lu, H.W. Sung. A genetically-encoded KillerRed protein as an intrinsically generated photosensitizer for photodynamic therapy. *Biomaterials*, 35, 500–508 (2014).

Lutzweiler, C., D. Razansky. Optoacoustic imaging and tomography: reconstruction approaches and outstanding challenges in image performance and quantification. *Sensors*, 13, 7345–7384 (2013).

Merčep, E., G. Jeng, S. Morscher, P.-C. Li, D. Razansky. Hybrid optoacoustic tomography and ultrasonography using concave arrays. *IEEE Transactions on UFFC*, 62(9), 1651–1661 (2015).

Mizrahi, A., J.C. Crowley, E. Shtoyerman, L.C. Katz. High-resolution in vivo imaging of hippocampal dendrites and spines. *Journal of Neuroscience*, 24(13), 3147–3151 (2004).

Nasiriavanaki, M., J. Xia, H.L. Wan, A.Q. Bauer, J.P. Culver, L.V. Wang. High-resolution photoacoustic tomography of resting-state functional connectivity in the mouse brain. *Proceedings of the National Academy of Sciences of the United States of America*, 111, 21–26 (2014).

Nienhaus, G.U. Chromophore photophysics and dynamics in fluorescent proteins of the GFP family. *European Biophysics Journal*, 44, S85–S85 (2015).

Ntziachristos, V., J. Ripoll, L.V. Wang, R. Weissleder. Looking and listening to light: the evolution of whole-body photonic imaging. *Nature Biotechnology*, 23, 313–320 (2005).

Ntziachristos, V. Going deeper than microscopy: the optical imaging frontier in biology. *Nature Methods*, 7, 603–614 (2010).

Razansky, D., M. Distel, C. Vinegoni, R. Ma, N. Perrimon, R.W. Köster, V. Ntziachristos. Multispectral optoacoustic tomography of deep-seated fluorescent proteins in-vivo. *Nature Photonics*, 3, 412–417 (2009).

Razansky, D., A. Buehler, V. Ntziachristos. Volumetric real-time multispectral optoacoustic tomography of biomarkers. *Nature Protocols*, 6, 1121–1129 (2011).

Razansky, D. Multispectral optoacoustic tomography - volumetric color hearing in real time. *IEEE Journal of Selected Topics in Quantum Electronics*, 18, 1234–1243 (2012).

Shcherbakova, D.M., M. Baloban, V.V. Verkhusha. Near-infrared fluorescent proteins engineered from bacterial phytochromes. *Current Opinion in Chemical Biology*, 27, 52–63 (2015).

Shu, X.K., A. Royant, M.Z. Lin, T.A. Aguilera, V. Lev-Ram, P.A. Steinbach, R.Y. Tsien. Mammalian expression of infrared fluorescent proteins engineered from a bacterial phytochrome. *Science*, 324, 804–807 (2009).

Song, L., C. Kim, K. Maslov, K.K. Shung, L.H.V. Wang. High-speed dynamic 3D photoacoustic imaging of sentinel lymph node in a murine model using an ultrasound array. *Medical Physics*, 36, 3724–3729 (2009).

Staley, J., P. Grogan, A.K. Samadi, H. Cui, M.S. Cohen, X. Yang. Growth of melanoma brain tumors monitored by photoacoustic microscopy. *Journal of Biomedical Optics*, 15(4), 040510 (2010).

Sun, J., K. Hynynen. The potential of transskull ultrasound therapy and surgery using the maximum available skull surface area. *Journal of the Acoustical Society of America*, 105(4), 2519–2527 (1999).

Tang, J.B., L. Xi, J.L. Zhou, H. Huang, T. Zhang, P.R. Carney, H.B. Jiang. Noninvasive high-speed photoacoustic tomography of cerebral hemodynamics in awake-moving rats. *Journal of Cerebral Blood Flow and Metabolism*, 35, 1224–1232 (2015).

Tzoumas, S., A. Nunes, I. Olefir, S. Stangl, P. Symvoulidis, S. Glasl, C. Bayer, G. Multhoff, V. Ntziachristos. Eigenspectra optoacoustic tomography achieves quantitative blood oxygenation imaging deep in tissues. *Nature Communications*, 7, 12121 (2016).

Wang, L.D., K. Maslov, L.H.V. Wang. Single-cell label-free photoacoustic flowoxigraphy in vivo. *Proceedings of the National Academy of Sciences of the United States of America*, 110, 5759–5764 (2013).

Wang, L.V. Tutorial on photoacoustic microscopy and computed tomography. *IEEE Journal of Selected Topics in Quantum Electronics*, 14, 171–179 (2008).

Wang, L.V. (Editor). *Photoacoustic Imaging and Spectroscopy*. CRC Press (2009).

Wang, X., Y. Pang, G. Ku, X. Xie, G. Stoica, L.V. Wang. Noninvasive laser-induced photoacoustic tomography for structural and functional in vivo imaging of the brain. *Nature Biotechnology*, 21(7), 803–806 (2003).

Wang, Y., K. Maslov, Y. Zhang, S. Hu, L.M. Yang, Y.N. Xia, J.A. Liu, L.H.V. Wang. Fiber-laser-based photoacoustic microscopy and melanoma cell detection. *Journal of Biomedical Optics*, 16, 011014 (2011).

Xu, Z., Q. Zhu, L.V. Wang. In vivo photoacoustic tomography of mouse cerebral edema induced by cold injury. *Journal of Biomedical Optics*, 16(6), 066020 (2011).

Yang, J.M., C. Favazza, R.M. Chen, J.J. Yao, X. Cai, K. Maslov, Q.F. Zhou, K.K. Shung, L.H.V. Wang. Simultaneous functional photoacoustic and ultrasonic endoscopy of internal organs in vivo. *Nature Medicine*, 18, 1297–1302 (2012).

Yang, S.H., D. Xing, Y.Q. Lao, D.W. Yang, L.M. Zeng, L.Z. Xiang, W.R. Chen. Noninvasive monitoring of traumatic brain injury and post-traumatic rehabilitation with laser-induced photoacoustic imaging. *Applied Physics Letters*, 90(24), 243902, (2007).

Yao, J., L.V. Wang. Sensitivity of photoacoustic microscopy. *Photoacoustics*, 2, 87–101 (2014).

Yao, J., L. Wang, J.-M. Yang, K.I. Maslov, T.T. Wong, L. Li, C.-H. Huang, J. Zou, L.V. Wang. High-speed label-free functional photoacoustic microscopy of mouse brain in action. *Nature Methods*, 12, 407–410 (2015).

Yao, J.J., J. Xia, K.I. Maslov, M. Nasiriavanaki, V. Tsytsarev, A.V. Demchenko, L.V. Wang. Noninvasive photoacoustic computed tomography of mouse brain metabolism in vivo. *NeuroImage*, 64, 257–266 (2013).

Zhang, E.Z., J.G. Laufer, R.B. Pedley, P.C. Beard. In vivo high-resolution 3D photoacoustic imaging of superficial vascular anatomy. *Physics in Medicine and Biology*, 54(4), 1035–1046 (2009).

Zhang, Q.Z., Z. Liu, P.R. Carney, Z. Yuan, H.X. Chen, S.N. Roper, H.B. Jiang. Non-invasive imaging of epileptic seizures in vivo using photoacoustic tomography. *Physics in Medicine and Biology*, 53(7), 1921–1931 (2008).

CHAPTER 7

Imaging Deep in the Brain with Wavefront Engineering

Roarke Horstmeyer, Maximillian Hoffmann, Haowen Ruan, Benjamin Judkewitz, and Changhuei Yang

CONTENTS

7.1 INTRODUCTION

Typically, we cannot see directly into or through biological tissue. The cellular structures comprising tissue are heterogeneous and exhibit a spatially varying refractive index. When light interacts with this varying refractive index it tends to scatter in many directions. This uncontrollable scattering subsequently causes tissue to appear turbid. The effect of scattering can be observed both with our own eyes, which cannot see through our own hands, for example, as well as with microscopes, which cannot form clear images of structures buried within most living organisms. The problem of optical scattering in tissue is particularly familiar within the field of neurophotonics, where it limits the depth at which the experimentalist can gain insights into activity inside the brain using optical tools.

Of course, neuroscientists have many experimental tools at their disposal and do not necessarily need to use light to probe the living brain. Electrophysiology, for example, is a common alternative. Electrical recording devices provide excellent temporal resolution and are commonly available now as multi-electrode arrays [1]. However, these methods tend to be invasive and cannot yet monitor the activity of large populations of neurons simultaneously and at subcellular resolution. It is also possible to use wavelengths outside of the optical spectrum to examine activity deep in the brain. For example, X-rays tend to scatter much less through heterogeneous tissue and thus offer high resolution in computed tomography. Alternatively, magnetic resonance imaging (MRI) remains an invaluable tool for medical diagnoses and functional MRI (fMRI) is also now a widespread technique for functional neuroimaging. Finally, photo-acoustic microscopy, which is based on the optical excitation of an ultrasound signal, can also image in vivo brain activity at high speeds [2].

Compared with light microscopy, however, these well-known techniques offer a limited ability to interact with and monitor the activity of individual cells or molecules. Thanks to a number of recent developments, a large suite of optical methods can now both probe and activate biochemical activity. These methods have become particularly useful within neuroscience. For example, is now common practice to tag cellular structures of interest, such as neurons, with genetically encoded fluorescent markers [3, 4]. The brightness of calcium-sensitive fluorescence can indicate the fast temporal dynamics of neuronal firing [5, 6], as can more recently developed voltage-sensitive fluorescent tags [7]. With such optical techniques, experimentalists can now monitor the activity of large, genetically defined populations of neurons *in vivo* and at subcellular resolution.

In addition, light can also be used to either stimulate or inhibit neural activity using optogenetics [8]. By focusing light on a small spot, it is possible to optogenetically excite activity between neurons at single-cell or subcellular resolution [9]. While it may be sufficient in some experiments to simply shine light on the entire brain to "globally" activate many neurons, having precise control over which neurons are locally activated or deactivated offers several benefits. First, residual heating and inefficiencies associated with high-energy light sources currently limit many optogenetic experiments [10]. Targeted focusing helps minimize these effects. Second, and perhaps more importantly, single-neuron control throughout a wide volume can untangle the many complex interactions within various neuronal circuits [9, 11]. The combination of optogenetic activation with imaging fluorescent excitation provides a flexible toolbox for the experimental neuroscientist to simultaneously interact with and measure many spatially resolved neurons at relatively high speeds [11, 12].

As mentioned above, however, light begins to scatter significantly as it enters deep into tissue. Thus, optical scattering still restricts many of the above advances in both probing and imaging neural activity to the outermost layers of the mammalian brain (see Figure 7.1). While techniques like tissue clearing, for example, may be used to minimize scattering to help resolve the entire brain at cellular resolution post-mortem [13, 14], these methods do not work in living tissue. As we will discuss in detail, two currently popular imaging methods, confocal microscopy and two-photon excitation microscopy, help to minimize the effects of optical scattering while imaging *in vivo* activity. Unfortunately, however, these techniques eventually face depth limits [15] due to their reliance on ballistic light for image formation.

Fortunately, there has been a large body of recent work in optics to help account for substantial optical scattering. In this chapter, we outline and discuss a variety of "wavefront

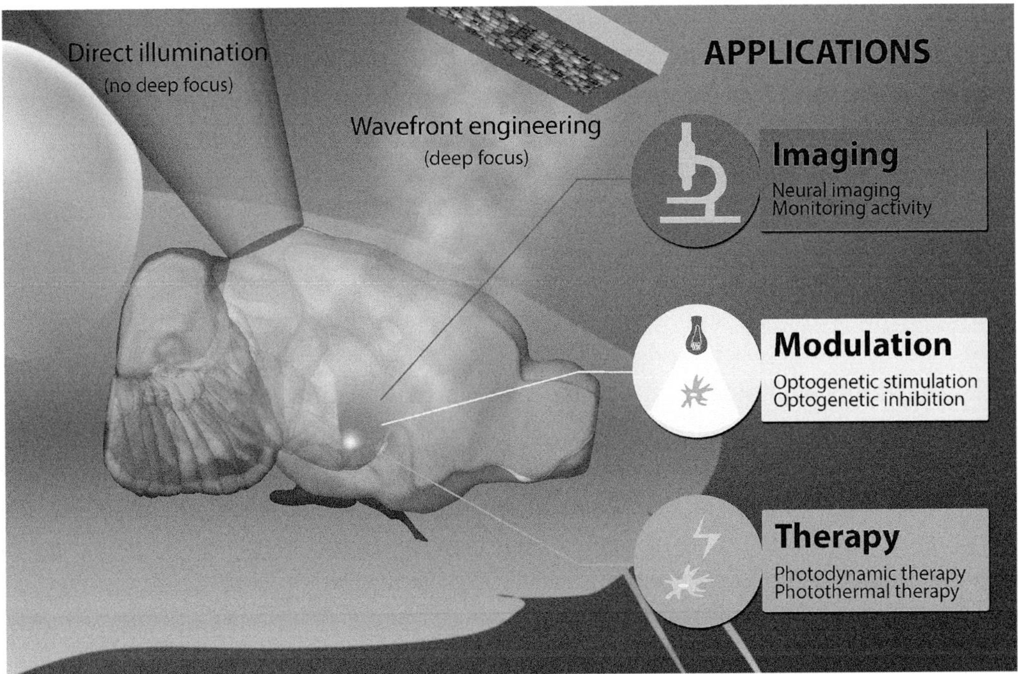

FIGURE 7.1 Wavefront engineering in the brain. Directly attempting to focus an unmodified beam of light (green) deep in the brain will fail due to scattering. By instead shaping the beam (red) into an appropriate pattern with wavefront engineering, it is possible to form a deep focus. Applications in neuroscience include extending the depth of imaging experiments, improving the accuracy and efficiency of optogenetic modulation, and enhancing treatments based on phototherapy.

engineering" tools that are available to extend the depth and accuracy of neurophotonic experiments within tissue. The outline here closely follows a recent review published on the subject [16], but focuses more on applications to experiments in neuroscience. The techniques we discuss offer the potential ability to focus light and form images much deeper in tissue (e.g. beneath 1 mm) than standard optical microscopes (including confocal and two-photon microscopes). With continued development, wavefront engineering will likely help monitor and control activity from larger brain volumes, and hopefully open up new areas of the *in vivo* brain for optical study in the future.

7.2 THE PROPAGATION OF LIGHT WITHIN TISSUE

Before describing the specifics of wavefront engineering, we first offer a brief review of optical scattering within tissue and the tools that are currently used to address it. For this discussion, it is helpful to break down the complex journey of a single photon through tissue into a discrete number of scattering events. Following this section, we will replace our photon viewpoint with a wave model.

In a standard neuroscience imaging experiment, one might illuminate a mouse brain with an epi-fluorescence microscope to visually examine its appearance at the cellular scale. When light from the microscope source reaches the brain, the incident photons will begin to scatter as they enter and progress through the tissue. We may characterize a homogeneous scattering medium by its scattering mean free path, l, which specifies the average distance travelled by a photon between two scattering events. A typical value for l in tissue is 100 μm [17]. Conventional microscopes imaging tissue samples thicker than l will also capture scattered photons, which deteriorate image quality. To overcome the negative effects of scattering, several optical techniques attempt to utilize only the non-scattered (i.e. ballistic) photons that pass through the sample. Confocal imaging is a well-known example that physically blocks scattered light. Optical coherence tomography uses interference to achieve a similar effect [18]. Finally, the multi-photon microscope is the third method that is in widespread use and is based on a nonlinear effect. By primarily measuring ballistic photons and "gating out" scattered light, these methods can image deeper than one scattering mean free path.

Unfortunately, the number of ballistic photons decays exponentially with tissue depth. Thus, while transformative over the past several decades, confocal and multi-photon microscopes eventually reach a depth limit, beyond which their images significantly degrade in quality. Previous analyses [15, 17, 19, 20] suggest that this tissue penetration upper limit occurs approximately at the transport mean free path (TMFP) depth in tissue, $l^* = l/(1-g) \approx 1$ mm [19, 21]. Here, we see the TMFP depth l^* depends on the mean free path l, but is also related to an anisotropy parameter, $g \approx 0.8$–0.95, which takes into account the forward scattering nature of tissue (Figure 7.1b). While several recent techniques improve ballistic photon collection and are challenging this 1 mm depth barrier, for example by using spatial field correlations [22], with combined spatial and coherence gating [23], using near-infrared light [24] or with three-photon excitation [21], many elastically scattered photons remain significantly beyond 1 mm. As the effects of tissue absorption are minimal until centimeter-scale depths [25], an ideal technique would account for, as opposed to block, such multiply scattered photons.

One typically refers to areas beyond the TMFP depth limit as the diffusive regime. There are several incoherent optical methods that attempt to form images here. Diffuse optical tomography [26, 27] and fluorescence molecular tomography [28] are two examples that can computationally detect the presence of macroscopic structures (e.g. tumors) well beneath several TMFP. However, their resolution is much lower than the above-mentioned gating microscopes. Furthermore, it deteriorates with imaging depth (Ntziachristos, 2010) and is unlikely to offer cellular-scale accuracy at millimeter depths, which is the goal of many neuroimaging experiments. Alternatively, one can directly image deep brain areas by inserting a fiber [29] or microendoscope into the brain, which maintains high spatial resolution for both imaging and focusing light [30]. While these insertion approaches can monitor activity in freely moving animals over long periods [31], they are invasive and are typically limited to a single and somewhat narrow field of view.

Unlike the incoherent imaging methods above, coherent light retains a significant amount of information after scattering many times through disordered material such as biological tissue [32, 33]. Instead of throwing these seemingly randomized photons away, a host of new technologies now capture and extract information from multiply scattered light. By measuring the spatial phase profile of a field exiting tissue, for example, it is possible to computationally "undo" the effects of scattering to maintain micron-scale resolution in a non-invasive manner. It is also possible to add an ideal wavefront shape to a beam of light to cause it focus through tissue to a micron-scale spot. Experiments often achieve these two related goals by using both what is known either as a wavefront-engineering or wavefront-shaping device, as well as some type of "guidestar" feedback from deep within the tissue. The remaining sections of this chapter explain how these two technologies combine to allow for the precise control of light within increasingly deep brain areas.

Designing the shape of optical fields is not a fundamentally new idea. Actively correcting random wavefront distortions is already a common practice in astronomy [34]. By measuring how the atmosphere perturbs light from a "guidestar" (i.e. an approximate point source of light), adaptive optics (AO) systems adjust and sharpen images within most ground-based telescopes. The variety of wavefront-engineering technologies that we discuss here share a common foundation with the principle of AO, but extend operation into the biophotonic setting, where heavy optical scattering dominates. Correspondingly, instead of removing low-order aberrations to clearly image "across" distortion, wavefront engineering aims for a generally more useful goal in biology: to send light "into" tissue to form a tight focus within.

Most recent wavefront-engineering methods for deep tissue imaging use one of two guidestar-based strategies. In this review, we classify these two strategies as using either a "feedback" or "conjugation" guidestar, and detail each in a separate section. Outside of the biological setting, an experimentalist can typically manipulate light on both sides of a scattering material, and thus does not usually need the assistance of a guidestar. Mosk et al. offer a thorough review of wavefront engineering through non-biological disordered media [35]. This review will focus on techniques that hold promise for controlling light inside *in vivo* brain tissue, where access is primarily limited to one side of the scattering medium (e.g. from a microscope positioned above the animal's head). Example guidestars that may find use in the brain include fluorescence markers [36, 37], nonlinear optical particles [38], multi-photon fluorescence [39–42], photo-acoustic feedback [43–46], ultrasound-modulated light [47–49], and microbubbles [50]. Before presenting details about these various guidestar strategies, we first review a simple mathematical model to explain how to measure and then correct for optical scattering in brain tissue using an SLM.

7.3 OPTICAL SCATTERING AND THE TRANSMISSION MATRIX

It is now convenient to restrict our attention to light described as a wave, which we assume is coherent, for simplicity (e.g. as from a laser). Returning once again to our simple neuroimaging experiment, let us assume that we use a coherent source to illuminate a particular area of the mouse brain. We may describe the incident monochromatic optical wave field as $u(\boldsymbol{r})$. As it enters the tissue, it will linearly propagate through the spatially varying tissue index of refraction, $n(\boldsymbol{r})$, following the scalar wave equation: $\nabla^2 u(\boldsymbol{r}) + k^2 n^2(\boldsymbol{r}) u(\boldsymbol{r}) = 0$. Here, k denotes wavevector. It is helpful to consider this optical field at two particular planes across space within the imaging experiment, as diagrammed in Figure 7.1. The first plane, which we refer to as the "input plane", is outside of the scattering medium. It contains a device to control the shape of the wavefront across this particular plane. Example devices that are commonly found in wavefront engineering experiments include spatial light modulators (SLMs), digital micro-mirror devices, or deformable mirrors. While each device has its particular advantages and disadvantages, they all offer the ability to modify the amplitude and/or phase of the optical field in a digitally controllable fashion at a discrete number of spatial locations. Given the discrete control N spatial degrees of freedom of the input optical field, u_a, it is useful to define it instead as a row vector $\boldsymbol{u_a}$ (denoted in bold) that contains N complex entries. A limited numerical aperture, or the diffraction limit, allows unambiguous discretization of u_a into a vector using the Nyquist–Shannon sampling theorem.

It is helpful to define a second "target" plane of interest at a particular depth L within the scattering tissue where one would like to form light into a tight focus. For example, this could be at a certain depth within a mouse or zebrafish brain, at a plane perpendicular to the optical axis, where focusing light is otherwise not possible with a standard microscope (see Figure 7.1). Similar to the optical field at the input plane, it is also helpful to spatially discretize the target field into M entries at $\lambda / 2n$ spacing. While both input and target fields typically exist along a two-dimensional plane in practice, we primarily consider each field along just one spatial dimension here for simple notation.

Once discretized, the forward-propagating input field vector $\boldsymbol{u_a}$ now connects to the optical field at the target plane, $\boldsymbol{u_b}$, through a matrix equation: $\boldsymbol{u_a T} = \boldsymbol{u_b}$. Here, we define scattering between the input and target planes, caused by an inhomogeneous refractive index in the wave equation, via multiplication with a transformation matrix $\boldsymbol{T}$ with $N \times M$ complex entries. If light at our embedded target plane is primarily forward traveling, as in the case of highly anisotropic scattering, then $\boldsymbol{T}$ is the well-known transmission matrix [51]. Deep within tissue, where scattering becomes isotropic, the total target field includes contributions from all scatterer regions. However, a linear relationship will still connect the total field at both planes. While not included here for simplicity, any reflected light may also be modeled within the $2N \times 2M$ scattering matrix [52, 53]. Finally, for graphical clarity we adopt the convention of multiplying $\boldsymbol{T}$ from the *left*, with a row vector, to denote light propagating *into* tissue (Figure 7.2a). This allows the variables to appear in the same order as the scattering process (e.g. propagation from $\boldsymbol{u_a}$ to $\boldsymbol{u_b}$ is left to right, matching the order of the above matrix multiplication). We will use column vectors to denote light propagation in the opposite direction, from the target plane back to the input plane (Figure 7.2b).

If a point source sequentially shifts across the input plane, then scattering will cause the target optical field to fluctuate. We mathematically express an input field containing a point source in its ith discrete location as, $\boldsymbol{u_a} = \delta_i$, where we define δ_i as the ith unit vector. For any discrete input location x_a, this particular input field has a one in its ith discrete entry, $\boldsymbol{u_a}\left[x_a = i\right] = 1$, and is zero otherwise: $\boldsymbol{u_a}\left[x_a \neq i\right] = 0$. The ith resulting target field is the matrix product, $\delta_i \boldsymbol{T} = \boldsymbol{t_i}$, where $\boldsymbol{t_i}$ denotes the ith transmission matrix row. Almost always, scattering results in a random field with a grainy appearance, called speckle. If the medium is sufficiently thick and turbid, then the embedded speckle field's spatial variations approach a correlation distance of $\lambda / 2n$ (that is, the coherent speckle becomes "fully developed" within the brain tissue [54]). In this limit, each discrete element of $\boldsymbol{t_i}$ approaches a complex random variable, with a circularly symmetric Gaussian distribution,

(a) Standard neuroimaging setup (b) Simplified diagram

SLM

Input Plane

Target Plane

Scattering Tissue

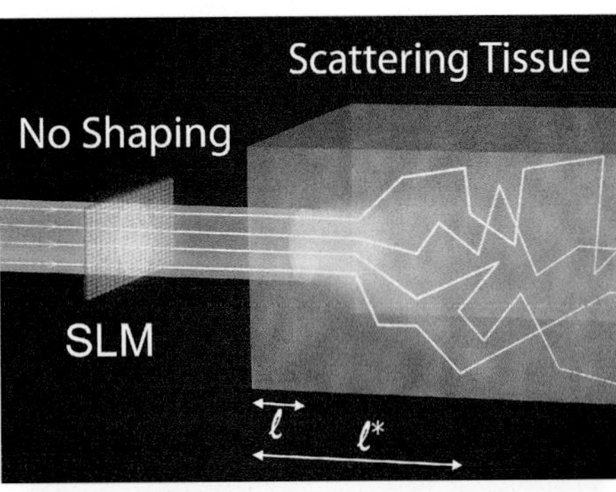

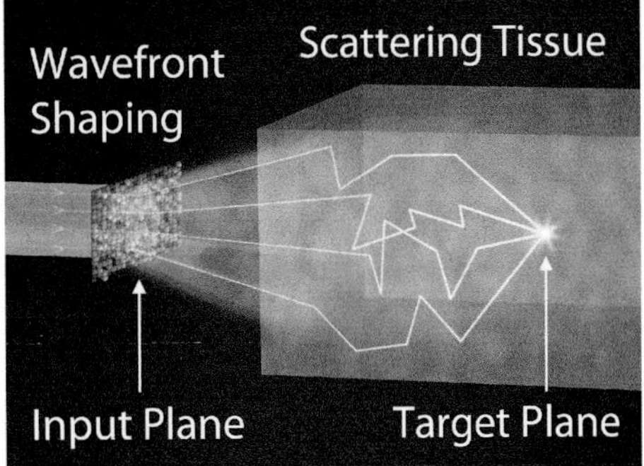

FIGURE 7.2 Principle of wavefront engineering. (a) Simplified schematic of a neurophotonics experiment using a wavefront-engineering device: a spatial light modulator (SLM) placed in the light beam path at an "input" plane. Light is then focused towards a "target" plane located within brain tissue. (b) Simplified diagram of the setup in (a), showing same input and target planes. (b, top) Unmodified coherent beam travels 1 mean free path (l) with minimal scattering into tissue. A fraction of beam directionality is preserved up to the transport mean free path length, l^*. (b, bottom) By appropriately engineering the wavefront at the input plane with the SLM, it is possible to form a focus within tissue beyond l^*. (Figure adapted from [16].)

which is not correlated with its neighbors. If the plane of interest is deep enough to satisfy $L \gg l$, then light is in the multiple scattering regime. Here, a small change in the input wavefront (e.g. a shift to input unit vector δ_{i+1}) will produce a nearly uncorrelated, yet still fully developed, target speckle field. At such depths, it thus becomes fair to approximate each element of the transmission matrix as an uncorrelated random variable.

Remaining sources of correlation within $\boldsymbol{T}$, caused for example by energy conservation and a finite thickness, are of both practical application and ongoing theoretical interest [55]. One form of correlation, termed the "memory effect" [56], helps computational imaging techniques recover objects hidden behind thin, scattering layers [57, 58], including layers of tissue [59]. However, these methods require a finite separation between the scatterer and object, and thus cannot directly focus or image to a plane embedded within tissue. The high anisotropy of tissue offers a second useful form of correlation that allows scanning across an embedded plane but over a limited range [60].

7.4 WAVEFRONT ENGINEERING USING FEEDBACK GUIDESTARS: SEQUENTIAL OPTIMIZATION

Our transmission matrix model informs us of the amplitude and phase of light at an embedded target plane, given an arbitrary input field. If the matrix $\boldsymbol{T}$ is known, then we can invert this scattering system to instead *create* any desired target field (e.g. a focus within tissue) using a specifically designed input field. We may express this input field solution mathematically with a matrix inverse: $\boldsymbol{u}_a = \boldsymbol{u}_b\boldsymbol{T}^{-1}$ (or a pseudo-inverse for rectangular $\boldsymbol{T}$). That is, we determine what wavefront shape we would like to have at the input plane, $\boldsymbol{u}_a$, from the desired field we would like to create at the embedded target plane (e.g. a focus at a particular neuron), through a matrix multiplication with $\boldsymbol{T}^{-1}$. If $\boldsymbol{T}$ is unitary, then we can equivalently apply a conjugate transpose to find $\boldsymbol{u}_a$, by computing $\boldsymbol{u}_a = \boldsymbol{u}_b\boldsymbol{T}^*$. While typically not an equivalent operation, conjugation closely approximates inversion when the scattering and detection system has minimal loss [61]. In practice, combining both matrix inversion and conjugation into one operator helps overcome experimental noise [62].

The above transmission matrix inversion process assumes that we can measure $\boldsymbol{T}$, which typically requires some sort of detector at the target plane. Let us assume for the moment that we are somehow able to place M light-sensitive detectors at the M discretized locations along the embedded target plane. Then, we could apply the following three steps to reverse optical scattering. First, scan through N different orthonormal input fields on the SLM and measure each resulting complex field at the target plane. For example, if each input field is the ith unit vector, δ_i, then each target field is the ith matrix row, $\boldsymbol{t}_i$. Next, computationally construct the transmission matrix $\boldsymbol{T}$ from the N measured target fields (i.e. place each $\boldsymbol{t}_i$ into the ith row of $\boldsymbol{T}$). Finally, given a desired target field $\boldsymbol{u}_b$, solve the above inverse matrix equation to determine the best input field $\boldsymbol{u}_a$ to form with the SLM at the input plane.

This way to measure and invert the transmission matrix is quite flexible. For example, it may deliver arbitrary patterns of light (e.g. an image of a flower) through highly disordered material [51, 62–65]. While an elegant solution to overcome optical scattering, measuring all of the target speckle fields $\boldsymbol{t}_i$ that make up $\boldsymbol{T}$ from *within* tissue is practically challenging. One experimental attempt at this used the photo-acoustic effect [43] (Figure 7.3a). Here, light at an embedded plane induced an ultrasound signal through localized thermal expansion [66]. This time-varying ultrasound response, detected with an external transducer, offered an indirect estimate of local light intensity along one scatterer dimension. Linearly translating the detector system, while scanning through SLM patterns at each unique location, built up a full transmission matrix (Figure 7.3b). Such a strategy may in principle help deliver arbitrary neural stimulation patterns [67] deeper within the brain, where multiple scattering would otherwise not allow direct projection of the desired pattern.

Often, neurophotonic experiments require just a single optical focus at a particular location of interest (e.g. to excite a particular neuron, or to form a tight focal spot that will later be scanned around). Given this relatively simple aim, it is often not necessary to measure the entire transmission matrix. Instead, if we consider a single fixed location $x_b = c$ for targeted focusing, then we can restrict our attention to just one value of the target field, $u_b\left[c\right]$, in our mathematical framework. This in turn allows us to only consider the cth

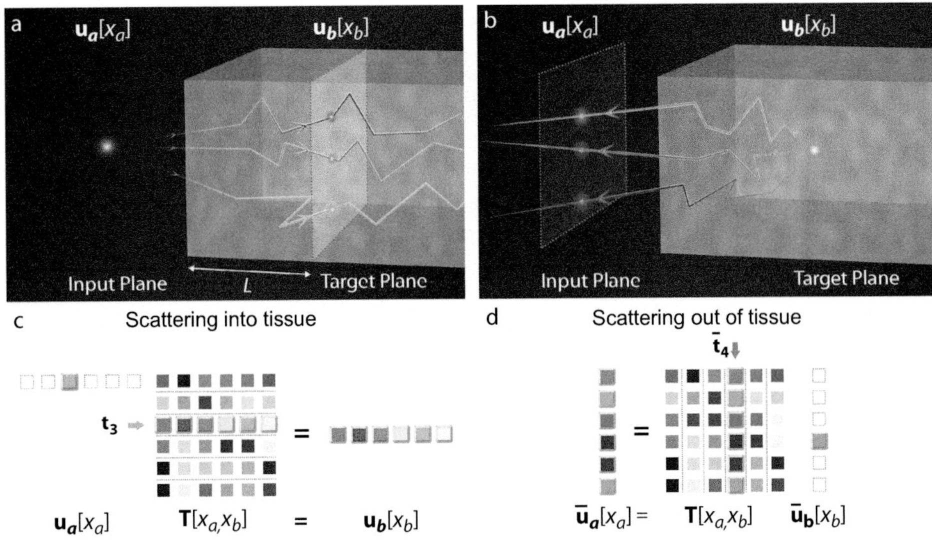

FIGURE 7.3 (a) Forward optical scattering a distance L into tissue (input to target plane). (b) Reverse optical scattering out of tissue (target to input plane). (c) Transmission matrix model for forward scattering. A discrete input point source at position 3 sets the vector on the left: $\boldsymbol{u_a}\left[x_a\right]=\delta_3$, the third unit vector. The target field shown on right is then $\boldsymbol{t}_3$, the third transmission matrix row. (d) Matrix model for scattering from an embedded guidestar point. Often, different matrices are used for scattering in and out of tissue. Here, we assume time-reversal symmetry and use the same matrix T for both types of scattering, but perform matrix multiplication from opposite sides. Assuming a point source at target plane location 4 sets $\bar{\boldsymbol{u}}_b\left[x_b\right]=\bar{\delta}_4$ (column vector) on the right. The input field to the left of the matrix then becomes $\bar{\boldsymbol{t}}_4$, the fourth transmission matrix column. (Figure adapted from [16].)

column of the transmission matrix, since its remaining values do not affect our single spot of interest. In other words, we need not multiply our input field with the entire transmission matrix to determine the target optical field at spot c, $u_b\left[c\right]$. Instead, we must only compute an inner product: $\boldsymbol{u_a}\bar{\boldsymbol{t}}_c = u_b\left[c\right]$, where $\bar{\boldsymbol{t}}_c$ is the cth transmission matrix column (see Figure 7.3c). We have used a bar here to identify it as a column vector, while we have written row vectors without a bar. We only need to know the N entries of the vector $\bar{\boldsymbol{t}}_c$ to fully map the scattering relationship between the input plane and the field at the cth location at the embedded target plane. Knowing the N values of $\bar{\boldsymbol{t}}_c$, we can now "design" what we would like the target field to be at point c by constructing an appropriately shaped input field, $\boldsymbol{u_a}$. For example, we may form a bright focus at the target location point $x_b = c$ by shaping the incident wavefront into the conjugate transpose of $\bar{\boldsymbol{t}}_c$. This becomes clear if we set $\boldsymbol{u_a} = \bar{\boldsymbol{t}}_c^*$ in the above inner product: $\bar{\boldsymbol{t}}_c^*\bar{\boldsymbol{t}}_c \approx 1$, the maximum possible normalized amplitude at target location c.

Just like our procedure to measure $\boldsymbol{T}$, one may also determine the values within $\bar{\boldsymbol{t}}_c$ by scanning through orthogonal SLM patterns. Now, however, one only needs to detect and measure light at the single target location c. The first demonstrations of wavefront-shaped focusing through disordered media used a single-pixel detector to obtain such measurements [32, 68]. Instead of cycling through orthogonal SLM patterns, it is also possible to use feedback between the single detector and the SLM (e.g. with a genetic or hill-climbing algorithm) [53, 69]. This encourages speedy maximization of delivered light intensity. With either approach, the ideal SLM-shaped input field remains the conjugated transmission matrix column vector, $\overline{t}_c^*$ [70], and the focal spot brightness increases linearly with the number of conjugated optical modes [32].

For many biological experiments, it is challenging to directly embed a physical detector (e.g. a photodiode) into tissue to measure $\overline{t}_c$. Similar to indirectly measuring the transmission matrix, one may use a signal that is correlated with localized light intensity to provide SLM feedback. We refer to such a mechanism as a "feedback guidestar", of which there are several varieties (Figure 7.4). One of the first demonstrations of overcoming scattering with an embedded feedback guidestar used a fluorescent bead inside a disordered material. A feedback algorithm connecting the fluorescence intensity (detected outside the material) with an SLM at the input plane could experimentally enhance fluorescent excitation by a factor of 20 or more [36, 71].

A fluorescent guidestar often requires invasive positioning and fixes light delivery to a single location. A moveable ultrasound focal spot, which frequency-shifts light [72] via index of refraction modulation and scatterer displacement, offers a non-invasive feedback

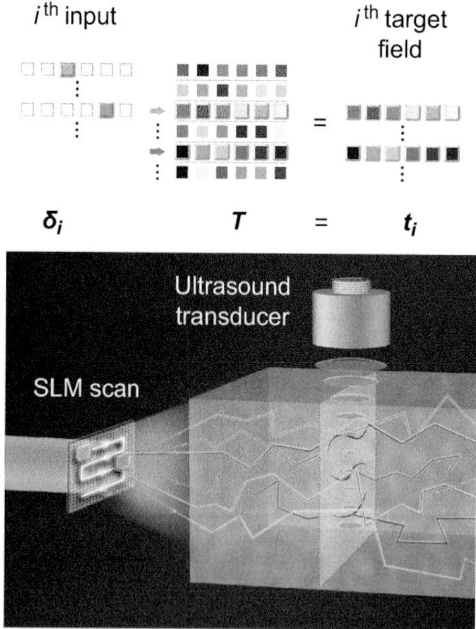

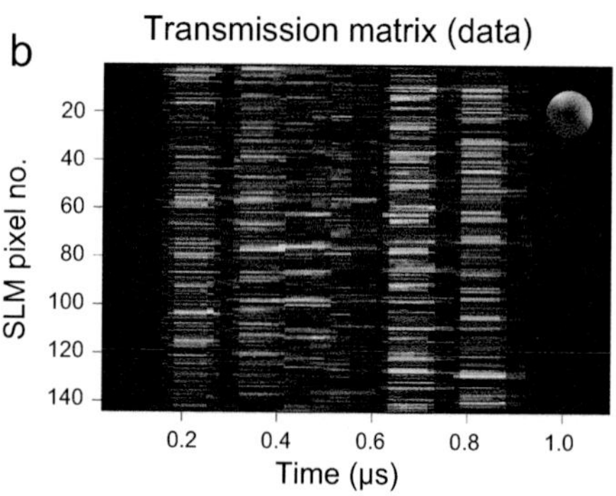

FIGURE 7.4 Measuring the optical transmission matrix within tissue. Scanning one transparent pixel across an input SLM and detecting each target field samples the rows of T. An external ultrasound transducer may be used, for example, to indirectly measure each target field from within tissue via the photo-acoustic effect. (b) Compiled together, these photo-acoustic measurements form the optical transmission matrix that connects the input and target planes [43]. (Figure adapted from [16]).

alternative [73]. Since ultrasound scatters weakly-in tissue, it provides a useful mechanism to extract optical information from an embedded target plane. Alternatively, instead of sending in ultrasound and detecting modulated light, an externally measured photo-acoustic signal also indicates the local strength of incident light, as mentioned earlier. Standard photo-acoustic "guidestar" feedback originates from a single spot and is limited to the acoustic resolution [44, 45]. Nonlinear photo-acoustic feedback [46] enables focusing somewhat close the optical resolution limit (~5–7 μm, Figure 7.4c).

It is also possible to create guidestar-like feedback by monitoring the two-photon fluorescence signal generated directly inside a scattering sample [39, 41, 74] (Figure 7.3e). Several techniques have also applied frequency multiplexing to speed up the measurement of the elements of $\overline{t}_c$ to achieve time-efficient wavefront correction [40, 42] with two photon feedback (Figure 7.4d). In a related manner, one may track the coherent interference of backscattered light from different sample layers [75, 76]. The above examples achieve feedback with optical mechanisms that are also shared by two other neuroimaging modalities: two-photon excitation microscopy and optical coherence tomography (OCT), respectively. Two-photon microscopy, in particular, is currently in widespread use in the neuroimaging community and there is a growing body of work applying adaptive optics (AO) techniques to improve its performance (see [77, 78] for recent examples). Due to the similarity between AO and wavefront engineering, it is helpful to briefly discuss their connections before proceeding to describe our second class of guidestar technique.

As noted in the introduction, AO is traditionally thought of as the imaging complement of wavefront engineering: the former passively removes aberrations after light has propagated from an object of interest, whereas the latter actively shapes light so that it can pass through an aberrative medium (e.g. to a focus). Early on, AO applied similar feedback algorithms to iteratively measure and correct image distortions caused by the eye's lens [79]. Later, AO microscopes applied fluorescence feedback to remove low-order aberrations from ballistic light [80]. Recent AO microscopes now also correct for complex image distortions caused by non-ballistic light [40, 81–83], which can produce images of excellent quality at moderate tissue depths. As with wavefront engineering, AO does not necessarily require a physically distinct feedback guidestar. Systems may apply feedback from the intrinsic two-photon [84] or fluorescent [85] light intensities that are detected at the image plane. Finally, in addition to aberration removal from the imaging path, AO microscopes also commonly correct for distortions within their illumination path [86]. This closely matches the principle of wavefront engineering, which almost always uses its corrective SLM in the illumination path.

While conceptually similar, four key differences help distinguish guidestar-based wavefront engineering from AO techniques. First, many AO setups focus light into a guidestar from the input side. Significant scattering precludes the ability to directly focus light into a guidestar within tissue, giving rise to our list of inventive feedback mechanisms in Table 7.1. Second, a wavefront correction map in AO can remove aberrations over a finite area, termed the isoplanatic patch [87]. The isoplanatic patch beneath 1 TFMP in tissue is limited to several square microns or less, limiting the relevance of one optimal input wavefront effectively to one specific target location. Third, since many AO setups

TABLE 7.1 List of Biological Tissue Guidestars for Feedback (Top) and Conjugation (Bottom)

	Guidestar mechanism	Ref	Min. spot size	Non- invasive?	Translate position?	PBR	Required time (sec.)	Coupling efficiency	One Side?
Feedback Guidestars	Fluorescence	[36, 71, 88]	1 μm	No	No	20–40	10^3	Low	No
	Ultrasound	[73	1 mm	Yes	Yes	10	10^3–10^4	Low	No
	Photo-acoustic	[44, 45]	35 μm	Yes	Yes	5–10	10^3	Moderate	No
	Nonlinear Photo-acoustic	[46]	5 μm	Yes	Yes	6000	10^3	Moderate	No
	Two-photon	[40, 41, 42]	1–5 μm	Yes*	Yes	20	10^2–10^3	Moderate	Yes
	Coherence gating	[75, 76]	1 μm	Yes	Yes	50	1–10	Moderate	Yes
Conj. Guidestars	Second harmonics	[38]	2 μm	No	No	400	1	Moderate	No
	Two-photon, AO	[89, 77]	1 μm	Yes	Yes	10	<.01	Moderate	Yes
	F-Sharp	[90]	1 μm	Yes	Yes	100	1	Moderate	Yes
	Fluorescence	[37]	1 μm	No	No	10–100	10^2	Low	No
	Ultrasound	[47–49]	30 μm	Yes	Yes	5	<.01	Low	Yes
	Iterative Ultrasound	[91–93]	10 μm	Yes	Yes	30	1–10	Low	Yes
	Variance Encoded Ultrasound (TROVE)	[94]	5 μm	Yes	Yes	100	10^3–10^4	Low	No
	Particle displacement (TRAP, TRACK)	[95, 96]	5–10 μm	Yes*	Yes	300	<1	High	No
	Micro-bubble	[50]	2 μm	Yes*	Yes	500	<1	High	No

Minimum spot size selected as minimum across all similar demonstrations. Peak-to-background ratio (PBR) selected as maximum across all similar demonstrations. Required time is an order-of-magnitude approximation for current demonstrations, but will likely drop in the future. Coupling efficiency p, defined as the percentage of incident optical power transferred into a detectable guidestar signal, is high if $p>0.5$, moderate if $0.5>p>0.01$, and low if $p<0.01$. (*) Denotes possible invasive insertion used in experiment, but not required in principle. "One side" column is yes, if technique demonstrated with illumination and detection on same scatterer side, as required for most *in vivo* experiments. (Table adapted from [16].)

typically correct distortions in the imaging arm, they often must operate with incoherent light, as opposed to the coherent light used in wavefront engineering setups. Finally, the number of parameters (and hence measurements) required to fit an AO aberration map is typically small. An AO transmission matrix contains in the order of $N \approx 100$ modes. This is suitable for devices like deformable mirrors, which are commonly found in AO setups and contain approximately the same number of independently controllable elements. In contrast, deep-tissue guidestars must accurately map and correct for the $N \approx 10^5$ or more independent optical modes in the diffusive regime [35]. This task typically requires many more measurements and many more degrees of freedom on the device used for wavefront engineering, which in turn calls for either an SLM or a digital micro-mirror device.

7.5 TIME-REVERSAL AND PHASE CONJUGATION

Feedback guidestar techniques measure the scattering response of light passing through brain tissue sequentially, by taking many measurements over time. Unfortunately, accurate focusing can require millions of unique measurements [35], and the scattering response of *in vivo* tissue can change on a sub-second time scale [97, 98]. To help overcome this time constraint, one would ideally like to measure the scattering response of tissue in a parallelized manner. By adopting the principle of time-reversal, or "optical phase conjugation" (OPC), it is possible to refocus light to a guidestar spot using a single snapshot measurement.

To briefly outline the principle of OPC, let us assume that an ideal guidestar emits a monochromatic optical wave with frequency ω. At any position $\boldsymbol{r}$ within the scattering tissue, we may express this emitted scalar field as, $u(\boldsymbol{r},t) = \mathrm{Re}\{A(\boldsymbol{r})\exp[i(\phi(\boldsymbol{r}) - \omega t)]\}$. Here, A specifies the field amplitude and the ϕ defines its spatially varying phase. Given propagation within a lossless medium, the wave equation remains time-invariant. For any forward propagating field $u(\boldsymbol{r},t)$ that is scattering out of the medium, there also exists a wave function $u' = u(\boldsymbol{r},-t)$ that will precisely retrace the path of u back through every scattering interaction to its original guidestar location. If the constant phase lines for u must satisfy $\phi(\boldsymbol{r}) = \omega t$, then the constant phase lines of u' must satisfy $\phi(\boldsymbol{r}) = -\omega t$. It is clear the phase lines of u' also satisfy the relation, $-\phi(\boldsymbol{r}) = \omega t$. The left-hand side of this equation represents the spatial phase conjugate of u. We can therefore "rewind" an arbitrary field u back into its original guidestar spot by conjugating its spatial phase, and then allowing it to continue to travel ahead in time.

To describe OPC within our linear algebra framework, we represent the light emerging *from* the target plane's guidestar as $\bar{\boldsymbol{u}}_b$, where the bar denotes a column vector. We assume time-reversal symmetry to express its transformation to the input plane as, $\bar{\boldsymbol{u}}_a = \boldsymbol{T}\bar{\boldsymbol{u}}_b$. Light propagating from the target to input plane, as a column vector, now multiplies into our transmission matrix from the right (Figure 7.2b, d). In practice, sample absorption and a limited detection aperture prevent collection of the complete guidestar field. Despite this challenge, the following phase conjugation strategy can still successfully focus light into tissue. First, create a discrete point source of light at a specific target location $x_b = c$ within the brain: $\bar{\boldsymbol{u}}_b = \bar{\delta}_c$, the cth unit column vector. Next, capture the resulting scattered field at the input plane: $\bar{u}_a = T\bar{\delta}_c = \bar{t}_c$, the cth column of T (Figure 7.4a). Finally, phase conjugate the detected field into $\bar{t}_c^*$, and send the conjugate transpose field back towards the target plane: $\bar{t}_c^* T = \delta_c + \varepsilon$. This conjugated field refocuses to a discrete spot at the guidestar origin (Figure 7.4b). Incomplete measurement and imperfect conjugation of $\bar{t}_c$ also typically introduces a finite background field, ε. Note this expression closely matches the final result of our previous "feedback guidestar" strategy. Here, OPC directly measures and conjugates the column vector $\bar{t}_c$, whereas feedback from the target plane determines the elements of $\bar{t}_c^*$ in sequence.

The first OPC experiment to focus light through thick biological tissue used a hologram to record and conjugate the scattered field [33]. Its demonstration matched results from early non-biological OPC experiments [99, 100]. Shortly thereafter, focusing through tissue was achieved with digital optical phase conjugation (DOPC), which interferometrically measures an optical field on a digital detector, and then creates its phase conjugate using a pixel-to-pixel matched SLM [101]. Here, unlike early phase conjugation work within ophthalmology [102], DOPC uses a modern SLM that contains several million individually addressable elements. Thus, DOPC may compensate for complex optical transformations that involve millions of propagation modes, such as diffusive scattering in tissue [35]. Furthermore, unlike analog holograms, DOPC can be digitally modified to correct for minor setup misalignments [103], for example, and can also in principle shape continuous and pulsed beams of arbitrary power.

7.6 WAVEFRONT ENGINEERING WITH CONJUGATION GUIDESTARS: PARALLEL DETECTION

Without direct access to either side of scattering tissue, for instance, during an *in vivo* neuroimaging experiment, OPC requires an effective "conjugation guidestar" to emit light for subsequent refocusing. Fluorescent excitation within the retina is one of the first conjugation guidestars applied in a biological context [102]. Here, a focused beam stimulates a small incoherent light source at the retina, which is externally measured to compensate for aberrations. Embedded fluorescent microspheres [104] and proteins [89, 105] act as similar conjugation beacons in AO microscopy. As noted above, it is also possible to excite two-photon fluorescence guidestars for improved AO imaging [39], including that of *in vivo* neural activity [77]. As noted earlier, since two-photon imaging is already a common neuroscience technique, the non-invasive formation of a two-photon fluorescence guidestar deep in tissue is an experimentally appealing goal. One recent technique termed F-Sharp achieves such a focus by interfering two beams [90], after which conjugation-based correction can significantly improve resolutions at depths up to 0.5 mm. Finally, it is also possible to use coherence gating to form an approximate guidestar [106], as also demonstrated within an OCT system [107]. Coherence gating offers another means to achieve conjugation-based correction non-invasively and has been demonstrated up to moderate tissue depths.

As the depth of the target plane increases into the diffusive regime, the distorted guidestar wavefront that one must detect begins to approach a fully developed speckle pattern. Just as in the case of feedback guidestars, the above list of AO techniques will typically not fully succeed in correcting such highly scattered fields. Instead, one may adopt one of a number of alternative techniques for deep-tissue conjugation-based focusing. We offer a summarized list of example deep-tissue conjugation guidestar mechanisms in Table 7.1.

The first guidestar for conjugation-based focusing through fully disordered media used the nonlinear second-harmonic generation signal from a nanoparticle [38]. Shortly thereafter, Vellekoop et al. conjugated light from a fluorescent marker embedded beneath 0.5 mm of tissue using an SLM [37] (Figure 7.5c). While their conjugated spot sizes may approach the diffraction limit, both nonlinear and fluorescent guidestar signals originate from a fixed, spatially confined spot at a different wavelength than excitation, and thus tend to require lengthy exposure times (≥1 second).

As discussed in the context of feedback guidestars, a confined ultrasound focus can frequency-shift light at a defined spot within *in vivo* tissue [108]. Several experimental setups can detect and phase conjugate the ultrasound-modulated wavefront scattered from this finite spot, thus creating an ultrasound conjugation guidestar [47–49]. Since ultrasound may be freely focused in different tissue areas, it offers an easily translatable guidestar (Figure 7.6a). It can also operate at millisecond time scales [109]. These two features offer significant promise for application within the *in vivo* brain. The resolution of early "time-reversed ultrasonically encoded" (TRUE) demonstrations was fixed by the size of the ultrasound focus (25–50 μm), which may be sufficient for some neuroimaging experiments. While increasing the ultrasound frequency shrinks this spot size, it also decreases

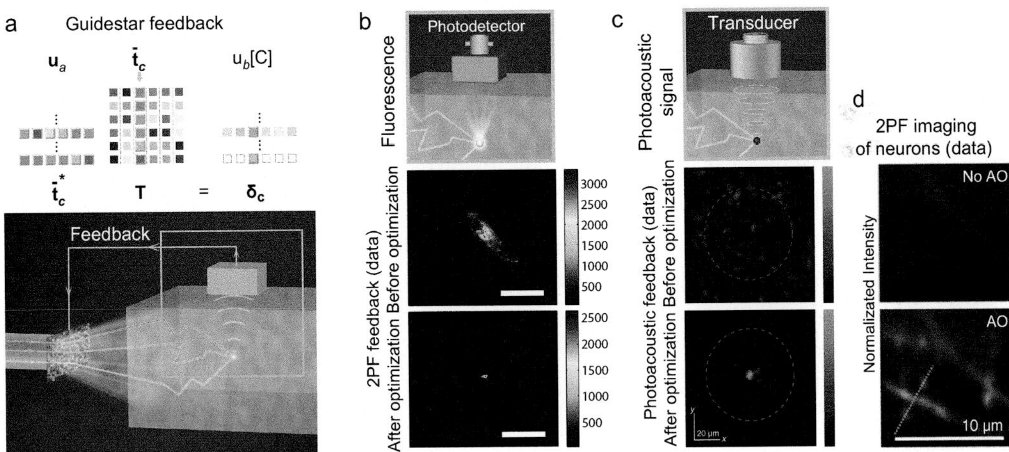

FIGURE 7.5 Feedback guidestars for wavefront engineering. (a) Feedback guidestar matrix model. The scattered field is measured from one discrete location within brain tissue, here $u_b[3]$. Feedback from this measurement helps determine the optimal shape to pattern the input wavefront with the SLM. (b) Photo-acoustic feedback measured via an ultrasound transducer optimizes light delivery to a tight focus [46]. (c) Using fluorescence feedback, such as two-photon fluorescence (2PF). After optimization, 2PF feedback significantly improves the ability to focus light through $L = 1$ mm of brain tissue to the brightest fluorescent spot [41]. (d) Two-photon fluorescence feedback can also be recorded in a multiplexed manner to increase signal and resolution from labeled neurons [42]. (Figure adapted from [16].)

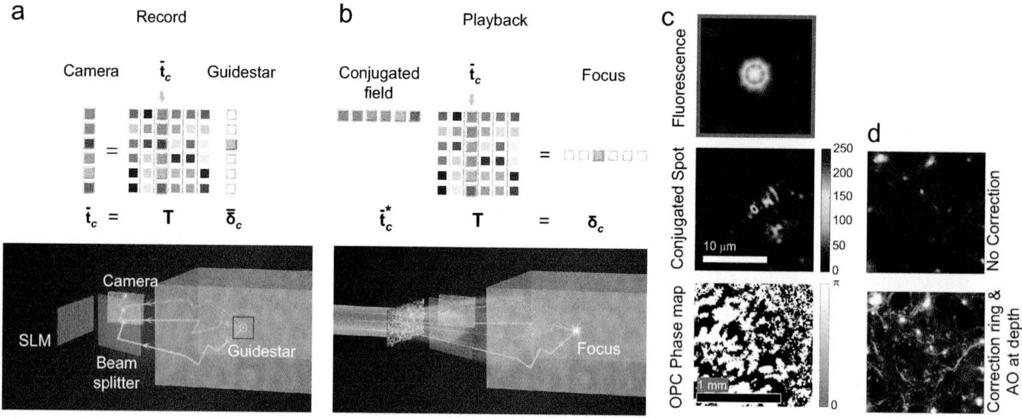

FIGURE 7.6 OPC with conjugation guidestars. (a) Matrix model and setup for detecting the field from an embedded guidestar. Light from the target field spot $\overline{\delta}_3$ forms the speckle field $\overline{t}_3$ at the input plane camera. (b) Matrix model and setup for conjugation guidestar focusing. SLM-shaping an incident wavefront into conjugate field $\overline{t}_3^*$ refocuses to δ_3. (c) Example experimental demonstration using a 0.2 μm fluorescent bead as a conjugation guidestar, showing the focal spot intensity and binary conjugated phase map [37]. (d) Example correction of *in vivo* neurons with a two-photon guidestar AO microscope before (top) and after (bottom) correction [77]. (Figure adapted from [16].)

the number of photons modulated from deep within the tissue, presenting a trade-off between resolution and depth [110]. A potential solution to this resolution trade-off is iterative TRUE [91–93], where a feedback loop between the conjugated and detected fields can experimentally shrink the focus spot size by a factor of three.

Instead of improving TRUE focusing resolution with iterative feedback, one may also adopt a statistical approach, termed TROVE [94]. Conceptually, the relatively large TRUE focus modulates many optical speckles. The modulated field emerging from the tissue is thus a superposition of weighted optical fields, each originating from a unique location within the ultrasound spot. Standard TRUE simply detects and conjugates this entire superposition. Alternatively, through a series of measurements and application of Gaussian statistics, TROVE computationally decomposes this superposition back into its individually weighted components. By conjugating the appropriate component, it is possible to refocus to a spot the size of one speckle (demonstrated so far: ~5 μm, Figure 7.6a bottom).

Even without ultrasonic frequency encoding, multiple measurements of a time-varying optical field can still lead to a sharp, phase conjugated focus within tissue. One simple strategy involves any small object that absorbs or scatters light at the target plane. First, a scattered field is measured when the target is present (u_0). Second, a different scattered field is measured after the target has moved away from the area of interest (u_1). Subtracting the second field from the first and conjugating the result (i.e. displaying $[u_0 - u_1]^*$ on the input plane SLM) will focus light to the target location. This concept has been recently demonstrated with targets that physically move beneath a scatterer [95, 96]. It is also possible to achieve a similar effect using microbubbles embedded in tissue, where the microbubbles are popped between each field measurement (Figure 7.6b) [50].

7.7 FUTURE EXPERIMENTAL DEVELOPMENT

Before finding widespread applications in neuroscience, guidestar techniques must first successfully operate within living tissue that is free to move (Figure 7.7). Any unknown change in scattering response between guidestar signal measurement and wavefront playback is currently a source of error. Even if an ideal wavefront is measured for OPC, its sharp focus will quickly spread into random speckle during prolonged *in vivo* conjugation [97]. Macroscopic effects, like heartbeat and blood flow, are a primary cause of movement across the brain in anesthetized animals. Studies indicate that an OPC system response time below 50 ms can overcome the error associated with movements in unconstrained tissue, while system response times may also relax to several seconds with immobilized (i.e., pinched) tissue [97, 98]. Thick brain slices also exhibit a similar multi-second decorrelation constant [111].

A select number of current setups, using either digital feedback [68] or nonlinear optical conjugation [112], operate within a 50 ms time window. Lock-in detection with TRUE focusing can now focus within less than a millisecond [109]. Further speed-up may be achieved by using an integrated detector and SLM [113], or with feedback algorithms that account for the temporal dynamics of the scattering medium [114, 115]. Finally, if the sample of interest can fit within an optical cavity, all-optical feedback could execute wavefront engineering at extremely fast time scales [116].

In parallel with increased speed, future wavefront engineering technologies hope to offer focusing depths that exceed one centimeter (i.e. operate in the macroscopic regime [17]). This will enable optical access to most regions of the in vivo rat brain, for example.

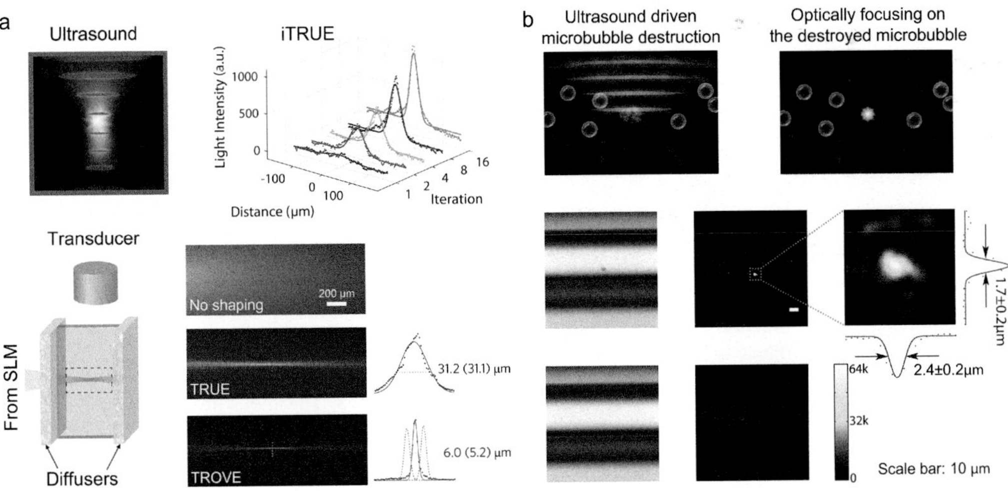

FIGURE 7.7 Additional conjugation guidestars. (a) Experimental demonstration of an ultrasound conjugation guidestar. Iterative iTRUE can sharpen the conjugated TRUE spot size by a factor of 3 [93], while TROVE reduces the focal spot width from 31 μm to 5 μm [94]. (b) Experimental demonstration of microbubbles as a conjugation guidestar mechanism [50]. The difference between the scattered optical fields measured before and after microbubble bursting enables focusing through a scattering slab to a micron-scale spot. (Figure adapted from [16].)

While optical absorption becomes a limiting factor in this regime, recent work demonstrates that OPC may focus through over 3 cm through tissue with safe light intensity levels [117]. Such depths have yet to be demonstrated with guidestar feedback. However, high-efficiency fluorescent emitters and nonlinear scatterers may help eventually realize this goal. Models also predict that ultrasound guidestars, which become attenuated at large depths, may currently extend to centimeter depth scales [110, 118].

Finally, most of the guidestar techniques reviewed here achieve refocusing with monochromatic light. The spectral and temporal dependence of optical scattering introduces additional degrees of freedom for wavefront control. Direct time-reversal of scattered pulses is well-studied with ultrasound [119] and may also lead to enhanced focusing in the optical regime. Several recent electromagnetic implementations show promising benefits from spectral/temporal control [120–124]. While few achieve this control from just one side of the scattering material (a prerequisite for its use in most neurophotonic experiments), it is possible that we will see this in future setups (Figure 7.8).

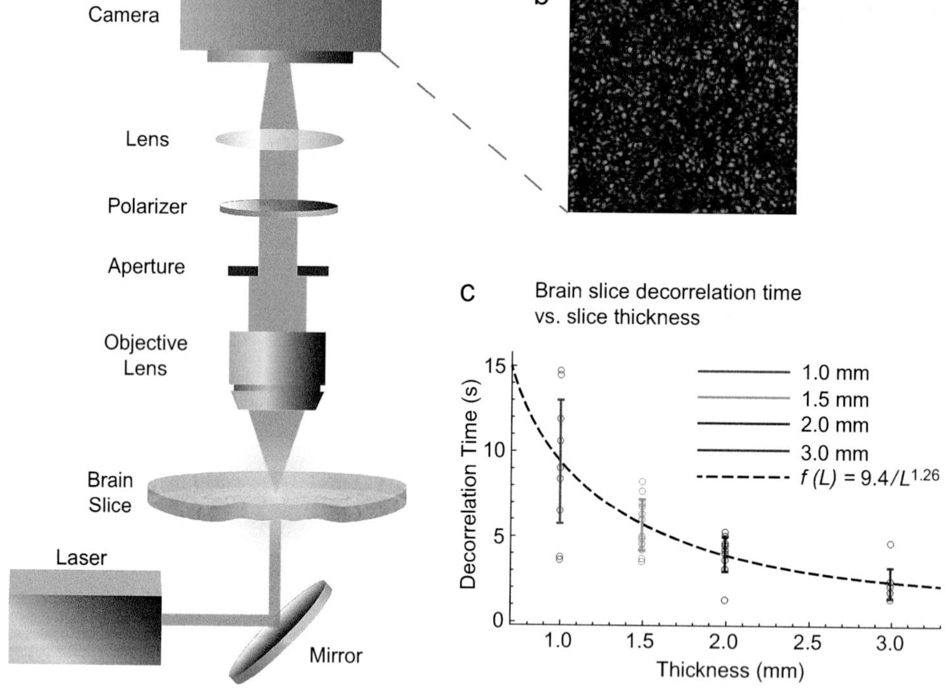

FIGURE 7.8 Decorrelation time of mouse brain tissue [111]. (a) Diagram of a decorrelation experiment, which measures how the optical field varies over time after passing through an *in vitro* acute brain slice between 1 and 3 mm thick. (b) Example scattered speckle intensity. (c) Average decorrelation time as a function of brain tissue thickness (i.e. depth), with a fit from diffusion theory. The scattered field changes in the order of several seconds, indicating that the average lifetime of a sharp guidestar-based focus inside *in vitro* tissue is approximately 1 sec. Due to bulk motion and blood flow, the required update time within *in vivo* is expected to be much less than 1 sec. (closer to 10–50 ms [98]). (Figure adapted from [111].)

7.8 APPLICATIONS OF GUIDESTARS IN NEUROSCIENCE

A variety of potential applications await both feedback and conjugation guidestar focusing inside of brain tissue. Many of these applications are related to two connected goals: light delivery and imaging. Here, we briefly review applications for each.

Most directly, wavefront engineering may help deliver more light to a single spot deep in the brain than previously possible. Photodynamic therapy (PDT) is one example application that may benefit from such targeted focusing. Here, a tight focus of light can photochemically activate drugs or biomolecules at a localized region within the brain, as opposed to a large area, in an efficient manner. Example treatments that might benefit cover a broad range, stretching from the removal of solid tumors to ophthalmic diseases [125]. Highly concentrated optical power can also directly ablate tissue [126]. This is typically useful for precise surgeries or in studies aiming to deactivate particular neurons, for example. In a related manner, focused light may also help induce targeted damage, such as photothrombotic stroke [127]. Wavefront engineering may help extend these capabilities beneath superficial brain layers and improve their accuracy and efficiency. Finally, wavefront engineering for particle manipulation is currently possible across highly scattering media [82, 128]. It may also eventually find use within thick tissue, and perhaps even within the *in vivo* brain.

As mentioned in Section 7.1: Introduction, the most likely widespread application of deep-tissue light delivery in neuroscience is for optogenetic excitation. When illuminated with sufficient light, optogenetic markers may activate or deactivate various physiological processes [8]. When these light-activated proteins are targeted to neurons, it is possible to activate or deactivate neuronal activity, which would ideally be possible at cellular-level precision across the entire brain. Initial demonstrations suggest that guidestar-based focusing can improve both the resolution and penetration depth of current optogenetic techniques [67, 83, 129, 130]. Future work will likely identify specific applications of this new and flexible optical toolset. For example, selective OPC excitation of neuronal circuits present in deep brain areas may eventually help uncover new interactions and functions.

The second application of wavefront engineering in neuroscience is to create clear, resolved images of neural activity. As already mentioned, a variety of information is available from monitoring the fluorescent excitation of neurons and other cells within the brain and nervous system. For example, it is now quite common to use genetically encoded calcium [3] or voltage [7] indicators to record neuronal firing. A guidestar-generated focus will locally excite fluorescence within a confined area deep in the brain. Scanning this focus around can then sweep out an image. In brain areas where it is possible to form an aberrated focus, wavefront engineering may directly improve the resolution and excitation efficiency (and thus frame rates) of current microscopes. Wavefront engineering has already been used to image ~0.4 mm into the mouse brain cortex using an ultrasound guidestar [131]. In principle, it may also be possible to monitor neuronal activity deeper in the brain, for example within the mouse hippocampus. The future "holy grail" of guidestar techniques in neurophotonics will likely be for the joint optogenetic excitation and fluorescent readout from *in vivo* neurons at arbitrary depths through the intact skull (e.g. to

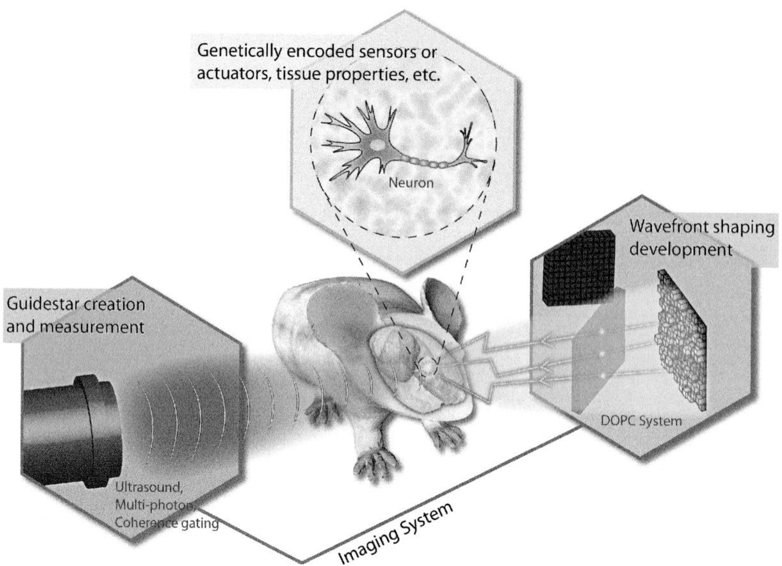

FIGURE 7.9 Three key areas for the future success of deep-tissue imaging and focusing in neuroscience. (a) Improved speed, sensitivity and pixel count of both light measurement devices (i.e. photodetectors and pixel arrays) and elements for wavefront engineering (i.e. SLMs and DMDs), as well as improved computational methods for deep focusing. (b) Available guidestar mechanisms for feedback from within the brain (see Table 7.1), with many new possibilities remaining. (c) Improved optogenetic tags and fluorescent markers (calcium and/or voltage-sensitive), among other biochemical tools. This latter category is currently changing the landscape of neurophotonics. Continued development here will inform future guidestar focusing systems.

extend the reach of current all-optical interrogation methods [132] to any location within an unperturbed mouse brain). Given the quick pace of development in technologies surrounding guidestar techniques (see Figure 7.9), it will not be surprising if this goal is realized in the near future.

In summary, wavefront engineering methods hold the unique promise for maintaining optical resolution beneath 1 mm of scattering tissue. These techniques may eventually help improve the resolution of optogenetic excitation and fluorescent readout for neurons deep inside the living brain. With more development, this technology may play an important role in uncovering the function of the brain at cellular resolution throughout its large volume.

REFERENCES

1. Spira, M.E. and A. Hai, Multi-electrode array technologies for neuroscience and cardiology. *Nature Nanotechnology*, 2013. **8**(2): p. 83–94.
2. Yao, J., et al., High-speed label-free functional photoacoustic microscopy of mouse brain in action. *Nature Methods*, 2015. **12**(5): p. 407–410.
3. Rose, T., et al., Putting a finishing touch on GECIs. *Frontiers in Molecular Neuroscience*, 2014. **7**: p. 88.
4. Chen, T.-W., et al., Ultrasensitive fluorescent proteins for imaging neuronal activity. *Nature*, 2013. **499**(7458): p. 295–300.

5. Svoboda, K., et al., In vivo dendritic calcium dynamics in neocortical pyramidal neurons. *Nature*, 1997. **385**: p. 161–165.
6. Ohki, K., et al., Functional imaging with cellular resolution reveals precise micro-architecture in visual cortex. *Nature*, 2005. **433**: p. 597–603.
7. Storace, D., et al., Toward better genetically encoded sensors of membrane potential. *Trends in Neurosciences*, 2016. **39**: p. 277–289.
8. Zhang, F., et al., Optogenetic interrogation of neural circuits: technology for probing mammalian brain structures. *Nature Protocols*, 2010. **5**(3): p. 439–456.
9. Packer, A.M., B. Roska, and M. Hausser, Targeting neurons and photons for optogenetics. *Nature Neuroscience*, 2013. **16**(7): p. 805–815.
10. Williams, J.C. and T. Denison, From optogenetic technologies to neuromodulation therapies. *Science Translational Medicine*, 2013. **5**(177): p. 177ps6.
11. Packer, A.M., et al., Simultaneous all-optical manipulation and recording of neural circuit activity with cellular resolution in vivo. *Nature Methods*, 2015. **12**: p. 140–146.
12. Rickgauer, J.P., K. Diesseroth, and D.W. Tank, Simultaneous cellular-resolution optical perturbation and imaging of place cell firing fields. *Nature Neuroscience*, 2014. **17**: p. 1816–1824.
13. Dodt, H.-U., et al., Ultramicroscopy: three-dimensional visualization of neuronal networks in the whole mouse brain. *Nature Methods*, 2007. **4**(4): p. 331–336.
14. Hama, H., et al., Scale: a chemical approach for fluorescence imaging and reconstruction of transparent mouse brain. *Nature Neuroscience*, 2011. **14**(11): p. 1481–1488.
15. Helmchen, F. and W. Denk, Deep tissue two-photon microscopy. *Nature Methods*, 2005. **2**(12): p. 932–940.
16. Horstmeyer, R., H. Ruan, and C. Yang, Guidestar-assisted wavefront-shaping methods for focusing light into biological tissue. *Nature Photonics*, 2015. **9**(9): p. 563–571.
17. Ntziachristos, V., Going deeper than microscopy: the optical imaging frontier in biology. *Nature Methods*, 2010. **7**(8): p. 603–611.
18. Huang, D., et al., Optical coherence tomography. *Science*, 1991. **254**(5035): p. 1178–1181.
19. Theer, P. and W. Denk, On the fundamental imaging-depth limit in two-photon microscopy. *Journal of the Optical Society of America A*, 2006. **23**(12): p. 3139–3149.
20. Oheim, M., et al., Two-photon microscopy in brain tissue: parameters influencing imaging depth. *Journal of Neuroscience Methods*, 2001. **111**: p. 29–37.
21. Horton, N.G., et al., In vivo three-photon microscopy of subcortical structures within an intact mouse brain. *Nature Photonics*, 2013. **7**(3): p. 205–209.
22. Kang, S., et al., Imaging deep within a scattering medium using collective accumulation of single-scattered waves. *Nature Photonics*, 2015. **9**(4): p. 253–258.
23. Matthews, T.E., et al., Deep tissue imaging using spectroscopic analysis of multiply scattered light. *Optica*, 2014. **1**(2): p. 105–111.
24. Hong, G., et al., Through-skull fluorescence imaging of the brain in a new near-infrared window. *Nature Photonics*, 2014. **8**(9): p. 723–730.
25. Weissleder, R. and V. Ntziachristos, Shedding light onto live molecular targets. *Nature Medicine*, 2003. **9**(1): p. 123–128.
26. den Outer, P.N., A. Lagendijk, and T.M. Nieuwenhuizen, Location of objects in multiple-scattering media. *Journal of the Optical Society of America A*, 1993. **10**(6): p. 1209–1218.
27. Arridge, S.R., Optical tomography in medical imaging. *Inverse Problems*, 1999. **15**(2): p. R41.
28. Ntziachristos, V., et al., Fluorescence molecular tomography resolves protease activity in vivo. *Nature Medicine*, 2002. **8**(7): p. 757–760.
29. Levene, M.J., et al., In vivo multiphoton microscopy of deep brain tissue. *Journal of Neurophysiology*, 2004. **91**(4): p. 1908–12.
30. Szabo, V., et al., Spatially selective holographic photoactivation and functional fluorescence imaging in freely behaving mice with a fiberscope. *Neuron*, 2014. **84**(6): p. 1157–1169.

31. Barretto, R.P.J., et al., Time-lapse imaging of disease progression in deep brain areas using fluorescence microendoscopy. *Nature Medicine*, 2011. **17**(2): p. 223–228.
32. Vellekoop, I.M. and A.P. Mosk, Focusing coherent light through opaque strongly scattering media. *Optics Letters*, 2007. **32**(16): p. 2309–2311.
33. Yaqoob, Z., et al., Optical phase conjugation for turbidity suppression in biological samples. *Nature Photonics*, 2008. **2**(2): p. 110–115.
34. Hardy, J.W., *Adaptive Optics for Astronomical Telescopes*. New York: Oxford University Press, 1998.
35. Mosk, A.P., et al., Controlling waves in space and time for imaging and focusing in complex media. *Nature Photonics*, 2012. **6**(5): p. 283–292.
36. Vellekoop, I.M., et al., Demixing light paths inside disordered metamaterials. *Optics Express*, 2008. **16**(1): p. 67–80.
37. Vellekoop, I.M., M. Cui, and C. Yang, Digital optical phase conjugation of fluorescence in turbid tissue. *Applied Physics Letters*, 2012. **101**(8): p. 081108.
38. Hsieh, C.L., et al., Digital phase conjugation of second harmonic radiation emitted by nanoparticles in turbid media. *Optics Express*, 2010. **18**(12): p. 12283–12290.
39. Aviles-Espinosa, R., et al., Measurement and correction of in vivo sample aberrations employing a nonlinear guide-star in two-photon excited fluorescence microscopy. *Biomedical Optics Express*, 2011a. **2**(11): p. 3135–3149.
40. Tang, J., R.N. Germain, and M. Cui, Superpenetration optical microscopy by iterative multiphoton adaptive compensation technique. *Proceedings of the National Academy of Sciences of the United States of America*, 2012. **109**(22): p. 8434–8439.
41. Katz, O., et al., Noninvasive nonlinear focusing and imaging through strongly scattering turbid layers. *Optica*, 2014a. **1**(3): p. 170–174.
42. Wang, C., et al., Multiplexed aberration measurement for deep tissue imaging in vivo. *Nature Methods*, 2014a. **11**: p. 1037–1040.
43. Chaigne, T., et al., Controlling light in scattering media non-invasively using the photoacoustic transmission matrix. *Nature Photonics*, 2014. **8**(1): p. 58–64.
44. Kong, F., et al., Photoacoustic-guided convergence of light through optically diffusive media. *Optics Letters*, 2011. **36**(11): p. 2053–2055.
45. Caravaca-Aguirre, A.M., et al., High contrast three-dimensional photoacoustic imaging through scattering media by localized optical fluence enhancement. *Optics Express*, 2013. **21**(22): p. 26671–26676.
46. Lai, P., et al., Photoacoustically guided wavefront shaping for enhanced optical focusing in scattering media. *Nature Photonics*, 2015. **9**(2): p. 126–132.
47. Xu, X., H. Liu, and L.V. Wang, Time-reversed ultrasonically encoded optical focusing into scattering media. *Nature Photonics*, 2011. **5**(3): p. 154–157.
48. Si, K., R. Fiolka, and M. Cui, Fluorescence imaging beyond the ballistic regime by ultrasound-pulse-guided digital phase conjugation. *Nature Photonics*, 2012a. **6**(10): p. 657–661.
49. Wang, Y.M., et al., Deep-tissue focal fluorescence imaging with digitally time-reversed ultrasound-encoded light. *Nature Communications*, 2012. **3**: p. 928.
50. Ruan, H., M. Jang, and C. Yang, Optical focusing inside scattering media with time-reversed ultrasound microbubble encoded light. *Nature Communications*, 2015. **6**: p. 8968.
51. Popoff, S.M., et al., Measuring the transmission matrix in optics: an approach to the study and control of light propagation in disordered media. *Physical Review Letters*, 2010a. **104**(10): p. 100601.
52. Beenakker, C.W.J., Random-matrix theory of quantum transport. *Reviews of Modern Physics*, 1997. **69**(3): p. 731–808.
53. Vellekoop, I.M., *Controlling the Propagation of Light in Disordered Scattering Media*. PhD thesis. University of Twente, Enschede, The Netherlands, 2008. Available at http://arxiv.org/abs/0807.1087.

54. Goodman, J.W., *Speckle Phenomena in Optics: Theory and Applications*. Roberts and Company, 2007. Bar-Ilan University, Ramat-Gan, Israel.
55. Wiersma, D.S., Disordered photonics. *Nature Photonics*, 2013. **7**(3): p. 188–196.
56. Feng, S., et al., Correlations and fluctuations of coherent wave transmission through disordered media. *Physical Review Letters*, 1988. **61**(7): p. 834–837.
57. Bertolotti, J., et al., Non-invasive imaging through opaque scattering layers. *Nature*, 2012. **491**(7423): p. 232–234.
58. Katz, O., et al., Non-invasive single-shot imaging through scattering layers and around corners via speckle correlations. *Nature Photonics*, 2014b. **8**(10): p. 784–790.
59. Yang, X., Y. Pu, and D. Psaltis, Imaging blood cells through scattering biological tissue using speckle scanning microscopy. *Optics Express*, 2014. **22**(3): p. 3405–3413.
60. Judkewitz, B., et al., Translation correlations in anisotropically scattering media. *Nature Physics*, 2015. **11**: p. 684–689.
61. Tanter, M., J.L. Thomas, and M. Fink, Time reversal and the inverse filter. *Journal of the Acoustical Society of America*, 2000. **108**: p. 223–234.
62. Popoff, S., et al., Image transmission through an opaque material. *Nature Communications*, 2010b. **1**: p. 81.
63. Kim, M., et al., Maximal energy transport through disordered media with the implementation of transmission eigenchannels. *Nature Photonics*, 2012. **6**(9): p. 581–585.
64. Cizmar, T. and K. Dholakia, Exploiting multimode waveguides for pure fibre-based imaging. *Nature Communications*, 2012. **3**: p. 1027.
65. Choi, Y., et al., Measurement of the time-resolved reflection matrix for enhancing light energy delivery into a scattering medium. *Physical Review Letters*, 2013. **111**(24): p. 243901.
66. Wang, L.V. and S. Hu, Photoacoustic tomography: in vivo imaging from organelles to organs. *Science*, 2012. **335**(6075): p. 1458–1462.
67. Papagiakoumou, E., et al., Functional patterned multiphoton excitation deep inside scattering tissue. *Nature Photonics*, 2013. **7**(4): p. 274–278.
68. Conkey, D.B., A.M. Caravaca-Aguirre, and R. Piestun, High-speed scattering medium characterization with application to focusing light through turbid media. *Optics Express*, 2012a. **20**(2): p. 1733–1740.
69. Conkey, D.B., et al., Genetic algorithm optimization for focusing through turbid media in noisy environments. *Optics Express*, 2012b. **20**(5): p. 4840–4849.
70. Vellekoop, I.M. and A.P. Mosk, Universal optimal transmission of light through disordered materials. *Physical Review Letters*, 2008. **101**(12): p. 120601.
71. Vellekoop, I.M. and C.M. Aegerter, Scattered light fluorescence microscopy: imaging through turbid layers. *Optics Letters*, 2010a. **35**(8): p. 1245–1247.
72. Leutz, W. and G. Maret, Ultrasonic modulation of multiply scattered light. *Physica B*, 1995. **204**(1–4): p. 14–19.
73. Tay, J.W., et al., Ultrasonically encoded wavefront shaping for focusing into random media. *Scientific Reports*, 2014. **4**: p. 3918.
74. Aviles-Espinosa, R., et al., Measurement and correction of in vivo sample aberrations employing a nonlinear guide-star in two-photon excited fluorescence microscopy. *Biomedical Optics Express*, 2011b. **2**: p. 3135–3149.
75. Fiolka, R., K. Si, and M. Cui, Complex wavefront corrections for deep tissue focusing using low coherence backscattered light. *Optics Express*, 2012. **20**(15): p. 16532–16543.
76. Jang, J., et al., Complex wavefront shaping for optimal depth-selective focusing in optical coherence tomography. *Optics Express*, 2013. **21**(3): p. 2890–2902.
77. Wang, K., et al., Direct wavefront sensing for high-resolution in vivo imaging in scattering tissue. *Nature Communications*, 2015. **6**: p. 1–6.
78. Kong, L., J. Tang, and M. Cui, In vivo volumetric imaging of biological dynamics in deep tissue via wavefront engineering. *Optics Express*, 2016a. **24**: p. 1214–1221.

79. Liang, J., D.R. Williams, and D.T. Miller, Supernormal vision and high-resolution retinal imaging through adaptive optics. *Journal of the Optical Society of America A*, 1997. **14**(11): p. 2884–2892.
80. Neil, M.A., et al., Adaptive aberration correction in a two-photon microscope. *Journal of Microscopy*, 2000. **200**: p. 105–108.
81. Ji, N., D.E. Milkie, and E. Betzig, Adaptive optics via pupil segmentation for high-resolution imaging in biological tissues. *Nature Methods*, 2010. **7**(2): p. 141–147.
82. Cizmar, T., M. Mazilu, and K. Dholakia, In situ wavefront correction and its application to micromanipulation. *Nature Photonics*, 2010. **4**(6): p. 388–394.
83. Kong, L. and M. Cui, In vivo fluorescence microscopy via iterative multi-photon adaptive compensation technique. *Optics Express*, 2014. **22**(20): p. 23786–23794.
84. Albert, O., et al., Smart microscope: an adaptive optics learning system for aberration correction in multiphoton confocal microscopy. *Optics Letters*, 2000. **25**(1): p. 52–54.
85. Booth, M.J., et al., Adaptive aberration correction in a confocal microscope. *Proceedings of the National Academy of Sciences of the United States of America*, 2002. **99**(9): p. 5788–5792.
86. Booth, M.J., Adaptive optical microscopy: the ongoing quest for a perfect image. *Light: Science and Applications*, 2014. **3**: p. e165.
87. Fried, D.L., Anisoplanatism in adaptive optics. *Journal of the Optical Society of America*, 1982. **72**(1): p. 52–61.
88. van Putten, E.G., A. Lagendijk, and A.P. Mosk, Optimal concentration of light in turbid materials. *Journal of the Optical Society of America*, 2011. **28**(5): p. 1200–1203.
89. Wang, K., et al., Rapid adaptive optical recovery of optimal resolution over large volumes. *Nature Methods*, 2014b. **11**(6): p. 625–628.
90. Papadopoulos, I.N., et al., Scattering compensation by focus scanning holographic aberration probing (F-SHARP). *Nature Photonics*, 2016. **11**(2): p. 116–123.
91. Si, K., R. Fiolka, and M. Cui, Breaking the spatial resolution barrier via iterative sound-light interaction in deep tissue microscopy. *Scientific Reports*, 2012b. 2: p. 748.
92. Suzuki, Y., et al., Continuous scanning of a time-reversed ultrasonically encoded optical focus by reflection-mode digital phase conjugation. *Optics Letters*, 2014. **39**(12): p. 3441–3444.
93. Ruan, H., et al., Iterative time-reversed ultrasonically encoded light focusing in backscattering mode. *Scientific Reports*, 2014. **4**: p. 7156.
94. Judkewitz, B., et al., Speckle-scale focusing in the diffusive regime with time reversal of variance-encoded light (TROVE). *Nature Photonics*, 2013. **7**(4): p. 300–305.
95. Zhou, E.H., et al., Focusing on moving targets through scattering samples. *Optica*, 2014. **1**(4): p. 227–232.
96. Ma, C., et al., Time-reversed adapted-perturbation (TRAP) optical focusing onto dynamic objects inside scattering media. *Nature Photonics*, 2014. **8**(12): p. 931–936.
97. Meng, C., E.J. McDowell, and C. Yang, An in vivo study of turbidity suppression by optical phase conjugation (TSOPC) on rabbit ear. *Optics Express*, 2010. **18**(1): p. 25–30.
98. Jang, M., et al., Relation between speckle decorrelation and optical phase conjugation (OPC)-based turbidity suppression through dynamic scattering media: a study on in vivo mouse skin. *Biomedical Optics Express*, 2015. **6**(1): p. 72–85.
99. Leith, E.N. and J. Upatnieks, Holographic imagery through diffusing media. *Journal of the Optical Society of America*, 1966. **56**(4): p. 523–523.
100. Goodman, J.W., et al., Wavefront-reconstruction imaging through random media. *Applied Physics Letters*, 1966. **8**(12): p. 311–313.
101. Cui, M. and C. Yang, Implementation of a digital optical phase conjugation system and its application to study the robustness of turbidity suppression by phase conjugation. *Optics Express*, 2010. **18**(4): p. 3444–3455.
102. Diaz Santana Haro, L. and J.C. Dainty, Single-pass measurements of the wave-front aberrations of the human eye by use of retinal lipofuscin autofluorescence. *Optics Letters*, 1999. **24**(1): p. 61–63.

103. Jang, M., et al., Method for auto-alignment of digital optical phase conjugation systems based on digital propagation. *Optics Express*, 2014a. **22**(12): p. 14054–14071.
104. Azucena, O., et al., Wavefront aberration measurements and corrections through thick tissue using fluorescent microsphere reference beacons. *Optics Express*, 2010. **18**(16): p. 17521–17532.
105. Tao, X., et al., Adaptive optics microscopy with direct wavefront sensing using fluorescent protein guide stars. *Optics Letters*, 2011. **36**(17): p. 3389–3391.
106. Rueckel, M., J.A. Mack-Bucher, and W. Denk, Adaptive wavefront correction in two-photon microscopy using coherence-gated wavefront sensing. *Proceedings of the National Academy of Sciences of the United States of America*, 2006. **103**(46): p. 17137–17142.
107. Hermann, B., et al., Adaptive-optics ultrahigh-resolution optical coherence tomography. *Optics Letters*, 2004. **29**(18): p. 2142–2144.
108. Lev, A. and B. Sfez, In vivo demonstration of the ultrasound-modulated light technique. *Journal of the Optical Society of America A*, 2003. **20**(12): p. 2347–2354.
109. Liu, Y., et al., Bit-efficient, sub-millisecond wavefront measurement using a lock-in camera for time-reversal based optical focusing inside scattering media. *Optics Letters*, 2016. **41**: p. 1321–1324.
110. Jang, M., et al., Model for estimating the penetration depth limit of the time-reversed ultrasonically encoded optical focusing technique. *Optics Express*, 2014b. **22**(5): p. 5787–5807.
111. Brake, J., M. Jang, and C. Yang, Analyzing the relationship between decorrelation time and tissue thickness in acute rat brain slices using multispeckle diffusing wave spectroscopy. *Journal of the Optical Society of America A*, 2016. **33**: p. 270–275.
112. Liu, Y., et al., Optical focusing deep inside dynamic scattering media with near-infrared time-reversed ultrasonically encoded (TRUE) light. *Nature Communications*, 2015. **6**: 5904.
113. Laforest, T., et al., A 4000 Hz CMOS image sensor with in-pixel processing for light measurement and modulation. IEEE 11th International New Circuits and Systems Conference (NEWCAS), 2013. p. 1–4.
114. Vellekoop, I.M. and C.M. Aegerter, Focusing light through living tissue. *Proceedings of SPIE*, 2010b. **7554**: p. 755430.
115. Stockbridge, C., et al., Focusing through dynamic scattering media. *Optics Express*, 2012. **20**(14): p. 15086–15092.
116. Nixon, M., et al., Real-time wavefront shaping through scattering media by all-optical feedback. *Nature Photonics*, 2013. **7**(11): p. 919–924.
117. Shen, Y., et al., Focusing light through biological tissue and tissue-mimicking phantoms up to 9.6 cm in thickness with digital optical phase conjugation. *Journal of Biomedical Optics*, 2016. **21**: p. 085001.
118. Hollmann, J.L., et al., Diffusion model for ultrasound-modulated light. *Journal of Biomedical Optics*, 2014. **19**(3): p. 035005.
119. Fink, M., Time reversed acoustics. *Physics Today*, 1997. **50**: p. 34–40.
120. Lerosey, G., et al., Focusing beyond the diffraction limit with far-field time reversal. *Science*, 2007. **315**(5815): p. 1120–1122.
121. Aulbach, J., et al., Control of light transmission through opaque scattering media in space and time. *Physical Review Letters*, 2011. **106**(10): p. 103901.
122. McCabe, D.J., et al., Spatio-temporal focusing of an ultrafast pulse through a multiply scattering medium. *Nature Communications*, 2011. **2**: p. 447.
123. Katz, O., et al., Focusing and compression of ultrashort pulses through scattering media. *Nature Photonics*, 2011. **5**(6): p. 372–377.
124. Paudel, H.P., et al., Focusing polychromatic light through strongly scattering media. *Optics Express*, 2013. **21**(14): p. 17299–17308.
125. Huang, Z., A review of progress in clinical photodynamic therapy. *Technology in Cancer Research and Treatment*, 2005. **4**(3): p. 283–293.

126. Vogel, A. and V. Venugopalan, Mechanisms of pulsed laser ablation of biological tissues. *Chemical Reviews*, 2003. **103**(2): p. 577–644.
127. Labat-gest, V. and S. Tomasi, Photothrombotic ischemia: a minimally invasive and reproducible photochemical cortical lesion model for mouse stroke studies. *Journal of Visualized Experiments*, **76**(e50370).
128. Volpe, G., et al., Speckle optical tweezers: micromanipulation with random light fields. *Optics Express*, 2014. **22**(15): p. 18159–18167.
129. Yoon, J., et al., Optogenetic signaling-pathway regulation through scattering skull using wavefront shaping. *ArXiv*, 2015a. **1502**.04826.
130. Yoon, J., et al., Optogenetic control of cell signaling pathway through scattering skull using wavefront shaping. *Scientific Reports*, 2015b. **5**: p. 13289.
131. Kong, L., J. Tang, and M. Cui, In vivo volumetric imaging of biological dynamics in deep tissue via wavefront engineering. *Optics Express*, 2016b. **24**(2): p. 1214–1221.
132. Emiliani, V., et al., All-optical interrogation of neural circuits. *The Journal of Neuroscience*, 2015. **35**(41): p. 13917–13926.

CHAPTER 8

Nanoscopic Imaging to Understand Synaptic Function

Daniel Choquet and Anne-Sophie Hafner

CONTENTS

8.1 INTRODUCTION

The adult brain contains over 100 billion neurons, each of which connects to thousands of neuronal partners (Garner et al., 2002). Understanding how this large number of cells interacts to produce a complex organism behavior is an intellectual challenge of the highest order. The neuronal dendritic arbor is optimized to receive a large number of information through inputs from neurons that can be located from micrometers to several centimeters away.

The term synapse comes from the Greek *synapsis* that means "conjunction" (the condition of being joined) and was introduced in 1897 by the English physiologists Michael Foster and Charles Sherrington (Foster and Sherrington, 1897; Tansey, 1997) to describe neuronal connections. At chemical synapses – the topic of this chapter – the plasma membrane of a presynaptic neuron is in close apposition (20–25 nm) with the membrane of the postsynaptic neuron. After more than a century of neuroscience research, there is now strong indications that information storage in the brain involves in large part experience-driven modifications of synaptic strength in addition to other parameters, such as circuit remodeling and changes in cell excitability.

Synaptic strength is the measure of the response of a given synapse to a given input and is determined by the complement of proteins expressed, also referred to as its "proteome", and their post-translational modifications. Because of their complex morphology, neurons are faced with the challenge of maintaining and regulating up to thousands of synapses as individual units. In fact, each synaptic connection can undergo up- or down-regulation of its synaptic strength depending on the intensity/frequency of presynaptic inputs it is receiving. In this chapter, we focus on the excitatory glutamatergic synapses for their ability to exhibit multiple forms of synaptic plasticity, stimulation-induced modulation of the synaptic strength.

The evidence that such changes in the properties of glutamatergic synapses are actually involved in memory has indeed kept growing during the last 45 years. While inhibitory synapses are almost exclusively shaft-type synapses (i.e. on the main part of the dendrites), glutamatergic synapses can be either on the shaft (on inhibitory neurons or during early development) or on special micrometer-sized protrusions on dendrites called "spines". In the adult brain, over 90% synapses formed into excitatory neurons are made on spines.

Dendritic spines were initially described in 1888 by the neuroanatomist Ramón y Cajal (Ramón y Cajal, 1888). They arise from the dendritic shaft of the postsynaptic neuron and receive inputs from glutamatergic presynaptic axon terminals. Spines exhibit a large variety of shapes and sizes that suggest an important functional diversity, morphological changes being tightly linked to biochemical reactions taking place in the isolated environment of the spine (reviewed in Sheng and Kim, 2011). Indeed, single spine remodeling occurs upon synaptic stimulation in the order of seconds to minutes, and *in vivo* imaging of adult mouse brain revealed that motor-learning tasks training can induce spine remodeling as well as new spine formation, for example, in the apical dendrites of layer 5 neurons in the motor cortex (reviewed in Rochefort and Konnerth, 2012). The fact that dendritic spines are so versatile and profoundly plastic is a strong argument for them being a key structure of memory encoding. However, for decades the optical tools applied to study spine morphology/plasticity were limited and largely not adapted to the investigation of such small structures as spines, objects of hundreds of nanometers (Figure 8.1A).

The emergence of super-resolution microscopy (SRM) in neuroscience is rapidly transforming our view of the synapse and especially of its protein organization. Ernst Abbe initially described the limit of resolution using light at the end of the 19th century (Abbe, 1873). Because light is a wave, a single emitter imaged through an objective within a microscope appears as a blurry spot known as the point-spread function (PSF). The spot size depends on the wavelength of the emitted light and the numerical aperture of the objective. The best resolution achievable on a microscope is the minimal distance necessary to resolve identical single objects.

During the last two decades, several groups have put in place different main strategies to "break" the diffraction-limit. This ensemble of techniques – coined superresolution microscopy – was awarded the Nobel prize in 2014 (Choquet, 2014). The first strategies are based on spatial modulation of fluorescence using patterned illumination. The related techniques include stimulated emission depletion (STED) and saturated structured illumination microscopy (SSIM). The second strategies take advantage of single-emitter imaging by separating in time the emission of single fluorophores within a sample. The related techniques include stochastic optical reconstruction microscopy (STORM) and

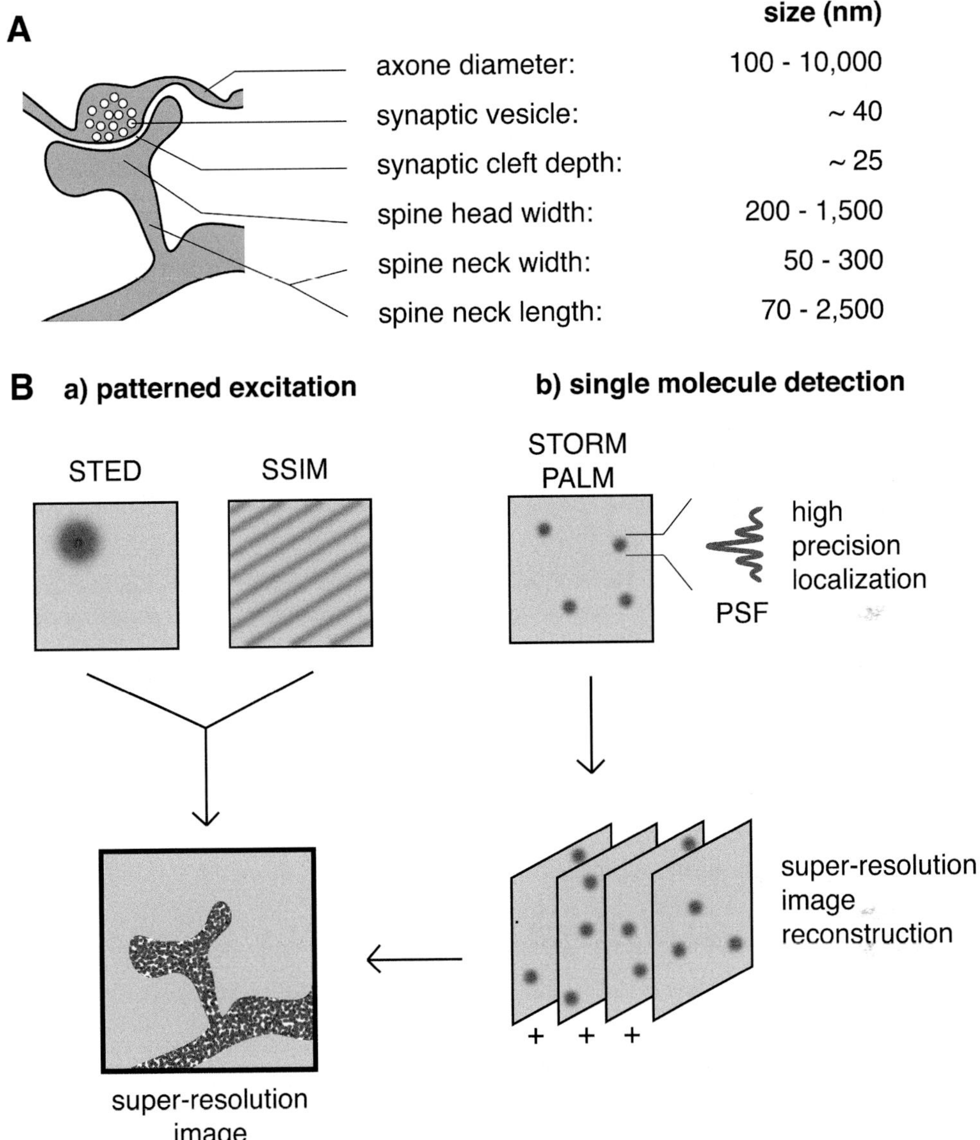

FIGURE 8.1 The synapse, a nanoscale object to study with superresolution methods. (A) Schematic representation of a dendritic spines with most of its characteristic (Bartol et al., 2015; Tønnesen et al., 2014). (B) In order to obtain a superresolved image of a sample there are two main strategies: a) use patterned illumination, or b) separate in time the emission of single fluorophores and reconstruct an image by compiling detections.

photo-activated localization microscopy (PALM). An additional, somewhat older, strategy to measure nanoscale distances between molecules which is not strictly recognized as a superresolution imaging method, uses the property of a fluorophore to transfer energy to a neighboring fluorophore within the nanometer range. This method is known as Foerster resonance energy transfer (FRET) (Figure 8.1B) (reviewed in Huang et al., 2009).

As the nanoscale organization of synaptic proteins is revealed by those optical methods, it implements our basic understanding of synaptic transmission and plasticity, two fundamental aspects of information processing and storage in the brain. In this chapter, we discuss the implications of SRM applied to neuroscience and the recent findings they generated, the new models for the regulation of synaptic transmission, and directions for future studies.

8.2 THE SYNAPSE: A COMPLEX SUB-DIFFRACTION LIMITED OBJECT

In the adult central nervous system, dendritic spines receive most glutamatergic excitatory inputs. Presynaptic axon terminals or boutons containing synaptic vesicles of ~40 nm in diameter are opposed and linked to those postsynaptic protrusions to form mature synapses. Dendritic spines are small subcellular organizations connected to the neuron dendritic shaft by a thin spine neck (length ~1 μm, diameter ~100 nm) that acts as a diffusion barrier and leads to a spine head of typically 200 nm to 1 μm in diameter. The spine head contains the postsynaptic density (PSD), a GigaDalton macromolecular structure of typically less than 500 nm^2 apposed intracellularly to the postsynaptic membrane and formed of a complex assembly of scaffold proteins responsible for anchoring numerous key players of synaptic transmission such as glutamate receptors, kinases, and phosphatases. The PSD also contains an assembly of adhesion proteins responsible for the adhesion between the post- and pre-synapse. The spine cytoplasm contains an impressive number of proteins both quantitatively and qualitatively. The dimensions of both pre- and post-synapse elements are in the range – or below – the diffraction limit (in the order of half the wavelength of the used light, i.e. 200–400 nm). Thus, for decades, the techniques of choice to study the morphological organization of synapses were electron microscopy or biochemical isolation. Light microscopy techniques harbor many advantages. For instance, compared to electron microscopy, it allows imaging of a large volume, a whole neuron, and the study of structural modifications over time with live-cell imaging. However, only the advent of superresolution microscopy techniques has allowed optical microscopy to deeply investigate synapse structure in live cells.

For the last century, spine morphology has been studied to investigate a possible correlation with its function. STED microscopy is the SRM technique of choice for this application. STED microscopy aims to improve the spatial resolution by reducing the diameter of the PSF beyond the diffraction limit (Hell and Wichmann, 1994). This is achieved by using two laser beams. The first one is a classic diffraction limited laser beam as in a regular confocal microscope. A second laser beam (known as the STED beam) is simultaneously applied to the sample to de-excite fluorescent molecules by stimulated emission at a wavelength that is longer than the fluorescence. By shaping the STED beam like a doughnut in the focal plane, it actively switches off the fluorescence on a circular rim around the PSF and thus only permits fluorescence to be emitted from the center of the doughnut (known as the null). So far, the fluorescence spot size has been reduced down to D10–20 nm, which is more than an order of magnitude smaller than what diffraction-limited light microscopy can achieve. In 2008, Valentin Nagerl and colleagues used STED microscopy to image dendritic spines in living tissue (Nagerl et al., 2008). For the first time, fine details such as the shape of the spine head or the width of the spine neck could be faithfully resolved and measured from YFP-positive CA1 pyramidal neurons in living organotypic hippocampal slices. STED microscopy permits the imaging of activity-driven structural changes of spines with unprecedented precision, revealing a diversity of shapes and dynamics that cannot be detected by established regular optical microscopy techniques. In 2014, two independent groups used STED combined with fluorescent recovery after photobleaching (FRAP) to investigate the correlation between spine dimensions (i.e. head volume, neck length, neck width, etc.) and diffusion properties in this subcellular compartment. Both research groups identified neck width as the most influential determinant of compartmentalization (Takasaki and Sabatini, 2014; Tønnesen et al., 2014). Thus, modulating spine neck width during plasticity might be the most efficient way to recruit and/or eliminate specific synaptic components.

Tissue preparation and image analysis are quite similar for STED microscopy and regular confocal microscopy. Thus, several groups have used this method to study distributions of immune-labeled presynaptic proteins in fixed-tissue preparations (Kittel et al., 2006; Willig et al., 2006). It is thought that the spacing between Ca^{2+} channels and vesicles at active zones is influencing the dynamic properties of synaptic transmission. For instance, super-resolution imaging of the protein Bruchpilot (BRP) at the neuromuscular junction in larval *Drosophila* revealed an unexpected donut shape co-localizing with presynaptic active zone markers (Kittel et al., 2006). Furthermore, in the mutant brp-/- *Drosophila*, the number of vesicles released per action potential is dramatically reduced, indicating a disrupted coupling between synaptic vesicles and active zones. Hence, BRP and its mammalian homologue CAST seem to ensure the tight proximity of synaptic vesicles and calcium channels essential to allow calcium influx-triggered neurotransmitter release. Calcium influx in the presynaptic terminal induces the SNARE-mediated fusion of docked synaptic vesicles with presynaptic membranes. After fusion, the mechanism underlying the recycling of vesicle proteins remains unknown. Also, STED imaging of synaptotagmin, a vesicle resident V-SNARE, showed that it remains concentrated in small clusters after exocytosis instead of being dispersed across the plasma membrane (Willig et al., 2006). This finding strongly suggests that synaptotagmin as well as other vesicle proteins might be rapidly endocytosed as clusters after fusion. Together those studies provide important insights into the molecular machinery involved in neurotransmitter release at the synapse.

Understanding synaptic function requires the determination of synaptic molecular components and their relative organization. As synapses are often considered as individual organelles, it is essential to their function that they sequester many of their key components. This is achieved in large part by scaffolding proteins. In fact, pre- and postsynapses express various types of scaffolding proteins providing binding sites for cytoplasmic and transmembrane proteins. On the presynaptic side, the main scaffolding proteins Bassoon, Munc13, and Piccolo have a pyramidal organization. They have been hypothesized to indirectly participate to drive synaptic vesicles to the presynaptic active zone. On the postsynaptic side, the anchoring of ionotropic glutamatergic receptors (i.e. AMPARs, NMDARs, Kainate receptors) in the postsynaptic density is essential to synaptic transmission. For instance, binding of the AMPAR auxiliary subunit stargazin to the main scaffolding protein of the postsynaptic density PSD-95 leads to AMPAR synaptic anchoring (Chen et al., 2000; Hafner et al., 2015; Opazo et al., 2010). Thus, on both sides of the synapse, scaffolding proteins play major roles that remain to be fully unraveled. With the aim of correlating synaptic structures with function, the laboratories of Xiaowei Zhuang and Catherine Dulac have developed a method and determined the relative position of ten key protein components of the glutamatergic synapse in the brain (Dani et al., 2010). By determining the position of each molecule with nanometer precision, STORM allows a high-resolution structural reconstruction of molecular assemblies. The authors systematically imaged a large number of synapses in various brain regions using Multicolor 3D STORM in order to map out the protein organization. Multicolor 3D STORM acquires its high resolution based on single-molecule imaging of photo-switchable fluorescent probes. As described above, the resolution of an optical system is based on its ability to differentiate the emission of two neighboring molecules. Hence, for molecules that give overlapping emission through a diffraction-limited system, STORM resolves these molecules by stochastically activating them at different times during image acquisition (Betzig et al., 2006; Hess et al., 2006; Rust et al., 2006). At any time, only a sparse optically resolvable subset of molecules is activated, allowing the images of these fluorophores to be readily separated from each other. As a result, the position of each individual fluorophore can be determined to a precision substantially beyond the diffraction-limited resolution. This procedure is repeated as many times as necessary to obtain the localizations of most fluorophores in the sample. A super-resolved image is subsequently reconstructed from these positions. For 3D STORM, a cylindrical lens is inserted in the detection path of the microscope to render the image of a single fluorophore elliptical. The lateral (x, y) and axial (z) coordinates of each fluorophore are determined from the centroid position and ellipticity of the image, respectively (Huang et al., 2008) (Figure 8.2).

Using this approach, one could determine the protein architecture of pre- and postsynaptic proteins with an unprecedented precision (Dani et al., 2010). The authors could also determine the orientation of synaptic proteins by taking advantage of epitopes located in different domains of the corresponding protein. They could show that the presynaptic scaffolding Bassoon and Piccolo are perpendicularly oriented with respect to the synaptic cleft, their N-terminal domain toward the postsynapse, whereas the postsynaptic scaffolding protein Homer1 seems to take a parallel orientation with respect to the postsynaptic density. Unraveling the relative positioning and orientation of proteins in the confined environment of the synapse gives rise to new models for protein interplays and regulation of synaptic function.

For decades, the postsynaptic density of glutamatergic synapses was viewed as a compact mesh of randomly organized proteins. However, this was challenged by electron microscopy tomography imaging of the main organizer of the postsynaptic density – the scaffolding protein PSD-95 – that appeared as highly organized vertical filaments, their N-terminal domain being inserted in the membrane through a double palmitoylation and their C-terminal into the spine cytoplasm (Chen et al., 2008, 2011). From there, other groups wondered if other postsynaptic components might exhibit nanoscale inter-molecular organizations. Fukata and colleagues observed, using 2-color STED, a subdomain organization

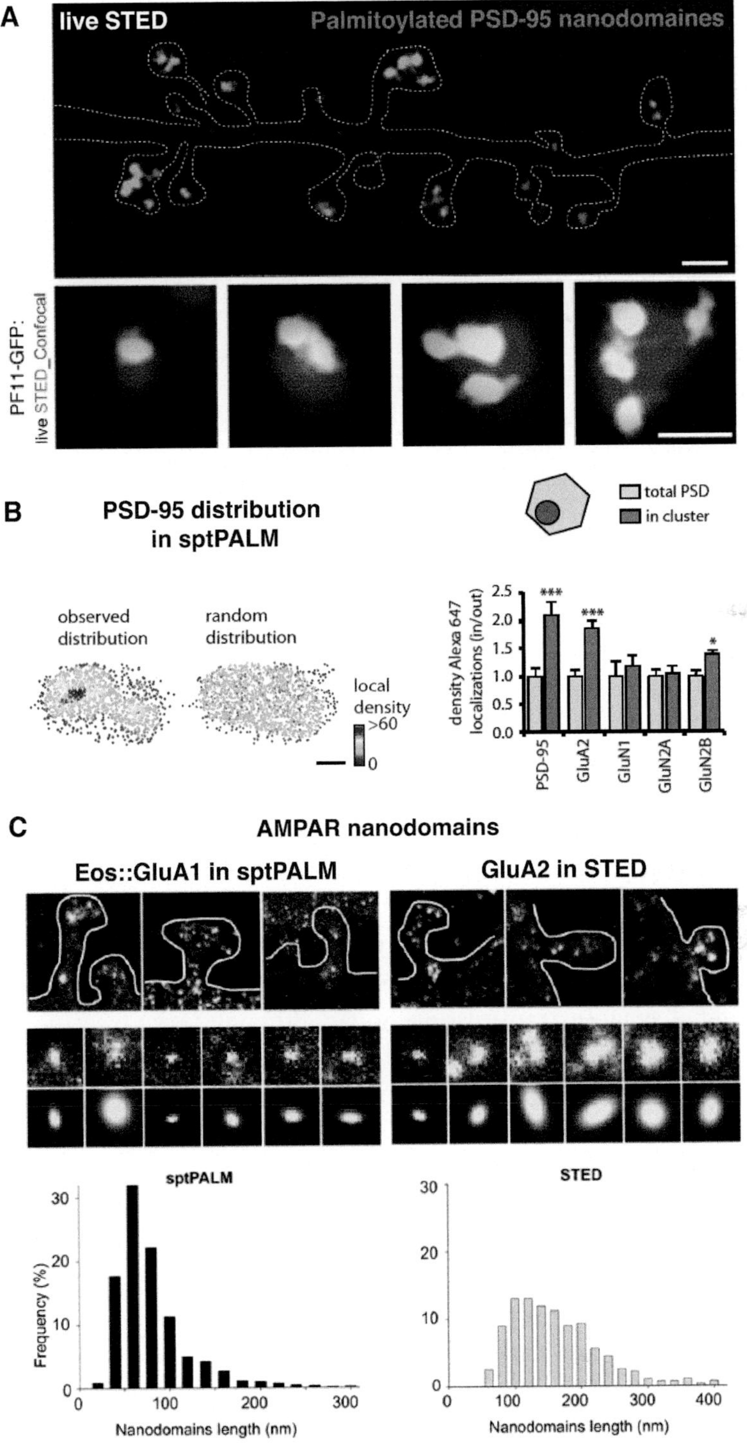

FIGURE 8.2 Breaking the diffraction limit to reveal nanoscale organization. Summarized results from three studies published in 2013 and revealing the organization in nanodomains of PSD-95 and AMPARs at synapses. (A) Live-cell imaging of palmitoylated PSD-95 using PF11-GFP by STED microscopy (green), but not by conventional confocal microscopy (red pseudocolor), efficiently

FIGURE 8.2 (CONTINUED)
detects multiple subspine clusters (1 to 4 clusters/spine) in neurons. Scale bars: (top) 1 μΓῗ; (bottom) 500 nm (adapted from Fukata et al., 2013). (B) (left) Single-molecule localization of PSD-95::Eos2. Individual molecules were color-coded according to their local density and a homogenous distribution was generated by randomly sampling equal numbers of localizations. Scale bar represents 100 nm. (right) Relative enrichment of PSD-95, GluA2, GluN1, GluN2A, and GluN2B localizations revealed with Alexa 647 inside and outside of subsynaptic PSD-95 clusters (adapted from MacGillavry et al., 2013). (C) Quantification of AMPAR nanodomain size and density by multiple superresolution light imaging techniques. Gallery of super-resolution images of spines imaged by sptPALM (Eos::GluA1) and STED (Surf-GluA2-ATTO647N) (top). Examples of AMPAR nanodomains imaged by sptPALM and STED and the corresponding fit using 2D anisotropic Gaussian functions (middle). Distributions of the principal (length) axis of nanodomains obtained by 2D anisotropic Gaussian fitting. sptPALM and STED display similar distributions centered at 80, 120, and 70 nm, respectively (adapted from Nair et al., 2013).

of PSD-95 between palmitoylated and non-palmitoylated forms (Fukata et al., 2013). These subdomains are maintained and regulated via the activity of local DHHC-type palmitoyl acyltransferases. The same year, Daniel Choquet's group used electron microscopy as well as several SRM techniques, including STED and STORM microscopy, to discover that AMPAR are organized in nanodomains of around 80 nm diameter, instead of being diffusively distributed in the PSD (Nair et al., 2013). Similar findings were reported in parallel in other labs (Fukata et al., 2013; Macgillavry et al., 2013). Since AMPARs have a low affinity for glutamate, their relative position with respect to presynaptic neurotransmitter release sites will influence postsynaptic responses to presynaptic signals. Recently, the Blanpied group discovered (Tang et al., 2016) that the pre-synaptic release events and machinery are aligned to postsynaptic AMPA receptors and scaffold proteins, introducing the concept of a nanocolumn for synaptic organization. Thus, SRM imaging of the postsynapse is revealing a novel layer of unexpected complexity to the regulation of postsynaptic processing of synaptic transmission. So far, although very powerful, the STED and STORM methods give no, or very limited, access to fast protein dynamics. Most proteins have a half-life in the order of several hours to several days. Thus, maintaining synaptic protein homeostasis is a challenge that each synapse must face by replacing constantly and dynamically its components.

8.3 MOBILITY OF SYNAPTIC COMPONENTS

In the last 40 years, our understanding of protein dynamics improved tremendously with the development of strategies allowing real-time imaging of protein diffusion. Initially, a method was developed to study protein and lipid dynamics at the cell surface, fluorescence recovery after photobleaching (FRAP) (Axelrod et al., 1976). For this method, protein or lipids of interest have to be specifically labeled with fluorophores. The core idea of FRAP is to photobleach a specific area of a living cell and analyze fluorescence recovery in the bleached region. Since the emergence of GFP-like variants fluorescent proteins, FRAP and related methods such as fluorescent loss in photobleaching (FLIP) or photoactivation, have been extensively used to study the population of proteins' diffusion properties, stability, and turnover. For a uniform population of molecules, measurements of average parameters versus an ensemble of unique elements should yield identical values. However, for most biological systems, populations of molecules do not display a homogenous behavior. For instance, synaptic PSD-95 molecules are highly stable whereas cytoplasmic PSD-95 are highly mobile (Sturgill et al., 2009).

By the end of 1980s, it became possible to track and analyze individual particles with a nanometer-level precision, through a video-microscopy method known as single particle tracking (SPT) (Geerts et al., 1987; Qian et al., 1991). During imaging, even though each particle appears as a diffraction limited spot, the lateral (x, y) coordinates of each particle can be determined by the center of the centroid, as in STORM superresolution image reconstruction. The trajectories of individual particles are reconstructed to extract the characteristics of each particle motion, the sequence of position frame by frame.

Initially, protein labeling was performed using labeled antibodies recognizing specifically extracellular epitopes on transmembrane proteins of interest. This form of SPT has been extensively used to characterize glutamate receptor dynamics at the nanoscale (Borgdorff and Choquet, 2002; Tardin et al., 2003; Groc et al., 2004; Bats et al., 2007; Heine et al., 2008). Together, those studies revealed that AMPARs and NMDARs: 1) diffuse at the neuron surface; 2) display three types of diffusion behaviors: free-diffusion, confined diffusion, or immobile; 3) alternate rapidly between these different behaviors at the single receptor level. The main limitation remained that only a small subset of the protein population could be labeled for a given experiment. The lack of population coverage was circumvolved in 2010 by the development of universal point-accumulation-for-imaging-in-nanoscale-tomography (uPAINT) (Giannone et al., 2010). The approach consists in recording short single-molecule trajectories that appear sequentially upon continuously labeling molecules of interest during the experiment. Despite this improvement, antibody-based SPT in living cells remains applicable exclusively to surface proteins or lipids.

To access the motion of individual intracellular proteins, the laboratories of Eric Betzig and Jennifer Lippincott-Schwartz developed the sptPALM technique combining SPT and photoactivated localization microscopy (PALM) (Betzig et al., 2006; Manley et al., 2008). PALM uses photoactivatable dyes for super-resolved localization of molecules in cells. By imaging fluorescent dyes in a sample in a non-time-correlated manner, one can reconstruct a super-resolved image, as in STORM microscopy. Thus, PALM imaging of fluorescent-protein chimeras reveals the organization of genetically expressed proteins with nanoscale precision, density of molecules being high enough to provide structural context. The protein of interest is genetically fused to EosFP, a photoconvertible fluorescent protein with a high photon count and a high contrast ratio between converted and unconverted forms. The addition of SPT to regular PALM allowed localization and tracking of many overlapping trajectories because the distance between fluorescent molecules at any time is greater than several times the width of their PSF. Notably, sptPALM probes the dynamics of a large number of molecules in a single cell. Using this method, Chazeau and colleagues in 2014 could investigate in details the dynamics of actin structures in spines that were originally described over three decades ago. In the 1980s, the Thomas Reese group described filamentous structures in both pre- and postsynapses using electron microscopy imaging on cerebellar cortex slices after rapid freezing and the freeze-etch technique of tissue preparation (Landis, 1983; Landis et al., 1988). Further studies confirmed that those filaments are actin-filaments and F-actin appears to be the driving force of spine morphological remodeling. In order to specifically determine the dynamics of postsynaptic branched F-actin networks, sptPALM was used to compare F- actin and Arp2/3 movements in dendritic spines of mature primary rat hippocampal cultured neurons (Chazeau et al., 2014). They fused actin and the ArpC5A subunit of the Arp2/3 complex fused to the photo-switchable mEOS2 fluorescent protein. They found the actin modulator, Arp2/3 and WAVE complex, to be regulated during spine remodeling and the PSD to play a critical organizing role in recruiting those branched F-actin regulators. In contrast, the growing F-actin (+) end is localized at the tips of membrane protrusions. It provides important insights on the molecular mechanisms behind spine remodeling typically associated with synaptic plasticity or spine morphology-related diseases such as the fragile X-syndrome.

SRM applied to molecular dynamics are powerful tools to dissect protein function and regulation through the spectrum of their diffusion properties. In cells, protein mobility is influenced by various factors including viscosity, volume geometry, and interactions with other molecules. In fact, most proteins have to interact with others in homo- or heteromeric macromolecular complexes to accomplish their function.

8.4 STUDYING THE SYNAPTIC INTERACTOME IN LIVE NEURONS

8.4.1 Monitoring Protein Interactions in Synapse (Figure 8.3)

The extensive unraveling of the molecular proteome of synapses has started about 15 years ago with the explosion in proteomics approaches (Bayes et al., 2011; Collins et al., 2006; Li et al., 2004; Sheng and Kim, 2011; Takamori et al., 2006). Many studies have reported the specific proteome of pre- or postsynaptic domains of central synapses that each contain over a thousand protein components. The postsynaptic proteome is composed of about 15% membrane proteins (i.e. ion channels, receptors, and adhesion proteins), the rest being a diverse range of signaling, structural scaffolding, and metabolic proteins. These various components are not randomly distributed in the synapse. On the contrary, they seem to be precisely organized in various subdomains and layers thanks to a complex set of multiplexed protein–protein and protein–lipid interactions that in large part dictate their function. This concept has led to the notion of a "synaptic molecular interactome". The general principles underlying the physical and functional organization of the various synaptic elements are now beginning to emerge and have been reviewed extensively (Grant, 2013; Li et al., 2010; Sheng and Kim, 2011). Some important features of these principles are that: 1) proteins are organized in a hierarchy of structures – individual proteins, simple complexes, and supercomplexes; 2) proteins are organized in domains and layers – membranous, juxtamembranous, and cytosolic; 3) this organization is highly dynamic and regulated by neuronal activity, with a very broad spectrum of exchange and turnover rates – from seconds to hours.

Importantly, it is becoming apparent that the nanoscale organization of synaptic components is a determinant of their functions. A good example is that of AMPA type glutamate receptors, whose relatively low affinity for glutamate imposes that they are localized in front of release sites in the presynaptic terminal to be fully activated (MacGillavry et al., 2011; Macgillavry et al., 2013; Nair et al., 2013; Tang et al., 2016). Another example is the recent demonstration that the synaptic cleft is organized on a nanoscale into sub-compartments marked by distinct trans-synaptic complexes (Perez de Arce et al., 2015). SynCAM 1 delineates the postsynaptic perimeter as determined by immunoelectron microscopy and superresolution imaging. One might expect this peripheral organization of SynCAM1 to have important functions, such as acting as a diffusion barrier between PSD and extra-PSD domains, either directly acting as obstacles to diffusion of through the zippering of the pre- and post-synaptic membranes that leads to a thinning of the cleft. Indeed, as several transmembrane proteins of the PSD have very tall extracellular domains (e.g. the AMPAR extracellular domain protrudes by nearly 15 nm), one might expect that a reduction of the cleft height acts as a diffusion barrier. At the spine level, a marked subcellular organization is also observed. Beyond the very peculiar shape of spines that is certainly dictated by yet to be identified specific molecular arrangements, specific functional subdomains have been identified. Of particular note is the endocytic zone (EZ) that lies a couple hundred nanometers lateral to the PSD and where receptor endocytosis is thought to occur (Blanpied et al., 2002; Lu et al., 2007; Petrini et al., 2009; Racz et al., 2004). It should, however, be mentioned that EZ are also abundantly present in the dendritic shaft and the respective contribution of shaft versus spine endocytosis has not been determined yet.

At developing but also at mature synapses, there is an intense activity dependent plasticity of synapse organization that underlies their functional plasticity. For example, processes of activity dependent long-term synaptic potentiation and depression have been shown to be largely mediated by rapid changes in AMPAR receptor number and composition at synapses (Anggono and Huganir, 2012; Choquet, 2010; Herring and Nicoll, 2016; Opazo et al., 2012). These processes of synaptic plasticity are actually accompanied by major changes in synaptic organization, starting by an increase in their volume and synaptic content, and the complex cascade of events that is associated with these morpho-functional changes still needs to be deciphered.

Notably, various diseases are associated with subtle to major changes in synapse organization, with some ultimately leading to synapse disappearance. Synaptic diseases – coined as synaptopathies – are now recognized as a major cause of brain disease. Mutations in over 200 genes result in disruption to the postsynaptic proteome causing over 130 brain diseases (Bayes et al., 2011). The varied manifestations of impaired perceptual processing, executive function, social interaction, communication, and/or intellectual ability in intellectual disability, autism spectrum disorder, and schizophrenia appear to emerge from altered neural microstructure, function, and/or wiring rather than gross changes in neuron number or morphology (Volk et al., 2015).

This dynamic nanoscale organization of synaptic elements, and physiological or abnormal regulation, relies on a complex multivalent set of interactions that just begins to be unraveled (i.e. the synaptic interactome). Understanding synapse function, and the rules that govern their activity-dependent morpho-functional plasticity in health and disease, will obviously require a full and extensive description and understanding of not only

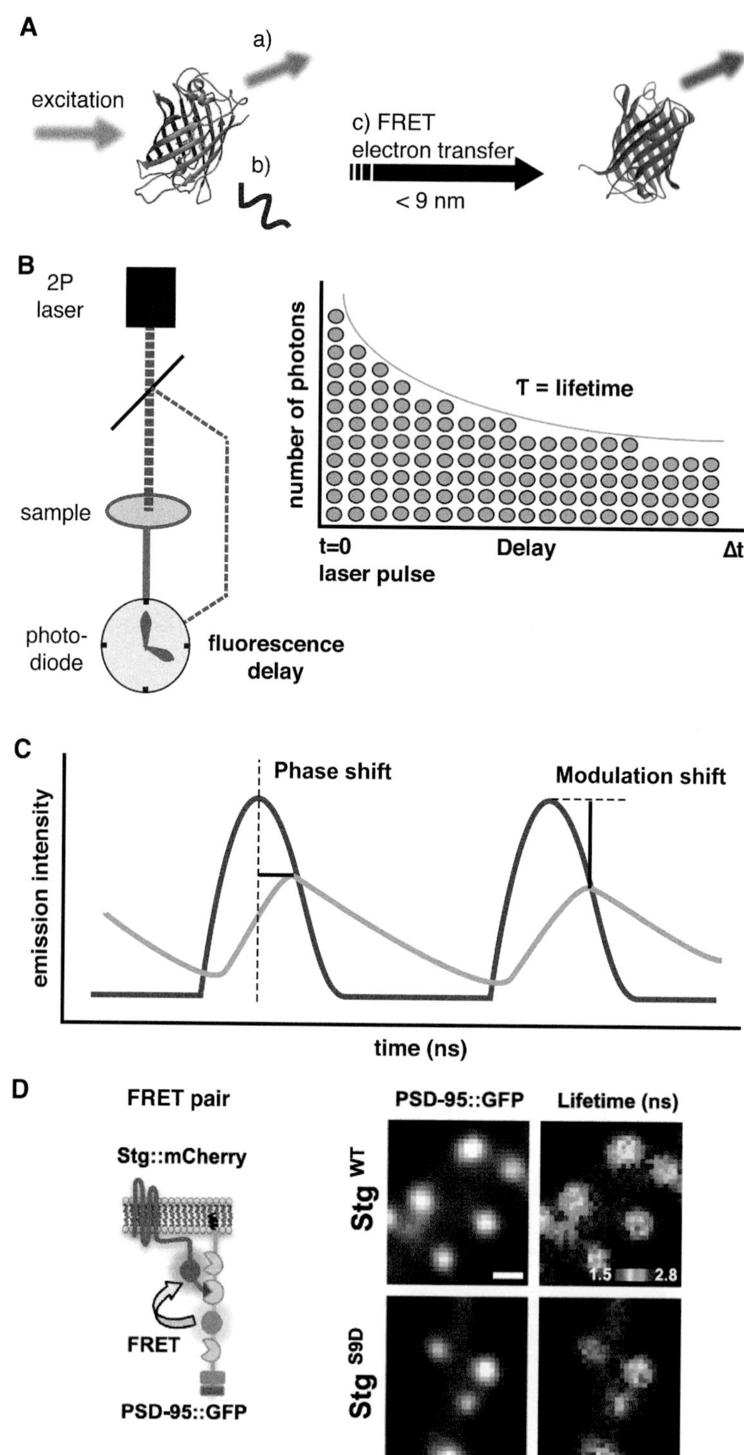

FIGURE 8.3 FRET-FLIM to study synaptic protein interactions. (A) Schematic representation of fluorophore de-excitation: a) non-radiative characterized by the emission of heat, b) radiative characterized by the emission of photons, c) FRET with an acceptor fluorophore. (B) Principle of time

FIGURE 8.3 (CONTINUED)
domain-based FLIM. The sample excited by a short laser pulse emits fluorescence with a certain lifetime characteristic for each fluorophore. In the presence of an acceptor fluorophore, the donor fluorophore will stay less time in the excited state therefore exhibiting a shorter lifetime. (C) Principle of frequency domain FLIM. The sample is excited at different intensities. The phase shift between the excitation (blue line) and the emission response (red lines) of the sample is correlated to the lifetime of the sample. Adapted from www.lambertinstruments.com. (D) Dendritic spines from cultured hippocampal neurons expressing PSD-95::GFP and STG::mCherry. Frequency domain FLIM was used to measure the efficiency of the binding between PSD-95 and STG in synapses. In neurons expressing STG carrying phosphomutations (STG S9D::mCherry) PSD-95::GFP lifetime is reduced revealing a stronger binding of the two proteins (adapted from Hafner et al., 2015).

synapse constituents but also of their interactions. Unfortunately, the tools at hand are scarce and endpoint measures extraordinarily limited, particularly for live cell measurements which are indispensable to study plasticity mechanisms. These endpoint measures are basically limited to protein localization and basic rules of interactions. As described above, a major leap forward has been recently made in our capacity for multiplex protein localization at high resolution thanks to the development of superresolution imaging compatible with live cell recordings. Largely limited to around 30 nm resolution in live cells, the SRM approaches are not yet ready to measure protein–protein interactions. Hence, measurement of protein interactions has been largely limited to either indirect inferences from modifications of their localizations upon given molecular or pharmacological manipulations, or based largely on bulk biochemical approaches which although powerful, can in no way capture the exquisite subtlety of regulations that occur at single spines. We are thus in desperate need of methods that would allow multiplex measurement of protein–protein interactions in live cells. While the ideal approach allowing extensive multiplexing is yet to be developed, recent years have seen an extensive use of a rather old approach, Fluorescence Resonance Energy Transfer (FRET), that is ideally suited to measure protein–protein, or protein–lipid interactions with sub-nanoscale resolution in live cells (Forster, 1993; Lakowicz, 2006; Yasuda, 2006). We will review some examples here where FRET has been used to build new biosensor reporters of various biochemical activities, or to measure functional regulations of protein–protein interactions in live neurons.

FRET is the process through which a donor fluorophore transfers its energy to an acceptor fluorophore. Fluorescence is a radiative process during which a molecule excited by light absorption undergoes de-excitation by light emission. This phenomenon is transient and its kinetic is dependent on the intrinsic properties of the fluorophore and the environment. FRET is a non-radiative process which requires a donor fluorophore in the excited state and an acceptor fluorophore. FRET only occurs when: (1) donor emission spectrum and acceptor absorption spectrum overlap; (2) the two fluorophores are in close proximity – 1–9 nm away; (3) the fluorophores are in a correct relative orientation. The intensity of FRET varies like the 6th power of the distance between two fluorophores and is thus exquisitely sensitive and perfectly adapted at the measure of intermolecular interactions. The revival of this technique largely developed in the 1970s arises from the now common use of fluorescence proteins to label and localize proteins in live cells. By labeling proteins of interest or the membrane with appropriate fluorophores, it is possible to study protein interactions in living cells, or relative position toward the plasma membrane with very good spatial (diffraction limited) and temporal (down to ~100 Hz) resolution.

Although FRET can be measured by multiple approaches (Yasuda, 2012), the method of choice is to monitor FRET through Fluorescence Lifetime Imaging Microscopy (FLIM). This non-invasive method allows the study of interactions independently of fluorophore concentration and photobleaching. The fluorescence lifetime that is typically around few nanoseconds is the average time gap between the excitation of the fluorophore and the emission of fluorescence. In the context of FRET, the presence of an acceptor fluorophore offers an extra pathway for de-excitation of the donor fluorophore, thus decreasing its fluorescence lifetime. There are two methods to measure fluorescence lifetime: (1) time domain – more sensitive and slower, and (2) frequency domain – less precise, but faster – based FRET-FLIM measurements. The time domain technique is also known as time-correlated single-photon counting (TCSPC). Using a picosecond pulsed laser, the time delay between the laser pulse and the arrival of the first detected photon emitted by the sample is measured (Figure 8.3B). By repeating this process for several seconds every tens of ns, a de-excitation curve can be reconstructed to access the fluorescence lifetime by estimating the curve decay.

This technique gives a very accurate estimation of fluorescence lifetime. However, since only one photon is detected during each round of measurements the reconstruction of a large area can take up to several minutes.

The frequency domain method varies rapidly the excitation light intensity and measures the corresponding phase shift in fluorescence emission using a CCD camera, allowing estimation of the donor fluorophore fluorescence lifetime per pixel (Figure 8.3C). This method allows a rapid "interaction" mapping of large areas of interest in space and time at around 1 Hz in a diffraction limited manner.

One of the first FRET-based sensors has been developed to measure intracellular calcium (Miyawaki et al., 1997). Since then, a plethora of sensors with improved sensitivity for calcium, but also able to measure various kinase activity have been produced. They either take advantage of changes in the kinase conformation upon activation to modify the distance between genetically fused donor and acceptor fluorescent proteins or consist of four linked components: a donor, a specific kinase target polypeptide, a phosphoamino-acid binding domain, and an acceptor (Ni et al., 2006). Phosphorylation of the kinase target polypeptide causes binding between the kinase target polypeptide and the phospho-amino-acid-binding domain, thereby altering the FRET efficiency between the donor and the acceptor. More recently, sensors for Ras, rho, rac, erk, and CaMKII activity optimized for FLIM have been developed and were shown to be useful for measuring the activity of these proteins in single dendritic spines in hippocampal slices (Harvey et al., 2008; Lee et al., 2009; Murakoshi et al., 2011; Nishiyama and Yasuda, 2015; Yasuda et al., 2006; Zhai et al., 2013). Interestingly, time-resolved imaging has been recently used to reveal heterogeneous landscapes of nanomolar Ca^{2+} in neurons and astroglia (Zheng et al., 2015).

FRET has also proven to be extremely useful to measure activity dependent changes in interaction between synaptic proteins or between proteins and the membrane.

We developed a FRET pair that reports the interaction between the c-terminus of the AMPAR auxiliary subunit stargazin and the PDZ domains of PSD-95 (Sainlos et al., 2011). Because the C-terminus of stargazin is the ligand that binds into the PDZ domain groove, it cannot be modified. We showed that an acceptor fluorescent protein inserted 22 amino-acids upstream of the last amino acid of stargazin efficiently FRET with a donor protein inserted next to the PDZ domains of PSD-95 and faithfully reported the interaction. We first used this assay to develop biomimetic divalent ligands for the acute disruption of synaptic AMPAR stabilization (Sainlos et al., 2011). We further extended the use of this FRET pair in the search for the mechanism of activity dependent AMPAR stabilization at the PSD. Previous work had established that the phosphorylation of a stretch of nine serines surrounded by arginins (the RS domain) way upstream of the stargazin c-terminus was instrumental in the bidirectional control of synaptic plasticity (Tomita et al., 2005), likely by regulating calcium induced stargazin anchoring to PSD95 and thus immobilizing AMPAR (Opazo et al., 2010). It was, however, quite mysterious how modifications in this stretch of serines could regulate downstream binding of stargazin C-terminus to PSD-95. Two points hinted us to the mechanism.

First, for a long time, PSD-95 was depicted in schematic diagrams as an elongated molecule parallel to the plasma membrane. This vision of the molecular organization of the PSD changed in 2008 after a publication by Thomas Reese's group who used EM tomography to reveal the perpendicular orientation of PSD-95 with respect to the plasma membrane within the PSD (Chen et al., 2008; Chen et al., 2011). At the synapse, PSD-95 seems to adopt an extended or "open" conformation with its N-terminus attached to the plasma membrane via palmitoylation and its C-terminus projecting into the cytoplasm. In this conformation, the three PDZ domains of PSD-95 are most likely to be located at various distances from the plasma membrane, PDZ1 being close to the membrane, PDZ2 in the middle, and PDZ3 deeper in the core of the PSD. This perpendicular orientation of PSD-95 prompted us to re-think the whole architecture of protein-protein interactions in the PSD.

Second, Tomita's group showed that the phosphorylation state of stargazin regulates the interaction of its CTD with negatively charged phospholipids (Sumioka et al., 2010). This interaction was confirmed by direct visualization using electron crystallography (Roberts et al., 2011). In absence of phosphorylation, the RS-domain is positively charged because of a stretch of seven arginine residues. Sumioka et al. showed that by modulating the plasma membrane phospholipid composition, they could actually change the amplitude of AMPAR synaptic current (Sumioka et al., 2010), preventing stargazin interaction with the plasma membrane enabling additional AMPAR synaptic trapping.

We thus postulated that PSD-95 orientation might dictate the dependence upon phosphorylation of the binding of stargazin to PSD-95 (Figure 8.3D). Using FRET imaging to monitor stargazin interaction with the plasma membrane and synaptic interaction with PSD-95, we found that modifying the effective length of the stargazin cytoplasmic tail regulates binding to PSD-95 by allowing access to deeper domains of PSD-95 (Hafner et al., 2015). Single particle tracking and electrophysiological approaches furthermore revealed that the length of the C-tail is the central determinant in the ability of stargazin to recruit AMPAR and control synaptic transmission. A similar approach was used to investigate PSD-95 interaction with NMDA receptors in dendritic spines showing that it is controlled by calpain, CaMKII, and Src family kinase (Dore et al., 2014). Other examples of FRET based studies of protein–protein interactions at the synapse include the study of the assembly, priming, and rearrangements of SNAREs (Degtyar et al., 2013; Takahashi et al., 2015). An interesting variation to FRET is the technology of BRET (Bioluminescence Energy Transfer) that has been elegantly applied to image the dynamic remodeling of scaffold interactions in dendritic spines controlled synaptic excitability (Goyet et al., 2016; Moutin et al., 2012).

8.5 CONCLUSION

The recent years have seen a major increase in our understanding of synapse nanoscale organization and related function thanks to the development and application of a variety of superresolution imaging approaches associated with functional readouts through electrophysiology or smart biosensors. However, most studies have so far been limited to measuring the properties of one or two synaptic component at a time. The complexity of interacting networks in the synapse will require a much higher level of multiplexing if we wish to understand how the synapse interactome relates to physiology. A related challenge that lies in front of us, and might be within reach, is the capacity to simultaneously localize with nanometer accuracy a series of proteins within a complex, ideally in live cells. This will provide invaluable information on the dynamic of multimolecular complexes that play such delicate roles in synapse function.

BIBLIOGRAPHY

Abbe, E. (1873). Contributions to the theory of the microscope and microscopic detection (Translated from German). *Arch Mikroskop Anat* 9, 413–468.

Anggono, V., and Huganir, R.L. (2012). Regulation of AMPA receptor trafficking and synaptic plasticity. *Curr Opin Neurobiol* 22, 461–469.

Bayes, A., van de Lagemaat, L.N., Collins, M.O., Croning, M.D., Whittle, I.R., Choudhary, J.S., and Grant, S.G. (2011). Characterization of the proteome, diseases and evolution of the human postsynaptic density. *Nat Neurosci* 14, 19–21.

Blanpied, T.A., Scott, D.B., and Ehlers, M.D. (2002). Dynamics and regulation of clathrin coats at specialized endocytic zones of dendrites and spines. *Neuron* 36, 435–449.

Choquet, D. (2010). Fast AMPAR trafficking for a high-frequency synaptic transmission. *Eur J Neurosci* 32, 250–260.

Choquet, D. (2014). The 2014 Nobel Prize in chemistry: a large-scale prize for achievements on the nanoscale. *Neuron* 84, 1116–1119.

Collins, M.O., Husi, H., Yu, L., Brandon, J.M., Anderson, C.N., Blackstock, W.P., Choudhary, J.S., and Grant, S.G. (2006). Molecular characterization and comparison of the components and multiprotein complexes in the postsynaptic proteome. *J Neurochem* 97 Suppl 1, 16–23.

Degtyar, V., Hafez, I.M., Bray, C., and Zucker, R.S. (2013). Dance of the SNAREs: assembly and rearrangements detected with FRET at neuronal synapses. *J Neurosci* 33, 5507–5523.

Dore, K., Labrecque, S., Tardif, C., and De Koninck, P. (2014). FRET-FLIM investigation of PSD95-NMDA receptor interaction in dendritic spines; control by calpain, CaMKII and Src family kinase. *PloS one* 9, e112170.

Förster, V.T. (1993). *Intermolecular Energy Migration and Fluorescence* (Translation of Förster, T., 1948). American Institute of Physics, New York edn.

Foster, M., and Sherrington, S.C. (1897). *A Textbook of Physiology*, 7th edn, part 3 edn. Macmillan, New York.

Fukata, Y., Dimitrov, A., Boncompain, G., Vielemeyer, O., Perez, F., and Fukata, M. (2013). Local palmitoylation cycles define activity-regulated postsynaptic subdomains. *J Cell Biol* 202, 145–161.

Goyet, E., Bouquier, N., Ollendorff, V., and Perroy, J. (2016). Fast and high resolution single-cell BRET imaging. *Sci Rep* 6, 28231.

Grant, S.G. (2013). SnapShot: organizational principles of the postsynaptic proteome. *Neuron* 80, 534 e531.

Hafner, A.S., Penn, A.C., Grillo-Bosch, D., Retailleau, N., Poujol, C., Philippat, A., Coussen, F., Sainlos, M., Opazo, P., and Choquet, D. (2015). Lengthening of the stargazin cytoplasmic tail increases synaptic transmission by promoting interaction to deeper domains of PSD-95. *Neuron* 86, 475–489.

Harvey, C.D., Yasuda, R., Zhong, H., and Svoboda, K. (2008). The spread of Ras activity triggered by activation of a single dendritic spine. *Science* 321, 136–140.

Herring, B.E., and Nicoll, R.A. (2016). Long-term potentiation: from CaMKII to AMPA receptor trafficking. *Annu Rev Physiol* 78, 351–365.

Lakowicz, J.R. (2006). *Principles of Fluorescence Spectroscopy*, 3rd ed. Plenum, New York edn.

Lee, S.J., Escobedo-Lozoya, Y., Szatmari, E.M., and Yasuda, R. (2009). Activation of CaMKII in single dendritic spines during long-term potentiation. *Nature* 458, 299–304.

Li, K.W., Hornshaw, M.P., Van Der Schors, R.C., Watson, R., Tate, S., Casetta, B., Jimenez, C.R., Gouwenberg, Y., Gundelfinger, E.D., Smalla, K.H., and Smit, A.B. (2004). Proteomics analysis of rat brain postsynaptic density. Implications of the diverse protein functional groups for the integration of synaptic physiology. *J Biol Chem* 279, 987–1002.

Li, K.W., Klemmer, P., and Smit, A.B. (2010). Interaction proteomics of synapse protein complexes. *Anal Bioanal Chem* 397, 3195–3202.

Lu, J., Helton, T.D., Blanpied, T.A., Racz, B., Newpher, T.M., Weinberg, R.J., and Ehlers, M.D. (2007). Postsynaptic positioning of endocytic zones and AMPA receptor cycling by physical coupling of dynamin-3 to Homer. *Neuron* 55, 874–889.

MacGillavry, H.D., Kerr, J.M., and Blanpied, T.A. (2011). Lateral organization of the postsynaptic density. *Mol Cell Neurosci* 48, 321–331.

MacGillavry, H.D., Song, Y., Raghavachari, S., and Blanpied, T.A. (2013). Nanoscale scaffolding domains within the postsynaptic density concentrate synaptic AMPA receptors. *Neuron* 78, 615–622.

Miyawaki, A., Llopis, J., Heim, R., McCaffery, J.M., Adams, J.A., Ikura, M., and Tsien, R.Y. (1997). Fluorescent indicators for Ca2+ based on green fluorescent proteins and calmodulin. *Nature* 388, 882–887.

Moutin, E., Raynaud, F., Roger, J., Pellegrino, E., Homburger, V., Bertaso, F., Ollendorff, V., Bockaert, J., Fagni, L., and Perroy, J. (2012). Dynamic remodeling of scaffold interactions in dendritic spines controls synaptic excitability. *J Cell Biol* 198, 251–263.

Murakoshi, H., Wang, H., and Yasuda, R. (2011). Local, persistent activation of Rho GTPases during plasticity of single dendritic spines. *Nature* 472, 100–104.

Nair, D., Hosy, E., Petersen, J.D., Constals, A., Giannone, G., Choquet, D., and Sibarita, J.B. (2013). Super-resolution imaging reveals that AMPA receptors inside synapses are dynamically organized in nanodomains regulated by PSD95. *J Neurosci* 33, 13204–13224.

Ni, Q., Titov, D.V., and Zhang, J. (2006). Analyzing protein kinase dynamics in living cells with FRET reporters. *Methods* 40, 279–286.

Nishiyama, J., and Yasuda, R. (2015). Biochemical computation for spine structural plasticity. *Neuron* 87, 63–75.

Opazo, P., Labrecque, S., Tigaret, C.M., Frouin, A., Wiseman, P.W., De Koninck, P., and Choquet, D. (2010). CaMKII triggers the diffusional trapping of surface AMPARs through phosphorylation of stargazin. *Neuron* 67, 239–252.

Opazo, P., Sainlos, M., and Choquet, D. (2012). Regulation of AMPA receptor surface diffusion by PSD-95 slots. *Curr Opin Neurobiol* 22, 453–460.

Perez de Arce, K., Schrod, N., Metzbower, S.W., Allgeyer, E., Kong, G.K., Tang, A.H., Krupp, A.J., Stein, V., Liu, X., Bewersdorf, J., and Blanpied, T.A. (2015). Topographic mapping of the synaptic cleft into adhesive nanodomains. *Neuron* 88, 1165–1172.

Petrini, E.M., Lu, J., Cognet, L., Lounis, B., Ehlers, M.D., and Choquet, D. (2009). Endocytic trafficking and recycling maintain a pool of mobile surface AMPA receptors required for synaptic potentiation. *Neuron* 63, 92–105.

Racz, B., Blanpied, T.A., Ehlers, M.D., and Weinberg, R.J. (2004). Lateral organization of endocytic machinery in dendritic spines. *Nat Neurosci* 7, 917–918.

Ramon y Cajal, S. (1888). Estructura de los centros nerviosos de las aves. *Rev Trim Histol Norm Pat* 1, 1–10.

Sainlos, M., Tigaret, C., Poujol, C., Olivier, N.B., Bard, L., Breillat, C., Thiolon, K., Choquet, D., and Imperiali, B. (2011). Biomimetic divalent ligands for the acute disruption of synaptic AMPAR stabilization. *Nat Chem Biol* 7, 81–91.

Sheng, M., and Kim, E. (2011). The postsynaptic organization of synapses. *Cold Spring Harb Perspect Biol* 3(12): a005678.

Sumioka, A., Yan, D., and Tomita, S. (2010). TARP phosphorylation regulates synaptic AMPA receptors through lipid bilayers. *Neuron* 66, 755–767.

Takahashi, N., Sawada, W., Noguchi, J., Watanabe, S., Ucar, H., Hayashi-Takagi, A., Yagishita, S., Ohno, M., Tokumaru, H., and Kasai, H. (2015). Two-photon fluorescence lifetime imaging of primed SNARE complexes in presynaptic terminals and beta cells. *Nat Commun* 6, 8531.

Takamori, S., Holt, M., Stenius, K., Lemke, E.A., Gronborg, M., Riedel, D., Urlaub, H., Schenck, S., Brugger, B., Ringler, P., and Müller, S.A. (2006). Molecular anatomy of a trafficking organelle. *Cell* 127, 831–846.

Tang, A.H., Chen, H., Li, T.P., Metzbower, S.R., MacGillavry, H.D., and Blanpied, T.A. (2016). A trans-synaptic nanocolumn aligns neurotransmitter release to receptors. *Nature 536, 210–214.*

Tomita, S., Stein, V., Stocker, T.J., Nicoll, R.A., and Bredt, D.S. (2005). Bidirectional synaptic plasticity regulated by phosphorylation of stargazin-like TARPs. *Neuron* 45, 269–277.

Volk, L., Chiu, S.L., Sharma, K., and Huganir, R.L. (2015). Glutamate synapses in human cognitive disorders. *Annu Rev Neurosci* 38, 127–149.

Yasuda, R. (2006). Imaging spatiotemporal dynamics of neuronal signaling using fluorescence resonance energy transfer and fluorescence lifetime imaging microscopy. *Curr Opin Neurobiol* 16, 551–561.

Yasuda, R. (2012). Imaging intracellular signaling using two-photon fluorescent lifetime imaging microscopy. *Cold Spring Harb Protoc* 2012, 1121–1128.

Yasuda, R., Harvey, C.D., Zhong, H., Sobczyk, A., van Aelst, L., and Svoboda, K. (2006). Supersensitive Ras activation in dendrites and spines revealed by two-photon fluorescence lifetime imaging. *Nat Neurosci* 9, 283–291.

Zhai, S., Ark, E.D., Parra-Bueno, P., and Yasuda, R. (2013). Long-distance integration of nuclear ERK signaling triggered by activation of a few dendritic spines. *Science* 342, 1107–1111.

Zheng, K., Bard, L., Reynolds, J.P., King, C., Jensen, T.P., Gourine, A.V., and Rusakov, D.A. (2015). Time-resolved imaging reveals heterogeneous landscapes of nanomolar Ca(2+) in neurons and astroglia. *Neuron* 88, 277–288.

CHAPTER 9

Chemical Clearing of Brains

Klaus Becker, Christian Hahn, Nina Jährling,
Marko Pende, Inna Sabdyusheva-Litschauer,
Saiedeh Saghafi, Martina Wanis, and Hans-Ulrich Dodt

CONTENTS

9.1 INTRODUCTION

The rapid progress of optical sectioning microscopy as confocal or light sheet microscopy in combination with vast advancements in 3D computational image processing brought a renaissance of chemical clearing techniques for biological tissues in the recent years and triggered the development of a still-growing number of novel tissue-clearing protocols. This chapter gives an overview of the most common chemical tissue clearing approaches that have been published up to 2015, as well as a brief description of these protocols.

The first attempts of chemical tissue clearing date back to the early 20th century, when H. Lundvall (1904) and Werner Spalteholz (1911) published pioneering studies about the clearing and brightening effect of some organic solvents on biological tissues. Spalteholz was the first who described a common physical principle underlying solvent-based tissue clearing: soaking biological tissues in a medium of similar refractive index as the cellular proteins (usually around 1.55–1.56 for dehydrated samples and about 1.45–1.49 for specimens that are not dehydrated) considerably diminishes intra-tissue variations of the refractive index (Figure 9.1). As a result, the average deviation of light beams in the sample is markedly reduced so that it becomes translucent, if light absorption by chromophores is not too intense. Although this simple theory still is assumed as basically valid, newer studies have found that further mechanisms may largely contribute to the clearing effect of distinct chemicals (Tuchin, 2005; Rylander et al., 2006; Genina et al., 2010; Zhu et al., 2013).

Motivated by his straightforward hypothesis, Spalteholz systematically tested various mixtures of organic solvents for their potential to clear isolated organs from humans and animals, e.g. the human heart. He finally established a mixture consisting of about 5 vol. parts methyl salicylate and 3 volume parts benzyl benzoate (or 3 vol. parts of methyl

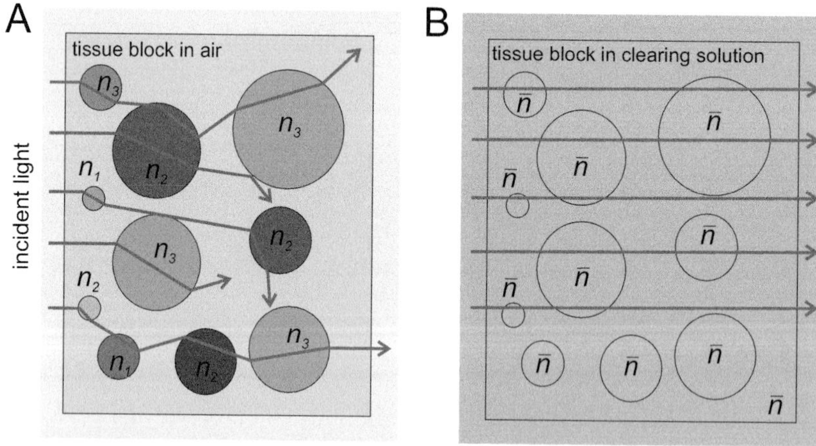

FIGURE 9.1 Principle of refractive index matching: (**A**) Biological tissues are not optically homogeneous, but exhibit transitions between regions of different refractive index (n_1, n_2, n_3, …). At these transition sites light is scattered making the tissue opaque. (**B**) By imbuing a sample in a medium, which approximately matches the average refractive index of the tissue, the refractive index variations are reduced and light scattering is minimized. As a consequence, the light transparency of the sample is enhanced.

salicylate and 1 vol. part of isosafrole, respectively) that renders many anatomical preparations transparent (Spalteholz, 1911). In the following years, he applied this solution to the commercial fabrication of so-called "Spalteholz preparations", which were sold worldwide to universities and anatomical institutions for educational purposes. However, the scientific impact of his preparation technique remained quite limited, apart from occasional applications in embryological and developmental studies (Keller and Dodt, 2012).

With the upcoming of modern computer-aided 3D-microscopy techniques, like confocal microscopy, optical tomography, and light sheet microscopy at the end of the last century, the number of possible applications for chemical specimen clearing began rapidly to grow. Whereas microscopy of large anatomical preparations was principally restricted to thin, virtually flat, sections before, these novel optical instruments brought true volumetric microscopy into reach. Since volumetric samples, due to their thickness, are usually opaque and hence prevent any kind of transmitted light microscopy, the idea of chemical tissue clearing rapidly found novel applications in 3D-microscopy. Presently, tissue clearing is a cutting-edge topic of research, a fact that's illustrated by the vast increase of publications dealing with specimen clearing during the last decade (Figure 9.2).

Chemical tissue clearing agents can be coarsely grouped into lipophilic solvents and water-based (hydrophilic) clearing cocktails. The first group is not mixable with water and thus requires careful tissue dehydration. Substances of the second group can be applied directly after fixation of the tissue. Lipophilic clearing agents usually penetrate biological tissues more easily than hydrophilic clearing substances, since they solubilize the lipids forming the cell membranes. Therefore, the clearing process with these substances is generally fast, rarely taking more than a few days even with large samples, e.g. entire brains from rats.

Contrarily, the clearing process obtained just by incubation in a refractive index matched hydrophilic medium usually is slow, taking up to several months or may even not be possible. Therefore, most hydrophilic clearing media include additions of surfactants

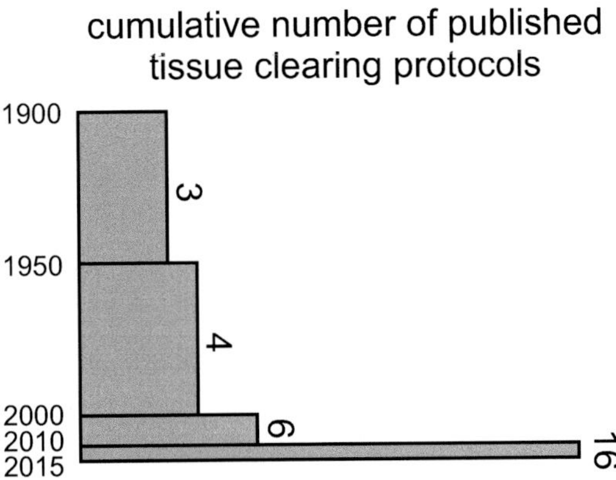

FIGURE 9.2 Tissue-clearing protocols published from 1900 up to now: the development of modern 3D microscopy techniques as confocal or light sheet microscopy brought a renaissance of interest in tissue clearing that is reflected in a vast increase of publications in the recent years.

to solubilize the cell membranes. Surfactants that have been successively applied for this purpose are sodium dodecyl sulphate (SDS) and Triton X100 (Hama et al., 2011; Chung et al., 2013; Susaki et al., 2014; Tomer et al., 2014). In work on CLARITY that has recently been published, electrical current is additionally applied to accelerate the evacuation of the emulsified membrane fragments in an electrophoresis-like process (Chung et al., 2013).

Different clearing approaches utilize hypoosmotic solutions of urea, which induce a diffusion mediated influx of water into biological tissues. The increased water content causes a decrease in density and average refractive index that is regarded as a major component in the clearing effect of urea solutions. Recently published clearing techniques that use this effect are Sca/e (Hama et al., 2011), FRUIT (Hou et al., 2015), or CUBIC (Susaki et al., 2014). A common, almost unwanted, side-effect of tissue hyperhydration by urea is swelling of the samples that can be up to twice their original size for the Sca/e clearing protocol. Such extremely hyperhydrated specimens change to a very soft and fragile state making them very difficult to handle. In some improved clearing protocols using urea which were recently published, it was attempted to counterbalance this effect by adding further hyperosmotic compounds to the clearing cocktail, or introducing further processing steps to reverse the swelling (Hama et al., 2015; Hou et al., 2015; Susaki et al., 2015). Presently, tissue clearing is in focus of research and presumably various novel approaches encompassing present limitations will come up in the near future.

9.2 TISSUE CLEARING PROTOCOLS

This paragraph provides an overview of common clearing approaches from the literature up to 2015. Appropriate specimens are, for example, entire mouse brains, mouse hippocampi, spinal cord, or mouse embryos. Some of the presented clearing approaches, e.g. 3DISCO (Becker et al., 2012; Ertürk et al., 2012) or CUBIC (Susaki et al., 2014), are even applicable for very large samples, such as whole adult mice or entire organs from rats and other mammalians. For each method that is addressed, a short protocol is provided which briefly describes the essential steps of the according procedures. Nevertheless, the publication in which the respective method was originally described should be consulted for more details and specific modifications and variants not addressed here. Table 9.1 provides a brief overview about the clearing approaches that are considered. It also names the original publication, where the method was firstly described, the applied chemicals, and the compatibility with GFP as the most prominent genetic fluorescence marker. It is further indicated whether the technique requires tissue dehydration before clearing.

9.2.1 Tissue Clearing Using Lipophilic Solvents

Solvent-based tissue clearing approaches mostly provide very good clearing results. Due to their excellent tissue penetration properties with dehydrated tissues they can clear even large samples such as entire brains or embryos within a few days. Tissue dehydration can be done by ethanol (Dodt et al., 2007), tetrahydrofuran (Becker et al., 2012), or *tert*-butanol (Schwarz et al., 2015). For fluorescent labeled specimens, the latter two are recommended since they affect the fluorescence of genetic fluorescence markers less (Becker et al., 2012; Schwarz et al., 2015). In order to obtain best possible clearing results, it is important to

TABLE 9.1 History of Chemical Tissue Clearing

year of publication	Acronym	Original publication	Reagents	requires dehydration	GFP preservation	Refractive index	volume change
1904		Lundvall, Anatomischer Anzeiger (Jena) 25:219–222	Dehydration in grades series of ethanol followed by incubation in **benzene** mixed with **peppermint oil** and **carbon disulphide**	+	?	1.5–1.6	shrinks
1911		Spalteholz, Über das Durchsichtigmachen von menschlichen und tierischen Präparaten. S. Hirzel, Leipzig.	Dehydration in grades series of ethanol followed by incubation in mixtures of **methyl salicylate** and **benzyl benzoate** or **isosafrol**	+	–	1.5–1.6	shrinks
1922		Drahn, Berliner tierärztliche Wochenschrift 38:97–100.	Dehydration in grades series of ethanol followed by incubation in solutions of **naphtalene** in **tetraline**	+	?	1.5–1.6	shrinks
1989	BABB, Murray's clear	Dent et al., Development 105(1):61–74.	Dehydration in ethanol. Incubation in 1 vol. part of **benzyl alcohol** + 2 vol. parts **benzylbenzoate**	+	(+)	1.55	shrinks
2001	FocusClear	Chiang et al., J Comp Neurol. 440(1):1–11	Mixture of unknown composition, probably containing **DMSO**, **diatrizoate acid**, **EDTA**, **glucamine**, **NADP**, **sodium diatrizoate** and **polyoxyalkalene derivatives**	–	+	1.47	none
2009		Tsai et al., J Neurosci 29(46), 14553–14570.	60% **sucrose** solution after membrane permeabilization with 2% (v/v) **Triton X-100**			1.44	shrinks
2011	Sca/e	Hama et al., Nat. Neurosci. 14(11) 1481–1488	Aqueous solutions of 4M **urea** containing 0.1% **Triton X-100** and varying fractions of **glycerol**.	–	+	1.38	expands
2012	3DISCO	Becker et al., PlosOne, Ertürk et al., Nat. Prot. 7(11)	Dehydration in peroxide free **tetrahydrofuran (THF)** followed by clearing in **peroxide free benzyl ether**.	+	+	1.56	shrinks
2013	ClearT	Kuwajima et al., Development 140: 1364–1368	Incubation of specimen in aqueous solutions of **formamide** ($Clear^{T}$) or mixtures of **formamide**, water and **polyethylene glycol** ($Clear^{T2}$)	–	+	1.44	none
2013	CLARITY	Chung et al., Nature 497: 332–337	Stabilization of tissue by **acrylamide hydrogel** infusion. Then removal of cell membranes via electrophoresis in a buffer solution containing **sodium dodecyl sulfate (SDS)**. Specimens obtain their final transparency by submersion in **FocusClear** tissue clearing medium or 80% **glycerol** ($n \sim 1.45$).	–	–	1.45	expands

(Continued)

TABLE 9.1 (CONTINUED) History of Chemical Tissue Clearing

year of publication	Acronym	Original publication	Reagents	requires dehydration	GFP preservation	Refractive index	volume change
2013	SeeDB	Ke et al., Nat. Neurosci 16:8 1154–1161	80% aqueous **fructose** solution containing 0.5% **alpha-thioglycerol** in order to prevent browning due to Maillard reaction.	–	+	1.48	shrinks
2014	CUBIC	Susaki et al., Cell 157: 1–14	Incubation in mixtures of an amino alcohol (**N,N,N',N'-tetrakis(2-hydroxypropyl)ethylene diamine**), **urea**, **sucrose** and **Triton X-100**.	–	+	1.38–1.48	expands
2015	TDE	Aoyagi et al., PLOS ONE, Doi:10.1371/journal.pone.0116280 Costantini et al., Scientific Reports, 5:9808	Clearing solution consisting of 30–97% **2,2'-thiodiethanol (TDE)** in **PBS**. Contains 0.5% **alpha-thioglycerol**	–	+	1.42–1.52	(shrinks)
2015	FRUIT	Hou et al., Frontiers in Neuroanat. 9(19)	Clearing in solutions containing **fructose** and **urea** in grades concentrations	–	+	1.48	none
2015	FluoBABB	Schwarz et al., PLOS ONE, Doi:10.371/journal.pone.0124650	Dehydration with tert-butanol. Then clearing with BABB that had been pH adjusted to 9.5 by adding **triethylamine**.	+	+	1.55	shrinks
2015	Sca/eS	Hama et al., Nat. Neurosci, doi: 10.1038/nn.4107	Clearing in a sequence of solutions containing sorbitol, **glycerol**, **urea**, **Triton X-100**, **Methyl-ß-cyclodextrin**, **gamma-cyclodextrin**, **N-acetyl-L-hydroxyproline**, and **dimethyl sulfoxide**.	–	+	1.334–1.447	none

The table provides an overview about tissue clearing approaches published form the early beginnings up to 2015. More detailed information about these techniques is provided in the text.

remove even minute traces of water from the sample, since they prevent the permeation of the clearing medium and cause opacities in the sample. This can be best achieved by adding a 3A molecular sieve (e.g. Sigma-Aldrich, order no. 208582) to the undiluted dehydration medium used in the final dehydration step at least one day before use.

As a side-effect, all dehydration agents cause a certain amount of shrinkage that can be up to 50% of the original tissue volume. Compared to this, the amount of shrinkage caused by the clearing medium is usually negligible. Too fast dehydration by applying too few intermediate concentration steps can lead to dehydration artefacts as cracks and ruptures in the tissue. Therefore, the number of steps should be sufficient to achieve a most gentle removal of water (50%, 70%, 80%, 90%, 96%, 2 × 100% ethanol or THF, one day each usually is a safe choice for a fixed mouse brain). Smaller samples, e.g. fixed mouse hippocampi or spinal cord, can be dehydrated using wider concentration steps and reduced incubation times (Ertürk et al., 2012).

It is a known drawback of solvent-based clearing agents that they can severely quench the fluorescence of genetic markers such as EGFP, GFP, YFP, or tomato. This often makes them of only limited use for genetically labeled specimens, especially if the expression rates of the genetic marker are low. However, some recent clearing approaches even achieve good fluorescence preservation with lipophilic clearing agents, e.g. 3DISCO (Becker et al., 2012; Ertürk et al., 2012) or FluoBABB (Schwarz et al., 2015).

9.2.1.1 *Tissue Clearing with Mixtures of Benzene and Carbon Disulphide* (Lundvall, 1904)

To our knowledge, Lundvall was the first who reported clearing of anatomical preparations. He dehydrated embryos of diverse vertebrates in ascending concentration series of ethanol and then transferred them into solutions of benzene containing increasing fractions of carbon disulphide. Finally, the specimens were stored in tightly sealed glass containers filled with a mixture of about 80% benzene and 20% carbon disulphide (Lundvall, 1904). Due to the intense rotten eggs-like smell of carbon disulphide and the severe toxicity and carcinogenicity of both substances, this early technique of tissue clearing soon was replaced by more manageable attempts.

9.2.1.2 *Tissue Clearing with Methyl Salicylate and Benzyl Benzoate/Isosafrole* (Spalteholz, 1911)

Werner Spalteholz (1911) published a more feasible method for rendering whole organs, embryos, and bones of human and animal origin transparent that became a standard method for clearing of anatomical demonstration samples for several decades. During the first half of the 20th century these "Spalteholz preparations" were distributed worldwide to various academic institutions for educational purposes. Since modern 3D microscopy techniques for imaging volumetric samples were missing at this time, the Spalteholz preparation technique to our knowledge was never applied in the context of microscopy. Later on, it was almost replaced by a mixture of benzyl alcohol and benzyl benzoate (BABB) that provides comparable clearing results, but generally performs faster (Dent et al., 1989; Klymkowsky and Hanken, 1991). As well as BABB, Spalteholz clearing quenches the fluorescence of genetic markers as GFP, but may be compatible with some immunostainings.

TABLE 9.2 Spalteholz Clearing

Specimen	methyl salicylate (w/w)	benzyl benzoate or isosafrole (w/w)
Human decalcified bone	5 (3)	3 (1)
Human muscles	1 (9)	1 (5)
Large human embryos	2 (18)	1 (5)
Small human embryos	3 (27)	1 (5)
Very small human embryos	5 (9)	1 (1)
Tissue from frogs and fish	3 (27)	1 (5)
Invertebrates	4 (36)	1 (5)

Mixing ratios of methyl salicylate and benzyl benzoate (or isosafrole, respectively) as suggested by Spalteholz (1911) for different specimen types.

Protocol for specimen clearing according to Spalteholz (1911).

- Fixate specimens in ethanol or 4% formaldehyde solution.
- If required, decalcify specimens, e.g. with diluted acetic acid.
- If required, bleach specimens with hydrogen peroxide containing 1% formaldehyde to reduce maceration of sensitive samples.
- Carefully rinse the specimens with water.
- Dehydrate specimen in an ascending series of ethanol concentrations.
- Transfer specimens into benzene for removing traces of alcohol. Change the benzene twice (caution benzene is carcinogenic).
- Transfer specimen in the Spalteholz solution with the appropriate refractive index according to Table 9.2.
- Remove residuals of benzene by gently applying vacuum (a vacuum which is too strong or increases too fast may damage the sample).
- Tightly seal the specimen container for long-term storage of the Spalteholz preparation.

9.2.1.3 Tissue Clearing with Tetralin/Naphthalene (Drahn, 1922)

About ten years after Spalteholz, Drahn (1922) described a mixture of the liquid aromatic hydrocarbon tetralin and about 25% naphthalene to clear embryos and decalcified bones dehydrated by ethanol. He found that this mixture exhibits a very fast tissue penetration, rendering an entire pig embryo of about 3 cm in size almost wholly transparent within only 8 hours. A practical drawback of the mixture is its extremely penetrating odor reminiscent of moth powder. Since it is also of severe toxicity and teratogenicity, its use should be limited to anatomical preparations that are permanently kept in tightly sealed glass containers. Handling of the mixture must be done under a good working fume hood. As the Spalteholz technique, this clearing approach to our knowledge was never applied in the context of microscopy. To our knowledge, there is no information about the compatibility with fluorescence markers or immunostainings available.

Protocol for clearing of anatomical samples with tetralin/naphthalene (Drahn, 1922).

- Add 33.33 g naphthalene to 100 g 1,2,3,4-tetrahydronaphthalene in a beaker glass under a well-functioning fume hood. Put the beaker on a magnetic stirrer and gently mix while gently heating until all tetralin has been dissolved. After cooling down the refractive index of the mixture should be approximately 1.561.
- Bleach specimens in hydrogen peroxide if required.
- Stepwise dehydrate samples in an ascending series of ethanol.
- Incubate the samples in the clearing mixture until they become clear. The specimen containers should be kept tightly sealed due to the intense odor and toxicity of the mixture.

9.2.1.4 *Tissue Clearing with Benzyl Alcohol/Benzyl Benzoate (BABB)* (Dent et al., 1989; Schwarz et al., 2015)

In 1989, Murray developed a clearing cocktail consisting of 1 vol. parts benzyl alcohol (BA) and 2 vol. parts benzyl benzoate (BB) that over years became a standard for clearing biological samples in the field of embryological and developmental research (Dent et al., 1989; Klymkowsky and Hanken, 1991). This mixture, which is commonly termed BABB or "Murray's clear", clears many kinds of samples excellently in a time range of a few hours or days. An entire juvenile mouse brain or mouse embryo is rendered virtually completely transparent within three days (Figure 9.3A–B). As with other lipophilic clearing agents, it is crucial that the samples are completely dehydrated before clearing. Minute residuals of water can be removed by brief incubation in an apolar solvent as hexane following the last dehydration step (Dodt et al., 2007). As with other organic solvents BABB severely quenches the fluorescence of genetic markers as GFP, if the incubation times are longer than a few hours or it contains peroxides. Nevertheless, BABB can be successfully applied to GFP expressing mouse brains and embryos, if the GFP expression rate is sufficiently high and incubation times are not too long (Figure 9.3C–D) (Dodt et al., 2007). However, newer clearing approaches with better GFP compatibility as 3DISCO generally provides a superior alternative for samples expressing genetic markers (Becker et al., 2012; Ertürk et al., 2012).

BABB that was exposed to oxygen and/or sunlight can contain significant amounts of peroxides. Since these peroxides destroy GFP fluorescence even in small traces >1 ppm, BABB should always be checked for peroxide contaminations, e.g. by Quantofix 25 test stripes (Sigma Aldrich, order No. Z249254). The test stripes are briefly dipped into the BABB and then rinsed under a water tap. The presence of peroxides is indicated by a blue coloration appearing after a few seconds. BABB contaminated with peroxides can be purified by column chromatography with activated basic aluminum oxide. A respective protocol that was developed for the purification of benzyl ether, but which can also be applied to BABB is described in Becker et al. (2012).

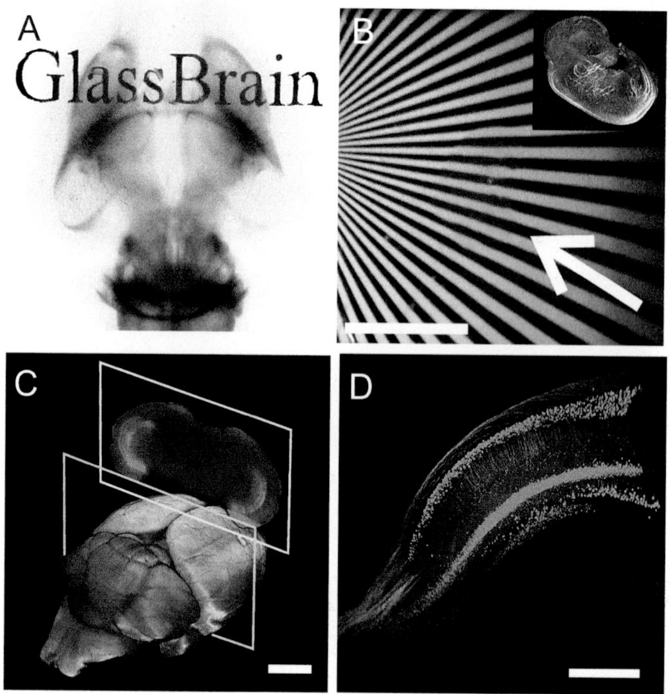

FIGURE 9.3 Clearing of mouse brains with BABB: (**A**) Juvenile mouse brain (about three weeks old) cleared with BABB in less than one week. The sample is so transparent that the letters of the word "glasbrain" can easily be read throughout the entire thickness of the samples (**B**) Mouse embryo E12.5 (arrow). The sample is placed on a Siemens star to demonstrate its transparency (scale bar 1 cm). Upper right corner: 3D-reconstruction of the embryo using ultramicroscopy. Nerve fibers were labeled by immunostaining. (**C**) Ultramicroscopy reconstruction of a GFP-expressing mouse brain cleared with BABB. Hippocampal pyramidal and granule cell layers are detectible in digital sections (length of scale bar 1 mm). (**D**) Ultramicroscopy reconstruction obtained from an excised hippocampus of a thy1-GFP-M transgenic mouse. Pyramidal neurons and granule cells exhibit an intense GFP-signal (length of scale bar ~100 μm). All samples were cleared according to Dodt et al. (2007).

Protocol for clearing GFP expressing mouse brains with ethanol/BABB according to Dodt et al. (2007).

- Deeply anaesthetize mice by intraperitoneal injection of pentobarbital (10 mg/kg).
- Transcardially perfuse them with 40 ml ice-cold 0.1 M PBS containing 1,000 units/ml heparin, followed by 80 ml 4% paraformaldehyde in 0.1 M ice-cold PBS (pH 7.4).*
- Dissect the brains from the scull and fix them in 4% paraformaldehyde for about 1h at 4°C.
- Dehydrate the brains using an ascending series of ethanol concentrations (30%, 50%, 70%, 80%, 90%, 96%, and twice in 100% EtOH for 1 d each).

* To improve fluorescence preservation, the pH of buffer and fixation solutions can be adjusted to a more basic value of about 8.5.

- Rinse the brains in 100% hexane for 1h to remove even small residuals of water.
- Incubate the brains in BABB (1 vol. part benzyl alcohol + 2 vol. parts benzyl benzoate, peroxide free) for at least two days until they become transparent.

Protocol for clearing of Mouse Spinal Cord using tetrahydrofuran (THF) and BABB (3DISCO)* (Ertürk et al., 2011).

- Deeply anaesthetize mice by intraperitoneal injection of pentobarbital (10 mg/kg).
- Transcardially perfuse them with 0.1 M phosphate buffer (pH 7.4) for 5–10 min, followed by 4% paraformaldehyde (pH 7.4†) in 0.1 M PB for 40 min at 3 ml/min.
- Dissect the spinal cords and post-fix them in 4% paraformaldehyde at 4°C overnight.
- Carefully remove the dura mater from the spinal cord and dissect it into 3–4 mm long segments.
- Incubate the samples in 50% THF for 30 min, 80% THF for 30 min, and twice in 100% THF for 30 min. THF should be checked for peroxides before use and cleaned, if necessary, by column chromatography as described in Becker et al. (2012).
- Rinse the samples in 100% dichloromethane for 20 min for defatting.
- Incubate the samples in peroxide-free BABB (1 vol. parts benzyl alcohol + 2 vol. parts benzyl benzoate) for 10–15 min until they are sufficiently transparent.

Recently, Schwarz et al. (2015) published a modified BABB clearing protocol for GFP expressing mouse brains (FluoBABB) that exhibits only minimal fluorescence quenching. It allows storing specimens for up to several months without significant loss of fluorescence. This was achieved by tissue dehydration with *tert*-butanol (tB) and adding triethylamine to the dehydration and clearing medium to shift the pH to a basic value of about 9.5.‡

Clearing protocol for GFP-expressing mouse brains using Tert-Butanol and pH-adjusted BABB§ (FluoBABB) according to Schwarz et al. (2015).

- Deeply anesthetize mice and transcardially perfuse them using an ice-cold solution of 4% paraformaldehyde in PBS.
- Remove the brains from the scull and post-fix them for about 20 h at 4°C.
- Wash the brains two times in PBS at 4°C.

* In a more recent version of the 3DISCO protocol (Ertürk et al., 2012) BABB was substituted by benzyl ether, which provides faster clearing and better fluorescence preservation (see Section 9.2.1.5)

† To improve fluorescence preservation, the pH of buffer and fixation solutions can be adjusted to a more basic value of about 8.5.

‡ The pH-adjustment of tert-butanol and BABB requires a pH-meter that is equipped with a special electrode for use with organic solvents (e.g. P/N 662-2837, Mettler-Toledo). A standard pH electrode would provide no reliable results and would be damaged (Galster, 1991).

§ Some further variants of this protocol (e.g. using 1-propanol instead of tert-butanol for dehydration to reduce tissue shrinkage) are described in Schwarz et al. (2015).

- Dehydrate the brains at 30 °C in an ascending concentration series of *tert*-butanol (e.g. Sigma-Aldrich, order-no. 471712) dissolved in distilled water (30%, 50%, 70%, 80%, 96%, and twice 100% for one day each). The pH[2] of each solution is adjusted to 9.5 by addition of triethylamine (e.g. Sigma-Aldrich, order-no. T0886).
- Incubate the specimens in BABB at 30°C (1 vol. parts benzyl alcohol + 2 vol. parts benzyl benzoate) that has been adjusted to a pH of 9.5 by addition of triethylamine (FluoBABB).
- Change the clearing solution after one day.
- Mount the samples in FluoBABB for microscopy inspection.
- For storage keep the samples in the clearing solution at 4°C.

9.2.1.5 *Tissue Clearing with Tetrahydrofuran and Benzyl Ether (3DISCO)* (Becker et al., 2012; Ertürk et al., 2012)

Benzyl ether (DBE) was first suggested as a substitute for BABB by Becker et al. (2012). Combined with tetrahydrofuran (THF) as a dehydration medium it offers better tissue transparency (Figure 9.4A) and improved fluorescence in mouse brains, embryos, and organs (Becker et al., 2012; Ertürk et al., 2012) (Figure 9.4B–C). Due to its lower viscosity, DBE more easily penetrates specimens, providing faster clearing results. Furthermore, DBE is clearly cheaper and significantly less toxic than BABB (Burdock and Ford, 1992).

It is important that DBE, which is used for clearing of fluorescence labeled samples is free from contaminations with peroxides, since peroxides efficiently quench fluorescence even in minute concentrations (>1 mg/L) (Alnuami et al., 2008). These ether peroxides are formed by exposure to oxygen. Under the influence of light, they react further to benzaldehyde (Eichel and Othmer, 1949), which is also a fluorescence quencher for GFP (Figure 9.4D). Benzaldehyde contaminations in DBE that had been stored for a prolonged time can be detected by a characteristic bitter-almond-like odor, or by using Brady's test for aldehydes (Brady and Elsmie, 1926). Depending on the manufacturer and the lot number, DBE can contain peroxide concentrations >1 mg/l at time of purchase, making it unsuitable for clearing GFP expressing samples. Peroxides can be easily detected, e.g. using Quantofix 25 test stripes (Sigma Aldrich Austria, Order No. Z249254). The test stripes are briefly dipped into the clearing solution and shortly rinsed under a water tap. Peroxides are indicated by a blue coloration appearing after a few seconds.

To remove peroxide contaminations in DBE and BABB absorption column chromatography with activated basic aluminum oxide (basic activated Brockman 1, Sigma-Aldrich, order-no. 199443) can be applied as described in Figure 9.4E or Becker et al. (2012). The generation of new peroxides in purified DBE can be slowed down by using brown bottles that are filled with an inert gas, such as argon.

Its compatibility with most immunostainings and the strong clearing effect even on very large samples makes 3DISCO a very good clearing approach for large immune-labeled samples. A modified version of 3DISCO, termed iDISCO, that was developed for immunostaining and clearing of mouse embryos was published by Renier et al. (2014).

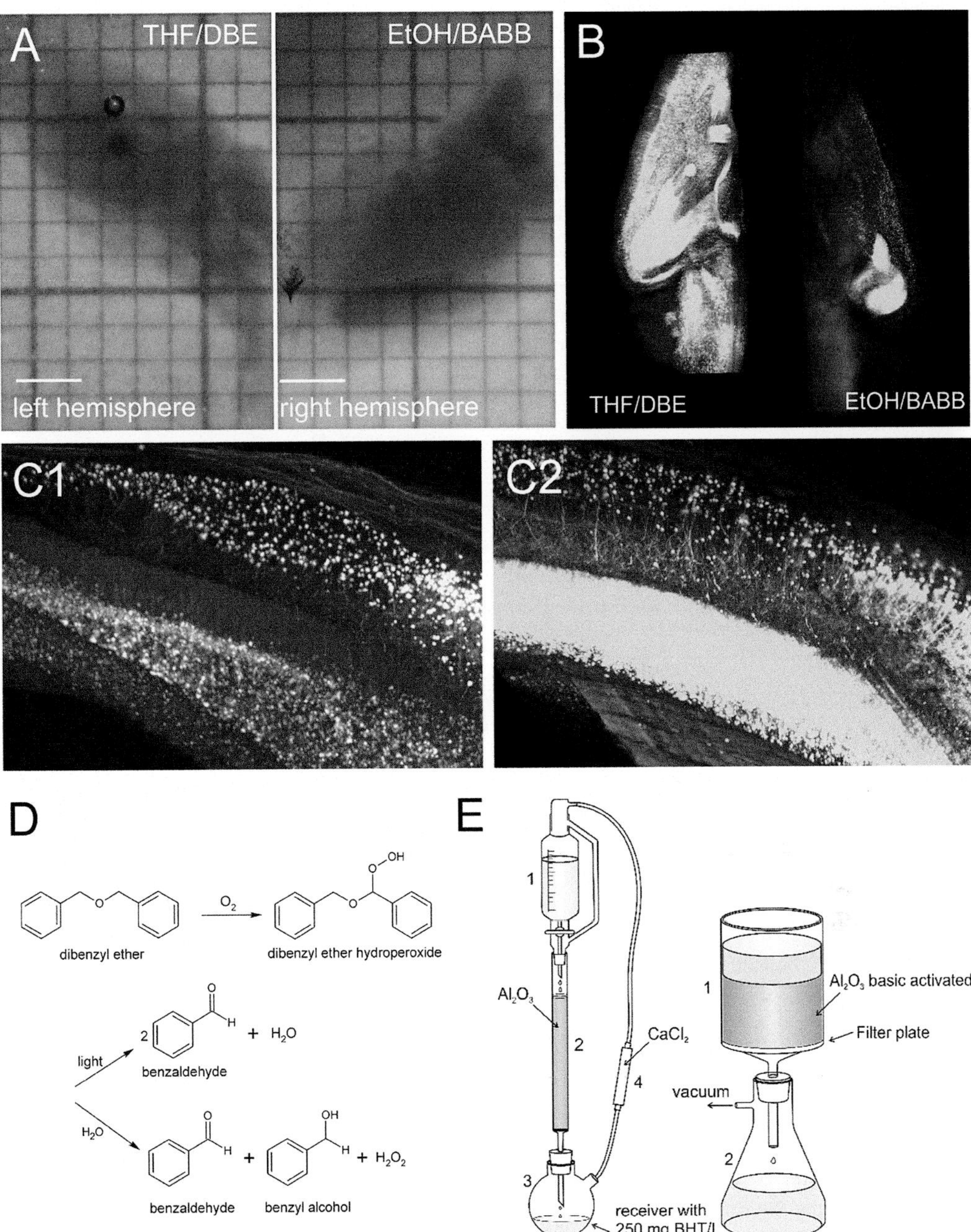

FIGURE 9.4 Clearing of mouse brains with THF and DBE (3DISCO): (**A**) Two differently cleared hemispheres from an adult mouse brain (postnatal day 101). Clearing with THF/DBE provides higher tissue transparency compared to EtOH/BABB. Incubation times in EtOH and THF or BABB and DBE, respectively were kept identical. Length of scale bar: 2 mm. (**B**) 3D reconstructions obtained from ultramicroscopy recordings of the two hemispheres depicted in (A). GFP-fluorescence is better preserved in the hemisphere treated with THF and DBE. (**C**) Ultramicroscopy recordings of dissected mouse hippocampi dehydrated and cleared with EtOH and BABB (**C1**) or 3DISCO (**C2**) confirming the better signal preservation with THF/DBE. (**D**) Chemical reaction of

FIGURE 9.4 (CONTINUED)
DBE with light and oxygen. To prevent DBE form forming peroxides and oxidation to benzaldehyde, which both are GFP-fluorescence quenchers, it should be protected from light and oxygen by storing in brown bottles filled with argon gas. (E) Apparatus for peroxide removal from THF **1**: Dropping funnel with pressure compensation. **2**: Chromatography column filled with basic activated aluminum oxide activity grade Brockman 1 **3**: Two necked round bottom flask **4**: Drying tube filled with calcium chloride (left). **B**: Apparatus for peroxide removal from DBE and BABB **1**: Filter unit with filter plate (16–40 μm pore size) **2**: Vacuum tight filtering flask (right). Figure modified from Becker et al. (2012). **Removal of peroxides in THF and DBE:** Peroxide removal in THF is done by column absorption chromatography with basic activated aluminum oxide activity grade Brockman I (Sigma-Aldrich, Austria, Order-No. 199443, about 250 g per liter THF) (Figure 9.4E left). The stabilizer, which is contained in commercially distributed THF is also removed during chromatography. This stabilizer is required to prevent the generation of dangerous amounts of peroxides by sunlight or exposure to oxygen. Therefore, it is essential by safety reasons to substitute it, e.g. by adding 250 mg/l butyl hydroxyl toluol (BHT) into the receiver flask (Sigma-Aldrich, Austria, Order-No. W218405). THF which is insufficiently stabilized can explode after prolonged exposure to oxygen and/or sunlight and may be dangerous to life! Due to their higher viscosity and boiling points, removal of peroxides in BABB and DBE can be done by vacuum filtering. The filter funnel is filled with ~250 g activated aluminum oxide per liter and suction is applied to the receiver flask (Figure 9.4E right).

*Clearing protocol for mouse organs using 3DISCO** (Ertürk et al., 2012).

- Deeply anesthetize mice and transcardially perfuse them using 0.1 M PBS for 5–10 min followed by 30–40 min perfusion of 4% PFA (w/v) in 0.1 PBS (pH 7.4)† at a rate of 3 ml/min. The perfusion time can be shortened as long as the blood is completely removed from the tissue.
- Dissect the organs of interest and postfix them in 4% PFA overnight. For small pieces of tissue (e.g. spinal cords), this step can be shortened or omitted.
- Remove the extra tissues (connective tissue, meninges, or dura mater) surrounding the organ in a PBS-filled dish using forceps and visual control under a binocular stereoscope.
- Perform dehydration of the samples according to Table 9.3. Place the vials on a turning wheel or shaker for optimal mixing. To optimize the clearing for tissues not mentioned in the table, follow the procedure for the tissue having the most similar size and composition.
- For some tissues, e.g. spinal cord, a brief intermediate incubation step in dichloromethane (DCM) following dehydration is useful. DCM is a strong fat dissolver that helps to dissolve fatty or heavily myelinated structures.

---

* There is also a previous version of the 3DISCO protocol using BABB instead of DBE for clearing (see Section 9.2.1.4).

† For GFP expressing animals, the pH may be increased to 8.0–8.5 by adding 1N NaOH for better fluorescence preservation.

TABLE 9.3 Clearing with THF and DBE (3DISCO)

Reagents (vol/vol)	Mammary gland, lymph node	Spinal cord, lung, spleen	Brain stem	Brain	Brain (long protocol)
50% THF	20 min	30 min	1 h	1 h	
70% THF	20 min	30 min	1 h[a]	1 h	
80% THF	20 min	30 min	1 h	1 h	
100% THF	3 × 20 min	3 × 30 min	2 × 1 h	1 h, overnight, 1 h	3 × 12 h
DCM	15 min	20 min	45 min	–	–
DBE	≥15 min	≥15 min	≥30 min	≥3 h	1–2 d

Incubation times for dehydration and clearing of mouse brain tissue and other organs provided by Ertürk et al. (2012).

[a] For small tissues the 70% THF step can be skipped to save time.

- Transfer the samples into peroxide-free DBE for clearing. Changing the DBE about two or three times improves the results (e.g. refresh the DBE for spinal cord every 10 min and every 12 h for mouse brains).
- Mount the samples in fresh peroxide-free DBE for imaging. Since the fluorescence degrades over time, image the samples as soon as possible after clearing.

9.2.2 Tissue Clearing Using Hydrophilic Compounds

Differently from organic solvents, hydrophilic clearing agents are applied without previous dehydration. This has the advantage that tissue shrinking artefacts are prevented and processing speed and throughput may increase. On the other hand, due to the long clearing times required with most hydrophilic clearing protocols, this time-saving effect is usually compensated. As an important benefit, hydrophilic clearing agents preserve the fluorescence of genetic markers excellently, even after prolonged storage.

However, for the achievable tissue transparency often is better with lipophilic solvents. This especially holds true for very large or heavily myelinated specimens, e.g. entire brains or other organs from rats.

9.2.2.1 *FocusClear* (Chiang, 2002)

FocusClear was first mentioned in a publication dealing with the mapping of brain neuropils in the cockroach *Diploptera punctata* (Chiang et al., 2001) and was soon patented under the US patent number 6,472,216 B1 (2002). It is usually applied as a clearing solution for small samples such as brains slices from mice, or entire insect brains (Chiang, 2002). According to the enclosed manufacturer's instructions, it can also be used for enhancing the light transparency of various other microscopic samples, such as tumors and even some kinds of plant tissues.

FocusClear is distributed by the Canadian company Cedarlane Labs. The composition of FocusClear is proprietary. According to the patent specification, it may contain dimethyl sulfoxide (DMSO), diatrizoate acid, EDTA, glucamine, NADP, sodium diatrizoate, and polyoxyalkalenes in unknown proportions. The essential component is presumably sodium diatrizoate (3,5-Bis(acetylamino)-2,4,6-triiodobenzoic acid, sodium salt,

CAS 737-31-5), which is a highly refractive, iodine containing, aromatic compound of distinct solubility in water (about 350 mg/ml H_2O). Sodium diatrizoate is also applied as a density gradient centrifugation medium and as a contrast agent in x-ray examinations.

The price of FocusClear is very high, currently about $180 per 5 ml vial, which approximately corresponds to the amount needed for clearing a whole mouse brain. A nearly saturated aqueous solution of sodium diatrizoate (~75%) has the same refractive index as FocusClear (n = 1.45) and may serve as a significantly cheaper substitute for many applications. Large objects such as entire mouse brains do not become sufficiently clear in FocusClear without previous processing, e.g. by CLARITY (Chung et al., 2013) or PACT (Yang et al., 2014). However, FocusClear may be a good choice for clearing small and thin samples such as brain slices up to a few hundred micrometer thickness or fly brains with low afford.

Clearing protocol for fly brains and brain slices using FocusClear.

- Fix samples using 4% formaldehyde or glutaraldehyde.
- Wash samples thoroughly after fixation.
- Transfer samples into a small amount of FocusClear solution. For an intact fly brain, about 100 µl, and for a mouse brain slice 200 µl are recommended.
- Incubate the samples until they become clear (about 10 min to 4 h, depending on the specimen size). For prolonged incubation the specimen chamber should be tightly sealed, e.g. with parafilm, to prevent evaporation.
- For microscopy inspection, mount the specimen in a fresh drop of FocusClear.

9.2.2.2 *Tissue Clearing with Urea* (Hama et al., 2011, 2015)

Urea (CH_4N_2O) is a relatively small organic molecule that easily penetrates cell membranes. There it induces an osmotically driven influx of water into the cytoplasm that causes hyperhydration and swelling of the tissue, which in turn reduces the refractive index difference between tissue and incubation medium. As a consequence, the tissue becomes more transparent due to less stray light generation. High concentrations of urea additionally lead to denaturation and increased hydration of proteins, which further lowers the refractive index (Richardson and Lichtman, 2015). The refractive index of a mouse brain that was hyperhydrated in a 4M solution of urea is about 1.38. This value is significantly lower than the refractive index of other tissue-clearing media designed for refractive index matching of proteins.

Hama et al. (2011) were the first to apply highly concentrated aqueous solutions of urea, with some additions of glycerol and the detergent Triton X-100, as a clearing medium for thick mouse brain sections and entire mouse brains. Hama et al. developed different variations of their clearing protocol, which they named with the acronyms Sca/eA2, Sca/eU2, and Sca/eB4. Sca/eA2 contains 4M urea and 10% (w/v) glycerol and 0.1% (w/v) Triton X-100, Sca/eU2 contains 4M urea, 30% (w/v) glycerol, and 0.1% (w/v) Triton X-100, and Sca/eB4 contains 8M urea and 0.1% (w/v) Triton X-100. According to the authors, these different versions have different clearing speed and cause different degrees of tissue expansion. For some time, Sca/eA2 was commercially distributed under the brand name "Scaleview-A2" by the Japanese microscope company Olympus, bundled with a specialized long working distance (WD) immersion objective (×25/1.0 at WD 4 mm; ×25/0.90 at WD 8 mm) that is corrected for the refractive index of Sca/eA2 (~1.38).

As a major drawback, specimens treated with Sca/e show a pronounced swelling that can be up to twice the original volume. Furthermore, the increased content of water gives the tissue a soft and fragile texture making the cleared samples difficult to handle. All variants of Sca/e described by Hama et al. (2011) require very long incubation times, e.g. about two weeks for a 13.5 days old mouse embryo. While the different modifications of Sca/e may sufficiently clear brain slices, mouse embryos and very young mouse brains (below postnatal day 14) in an acceptable time, myelinated structures do not become transparent. We even failed to clear (almost fat free) porcine muscle samples of about 5–10 mm edge length by incubation in Sca/e A2 for more than ten weeks using the protocol provided by Hama et al. 2011. In a later publication from Hama et al. (2015), it is pointed out that the method additionally requires the application of freeze-thaw cycles to make the cell membranes more permeable, a crucial point that is not clearly addressed in Hama et al. (2011).

Recently, Hama et al. presented an improved version of the Sca/e clearing technique for entire mouse brains, termed Sca/eS. As stated by the authors, it avoids excessive tissue swelling by combining urea with the sugar alcohol sorbitol (Hama et al., 2015). Additions of 1 mM methyl-ß-cyclodextrin (e.g. Sigma-Aldrich, order-no. 332615), 1 mM γ-cyclodextrin (e.g. Sigma-Aldrich, order-no. C4892), and 1% N-acetyl-L-hydroxyproline (e.g. Sigma-Aldrich, order-no. 441562) to the urea/sorbitol solution were made to extract cholesterol from the cell membranes and to loosen collagen structures. The method, which involves subsequent incubation steps in a total of six different solutions, was optimized for entire mouse brains. In these solutions the samples first exhibit swelling that later on in the clearing process is counterbalanced by shrinkage so that the original volume of the mouse brain is approximately maintained.

Clearing protocol for mouse brain slices, juvenile entire mouse brains and embryos using Sca/E A2 (Hama et al., 2011).

- Fix mouse brain slices in 4% formaldehyde solution.
- Prepare a 4M solution of urea and add 10% (w/v) glycerol and 0.1% (w/v) Triton X-100. The mixture should have a pH of 7.7 and a refractive index of 1.382 at 589 nm wavelength.
- Incubate the samples in this solution, until they are sufficiently transparent. According to Hama et al. (2011) this takes about 48 h for a 1.5 mm thick mouse brain slice and about 14 days for an entire brain of a mouse at p15. The authors further report a linear size expansion of about 1.25 for an entire mouse brain (Hama et al., 2011).
- Mount the samples in the clearing solution for microscopy inspection.

Clearing of entire mouse brains using the improved protocol Sca/eS (Hama et al., 2015).

- Fix mouse brains in 4% formaldehyde solution.
- Prepare the clearing solutions S0, S1, S2, S3, and S4 according to Table 9.4.
- Incubate the mouse brains in the solutions S1–S3 for 12 h each at 37°C.
- Wash the samples for at least 6 h in PBS.
- Transfer the samples into solution S4 for 12 h at 37°C.
- Mount the cleared samples at room temperature in fresh solution S4 for microscopy. The refractive index of solution S4 at 589 nm is $n = 1.437$.

According to Hama et al. (2015), this protocol was successfully applied for clearing even aged mouse brains from an Alzheimer's disease model. As an important advantage of their clearing method, the authors point out the excellent preservation of the ultrastructure of the tissue and the high compatibility with genetic markers as GFP. The cleared samples are reported to preserve a soft, but not too fragile, texture so that they can easily be handled or cut using a scalpel (Hama et al., 2015).

9.2.2.3 Tissue Clearing Using Formamide (Clear^T, Kuwajima et al. 2013)

Kuwajima et al. (2013) suggested aqueous solutions of formamide for clearing mouse embryos, as well as thick mouse brain slices and termed their tissue clearing approach Clear^T. Clear^T alone is not compatible with GFP expressing samples since formamide diminishes the fluorescence of GFP. Therefore, the authors developed a modified more GFP compatible protocol, which they named Clear^{T2}. In Clear^{T2}, 10–20% of the water is substituted by polyethylene glycol (PEG, molecular weight 8,000 g/mol), which was found to stabilize the fluorescence of GFP in the presence of formamide. The clearing capability of Clear^{T2} is reported as lower compared to Clear^T, probably due to the higher viscosity of the solution.

TABLE 9.4 Clearing of Whole Mouse Brains with Sca/eS

Ingredients	**S0**	**S1**	**S2**	**S3**	**deSca/ing**	**S4**
D-(-)-sorbitol (w/v)%	20	20	27	36.4		40
Glycerol (w/v)%	5	10		2.7		10
Urea (M)		4	2.7			4
Triton X-100 (w/v)%		0.2	0.1			0.2
Methyl-ß-cyclodextrin (mM)	1					
ɣ-Cyclodextrin (mM)	1					
N-acetyl-L-hydroxyproline (w/v)%	1					
Dimethyl sulfoxide (v/v)%	3		8.3	9.1		15–25

Composition of the solutions S0–S4 for sequential incubation of mouse brains according to Hama et al. (2015). All compounds are dissolved in 1 × PBS.

Kuwajima et al. demonstrate that Clear$^{T/T2}$ can render 14.5 days old mouse embryos transparent. They further present images of 800 μm thick slices from GFP expressing mouse brains and embryos rendered transparent with Clear^{T2} which maintain fluorescence. As a special advantage of their clearing protocol, Kuwajima et al. point out that Clear$^{T/T2}$ produces nearly no tissue swelling or shrinkage. Since Clear^{T} as well as Clear^{T2} is free from detergents, the fluorescence of important lipophilic tracers as DiI (1,1'-dioctadecyl-3,3,3',3'-tetramethylindocarbocyanine perchlorate) is also preserved. Due to their limited tissue penetration and fat dissolving capability, Clear$^{T/T2}$ are not suitable for clearing heavily myelinated samples, e.g. entire adult murine brains.

Clearing Protocol for mouse embryos, juvenile entire mouse brains and brain slices using Clear$^{T/T2}$ (Kuwajima et al., 2013).

- Fix mouse brain slices in 4% formaldehyde solution.
- For Clear^{T} prepare solutions of 20%, 40%, 80%, and 95% formamide in water. For clearing GFP expressing samples with Clear^{T2} prepare formamide solutions containing 25% formamide + 10% PEG (MW 8000 g/mol) (e.g. Sigma-Aldrich order-no. 89510), and 50% formamide + 20% PEG.
- Incubate the samples in the clearing solutions according to Table 9.5 (Clear^{T}) or Table 9.6 (Clear^{T2}).
- For microscopy inspection mount the sample in a small volume of the last clearing medium.

TABLE 9.5 Tissue Clearing with Clear^{T}

Solution (v/v)	**Whole embryos or heads**	**Whole dissected brains**	**Half embryonic brains**	**Sections (20–1000 μm)**
20% formamide	30 min	30 min	30 min	5 min
40% formamide	30 min	30 min	30 min	5 min
80% formamide	2 hours	2 hours	2 hours	5 min
95% formamide	30 min	30 min	30 min	5 min
95% formamide	5–16 hours	Overnight, 2 days	4 hours	15 min

Incubation scheme for mouse brains and embryos according to Kuwajima et al. (2013). Formamide (>99%) is diluted in × PBS. The H 7.5 is adjusted to 7.5 with NaOH.

TABLE 9.6 Tissue Clearing with Clear^{T2}

Solution (w/vol)	**Embryonic heads or brains**	**Sections (20–1,000 μm)**
25% formamide + 10% PEG (w/v)	1 hour	10 min
50% formamide + 20% PEG (w/v)	1 hour	5 min
50% formamide + 20% PEG (w/v)	5–16 hours	15–60 min

Incubation scheme for GFP expressing tissues according to Kuwajima et al. (2013). Polyethylene glycol (molecular weight 8,000 g/mol) is dissolved in 1 × PBS before adding formamide. The pH is adjusted to 7.5 with NaOH.

9.2.2.4 *Tissue Clearing by Electrically Accelerated Membrane Lipid Extraction (CLARITY)* (Chung et al. 2013)

CLARITY is an innovative tissue clearing approach presented by Chung et al. (2013). CLARITY can be applied on samples up to several cm in size, e.g. entire brains and whole organs from mice, or even rats. With some experience CLARITY exhibits excellent clearing results and preserves GFP fluorescence as well as most immunostainings (Figure 9.5A–D). A novelty of the CLARITY approach is the mechanical stabilization of the tissue that is achieved by embedding the sample in a hydrogel matrix before clearing. As a second innovation, electrical current is used to facilitate lipid extraction from the cell membranes. During the lipid extraction process, the hydrogel matrix reliably prevents structural deformation and major loss of cellular proteins in the sample (Chung et al., 2013).

Although, the method is less than three years old, already numerous modifications and improvements of the original CLARITY protocol have been described (e.g. Tomer et al., 2014; Yang et al., 2014; Li et al., 2015; or Treweek et al., 2015). Yang et al. (2014) developed a simplified version of the protocol, which works without the application of an electric field (passive clarity technique PACT). A comparable protocol also omitting electrophoresis is described in Tomer et al. (2014).

A recent study from Epp et al. (2015) comparing electrophoresis-assisted and PACT found that optimal tissue clearing results cannot be obtained by passive clearing alone, if the samples are bigger than a few mm^3, e.g. entire brains from mice. The publication also addresses certain issues with the reproducibility of the CLARITY technique, which emerged after the first publication of the method in the year 2013, where it was found difficult by many researchers to implement CLARITY in their lab. To facilitate the accessibility of CLARITY, Epp et al. (2015) published an optimized version of the original CLARITY protocol providing more detailed descriptions of technical details, such as optimal hydrogel composition, clearing times, and temperature. Although this protocol was optimized for mouse brains it can be adapted to other samples as well.

A highly promising property of gel-stabilized tissue already demonstrated by Chung et al. (2013) is the option of infiltrating antibodies for fluorescence staining very deep into the tissue, since there are no cell membranes anymore that could act as barriers. Nevertheless, a diffusion driven, passive permeation of antibodies into the hydrogel matrix unexpectedly turned out to be notoriously slow, making such an approach almost untenable, if multiple primary and secondary antibodies were to be sequentially infiltrated into the tissue. Li et al. (2015) recently found a promising solution for this problem with their important discovery that the permeation speed of antibodies into the hydrogel matrix can be dramatically increased by a factor of more than 800, if an electrical field, similar to the one that is used for lipid extraction, is applied.

9.2.2.4.1 Clearing Samples with CLARITY The demands of CLARITY on the lab instrumentation are relatively high and establishing the technique can be somewhat tricky before obtaining the first satisfying results. Due to the complex technical instrumentation required (Figure 9.5E), only a principal description of the required setup is provided here.

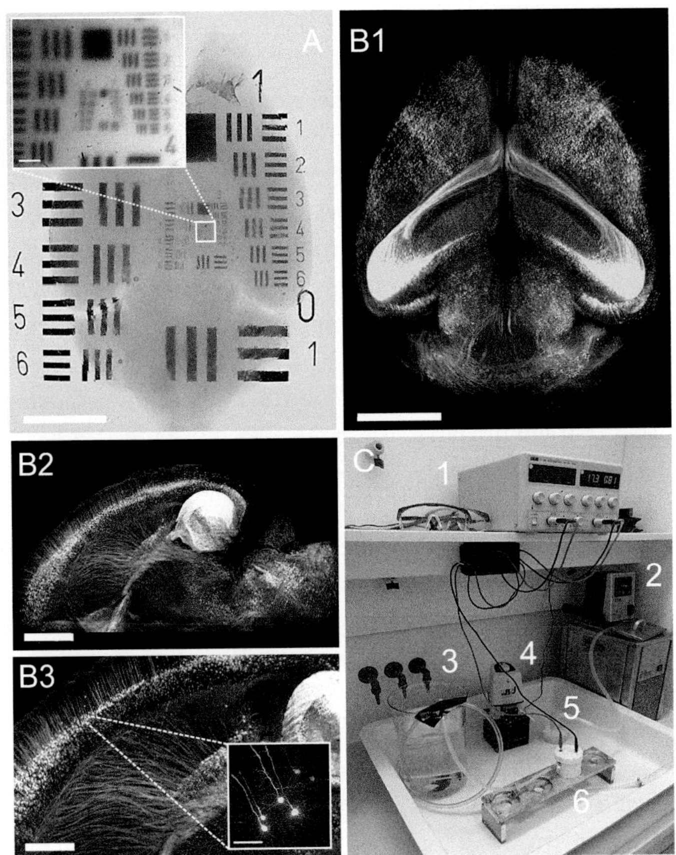

FIGURE 9.5 Clearing of mouse brains with CLARITY: (**A**) Brain after treatment with CLARITY placed on an USAF resolution test chart. The transparency is good enough to dissolve structures on a micron scale through the whole brain. Length of scale bars: 5 mm and 100 μm. (**B**) The images in B1-B3 show ultramicroscopy reconstructions of the mouse brain depicted in (A). All images were taken with a 2×, N.A 0.14 objective corrected with custom made correction optics and 0.63× demagnifier. The thin light sheet was generated with aspherical optics (Saghafi et al., 2014). Length of scale bars: B1: 5 mm, B2: 2 mm, B3: 1 mm and 100 μm. (**C**) Experimental setup for CLARITY. 1. Constant current power supply. 2. Water bath with thermostat. 3. Reservoir with SDS buffer solution. 4. Chemical resistant membrane pump. 5. Particle filter. 6. Specimen chamber.

Detailed examples how to implement the technique were published, e.g. by Treweek et al. (2015) or Epp et al. (2015).

A setup for CLARITY comprises a specimen chamber containing the sample to be cleared and two platinum electrodes for electric field generation, a reservoir for the SDS-buffer solution, and a pump for circulating the SDS lipid extraction buffer in a closed circuit. A water bath equipped with a cooling thermostat is recommended to enable cooling of the system if necessary, since the heat production due to the electric current in the specimen chamber can be quite high. A constant current voltage source (or a commercial electrophoresis power supply) with one or more outlets of at least 75W is required for powering the system (Figure 9.5E).

In all variations of the CLARITY protocol the tissue is mechanically stabilized via an acrylamide hydrogel mesh. Therefore, the animal is perfused with a cocktail containing the monomers acrylamide and bis-acrylamide, as well as formaldehyde. Also, a polymerization starter has to be contained in the mixture. After dissection of the organs of interest, the acrylamide monomers are thermally polymerized to an elastic hydrogel at 37°C. During the polymerization process, the acryl amide monomers become cross-linked to proteins, nucleic acids, and other small, low molecular weight cellular tissue compounds, forming a hybrid, mechanically stable, molecular mesh. Importantly, the lipids constituting the cell membranes do not become integrated in this mesh, facilitating their later extraction (Chung et al., 2013).

In the next step the membrane lipids are solubilized by the anionic detergent SDS, forming highly negatively charged micelles. These micelles move towards to the anode and are extracted from the tissue. Importantly, the polymerized hydrogel preserves even fine protein structures such as membrane-localized proteins, synapses, and spines from displacement and structural alterations during lipid extraction (Chung et al., 2013).

Finally, the specimens are carefully washed to completely remove the detergent and then incubated in an aqueous clearing medium such as FocusClear (CellExplorer Labs, Taiwan), or 80% glycerol solution with a refractive index of about 1.45 (Chung et al., 2013). Since the lipid membranes were removed, the refractive-matching medium rapidly saturates the tissue, thereby making the tissue transparent via refractive index matching (Figure 9.1). Since FocusClear (see Section 9.2.2.1) is extremely expensive in the amounts needed for brains and other whole mount samples, Yang et al. (2014) suggest an equivalent substitute consisting of 88% (w/v) of the nonionic density gradient centrifugation medium Histodenz (5-(N-2,3-Dihydroxypropylacetamido)-2,4,6-triiodo-N,N′-bis(2,3-dihydroxypropyl) isophthalamide, Sigma Aldrich, order-no. D2158) and 0.1% Tween-20 (Sigma-Aldrich, order-no. P1379) in PBS, which they termed RIMS. 0.01% sodium azide (Sigma-Aldrich, order-no. 71290) is added to prevent bacterial contamination of RIMS. The pH is adjusted with NaOH to a value of 7.5.

Since 2015, a ready-made, commercial CLARITY system is also available under the brand name sCLARITY. It comprises ready-made devices for electrophoresis-driven lipid extraction and tissue gel hybridization, as well as equipment for refractive index matching and specimen mounting. According to the manufacturer (Company Live Cell Instrument (LCI), Seoul, Korea) sCLARITY comprises several patent pending improvements making the system faster and more convenient to use. The feasibility of this machine has still to be proven by significant publications.

9.2.2.5 *Tissue Clearing with Fructose Solutions* (Ke et al., 2013; Hou et al. 2015)

Incubation in solutions of ascending concentrations of fructose was proposed under the acronym SeeDB (See Deep Brain) by Ke et al. (2013) for clearing juvenile mouse brains and brain hemispheres from adult mice. Fructose is a mono saccharide that is highly soluble in water (~3750 g/L at 20°C). An 80% fructose solution (w/w) exhibits a refractive index of about 1.49, which is higher than the refractive index of other water-based tissue clearing substances such as 4M solutions of urea (1.39) or 60% sucrose (1.43). The authors found that high-concentration fructose solutions, differently to sucrose, do not cause relevant tissue shrinkage or expansion (Tsai et al., 2009). However, prolonged incubation in fructose solutions (>7 days), can cause browning and auto-fluorescence accumulation due to the Maillard reaction among the fructose molecules and amino acids from sample proteins. As a solution, Ke et al. (2013) found out that this effect can be prevented by adding a reductive such as α-thioglycerol or ß-mercaptoethanol to the fructose solutions.

As with other detergent-free clearing approaches based on hydrophilic substances, this method suffers from an insufficient permeation of the clearing medium across the cell membranes. In case of SeeDB, this drawback is even increased by the high viscosity of concentrated fructose solutions. Nevertheless, this method may be sufficient for, e.g. thin specimens as murine brain slices, if the requests on transparency are not too high. However, larger samples as mouse brains do not become clear with SeeDB. This is already obvious by the images presented in Ke et al. (2013, e.g. supplemental Figure 9.3).

Protocol for clearing murine brain slices, juvenile mouse brains or mouse embryos with SeeDB (Ke et al., 2013).

- Fix the samples in 4% formaldehyde solution.
- Sequentially incubate the samples in fructose solutions of increasing concentrations (20–30 ml each) of 20%, 40%, and 60% (w/v) at 25°C for 4–8 h each. To prevent browning add 0.5% α-thioglycerol (Sigma-Aldrich, e.g. order-no. M1753) to each fructose concentration. Gently shake or rotate the specimen tubes for better mixing during dehydration.
- Incubate samples in 80% and 100% (w/v) fructose containing 0.5% α-thioglycerol for 12h each.
- Finally, transfer the samples in 80.2% (w/w) fructose solution containing 0.5% α-thioglycerol (SeeDB) for further 24 h.
- Mount the samples in SeeDB for microscopy inspection.

Another tissue clearing approach using fructose for refractive index matching, which tries to compass the drawbacks of SeeDB as its high viscosity was published by Hou et al. (2015). Their clearing cocktail is termed FRUIT, and in addition to fructose, it contains urea. The authors report no tissue swelling as observed with pure urea solutions. They further report a 97% lower viscosity compared to SeeDB. Due to this low viscosity, FRUIT can even be applied via arterial perfusion (Ke et al., 2013).

Contrarily to other clearing solutions containing detergents (e.g. Sca/e or CLARITY) FRUIT preserves the fluorescence of lipophilic tracers as DiI (1,1'-dioctadecyl-3,3,3',3'-tetramethylindocarbocyanine perchlorate) since it, as with SeeDB, contains no detergents (Hou et al., 2015). Although FRUIT produces better clearing results compared to SeeDB, the obtainable tissue transparency is not as good as, for example, with CLARITY (Chung et al., 2013), CUBIC (Susaki et al., 2014), or organic solvent-based clearing approaches such as 3DISCO (Ertürk et al., 2012).

TABLE 9.7 Clearing of Brains with FRUIT

Solution	A	B	C	D	E
% (w/v) fructose	35	40	60	80	100
% (w/v) urea	16.8	19.2	28.8	38.4	48.0[a]
% (w/v) α-thioglycerol	0.5	0.5	0.5	0.5	0.5

Composition of the clearing solutions A–E.

[a] If the whole amount of urea cannot be dissolved, add urea until the solution becomes saturated.

Protocol for clearing mouse brains with fruit (Hou et al., 2015).

- Prepare the FRUIT solutions according to Table 9.7 (20–30 ml per mouse brain).
- Deeply anesthetize mice with pentobarbital and transcardially perfuse them with 1 × PBS followed by 4% PFA.
- Excise the brains and postfix them in 4% PFA at 4°C overnight.
- Serially incubate the brains in solutions A–C for 8 h each.
- Incubate the brains in solution D for 12 h.
- Incubate the brains in solution E for 24 h.
- Mount samples in solution E for microscopy.
- Samples can be stored at 4°C in solution E over two months.

9.2.2.6 *Tissue Clearing Using Amino Alcohols* (CUBIC, Susaki et al. 2014)

Starting from the urea-based clearing solution Sca/eA2 (Hama et al., 2011), Susaki et al. (2014) developed an improved protocol, which they termed CUBIC. It comprises sequential incubation steps in two clearing solutions (CUBIC 1 and 2) which contain, beside urea and the anionic detergent Triton X-100, supplements of the aminoalcohols Quadrol (N,N,N',N'-tetrakis(2-hydroxypropyl)ethylenediamine) (CUBIC 1), or triethanolamine (CUBIC 2).

It was shown in screening experiments that both aminoalcohols considerably enhance tissue transparency, probably by solvating anionic phospholipids from the tissue (Susaki et al., 2014). CUBIC 1 clears most regions of an entire mouse brain. Subsequent incubation in CUBIC 2 further enhances the transparency of myelin-rich structures deep in the brain that were not sufficiently rendered transparent with CUBIC 1. The refractive index of CUBIC 2 (1.48–1.49) is higher compared to CUBIC 1 (1.45–1.46), which further improves the transparency of the tissue. Therefore, Cubic 2 is also recommended as the mounting medium for microscopy inspection (Susaki et al., 2014).

CUBIC 1 can be directly used for intracardial perfusion. Interestingly, Tainaka et al. (2014) found that during perfusion the aminoalcohol Quadrol, which is contained in CUBIC 1, markedly decolorizes the blood by eluting the heme chromophore from the hemoglobin. Although, such a bleaching effect could also be achieved, e.g. using Dent's bleach containing hydrogen peroxide (Klymkowsky and Hanken, 1991), there had always been the drawback that peroxides essentially destroy the fluorescence of genetic markers as GFP (Alnuami et al., 2008). However, Quadrol decolorizes blood without affecting the fluorescence of genetic markers, exhibiting its bleaching properties already at moderately basic pH conditions (around pH 10) that are still within the optimal range for GFP fluorescence (Tainaka et al., 2014).

Clearing protocol for entire mouse brains with CUBIC (Susaki et al., 2014).

- Dissect the brain and fix it in 4% formaldehyde solution.
- Prepare the following solutions:
- CUBIC reagent 1: solve 25 % urea (w/w), 25% Quadrol (w/w) (N,N,N',N'-tetrakis(2-hydroxypropyl)ethylenediamine, Sigma-Aldrich, order-no: 122262), and 15% Triton X-100 (w/w) in water. Sucrose solution: solve 20% sucrose (w/v) in PBS. CUBIC reagent 2: solve 50% sucrose (w/w), 25% urea (w/w), 10% (w/w) Triethanolamine, Sigma-Aldrich, order-no. 90279), and 0.1% Triton X-100 (v/v) in water.
- Immerse the brain in ~10 g of CUBIC reagent 1 for 3 days at 37°C while gently shaking.
- Renew the clearing solution and incubate the brain for further 3–4 days at 37°C.
- Wash the brain several times in PBS at room temperature while gently shaking.
- Immerse the brain in 20% sucrose dissolved in PBS (degas before use).
- Incubate the brain for 3–7 days in CUBIC 2 reagent (~10 g per brain).
- Mount the sample in CUBIC 2 for microscopy.
- After microscopy the brain should be washed again with PBS, immersed in 20% sucrose/PBS in which it can be stored at −80°.

9.2.2.7 *Tissue Clearing with Thiodiethanol (TDE)* (Aoyagi et al., 2015; Costantini et al., 2015)

2,2'-thiodiethanol was firstly introduced into histology by Staudt et al. (2007) as a highly refractive mounting medium for cultured cells that is mixable with water in any proportions. It allows an easy adjustment of the refractive index by simple dilution with water in the range from of 1.33 (100% water) to 1.52 (100% TDE). Two recent publications (Aoyagi et al., 2015; Costantini et al., 2015) also propose TDE as a cheap and straightforward clearing medium for murine brain slices. The fixed slices are either directly incubated in 60% aqueous TDE solution (Aoyagi et al., 2015) or in differently concentrated TDE solutions up to 80% to best possibly match the refractive index correction of the available objectives (Costantini et al., 2015). Generally, the obtainable tissue transparency increases with higher TDE concentrations up to 80% provided that suitable objectives are available. Costantini et al. (2015) further recommend a 63% solution of TDE (n = 1.45) as an inexpensive substitute for FocusClear as the final refractive matching medium in CLARITY (Chung et al., 2013).

While solutions containing 97% TDE markedly quench the fluorescence of GFP, Costantini et al. (2015), Aoyagi et al. (2015), and Staudt et al. (2007) found no pronounced quenching effect for TDE concentrations of up to about 80%, which corresponds to a refractive index of $n = 1.48$. Interestingly, Staudt et al. (2007) observed an enhancing effect of TDE on the fluorescence of the cyanine dye Cy3 that is frequently used as fluorochrome with immunostainings, by a factor of more than four-fold compared to samples mounted in PBS.

As with most other hydrophilic clearing approaches not utilizing detergents to solubilize cell membranes such as CLARITY or CUBIC, the transparency that can be achieved by sole immersion in TDE is quite limited. TDE solutions alone do not clear, e.g. whole brains that were not pre-treated with detergents, e.g. in CLARITY or PAC (see Section 9.2.2.4). However, TDE may be useful as a rapid, cheap, and simple method to clear brain slices of a few hundred micrometer thickness without much effort. Aoyagi et al. (2015) report maximal penetration depths in TDE-cleared brain slices of up 2 mm with two-photo microscopy, which may be sufficient for various applications.

Protocol for clearing brain slices with TDE (Aoyagi et al., 2015; Costantini et al., 2015).

- Prepare aqueous TDE solutions (e.g. Sigma-Aldrich, order-no. 166782) of the required refractive index (not more than 80% (v/v) if preservation of GFP fluorescence is required) by dilution with water or PBS. Since the relation between the concentration (v/v) of TDE and the refractive index is linear (Staudt et al., 2007) the concentration C (in VOL%) that is required to match a refractive index n can be simply determined by the expression $C = (n - 1.33)/0.0019$.
- Incubate the brain slices in a small volume of TDE solution for about 1 h until they are clear enough for microscopy inspection. Clearing with TDE introduces a moderate tissue shrinkage of about 10%, which is without significant anisotropic distortion (Costantini et al., 2015).
- Mount the sample in the TDE solution for microscopy.

9.3 DISCUSSION

Chemical tissue clearing is a field of research which is presently undergoing a fast evolution. Therefore, it can be expected that a lot of further tissue clearing approaches will be continuously developed in the next years. While small samples in the size range of up to 1 cm, e.g. mouse brains, can presently be made nearly fully transparent with recent approaches such as 3DISCO (Ertürk et al., 2012), CLARITY (Chung et al., 2013), or CUBIC (Susaki et al., 2014) and with excellent preservation of fluorescence markers, there is still room for further improvements in clearing of whole-body samples from entire mice and other experimental animals. Although recently, promising steps in this respect were made, e.g. by Tainaka et al. (2014), the presently achievable transparency, e.g. of entire adult mice, is still not good enough to resolve all structures of interest through the whole thickness of these samples.

Microscopy of such highly transparent whole mount samples would further require the availability of refractive index matched microscope objectives of sufficient working distance and high numerical aperture in combination with low magnifications (Dodt et al., 2015). A further yet unresolved issue is the clearing of bone (Susaki and Ueda, 2016). Although, it was recently possible to clear longitudinally bisected tibias from 8–12 weeks old mice using BABB or 3DISCO (Acar et al., 2015), very large samples such as entire animals would require adequate decalcification prior to clearing. A promising approach for decalcification that (in contrast to well-known approaches such as incubation in diluted hydrochloric or formic acid) does not affect fluorescence markers such as GFP, may be the application of chelators as EDTA. Combined with a modified passive CLARITY technique (PACT-deCAL), EDTA was already successfully applied for bone clearing by Treweek et al. (2015).

Secondary to light scattering, light absorption at chromophores, such as hemoglobin, bilirubin, melatonin, and others is an obstacle that cannot be addressed by reduction of light scattering. Although a harsh bleaching approach, e.g. with hydrogen peroxide, or perchloric acid would be possible to decolorize samples, it would destroy most fluorescence markers, making the samples worthless for fluorescence microscopy, except inspections in autofluorescent light. As a step in this direction, the problem of a fluorescence-friendly removal of hemoglobin was already successfully addressed by Tainaka et al. (2014). However, the removal of other pigmentations disturbing the clearing process, e.g. melatonin, is an unsolved challenge (Susaki and Ueda, 2016) that still is awaiting a solution.

REFERENCES

Acar M, Kocherlakota KS, Murphy MM, Peyer JG, Oguro H, Inra CN, Jaiyeola C, Zhao Z, Luby-Phelps K, Morrison SJ (2015) Deep imaging of bone marrow shows non-dividing stem cells are mainly perisinusoidal. *Nature* 526:126–130.

Alnuami AA, Zeedi B, Qadri SM, Ashraf SS (2008) Oxyradical-induced GFP damage and loss of fluorescence. *International Journal of Biological Macromolecules* 43:182–186.

Aoyagi Y, Kawakami R, Osanai H, Hibi T, Nemoto T (2015) A rapid optical clearing protocol using 2,2′-thiodiethanol for microscopic observation of fixed mouse brain. *PLoS One* 10:e0116280.

Becker K, Jährling N, Saghafi S, Weiler R, Dodt H-UU, Jahrling N, Saghafi S, Weiler R, Dodt H-UU (2012) Chemical clearing and dehydration of GFP expressing mouse brains. *PLoS One* 7:e33916.

Brady OL, Elsmie GV (1926) The use of 2:4-dinitrophenylhydrazine as a reagent for aldehydes and ketones. *Analyst* 51:77–78.

Burdock GA, Ford RA (1992) Safety evaluation of dibenzyl ether. *Food and Chemical Toxicology* 30:559–566.

Chiang A (2002) Aqueous tissue clearing solution. *Patent No.: US 6,472,216 B1.*

Chiang AS, Liu YC, Chiu SL, Hu SH, Huang CY, Hsieh CH (2001) Three-dimensional mapping of brain neuropils in the cockroach, *Diploptera punctata. The Journal of Comparative Neurology* 440:1–11.

Chung K, Wallace J, Kim S-Y, Kalyanasundaram S, Andalman AS, Davidson TJ, Mirzabekov JJ, Zalocusky KA, Mattis J, Denisin AK, Pak S, Bernstein H, Ramakrishnan C, Grosenick L, Gradinaru V, Deisseroth K (2013) Structural and molecular interrogation of intact biological systems. *Nature* 497:332–337.

Costantini I, Ghobril J-P, Di Giovanna AP, Mascaro ALA, Silvestri L, Müllenbroich MC, Onofri L, Conti V, Vanzi F, Sacconi L, Guerrini R, Markram H, Iannello G, Pavone FS (2015) A versatile clearing agent for multi-modal brain imaging. *Scientific Reports* 5:9808.

Dent JA, Polson AG, Klymkowsky MW (1989) A whole-mount immunocytochemical analysis of the expression of the intermediate filament protein vimentin in Xenopus. *Development* 105:61–74.

Dodt H-U, Leischner U, Schierloh A, Jährling N, Mauch CP, Deininger K, Deussing JM, Eder M, Zieglgänsberger W, Becker K (2007) Ultramicroscopy: three-dimensional visualization of neuronal networks in the whole mouse brain. *Nature Methods* 4:331–336.

Dodt H-U, Saghafi S, Becker K, Jährling N, Hahn C, Pende M, Wanis M, Niendorf A (2015) Ultramicroscopy: development and outlook. *Neurophotonics* 2:041407.

Drahn F (1922) Ein neues Durchtränkungsmittel für histologische und anatomische Objekte. *Berlin tierärztliche Wochenschr* 38:97–100.

Eichel FG, Othmer DF (1949) Benzaldehyde by autoxidation by dibenzyl ether. *Industrial and Engineering Chemistry Research* 41:2623–2626.

Epp JR, Niibori Y, Hsiang HL, Mercaldo V, Deisseroth K, Josselyn SA, Frankland PW (2015) Optimization of CLARITY for clearing whole-brain and other intact organs. *eNeuro* 2:1–15.

Ertürk A, Becker K, Jährling N, Mauch CP, Hojer CD, Egen JG, Hellal F, Bradke F, Sheng M, Dodt H-U (2012) Three-dimensional imaging of solvent-cleared organs using 3DISCO. *Nature Protocols* 7:1983–1995.

Ertürk A, Mauch CP, Hellal F, Förstner F, Keck T, Becker K, Jährling N, Steffens H, Richter M, Hübener M, Kramer E, Kirchhoff F, Dodt HU, Bradke F (2011) Three-dimensional imaging of the unsectioned adult spinal cord to assess axon regeneration and glial responses after injury. *Nature Medicine* 18:166–171.

Galster H (1991) *pH Measurement. Fundamentals, Methods, Applications, Instrumentation.* Weinheim, Germany: Wileh-VCH.

Genina EA, Bashkatov AN, Tuchin VV (2010) Tissue optical immersion clearing. *Expert Review of Medical Devices* 7:825–842.

Hama H, Hioki H, Namiki K, Hoshida T, Kurokawa H, Ishidate F, Kaneko T, Akagi T, Saito T, Saido T, Miyawaki A (2015) ScaleS: an optical clearing palette for biological imaging. *Nature Neuroscience* 18:1518–1529.

Hama H, Kurokawa H, Kawano H, Ando R, Shimogori T, Noda H, Fukami K, Sakaue-Sawano A, Miyawaki A (2011) Scale: a chemical approach for fluorescence imaging and reconstruction of transparent mouse brain. *Nature Neuroscience* 14:1481–1488.

Hou B, Zhang D, Zhao S, Wei M, Yang Z, Wang S, Wang J, Zhang X, Liu B, Fan L, Li Y, Qiu Z, Zhang C, Jiang T (2015) Scalable and DiI-compatible optical clearance of the mammalian brain. *Frontiers in Neuroanatomy* 9:19.

Ke M-T, Fujimoto S, Imai T (2013) SeeDB: a simple and morphology-preserving optical clearing agent for neuronal circuit reconstruction. *Nature Neuroscience* 16:1154–1161.

Keller PJ, Dodt HU (2012) Light sheet microscopy of living or cleared specimens. *Current Opinion in Neurobiology* 22:138–143.

Klymkowsky MW, Hanken J (1991) Whole-mount staining of Xenopus and other vertebrates. *Methods in Cell Biology* 36:419–441.

Kuwajima T, Sitko AA, Bhansali P, Jurgens C, Guido W, Mason C (2013) ClearT: a detergent- and solvent-free clearing method for neuronal and non-neuronal tissue. *Development* 140:1364–1368.

Li J, Czajkowsky DM, Li X, Shao Z (2015) Fast immuno-labeling by electrophoretically driven infiltration for intact tissue imaging. *Scientific Reports* 5:10640.

Lundvall H (1904) Über demonstration embryonaler Knorpelskelette. *Anatomischer Anzeiger* 25:219–222.

Renier N, Wu Z, Simon DJ, Yang J, Ariel P, Tessier-Lavigne M (2014) IDISCO: a simple, rapid method to immunolabel large tissue samples for volume imaging. *Cell* 159:896–910.

Richardson DS, Lichtman JW (2015) Clarifying tissue clearing. *Cell* 162:246–257.

Rylander CG, Stumpp OF, Milner TE, Kemp NJ, Mendenhall JM, Diller KR, Welch AJ (2006) Dehydration mechanism of optical clearing in tissue. *Journal of Biomedical Optics* 11:041117.

Saghafi S, Becker K, Hahn C, Dodt HU (2014) 3D-ultramicroscopy utilizing aspheric optics. *Journal of Biophotonics* 7:117–125.

Schwarz MK, Scherbarth A, Sprengel R, Engelhardt J, Theer P, Giese G (2015) Fluorescent-protein stabilization and high-resolution imaging of cleared, intact mouse brains. *PLoS One* 10:e0124650.

Spalteholz W (1911) Über das Durchsichtigmachen von menschlichen und tierischen Präparaten. 1911:48.

Staudt T, Lang MC, Medda R, Engelhardt J, Hell SW (2007) 2,2'-thiodiethanol: a new water soluble mounting medium for high resolution optical microscopy. *Microscopy Research and Technique* 70:1–9.

Susaki EA, Tainaka K, Perrin D, Kishino F, Tawara T, Watanabe TM, Yokoyama C, Onoe H, Eguchi M, Yamaguchi S, Abe T, Kiyonari H, Shimizu Y, Miyawaki A, Yokota H, Ueda HR (2014) Whole-brain imaging with single-cell resolution using chemical cocktails and computational analysis. *Cell* 157:726–739.

Susaki EA, Tainaka K, Perrin D, Yukinaga H, Kuno A, Ueda HR (2015) Advanced CUBIC protocols for whole-brain and whole-body clearing and imaging. *Nature Protocols* 10:1709–1727.

Susaki EA, Ueda HR (2016) Whole-body and whole-organ clearing and imaging techniques with single-cell resolution: toward organism-level systems biology in mammals. *Cell Chemical Biology* 23:137–157.

Tainaka K, Kubota SI, Suyama TQ, Susaki EA, Perrin D, Ukai-Tadenuma M, Ukai H, Ueda HR (2014) Whole-body imaging with single-cell resolution by tissue decolorization. *Cell* 159:911–924.

Tomer R, Ye L, Hsueh B, Deisseroth K (2014) Advanced CLARITY for rapid and high-resolution imaging of intact tissues. *Nature Protocols* 9:1682–1697.

Treweek JB, Chan KY, Flytzanis NC, Yang B, Deverman BE, Greenbaum A, Lignell A, Xiao C, Cai L, Ladinsky MS, Bjorkman PJ, Fowlkes CC, Gradinaru V (2015) Whole-body tissue stabilization and selective extractions via tissue-hydrogel hybrids for high-resolution intact circuit mapping and phenotyping. *Nature Protocols* 10:1860–1896.

Tsai PS, Kaufhold JP, Blinder P, Friedman B, Drew PJ, Karten HJ, Lyden PD, Kleinfeld D (2009) Correlations of neuronal and microvascular densities in murine cortex revealed by direct counting and colocalization of nuclei and vessels. *Journal of Neuroscience* 29:14553–14570.

Tuchin VV (2005) Optical clearing of tissues and blood using the immersion method. *Journal of Physics D: Applied Physics* 38:2497–2518.

Yang B, Treweek JBB, Kulkarni RPP, Deverman BEE, Chen C-KK, Lubeck E, Shah S, Cai L, Gradinaru V (2014) Single-cell phenotyping within transparent intact tissue through whole-body clearing. *Cell* 158:945–958.

Zhu D, Larin KV, Luo Q, Tuchin VV (2013) Recent progress in tissue optical clearing. *Laser and Photonics Reviews* 7:732–757.

CHAPTER 10

Advanced Light-Sheet Microscopy to Explore Brain Structure on an Organ-Wide Scale

Ludovico Silvestri and Francesco S. Pavone

CONTENTS

Light-sheet microscopy has emerged as the method of choice to perform brain studies the at whole-organ scale. Indeed, this technique achieves high three-dimensional resolution at unprecedented volumetric throughput. In this chapter, we review the basics of the technology and the most recent developments and applications for whole-brain structural imaging.

10.1 INTRODUCTION

The brain is an extremely complex organ, whose structure and function spans many orders of magnitude. Indeed, single neurons often span the entire encephalon – a cm scale in rodents, a dm scale in humans – but with ramifications that are usually micrometric or sub-micrometric in thickness (Lichtman and Denk, 2011). On the other hand, brain activity relies on the transmission of electrical signals along neurons and chemical exchange at synapses, with time scales in the ms range. Investigating the subtle interaction between brain structure and function thus requires the ability to map large volumes with high spatial resolution and substantial speed.

Strictly speaking, no single imaging technology is able to cover all the relevant spatio-temporal scales (Allegra Mascaro et al., 2015). However, in the last years light-sheet fluorescence microscopy (LSFM) has demonstrated its potential to provide a brain-wide, yet cellular-resolution view of the entire brain in several animal models (Huisken and Stainier, 2009; Keller and Dodt, 2012; Keller and Ahrens, 2015; Keller et al., 2015). The high throughput of this method allows not only to investigate large-scale neuronal activity with unprecedented temporal resolution, but finds also an important application in charting brain cytoarchitecture in a scalable way. Indeed, albeit imaging speed is not essential in principle for *ex vivo* specimens, it is fundamental from a practical perspective to analyze a significant number of samples on a reasonable timescale. For instance, no complete reconstruction of a murine brain (which has a volume of approximately 1 cubic cm) has been produced hitherto with confocal or two-photon scanning microscopies, while such samples are routinely imaged in less than one day with LSFM. Furthermore, the recent development of methods that increase effective spatial resolution by physically expanding the sample under observation (Chen et al., 2015) needs to be complemented by high-speed, scalable imaging methods. In this respect, LSFM has been demonstrated to be the technique of election, producing large-scale reconstruction of nervous tissue at unprecedented resolution (Migliori et al., 2018; Gao et al., 2019).

Light-sheet microscopy was invented in the early 20th century by the German physicist Heinrich Siedentopf and the physical chemist Richard Zsigmondy (Siedentopf and Zsigmondy, 1902), who introduced a planar illumination system into a standard microscope (Figure 10.1a), bringing the concept of dark-field oblique illumination to its limit. The "ultramicroscope" was used by Zsigmondy to study colloidal suspensions, and his achievements in the field were awarded with the Nobel Prize in Chemistry in 1925. Decades after Siedentopf and Zsigmondy's pioneering work, planar illumination was coupled with fluorescence imaging of biological specimens by Voie and colleagues in 1993 (Voie et al., 1993). Eleven years later, Ernst Stelzer's group in Heidelberg started a more systematic development of light-sheet fluorescence microscopy, unraveling its inherent potential in the life sciences (Huisken et al., 2004). From that moment on, the popularity of planar illumination has steadily increased, and currently many labs around the world either built or bought a light-sheet microscope.

10.2 LIGHT-SHEET FLUORESCENCE MICROSCOPY

The principle of LSFM is fairly simple: the sample is illuminated from the side with a thin sheet of light, and fluorescence is detected along an axis perpendicular to the illumination plane (Fig. 10.1b). By placing the sheet of light in the focal plane of the detection objective, only in-focus fluorophores are excited, and no out-of-focus fluorescence is generated inside the sample. Selective planar illumination thus allows optical sectioning while maintaining a wide-field detection scheme. In contradistinction to confocal and two-photon microscopies, LSFM is therefore extremely fast as an entire plane is imaged at once.

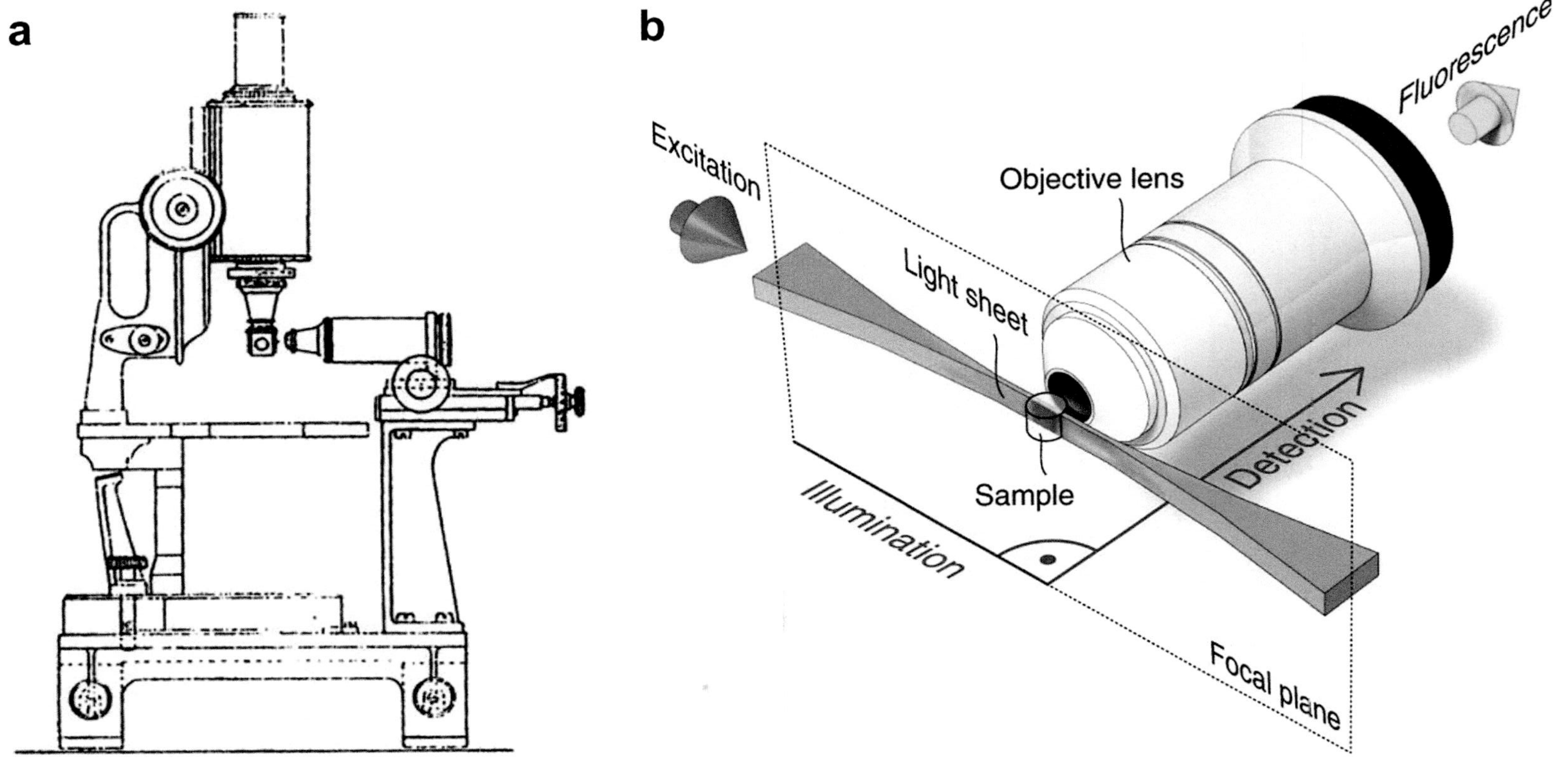

FIGURE 10.1 Light-sheet fluorescence microscopy. (a) A drawing of the first "ultramicroscope" developed by Siedentopf and Zsigmondy, from the original paper (Siedentopf and Zsigmondy, 1902). (b) The operating principle of LSFM, adapted with permission from Huisken and Stainier (2009).

The excitation sheet of light can be produced either using a cylindrical lens (Huisken et al., 2004) or virtually, by scanning a laser line inside the specimen (Keller et al., 2008). The first method is simpler to implement, and has the benefit of keeping peak illumination intensity low (Power and Huisken, 2017). On the other hand, beam scanning can be coupled with confocal line detection, increasing image contrast by 100% or more (Silvestri et al., 2012). Confocal LSFM is nowadays increasingly popular as it can be easily implemented by taking advantage of the rolling shutter of sCMOS cameras (Baumgart and Kubitscheck, 2012).

The lateral and axial resolution of an LSFM apparatus are limited by the detection and illumination optics, respectively. In detail, the smallest resolvable distance in the plane defined by the illumination light sheet is given by the traditional Abbe's formula for wide-field microscopes:

$$d_{xy} = \frac{0.61\lambda}{NA_{det}}$$

where λ is the wavelength of fluorescence emission, and NA_{det} is the numerical aperture of the detection objective. On the other hand, the axial resolution of the microscope is given by the thickness of the illumination volume. In most implementations, the sheet of light is obtained by focusing a Gaussian beam (either in 1D or 2D). From the optics of Gaussian beams, the $1/e^2$ waist thickness is:

$$d_z = \frac{\lambda}{\pi\, NA_{ill}}$$

with NA_{ill} indicating the $1/e^2$ numerical aperture of the excitation beam. The available field of view (FOV) is usually identified with the confocal parameter of the beam, i.e. the region around the waist where the beam radius is constant within a factor $\sqrt{2}$:

$$FOV = \frac{2\pi\, d_z^2}{\lambda}$$

Since the FOV is proportional to the axial resolution, imaging a larger field of view is at the cost of poorer optical sectioning. In order to mitigate this trade-off, the use of non-Gaussian beams has been proposed. In particular, scanned Bessel (Fahrbach et al., 2010) or Airy (Vettenburg et al., 2014) beams are able to maintain the same thickness for longer distances and have been used to improve axial resolution in LSFM (Planchon et al., 2011). 2D beam shaping has been used as well (Saghafi et al., 2018).

Although the spatial resolution of LSFM might be poorer than that of two-photon or confocal microscopy (especially axially), this technique affords a data throughput orders of magnitude larger than point-scanning 3D microscopy methods. Indeed, the volume rate (VR) of a standard LSFM is given by:

$$VR_{LSFM} = \#_{vox}\, v_{vox}\, FR$$

where $\#_{vox}$ is the number of voxels acquired by the camera at each exposure, FR is the camera frame rate, and v_{vox} is the voxel volume. On the other hand, the imaging speed of a point-scanning microscope is just the ratio between the voxel size v_{vox} and the dwell time t_{dwell}:

$$VR_{scan} = \frac{v_{vox}}{t_{dwell}}$$

A direct comparison of the last two formulas show a volumetric speedup by a factor $\left(\#_{vox}\, t_{dwell}\, FR\right)$. With typical values $\#_{vox} = 2048 \times 2048$, $t_{dwell} = 5\,\mu s$, $FR = 50\,Hz$, the net speedup is 1000×. In practice, LSFM is capable of obtaining micron-resolution datasets with a volumetric rate of a couple of about 10 mm^3/min, opening completely new possibilities for structural and functional imaging of entire brains.

10.3 WHOLE-BRAIN STRUCTURAL IMAGING WITH LIGHT-SHEET MICROSCOPY

The high volumetric rate of LSFM makes it an ideal technique to reconstruct macroscopic specimens with microscopic resolution. However, this requires the sample itself to be transparent, in order for the light to penetrate several millimeters (or even centimeters) inside the specimen. Dodt and colleagues were the first to report the combination of LSFM with chemical clearing of brain samples (Dodt et al., 2007). They modified the diaphanization protocol described by Spalteholz in 1911 (Spalteholz, 1914), and still used in anatomical investigation, to better preserve endogenous fluorescence of mouse brain samples labeled with GFP (Figure 10.2a, b). Following this pioneering work, a plethora of clearing methods have been described. For a thorough survey of tissue clearing technologies, we

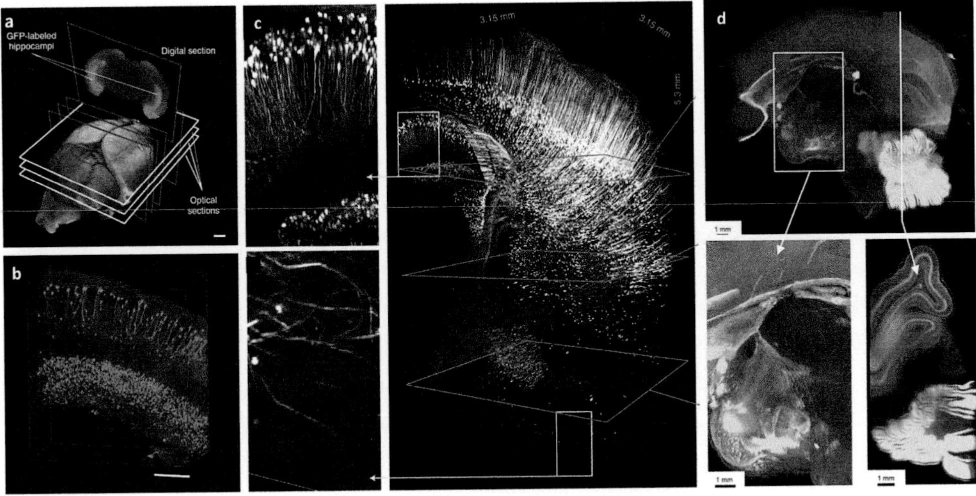

FIGURE 10.2 LSFM of cleared specimens. (a) First reconstruction of a whole mouse brain with LSFM, by Dodt and co-authors, and (b) 3D rendering of an excised hippocampus. Reproduced with permission from Dodt et al. (2007). (c) High-resolution reconstruction of a thy1-GFP-M transgenic mouse brain, adapted with permission from Tomer et al. (2014). (d) 3D imaging of a cleared unsectioned marmoset brain, adapted with permission from Susaki et al. (2014).

refer the reader to several recent reviews (Richardson and Lichtman, 2015; Silvestri et al., 2016; Susaki and Ueda, 2016). Here, we only notice that the main challenges in the field are the achievement of high and uniform transparency, high preservation of endogenous fluorescence, speed, safety, and reproducibility (Silvestri et al., 2016).

From an imaging perspective, it should be noted that clearing solutions usually show high refractive indices (sometimes even higher than that of microscope immersion oil) and thus require specialized immersion objectives to avoid the introduction of heavy aberrations (Silvestri et al., 2014). Furthermore, the sample itself might introduce defocus and other low-order aberrations. It is therefore important to introduce some adaptive correction of the microscope parameters in order to maintain high imaging quality across the entire specimen (Tomer et al., 2014; Silvestri et al., 2017) (Figure 10.2c).

High-resolution LSFM of cleared mouse brain have shown the possibility of reconstructing neuronal arborizations and fine neurites in transgenic animals which have been previously imaged *in vivo* with two-photon fluorescence microscopy (Silvestri et al., 2014). Schwartz and colleagues demonstrated the possibility of following single long-range axonal projections exploiting the high fluorescence levels obtained with viral transfection (Schwarz et al., 2015). Despite these results, the application of high-resolution LSFM has been hitherto limited to proof-of-principle studies rather than real biological problems, as the extremely large amounts of data collected with this technique poses unprecedented challenges in data analysis (see Section 10.5 below).

At the time of writing this chapter, the main biological findings obtained with planar illumination rely on low-resolution LSFM. Indeed, as a voxel size of a few microns is barely enough to recognize single cell bodies, quantification of the abundance of fluorescent neurons have been a particularly fruitful application of light-sheet microscopy of cleared mouse brains. Depending on the labeling strategy, this approach can be used to address different questions.

Trans-synaptic retrograde tracers have been successfully employed to map structural connectivity between selected cell types across the entire encephalon. For instance, Niedworok and co-authors reported a qualitative map of presynaptic inputs to olfactory bulbs (Niedworok et al., 2012). A more systematic study, with proper registration of the images to a reference space and quantitative cell detection, has found distinct anatomical groups of dopamine projection neurons in the ventral tegmental area (Menegas et al., 2015). A similar approach has also been used to investigate the connectivity of human stem cells implanted in the mouse brain (Doerr et al., 2017).

Single-cell mapping in entire cleared mouse brain is particularly suited to investigate whole-brain behavior-specific neuronal activation by looking at the expression of immediate early genes. Renier and colleagues used immunohistochemistry (based on the iDISCO protocol: Renier et al., 2014) to map activation in explorative tasks and in parental behaviors (Renier et al., 2016). Ye and co-authors used a similar approach, although based on the use of a transgenic animal rather than immunolabeling, to study whole-brain effects of appetitive versus aversive stimuli (Ye et al., 2016). This latter study was also complemented by tracing of activation-specific axonal bundles, and by optogenetic stimulation of a subset of neurons defined by their activation, leading to an exquisite global view of the brain-wide circuits involved in punishment and reward. Activation mapping using immediate early gene expression has also been performed by Susaki and colleagues (Susaki et al., 2014). The same authors also applied LSFM to image the entire marmoset brain, opening the possibility of studying primate nervous system at an organ-wide level, but still with cellular resolution (Figure 10.2d).

Finally, LSFM of cleared brain samples has been employed to map structures different from single cells, like blood vessels (Jahrling et al., 2009; Di Giovanna et al., 2017; Lugo-Hernandez et al., 2017) and amyloid plaques (Jahrling et al., 2015; Liebmann et al., 2016), providing again a valuable three-dimensional view of the entire mouse brain or of portions of human brain (Liebmann et al., 2016).

10.4 CHALLENGES IN DATA ANALYSIS

Light-sheet fluorescence microscopy is revolutionizing life sciences as it enables 3D high-resolution imaging with unprecedented throughput. However, this resolution will be incomplete until we have robust, reliable, and user-friendly methods to analyze the enormous amounts of data generated by this technique. Indeed, the experimental bottleneck has shifted from image acquisition to image processing, and this is probably the reason why most of the main biological findings obtained using LSFM exploit low-magnification systems (see Section 10.3), which are far from the diffraction limits of optical microscopy.

Actually, several efforts have been made by the scientific community to develop tools for handling and analyzing images datasets in the TeraByte range, which are easily produced in a single day by a light-sheet apparatus. Given the data size, even the basic building blocks need to be rethought. For instance, a new file format allowing efficient access to a compressed version of the imaging dataset have been proposed in (Amat et al., 2015). Also, the stitching of the multiple tiles that cover a macroscopic volume requires specialized tools (Bria and Iannello, 2012). Visualization and manual annotation of images is another challenging task, which can be addressed using a multi-resolution strategy like that described in Bria et al. (2016).

Although visual inspection of the data is still a mandatory step towards their interpretation, nonetheless the extra-large images typical of LSFM need to be analyzed in a completely automated fashion. Frasconi and co-authors described BrainCellFinder, a method based on deep neural networks to localize cells in large images with accuracy comparable to human annotation (Frasconi et al., 2014) (Figure 10.3a). BrainCellFinder has enabled a thorough analysis of the spatial distribution of Purkinje cells in the mouse cerebellum (Silvestri et al., 2015) (Figure 10.3b). On a wider scale, Renier and co-authors developed ClearMap, a tool for brain-wide analysis of cells labeled through immunohistochemistry (Renier et al., 2016) (Figure 10.3c, d). This software includes both a 3D registration step to align the acquired volume onto a reference atlas, and a cell detection step based on image filtering and peak detection. ClearMap enables quantitative comparison of brain activation – measured through density of c-fos-positive cells – between different subjects (Figure 10.3c) and across various brain areas (Figure 10.3d).

Although several proof-of-principles have been achieved for LSFM data analysis, the developed tools seem not to be mature enough to be used outside the hands of their

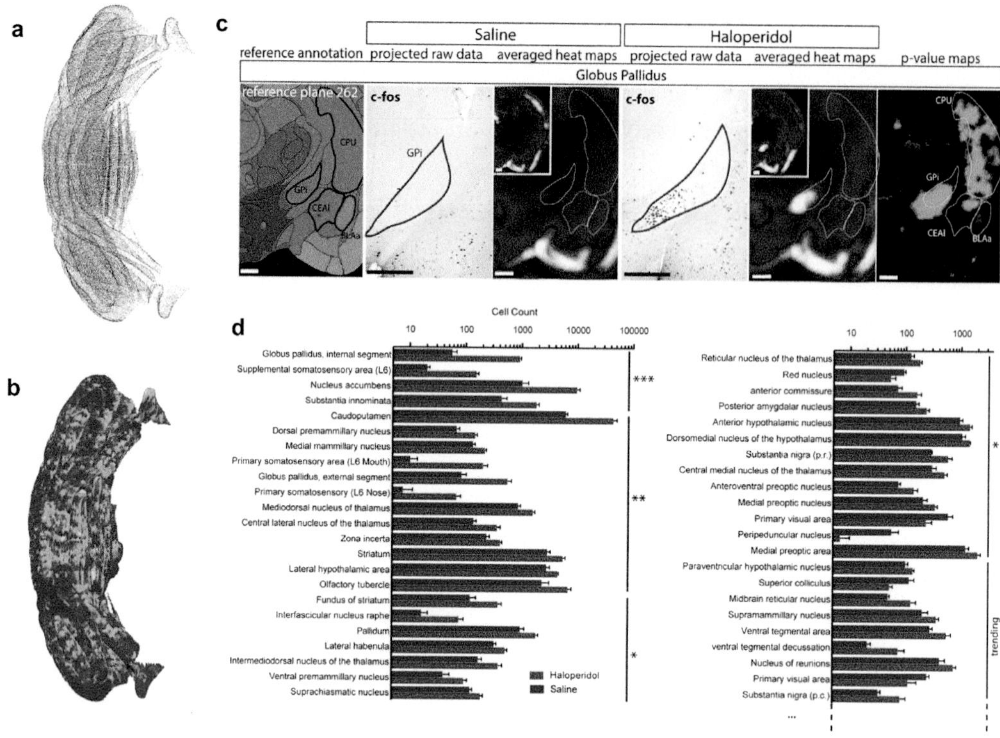

FIGURE 10.3 Data analysis challenges related to LSFM high-throughput imaging. (a) All Purkinje cells in the mouse cerebellum localized with BrainCellFinder, and visualized as a point cloud. (b) From the dataset in (a), a heatmap of the inter-cellular gaps size (blue = small gaps, red = large gaps). (a) and (b) taken with permission from Silvestri et al. (2015). (c) Example of comparative analysis of brain activation using ClearMap. The same region (Globus Pallidus) is inspected in animals treated with saline or with Haloperidol. In both cases, representative raw data from single mice and averaged heat maps from multiple animals are reported, together with a statistical map representing the p-value of local cell densities between the two groups (green indicates higher density in Haloperidol-treated subjects, whilst red is higher density in saline-treated mice). (d) Counts of activated cells for both experimental groups, across different brain areas. (c) and (d) adapted with permission from Renier et al. (2016).

developers. However, we are confident that in the next few years these instruments will become more and more reliable and flexible, eventually releasing all the power of light-sheet based techniques and completing the ongoing transition that is bringing life sciences from two dimensions to three (Keller et al., 2006).

ACKNOWLEDGMENTS

The authors have received financial support from the European Union's H2020 research and innovation program under grant agreements No. 720270 (Human Brain Project) and 654148 (Laserlab-Europe), and from the EU program H2020 EXCELLENT SCIENCE – European Research Council (ERC) under grant agreement ID n. 692943 (BrainBIT). The authors have also been supported by the Italian Ministry for Education, University, and

Research in the framework of the Flagship Project NanoMAX and of Eurobioimaging Italian Nodes (ESFRI research infrastructure), and by "Ente Cassa di Risparmio di Firenze" (private foundation).

REFERENCES

Allegra Mascaro, A. L., L. Silvestri, L. Sacconi and F. S. Pavone (2015). "Towards a comprehensive understanding of brain machinery by correlative microscopy." *Journal of Biomedical Optics* **20**(6): 61105.

Amat, F., B. Hockendorf, Y. Wan, W. C. Lemon, K. McDole and P. J. Keller (2015). "Efficient processing and analysis of large-scale light-sheet microscopy data." *Nature Protocols* **10**(11): 1679–1696.

Baumgart, E. and U. Kubitscheck (2012). "Scanned light sheet microscopy with confocal slit detection." *Optics Express* **20**(19): 21805–21814.

Bria, A. and G. Iannello (2012). "TeraStitcher - a tool for fast automatic 3D-stitching of teravoxel-sized microscopy images." *BMC Bioinformatics* **13**: 316.

Bria, A., G. Iannello, L. Onofri and H. Peng (2016). "TeraFly: real-time three-dimensional visualization and annotation of terabytes of multidimensional volumetric images." *Nature Methods* **13**(3): 192–194.

Chen, F., P. W. Tillberg and E. S. Boyden (2015). "Optical imaging. Expansion microscopy." *Science* **347**(6221): 543–548.

Di Giovanna, A. P., A. Tibo, L. Silvestri, M. C. Muellenbroich, I. Costantini, A. L. Allegra Mascaro, L. Sacconi, P. Frasconi and F. S. Pavone (2018). "Whole-brain vasculature reconstruction at the single capillary level." Scientific Reports 8(1): 12573.

Dodt, H. U., U. Leischner, A. Schierloh, N. Jahrling, C. P. Mauch, K. Deininger, J. M. Deussing, M. Eder, W. Zieglgansberger and K. Becker (2007). "Ultramicroscopy: three-dimensional visualization of neuronal networks in the whole mouse brain." *Nature Methods* **4**(4): 331–336.

Doerr, J., M. K. Schwarz, D. Wiedermann, A. Leinhaas, A. Jakobs, F. Schloen, I. Schwarz, M. Diedenhofen, N. C. Braun, P. Koch, D. A. Peterson, U. Kubitscheck, M. Hoehn and O. Brustle (2017). "Whole-brain 3D mapping of human neural transplant innervation." *Nature Communications* **8**: 14162.

Fahrbach, F. O., P. Simon and A. Rohrbach (2010). "Microscopy with self-reconstructing beams." *Nature Photonics* **4**(11): 780–785.

Frasconi, P., L. Silvestri, P. Soda, R. Cortini, F. S. Pavone and G. Iannello (2014). "Large-scale automated identification of mouse brain cells in confocal light sheet microscopy images." *Bioinformatics* **30**(17): i587–i593.

Gao, R., S. M. Asano, S. Upadhyayula, I. Pisarev, D. E. Milkie, T. L. Liu, V. Singh, A. Graves, G. H. Huynh, Y. Zhao, J. Bogovic, J. Colonell, C. M. Ott, C. Zugates, S. Tappan, A. Rodriguez, K. R. Mosaliganti, S. H. Sheu, H. A. Pasolli, S. Pang, C. S. Xu, S. G. Megason, H. Hess, J. Lippincott-Schwartz, A. Hantman, G. M. Rubin, T. Kirchhausen, S. Saalfeld, Y. Aso, E. S. Boyden and E. Betzig (2019). "Cortical column and whole-brain imaging with molecular contrast and nanoscale resolution." *Science* **363**(6424).

Huisken, J. and D. Y. Stainier (2009). "Selective plane illumination microscopy techniques in developmental biology." *Development* **136**(12): 1963–1975.

Huisken, J., J. Swoger, F. Del Bene, J. Wittbrodt and E. H. K. Stelzer (2004). "Optical sectioning deep inside live embryos by selective plane illumination microscopy." *Science* **305**(5686): 1007–1009.

Jahrling, N., K. Becker and H. U. Dodt (2009). "3D-reconstruction of blood vessels by ultramicroscopy." *Organogenesis* **5**(4): 227–230.

Jahrling, N., K. Becker, B. M. Wegenast-Braun, S. A. Grathwohl, M. Jucker and H. U. Dodt (2015). "Cerebral beta-amyloidosis in mice investigated by ultramicroscopy." *PLoS One* **10**(5): e0125418.

Keller, P. J. and M. B. Ahrens (2015). "Visualizing whole-brain activity and development at the single-cell level using light-sheet microscopy." *Neuron* **85**(3): 462–483.

Keller, P. J., M. B. Ahrens and J. Freeman (2015). "Light-sheet imaging for systems neuroscience." *Nature Methods* **12**(1): 27–29.

Keller, P. J. and H. U. Dodt (2012). "Light sheet microscopy of living or cleared specimens." *Current Opinion in Neurobiology* **22**(1): 138–143.

Keller, P. J., F. Pampaloni and E. H. Stelzer (2006). "Life sciences require the third dimension." *Current Opinion in Cell Biology* **18**(1): 117–124.

Keller, P. J., A. D. Schmidt, J. Wittbrodt and E. H. K. Stelzer (2008). "Reconstruction of zebrafish early embryonic development by scanned light sheet microscopy." *Science* **322**(5904): 1065–1069.

Lichtman, J. W. and W. Denk (2011). "The big and the small: challenges of imaging the brain's circuits." *Science* **334**(6056): 618–623.

Liebmann, T., N. Renier, K. Bettayeb, P. Greengard, M. Tessier-Lavigne and M. Flajolet (2016). "Three-dimensional study of Alzheimer's disease hallmarks using the iDISCO clearing method." *Cell Reports* **16**(4): 1138–1152.

Lugo-Hernandez, E., A. Squire, N. Hagemann, A. Brenzel, M. Sardari, J. Schlechter, E. H. Sanchez-Mendoza, M. Gunzer, A. Faissner and D. M. Hermann (2017). "3D visualization and quantification of microvessels in the whole ischemic mouse brain using solvent-based clearing and light sheet microscopy." *Journal of Cerebral Blood Flow and Metabolism* **37**(10): 3355–3367.

Menegas, W., J. F. Bergan, S. K. Ogawa, Y. Isogai, K. Umadevi Venkataraju, P. Osten, N. Uchida and M. Watabe-Uchida (2015). "Dopamine neurons projecting to the posterior striatum form an anatomically distinct subclass." *Elife* **4**: e10032.

Migliori, B., M. S. Datta, C. Dupre, M. C. Apak, S. Asano, R. Gao, E. S. Boyden, O. Hermanson, R. Yuste and R. Tomer (2018). "Light sheet theta microscopy for rapid high-resolution imaging of large biological samples." *BMC Biology* **16**(1): 57.

Niedworok, C. J., I. Schwarz, J. Ledderose, G. Giese, K. K. Conzelmann and M. K. Schwarz (2012). "Charting monosynaptic connectivity maps by two-color light-sheet fluorescence microscopy." *Cell Reports* **2**(5): 1375–1386.

Planchon, T. A., L. Gao, D. E. Milkie, M. W. Davidson, J. A. Galbraith, C. G. Galbraith and E. Betzig (2011). "Rapid three-dimensional isotropic imaging of living cells using Bessel beam plane illumination." *Nature Methods* **8**(5): 417–423.

Power, R. M. and J. Huisken (2017). "A guide to light-sheet fluorescence microscopy for multiscale imaging." *Nature Methods* **14**(4): 360–373.

Renier, N., E. L. Adams, C. Kirst, Z. Wu, R. Azevedo, J. Kohl, A. E. Autry, L. Kadiri, K. Umadevi Venkataraju, Y. Zhou, V. X. Wang, C. Y. Tang, O. Olsen, C. Dulac, P. Osten and M. Tessier-Lavigne (2016). "Mapping of brain activity by automated volume analysis of immediate early genes." *Cell* **165**(7): 1789–1802.

Renier, N., Z. H. Wu, D. J. Simon, J. Yang, P. Ariel and M. Tessier-Lavigne (2014). "iDISCO: a simple, rapid method to immunolabel large tissue samples for volume imaging." *Cell* **159**(4): 896–910.

Richardson, D. S. and J. W. Lichtman (2015). "Clarifying tissue clearing." *Cell* **162**(2): 246–257.

Saghafi, S., N. Haghi-Danaloo, K. Becker, I. Sabdyusheva, M. Foroughipour, C. Hahn, M. Pende, M. Wanis, M. Bergmann, J. Stift, B. Hegedus, B. Dome and H. U. Dodt (2018). "Reshaping a multimode laser beam into a constructed Gaussian beam for generating a thin light sheet." *Journal of Biophotonics* **11**(6): e201700213.

Schwarz, M. K., A. Scherbarth, R. Sprengel, J. Engelhardt, P. Theer and G. Giese (2015). "Fluorescent-protein stabilization and high resolution imaging of cleared, intact mouse brains." *Plos One* **10**(5): e0124650.

Siedentopf, H. and R. Zsigmondy (1902). "Uber sichtbarmachung und größenbestimmung ultramikoskopischer teilchen, mit besonderer anwendung auf goldrubingläser." *Annalen der Physik* **315**(1): 1–39.

Silvestri, L., A. L. Allegra Mascaro, I. Costantini, L. Sacconi and F. S. Pavone (2014). "Correlative two-photon and light sheet microscopy." *Methods* **66**(2): 268–272.

Silvestri, L., A. Bria, L. Sacconi, G. Iannello and F. S. Pavone (2012). "Confocal light sheet microscopy: micron-scale neuroanatomy of the entire mouse brain." *Optics Express* **20**(18): 20582–20598.

Silvestri, L., I. Costantini, L. Sacconi and F. S. Pavone (2016). "Clearing of fixed tissue: a review from a microscopist's perspective." *Journal of Biomedical Optics* **21**(8): 081205.

Silvestri, L., M. C. Muellenbroich, I. Costantini, A. P. Di Giovanna, L. Sacconi and F. S. Pavone (2017). "RAPID: real-time image-based autofocus for all wide-field optical microscopy systems." *bioRxiv*.doi: https://doi.org/10.1101/170555

Silvestri, L., M. Paciscopi, P. Soda, F. Biamonte, G. Iannello, P. Frasconi and F. S. Pavone (2015). "Quantitative neuroanatomy of all Purkinje cells with light sheet microscopy and high-throughput image analysis." *Frontiers in Neuroanatomy* **9**: 68.

Silvestri, L., L. Sacconi and F. S. Pavone (2014). "Correcting spherical aberrations in confocal light sheet microscopy: a theoretical study." *Microscopy Research and Technique* **77**(7): 483–491.

Spalteholz, W. (1914). *Über das durchsichtigmachen von menschlichen und tierischen präpareten und seine theoretischen bedingungen*. n.p.

Susaki, E. A., K. Tainaka, D. Perrin, F. Kishino, T. Tawara, T. M. Watanabe, C. Yokoyama, H. Onoe, M. Eguchi, S. Yamaguchi, T. Abe, H. Kiyonari, Y. Shimizu, A. Miyawaki, H. Yokota and H. R. Ueda (2014). "Whole-brain imaging with single-cell resolution using chemical cocktails and computational analysis." *Cell* **157**(3): 726–739.

Susaki, E. A. and H. R. Ueda (2016). "Whole-body and whole-organ clearing and imaging techniques with single-cell resolution: toward organism-level systems biology in mammals." *Cell Chemical Biology* **23**(1): 137–157.

Tomer, R., L. Ye, B. Hsueh and K. Deisseroth (2014). "Advanced CLARITY for rapid and high-resolution imaging of intact tissues." *Nature Protocols* **9**(7): 1682–1697.

Vettenburg, T., H. I. Dalgarno, J. Nylk, C. Coll-Llado, D. E. Ferrier, T. Cizmar, F. J. Gunn-Moore and K. Dholakia (2014). "Light-sheet microscopy using an Airy beam." *Nature Methods* **11**(5): 541–544.

Voie, A. H., D. H. Burns and F. A. Spelman (1993). "Orthogonal-plane fluorescence optical sectioning - 3-dimensional imaging of macroscopic biological specimens." *Journal of Microscopy-Oxford* **170**: 229–236.

Ye, L., W. E. Allen, K. R. Thompson, Q. Tian, B. Hsueh, C. Ramakrishnan, A. C. Wang, J. H. Jennings, A. Adhikari, C. H. Halpern, I. B. Witten, A. L. Barth, L. Luo, J. A. McNab and K. Deisseroth (2016). "Wiring and molecular features of prefrontal ensembles representing distinct experiences." *Cell* **165**(7): 1776–1788.

II

Neurophotonic Control and Perturbation

CHAPTER 11

Optogenetic Modulation of Neural Circuits

Mathias Mahn, Oded Klavir, and Ofer Yizhar

CONTENTS

11.1 INTRODUCTION

Optogenetic technology uses light to control or modulate neuronal function, providing a means of interacting with neurons that is unchallenged in its temporal resolution, spatial resolution, and cell type specificity. First applied in mammalian neurons in 2005 (Boyden et al., 2005; Li et al., 2005), this approach comprises a single-component, genetically encoded system to activate, inhibit, or otherwise modulate the activity of neurons with light. The actuators used in optogenetic technology, essentially converting electromagnetic energy to changes in neuronal excitability, are light-sensitive proteins that serve as ion channels, pumps, or biochemical pathway modulators. These tools, typically proteins of the microbial rhodopsin family (Yizhar et al., 2011a; Zhang et al., 2011), are genetically encoded single-component actuators, which allows for selective targeting to specific cell types and therefore interrogation of individual components of highly complex neural systems. Once the targeted neuronal population expresses the genetically encoded optogenetic tool, its function can be controlled with light. In contrast with pharmacological manipulations, which suffer from diffusion of the active compound and poor temporal control, light is not subject to diffusion and can be delivered with millisecond precision. This provides exquisite experimental control that cannot be matched by other genetically encoded approaches, such as pharmacological manipulation of receptors or another genetically encoded system: designer receptors exclusively activated by designer drugs (DREADDs; Armbruster et al., 2007; Vardy et al., 2015).

Optogenetics has grown rapidly, and information about the intricate details of this methodology would easily fill several books. This chapter aims to serve as a concise introduction and reference guide to the field and its state-of-the-art tools, to enable the reader to plan optogenetic experiments including the appropriate controls while being aware of possible caveats. Furthermore, we provide references to both original research articles and reviews in order to allow the reader to easily locate appropriate literature sources for in-depth reading.

11.2 GENETICALLY ENCODED TOOLS FOR OPTOGENETIC MODULATION

Optogenetic actuators are light-sensitive proteins that have been engineered from their naturally occurring forms and adapted for expression in heterologous systems in which they can allow light-based manipulation of cellular function. Most of these optogenetic actuators are not intrinsically light-sensitive, and therefore need to bind a chromophore as a cofactor. The biophysical properties of this cofactor correlate with the energy and thereby the wavelength of light necessary for inducing state transitions in the actuator protein (Hellingwerf et al., 1996). The commonly used optogenetic tools are currently based on three classes of such cofactors: polyenes, aromatics, and tetrapyrroles. Rhodopsins, which bind the polyene retinal, are the most abundantly used class in optogenetic applications. Light-gated ion channels and pumps, as well as G-protein coupled opsins all belong to this class and will be discussed in detail. Flavin-based chromophore-binding photoreceptors constitute another large class of proteins including phototropins (Christie, 2007), cryptochromes (Chaves et al., 2011), and blue light-using FAD-photoreceptors (BLUF, (Masuda, 2013)), all of which utilize the aromatic cofactor flavin. The aromatic cofactor p-coumaric acid serves as a chromophore for the xanthopsin family of photoreceptors (Kort et al., 1996). The main light-sensitive proteins from the family of tetrapyrrole binding proteins currently utilized for optogenetic approaches are phytochromes (Quail, 2002) which bind bilins. The only classes of proteins that do not require cofactors for activation are the GFP-like protein Dronpa (Ando et al., 2004) and the ultraviolet light receptor UVR8 (Brown et al., 2009). Numerous tools have been designed using these diverse naturally light-sensitive proteins. However, the most well-known and abundantly used optogenetic actuators belong to the family of microbial rhodopsins, and these are discussed in greater detail in this chapter.

11.2.1 Tools for Optogenetic Excitation

Channelrhodopsin-2 (ChR2) is a light-gated cation channel originally isolated from the alga *Chlamydomonas reinhardtii* (Nagel et al., 2002; Nagel et al., 2003). ChR2 is perhaps the most well-known optogenetic tool. Activation of ChR2 with blue light (460 nm peak absorption, 1.1 mW mm^{-2} half maximal activation light power density (EC50), 1.1 pS single channel conductance) triggers precisely timed photocurrents that decay with a time constant of ~10 ms at 25°C (Lin et al., 2009). ChR2 was first applied in 2005 to cultured hippocampal neurons and shown to allow millisecond-resolution control of spiking activity using blue light illumination (Boyden et al., 2005). Numerous studies ensued, utilizing ChR2 and other microbial rhodopsins in a variety of neural circuits and animal models (Fenno et al., 2011). Specific experimental needs have driven further protein engineering efforts which led to the expansion of a diverse tool palette for optogenetic excitation. For instance, fast channel closing kinetics ($_{\tau off}$) are necessary to allow for high temporal precision and the reliable driving of high firing rates. To efficiently recruit cells in a large brain volume, the operational light sensitivity, single channel conductance, and expression level should be maximized. Furthermore, to control two different neuronal populations independently, or to combine optogenetic control with imaging of fluorescent reporters, opsins with shifted excitation wavelengths were designed. These efforts resulted in opsin variants with altered spectral sensitivity, kinetic properties, operational light sensitivity, and single channel conductance. Below is a brief summary of the major classes of ChRs and some of the considerations that can guide the choice of tool. For a more in-depth perspective, we refer the reader to Wietek and Prigge (2016).

The blue light-sensitive channelrhodopsin ChR2 has been used extensively in optogenetic experiments. Although the wild-type ChR2 in itself can allow neuronal excitation (Boyden et al., 2005), the most commonly used variant has been ChR2(H134R), in which a single amino acid replacement led to increased photocurrents and improved peak/steady-state ratio, effectively improving the stability of optogenetic excitation during longer pulse trains (Nagel et al., 2005). Additional mutagenesis work yielded several optimized variants, including the ultra-fast ChETA variants (Gunaydin et al., 2010), the slow variants (also termed step-function opsins (Berndt et al., 2009), and the enhanced T159C mutant (Berndt et al., 2011) which shows stronger photocurrents, presumably due to improved chromophore binding (Ullrich et al., 2013). Generally speaking, these mutated ChRs, based on ChR2, vary widely in kinetics, but preserve the spectral properties of ChR2. Variants with slower kinetics are effectively more light-sensitive (Berndt et al., 2009; Yizhar et al., 2011b) since more channels remain in the open state during an illumination pulse (see Section 11.4.1 Light Requirements for Efficient Optogenetic Modulation for further discussion). There are several exceptions to this rule, including the fast ChR Chronos (Klapoetke et al., 2014), which seems to allow both fast (τ_{off} < 5 ms) and light-sensitive (0.3 mW mm^{-2} light power for half maximal activation for Chronos) optogenetic excitation, and the naturally occurring anion-conducting channelrhodopsins of *Guillardia theta* GtACR1 and GtACR2, which possess a higher single-channel conductance (Govorunova et al., 2015a).

11.2.2 Multi-Wavelength Approaches

One of the major advantages of light-based approaches is the potential for multiplexed interrogation of the experimental system, whether through combination of fluorescence imaging with optogenetic manipulations, or through the use of multiple optogenetic actuators with distinct spectral sensitivities. Such spectral multiplexing can be applied, for example, by utilizing a green activity indicator such as GCaMP (Chen et al., 2013) with a red-shifted optogenetic tool (Zhang et al., 2008; Yizhar et al., 2011c; Prigge et al., 2012; Lin et al., 2013; Klapoetke et al., 2014), or vice versa. However, the action spectra of red- and blue-shifted rhodopsins are somewhat overlapping, and spectral overlap also poses constraints on combined optogenetic-imaging experiments. Although such multi-wavelength experiments are feasible, they need to be properly calibrated and controlled.

The development of red-shifted rhodopsins was motivated by several potential applications. These include improved tissue penetration (see Section 11.4.2 Optical Properties of Brain Tissue) and multiplexed photostimulation of two or more neuronal populations (Yizhar et al., 2011c; Prigge et al., 2012; Klapoetke et al., 2014). VChR1, a red-shifted cation channel from the alga *Volvox carteri* (Zhang et al., 2008) was first used as a basis for engineering of the high-efficiency red-shifted channelrhodopsin variants C1V1 and ReaChR (Yizhar et al., 2011c; Lin et al., 2013) that allow excitation of neurons with red-shifted light. ChRimson (Klapoetke et al., 2014), a naturally occurring ChR variant isolated from *Chlamydomonas noctigama*, shows the most red-shifted action spectrum, with peak absorption at ~590 nm, while the chimeric opsins C1V1 and ReaChR both maximally absorb at ~540 nm. Another important application of red-shifted channelrhodopsins is the combination of optogenetic modulation with activity reporter imaging. Using single photon whole field excitation, this can be achieved by using a green fluorescent indicator in combination with a red light sensitive opsin (Akerboom et al., 2013; Chen et al., 2013), or vice versa (Hochbaum et al., 2014; Dana et al., 2016).

Importantly, the current palette of red-shifted excitation channelrhodopsins all respond quite robustly to blue light (Wietek and Prigge, 2016), imposing limiting constraints on dual-channel excitation experiments, primarily on the blue side of the visual spectrum. In contrast, ChR2 and other blue light-activated ChRs show very low response to light at wavelengths above 590 nm, and therefore separation on this side of the visual spectrum is more complete. Separation on the red side of the spectrum could be achieved by using a red-shifted ChR variant with an operational light sensitivity significantly lower than the blue opsin, allowing for separation on the basis of irradiance combined with wavelength

(Klapoetke et al., 2014). When considering such dual/multi-wavelength experiments, it is important to consider several sources of variability, including the non-linear nature of neuronal excitability, the diversity of opsin expression levels across cells and the diversity of cellular excitability across the population. While the action spectrum of a ChR might appear to show as little as 20% absorption at a particular spectral band, the probability of action potential firing in response to light delivery at this wavelength will depend on the expression level of the opsin molecules and might be as efficient as the excitation induced by stimulation at the peak wavelength under some conditions. This constraint might be overcome by additional improvement upon current red-shifted activity sensors (Gong et al., 2014; Dana et al., 2016; Lou et al., 2016). However, red calcium indicators can show a calcium-independent increase in fluorescence upon blue illumination, an artifact that was only recently overcome with the development of jRCaMPs (Dana et al., 2016) allowing for non-interfering single photon stimulation and imaging of a calcium indicator. Another possibility is to combine two-photon indicator imaging with single or two-photon opsin activation (Packer et al., 2015). While two-photon scanning leads to a reduction of red-shifted opsin activation during indicator imaging compared to single photon approaches, also here crosstalk might not be negligible and potentially warrants a shift to a combination of a red-shifted activity indicator with a blue light-sensitive opsin (Forli et al., 2018).

11.2.3 Tools for Optogenetic Inhibition

Optogenetic inhibition was first accomplished using the light-driven chloride pump halorhodopsin (HR) from *Natronomonas pharaonis* (Scharf and Engelhard, 1994; Zhang et al., 2007). The red-shifted action spectrum of this microbial rhodopsin allows multiplexed excitation and inhibition when combined with the blue light-sensitive ChRs. Activation of the proton-pumping bacteriorhodopsin (Gradinaru et al., 2010) and archaerhodopsin (Chow et al., 2010) can also hyperpolarize neurons, effectively inhibiting them due to the outward-directed movement of protons during illumination of these tools. Additional engineering contributed to improved photocurrent amplitudes (and thereby more efficient inhibition) mediated by improved membrane targeting of these proteins, which originate in prokaryotes and therefore initially failed to express robustly in mammalian neurons (Gradinaru et al., 2010; Mattis et al., 2011). The membrane targeting-enhanced versions of halorhodopsin (eNpHR3.0) and archaerhodopsin (eArch3.0) have therefore become the most widely used tools for optogenetic silencing of neurons, and numerous studies have utilized these tools for temporally precise and reversible optogenetic inhibition (Fenno et al., 2011). As with the excitatory ChRs, the anion-pumping HR displays a pronounced peak upon initial illumination, which then decays to a steady-state photocurrent, a behavior that should be taken into account when designing long-term inhibition experiments (Mattis et al., 2011; Goshen et al., 2011). Importantly, it has been shown that ion-pumping microbial rhodopsins can shift the concentrations of intracellular ions to non-physiological levels. In the case of halorhodopsin, this can lead to accumulation of chloride in the neuron, inducing changes in the reversal potential of GABAergic synapses (Raimondo et al., 2012) while archaerhodopsin was shown to increase the intracellular pH, inducing action potential independent Ca^{2+} influx and elevating spontaneous vesicle release in presynaptic terminals (Mahn et al., 2016).

Given the limitations of ion-pumping microbial rhodopsins, several groups have developed anion-conducting rhodopsins (ACRs) that rely on a passive channel mechanism rather than active pumping of ions across the membrane. Channels have two major functional differences compared to pumps: they can conduct multiple ions during each photoreaction cycle, thereby increasing the photocurrent yield per photon (which increases the effective light-sensitivity of the manipulated neurons), and they conduct ions according to their natural reversal potential, which could at least partially avoid non-physiological changes in ion concentration gradients. Originally, two variants of ACRs were engineered, based on the wild-type ChR2 sequence (Wietek et al., 2014) or the chimeric C1C2 channelrhodopsin (Berndt et al., 2014). Both of these channels, engineered by substituting positively charged amino acids within the inner pore region of the channel, were later optimized further to yield more chloride-selective variants (Wietek et al., 2015; Berndt et al., 2016). The resulting channels were shown to allow optogenetic inhibition with blue light, *in vitro* and *in vivo*. In addition, a novel class of chloride-conducting channelrhodopsins was discovered in the cryptophyte group of algae (Govorunova et al., 2015a). These algae, *Guillardia theta*, express natural anion-conducting channelrhodopsins (GtACR1 and GtACR2) that show high single-channel conductance and high chloride-selectivity. Searching for proteins with sequence similarities to GtACRs resulted in the rapid discovery of additional ACRs (Govorunova et al., 2015b; Wietek et al., 2016; Govorunova et al., 2017; Govorunova et al., 2018). While these ACRs show low sequence similarity to that of ChRs from *Chlamydomonas*, the overall structure is similar, and key amino acid positions are somewhat conserved, allowing utilization of known mutations for manipulating the kinetic properties of chloride-conducting ChRs (Sineshchekov et al., 2015; Wietek et al., 2015; Berndt et al., 2016; Wietek et al., 2017; Govorunova et al., 2017). While this new class of optogenetic tools seems to be a valuable addition to the already diverse toolset, one must keep in mind that chloride is dynamically regulated in neurons. Gamma-Aminobutyric acid (GABA), the ligand of endogenous chloride channels ($GABA_AR$), is thought to exert excitatory effects in developing neurons (Ben-Ari et al., 1989; Ben-Ari, 2002; Ben-Ari, 2014); however, this theory is under active debate (Minlebaev et al., 2007; Bregestovski and Bernard, 2012; Kirmse et al., 2015; Valeeva et al., 2016). Excitatory GABA effects were also observed in some neurons in the adult brain (Haam et al., 2012) and during disease manifestation (Kahle et al., 2008). Furthermore, there is evidence of a depolarizing chloride reversal potential in the axon (Pugh and Jahr, 2011; Pugh and Jahr, 2013; Price and Trussell, 2006; Takkala et al., 2016; Mahn et al., 2016; Messier et al., 2018). We recently ameliorated the effect of axonal excitation by reducing the amount of axonal ACRs by fusing the opsin to a trafficking sequence that localizes the light-gated ion channels to the soma and proximal dendrites. This led to the generation of stGtACR1 and stGtACR2, allowing for efficient neuronal inhibition at low light power and with reduced light onset excitation (Mahn et al., 2018). Nevertheless, the effect of chloride-conducting ChRs on the neuronal networks may therefore be complex and the effect of light-gated chloride channel activation should be validated using electrophysiology or other measurements of neuronal function for each experimental configuration.

11.2.4 Tools for Optogenetic Modulation of Signal Transduction

11.2.4.1 G-Protein Coupled Opsins

In contrast with ion-channels or ion-pumps, this class of photoreceptors is coupled to intracellular signaling cascades. Indeed, one of the first successful attempts to sensitize neurons to light was to express the *Drosophila* photoreceptor genes Rh1, arrestin-2, and the G protein alpha subunit in cultured hippocampal neurons (Zemelman et al., 2002). The simplicity and robustness of channelrhodopsins and microbial ion-pumps have made them the tools of choice for direct control of action potential firing. However, G-protein coupled optogenetic tools provide an opportunity to either probe the effect of signaling cascade modulation on neuronal function, or to hijack the second messenger system to influence neuronal dynamics, in a spatially and temporally specific manner. Moreover, these tools might be better suited for temporally defined modulation of glial cells, due to the important role of the G-protein signaling pathways in the natural activity of these cells (Bradley and Challiss, 2012).

While the mechanism of rhodopsin signaling in retinal photoreceptors are well established (Palczewski, 2006), their activity in heterologous systems (such as central nervous system neurons) relies on coupling to native G-protein signaling cascades. In retinal rod cells, vertebrate rhodopsin couples to the G protein transducin (G_t). While transducin is not expressed in neurons, its alpha subunit belongs to the G_i subfamily (Simon et al., 1991). When expressed in neurons, vertebrate rhodopsin can couple to other $G_{i/o}$ family proteins (Li et al., 2005), thereby allowing for light-gated activation of the $G_{i/o}$ pathway. One limitation to these tools arises from the fact that rod and cone opsins possess a thermally unstable photoproduct, meaning that the light-absorbing chromophore retinal is released upon activation. This effect, known as bleaching, means that the opsin must bind a new cis-retinal before it can be reactivated. The light-induced signaling responses of vertebrate rhodopsin therefore adapt during repetitive stimulation – a limitation that was eased using cone short or long wavelength opsins (Masseck et al., 2014). However, light-induced adaptation still imposes a limit on the magnitude and reproducibility of vertebrate rod and cone opsin activation (Bailes et al., 2012). While *in vivo*, sufficient amounts of cis-retinal are present in mammalian tissues, the need to bind cis-retinal after every photoactivation necessitates the addition of cis-retinal to the medium for *in vitro* applications (Masseck et al., 2014). A second potential concern regarding the use of these proteins for repeated light-stimulation is the potential internalization of membrane-resident rhodopsins following activation (Siuda et al., 2015a).

G-protein coupled receptors (GPCRs) have a highly conserved seven transmembrane helix structure. This similarity led to the idea that it might be possible to switch the cytoplasmic domains from a light-gated GPCR with those of a ligand-gated GPCR, thereby generating a chimeric light-gated GPCR with the intracellular coupling of the ligand-gated GPCR. Indeed, this approach was successfully employed (Kim et al., 2005) and seems to be generalizable to all major G-protein signaling pathways (G_α, $G_{i/o}$ and G_s) by the generation of chimeric proteins replacing the cytoplasmic loops of rhodopsin (Airan et al., 2009; Siuda et al., 2015b). Currently, most of these chimeric light-gated GPCRs, also referred to as OptoXRs (Airan et al., 2009), are based on vertebrate rhodopsin, which – as noted above – necessitates the binding of a new cis-retinal after every photocycle. Several other photoreceptor types possess an intrinsic photocycle, which enables the receptors to restore cis-retinal by photo-regeneration (Byk et al., 1993), alleviating the necessity to bind new cis-retinal after every photocycle. Utilization of such photoreceptors with intrinsic photocycles, or the generation of chimeras based on unique naturally occurring variants, could lead to highly efficient optogenetic tools for the investigation of GPCR signaling and for manipulating neural circuit function (Bailes et al., 2012; Karunarathne et al., 2013; McGregor et al., 2016).

11.2.4.2 Engineered Tools Based on Light-Gated Protein–Protein Interactions

Naturally occurring photoreceptors in plants and microorganisms have often evolved in a modular fashion, with a highly conserved light sensing domain fused to an effector domain (Han et al., 2004; Green, 2004; Glantz et al., 2016). In these proteins, light absorption leads to a conformational change that in turn influences the activity of the effector domain. Such conformational changes can lead to unmasking of an active region of the effector domain or to an increased affinity to another protein and thereby dimerization (Nash et al., 2011) or oligomerization (Bugaj et al., 2013). Several such naturally occurring proteins have been utilized to engineer optogenetic tools based on this light-response mechanism to achieve light-induced protein–protein interaction.

Plant, bacterial, and fungal proteins utilize light-oxygen-voltage-sensing (LOV) domains (Christie et al., 1999) as light sensor (Herrou and Crosson, 2011; Losi and Gärtner, 2012; Christie et al., 2015). Their small size (~120 amino acids) and flavin-based cofactors, which are abundantly available in animal tissues, make these blue-light sensors especially suitable for optogenetic approaches. Several studies successfully utilized LOV domains fused to effector proteins to accomplish light-based control of subcellular protein concentrations (Guntas et al., 2015; Kawano et al., 2015; Wang et al., 2016; Niopek et al., 2016), gene expression (Wang et al., 2012; Ma et al., 2013), DNA manipulation (Kawano et al., 2016), calcium signaling (Pham et al., 2011), or protein stability (Bonger et al., 2014). Different LOV domain proteins were engineered to achieve changes in the dimerization state of the protein upon illumination. Vivid (VVD) is one such protein, which homodimerizes upon light absorption (Zoltowski et al., 2007). This dimerization is well-characterized and mutants spanning a wide range of inactivation kinetics are available (Zoltowski et al., 2009). A heterodimerizing version of VVD has been generated (Kawano et al., 2015), and was utilized to reconstitute a split Cre recombinase enzyme, leading to the generation of an efficient light-activatable Cre-dependent expression system (Kawano et al., 2016).

The second most frequently utilized type of proteins are cryptochromes (CRYs). Several CRY-based oligomerization and heterodimerization tools for light-gated protein interactions and transcription control were described (Kennedy et al., 2010; Taslimi et al., 2016). CRY-based tools are used less frequently than LOV domain proteins mostly due to their size. Their flavin adenine dinucleotide (FAD) cofactor binding domain alone is over four times larger than the entire LOV domain (~500 vs ~120 amino acids). The blue light sensor using FAD (BLUF) domain is another flavin-binding domain, similar to CRY or LOV domains (Masuda, 2013). BLUF-coupled photoactivated nucleotidyl cyclases, like naturally occurring light-gated andenylyl cyclases (Stierl et al., 2011; Penzkofer et al., 2015) or the light-gated guanylyl cyclase BlgC (Ryu et al., 2010; Kim et al., 2015), are thus far the most promising additions to the optogenetic toolbox. However, recently described naturally occurring rhodopsin-based guanylyl cyclases show lower dark activity and higher specificity (Gao et al., 2015).

Phytochromes, belonging to the tetrapyrrole binding class of light sensitive proteins, share common domains consisting of a photosensory core module (PCM) and an effector module. Their red/far-red light absorption (Li et al., 2011) makes them interesting for optogenetic tool development as longer light wavelengths allow for deeper tissue penetration or dual wavelength experiments (see Section 11.4.2 Optical Properties of Brain Tissue). The sensitivity to longer wavelengths is conveyed by bilin-based co-factors (Rockwell et al., 2006). Bacterial phytochromes utilize biliverdin (Bhoo et al., 2001), which is available in mammalian tissues. Plant and cyanobacterial phytochromes, however, depend on phytochromobilin and phycobilins, which are not endogenously available in mammalian cells or tissues and therefore need to be substituted.

In contrast to the light sensors discussed so far, two unique proteins have been utilized that do not necessitate an additional cofactor since their light-dependent functions are encoded by their protein structure. In the UV-absorbing protein UVR8, light is absorbed by specific tryptophans that act as a chromophore, making it sensitive only to very energy-rich irradiation with a peak at 28 nm (Brown et al., 2009). While absorption in the UV range can be a limit for *in vivo* applications due to limited penetration depth and tissue heating, it does allow for good spectral separation and thereby combination with fluorophores or other light gated tools. UVR8 fusion proteins were engineered to render cellular functions such as gene expression (Crefcoeur et al., 2013) or protein secretion (Chen et al., 2013) light controllable. Another cofactor-independent protein is Dronpa, a reversible photoswitching tool based on a coral GFP-like fluorescent protein (Ando et al., 2004; Habuchi et al., 2005). Dronpa was originally described for reversible protein highlighting as it can be reversibly switched from its fluorescent state to a dark state and back by illumination with 488 nm and 405 nm light, respectively. In the following, tetrameric and heterodimeric mutants of Dronpa that dissociate upon illumination with 488 nm light (Zhou et al., 2012) were developed and have been used to engineer several optogenetic tools (Zhou et al., 2012; Zhou et al., 2017). A useful feature of this system is that its switching state can be read out by exciting fluorescence with weak 488 nm light, albeit not without risking some activation. Furthermore, while in the fluorescent state, the fluorescence can be used to localize the tool, thereby alleviating the need to fuse it with a fluorophore.

In summary, a wide palette of optogenetic tools is available for neuroscience applications. While these tools allow a vast array of neural circuit manipulations, the choice of tool depends on many factors, ranging from the organism in which manipulation is performed to the temporal precision of the manipulation and the tissue volume that needs to be targeted.

11.3 GENE TARGETING APPROACHES FOR OPTOGENETICS

Genetic targeting is the main instrument of cell type specificity in optogenetics, made possible by the single-component nature of most commonly used optogenetic tools.

Several targeting approaches have been developed for delivering opsin genes to defined cells. Ideally, the targeting approach used to express any optogenetic tool should result in highly specific expression in the targeted neuron population, provide reliable modulation of neuronal excitability with moderate light power, and avoid cellular toxicity due to overexpression. The targeting approach should be selected to optimize these parameters, and its impact on the targeted (and non-targeted) neurons should be carefully examined and controlled for.

Current gene targeting approaches are designed to capitalize on: (i) cell type-specific expression of unique promoter elements; (ii) local or long-range synaptic connectivity; or (iii) expression of activity-dependent immediate-early genes. These three approaches are described below, along with the considerations that need to be made in choosing the optimal targeting approach.

11.3.1 Promoter-Based Targeting

Certain neuronal subtypes have been shown to express unique genes or combinations of genes involved, for example, in the specification of the neurotransmitter released by these cells, their electrophysiological properties, receptor expression profiles, and other molecular identifiers. In some cases, including, for example, parvalbumin-expressing GABAergic neurons, the expression of a unique gene in the targeted population allows the use of a specific promoter sequence for targeting these neurons. If the promoter sequence is short enough, it can be used directly, in a viral vector (Dimidschstein et al., 2016). In cases where the regulatory sequences (promoters and enhancers) are too large to package into a viral vector, specific expression can be obtained using a Cre driver line combined with a Cre-dependent viral vector (Sohal et al., 2009). A transgenic mouse is one that is genetically engineered with a DNA sequence often randomly inserted into its genome. For optogenetic applications, an opsin gene coupled to a fluorescent protein gene can be inserted into the mouse genome, coupled with upstream regulatory sequences that confer cell type selectivity. This approach can allow selective expression in neurons expressing the gene driven by the specific promoter, throughout the brain and other tissues (Arenkiel et al., 2007; Zhao et al., 2011). The main benefit of this approach is that once specific expression is confirmed, it only requires the placement of an optical fiber or other light-delivery device in order to obtain optogenetic control of the targeted population (Lim et al., 2012; Guo et al., 2014; Allen et al., 2017). Since expression is brain-wide, one should consider the potential confounds arising from off-target effects resulting from optogenetic modulation of distal neurons that send axonal collaterals or dendritic branches to the illuminated area.

To circumvent some of the drawbacks of brain-wide expression of optogenetic tools, a more localized approach involves the use of viral vectors for region-specific gene targeting. This approach benefits from localized expression patterns of viral vectors and is more versatile and cost-effective compared with the transgenic approach. Viral vectors are engineered viruses, genetically modified to remove pathogenic gene sequences and, most typically, prevent additional replication and transduction cycles that are part of the natural life-cycle of wild-type viruses (Verma and Weitzman, 2005). Viral vectors used for optogenetic experiments are typically derived from the adeno-associated virus (AAV) and lentivirus (LV) families. AAV-based vectors are the most commonly used, as they are less immunogenic, thereby minimizing immune activation upon injection into brain tissue, and since they do not broadly integrate into the host genome but rather remain inside the nucleus as episomes (Duan et al., 1998). Despite the limited payload size of viral vectors (Grieger and Samulski, 2005, for AAV), some minimal promoter fragments can be used that retain cell-type specificity (Yizhar et al., 2011b). In such cases, the viral vector can be used in non-transgenic animals and achieve specific expression, e.g. in cortical excitatory neurons (CaMKII, Dittgen et al., 2004), hypothalamic hypocretin-secreting neurons (Adamantidis et al., 2007), oxytocinergic neurons (Knobloch et al., 2012), or astrocytes (Jakobsson et al., 2003; Gradinaru et al., 2009).

In most cases, however, a minimal promoter fragment cannot attain sufficient specificity, in which case combinatorial transgenic/viral approaches can be used for robust cell type-specific expression. For example, parvalbumin (PV) neurons can be targeted by injecting a Cre-dependent viral vector into a knock-in mouse expressing the Cre recombinase specifically in PV neurons (Sohal et al., 2009; Cardin et al., 2009). This approach allows combinatorial use of numerous types of viral vectors and mouse lines, and is the most popular approach for anatomically defined and cell type-specific expression of optogenetic and other genes (e.g. genetically encoded calcium or voltage indicators; see Chapter 5 by Chenchen Song and Thomas Knöpfel). A growing number of Cre driver lines facilitate this approach (Taniguchi et al., 2011), and some rat Cre driver lines are also available (Witten et al., 2011). Cre driver mouse lines can also be crossed with conditional opsin-expressing lines, which express opsin genes within a Cre-dependent cassette located in a ubiquitously expressed genomic locus. This approach can achieve brain-wide expression in a specific neuronal population (Madisen et al., 2012; Harris et al., 2014) and avoids the need to inject viral vectors. There are, however, two main drawbacks of this method compared with the combinatorial Cre-based viral approach: (i) the copy numbers of opsin genes attained using the viral vector approach are typically much higher than using the transgenic approach, due to the multiplicity of infection that occurs when injecting viral vectors, leading to higher amplitude photocurrents and, presumably, more robust optogenetic modulation; (ii) since the Cre-dependent expression of the opsin gene is non-reversible, opsin expression in the adult animal encompasses all neurons in which the Cre driving promoter was active, throughout embryonic and postnatal development. For example, neurons that expressed the promoter during day E12 of embryonic development would remain opsin-expressing even though they no longer express the promoter in the adult stage.

11.3.2 Connectivity-Based Targeting

In neural circuits, connectivity and function are thought to be tightly linked. Neurons projecting to defined long-range targets have been shown to possess defined functional properties (e.g. Herry et al., 2008), and dissecting these roles has been a major application of optogenetic tools (Sjulson et al., 2016). One method for determining synaptic connectivity between two defined neuronal populations is to express a channelrhodopsin gene in one, record from another and stimulate the expressing population of neurons or its projections with light (Petreanu et al., 2007; Warden et al., 2012; Scott et al., 2015;

Klavir et al., 2017). The major advantage of this method is that it provides a direct measure of functional synaptic connectivity between a ChR-expressing neuronal population and a second target population, either within the local circuit (Adesnik and Scanziani, 2010) or in a distant brain region (Melzer et al., 2012). While this approach is highly useful in the slice preparation as axons remain functional even when severed during the slicing procedure (Petreanu et al., 2007), *in vivo* experiments with this approach require greater care since light stimulation of intact axons might activate axons of passage that do not terminate at the illuminated region (Llewellyn et al., 2010). To establish the functional role of a particular population of neurons that projects to a defined brain target, several techniques have been developed that utilize a unique set of viral vectors that can retrogradely transduce neurons based on their local or long-range connectivity.

Herpes simplex vectors: these vectors are derived from the herpes virus, which can effectively transduce axonal terminals and use the cellular machinery to introduce its genetic material into the neuronal nucleus (Smith, 2012). Using engineered herpes simplex virus (HSV) vectors, it is possible to direct the expression of a gene to a population of neurons that forms a long-range axonal projection to a defined brain region (Lima et al., 2009). A HSV vector driving the Cre recombinase can be combined with either a conditional mouse line carrying the opsin gene (Madisen et al., 2012; Harris et al., 2014), or a Cre-dependent viral vector injected at the region containing the neuronal cell bodies (Znamenskiy and Zador, 2013). This method was successfully used by several labs (Senn et al., 2014; Lima et al., 2009). However, HSV-transduced neurons tend to become unhealthy after several weeks of expression and this needs to be taken into account in the experimental design (Ugolini, 1992; Ugolini, 2010).

Canine adenovirus vectors (CAV): the canine adenovirus possesses retrograde transduction properties similar to those of HSV (Kremer et al., 2000). It can be injected at the target region and retrogradely transduce the neurons projecting to this region. Cell health has been reported to be better with CAV than with HSV, but this should be tested and controlled for due to the potential variation in the sensitivity of different neurons to expression of this vector. A recent study has reported a fundamental difference between HSV and CAV, demonstrating that while HSV can only transduce synaptic terminals, CAV might also transduce axons of passage (Schwarz et al., 2015). This is obviously an important consideration when designing and interpreting such experiments. Furthermore, CAV and HSV have been reported to possess somewhat distinct tropism for axonal terminal populations, leading to incomplete overlap between the cell populations retrogradely labeled by these two vector types through an unknown mechanism (Senn et al., 2014).

Retrograde AAV: the retrograde-transducing AAV was developed using a directed evolution approach from a library of AAV capsids (Tervo et al., 2016), yielding an AAV vector that can be used in a similar manner to that described above for HSV and CAV. The main advantage of this vector is that AAVs are known to be minimally toxic to neurons, and can provide robust, lasting expression of transgenes. Like HSV and CAV, this vector can be used to either drive the expression of the opsin gene directly, or of a Cre recombinase combined with Cre-dependent expression of an opsin gene at an upstream location.

Rabies virus targeting: a major advantage of the rabies virus is that it allows the expression of a transgene through trans-synaptic targeting from a defined neuronal population. The rabies virus undergoes trans-synaptic transport by utilizing pre- and post-synaptic molecules, transporting its genome across specific synaptic connections (Ugolini, 1995). While the wild-type rabies virus can perform this process repeatedly, replicating in the host cells and thus traversing across multiple synaptic layers from the primary population, the engineered variants cannot (Wickersham et al., 2007a). Using these engineered rabies vectors, researchers have mapped the whole-brain inputs onto defined neuronal populations (Ogawa et al., 2014), and onto single cells in the mammalian cortex (Wickersham et al., 2007b; Wertz et al., 2015). The advantage of the rabies system over those described above is its exquisite trans-synaptic specificity. However, until recently neurons expressing the rabies-derived transgenes showed limited viability within 7–14 days from transduction, thus severely limiting the experimental time window. Ciabatti et al. overcame this limitation by engineering a self-inactivating rabies virus (SiR) which does not cause long-term cytotoxicity and outperforms the current retro-AAV (Tervo et al., 2016) in labeling subcortical structures (Ciabatti et al., 2017).

11.4 LIGHT DELIVERY

Optogenetic technology allows researchers to induce light-dependent changes in the activity of genetically defined neuron populations. The induced effect depends on the opsin being expressed and the properties of light delivered at the site of action. For a given opsin and expression system, the parameters controlled by the experimenter during the experiment are the wavelength, the photon flux, and the temporal pattern of illumination. Optimization of these parameters for efficient optogenetic control is discussed in Section 11.4.1.

To design optimal optogenetic experiments, one needs to calculate, or measure the photon flux at the site containing the opsin-expressing cells. The change in radiant flux density and distribution with distance from the source is strongly dependent on the optical properties of the medium. For example, air very weakly absorbs light in the visible spectrum. Therefore, light power measured directly in free space by a power meter is dramatically different from incident power at the same distance from the light source in the intact brain. In brain tissue, the distribution of light from a source is dictated by scattering and absorption, which are both wavelength-dependent and can also vary between brain regions. To deliver the required photon flux at the target, it is therefore crucial to consider the attenuation introduced by the optical properties of the tissue (discussed in Sections 11.4.2–11.4.3). Delivery of light into brain tissue can involve significant heating and potentially lead to undesired artifacts. We discuss heating-related considerations in Section 11.4.4 and provide some examples for commonly used light-delivery schemes in Sections 11.4.5–11.4.7.

11.4.1 Light Requirements for Efficient Optogenetic Modulation

In the case of optogenetic excitation of neurons, the excitatory light-gated tool often needs to induce a photocurrent that is sufficiently large to depolarize the targeted cells beyond their action potential thresholds. The magnitude of the required photocurrent depends on many factors, including the input resistance of the given cell, its action potential threshold and its maximal firing rate, which together determine the neuron's ability to reliably respond at the required frequency throughout the desired stimulation period. This is therefore a highly cell type-dependent measure and, for *in vivo* experiments, also depends on the behavioral state of the animal due to changing levels of neuromodulators and tonic inhibition of the targeted neuronal population (Lee and Maguire, 2014). Light power is typically not the limiting factor when a small tissue volume is targeted. However, as the tissue volume increases, the peak light power required to achieve efficient activation 100 μm or more away from the center of illumination can become a limiting factor, especially when using flat-cleaved optical fibers. Light sensitivity is quite similar among ChRs, as it is an intrinsic property of the retinal chromophore, which is common to all ChRs. However, the kinetic properties of their photocurrents can vary greatly and while some channels are extremely fast, closing with a time constant (τ_{off}) of several milliseconds (Gunaydin et al., 2010; Mattis et al., 2011), others are extremely slow, with τ_{off} of seconds to minutes (Berndt et al., 2009; Yizhar et al., 2011c). For a given ChR, increasing the light pulse duration will monotonically reduce the required irradiance, if the light pulse is shorter than the time needed for channels to complete a photocycle and close again (Figure 11.1a). The optimal light pulse duration and irradiance that minimize the amount of energy delivered to the tissue (radiant exposure) while inducing the required photocurrent can be estimated based on the kinetic parameters of the ChR variant in use (Figure 11.1b, c). This also means, that for a light stimulus that is longer than a given channel's τ_{off}, the closing kinetics and the effective irradiance for 50% activation are inversely proportional (Mattis et al., 2011). This is due to the shorter time in which open channels can accumulate for an opsin with fast closing kinetics. Slower opsins therefore allow for the use of lower irradiance combined with longer light pulses, effectively reducing tissue heating due to the improved heat dissipation with longer, lower radiant flux density pulses compared with short, high

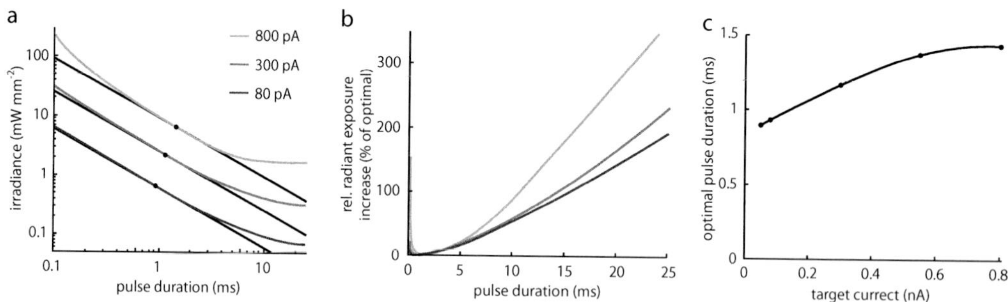

FIGURE 11.1 Optimal light pulse duration for ChR2 activation. Simulations of the necessary irradiance and light pulse duration to evoke 80, 300, and 800 pA current in a ChR2 expressing cell based on a six-state functional Markov model of opsin kinetics (PyRho; Evans, 2016; Grossman, 2013). (a) Necessary irradiance to induce the target current for a given light pulse duration. Black lines indicate equi-radiant exposure fits through the pulse duration resulting in minimal radiant exposure for comparison (black circles; $y = a \cdot x^{-1}$). (b) Relative increase in radiant expose compared to the optimal pulse duration. (c) Dependence of the optimal pulse duration on the target current.

radiant flux density ones. However, a longer open-state results in lasting depolarization and thereby reduced temporal control over action potential generation, even for brief illumination pulses. Also, the maximal inducible firing rates are lower, due to accumulation of refractory voltage gated sodium channels (Lin, 2011). The optimal tradeoff would therefore be to match the excitatory opsin to the maximal firing rate that needs to be reliably evoked experimentally, and then design the illumination pulse width to match the kinetics of this opsin. This allows minimization of the irradiance needed for efficient activation of a given cell population.

Optogenetic inhibition typically imposes more stringent constraints, due to the need for continuous modulation of the target cells. With some inhibitory optogenetic tools (e.g. halorhodopsin and archaerhodopsin), continuous illumination is required in order to achieve sustained silencing (Chow et al., 2010; Zhang et al., 2007; Gradinaru et al., 2010). However, newer inhibitory optogenetic tools such as soma targeted ACRs (Mahn et al., 2018; Messier et al., 2018) allow for reduced irradiance and step-function ACRs (Sineshchekov et al., 2015; Wietek et al., 2015; Berndt et al., 2016) allow for intermittent illumination (see Section 11.2.2 Tools for Optogenetic Inhibition). Regardless, light propagation and heating are major caveats that need to be considered when designing both optogenetic excitation and inhibition experiments (Wiegert et al., 2017).

11.4.2 Optical Properties of Brain Tissue

The probability of an opsin molecule switching from its dark-adapted, inactive state to its active state depends on the photon density and their wavelength. Figure 11.2 introduces radiometric quantities that are used to perform light power calculations. In optogenetic experiments, one needs to estimate the spectral flux density (Figure 11.2a, b) that should be emitted from a surface (such as the tip of an optical fiber; Figure 11.2c) to achieve the minimal necessary irradiance at another surface, such as the membrane of a channelrhodopsin-expressing neuron. As a light beam travels through a medium, its radiant flux density is attenuated by absorption and redistributed by scattering (Figure 11.2d, e). Light propagation in a turbid medium such as brain tissue is strongly affected by both of these phenomena (Azimipour et al., 2014; Gysbrechts et al., 2016). An important point to consider is that the parameters describing light transmission differ greatly between biological tissues and specifically in brain tissue, light transmission also varies substantially between different brain regions (Al-Juboori et al., 2013). *In vitro* measurements do not allow for the faithful estimation of blood absorption, and the absorption coefficient of brain tissue was estimated to typically be much smaller than the scattering coefficient and neglected. However, hemoglobin in red blood cells possesses a very high absorption coefficient in the visible range (Figure 11.3a) and likely leads to a significant increase in the absorption coefficient *in vivo*. According to Chugh et al. (2009), the small vessel cerebral blood volume, excluding major vessels, in mouse brain ranges from 2.8 to 5.9% of the tissue volume, with a brain-wide average of 4.4%. Using these values, absorption lengths can be estimated for light at different wavelengths (Figure 11.3a). According to this estimate, absorption will cause the spectral flux density at the peak excitation wavelength of ChR2 (470 nm) to attenuate by 60% within 1 mm in striatum and by 86% in the olfactory bulb. Moreover, if a

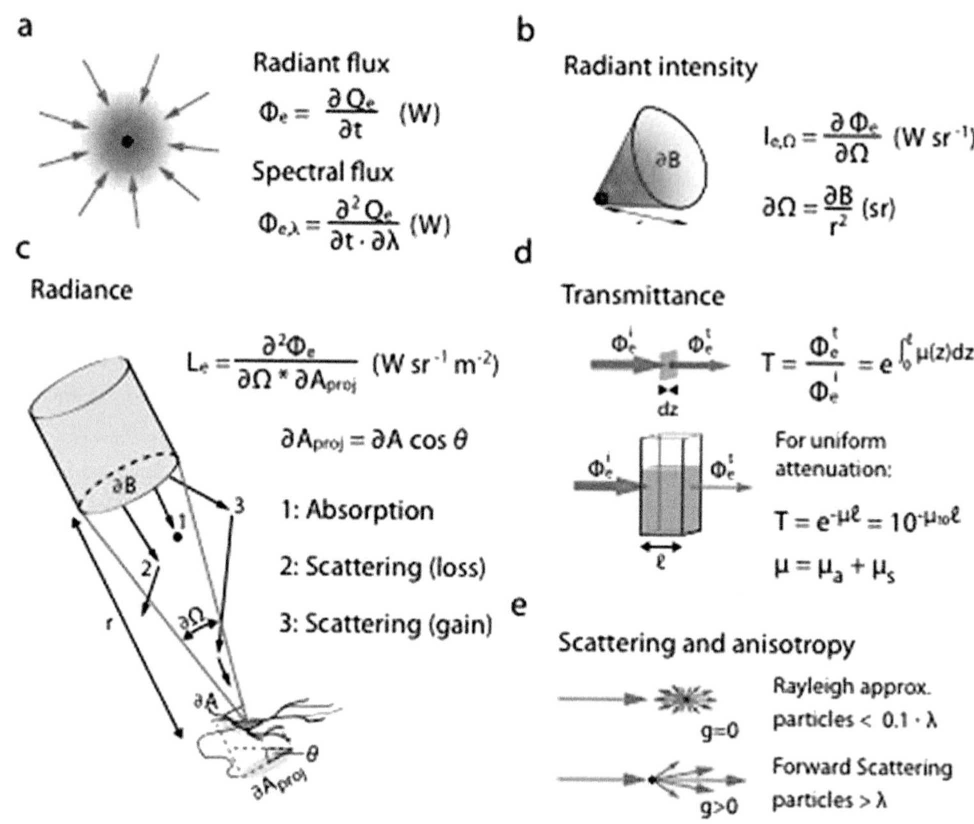

FIGURE 11.2 Radiometric quantities. (a) In radiometry the energy of electromagnetic waves (radiant energy: Q_e) per unit time is referred to as radiant flux (Φ_e). When a quantity is given within a wavelength range, the word "radiant" is replaced by "spectral", as for instance in spectral flux ($\Phi_{e,\lambda}$). (b) The radiant flux received or emitted from a point source within a range of directions (per solid angle) is called radiant intensity ($I_{e,\Omega}$). The radiant flux emitted, reflected, transmitted, or received per unit area is called radiant flux density ($E_{e,\lambda}$). The radiant flux density received per unit area is also referred to as irradiance. (c) Radiance is the radiant intensity received by a given surface. It depends on the radiant flux emitted in the relevant direction, the size of the two surfaces, their angle towards each other (projected area), their distance as well as characteristics of the light propagation through the medium. (d) When measuring the relative reduction in radiant flux (transmittance) of a narrow beam passing through a thin tissue section, light absorption as well as light scattering away from the original beam path can reduce the measured transmittance. The resulting transmittance is described by the Beer-Lambert law and is dependent on the attenuation coefficient (μ, also referred to as transport coefficient). The attenuation coefficient is a combination of the absorption coefficient (μ_a) and the scattering coefficient (μ_s). In chemistry, the decadic attenuation coefficient (μ_{10}) of a given substance is normally reported. (e) Scattering events at particles much smaller than the wavelength of the interacting photon are scattered forward and backwards with equal probabilities, resulting in the average of the scattering angle to be close to 0. When light interacts with particles bigger than the photon's wavelength, scattering mostly occurs in forward directions.

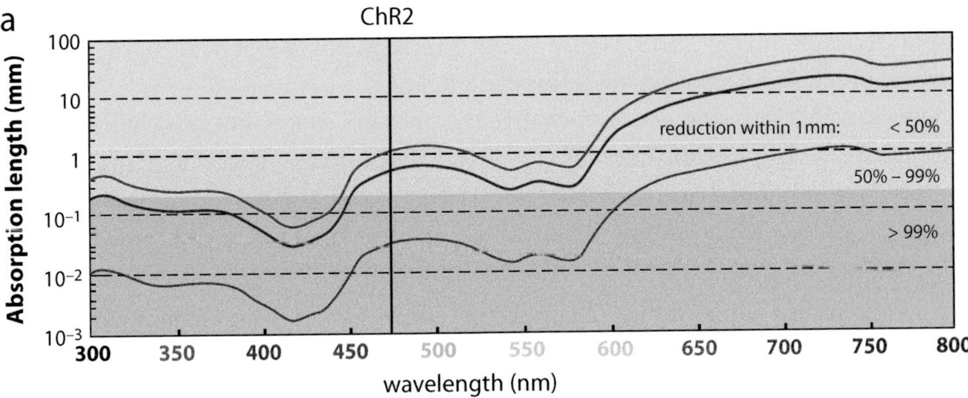

— striatum (estimated as: 62% Water, **2.8% blood** (62% O2 saturated))

— olf. bulb (estimated as: 62% Water, **5.9% blood** (62% O2saturated))

— **blood** (62% oxygen saturation)

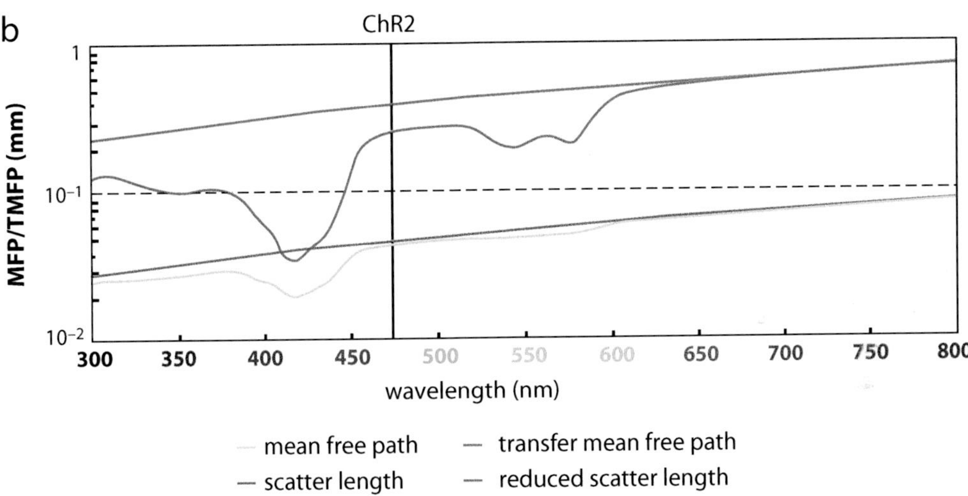

FIGURE 11.3 Estimated absorption length and mean free path in brain tissue. (a) Absorption length is the inverse of the absorption coefficient and can be interpreted as the distance a light beam can travel through a medium before reducing to 1/e times its initial radiant flux density. Absorption length of 62% oxygenated blood was estimated from the hemoglobin absorption coefficient (omlc.org). Estimates of the absorption length of the two brain areas with lowest and highest reported blood volumes (Chugh et al., 2009) are depicted. (b) The inverse of the attenuation coefficient is referred to as attenuation length or mean free path (MFP) and can be interpreted as the average distance traveled by a photon before it is absorbed or scattered. MFP based on the scattering coefficient estimated as $2.37\cdot(\lambda/500\text{nm})$-1.15 with or without considering absorption depicted in orange and red, respectively (Liu et al., 2015). The transfer mean path takes the anisotropy factor (g = 0.8803) into consideration: $\mu_s' = \mu_s\cdot(1\text{-}g)$.

large blood vessel passes though the beam path, it will dramatically attenuate the transmitted light (Azimipour et al., 2015, Figure 11.3a). Similarly, blood that has coagulated around an optical implant can strongly absorb applied light.

The process of scattering does not change the photon's energy but can change its direction (Figure 11.2e). Scattering therefore causes a light beam to disperse, leading to a reduction in the radiant flux density. In tissue, scattering is due to interaction of photons with particles of various sizes ranging from the phospholipids of cell membranes over cellular organelles like mitochondria to entire cell bodies (Beauvoit et al., 1995). A scattering event can often be approximated by the change in the direction of a photon due to interaction with a homogenous sphere of a given size, which is described by the Mie solution to the Maxwell equation (Stratton, 1941). Mie solutions for calculation of the scattering phase function are expressed in terms of infinite series. For spheres much larger and much smaller than the wavelength of the light, the exact Mie solution is typically analytically approximated. For particle sizes up to about 10% of the wavelength, the Rayleigh approximation is valid and scattering events for such small particles are therefore often referred to as Rayleigh scattering (Figure 11.2e). The intensity of Rayleigh scattering is identical in forward and reverse directions and increases rapidly with increasing particle size or decreasing wavelength. Scattering by particles with roughly the same size as the wavelength is typically referred to as Mie scattering. The probability of Mie forward-scattering increases with increasing particle size-to-wavelength ratio. The average of the cosine of the scattering angle in this function is referred to as the anisotropy coefficient (g). It ranges from 1 (only forward scattering) to –1 (only backward scattering), with an anisotropy coefficient value of zero corresponding to isotropic scattering. Scattering phase functions with strong forward scattering can be approximated by the Henyey–Greenstein phase function (Henyey and Greenstein, 1941), which is fully parametrized by g. The anisotropy coefficient of brain tissue is typically greater than 0.8, meaning that it is dominated by forward scattering and a single scattering event is unlikely to completely change the direction of a photon. The distance at which a light beam becomes diffuse is therefore a multiple scattering process and can be estimated by taking g into consideration. The reduced scattering coefficient is defined as: $\mu_s' = \mu_s\ (1 - g)$. If only forward scattering occurs ($g = 1$), scattering has no effect on the direction of a light beam, while a single scattering event of each photon is sufficient to disperse a light beam during isotropic scattering ($g = 0$). This distance after which the light beam fraction continuing to travel along its original trajectory is reduced to ~37 % is called the transfer mean free path (TMFP; Figure 11.3b).

Absorption as well as scattering and therefore also the penetration depth into the brain depend strongly on the wavelength of light. Within the visible spectrum, longer wavelengths penetrate deeper. A wavelength range with favorable light penetration characteristics is referred to as an "optical window". The classical optical window used for brain imaging is in the range of 650–950 nm, resulting from the dramatic reduction in hemoglobin absorption beyond 650 nm (Figure 11.3). Opsins activated with wavelengths above 600 nm can therefore be advantageous when large brain regions are targeted for optogenetics (Gysbrechts et al., 2016). For a more in-depth review on the optical properties of biological tissues, we refer the reader to Jacques (2013).

11.4.3 Irradiance Estimation

Mathematically, the propagation of light in a medium is described by the radiation transport equation (RTE, Ishimaru, 1977; Wang and Wu, 2009). Solving this equation allows a calculation of the radiant flux at any point for a given light source configuration. The RTE does not consider the wave properties of light; therefore, phenomena such as diffraction and interference, which also occur when a coherent light source is used, are not considered. Analytical solutions to the transport equation only exist for a few specific cases, often necessitating approximations by numerical methods. In general, analytical solutions as well as numerical methods require detailed information about the optical properties of brain tissue, which can vary considerably between brain areas, so implementations of this fundamental approach are still rare. Early work on estimating optogenetic light transport used the Kubelka–Munk model (KM, Kubelka, 1948; Vo-Dinh, 2003; Aravanis et al., 2007), which was originally designed to model light reflectance from opaque surfaces. The KM model can be used to calculate a rough estimate of light propagation by assuming isotropic scattering and isotropic illumination (Neuman and Edström, 2010). However, the isotropic scattering assumption is only valid after the light has passed through the transfer mean free path. In optogenetic experiments, however, light is often delivered from a small fiber tip, in stark contrast to the assumed illumination with diffuse light, potentially leading to imprecise irradiance estimates especially close to, as well as lateral to, the fiber-brain interface. More appropriate modeling approaches are based on Monte Carlo simulations of light transport (MC, Wilson and Adam, 1983; Wang et al., 1995; Pifferi et al., 1998; Bernstein et al., 2008; Chow et al., 2010; Fang, 2010; Kahn et al., 2011; Yona et al., 2016) where the propagation of ensembles of single photons is modeled. This method allows integrating local tissue inhomogeneity (Liu et al., 2015) and can be adapted to various light delivery paradigms. To generate reliable irradiance estimations, the path of many photons must be simulated, making it a computationally costly and therefore time-consuming process. Yona et al. (2016) recently described an analytical approach to estimate the light propagation from an optical fiber (Figure 11.4a, b) based on the beam spread function (BSF) method (McLean et al., 1998). Several software tools have recently been described which allow users to estimate light propagation in a given experimental paradigm, and can greatly aid in the design of optogenetic experiments (Liu et al., 2015; Stujenske et al., 2015; Yona et al., 2016).

11.4.4 Heating

Brain tissue heating is a direct result of light absorption and a great concern when using optogenetics, not only due to the damage it may cause but also due to the possibility of heat-induced neuronal activation (Stujenske et al., 2015; Rungta et al., 2017). Optogenetic experiments, particularly involving inhibitory optogenetic tools, often require prolonged illumination of neural tissue which may cause temperature changes. While some brain systems, specifically in the hypothalamus, are directly involved with regulating thermal homeostasis and contain neurons that are highly sensitive to temperature changes (Tabarean et al., 2005), there are also reports of changes in the activity of other neural systems such as the hippocampus (Thompson et al., 1985; Kim and Connors, 2012), and

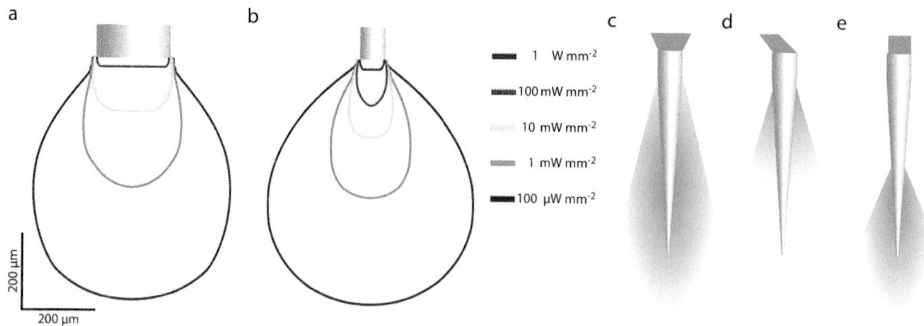

FIGURE 11.4 Light delivery to deep brain targets. (a, b) Estimation of the radiant flux density due to light propagation from a flat-cleaved optical fiber, based on Yona et al. (2016) using light transfer parameters as shown in Figure 11.3. Contour lines for 5 mW of 470 nm light exiting a 200 μm (a) or 62 μm (b) diameter fiber with a numerical aperture of 0.39. Radiant flux density at the fiber tip is 160 mW mm^{-2} and 1.66 W mm^{-2} for the 200 μm and 62 μm diameter fiber, respectively. Radiant flux density differs strongly only close to the fiber tip. (c–d) Schematic of light propagation from a tapered fiber (Pisanello et al., 2017). Maximal radiant flux densities can be reduced by increasing the emitting surface of the optic fiber. This approach allows for efficient opsin activation in large tissue volumes. Furthermore, light exit height along the taper can be controlled by the light input angle to the fiber. Light entering the fiber at an angle exits early along the taper (d), while light entering the fiber orthogonal to the surface exits the fiber closer towards the end of the taper (e).

in neuronal firing rates (Reig et al., 2010) due to thermal changes. In fact, it has been demonstrated that the commonly used 5 mW and 10 mW radiant flux at 532 nm delivered through a 200 μm diameter optical fiber are sufficient to elicit an increase in firing rates of around 30% and 40%, respectively, in wild-type mice which do not express any opsin (Stujenske et al., 2015). Heat will be generated wherever light is absorbed, proportionately to the radiant flux density and dissipate into the outlying non-illuminated tissue by a combination of heat diffusion and convective transport (Yizhar et al., 2011b). Absorption coefficients vary by wavelength and are approximately ten times higher for 475 nm than 630 nm (Figure 11.3), therefore the effect of heating varies markedly between different wavelengths. Wavelengths that are more strongly absorbed heat a smaller brain volume, leading to a larger temperature increase at the same irradiance. Taking this into consideration, Stujenske et al (2015) developed a model to predict heat changes according to light power, wavelength, and optical fiber size. The model predicts that for continuous illumination at 10 mW, for example, a temperature change of 1–4°C will occur across a large volume of tissue. This temperature change is sufficient to induce physiological and behavioral changes (Thompson et al., 1985; Moser et al., 1993) and to significantly alter blood flow (Rungta et al., 2017) and local neuronal firing rates (Stujenske et al., 2015) in naïve animals.

These findings highlight two major requirements when planning optogenetic experiments – one is to consider tissue heating and the other is to use opsin-free light-stimulated controls. There are several factors which could be considered in order to reduce the heating effects: the first is fiber diameter – per a given light power, a larger fiber diameter reduces the induced heat at the fiber tip due to an effective reduction in radiant flux density (Figure 11.4; Chuong et al., 2014). The second factor is wavelength – as mentioned above, absorption is higher for shorter wavelengths and therefore peak temperature increase is predicted to be two-fold lower for 593 nm light as compared with 532 nm at the same radiant flux density. Therefore, if prolonged stimulation is needed, it might be advisable to illuminate at wavelengths above the action spectrum peak, as even with the higher light power required at the suboptimal excitation wavelength, the lower tissue absorption can lead to a lower peak temperature increase. Another factor is the duty cycle – pulsing the illumination instead of continuous irradiation reduces the induced heat dramatically (Stujenske et al., 2015). When reporting on optogenetic experiments, irradiance (or light intensity, in watts per unit area) is commonly used instead of net photon flux, which allows a direct understanding of the amount of energy delivered to the tissue, the critical parameter for potential side-effects like tissue heating.

11.4.5 Surface Targets

Optogenetics is readily applicable to light-accessible preparations such as cultured neurons, brain slices, transparent organisms such as zebrafish larvae, or to the cortical surface of rodents, allowing for extensive flexibility in light delivery. As long as the target population is close to the air-tissue interface (as with cortical neurons or in smaller organisms), light can be delivered in a manner that can be rapidly altered in both the spatial and temporal domains (Guo et al., 2014; Shipley et al., 2014; Mohammad et al., 2017). One way to achieve spatially restricted illumination is through the use of two-dimensional LED arrays (Grossman et al., 2010). The main downside of this system is the limited spatial resolution due to the minimal LED size. Light patterning with greater resolution can be attained using digital micro-mirror devices (DMD) or through wavefront modulation by a liquid crystal spatial light modulator (SLM). The DMD system uses a large array of independently switchable micro-mirrors. Multiple wavelengths can be delivered using a high-power broad-band light source filtered to the desired wavelengths and modulated through the DMD to create the desired pattern (Pashaie et al., 2014). The ability of the system to project multi-colored patterned images and the individual modulation of single mirrors at high frequency (up to several kHz) allows for the system to employ both spatially and temporally patterned photostimulation of the specimen. The light source for this type of system can be a LED array (Sakai et al., 2013) or lasers. Using a combination of dichroic mirrors for laser beam combination and acousto-optic tunable filters furthermore allows for rapid wavelength switching (Brinks et al., 2016). One drawback of DMDs is that they require high-power light sources, as the patterning is created by discarding the light from the unwanted areas. An alternative method to the DMD is the SLM, which is discussed in detail in Chapter 3. While SLMs allow greater spatial control and achieve high local intensities, their disadvantage compared with the DMD is their lower temporal resolution (Brinks et al., 2016) and a remaining fraction of undiffracted light, which can cause a focal excitation point in the specimen (Hernandez et al., 2014). When used with single-photon excitation, these approaches have limited axial resolution. For better spatial specificity and for deeper penetration into the tissue, using an SLM for holographic light patterning stimulation combined with a two-photon system has been applied for optogenetic stimulation (Oron et al., 2012). Two-photon optogenetics has been applied successfully in different surface targets from slices (Papagiakoumou et al., 2010) to superficial cortical layers *in vivo* (Prakash et al., 2012; Packer et al., 2012; Rickgauer et al., 2014; Packer et al., 2015; Chaigneau et al., 2016).

11.4.6 Deep Brain Targets

Targeting deep brain neural circuits with optogenetics *in vivo* presents several unique challenges. Light needs to reach the target with sufficient irradiance to induce opsin activation in a sufficient number of neurons. However, guiding the light into the structure should be done with minimum damage to the tissue. In behaving animals, stimulation should also be conducted with minimal disruption to the measured behavior. Deep brain optogenetic stimulation is typically carried out using a multimode optical fiber, guiding the light from the source to the deep target. In anesthetized animals, the fiber can be directed to the target using stereotactic arms. For chronic experiments, fibers are permanently implanted by attaching a fiberoptic implant to the skull using dental cement (Mahn et al., 2013). The dimensions of the fiber and its optical properties can dramatically influence the profile of the light reaching the brain (Yizhar et al., 2011b). The commonly used step-index fiber consists of two layers. A core layer is enveloped by a cladding layer with a higher refractive index than the core. The interface between the core and cladding determines the numerical aperture (NA), which is the maximum angle of incidence at which a light ray can still propagate through the fiber. This sets the maximum acceptance angle for incoming light and maximum exit angle of beams exiting the fiber. The exit angle sets the shape and size of the tissue volume affected by the light as the propagation of light within the transfer mean free path depends on the geometric spread of the light (Yizhar et al., 2011b; Yona et al., 2016). The main drawback of flat surface optical fibers is the high radiant flux density necessary at the fiber tip to reach sufficient irradiance within the targeted volume (Figure 11.4a, b). This problem decreases with increasing optical fiber diameter. However, wider fibers also cause more tissue damage. This tradeoff can be overcome with a tapered optical fiber, which can be used to flexibly illuminate a large brain volume (Figure 11.4c–e; Pisanello et al., 2017). Patterned light delivery in deep brain areas is possible with the mentioned approaches for surface targets in combination with removal of the covering brain tissue or implantation of a graded-index lens (Jung and Schnitzer, 2003; Levene et al., 2004). Also see Chapter 14 for further discussion.

11.4.7 Integrating Optogenetic Control with Electrophysiology

Good practice requires observing and measuring the neuronal outcome of optogenetic manipulations. This is commonly done either by immunohistochemical validation by staining for immediate early genes such as c-Fos (Gradinaru et al., 2009) or by integrating electrophysiological probes for *in vivo* recordings with a light guide (Anikeeva et al., 2011; Stark et al., 2012; Kvitsiani et al., 2013). Electrophysiological probes typically consist of an array of individual electrodes, stereotrodes (two intertwined microwires) or tetrodes (four intertwined microwires) for isolation of single-unit action potential waveforms. The integrated device is referred to as “optrode” or “optetrode” (Anikeeva et al., 2011), and allows for illumination of the area in which the electrodes are located while recording from light-modulated neurons (Pashaie et al., 2014). The spatial targeting, however, is far from perfect due to the random spatial mismatch of the wires. The bundle of wires can be replaced with a silicon multitrode design combined with an optical fiber (Royer et al., 2010). This allows for better spatial control and might reduce tissue damage. It is possible to combine an optical fiber into a Utah-style multi-electrode array (MEA) by replacing one of the electrodes in the array with an opening for the fiber tip (Zhang et al., 2009). This improves spatial targeting but only to the electrodes in the direct vicinity of the fiber, while all other electrodes in the array would record neurons farther away and thus not directly affected by the illumination. The problem of tissue damage and spatial mismatch is exacerbated when the research question calls for multiple light delivery and recording locations. These issues are elegantly resolved utilizing the optically transparent and highly conductive zinc oxide (ZnO) crystal in the design of a "micro-optoelectrode array" (MOA) for simultaneous multichannel optical stimulation and electrical recording (Lee et al., 2015). The MOA geometry is like that of the Utah-array, yet due to the features of the wide-bandgap transparent semiconductor – ZnO, each tip is both a light guide and a microelectrode. The MOA therefore allows simultaneous microscale electrophysiological recording in full spatial registry with photoexcitation (Lee et al., 2015). Yet the immediate proximity also creates light artifacts in the recordings, specifically in the lower-frequency local-field potential (LFP) band.

Recently, several new technologies have been developed to allow multi-modal manipulation and recording from the mammalian brain. One such approach combines different polymers, composite materials and low melting-temperature metals to create a single fiber probe using a thermal drawing process (TDP), a method also applied to glasses during fiber production (Canales et al., 2015). These multi-modal fibers can be used for simultaneous optical stimulation, drug delivery, and electrophysiological recordings *in vivo*. The TDP fabrication packs all these functions into a single flexible and small diameter probe (70–700 μm) and can be reduced further with selective etching of sacrificial cladding (Canales et al., 2015). One potential drawback associated with the use of physical tethers for light and drug delivery is that it could physically restrict the animal, interfering with its natural movement. Several wireless systems have been developed for *in vivo* light stimulation. These normally include LEDs or micro-LEDs that can be remotely controlled and powered by a battery (Park et al., 2015) or through near-field wireless power (Shin et al., 2017). Recently, new wireless devices were described which include an integrated light and fluid delivery system to allow combined pharmacological and optogenetic manipulation (Jeong et al., 2015). While these novel wireless systems permit experimental designs which were previously impossible, they are significantly more complex to utilize than the more conventional fiberoptic-based optogenetic techniques. Future development, along with commercialization of such solutions, will undoubtedly make these wireless solutions more broadly applicable.

11.5 EXPERIMENTAL PARADIGM

The widespread adoption of optogenetic technology in the field of neuroscience is predominantly ascribable to the ease of use and the robustness of microbial rhodopsins. With opsin-encoding viruses, mouse lines, and affordable light delivery systems widely available, optogenetic techniques have become commonplace in neuroscience laboratories. While allowing for a wide range of experimental paradigms, the variety of optogenetic tools presents a challenge in deciding on the most appropriate tool and paradigm, and understanding the experimental constraints of the chosen approach. The efficacy of optogenetic manipulation is dependent on many factors, including the type of opsin used, its expression level, the intrinsic neuronal firing properties of the targeted population, the targeted cell compartment, and the effective light power at this compartment. Therefore, the effect of any optogenetic modulation needs to be carefully characterized to allow for the interpretation of behavioral experiments.

11.5.1 Necessary Controls

The full list of necessary controls varies between experiments and depends on the specific technical details of the planned experiment. Here, we propose a minimal set of controls that should be performed when optogenetics is employed.

Viral spread and cell viability: if viruses are used as transgene delivery vectors, viral vector spread and expression level need to be evaluated. Many opsin viral vectors were designed to co-express a fluorophore. Standard histological methods can be used to visualize the virus expression level and area. Especially when the brain structure of interest is challenging to target due to its size or location, characterizing viral expression for every experimental animal can be used to correlate the variability in the behavioral effect to variability in expression area and, for instance, optical fiber placement. Even when an experiment is planned based on published work demonstrating robust optogenetic modulation of a defined set of neurons, the paradigm should be validated in each new experiment due to the potential variability of viral vector titers, optical hardware, and mouse strain.

Validation of optogenetic manipulation: it is crucial to measure the change in firing rate at least in a subset of animals by extracellular or intracellular recordings, *in vivo* or *in vitro*. High-frequency light pulse trains or constant illumination of an excitatory expressing pyramidal neuron can, for instance, lead to depolarization block, effectively reducing rather than increasing its firing rate (Herman et al., 2014). Whether such depolarization block occurs and to what extent is hard to predict, as it depends on many experiment- and setup-specific parameters which can greatly vary between laboratories. Electrophysiological characterization of the optogenetic manipulation should therefore be performed in every experiment, rather than relying on previous work.

Cell health: virus expression can impact cell health or change the electrophysiological properties of the targeted neurons. It is therefore necessary to include a control group injected with titer matched virus that expresses a control transgene. Researchers often use a virus encoding the same fluorophore that is co-expressed with the opsin. This control group can be used to evaluate direct effects of the virus injection surgery and potential phototoxic or heating effects due to the light delivery paradigm. Strong opsin expression has been reported to affect cell physiology (Miyashita et al., 2013). It is therefore also advisable to include an opsin expressing group where no light is applied. In case the experiment allows for multiple repeats of the same manipulation, the light and no light conditions can be tested in the same group (counterbalanced), which results in a within-animal control. Ferenczi et al. (2016) showed that opsin activation in a given neuronal population induces a slow, non-synaptic current in the same direction in nearby non-expressing neurons. Other potential effects that should be considered are transient and long-term changes induced at light onset or light termination (Raimondo et al., 2012; Mahn et al., 2016).

11.6 OUTLOOK

Since the first time a light-gated ion channel was expressed in neurons and successfully used to reliably stimulate these neurons with light, the use of optogenetics has vastly expanded and hence matured to a well-established toolkit for the investigation of brain function. While there is still room for improvement, for instance, in better spectral separation between different optogenetic tools or in refined approaches for synaptic terminal inhibition, optogenetic tools are often not the limiting factor for experimental approaches when the appropriate tool is chosen. Driven by the advances in genetically encoded sensors and actuators, gene delivery methods are constantly advancing allowing for targeting cell populations with ever-increasing specificity. Translational approaches, on the other hand, require less invasive methodologies. For this purpose, light in the visible or near-infrared spectrum is not ideal for opsin activation as tissue penetration depth is too limiting. Activation of genetically encoded tools with electromagnetic waves of other frequencies that allow for high spatial and temporal resolution in addition to good tissue penetration are therefore desirable. Current research efforts are directed at using low-frequency radio waves or magnetic fields (Stanley et al., 2015; Wheeler et al., 2016; Stanley et al., 2016), but the feasibility of these approaches is a topic of active debate (Meister, 2016).

REFERENCES

Adamantidis, A. R. et al., 2007. Neural substrates of awakening probed with optogenetic control of hypocretin neurons. *Nature*, 450(7168), pp. 420–424.

Adesnik, H. & Scanziani, M., 2010. Lateral competition for cortical space by layer-specific horizontal circuits. *Nature*, 464(7292), pp. 1155–1160.

Airan, R. D. et al., 2009. Temporally precise in vivo control of intracellular signalling. *Nature*, 458(7241), pp. 1025–1029.

Akerboom, J. et al., 2013. Genetically encoded calcium indicators for multi-color neural activity imaging and combination with optogenetics. *Frontiers in Molecular Neuroscience*, 6, p.2.

Al-Juboori, S. I. et al., 2013. Light scattering properties vary across different regions of the adult mouse brain. *PloS One*, 8(7), p. e67626.

Allen, W. E. et al., 2017. Global representations of goal-directed behavior in distinct cell types of mouse neocortex. *Neuron*, 94(4), pp. 891–907.e6.

Ando, R., Mizuno, H. & Miyawaki, A., 2004. Regulated fast nucleocytoplasmic shuttling observed by reversible protein highlighting. *Science (New York, N.Y.)*, 306(5700), pp. 1370–1373.

Anikeeva, P. et al., 2011. Optetrode: a multichannel readout for optogenetic control in freely moving mice. *Nature Neuroscience*, 15(1), pp. 163–170.

Aravanis, A. M. et al., 2007. An optical neural interface: in vivo control of rodent motor cortex with integrated fiberoptic and optogenetic technology. *Journal of Neural Engineering*, 4(3), pp. S143–S156.

Arenkiel, B. R. et al., 2007. In vivo light-induced activation of neural circuitry in transgenic mice expressing channelrhodopsin-2. *Neuron*, 54(2), pp. 205–218.

Armbruster, B. N. et al., 2007. Evolving the lock to fit the key to create a family of G protein-coupled receptors potently activated by an inert ligand. *Proceedings of the National Academy of Sciences of the United States of America*, 104(12), pp. 5163–5168.

Azimipour, M., Atry, F. & Pashaie, R., 2015. Effect of blood vessels on light distribution in optogenetic stimulation of cortex. *Optics Letters*, 40(10), pp. 2173–2176.

Azimipour, M. et al., 2014. Extraction of optical properties and prediction of light distribution in rat brain tissue. *Journal of Biomedical Optics*, 19(7), p. 75001.

Bailes, H. J., Zhuang, L.-Y. & Lucas, R. J., 2012. Reproducible and sustained regulation of Gαs signalling using a metazoan opsin as an optogenetic tool. *PloS One*, 7(1), p. e30774.

Beauvoit, B. et al., 1995. Correlation between the light scattering and the mitochondrial content of normal tissues and transplantable rodent tumors. *Analytical Biochemistry*, 226(1), pp. 167–174.

Ben-Ari, Y., 2002. Excitatory actions of gaba during development: the nature of the nurture. *Nature Reviews. Neuroscience*, 3(9), pp. 728–739.

Ben-Ari, Y., 2014. The GABA excitatory/inhibitory developmental sequence: a personal journey. *Neuroscience*, 279, pp. 187–219.

Ben-Ari, Y., Cherubini, E., Corradetti, R. & Gaiarsa, J. L., 1989. Giant synaptic potentials in immature rat CA3 hippocampal neurones. *The Journal of Physiology*, 416, pp. 303–325.

Berndt, A., Lee, S. Y., Ramakrishnan, C. & Deisseroth, K., 2014. Structure-guided transformation of channelrhodopsin into a light-activated chloride channel. *Science (New York, N.Y.)*, 344(6182), pp. 420–424.

Berndt, A. et al., 2009. Bi-stable neural state switches. *Nature Neuroscience*, 12(2), pp. 229–234.

Berndt, A. et al., 2011. High-efficiency channelrhodopsins for fast neuronal stimulation at low light levels. *Proceedings of the National Academy of Sciences of the United States of America*, 108(18), pp. 7595–7600.

Berndt, A. et al., 2016. Structural foundations of optogenetics: determinants of channelrhodopsin ion selectivity. *Proceedings of the National Academy of Sciences of the United States of America*, 113(4), pp. 822–829.

Bernstein, J. G. et al., 2008. Prosthetic systems for therapeutic optical activation and silencing of genetically-targeted neurons. *Proceedings of SPIE*, 6854, p. 68540H.

Bhoo, S. H. et al., 2001. Bacteriophytochromes are photochromic histidine kinases using a biliverdin chromophore. *Nature*, 414(6865), pp. 776–779.

Bonger, K. M. et al., 2014. General method for regulating protein stability with light. *ACS Chemical Biology*, 9(1), pp. 111–115.

Boyden, E. S. et al., 2005. Millisecond-timescale, genetically targeted optical control of neural activity. *Nature Neuroscience*, 8(9), pp. 1263–1268.

Bradley, S. J. & Challiss, R. A. J., 2012. G protein-coupled receptor signalling in astrocytes in health and disease: a focus on metabotropic glutamate receptors. *Biochemical Pharmacology*, 84(3), pp. 249–259.

Bregestovski, P. & Bernard, C., 2012. Excitatory GABA: how a correct observation may turn out to be an experimental artifact. *Frontiers in Pharmacology*, 3, p. 65.

Brinks, D., Adam, Y., Kheifets, S. & Cohen, A. E., 2016. Painting with rainbows: patterning light in space, time, and wavelength for multiphoton optogenetic sensing and control. *Accounts of Chemical Research*, 49(11), pp. 2518–2526.

Brown, B. A., Headland, L. R. & Jenkins, G. I., 2009. UV-B action spectrum for UVR8-mediated HY5 transcript accumulation in Arabidopsis. *Photochemistry and Photobiology*, 85(5), pp. 1147–1155.

Bugaj, L. J. et al., 2013. Optogenetic protein clustering and signaling activation in mammalian cells. *Nature Methods*, 10(3), pp. 249–252.

Byk, T. et al., 1993. Regulatory arrestin cycle secures the fidelity and maintenance of the fly photoreceptor cell. *Proceedings of the National Academy of Sciences of the United States of America*, 90(5), pp. 1907–1911.

Canales, A. et al., 2015. Multifunctional fibers for simultaneous optical, electrical and chemical interrogation of neural circuits in vivo. *Nature Biotechnology*, 33(3), pp. 277–284.

Cardin, J. A. et al., 2009. Driving fast-spiking cells induces gamma rhythm and controls sensory responses. *Nature*, 459(7247), pp. 663–667.

Chaigneau, E. et al., 2016. Two-photon holographic stimulation of ReaChR. *Frontiers in Cellular Neuroscience*, 10, p. 234.

Chaves, I. et al., 2011. The cryptochromes: blue light photoreceptors in plants and animals. *Annual Review of Plant Biology*, 62, pp. 335–364.

Chen, T.-W. et al., 2013. Ultrasensitive fluorescent proteins for imaging neuronal activity. *Nature*, 499, pp. 295–300.

Chow, B. Y. et al., 2010. High-performance genetically targetable optical neural silencing by light-driven proton pumps. *Nature*, 463(7277), pp. 98–102.

Christie, J. M., 2007. Phototropin blue-light receptors. *Annual Review of Plant Biology*, 58, pp. 21–45.

Christie, J. M., Blackwood, L., Petersen, J. & Sullivan, S., 2015. Plant flavoprotein photoreceptors. *Plant and Cell Physiology*, 56(3), pp. 401–413.

Christie, J. M. et al., 1999. LOV (light, oxygen, or voltage) domains of the blue-light photoreceptor phototropin (nph1): binding sites for the chromophore flavin mononucleotide. *Proceedings of the National Academy of Sciences of the United States of America*, 96(15), pp. 8779–8783.

Chugh, B. P. et al., 2009. Measurement of cerebral blood volume in mouse brain regions using micro-computed tomography. *NeuroImage*, 47(4), pp. 1312–1318.

Chuong, A. S. et al., 2014. Noninvasive optical inhibition with a red-shifted microbial rhodopsin. *Nature Neuroscience*, 17(8), pp. 1123–1129.

Ciabatti, E. et al., 2017. Life-long genetic and functional access to neural circuits using self-inactivating rabies virus. *Cell*, 170(2), pp. 382–392.e14.

Crefcoeur, R. P., Yin, R., Ulm, R. & Halazonetis, T. D., 2013. Ultraviolet-B-mediated induction of protein-protein interactions in mammalian cells. *Nature Communications*, 4, p. 1779.

Dana, H. et al., 2016. Sensitive red protein calcium indicators for imaging neural activity. *eLife*, 5.

Dimidschstein, J. et al., 2016. A viral strategy for targeting and manipulating interneurons across vertebrate species. *Nature Neuroscience*, 19(12), pp. 1743–1749.

Dittgen, T. et al., 2004. Lentivirus-based genetic manipulations of cortical neurons and their optical and electrophysiological monitoring in vivo. *Proceedings of the National Academy of Sciences of the United States of America*, 101(52), pp. 18206–18211.

Duan, D. et al., 1998. Circular intermediates of recombinant adeno-associated virus have defined structural characteristics responsible for long-term episomal persistence in muscle tissue. *Journal of Virology*, 72(11), pp. 8568–8577.

Fang, Q., 2010. Mesh-based Monte Carlo method using fast ray-tracing in Plücker coordinates. *Biomedical Optics Express*, 1, pp. 165–175.

Fenno, L., Yizhar, O. & Deisseroth, K., 2011. The development and application of optogenetics. *Annual Review of Neuroscience*, 34, pp. 389–412.

Ferenczi, E. A. et al., 2016. Optogenetic approaches addressing extracellular modulation of neural excitability. *Scientific Reports*, 6, p. 23947.

Forli, A. et al., 2018. Two-photon bidirectional control and imaging of neuronal excitability with high spatial resolution in vivo. *Cell Reports*, 22, pp. 3087–3098.

Gao, S. et al., 2015. Optogenetic manipulation of cGMP in cells and animals by the tightly light-regulated guanylyl-cyclase opsin CyclOp. *Nature Communications*, 6, p. 8046.

Glantz, S. T. et al., 2016. Functional and topological diversity of LOV domain photoreceptors. *Proceedings of the National Academy of Sciences of the United States of America*, 113(11), pp. E1442–E1451.

Gong, Y., Wagner, M. J., Zhong Li, J. & Schnitzer, M. J., 2014. Imaging neural spiking in brain tissue using FRET-opsin protein voltage sensors. *Nature Communications*, 5, p. 3674.

Goshen, I. et al., 2011. Dynamics of retrieval strategies for remote memories. *Cell*, 147(3), pp. 678–689.

Govorunova, E. G., Sineshchekov, O. A. & Spudich, J. L., 2015b. Proteomonas sulcata ACR1: a fast anion channelrhodopsin. *Photochemistry and Photobiology*, 92(2), pp. 257–263.

Govorunova, E. G. et al., 2015a. NEUROSCIENCE. Natural light-gated anion channels: a family of microbial rhodopsins for advanced optogenetics. *Science (New York, N.Y.)*, 349(6248), pp. 647–650.

Govorunova, E. G. et al., 2017. The expanding family of natural anion channelrhodopsins reveals large variations in kinetics, conductance and spectral sensitivity. *Scientific Reports*, 7, p. 43358.

Govorunova, E. G. et al., 2018. Extending the time domain of neuronal silencing with cryptophyte anion channelrhodopsins. *eNeuro*, 5, pp. ENEURO.0174-18.2018.

Gradinaru, V. et al., 2009. Optical deconstruction of parkinsonian neural circuitry. *Science (New York, N.Y.)*, 324(5925), pp. 354–359.

Gradinaru, V. et al., 2010. Molecular and cellular approaches for diversifying and extending optogenetics. *Cell*, 141(1), pp. 154–165.

Green, C. B., 2004. Cryptochromes: tail-ored for distinct functions. *Current Biology*, 14(19), pp. R847–R849.

Grieger, J. C. & Samulski, R. J., 2005. Packaging capacity of adeno-associated virus serotypes: impact of larger genomes on infectivity and postentry steps. *Journal of Virology*, 79(15), pp. 9933–9944.

Grossman, N. et al., 2010. Multi-site optical excitation using ChR2 and micro-LED array. *Journal of Neural Engineering*, 7(1), p. 16004.

Gunaydin, L. A. et al., 2010. Ultrafast optogenetic control. *Nature Neuroscience*, 13(3), pp. 387–392.

Guntas, G. et al., 2015. Engineering an improved light-induced dimer (iLID) for controlling the localization and activity of signaling proteins. *Proceedings of the National Academy of Sciences of the United States of America*, 112(1), pp. 112–117.

Guo, Z. V. et al., 2014. Flow of cortical activity underlying a tactile decision in mice. *Neuron*, 81(1), pp. 179–194.

Gysbrechts, B. et al., 2016. Light distribution and thermal effects in the rat brain under optogenetic stimulation. *Journal of Biophotonics*, 9(6), pp. 576–585.

Haam, J. et al., 2012. GABA is excitatory in adult vasopressinergic neuroendocrine cells. *The Journal of Neuroscience*, 32(2), pp. 572–582.

Habuchi, S. et al., 2005. Reversible single-molecule photoswitching in the GFP-like fluorescent protein Dronpa. *Proceedings of the National Academy of Sciences of the United States of America*, 102(27), pp. 9511–9516.

Han, Y., Braatsch, S., Osterloh, L. & Klug, G., 2004. A eukaryotic BLUF domain mediates light-dependent gene expression in the purple bacterium Rhodobacter sphaeroides 2.4.1. *Proceedings of the National Academy of Sciences of the United States of America*, 101(33), pp. 12306–12311.

Harris, J. A. et al., 2014. Anatomical characterization of Cre driver mice for neural circuit mapping and manipulation. *Frontiers in Neural Circuits*, 8, p. 76.

Hellingwerf, K. J., Hoff, W. D. & Crielaard, W., 1996. Photobiology of microorganisms: how photosensors catch a photon to initialize signalling. *Molecular Microbiology*, 21, pp. 683–693.

Henyey, L. G. & Greenstein, J. L., 1941. Diffuse radiation in the galaxy. *The Astrophysical Journal*, 93, pp. 70–83.

Herman, A. M. et al., 2014. Cell type-specific and time-dependent light exposure contribute to silencing in neurons expressing channelrhodopsin-2. *eLife*, 3, p. e01481.

Hernandez, O., Guillon, M., Papagiakoumou, E. & Emiliani, V., 2014. Zero-order suppression for two-photon holographic excitation. *Optics Letters*, 39(20), pp. 5953–5956.

Herrou, J. & Crosson, S., 2011. Function, structure and mechanism of bacterial photosensory LOV proteins. *Nature Reviews*, 9(10), pp. 713–723.

Herry, C. et al., 2008. Switching on and off fear by distinct neuronal circuits. *Nature*, 454, pp. 600–606.

Hochbaum, D. R. et al., 2014. All-optical electrophysiology in mammalian neurons using engineered microbial rhodopsins. *Nature Methods*, 11(8), pp. 825–833.

Ishimaru, A., 1977. Theory and application of wave propagation and scattering in random media. *Proceedings of the IEEE*, 65, pp. 1030–1061.

Jacques, S. L., 2013. Optical properties of biological tissues: a review. *Physics in Medicine and Biology*, 58(11), pp. R37–R61.

Jakobsson, J. et al., 2003. Targeted transgene expression in rat brain using lentiviral vectors. *Journal of Neuroscience Research*, 73(6), pp. 876–885.

Jeong, J.-W. et al., 2015. Wireless optofluidic systems for programmable in vivo pharmacology and optogenetics. *Cell*, 162(3), pp. 662–674.

Jung, J. C. & Schnitzer, M. J., 2003. Multiphoton endoscopy. *Optics Letters*, 28(11), pp. 902–904.

Kahle, K. T. et al., 2008. Roles of the cation-chloride cotransporters in neurological disease. *Nature Clinical Practice. Neurology*, 4(9), pp. 490–503.

Kahn, I. et al., 2011. Characterization of the functional MRI response temporal linearity via optical control of neocortical pyramidal neurons. *The Journal of Neuroscience*, 31(42), pp. 15086–15091.

Karunarathne, W. K. A., Giri, L., Kalyanaraman, V. & Gautam, N., 2013. Optically triggering spatiotemporally confined GPCR activity in a cell and programming neurite initiation and extension. *Proceedings of the National Academy of Sciences of the United States of America*, 110(17), pp. E1565–E1574.

Kawano, F., Okazaki, R., Yazawa, M. & Sato, M., 2016. A photoactivatable Cre-loxP recombination system for optogenetic genome engineering. *Nature Chemical Biology*, 12(12), pp. 1059–1064.

Kawano, F., Suzuki, H., Furuya, A. & Sato, M., 2015. Engineered pairs of distinct photoswitches for optogenetic control of cellular proteins. *Nature Communications*, 6, p. 6256.

Kennedy, M. J. et al., 2010. Rapid blue-light-mediated induction of protein interactions in living cells. *Nature Methods*, 7(12), pp. 973–975.

Kim, J. A. & Connors, B. W., 2012. High temperatures alter physiological properties of pyramidal cells and inhibitory interneurons in hippocampus. *Frontiers in Cellular Neuroscience*, 6, p. 27.

Kim, J.-M. et al., 2005. Light-driven activation of beta 2-adrenergic receptor signaling by a chimeric rhodopsin containing the beta 2-adrenergic receptor cytoplasmic loops. *Biochemistry*, 44(7), pp. 2284–2292.

Kim, T., Folcher, M., Doaud-El Baba, M. & Fussenegger, M., 2015. A synthetic erectile optogenetic stimulator enabling blue-light-inducible penile erection. *Angewandte Chemie (International ed. in English)*, 54(20), pp. 5933–5938.

Kirmse, K. et al., 2015. GABA depolarizes immature neurons and inhibits network activity in the neonatal neocortex in vivo. *Nature Communications*, 6, p. 7750.

Klapoetke, N. C. et al., 2014. Independent optical excitation of distinct neural populations. *Nature Methods*, 11(3), pp. 338–346.

Klavir, O. et al., 2017. Manipulating fear associations via optogenetic modulation of amygdala inputs to prefrontal cortex. *Nature Neuroscience*, 20(6), pp. 836–844.

Knobloch, H. S. et al., 2012. Evoked axonal oxytocin release in the central amygdala attenuates fear response. *Neuron*, 73(3), pp. 553–566.

Kort, R. et al., 1996. The xanthopsins: a new family of eubacterial blue-light photoreceptors. *The EMBO Journal*, 15(13), pp. 3209–3218.

Kremer, E. J., Boutin, S., Chillon, M. & Danos, O., 2000. Canine adenovirus vectors: an alternative for adenovirus-mediated gene transfer. *Journal of Virology*, 74(1), pp. 505–512.

Kubelka, P., 1948. New contributions to the optics of intensely light-scattering materials. *Journal of the Optical Society of America*, 38(5), pp. 448–457.

Kvitsiani, D. et al., 2013. Distinct behavioural and network correlates of two interneuron types in prefrontal cortex. *Nature*, 498(7454), pp. 363–366.

Lee, J., Ozden, I., Song, Y.-K. & Nurmikko, A. V., 2015. Transparent intracortical microprobe array for simultaneous spatiotemporal optical stimulation and multichannel electrical recording. *Nature Methods*, 12(12), pp. 1157–1162.

Lee, V. & Maguire, J., 2014. The impact of tonic GABAA receptor-mediated inhibition on neuronal excitability varies across brain region and cell type. *Frontiers in Neural Circuits*, 8, p. 3.

Levene, M. J. et al., 2004. In vivo multiphoton microscopy of deep brain tissue. *Journal of Neurophysiology*, 91(4), pp. 1908–1912.

Li, J., Li, G., Wang, H. & Wang Deng, X., 2011. Phytochrome signaling mechanisms. *The Arabidopsis Book*, 9, p. e0148.

Li, X. et al., 2005. Fast noninvasive activation and inhibition of neural and network activity by vertebrate rhodopsin and green algae channelrhodopsin. *Proceedings of the National Academy of Sciences of the United States of America*, 102(49), pp. 17816–17821.

Lim, D. H. et al., 2012. In vivo large-scale cortical mapping using channelrhodopsin-2 stimulation in transgenic mice reveals asymmetric and reciprocal relationships between cortical areas. *Frontiers in Neural Circuits*, 6, p. 11.

Lima, S. Q., Hromádka, T., Znamenskiy, P. & Zador, A. M., 2009. PINP: a new method of tagging neuronal populations for identification during in vivo electrophysiological recording. *PloS One*, 4(7), p. e6099.

Lin, J. Y., 2011. A userś guide to channelrhodopsin variants: features, limitations and future developments. *Experimental Physiology*, 96(1), pp. 19–25.

Lin, J. Y., Lin, M. Z., Steinbach, P. & Tsien, R. Y., 2009. Characterization of engineered channelrhodopsin variants with improved properties and kinetics. *Biophysical Journal*, 96, pp. 1803–1814.

Lin, J. Y. et al., 2013. ReaChR: a red-shifted variant of channelrhodopsin enables deep transcranial optogenetic excitation. *Nature Neuroscience*, 16(10), pp. 1499–1508.

Liu, Y. et al., 2015. OptogenSIM: a 3D Monte Carlo simulation platform for light delivery design in optogenetics. *Biomedical Optics Express*, 6(12), pp. 4859–4870.

Llewellyn, M. E., Thompson, K. R., Deisseroth, K. & Delp, S. L., 2010. Orderly recruitment of motor units under optical control in vivo. *Nature Medicine*, 16(10), pp. 1161–1165.

Losi, A. & Gärtner, W., 2012. The evolution of flavin-binding photoreceptors: an ancient chromophore serving trendy blue-light sensors. *Annual Review of Plant Biology*, 63, pp. 49–72.

Lou, S. et al., 2016. Genetically targeted all-optical electrophysiology with a transgenic Cre-dependent optopatch mouse. *The Journal of Neuroscience*, 36(43), pp. 11059–11073.

Ma, Z. et al., 2013. Fine tuning the LightOn light-switchable transgene expression system. *Biochemical and Biophysical Research Communications*, 440(3), pp. 419–423.

Madisen, L. et al., 2012. A toolbox of Cre-dependent optogenetic transgenic mice for light-induced activation and silencing. *Nature Neuroscience*, 15(5), pp. 793–802.

Mahn, M., Ron, S. & Yizhar, O., 2013. Viral vector-based techniques for optogenetic modulation in vivo. In: *Neuromethods*. s.l.: Humana Press, pp. 289–310.

Mahn, M. et al., 2016. Biophysical constraints of optogenetic inhibition at presynaptic terminals. *Nature Neuroscience*, 19(4), pp. 554–556.

Mahn, M. et al., 2018. High-efficiency optogenetic silencing with soma-targeted anion-conducting channelrhodopsins. *Nature Communications*, 9(1), p. 4125.

Masseck, O. A. et al., 2014. Vertebrate cone opsins enable sustained and highly sensitive rapid control of Gi/o signaling in anxiety circuitry. *Neuron*, 81(6), pp. 1263–1273.

Masuda, S., 2013. Light detection and signal transduction in the BLUF photoreceptors. *Plant and Cell Physiology*, 54(2), pp. 171–179.

Mattis, J. et al., 2011. Principles for applying optogenetic tools derived from direct comparative analysis of microbial opsins. *Nature Methods*, 9(2), pp. 159–172.

McGregor, K. M., Bécamel, C., Marin, P. & Andrade, R., 2016. Using melanopsin to study G protein signaling in cortical neurons. *Journal of Neurophysiology*, 116(3), pp. 1082–1092.

McLean, J. W., Freeman, J. D. & Walker, R. E., 1998. Beam spread function with time dispersion. *Applied Optics*, 37(21), pp. 4701–4711.

Meister, M., 2016. Physical limits to magnetogenetics. *eLife*, 5, p. e17210.

Melzer, S. et al., 2012. Long-range-projecting GABAergic neurons modulate inhibition in hippocampus and entorhinal cortex. *Science (New York, N.Y.)*, 335(6075), pp. 1506–1510.

Messier, J. E., Chen, H., Cai, Z.-L. & Xue, M., 2018. Targeting light-gated chloride channels to neuronal somatodendritic domain reduces their excitatory effect in the axon. *eLife*, 7, p. e38506.

Minlebaev, M., Ben-Ari, Y. & Khazipov, R., 2007. Network mechanisms of spindle-burst oscillations in the neonatal rat barrel cortex in vivo. *Journal of Neurophysiology*, 97(1), pp. 692–700.

Miyashita, T. et al., 2013. Long-term channelrhodopsin-2 (ChR2) expression can induce abnormal axonal morphology and targeting in cerebral cortex. *Frontiers in Neural Circuits*, 7, p. 8.

Mohammad, F. et al., 2017. Optogenetic inhibition of behavior with anion channelrhodopsins. *Nature Methods*, 14(3), pp. 271–274.

Moser, E., Mathiesen, I. & Andersen, P., 1993. Association between brain temperature and dentate field potentials in exploring and swimming rats. *Science (New York, N.Y.)*, 259(5099), pp. 1324–1326.

Nagel, G. et al., 2002. Channelrhodopsin-1: a light-gated proton channel in green algae. *Science (New York, N.Y.)*, 296(5577), pp. 2395–2398.

Nagel, G. et al., 2003. Channelrhodopsin-2, a directly light-gated cation-selective membrane channel. *Proceedings of the National Academy of Sciences of the United States of America*, 100(24), pp. 13940–13945.

Nagel, G. et al., 2005. Light activation of channelrhodopsin-2 in excitable cells of *Caenorhabditis elegans* triggers rapid behavioral responses. *Current Biology*, 15(24), pp. 2279–2284.

Nash, A. I. et al., 2011. Structural basis of photosensitivity in a bacterial light-oxygen-voltage/helix-turn-helix (LOV-HTH) DNA-binding protein. *Proceedings of the National Academy of Sciences of the United States of America*, 108(23), pp. 9449–9454.

Neuman, M. & Edström, P., 2010. Anisotropic reflectance from turbid media. I. Theory. *Journal of the Optical Society of America A*, 27(5), pp. 1032–1039.

Niopek, D. et al., 2016. Optogenetic control of nuclear protein export. *Nature Communications*, 7, p. 10624.

Ogawa, S. K. et al., 2014. Organization of monosynaptic inputs to the serotonin and dopamine neuromodulatory systems. *Cell Reports*, 8(4), pp. 1105–1118.

Oron, D., Papagiakoumou, E., Anselmi, F. & Emiliani, V., 2012. Two-photon optogenetics. *Progress in Brain Research*, 196, pp. 119–143.

Packer, A. M., Russell, L. E., Dalgleish, H. W. P. & Häusser, M., 2015. Simultaneous all-optical manipulation and recording of neural circuit activity with cellular resolution in vivo. *Nature Methods*, 12(2), pp. 140–146.

Packer, A. M. et al., 2012. Two-photon optogenetics of dendritic spines and neural circuits. *Nature Methods*, 9(12), pp. 1202–1205.

Palczewski, K., 2006. G protein-coupled receptor rhodopsin. *Annual Review of Biochemistry*, 75, pp. 743–767.

Papagiakoumou, E. et al., 2010. Scanless two-photon excitation of channelrhodopsin-2. *Nature Methods*, 7(10), pp. 848–854.

Park, S. I. et al., 2015. Soft, stretchable, fully implantable miniaturized optoelectronic systems for wireless optogenetics. *Nature Biotechnology*, 33(12), pp. 1280–1286.

Pashaie, R. et al., 2014. Optogenetic brain interfaces. *IEEE Reviews in Biomedical Engineering*, 7, pp. 3–30.

Penzkofer, A., Tanwar, M., Veetil, S. K. & Kateriya, S., 2015. Photo-dynamics of photoactivated adenylyl cyclase TpPAC from the spirochete bacterium *Turneriella parva* strain H(T). *Journal of Photochemistry and Photobiology B, Biology*, 153, pp. 90–102.

Petreanu, L., Huber, D., Sobczyk, A. & Svoboda, K., 2007. Channelrhodopsin-2-assisted circuit mapping of long-range callosal projections. *Nature Neuroscience*, 10(5), pp. 663–668.

Pham, E., Mills, E. & Truong, K., 2011. A synthetic photoactivated protein to generate local or global Ca(2+) signals. *Chemistry and Biology*, 18(7), pp. 880–890.

Pifferi, A., Taroni, P., Valentini, G. & Andersson-Engels, S., 1998. Real-time method for fitting time-resolved reflectance and transmittance measurements with a monte carlo model. *Applied Optics*, 37(13), pp. 2774–2780.

Pisanello, F. et al., 2017. Dynamic illumination of spatially restricted or large brain volumes via a single tapered optical fiber. *Nature Neuroscience*, 20(8), pp. 1180–1188.

Prakash, R. et al., 2012. Two-photon optogenetic toolbox for fast inhibition, excitation and bistable modulation. *Nature Methods*, 9(12), pp. 1171–1179.

Price, G. D. & Trussell, L. O., 2006. Estimate of the chloride concentration in a central glutamatergic terminal: a gramicidin perforated-patch study on the calyx of Held. *The Journal of Neuroscience*, 26(44), pp. 11432–11436.

Prigge, M. et al., 2012. Color-tuned channelrhodopsins for multiwavelength optogenetics. *The Journal of Biological Chemistry*, 287(38), pp. 31804–31812.

Pugh, J. R. & Jahr, C. E., 2011. Axonal GABAA receptors increase cerebellar granule cell excitability and synaptic activity. *The Journal of Neuroscience*, 31(2), pp. 565–574.

Pugh, J. R. & Jahr, C. E., 2013. Activation of axonal receptors by GABA spillover increases somatic firing. *The Journal of Neuroscience*, 33(43), pp. 16924–16929.

Quail, P. H., 2002. Phytochrome photosensory signalling networks. *Nature Reviews. Molecular Cell Biology*, 3(2), pp. 85–93.

Raimondo, J. V., Kay, L., Ellender, T. J. & Akerman, C. J., 2012. Optogenetic silencing strategies differ in their effects on inhibitory synaptic transmission. *Nature Neuroscience*, 15(8), pp. 1102–1104.

Reig, R. et al., 2010. Temperature modulation of slow and fast cortical rhythms. *Journal of Neurophysiology*, 103(3), pp. 1253–1261.

Rickgauer, J. P., Deisseroth, K. & Tank, D. W., 2014. Simultaneous cellular-resolution optical perturbation and imaging of place cell firing fields. *Nature Neuroscience*, 17(12), pp. 1816–1824.

Rockwell, N. C., Su, Y.-S. & Lagarias, J. C., 2006. Phytochrome structure and signaling mechanisms. *Annual Review of Plant Biology*, 57, pp. 837–858.

Royer, S. et al., 2010. Multi-array silicon probes with integrated optical fibers: light-assisted perturbation and recording of local neural circuits in the behaving animal. *The European Journal of Neuroscience*, 31(12), pp. 2279–2291.

Rungta, R. L. et al., 2017. Light controls cerebral blood flow in naive animals. *Nature Communications*, 8, p. 14191.

Ryu, M.-H., Moskvin, O. V., Siltberg-Liberles, J. & Gomelsky, M., 2010. Natural and engineered photoactivated nucleotidyl cyclases for optogenetic applications. *The Journal of Biological Chemistry*, 285(53), pp. 41501–41508.

Sakai, S., Ueno, K., Ishizuka, T. & Yawo, H., 2013. Parallel and patterned optogenetic manipulation of neurons in the brain slice using a DMD-based projector. *Neuroscience Research*, 75(1), pp. 59–64.

Scharf, B. & Engelhard, M., 1994. Blue halorhodopsin from *Natronobacterium pharaonis*: wavelength regulation by anions. *Biochemistry*, 33(21), pp. 6387–6393.

Schwarz, L. A. et al., 2015. Viral-genetic tracing of the input-output organization of a central noradrenaline circuit. *Nature*, 524(7563), pp. 88–92.

Scott, N., Prigge, M., Yizhar, O. & Kimchi, T., 2015. A sexually dimorphic hypothalamic circuit controls maternal care and oxytocin secretion. *Nature*, 525(7570), pp. 519–522.

Senn, V. et al., 2014. Long-range connectivity defines behavioral specificity of amygdala neurons. *Neuron*, 81(2), pp. 428–437.

Shin, G. et al., 2017. Flexible near-field wireless optoelectronics as subdermal implants for broad applications in optogenetics. *Neuron*, 93(3), pp. 509–521.e3.

Shipley, F. B., Clark, C. M., Alkema, M. J. & Leifer, A. M., 2014. Simultaneous optogenetic manipulation and calcium imaging in freely moving *C. elegans*. *Frontiers in Neural Circuits*, 8, p. 28.

Simon, M. I., Strathmann, M. P. & Gautam, N., 1991. Diversity of G proteins in signal transduction. *Science (New York, N.Y.)*, 252(5007), pp. 802–808.

Sineshchekov, O. A., Govorunova, E. G., Li, H. & Spudich, J. L., 2015. Gating mechanisms of a natural anion channelrhodopsin. *Proceedings of the National Academy of Sciences of the United States of America*, 112(46), pp. 14236–14241.

Siuda, E. R. et al., 2015a. Optodynamic simulation of β-adrenergic receptor signalling. *Nature Communications*, 6, p. 8480.

Siuda, E. R. et al., 2015b. Spatiotemporal control of opioid signaling and behavior. *Neuron*, 86(4), pp. 923–935.

Sjulson, L., Cassataro, D., DasGupta, S. & Miesenböck, G., 2016. Cell-specific targeting of genetically encoded tools for neuroscience. *Annual Review of Genetics*, 50, pp. 571–594.

Smith, G., 2012. Herpesvirus transport to the nervous system and back again. *Annual Review of Microbiology*, 66, pp. 153–176.

Sohal, V. S., Zhang, F., Yizhar, O. & Deisseroth, K., 2009. Parvalbumin neurons and gamma rhythms enhance cortical circuit performance. *Nature*, 459(7247), pp. 698–702.

Stanley, S. A. et al., 2015. Remote regulation of glucose homeostasis in mice using genetically encoded nanoparticles. *Nature Medicine*, 21(1), pp. 92–98.

Stanley, S. A. et al., 2016. Bidirectional electromagnetic control of the hypothalamus regulates feeding and metabolism. *Nature*, 531(7596), pp. 647–650.

Stark, E., Koos, T. & Buzsáki, G., 2012. Diode probes for spatiotemporal optical control of multiple neurons in freely moving animals. *Journal of Neurophysiology*, 108(1), pp. 349–363.

Stierl, M. et al., 2011. Light modulation of cellular cAMP by a small bacterial photoactivated adenylyl cyclase, bPAC, of the soil bacterium Beggiatoa. *The Journal of Biological Chemistry*, 286(2), pp. 1181–1188.

Stratton, J. A., 1941. *Electromagnetic Theory*. Mcgraw Hill Book Company.

Stujenske, J. M., Spellman, T. & Gordon, J. A., 2015. Modeling the spatiotemporal dynamics of light and heat propagation for in vivo optogenetics. *Cell Reports*, 12(3), pp. 525–534.

Tabarean, I. V. et al., 2005. Electrophysiological properties and thermosensitivity of mouse preoptic and anterior hypothalamic neurons in culture. *Neuroscience*, 135(2), pp. 433–449.

Takkala, P., Zhu, Y. & Prescott, S. A., 2016. Combined changes in chloride regulation and neuronal excitability enable primary afferent depolarization to elicit spiking without compromising its inhibitory effects. *PLoS Computational Biology*, 12(11), p. e1005215.

Taniguchi, H. et al., 2011. A resource of Cre driver lines for genetic targeting of GABAergic neurons in cerebral cortex. *Neuron*, 71(6), pp. 995–1013.

Taslimi, A. et al., 2016. Optimized second-generation CRY2-CIB dimerizers and photoactivatable Cre recombinase. *Nature Chemical Biology*, 12(6), pp. 425–430.

Tervo, D. G. R. et al., 2016. A designer AAV variant permits efficient retrograde access to projection neurons. *Neuron*, 92(2), pp. 372–382.

Thompson, S. M., Masukawa, L. M. & Prince, D. A., 1985. Temperature dependence of intrinsic membrane properties and synaptic potentials in hippocampal CA1 neurons in vitro. *The Journal of Neuroscience*, 5(3), pp. 817–824.

Ugolini, G., 1992. Transneuronal transfer of herpes simplex virus type 1 (HSV 1) from mixed limb nerves to the CNS. I. Sequence of transfer from sensory, motor, and sympathetic nerve fibres to the spinal cord. *The Journal of Comparative Neurology*, 326(4), pp. 527–548.

Ugolini, G., 1995. Specificity of rabies virus as a transneuronal tracer of motor networks: transfer from hypoglossal motoneurons to connected second-order and higher order central nervous system cell groups. *The Journal of Comparative Neurology*, 356(3), pp. 457–480.

Ugolini, G., 2010. Advances in viral transneuronal tracing. *Journal of Neuroscience Methods*, 194(1), pp. 2–20.

Ullrich, S., Gueta, R. & Nagel, G., 2013. Degradation of channelopsin-2 in the absence of retinal and degradation resistance in certain mutants. *Biological Chemistry*, 394(2), pp. 271–280.

Valeeva, G. et al., 2016. An optogenetic approach for investigation of excitatory and inhibitory network GABA actions in mice expressing channelrhodopsin-2 in GABAergic neurons. *The Journal of Neuroscience*, 36(22), pp. 5961–5973.

Vardy, E. et al., 2015. A new DREADD facilitates the multiplexed chemogenetic interrogation of behavior. *Neuron*, 86, pp. 936–946.

Verma, I. M. & Weitzman, M. D., 2005. Gene therapy: twenty-first century medicine. *Annual Review of Biochemistry*, 74, pp. 711–738.

Vo-Dinh, T., 2003. Basic instrumentation in photonics. *Biomedical Photonics Handbook*, Oak Ridge National Laboratory: Oak Ridge, TN.

Wang, H. et al., 2016. LOVTRAP: an optogenetic system for photoinduced protein dissociation. *Nature Methods*, 13(9), pp. 755–758.

Wang, L., Jacques, S. L. & Zheng, L., 1995. MCML--Monte Carlo modeling of light transport in multi-layered tissues. *Computer Methods and Programs in Biomedicine*, 47(2), pp. 131–146.

Wang, L. V. & Wu, H.-I., 2009. *Biomedical Optics*. s.l.:John Wiley & Sons, Inc.

Wang, X., Chen, X. & Yang, Y., 2012. Spatiotemporal control of gene expression by a light-switchable transgene system. *Nature Methods*, 9(3), pp. 266–269.

Warden, M. R. et al., 2012. A prefrontal cortex-brainstem neuronal projection that controls response to behavioural challenge. *Nature*, 492(7429), pp. 428–432.

Wertz, A. et al., 2015. PRESYNAPTIC NETWORKS. Single-cell-initiated monosynaptic tracing reveals layer-specific cortical network modules. *Science (New York, N.Y.)*, 349(6243), pp. 70–74.

Wheeler, M. A. et al., 2016. Genetically targeted magnetic control of the nervous system. *Nature Neuroscience*, 19(5), pp. 756–761.

Wickersham, I. R., Finke, S., Conzelmann, K.-K. & Callaway, E. M., 2007b. Retrograde neuronal tracing with a deletion-mutant rabies virus. *Nature Methods*, 4(1), pp. 47–49.

Wickersham, I. R. et al., 2007a. Monosynaptic restriction of transsynaptic tracing from single, genetically targeted neurons. *Neuron*, 53(5), pp. 639–647.

Wiegert, J. S. et al., 2017. Silencing neurons: tools, applications, and experimental constraints. *Neuron*, 95, pp. 504–529.

Wietek, J., Broser, M., Krause, B. S. & Hegemann, P., 2016. Identification of a natural green light absorbing chloride conducting channelrhodopsin from *Proteomonas sulcata*. *Journal of Biological Chemistry*, 291, pp. 4121–4127.

Wietek, J. & Prigge, M., 2016. Enhancing channelrhodopsins: an overview. *Methods in Molecular Biology (Clifton, N.J.)*, 1408, pp. 141–165.

Wietek, J. et al., 2014. Conversion of channelrhodopsin into a light-gated chloride channel. *Science (New York, N.Y.)*, 344(6182), pp. 409–412.

Wietek, J. et al., 2015. An improved chloride-conducting channelrhodopsin for light-induced inhibition of neuronal activity in vivo. *Scientific Reports*, 5, p. 14807.

Wietek, J. et al., 2017. Anion-conducting channelrhodopsins with tuned spectra and modified kinetics engineered for optogenetic manipulation of behavior. *Scientific Reports*, 7, p. 14957.

Wilson, B. C. & Adam, G., 1983. A Monte Carlo model for the absorption and flux distributions of light in tissue. *Medical Physics*, 10(6), pp. 824–830.

Witten, I. B. et al., 2011. Recombinase-driver rat lines: tools, techniques, and optogenetic application to dopamine-mediated reinforcement. *Neuron*, 72(5), pp. 721–733.

Yizhar, O. et al., 2011a. Microbial opsins: a family of single-component tools for optical control of neural activity. *Cold Spring Harbor Protocols*, 2011(3), p. top102.

Yizhar, O. et al., 2011b. Optogenetics in neural systems. *Neuron*, 71(1), pp. 9–34.

Yizhar, O. et al., 2011c. Neocortical excitation/inhibition balance in information processing and social dysfunction. *Nature*, 477(7363), pp. 171–178.

Yona, G., Meitav, N., Kahn, I. & Shoham, S., 2016. Realistic numerical and analytical modeling of light scattering in brain tissue for optogenetic applications. *eNeuro*, 3, pp. ENEURO--0059.

Zemelman, B. V., Lee, G. A., Ng, M. & Miesenböck, G., 2002. Selective photostimulation of genetically chARGed neurons. *Neuron*, 33(1), pp. 15–22.

Zhang, F. et al., 2007. Multimodal fast optical interrogation of neural circuitry. *Nature*, 446(7136), pp. 633–639.

Zhang, F. et al., 2008. Red-shifted optogenetic excitation: a tool for fast neural control derived from *Volvox carteri*. *Nature Neuroscience*, 11(6), pp. 631–633.

Zhang, F. et al., 2011. The microbial opsin family of optogenetic tools. *Cell*, 147(7), pp. 1446–1457.

Zhang, J. et al., 2009. Integrated device for optical stimulation and spatiotemporal electrical recording of neural activity in light-sensitized brain tissue. *Journal of Neural Engineering*, 6(5), p. 055007.

Zhao, S. et al., 2011. Cell type-specific channelrhodopsin-2 transgenic mice for optogenetic dissection of neural circuitry function. *Nature Methods*, 8(9), pp. 745–752.

Zhou, X. X., Chung, H. K., Lam, A. J. & Lin, M. Z., 2012. Optical control of protein activity by fluorescent protein domains. *Science (New York, N.Y.)*, 338(6108), pp. 810–814.

Zhou, X. X. et al., 2017. Optical control of cell signaling by single-chain photoswitchable kinases. *Science (New York, N.Y.)*, 355(6327), pp. 836–842.

Znamenskiy, P. & Zador, A. M., 2013. Corticostriatal neurons in auditory cortex drive decisions during auditory discrimination. *Nature*, 497(7450), pp. 482–485.

Zoltowski, B. D., Vaccaro, B. & Crane, B. R., 2009. Mechanism-based tuning of a LOV domain photoreceptor. *Nature Chemical Biology*, 5(11), pp. 827–834.

Zoltowski, B. D. et al., 2007. Conformational switching in the fungal light sensor Vivid. *Science (New York, N.Y.)*, 316(5827), pp. 1054–1057.

CHAPTER 12

Molecular Photoswitches for Synthetic Optogenetics

Shai Berlin and Ehud Y. Isacoff

CONTENTS

12.1 INTRODUCTION

Light is an instrumental tool for visualizing biology (Ramon, 1952; Ji et al., 2008). However, light can also be used to trigger activity in cells, through the activation of different classes of light-absorbing receptors – photoreceptors proteins (henceforth noted as photoreceptors) – such as rhodopsin and phytochromes, for example (Moglich et al., 2010). Photoreceptors, by means of their chromophores, relay photon-absorption to activation of particular intracellular signaling mechanisms. Whereas photoreceptors express at exclusive cells or organisms (e.g. retina or photosynthetic organisms), their ectopic expression in other cell types can, correspondingly, render the transfected cell responsive to light. This realization has led to the repurposing of photoreceptors into light-activated tools.

One of the first examples of this involved the expression of Arrestin, Rhodopsin and a G-protein to excite, or "chARGe" (i.e. elicit action potentials), neurons by light illumination (Zemelman et al., 2002). This and other genetically encoded optical actuators – coined optogenetic tools (Deisseroth et al., 2006; Deisseroth, 2015) – have now become widely used, by neuroscientists in particular, granting non-invasive photocontrol over cellular activity. Remarkably, with only six classes of photoreceptors, a large array of optogenetic tools has emerged (Boyden et al., 2005; Li et al., 2005; Schroder-Lang et al., 2007; Airan et al., 2009; Levskaya et al., 2009; Hegemann and Nagel, 2013; Konermann et al., 2013; Adamantidis et al., 2015; Deisseroth, 2015; Nihongaki et al., 2015a, 2015b; Polstein and Gersbach, 2015; Scheib et al., 2015). Notably, most of these photoreceptors are not native to mammalian cells or neurons. As a result, ectopic expression of optogenetic tools, particularly in neurons, can prove difficult owing to poor membrane trafficking or may induce adverse effects (e.g. Yizhar et al., 2011; Mahn et al., 2016; Perny et al., 2016; Malyshev et al., 2017). Thus, to study neuronal physiology, it should prove useful if we were able to photocontrol ion channels and receptors that are native to neurons. Unfortunately, most neuronal membrane proteins do not harbor chromophores and hence are not photoreceptors, and cannot be manipulated by light. To address this, it was conceived that *blind* proteins (i.e. non-photoreceptors) could be fashioned to be light-sensitive should they be tagged with synthetic chromophores – photoswitches (e.g. Figure 12.1a, c). These would mimic the actions of chromophores by 1) interacting with a particular protein, 2) absorbing photons, to then 3) modulate its activity (Figure 12.2).

The first demonstration of the use of the technique to control the function of a non-photoreceptor protein involved a photoswitch named *p*-azophenyldiphenylcarbamyl chloride (PAPC) (Kaufman et al., 1968). PAPC was designed to photo-regulate the soluble enzyme chymotrypsin. To this end, PAPC was assembled from a light-sensitive core and an enzyme-inactivating drug (viz. antagonist). During illumination, the core would toggle (i.e. isomerize, see Section 12.2.1) between two geometrically different isomeric forms (-*trans* and -*cis*, e.g. Figure 12.1a, c). This isomerization would then change the structure of the entire molecule, affecting the pharmacological properties of the antagonist, thereby regulating the enzyme's activity. The ingenuity and simplicity of the approach were quickly adopted by the same group, and others, to engineer several other photoswitchable enzymes (Bieth et al., 1969; Bieth et al., 1970, but see detailed list in Hug, 1978), and not too long after ion channels, explicitly nicotinic Acetylcholine receptors (nAChR) (Bieth et al., 1969; Deal et al., 1969; Bieth et al., 1970; Bartels et al., 1971; Bieth et al., 1973; Nass et al., 1978; Lester et al., 1979, 1980). Though revolutionary at the time, the approach was dormant for many years, but its potential was newly appreciated by contemporary neuroscientists, chemists, and biophysicists, leading to newer developments and, importantly, disseminating the technique outside the walls of specialized laboratories (Banghart et al., 2004; Folgering et al., 2004; Kocer et al., 2005; Brownlee, 2006). Consequently, over 45 different light-gated neuronal ion channels and receptors, not to mention photocontrol over molecules such as DNA (e.g. Kamei et al., 2007, but see Beharry and Woolley, 2011), have been engineered in very rapid succession (Szobota and Isacoff, 2010; Kramer et al., 2013; Berlin and Isacoff, 2017). This chapter

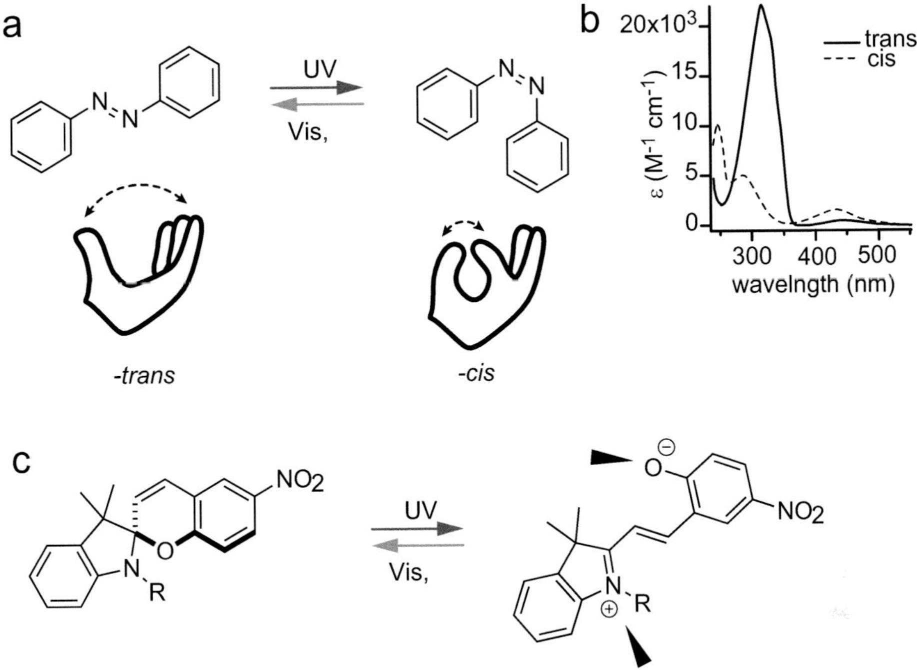

FIGURE 12.1 Light-dependent isomerization of photoswitchable-cores. **(a)** Molecular structure of *-trans* and *-cis* isomers of azobenzene. Near-UV light (violet) induces isomerization to *-cis* (pinched fingers), whereas green light reverses the reaction back to *-trans* (extended fingers). **(b)**, Absorption spectra of *-trans* and *-cis* isomers of azobenzene in ethanol. **(c)**. Molecular structure Spiropyran undergoing heterolytic cleavage upon UV irradiation (360–370 nm). This isomerization is accompanied by a very large change in polarity (arrowheads) and is reversible both thermally and photochemically, by visible light illumination (>460 nm) (panel b adapted from Szymanski et al., 2013).

will first review the design and modes of actions of photoswitches, as well as discuss several emerging light-gated tools with a focus on light-gated ion channels, for the study of neuronal function.

12.2 THE PHOTOSWITCHABLE-CORE

The design of a light-gated channel is generally preceded by the design of the photoswitch. This firstly entails selecting a photoswitchable-core. Photoswitches, akin to chromophores, are used for their ability to adopt, typically, two distinct conformations (e.g. *-cis* and *-trans* isomers, Figure 12.1 and see below) following the absorption of light. Ideally, the photoswitch is "inactive" (or *off*) in one form, and becomes "active" (*on*) when switched to its second isomeric form by light. This allows rapid and reversible toggling between the *off* and *on* states; enabling or disabling the effect of the photoswitch. The mechanism of action of photoswitches is very diverse, significantly diverging from that of naturally occurring chromophores. Briefly, chromophores are deeply embedded within the protein and there form a complex web of interactions. This enables even the most subtle change in the chromophore's geometry to alter the network of photoreceptor-chromophore contacts, inducing changes in function of the photoreceptor (Palczewski et al., 2000; Teller et al., 2001). Photoswitches, on the other hand, do not need to interact so elaborately with their protein target, significantly easing their design and implementation to alter protein function, detailed here as simple hand movements that drive changes in protein function (Figure 12.1a and Figure 12.2). In fact, some photoswitches do not even bind to their protein target, rather affect its nearby environment, such as induction of lateral tension in the membrane to open mechanosensitive channels (Folgering et al., 2004 and see below). Thus, the type of effect will significantly depend on the type of light-dependent isomerization of the photoswitchable-core and its location with regard to its protein target.

Photoswitches can undergo changes in geometry (i.e. *trans-to-cis* isomerization, Figure 12.1a), interconversions from open-to-closed states (Figure 12.1c), changes in polarity, changes in charge distribution, etc. There are several light-sensitive chemical groups that can be used as cores for photoswitches, such as azobenzenes, stilbenes, spiropyrans, diarylethenes, hemithioindigos, etc. (for a review, see Szymanski et al., 2013). However, most of these chemical groups are not commonly used in the synthesis of photoswitches due to unfavorable properties, with the exception of azobenzenes.

12.2.1 Azobenzenes

Azobenzenes are one of the most studied photochromic molecules, since their discovery in the early 1800s (Figure 12.1a) (Demselben, 1834; Noble, 1856; Hamon et al., 2009). In the dark at equilibrium, the azobenzene is found in its stable, -*trans* form; however, a mixture of the two structurally distinct isomers was obtained by the action of sunlight (Hartley, 1937). Later, Fischer and colleagues showed that irradiation with 365 nm produces a solution containing predominantly the -*cis* isomer, though not exclusively (Fischer et al., 1955), because the -*trans* and the -*cis* isomers display overlapping absorption spectra (Figure 12.1b). Under optimal excitations a maximum of ~80% -*cis* or ~95% -*trans* photostationary states can be obtained (Beharry and Woolley, 2011). Whereas the isomerization between -*trans* to -*cis* occurs rapidly (picoseconds: Sporlein et al., 2002) following illumination with 320–365 nm, the spontaneous reverting back to -*trans* is slow, even though the -*cis* isomer is less thermally stable. In fact, unmodified azobenzene in -*cis* will persist for days in the dark, until undergoing thermal relaxation to regenerate the -*trans* form. However, -*trans* could be obtained instantaneously by irradiation with blue/green light (typically 510 nm). From a photoswitch point of view, the ability of the azobenzene to remain in either -*trans* or -*cis* in the dark (i.e. bistability) is very beneficial as it enables the switch to remain in either forms (active or inactive) in the dark, without requiring persistent illumination. However, this bistability feature and its duration can be modified (i.e. accelerated relaxation to-*trans*) by decorating the molecule with additional chemical groups across the azobenzene molecule (Kienzler et al., 2013; Dong et al., 2015b). Lastly, the ease of synthesis, low rate of photobleaching (Szymanski et al., 2013), and lack of long-lived excited states or reactive intermediates (Bartels et al., 1971) make azobenzenes extremely attractive for inclusion in photoswitches.

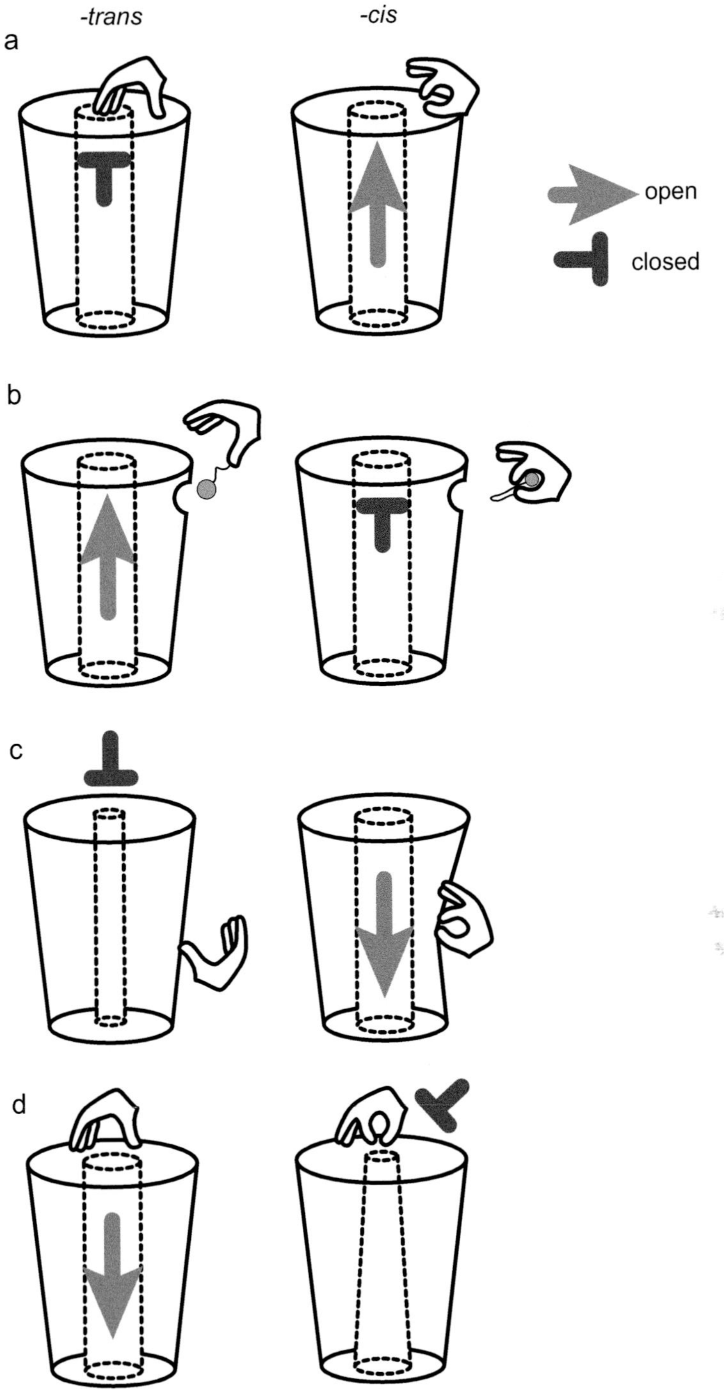

FIGURE 12.2 Protein photoswitching. Examples of photoswitching mechanisms. **(a)** A tethered photoswitch (thumb denotes site of conjugation to protein) blocks the pore of the channel when in *-trans*, but the block is relieved when the photoswitch is isomerized to *-cis*. **(b)** Reversibly uncaging a soluble (PCL) drug. **(c)** Photoswitching within the membrane. **(d)** A doubly-conjugated photoswitch to control movement of the pore-Photoswitchable tweezers.

12.2.2 Spiropyrans

Another group of light-sensitive molecules are spiropyrans. These undergo reversible heterolytic cleavage around the C_{spiro}–O bond following UV absorption (~360 nm) (Figure 12.1c). This type of interconversion (or cyclization) from closed-to-open form is accompanied by a large polarity change (and fluorescence emission (Marriott et al., 2008, Figure 12.1c, arrowheads). This process too can be reversed by thermal relaxation or absorption of longer wavelengths (>460 nm) (Szymanski et al., 2013). Spiropyrans have been incorporated in several interesting photoswitches to gate a bacterial ion channel (Kocer et al., 2005), to inhibit protein–protein interaction (Sakata et al., 2005), and even to bind DNA, in a light-dependent manner (Andersson et al., 2008). However, as noted above, this core is not as widely used as azobenzenes, and therefore will not be discussed in further depth in this chapter.

12.3 ACTIVE HEADGROUP

The next step in designing a photoswitch encompasses choosing selective and active headgroups to be appended to the core. As azobenzenes remain inert in both states, the mode of action of the photoswitch will typically be dictated by the added chemical headgroups located on either of its sides (Figure 12.3a). In these cases, the photoswitchable-core serves

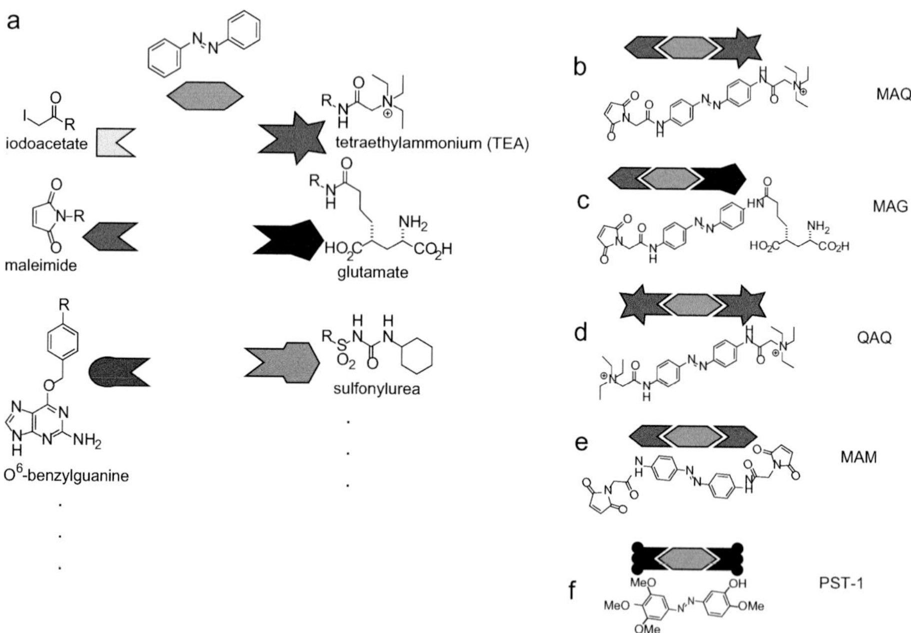

FIGURE 12.3 Building photoswitches. **(a)** Different chemical groups can be added on both sized of the azobenzene photoswitchable-core (top) to create new photoswitches (left lane – conjugating moieties, right – pharmacophores). **(b–f)** Examples of azobenzene-based photoswitches. Mixing and matching conjugating and pharmacophore chemical groups yields MAQ **(b)** (Banghart et al., 2004), MAG **(c)** (Volgraf et al., 2006), QAQ **(d)** (Mourot et al., 2012), MAM **(e)** (Habermacher et al., 2016) and PST1 **(f)** (Borowiak et al., 2015).

as means to: distance or draw closer the active headgroup to the protein (Figure 12.2a), act as a cage that could be reversibly undone to expose a drug (Figure 12.2b), induce changes within the membrane (Figure 12.2c), mechanically affect pore opening (Figure 12.2d), and in more complex cases, isomerization could be used to piece together a fragmented pharmacophore, to name a few.

With respect to ion channels and receptors, active headgroups often involve pharmacophores, such as blockers, agonists, antagonists, allosteric modulators, potentiators, etc. (see below). Choosing the right pharmacophore often benefits from exploring traditional pharmacology with its rich repertoire of compounds with extensive characterizations, some in clinical use, with analogous structures to photoswitchable cores, and, importantly, many commercially available.

Examples of specific agonists used in the design of different photoswitches include GluAzo; a light-switchable agonist for GluK receptors (Volgraf et al., 2007), based on the selective GluK1-receptor agonist LY339434 (Pedregal et al., 2000). The ATA-3 photoswitch is based on the ligand AMPA itself (i.e. 2-amino-3-(5-methyl-3-hydroxyisoxazol-4-yl)propanoic acid) and thereby is specific for AMPA-Receptors (GluARs) (Stawski et al., 2012)). ATG is a photoswitch based on glutamate to specifically activates NMDARs (GluNRs) (Laprell et al., 2015), whereas another glutamate-based photoswitch, MAG, may be used to activate Kainate receptors (GluKRs) (Figure 12.3a, c) (Volgraf et al., 2006), metabotropic GluRs (mGluRs) (Levitz et al., 2013), as well as GluNRs (Berlin et al., 2016).

An exemplary pharmacophore incorporated in numerous photoswitches is the potassium channel blocker TEA (tetraethylammonium) (Armstrong, 1971) (Figure 12.3a), belonging to the quaternary-ammonium family of drugs, referred to as QA hereafter. QA has been incorporated into MAQ, AAQ, QAQ, DENAQ, PhENAQ, BzAQ, etc. (Figure 12.3b, d) (Banghart et al., 2004; Chambers et al., 2006; Fortin et al., 2008; Banghart et al., 2009; Fortin et al., 2011; Mourot et al., 2011; Sandoz et al., 2011; Mourot et al., 2012). Despite the similarity among these photoswitches, their mechanism of actions can differ from one another. For example, whereas AAQ and PhENAQ are designed to enter the cell and block the channel from its inner vestibule (Banghart et al., 2009; Tochitsky et al., 2014), MAQ and QAQ block the channel from the outside in a light-dependent manner (Banghart et al., 2004; Sandoz et al., 2011; Mourot et al., 2013).

Other examples of interesting pharmacophores used in photoswitches include an anesthetic drug propofol (Stewart et al., 2011) in the design of AP2; a photoswitchable potentiator of $GABA_A$-receptors (Stein et al., 2012). A clinically relevant drug combretastatin A-4 (Pettit et al., 1989) has been incorporated in a photoswitch designed to inhibit microtubule dynamics, denoted PST-1 (Figure 12.3f) (Borowiak et al., 2015) and a potassium-sparing diuretic drug, amiloride, was used in a photoamiloride switch called PA1 (Schonberger et al., 2014).

An emerging class of photoswitches specific for mGluRs involves allosteric compounds. For example, the Gorostiza group searched chemical catalogues looking for chemical and structural homology with that of azobenzenes. They have identified VU0415374 to create Alloswitch-1; the first light-controlled negative allosteric modulator (NAM) for mGluR5 (Pittolo et al., 2014). Shortly after, they identified another NAM (VU0155041), to create

OptoGluNAM4.1, specifically targeting mGluR4 (Rovira et al., 2016). As noted above, in contrast to photoreceptors and their chromophores, these examples show the relative ease in creating target-specific photoswitches.

Aside pharmacophores, photoswitches can also include tethering moieties as the active headgroups. When two of these headgroups span the azobenzene, they enable bridging between two sites of the protein (Figure 12.2d and Figure 12.3e). Then isomerization of the photoswitch can exert mechanical forces to modulate protein's structure and activity, bearing a resemblance to a forceps motion (photoswitchable tweezers: see section 12.4.2.1).

Notably, though the mechanism of action of photoswitches can usually be predicted based on its chemical headgroup (e.g. an agonist is likely to open a channel), it is not always the case! For example, the action of the MAM-6 photoswitch consisting of the $GABA_A$-receptor agonist muscimol, unexpectedly yields photo-antagonism (Lin et al., 2014). A similar case has been reported for a light-gated glutamate receptor of the NMDA kind (LiGluN). Though GluNRs are activated by glutamate, a particular LiGluN variant used with a glutamate based-photoswitch, L-MAG (maleimide-azobenzene-glutamate), counterintuitively induces photo-antagonism, likely by obstructing clamshell closure. However, the same photoswitch can induce channel opening when tethered to another residue atop the ligand binding domain (LBD) of the GluN subunit (Berlin et al., 2016). These examples highlight the variety of responses that can be obtained by a single photoswitch.

12.4 PHOTOCHROMIC LIGANDS (PCL) VERSUS PHOTOTETHERED LIGANDS (PTL)

There are several types of photoswitches, namely freely diffusible (denoted PhotoChromic Ligand; PCL) or chemically conjugated to the protein of interest (denoted PhotoTethered Ligand; PTL) (e.g. Figure 12.2a vs 12.2b).

12.4.1 Photochromic Ligands (PCL)

PCLs are soluble, non-tethered photoswitches that, analogous to caged compounds (Ellis-Davies, 2007), can interact or effect the protein of interest likely in one isomeric form (either in *-trans* or *-cis*) following illumination. However, unlike caged-compounds that undergo irreversible photo-destruction, PCLs can be readily isomerized back and forth between isomers (i.e. turned *off*) by light; making this completely reversible! Another unique advantage of PCLs is that they target endogenous proteins, therefore do not require their protein target to be genetically modified and overexpressed, as in the case of PTLs (see below). This enables working under physiological conditions, and allows the user to begin the experiment only a few minutes after the PCL has been applied onto the preparation. However, the unrestricted diffusion of PCLs can be a limitation, as light-activated PCLs will diffuse outside the region of illumination to activate, inadvertently, receptors located in other regions. One way to bypass this is to apply broad illumination over the entire preparation with wavelengths that would turn *off* the escaping PCL molecules. Although seemingly impractical, broad illumination of the preparation is used with PCLs that are "active" in the thermally relaxed *-trans* form (denoted *trans*-agonists). This is done to prevent tonic activation of the receptors during the application of the PCLs, as is the case with

GluAzo (Volgraf et al., 2007) or ATA-3 (Stawski et al., 2012), for example. Nevertheless, a *cis*-agonist PCL is preferable, as is the case of the ATG photoswitch activating GluNRs (Laprell et al., 2015). Lastly, an elegant way to ensure the cessation of a *cis*-agonist PCL's activity once it diffuses outside the region of illumination is to speed up its spontaneous, thermal relaxation back to -*trans* (i.e. off). This can be achieved by red-shifting the absorption spectrum of the photoswitch, yielding a fast-relaxing azobenzene (Kienzler et al., 2013; Dong et al., 2015b).

Notably, though most PCLs are membrane impermeable, several have been described as exerting their effect from the intracellular side. This is thought to occur by infiltration of the photoswitch to the cytoplasm by passively crossing the lipid bilayer (namely, lipophilic/hydrophobic photoswitches) (Taglialatela et al., 1991; Banghart et al., 2009). Examples of bilayer-permeant photoswitches include the Acrylamide-Azobenzene-Quaternary ammonium (AAQ) photoswitch with its hydrophobic acrylamide group or JB253, a photoswitchable sulfonylurea (Banghart et al., 2009; Mourot et al., 2013; Broichhagen et al., 2014). Interestingly, charged photoswitches may cross the membrane by, presumably, permeating through endogenous channels, such as TRPV1 or P2X receptors (Mourot et al., 2012). Though the mechanisms of the latter are not fully understood, TRPV1 and P2X are suspected to undergo pore-dilation resulting in non-specific conduction of larger molecules (Virginio et al., 1999; Chung et al., 2008, but also see Bean, 2015; Li et al., 2015) the photoswitch in this case. Notably, photoswitches that permeate intracellularly via TRPV1 for example, provide several advantages: 1) to probe for the presence of the carrier channel, to subsequently 2) photo-block other endogenous channels and lastly, 3) though photoswitches are not genetically encoded, these can be confined to specific cells, such as cells exclusively expressing TRPV1 channels. Furthermore, cellular specificity can be obtained by varying the activity of the carrier channels, so that cells with TRPV1-hyperactivity would incorporate the photoswitch more readily (Mourot et al., 2013).

Despite these clear advantages, it is very hard to design PCLs with absolute target specificity. This is especially true when the photoswitches are used at high concentrations. In fact, this critical feature is shared among many conventional pharmacological or caged-compounds (Kramer et al., 2013).

12.4.2 Phototethered Ligands (PTL)

PTLs, on the other hand, are tethered (or bioconjugated) photoswitches (Figure 12.1a, c, and d). These are designed to irreversibly bind to a specific region of a protein via a chemical handle. This commonly requires modification of the protein of interest to include a specific genetic tether (see Section 12.5). This tethering presents several advantages over PCLs: immobilized PTLs cannot diffuse outside the illuminated region, can be used at lower concentrations to minimize off-target effects and can target a defined cellular population by limiting expression of the modified receptor with the use cell-specific promoters. Another key feature of PTLs is gaining a very high local concentration of the active head-group (~mM) (Volgraf et al., 2006), giving a competitive advantage to the photoswitch over other soluble and/or endogenous ligands. This feature is particularly useful when using channels with low affinity towards a particular ligand or blocker. For example, though some potassium channels exhibit weak sensitivity towards QA, these channels can still be efficiently photo-blocked by a QA-based photoswitch (e.g. MAQ: Banghart et al., 2004), owing to its high effective concentration near its binding site (Sandoz and Levitz, 2013). This also enables the user to express low-affinity receptors that are insensitive to native neurotransmission, but that can readily respond to the PTL and thus remain under photocontrol (Levitz et al., 2016).

Nonetheless, as most PTLs require an engineered receptor to be ectopically overexpressed, it is important to bear in mind that overexpression of the latter may lead to altered changes in receptor amounts, distribution, signaling promiscuity or unwarranted changes in cellular excitability. In addition, mutations in the proteins (e.g. in the LBD) may drastically affect channel function, such as ligand affinity (Kalbaugh et al., 2004; Chen et al., 2005). These command detailed functional analysis of the mutated receptors.

12.4.2.1 Photoswitchable Tweezer (PX)

A unique subclass of PTLs is the photoswitchable tweezer (PX) that contains two tethering moieties on both sides of the azobenzene-core (Figure 12.1d and Figure 12.3e). Doubly-tethered PXs function as photoswitchable crosslinkers, used to physically manipulate the structure of a molecule (e.g. DNA (Asanuma et al., 1999; Biswas and Burghardt, 2014)), peptide (Kumita et al., 2000), or an ion channel (Browne et al., 2014; Habermacher et al., 2016). This change in structure is obtained by the forces generated by the isomerization of the photoswitch that pulls when in *cis*- (or pushes, in the case of *trans*-isomerization) on regions it is conjugated to. Recently, with regard to neurons, this methodology has been applied to link two transmembrane segments of the P2X2 receptor. Photoisomerization of the switch induced structural rearrangements in the receptor to open it, mimicking the iris-like motion of the pore during channel opening (Habermacher et al., 2016).

12.4.2.2 Targeted Covalent Photoswitches (TCP)

A new addition to the PTL family of photoswitches consists of Targeted Covalent Photoswitches (TCP) (Izquierdo-Serra et al., 2016). TCPs resemble PTLs because they too contain a ligand, a photoswitchable core, and a tethering moiety to immobilize the photoswitch to the protein at a selected site. However, instead of the typical biotethers used for conjugating PTLs onto the protein (e.g. maleimide, Figure 12.3a, green and see Section 12.5), TCPs contain short-lived, reactive anchoring groups containing strong electrophilic moieties, such as epoxide or *N*-hydroxysuccinimide esters. These electrophiles can react with reactive amines and hydroxyl groups found in multiple amino acids. Thus, whereas PTL require the modification of the protein (e.g. cysteine mutagenesis), TCPs are intended to conjugate residues in wild-type channels or receptors. Luckily, whereas cysteines are very rare in proteins (see below), amino acids bearing nucleophilic groups that can react with electrophilic esters are not (e.g. lysines, serines, threonines, and tyrosines).

TCPs are likely to undergo attachment by affinity labeling (Wofsy et al., 1962), requiring the ligand of the photoswitch to first dock within its binding domain, so as to provide sufficient time for the anchoring reaction to occur. This implies that during labeling, the receptors may be activated as a result and, if the active headgroup is an agonist, may potentially lead to excessive activation of the receptors.

12.5 BIOTETHERS

To tether PTLs or PXs, several genetically encoded reactive tags can be inserted within the protein of interest (Spicer and Davis, 2014). Briefly, unnatural amino acid (UAA) incorporations enable to introduce unique orthogonal amino acids with residues that can undergo selective, abiotic chemical reactions with synthetic photoswitches. Whereas these broaden the palette of possible chemoselective reactions (Saxon and Bertozzi, 2000; Lang and Chin, 2014), their incorporation requires special handling, accessory components (e.g. tRNA-synthetases and suppressor tRNA: Young and Schultz, 2010), which are not typically accessible to most labs. In addition, the UAA incorporation process may yield lower protein levels, which may be insufficient to obtain a measurable photo-response. The most significant limitation of using UAAs is that many of the reactions are poorly carried out under physiological conditions (e.g. pH and temperature: Lang and Chin, 2014). Another method embraces the fusion of auxiliary proteins (e.g. SNAP and CLIP-tags: Keppler et al., 2003; Gautier et al., 2008; Reymond et al., 2011). This technique has only recently been used for tethering a benzylguanine photoswitch to a G-protein coupled receptor (Figure 12.3a) (Broichhagen et al., 2015a). Typically, the fusion of protein tags is easier to incorporate than UAAs; however, these proteins (enzymes in the case of SNAP and CLIP) are large (~150 a.a.) and may prove challenging to introduce in receptors or channels without affecting structure and function, particularly when introduced in important or sensitive domains of the protein, such as the LBD.

Single amino acid substitution, in our opinion, remains the most practical choice. This procedure does not require any specialist techniques (as little as primers and a thermocycler) and can be easily incorporated (almost) anywhere within the protein. Of the 20 canonical amino acids, the thiol of cysteine acts as a unique handle, easily exploitable by electrophiles reagents under physiological conditions, such as maleimides (Spicer and Davis, 2014). The maleimide-cysteine reaction fulfills most of the requirements for it to be used with living cells. First, the speed of the reaction between the tag (cysteine) and its substrate (maleimide-derivatized photoswitch) is fast (~minutes). Second, though maleimides can interact with any water-accessible cysteines, these are naturally of low abundance in proteins (<2%: Fodje and Al-Karadaghi, 2002), thus making this reaction protein- and site-selective. Lastly, the reaction is non-toxic.

12.6 MODERN EXAMPLES OF PHOTOSWITCHABLE PROTEINS

One of the first light-gated channels introduced, during the second coming of synthetic optogenetics was SPARK – a Synthetic Photoisomerizable Azobenzene-Regulated K^+-channel (Banghart et al., 2004). SPARK emerged at a time when molecular structures of ion channels and membrane proteins were becoming more and more available (Clapham, 2003). This information enabled the authors to devise a unique photo-regulated blocking mechanism by an azobenzene-based photoswitch. SPARK is a modified *Shaker* channel, blocked by QA (MacKinnon and Yellen, 1990; Blaustein et al., 2000). To confer light-sensitivity to this channel, Banghart et al. designed a PTL consisting of maleimide, azobenzene, and QA named MAL-AZO-QA (or MAQ, Figure 12.3b) (Banghart et al., 2004). They used the available molecular and structural information of similar channels (Doyle et al., 1998; Jiang et al., 2002) to find a tethering site on Shaker (E422C mutant) that would enable the QA to perfectly reach its binding site when the azobenzene was in *-trans*, but that this block would be alleviated when the photoswitch was shortened following light-absorption and isomerization to *-cis* (e.g. Figure 12.2a). The photoswitching occurred fast, displayed complete reversibility and could be repeated many times without any apparent exhaustion. The authors used SPARK to control neuronal excitability, so that during green light illumination (*trans*-MAQ) the potassium channel would be blocked, allowing the neuron to fire action potentials. Conversely, the neuron was silenced when MAQ was isomerized to *-cis*. Notably, the mechanism described here for the photo-block of SPARK does not diverge significantly from the original concept used to photo-regulate nAChRs, demonstrated almost 50 years ago by the Erlanger group (Kaufman et al., 1968; Bartels et al., 1971). Equally, this nearing-distancing concept was heavily relied on for the engineering of additional photoswitchable ion channels and receptors, such as photoswitchable glutamate-receptors, with glutamate drawn near or far from the LBD by light (Volgraf et al., 2006; Szobota et al., 2007; Wang et al., 2007; Baier and Scott, 2009; Numano et al., 2009; Caporale et al., 2011; Li et al., 2012; Gaub et al., 2014; Carroll et al., 2015; Berlin et al., 2016; Izquierdo-Serra et al., 2016; Levitz et al., 2016) or photoswitchable K^+-channels analogously regulated by QA-based photoswitches (Chambers et al., 2006; Fortin et al., 2011; Sandoz et al., 2011), a P2X receptor (Lemoine et al., 2013), not to mention $GABA_A$ receptors (Stein et al., 2012; Yue et al., 2012; Raimondo et al., 2016). A slight variation of this mode is the action of PXs (see Section 12.4.2.1) where the nearing-distancing force of azobenzene is applied to move domains in the protein, rather than a ligand (Browne et al., 2014; Habermacher et al., 2016).

Early on, in parallel to the appearance of SPARK, Bert Poolman's group developed a unique PCL to gate another channel; a bacterial mechanosensitive channel of large conductance (MscL) (Folgering et al., 2004). To this end, the group synthesized a photoswitch consisting of double-tailed phosphate amphiphiles containing an azobenzene core, denoted 4-Azo-5P. This switch was incorporated within the leaflets of a membrane, so that when isomerized from *-trans* to *-cis*, the bent form of the molecule would exert lateral pressure in the membrane. This pressure could be sensed by the channel MscL, also incorporated within the membrane, causing it to open (Figure 12.2c). Nowadays, this mode of action was extended to photoswitch, for example, membrane-associated enzymes, lipid rafts, and TRPV1 channels (Frank et al., 2015; Frank et al., 2016a; Frank et al., 2016b).

In 2005, the Feringa group reported their work on the photo-activation of MscL, albeit by a different approach (Kocer et al., 2005). Though mechanosensitive, the MscL channel also exhibits spontaneous opening, when polar or charged amino acids are introduced next to the pore at position G22 (Yoshimura et al., 1999). Koçer and colleagues took advantage of this phenomenon and synthesized a spiropyran-based PTL. As noted above (see Section 12.2.2), the light-mediated interconversion of spiropyran induces a large polarity change in the molecule. The photoswitch was accurately positioned by its tethering to G22C, via an iodoacetate biotether. Then, illumination induced charge movement and channel opening (Kocer et al., 2005). Nevertheless, unlike the above cases, the UV-induced channel opening displayed very long lags (~2 min), likely resulting from the slow gating or sensing of the channel of changes in charge. Therefore, though initially attractive, this approach did not gain much popularity, judging by the lack of photoswitchable proteins employing this technique. Together, these demonstrations clearly show the advantages in the incorporation of azobenezenes in photoswitches. Indeed, the latter are now used in over 45 different examples of photoswitchable ion channels and receptors.

12.7 NEXT STEPS TOWARDS NEW PHOTOSWITCHABLE RECEPTORS AND CHANNEL

Despite the growing popularity and constant improvements of the synthetic optogenetics method, there are still several issues to be addressed before its widespread use. One issue is photoswitch solubility, owing to the lipophilic nature of the azobenzene molecule. To this end, to increase polarity and solubility of the azobenzene, synthetic substitutions around the photoswitchable-core can make it more water-friendly, though possibly affecting bistability (Zhang et al., 2003; Broichhagen et al., 2015b). However, in some instances this drawback is embraced for the design of lipophilic photoswitches to be readily incorporated within the membrane to affect receptors, channels or enzymes (Broichhagen et al., 2015b; Frank et al., 2016a, 2016b), as is the case with the MscL channel (see above, and Folgering et al., 2004).

Another step taken forward was to lessen the burden and potential cellular damage of near-UV illumination for inducing *trans*-to-*cis* isomerization. This was realized by significantly red-shifting the absorption of photoswitches (Kienzler et al., 2013; Samanta et al., 2013; Rullo et al., 2014; Dong et al., 2015a, 2015b; Konrad et al., 2016). In addition to shifting their one-photon absorbance, it also improved their compatibility with two-photo-excitation (2PE) (e.g. Carroll et al., 2015), presenting better means for *in vivo* activation of the photoswitches. To date, there are several examples of the use of 2PE with azobenezene-based photoswitches for the activation of glutamate receptors (Izquierdo-Serra et al., 2014; Carroll et al., 2015; Gascon-Moya et al., 2015; Laprell et al., 2015). However, the field is far from mature at this point, significantly lagging behind other optical tools, namely caged-compounds and optogenetics (Oron et al., 2012; Olson et al., 2013).

Lastly, whereas the field is currently booming with new photoswitches, only a handful have been applied *in vivo* in mammalian brains. In fact, to the best of our knowledge, but a few groups have succeeded in showing the ability of PTLs and modified receptors to modulate neuronal function in live rodent brains (Lin et al., 2015; Levitz et al., 2016). Despite the latter, there are ample *in vivo* examples of photoswitches applied in other model preparations, such as blood, rodent muscle tissue, *C. elegans*, zebrafish, and flies, to name but a few (Janovjak et al., 2010; Beharry et al., 2011; Warp et al., 2012; Kauwe and Isacoff, 2013; Levitz et al., 2013; Borowiak et al., 2015; Dong et al., 2015b; Berlin et al., 2016). Importantly, as this method requires photoswitch and light-delivery, and in several cases, expression of a modified receptor, it is not surprising that the visual system has been a major platform for the method. In fact, there are multiple reports using the method to restore visual function

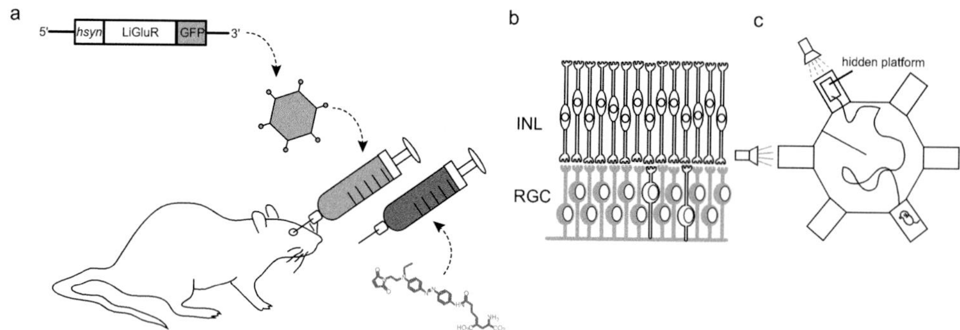

FIGURE 12.4 LiGluR expression in RGCs restores light responses in blind mice. **(a)** Intravitreal delivery of LiGluR and MAG_{460}. Schematics of the GFP-tagged LiGluR DNA encapsulated within an AAV virus and injected to the retina of blind mice. Following several weeks of expression of the LiGluR in the retinas (>4 weeks), an additional injection delivers the MAG_{460}-photoswitch (magenta). **(b)** Restriction of LiGluR expression in retinal ganglion cells (RGCs) is obtained by an AAV vector combining the human synapsin promoter (*hsyn*) and the AAV 2/2 capsid (inner nuclear layer, INL). **(c)** Water maze behavioral test. Mice that underwent dual injection (LiGluR and MAG_{460}) are able to find the correct platform by distinguishing between different temporal patterns of light projected at different locations of the maze. (Adapted from (Gaub et al., 2014.)

in blind mice and dogs (Polosukhina et al., 2012; Gaub et al., 2014; Tochitsky et al., 2014; Tochitsky and Kramer, 2015). For example, Gaub et al. have used intravitreal delivery of viruses encoding LiGluR (under specific neuronal promoters, e.g. Hsyn, Figure 12.4a, green syringe) to blind mice. Several weeks after, they delivered a red-shifted MAG photoswitch (MAG_{460}) to conjugate LiGluR-expressing RGCs (Figure 12.4a, magenta, Figure 12.4b). Lastly, with behavioral tests they were able to confirm that the LiGluR-treated blind mice exhibited light-avoidance behavior as their *wt*-littermates, as well as distinguishing between temporal patterns of light in a water maze test (Figure 12.4c). Notably, the molecular startegy used here proved successful in retinas originating from a canine model of retinal blindness (Gaub et al., 2014). Together, though at early-stages, these examples show the significant progress made towards the implementation of the method *in vivo*, and we anticipate many more examples in the very near future.

12.8 DISCUSSION

Synthetic optogenetics is a powerful technique for the control and study of native neuronal ion channels and receptors, *in vitro* and *in vivo*, with the use of light. It involves application of synthetic photoswitches that are designed to specifically interact with native or lightly modified receptors (cysteine-mutagenesis). Isomerization of the photoswitches, by two distinct wavelengths (e.g. near-UV and green) or by a single wavelength and thermal relaxation, ultimately affects protein function. In this chapter, we outline several key turning points in the development of this approach, and focus on the essential building-blocks required to make it work, namely photoswitchable-cores, pharmacophores, tethering groups, and protein engineering techniques. We also describe several examples of photoswitches and discuss the modes of action by which they render a multitude of neuronal membrane proteins photoswitchable (see Figure 12.1). We would like to emphasize, however, that despite our efforts to provide a broad overview of the subject, this chapter outlines but a fraction of the photoswitchable proteins, peptides, and molecules that are available to date. Thus, we encourage users to explore the literature for other examples that may involve their protein/molecule of choice. If it is missing in the repertoire of synthetic optogenetic tools, then we hope this to serve as motivation to engineer it!

The field of synthetic optogenetics is progressing owing to fruitful collaborations between chemists, physicists, and biologists (Brownlee, 2006). We therefore expect to see many more improvements in the technique such as better multiphoton-absorbing photoswitches and better labeling schemes to obtain more homogenous receptor populations, as well as new photoswitching mechanisms. In particular, we imagine next generation photoswitches to undergo through-skull photoactivation, as well as acoustic or magnetic manipulations. Together, we see a "bright" future for synthetic optogenetics.

REFERENCES

Adamantidis, A., Arber, S., Bains, J.S., Bamberg, E., Bonci, A., Buzsaki, G., Cardin, J.A., Costa, R.M., Dan, Y., Goda, Y., Graybiel, A.M., Hausser, M., Hegemann, P., Huguenard, J.R., Insel, T.R., Janak, P.H., Johnston, D., Josselyn, S.A., Koch, C., Kreitzer, A.C., Luscher, C., Malenka, R.C., Miesenbock, G., Nagel, G., Roska, B., Schnitzer, M.J., Shenoy, K.V., Soltesz, I., Sternson, S.M., Tsien, R.W., Tsien, R.Y., Turrigiano, G.G., Tye, K.M. & Wilson, R.I. (2015) Optogenetics: 10 years after ChR2 in neurons--views from the community. *Nat Neurosci*, **18**, 1202–1212.

Airan, R.D., Thompson, K.R., Fenno, L.E., Bernstein, H. & Deisseroth, K. (2009) Temporally precise in vivo control of intracellular signalling. *Nature*, **458**, 1025–1029.

Andersson, J., Li, S., Lincoln, P. & Andreasson, J. (2008) Photoswitched DNA-binding of a photochromic spiropyran. *J Am Chem Soc*, **130**, 11836–11837.

Armstrong, C.M. (1971) Interaction of tetraethylammonium ion derivatives with the potassium channels of giant axons. *J Gen Physiol*, **58**, 413–437.

Asanuma, H., Ito, T., Yoshida, T., Liang, X. & Komiyama, M. (1999) Photoregulation of the formation and dissociation of a DNA duplex by using the cis-trans isomerization of azobenzene. *Angew Chem Int Ed Engl*, **38**, 2393–2395.

Baier, H. & Scott, E.K. (2009) Genetic and optical targeting of neural circuits and behavior--zebrafish in the spotlight. *Curr Opin Neurobiol*, **19**, 553–560.

Banghart, M., Borges, K., Isacoff, E., Trauner, D. & Kramer, R.H. (2004) Light-activated ion channels for remote control of neuronal firing. *Nat Neurosci*, **7**, 1381–1386.

Banghart, M.R., Mourot, A., Fortin, D.L., Yao, J.Z., Kramer, R.H. & Trauner, D. (2009) Photochromic blockers of voltage-gated potassium channels. *Angew Chem Int Ed Engl*, **48**, 9097–9101.

Bartels, E., Wassermann, N.H. & Erlanger, B.F. (1971) Photochromic activators of the acetylcholine receptor. *Proc Natl Acad Sci U S A*, **68**, 1820–1823.

Bean, B.P. (2015) Pore dilation reconsidered. *Nat Neurosci*, **18**, 1534–1535.

Beharry, A.A., Wong, L., Tropepe, V. & Woolley, G.A. (2011) Fluorescence imaging of azobenzene photoswitching in vivo. *Angew Chem Int Ed Engl*, **50**, 1325–1327.

Beharry, A.A. & Woolley, G.A. (2011) Azobenzene photoswitches for biomolecules. *Chem Soc Rev*, **40**, 4422–4437.

Berlin, S. & Isacoff, E.Y. (2017) Optical control of glutamate receptors of the NMDA-kind in mammalian neurons, with the use of photoswitchable ligands. In *Biochemical Approaches to Glutamatergic Neurotransmission*, Parrot, S., Denoroy L (eds), Springer: Totowa, NJ.

Berlin, S., Szobota, S., Reiner, A., Carroll, E.C., Kienzler, M.A., Guyon, A., Xiao, T., Tauner, D. & Isacoff, E.Y. (2016) A family of photoswitchable NMDA receptors. *eLife*, **5**: e12040.

Bieth, J., Vratsanos, S.M., Wassermann, N. & Erlanger, B.F. (1969) Photoregulation of biological activity by photocromic reagents. II. Inhibitors of acetylcholinesterase. *Proc Natl Acad Sci U S A*, **64**, 1103–1106.

Bieth, J., Vratsanos, S.M., Wassermann, N.H., Cooper, A.G. & Erlanger, B.F. (1973) Photoregulation of biological activity by photochromic reagents. Inactivators of acetylcholinesterase. *Biochemistry*, **12**, 3023–3027.

Bieth, J., Wassermann, N., Vratsanos, S.M. & Erlanger, B.F. (1970) Photoregulation of biological activity by photochromic reagents, IV. A model for diurnal variation of enzymic activity. *Proc Natl Acad Sci U S A*, **66**, 850–854.

Biswas, M. & Burghardt, I. (2014) Azobenzene photoisomerization-induced destabilization of B-DNA. *Biophys J*, **107**, 932–940.

Blaustein, R.O., Cole, P.A., Williams, C. & Miller, C. (2000) Tethered blockers as molecular 'tape measures' for a voltage-gated K+ channel. *Nat Struct Biol*, **7**, 309–311.

Borowiak, M., Nahaboo, W., Reynders, M., Nekolla, K., Jalinot, P., Hasserodt, J., Rehberg, M., Delattre, M., Zahler, S., Vollmar, A., Trauner, D. & Thorn-Seshold, O. (2015) Photoswitchable inhibitors of microtubule dynamics optically control mitosis and cell death. *Cell*, **162**, 403–411.

Boyden, E.S., Zhang, F., Bamberg, E., Nagel, G. & Deisseroth, K. (2005) Millisecond-timescale, genetically targeted optical control of neural activity. *Nat Neurosci*, **8**, 1263–1268.

Broichhagen, J., Damijonaitis, A., Levitz, J., Sokol, K.R., Leippe, P., Konrad, D., Isacoff, E.Y. & Trauner, D. (2015a) Orthogonal optical control of a G protein-coupled receptor with a SNAP-tethered photochromic ligand. *ACS Cent Sci*, **1**, 383–393.

Broichhagen, J., Frank, J.A. & Trauner, D. (2015b) A roadmap to success in photopharmacology. *Acc Chem Res*, **48**, 1947–1960.

Broichhagen, J., Schonberger, M., Cork, S.C., Frank, J.A., Marchetti, P., Bugliani, M., Shapiro, A.M., Trapp, S., Rutter, G.A., Hodson, D.J. & Trauner, D. (2014) Optical control of insulin release using a photoswitchable sulfonylurea. *Nat Commun*, **5**, 5116.

Browne, L.E., Nunes, J.P., Sim, J.A., Chudasama, V., Bragg, L., Caddick, S. & North, R.A. (2014) Optical control of trimeric P2X receptors and acid-sensing ion channels. *Proc Natl Acad Sci U S A*, **111**, 521–526.

Brownlee, C. (2006) Gateways to collaboration. *ACS Chem Biol*, **1**, 10–13.

Caporale, N., Kolstad, K.D., Lee, T., Tochitsky, I., Dalkara, D., Trauner, D., Kramer, R., Dan, Y., Isacoff, E.Y. & Flannery, J.G. (2011) LiGluR restores visual responses in rodent models of inherited blindness. *Mol Ther*, **19**, 1212–1219.

Carroll, E.C., Berlin, S., Levitz, J., Kienzler, M.A., Yuan, Z., Madsen, D., Larsen, D.S. & Isacoff, E.Y. (2015) Two-photon brightness of azobenzene photoswitches designed for glutamate receptor optogenetics. *Proc Natl Acad Sci U S A*, **112**, E776–E785.

Chambers, J.J., Banghart, M.R., Trauner, D. & Kramer, R.H. (2006) Light-induced depolarization of neurons using a modified Shaker K(+) channel and a molecular photoswitch. *J Neurophysiol*, **96**, 2792–2796.

Chen, P.E., Geballe, M.T., Stansfeld, P.J., Johnston, A.R., Yuan, H., Jacob, A.L., Snyder, J.P., Traynelis, S.F. & Wyllie, D.J. (2005) Structural features of the glutamate binding site in recombinant NR1/NR2A N-methyl-D-aspartate receptors determined by site-directed mutagenesis and molecular modeling. *Mol Pharmacol*, **67**, 1470–1484.

Chung, M.K., Guler, A.D. & Caterina, M.J. (2008) TRPV1 shows dynamic ionic selectivity during agonist stimulation. *Nat Neurosci*, **11**, 555–564.

Clapham, D.E. (2003) Symmetry, selectivity, and the 2003 Nobel Prize. *Cell*, **115**, 641–646.

Deal, W.J., Erlanger, B.F. & Nachmansohn, D. (1969) Photoregulation of biological activity by photochromic reagents. 3. Photoregulation of bioelectricity by acetylcholine receptor inhibitors. *Proc Natl Acad Sci U S A*, **64**, 1230–1234.

Deisseroth, K. (2015) Optogenetics: 10 years of microbial opsins in neuroscience. *Nat Neurosci*, **18**, 1213–1225.

Deisseroth, K., Feng, G., Majewska, A.K., Miesenbock, G., Ting, A. & Schnitzer, M.J. (2006) Next-generation optical technologies for illuminating genetically targeted brain circuits. *J Neurosci*, **26**, 10380–10386.

Demselben (1834) Ueber das Stickstoffbenzid. *Annalen der Pharmacie*, **12**, 311–314.

Dong, M., Babalhavaeji, A., Hansen, M.J., Kalman, L. & Woolley, G.A. (2015a) Red, far-red, and near infrared photoswitches based on azonium ions. *Chem Commun (Camb)*, **51**, 12981–12984.

Dong, M., Babalhavaeji, A., Samanta, S., Beharry, A.A. & Woolley, G.A. (2015b) Red-shifting azobenzene photoswitches for in vivo use. *Acc Chem Res*, **48**, 2662–2670.

Doyle, D.A., Morais Cabral, J., Pfuetzner, R.A., Kuo, A., Gulbis, J.M., Cohen, S.L., Chait, B.T. & MacKinnon, R. (1998) The structure of the potassium channel: molecular basis of K+ conduction and selectivity. *Science*, **280**, 69–77.

Ellis-Davies, G.C. (2007) Caged compounds: photorelease technology for control of cellular chemistry and physiology. *Nat Methods*, **4**, 619–628.

Fischer, E., Frankel, M. & Wolovsky, R. (1955) Wavelength dependence of photoisomerization equilibria in azocompounds. *J Chem Phys*, **23**, 1367.

Fodje, M.N. & Al-Karadaghi, S. (2002) Occurrence, conformational features and amino acid propensities for the pi-helix. *Protein Eng*, **15**, 353–358.

Folgering, J.H., Kuiper, J.M., de Vries, A.H., Engberts, J.B. & Poolman, B. (2004) Lipid-mediated light activation of a mechanosensitive channel of large conductance. *Langmuir*, **20**, 6985–6987.

Fortin, D.L., Banghart, M.R., Dunn, T.W., Borges, K., Wagenaar, D.A., Gaudry, Q., Karakossian, M.H., Otis, T.S., Kristan, W.B., Trauner, D. & Kramer, R.H. (2008) Photochemical control of endogenous ion channels and cellular excitability. *Nat Methods*, **5**, 331–338.

Fortin, D.L., Dunn, T.W., Fedorchak, A., Allen, D., Montpetit, R., Banghart, M.R., Trauner, D., Adelman, J.P. & Kramer, R.H. (2011) Optogenetic photochemical control of designer K+ channels in mammalian neurons. *J Neurophysiol*, **106**, 488–496.

Frank, J.A., Franquelim, H.G., Schwille, P. & Trauner, D. (2016a) Optical Control of Lipid Rafts with Photoswitchable Ceramides. *J Am Chem Soc*, **138**, 12981–12986.

Frank, J.A., Yushchenko, D.A., Hodson, D.J., Lipstein, N., Nagpal, J., Rutter, G.A., Rhee, J.S., Gottschalk, A., Brose, N., Schultz, C. & Trauner, D. (2016b) Photoswitchable diacylglycerols enable optical control of protein kinase C. *Nat Chem Biol*, **12**, 755–762.

Frank, J.A., Moroni, M., Moshourab, R., Sumser, M., Lewin, G.R. & Trauner, D. (2015) Photoswitchable fatty acids enable optical control of TRPV1. *Nat Commun*, **6**, 7118.

Gascon-Moya, M., Pejoan, A., Izquierdo-Serra, M., Pittolo, S., Cabre, G., Hernando, J., Alibes, R., Gorostiza, P. & Busque, F. (2015) An optimized glutamate receptor photoswitch with sensitized azobenzene isomerization. *J Org Chem*, **80**, 9915–9925.

Gaub, B.M., Berry, M.H., Holt, A.E., Reiner, A., Kienzler, M.A., Dolgova, N., Nikonov, S., Aguirre, G.D., Beltran, W.A., Flannery, J.G. & Isacoff, E.Y. (2014) Restoration of visual function by expression of a light-gated mammalian ion channel in retinal ganglion cells or ON-bipolar cells. *Proc Natl Acad Sci U S A*, **111**, E5574–5583.

Gautier, A., Juillerat, A., Heinis, C., Correa, I.R., Jr., Kindermann, M., Beaufils, F. & Johnsson, K. (2008) An engineered protein tag for multiprotein labeling in living cells. *Chem Biol*, **15**, 128–136.

Habermacher, C., Martz, A., Calimet, N., Lemoine, D., Peverini, L., Specht, A., Cecchini, M. & Grutter, T. (2016) Photo-switchable tweezers illuminate pore-opening motions of an ATP-gated P2X ion channel. *eLife*, **5**, e11050.

Hamon, F., Djedaini-Pilard, F., Barbot, F. & Len, C. (2009) Azobenzenes—synthesis and carbohydrate applications. *Tetrahedron*, **65**, 10105–10123.

Hartley, G.S. (1937) The cis-form of azobenzene. *Nature*, **140**, 281.

Hegemann, P. & Nagel, G. (2013) From channelrhodopsins to optogenetics. *EMBO Mol Med*, **5**, 173–176.

Hug, D.H. (1978) The activation of enzymes with light. In Smith, K.C. (ed) *Photochemical and Photobiological Reviews*: *Volume 3*. Springer US, Boston, MA, pp. 1–33.

Izquierdo-Serra, M., Bautista-Barrufet, A., Trapero, A., Garrido-Charles, A., Diaz-Tahoces, A., Camarero, N., Pittolo, S., Valbuena, S., Perez-Jimenez, A., Gay, M., Garcia-Moll, A., Rodriguez-Escrich, C., Lerma, J., de la Villa, P., Fernandez, E., Pericas, M.A., Llebaria, A. & Gorostiza, P. (2016) Optical control of endogenous receptors and cellular excitability using targeted covalent photoswitches. *Nat Commun*, **7**, 12221.

Izquierdo-Serra, M., Gascon-Moya, M., Hirtz, J.J., Pittolo, S., Poskanzer, K.E., Ferrer, E., Alibes, R., Busque, F., Yuste, R., Hernando, J. & Gorostiza, P. (2014) Two-photon neuronal and astrocytic stimulation with azobenzene-based photoswitches. *J Am Chem Soc*, **136**, 8693–8701.

Janovjak, H., Szobota, S., Wyart, C., Trauner, D. & Isacoff, E.Y. (2010) A light-gated, potassium-selective glutamate receptor for the optical inhibition of neuronal firing. *Nat Neurosci*, **13**, 1027–1032.

Ji, N., Shroff, H., Zhong, H. & Betzig, E. (2008) Advances in the speed and resolution of light microscopy. *Curr Opin Neurobiol*, **18**, 605–616.

Jiang, Y., Lee, A., Chen, J., Cadene, M., Chait, B.T. & MacKinnon, R. (2002) The open pore conformation of potassium channels. *Nature*, **417**, 523–526.

Kalbaugh, T.L., VanDongen, H.M. & VanDongen, A.M. (2004) Ligand-binding residues integrate affinity and efficacy in the NMDA receptor. *Mol Pharmacol*, **66**, 209–219.

Kamei, T., Kudo, M., Akiyama, H., Wada, M., Nagasawa, J.i., Funahashi, M., Tamaoki, N. & Uyeda, T.Q.P. (2007) Visible-light photoresponsivity of a 4-(dimethylamino)azobenzene unit incorporated into single-stranded DNA: demonstration of a large spectral change accompanying isomerization in DMSO and detection of rapid (Z)-to-(E) isomerization in aqueous solution. *European J Org Chem*, **2007**, 1846–1853.

Kaufman, H., Vratsanos, S.M. & Erlanger, B.F. (1968) Photoregulation of an enzymic process by means of a light-sensitive ligand. *Science*, **162**, 1487–1489.

Kauwe, G. & Isacoff, E.Y. (2013) Rapid feedback regulation of synaptic efficacy during high-frequency activity at the Drosophila larval neuromuscular junction. *Proc Natl Acad Sci U S A*, **110**, 9142–9147.

Keppler, A., Gendreizig, S., Gronemeyer, T., Pick, H., Vogel, H. & Johnsson, K. (2003) A general method for the covalent labeling of fusion proteins with small molecules in vivo. *Nat Biotechnol*, **21**, 86–89.

Kienzler, M.A., Reiner, A., Trautman, E., Yoo, S., Trauner, D. & Isacoff, E.Y. (2013) A red-shifted, fast-relaxing azobenzene photoswitch for visible light control of an ionotropic glutamate receptor. *J Am Chem Soc*, **135**, 17683–17686.

Kocer, A., Walko, M., Meijberg, W. & Feringa, B.L. (2005) A light-actuated nanovalve derived from a channel protein. *Science*, **309**, 755–758.

Konermann, S., Brigham, M.D., Trevino, A.E., Hsu, P.D., Heidenreich, M., Cong, L., Platt, R.J., Scott, D.A., Church, G.M. & Zhang, F. (2013) Optical control of mammalian endogenous transcription and epigenetic states. *Nature*, **500**, 472–476.

Konrad, D.B., Frank, J.A. & Trauner, D. (2016) Synthesis of redshifted azobenzene photoswitches by late-stage functionalization. *Chemistry*, **22**, 4364–4368.

Kramer, R.H., Mourot, A. & Adesnik, H. (2013) Optogenetic pharmacology for control of native neuronal signaling proteins. *Nat Neurosci*, **16**, 816–823.

Kumita, J.R., Smart, O.S. & Woolley, G.A. (2000) Photo-control of helix content in a short peptide. *Proc Natl Acad Sci U S A*, **97**, 3803–3808.

Lang, K. & Chin, J.W. (2014) Cellular incorporation of unnatural amino acids and bioorthogonal labeling of proteins. *Chem Rev*, **114**, 4764–4806.

Laprell, L., Repak, E., Franckevicius, V., Hartrampf, F., Terhag, J., Hollmann, M., Sumser, M., Rebola, N., DiGregorio, D.A. & Trauner, D. (2015) Optical control of NMDA receptors with a diffusible photoswitch. *Nat Commun*, **6**, 8076.

Lemoine, D., Habermacher, C., Martz, A., Mery, P.F., Bouquier, N., Diverchy, F., Taly, A., Rassendren, F., Specht, A. & Grutter, T. (2013) Optical control of an ion channel gate. *Proc Natl Acad Sci U S A*, **110**, 20813–20818.

Lester, H.A., Krouse, M.E., Nass, M.M., Wassermann, N.H. & Erlanger, B.F. (1979) Light-activated drug confirms a mechanism of ion channel blockade. *Nature*, **280**, 509–510.

Lester, H.A., Krouse, M.E., Nass, M.M., Wassermann, N.H. & Erlanger, B.F. (1980) A covalently bound photoisomerizable agonist: comparison with reversibly bound agonists at electrophorus electroplaques. *J Gen Physiol*, **75**, 207–232.

Levitz, J., Pantoja, C., Gaub, B., Janovjak, H., Reiner, A., Hoagland, A., Schoppik, D., Kane, B., Stawski, P., Schier, A.F., Trauner, D. & Isacoff, E.Y. (2013) Optical control of metabotropic glutamate receptors. *Nat Neurosci*, **16**, 507–516.

Levitz, J., Popescu, A.T., Reiner, A. & Isacoff, E.Y. (2016) A toolkit for orthogonal and in vivo optical manipulation of ionotropic glutamate receptors. *Front Mol Neurosci*, **9**, 2.

Levskaya, A., Weiner, O.D., Lim, W.A. & Voigt, C.A. (2009) Spatiotemporal control of cell signalling using a light-switchable protein interaction. *Nature*, **461**, 997–1001.

Li, D., Herault, K., Isacoff, E.Y., Oheim, M. & Ropert, N. (2012) Optogenetic activation of LiGluR-expressing astrocytes evokes anion channel-mediated glutamate release. *J Physiol*, **590**, 855–873.

Li, M., Toombes, G.E., Silberberg, S.D. & Swartz, K.J. (2015) Physical basis of apparent pore dilation of ATP-activated P2X receptor channels. *Nat Neurosci*, **18**, 1577–1583.

Li, X., Gutierrez, D.V., Hanson, M.G., Han, J., Mark, M.D., Chiel, H., Hegemann, P., Landmesser, L.T. & Herlitze, S. (2005) Fast noninvasive activation and inhibition of neural and network activity by vertebrate rhodopsin and green algae channelrhodopsin. *Proc Natl Acad Sci U S A*, **102**, 17816–17821.

Lin, W.C., Davenport, C.M., Mourot, A., Vytla, D., Smith, C.M., Medeiros, K.A., Chambers, J.J. & Kramer, R.H. (2014) Engineering a light-regulated GABAA receptor for optical control of neural inhibition. *ACS Chem Biol*, **9**, 1414–1419.

Lin, W.C., Tsai, M.C., Davenport, C.M., Smith, C.M., Veit, J., Wilson, N.M., Adesnik, H. & Kramer, R.H. (2015) A comprehensive optogenetic pharmacology toolkit for in vivo control of GABA(A) receptors and synaptic inhibition. *Neuron*, **88**, 879–891.

MacKinnon, R. & Yellen, G. (1990) Mutations affecting TEA blockade and ion permeation in voltage-activated K+ channels. *Science*, **250**, 276–279.

Mahn, M., Prigge, M., Ron, S., Levy, R. & Yizhar, O. (2016) Biophysical constraints of optogenetic inhibition at presynaptic terminals. *Nat Neurosci*, **19**, 554–556.

Malyshev, A.Y., Roshchin, M.V., Smirnova, G.R., Dolgikh, D.A., Balaban, P.M. & Ostrovsky, M.A. (2017) Chloride conducting light activated channel GtACR2 can produce both cessation of firing and generation of action potentials in cortical neurons in response to light. *Neurosci Lett*, **640**, 76–80.

Marriott, G., Mao, S., Sakata, T., Ran, J., Jackson, D.K., Petchprayoon, C., Gomez, T.J., Warp, E., Tulyathan, O., Aaron, H.L., Isacoff, E.Y. & Yan, Y. (2008) Optical lock-in detection imaging microscopy for contrast-enhanced imaging in living cells. *Proc Natl Acad Sci U S A*, **105**, 17789–17794.

Moglich, A., Yang, X., Ayers, R.A. & Moffat, K. (2010) Structure and function of plant photoreceptors. *Annu Rev Plant Biol*, **61**, 21–47.

Mourot, A., Fehrentz, T., Le Feuvre, Y., Smith, C.M., Herold, C., Dalkara, D., Nagy, F., Trauner, D. & Kramer, R.H. (2012) Rapid optical control of nociception with an ion-channel photoswitch. *Nat Methods*, **9**, 396–402.

Mourot, A., Kienzler, M.A., Banghart, M.R., Fehrentz, T., Huber, F.M., Stein, M., Kramer, R.H. & Trauner, D. (2011) Tuning photochromic ion channel blockers. *ACS Chem Neurosci*, **2**, 536–543.

Mourot, A., Tochitsky, I. & Kramer, R.H. (2013) Light at the end of the channel: optical manipulation of intrinsic neuronal excitability with chemical photoswitches. *Front Mol Neurosci*, **6**, 5.

Nass, M.M., Lester, H.A. & Krouse, M.E. (1978) Response of acetylcholine receptors to photoisomerizations of bound agonist molecules. *Biophys J*, **24**, 135–160.

Nihongaki, Y., Kawano, F., Nakajima, T. & Sato, M. (2015a) Photoactivatable CRISPR-Cas9 for optogenetic genome editing. *Nat Biotechnol*, **33**, 755–760.

Nihongaki, Y., Yamamoto, S., Kawano, F., Suzuki, H. & Sato, M. (2015b) CRISPR-Cas9-based photoactivatable transcription system. *Chem Biol*, **22**, 169–174.

Noble, A. (1856) Justus Liebigs. *Ann Chem*, **98**, 253–256.

Numano, R., Szobota, S., Lau, A.Y., Gorostiza, P., Volgraf, M., Roux, B., Trauner, D. & Isacoff, E.Y. (2009) Nanosculpting reversed wavelength sensitivity into a photoswitchable iGluR. *Proc Natl Acad Sci U S A*, **106**, 6814–6819.

Olson, J.P., Kwon, H.B., Takasaki, K.T., Chiu, C.Q., Higley, M.J., Sabatini, B.L. & Ellis-Davies, G.C. (2013) Optically selective two-photon uncaging of glutamate at 900 nm. *J Am Chem Soc*, **135**, 5954–5957.

Oron, D., Papagiakoumou, E., Anselmi, F. & Emiliani, V. (2012) Two-photon optogenetics. *Prog Brain Res*, **196**, 119–143.

Palczewski, K., Kumasaka, T., Hori, T., Behnke, C.A., Motoshima, H., Fox, B.A., Le Trong, I., Teller, D.C., Okada, T., Stenkamp, R.E., Yamamoto, M. & Miyano, M. (2000) Crystal structure of rhodopsin: a G protein-coupled receptor. *Science*, **289**, 739–745.

Pedregal, C., Collado, I., Escribano, A., Ezquerra, J., Dominguez, C., Mateo, A.I., Rubio, A., Baker, S.R., Goldsworthy, J., Kamboj, R.K., Ballyk, B.A., Hoo, K. & Bleakman, D. (2000) 4-Alkyl- and 4-cinnamylglutamic acid analogues are potent GluR5 kainate receptor agonists. *J Med Chem*, **43**, 1958–1968.

Perny, M., Muri, L., Dawson, H. & Kleinlogel, S. (2016) Chronic activation of the D156A point mutant of channelrhodopsin-2 signals apoptotic cell death: the good and the bad. *Cell Death Dis*, **7**, e2447.

Pettit, G.R., Singh, S.B., Hamel, E., Lin, C.M., Alberts, D.S. & Garcia-Kendall, D. (1989) Isolation and structure of the strong cell growth and tubulin inhibitor combretastatin A-4. *Experientia*, **45**, 209–211.

Pittolo, S., Gomez-Santacana, X., Eckelt, K., Rovira, X., Dalton, J., Goudet, C., Pin, J.P., Llobet, A., Giraldo, J., Llebaria, A. & Gorostiza, P. (2014) An allosteric modulator to control endogenous G protein-coupled receptors with light. *Nat Chem Biol*, **10**, 813–815.

Polosukhina, A., Litt, J., Tochitsky, I., Nemargut, J., Sychev, Y., De Kouchkovsky, I., Huang, T., Borges, K., Trauner, D., Van Gelder, R.N. & Kramer, R.H. (2012) Photochemical restoration of visual responses in blind mice. *Neuron*, **75**, 271–282.

Polstein, L.R. & Gersbach, C.A. (2015) A light-inducible CRISPR-Cas9 system for control of endogenous gene activation. *Nat Chem Biol*, **11**, 198–200.

Raimondo, J.V., Richards, B.A. & Woodin, M.A. (2016) Neuronal chloride and excitability - the big impact of small changes. *Curr Opin Neurobiol*, **43**, 35–42.

Ramon, Y.C.S. (1952) Structure and connections of neurons. *Bull Los Angel Neuro Soc*, **17**, 5–46.

Reymond, L., Lukinavicius, G., Umezawa, K., Maurel, D., Brun, M.A., Masharina, A., Bojkowska, K., Mollwitz, B., Schena, A., Griss, R. & Johnsson, K. (2011) Visualizing biochemical activities in living cells through chemistry. *Chimia (Aarau)*, **65**, 868–871.

Rovira, X., Trapero, A., Pittolo, S., Zussy, C., Faucherre, A., Jopling, C., Giraldo, J., Pin, J.P., Gorostiza, P., Goudet, C. & Llebaria, A. (2016) OptoGluNAM4.1, a photoswitchable allosteric antagonist for real-time control of mGlu4 receptor activity. *Cell Chem Biol*, **23**, 929–934.

Rullo, A., Reiner, A., Reiter, A., Trauner, D., Isacoff, E.Y. & Woolley, G.A. (2014) Long wavelength optical control of glutamate receptor ion channels using a tetra-ortho-substituted azobenzene derivative. *Chem Commun (Camb)*, **50**, 14613–14615.

Sakata, T., Yan, Y. & Marriott, G. (2005) Optical switching of dipolar interactions on proteins. *Proc Natl Acad Sci U S A*, **102**, 4759–4764.

Samanta, S., Beharry, A.A., Sadovski, O., McCormick, T.M., Babalhavaeji, A., Tropepe, V. & Woolley, G.A. (2013) Photoswitching azo compounds in vivo with red light. *J Am Chem Soc*, **135**, 9777–9784.

Sandoz, G., Bell, S.C. & Isacoff, E.Y. (2011) Optical probing of a dynamic membrane interaction that regulates the TREK1 channel. *Proc Natl Acad Sci U S A*, **108**, 2605–2610.

Sandoz, G. & Levitz, J. (2013) Optogenetic techniques for the study of native potassium channels. *Front Mol Neurosci*, **6**, 6.

Saxon, E. & Bertozzi, C.R. (2000) Cell surface engineering by a modified Staudinger reaction. *Science*, **287**, 2007–2010.

Scheib, U., Stehfest, K., Gee, C.E., Korschen, H.G., Fudim, R., Oertner, T.G. & Hegemann, P. (2015) The rhodopsin-guanylyl cyclase of the aquatic fungus *Blastocladiella emersonii* enables fast optical control of cGMP signaling. *Sci Signal*, **8**, rs8.

Schonberger, M., Althaus, M., Fronius, M., Clauss, W. & Trauner, D. (2014) Controlling epithelial sodium channels with light using photoswitchable amilorides. *Nat Chem*, **6**, 712–719.

Schroder-Lang, S., Schwarzel, M., Seifert, R., Strunker, T., Kateriya, S., Looser, J., Watanabe, M., Kaupp, U.B., Hegemann, P. & Nagel, G. (2007) Fast manipulation of cellular cAMP level by light in vivo. *Nat Methods*, **4**, 39–42.

Spicer, C.D. & Davis, B.G. (2014) Selective chemical protein modification. *Nat Commun*, **5**, 4740.

Sporlein, S., Carstens, H., Satzger, H., Renner, C., Behrendt, R., Moroder, L., Tavan, P., Zinth, W. & Wachtveitl, J. (2002) Ultrafast spectroscopy reveals subnanosecond peptide conformational dynamics and validates molecular dynamics simulation. *Proc Natl Acad Sci U S A*, **99**, 7998–8002.

Stawski, P., Sumser, M. & Trauner, D. (2012) A photochromic agonist of AMPA receptors. *Angew Chem Int Ed Engl*, **51**, 5748–5751.

Stein, M., Middendorp, S.J., Carta, V., Pejo, E., Raines, D.E., Forman, S.A., Sigel, E. & Trauner, D. (2012) Azo-propofols: photochromic potentiators of GABA(A) receptors. *Angew Chem Int Ed Engl*, **51**, 10500–10504.

Stewart, D.S., Savechenkov, P.Y., Dostalova, Z., Chiara, D.C., Ge, R., Raines, D.E., Cohen, J.B., Forman, S.A., Bruzik, K.S. & Miller, K.W. (2011) p-(4-Azipentyl)propofol: a potent photoreactive general anesthetic derivative of propofol. *J Med Chem*, **54**, 8124–8135.

Szobota, S., Gorostiza, P., Del Bene, F., Wyart, C., Fortin, D.L., Kolstad, K.D., Tulyathan, O., Volgraf, M., Numano, R., Aaron, H.L., Scott, E.K., Kramer, R.H., Flannery, J., Baier, H., Trauner, D. & Isacoff, E.Y. (2007) Remote control of neuronal activity with a light-gated glutamate receptor. *Neuron*, **54**, 535–545.

Szobota, S. & Isacoff, E.Y. (2010) Optical control of neuronal activity. *Annu Rev Biophys*, **39**, 329–348.

Szymanski, W., Beierle, J.M., Kistemaker, H.A., Velema, W.A. & Feringa, B.L. (2013) Reversible photocontrol of biological systems by the incorporation of molecular photoswitches. *Chem Rev*, **113**, 6114–6178.

Taglialatela, M., Vandongen, A.M., Drewe, J.A., Joho, R.H., Brown, A.M. & Kirsch, G.E. (1991) Patterns of internal and external tetraethylammonium block in four homologous K+ channels. *Mol Pharmacol*, **40**, 299–307.

Teller, D.C., Okada, T., Behnke, C.A., Palczewski, K. & Stenkamp, R.E. (2001) Advances in determination of a high-resolution three-dimensional structure of rhodopsin, a model of G-protein-coupled receptors (GPCRs). *Biochemistry*, **40**, 7761–7772.

Tochitsky, I. & Kramer, R.H. (2015) Optopharmacological tools for restoring visual function in degenerative retinal diseases. *Curr Opin Neurobiol*, **34**, 74–78.

Tochitsky, I., Polosukhina, A., Degtyar, V.E., Gallerani, N., Smith, C.M., Friedman, A., Van Gelder, R.N., Trauner, D., Kaufer, D. & Kramer, R.H. (2014) Restoring visual function to blind mice with a photoswitch that exploits electrophysiological remodeling of retinal ganglion cells. *Neuron*, **81**, 800–813.

Virginio, C., MacKenzie, A., Rassendren, F.A., North, R.A. & Surprenant, A. (1999) Pore dilation of neuronal P2X receptor channels. *Nat Neurosci*, **2**, 315–321.

Volgraf, M., Gorostiza, P., Numano, R., Kramer, R.H., Isacoff, E.Y. & Trauner, D. (2006) Allosteric control of an ionotropic glutamate receptor with an optical switch. *Nat Chem Biol*, **2**, 47–52.

Volgraf, M., Gorostiza, P., Szobota, S., Helix, M.R., Isacoff, E.Y. & Trauner, D. (2007) Reversibly caged glutamate: a photochromic agonist of ionotropic glutamate receptors. *J Am Chem Soc*, **129**, 260–261.

Wang, S., Szobota, S., Wang, Y., Volgraf, M., Liu, Z., Sun, C., Trauner, D., Isacoff, E.Y. & Zhang, X. (2007) All optical interface for parallel, remote, and spatiotemporal control of neuronal activity. *Nano Lett*, **7**, 3859–3863.

Warp, E., Agarwal, G., Wyart, C., Friedmann, D., Oldfield, C.S., Conner, A., Del Bene, F., Arrenberg, A.B., Baier, H. & Isacoff, E.Y. (2012) Emergence of patterned activity in the developing zebrafish spinal cord. *Curr Biol*, **22**, 93–102.

Wofsy, L., Metzger, H. & Singer, S.J. (1962) Affinity labeling-a general method for labeling the active sites of antibody and enzyme molecules. *Biochemistry*, **1**, 1031–1039.

Yizhar, O., Fenno, L.E., Davidson, T.J., Mogri, M. & Deisseroth, K. (2011) Optogenetics in neural systems. *Neuron*, **71**, 9–34.

Yoshimura, K., Batiza, A., Schroeder, M., Blount, P. & Kung, C. (1999) Hydrophilicity of a single residue within MscL correlates with increased channel mechanosensitivity. *Biophys J*, **77**, 1960–1972.

Young, T.S. & Schultz, P.G. (2010) Beyond the canonical 20 amino acids: expanding the genetic lexicon. *J Biol Chem*, **285**, 11039–11044.

Yue, L., Pawlowski, M., Dellal, S.S., Xie, A., Feng, F., Otis, T.S., Bruzik, K.S., Qian, H. & Pepperberg, D.R. (2012) Robust photoregulation of GABA(A) receptors by allosteric modulation with a propofol analogue. *Nat Commun*, **3**, 1095.

Zemelman, B.V., Lee, G.A., Ng, M. & Miesenbock, G. (2002) Selective photostimulation of genetically chARGed neurons. *Neuron*, **33**, 15–22.

Zhang, Z., Burns, D.C., Kumita, J.R., Smart, O.S. & Woolley, G.A. (2003) A water-soluble azobenzene cross-linker for photocontrol of peptide conformation. *Bioconjug Chem*, **14**, 824–829.

CHAPTER 13

Applications of Nanoparticles for Optical Modulation of Neuronal Behavior

Chiara Paviolo, Shaun Gietman, Daniela Duc, Simon E. Moulton, and Paul R. Stoddart

CONTENTS

13.1 INTRODUCTION

The nervous system is essential to the functional transmission and processing of information within the human body. It consists of two main parts: the central nervous system (CNS), which includes the brain and spinal cord, and the peripheral nervous system (PNS) that comprises all other neural tissues in the body [1]. The basic unit of the nervous system is the neuron, a specialized cell that is capable of receiving and sending electrical signals on millisecond time-scales [2]. Neurons form a complex electrical network throughout the body that underpins perception, motor control, organ communication, and the maintenance of all physiological functions. Under pathological conditions, neuronal pathways can be partially or totally disrupted, resulting in the loss of electrical transmission. Clinical therapies to restore damaged neuronal networks range from bridging axonal gap connections (currently over distances less than 25 mm [3]), to neural prostheses and neural interfaces for non-treatable conditions (e.g. neuro-degenerative diseases or spinal cord injuries).

In this context, nanomaterials are expected to provide new opportunities for improvements in cell-based or immunological therapies [1, 4–7]. Due to their small size, nanotechnology-based devices can interact with biological systems at the molecular level, with a high degree of spatial and temporal specificity. They can penetrate the blood-brain barrier and deliver specific therapeutic agents, probes, or biological materials to targeted cells and tissues [8]. The availability of new experimental techniques and tools also allows complex biological processes to be monitored in real time at the single cell level. When coupled with photonic techniques, optically active nanoparticles provide a relatively non-invasive means to mediate cellular interventions with high spatial and temporal precision [4, 9].

The use of optically targeted nanoparticles in neuroscience has increased considerably over the past decade and the focus of this chapter is to provide a critical perspective on the use of NPs to optically modulate the activity of neuronal tissue. Plasmonic gold nanoparticles (Au NPs) have attracted particular attention, as they can be bio-conjugated for cell-specific targeting, can be delivered by injection, and match the dimensions of subcellular components, such as cell receptors and ion channels [4, 6]. In the context of stimulation and modulation of neural activity, Au NPs have already been successfully employed for several applications, including enhancement of neurite outgrowth [10, 11], modulation of intracellular calcium signaling [12, 13], neuron depolarization [14, 15], and suppression of neuronal activity [16].

The mechanism of Au NPs is generally based on photothermal and photomechanical interactions, but these effects, together with photoelectric and photochemical interactions, can be achieved with a range of other nanomaterials that will also be discussed here. However, given the focus on optically active nanoparticles for neuromodulation, this chapter will not address other important areas of related research, such as the use of light-absorbing microparticles [17], radio-frequency magnetic-field heating, or mechanical stimulation via superparamagnetic ferrite nanoparticles [18, 19], magneto-electric transduction via NPs [20], and acousto-electric interactions driven by ultrasound on piezoelectric nanomaterials [21, 22]. Similarly, the topics of gene-therapy and cellular uptake and neural toxicity of NPs have been extensively discussed in other recent publications [1, 23, 24] and therefore are not included here.

13.2 PROPERTIES OF NANOPARTICLES

Nanoparticles are generally considered to fall in the size range from 1 to 100 nm. The significant interest in nanoscale materials over the past few decades is due to (a) the relatively high surface-to-volume ratio, (b) size-related properties that are significantly different to the bulk material, and (c) length scales that are comparable with many large biomolecules and cellular structures. Nanoparticles can be fabricated from a wide variety of materials, including polymers, phospholipids, inorganics, and metals. This versatility has allowed for the development of nanoparticles with different shapes, sizes, surface chemistries, and cargoes (Petros and DeSimone, 2010). In the context of optically active materials, a range of effects have been exploited for neuronal modulation, based on photothermal, photomechanical, photoelectric, and photochemical energy transduction pathways. The following discussion is organized in terms of these specific energy transduction mechanisms, although in practice it may be found that multiple pathways are active.

13.2.1 Photothermal Effects

A wide range of nanoparticles have been utilized for their photothermal properties, with the different families of heating nanoparticles described in the review by Jaque et al. [26]. Au NPs have received particular attention as a photothermal material, owing to their unique properties that include optical response, chemical and physical stability, relatively low toxicity and wide range of possible surface functionalizations [23, 27]. For example, functionalization with specific ligands allows cellular and molecular specificity, which enables the interaction with target cells and tissues in controlled ways. Thus, Au NPs have been engineered to bind to voltage-gated sodium channels, transient receptor potential vanilloid member 1 (TRPV1) channels, and P2X3 receptor ion channels in dorsal root ganglion neurons [28].

The many applications of Au NPs in medicine and biology are closely related to their unique optical properties. When Au NPs are perturbed by an external light field in the visible or near-infrared (NIR) domain, the conduction electrons move away from their equilibrium position, creating a resonant coherent oscillation called the localized surface plasmon resonance (LSPR) [29]. The excited conduction electrons rapidly lose energy through collisions with the metal lattice and the resulting thermal energy is then transferred into the surrounding medium [30]. The precise position of the LSPR peak wavelength depends on the particle morphology, interparticle distance, and the refractive index of the surrounding medium [31]. For many biological applications, the plasmon absorption peak is selected to match the transparency window of biological tissues (600–1200 nm), meaning that nanorods (NRs), nanoshells, nanostars, and nanocages are the most suitable morphologies [4]. Despite this, to date only NRs and nanospheres (NSs) have been used for modulation of neuronal activity (see Table 13.1). Au NRs have proven to be particularly useful, as their resonance wavelength can be tuned by modification of the NR aspect ratio [32].

Au NPs also have several attractive features as "high precision" photothermal agents for *in vivo* neural modulation. The details of the photothermal energy conversion process are somewhat dependent on the characteristics of the light field (pulse length and peak energy) and the length scale on which the heating is considered [9]. As a result of their very small size relative to mammalian cells, Au NPs are most effective in heating their immediate environment. This allows the overall heat delivery to be reduced, as long as the particles are strategically positioned close to the target cell or receptor. It also leads to a reduction in the diffusion path length for cooling. Consequently, Au NP photothermal modulation acts on sub-millisecond timescales, which is critical for temporally precise stimulation of neuronal activity. Moreover, accurate targeting of NPs to the neurons, together with removal of excess particles by the circulation of interstitial fluids, allows off-target environmental heating to be minimized. These properties are likely to be critically important for avoiding

TABLE 13.1 Summary of NP Characteristics for Modulation of Neural Activity

NP	Size	Irradiation	Functionalization	Mechanism	Applications	Observed effects
Au Nanorods (NRs)	48.6 × 13.8 nm	780 nm	Poly(4-styrenesulfonic acid), silica	Photothermal	Peripheral nerve regeneration	Increased neurite length [10]
Au Nanospheres (NSs)	40 nm	–	Polyethylene glycol (PEG)	Mechanical	Peripheral nerve regeneration	Hind limb motor recovery, attenuation of microglial response, enhanced motor neuron protection, increased remyelination [11]
Au NSs	8.6 nm	–	Manganese-doped	Mechanical	Peripheral nerve regeneration	Increased neurite length [33]
Au NSs	10 nm	–	–	Mechanical	Integration into nerve conduits	Increased neurite length [34]
Au NSs	2–22 nm	–	–	Mechanical	Integration into nerve conduits	Promote adhesion and proliferation of Schwann cells [35]
Au NSs	5 nm	–	Chitosan	Mechanical	Integration into nerve conduits	Regeneration of the sciatic nerve [36]
Au NRs	Aspect ratio 3.4	780 nm	Silica	Photothermal	Modulation of electrical activity	Action potentials in primary auditory neurons [15]
Au NRs	80.4 × 15.3 nm	977 nm	–	Photothermal	Modulation of electrical activity	Action potentials in rat sciatic nerves *in vivo* [14]
Au NRs	71.3 × 18.5 nm	785 nm	Amine-terminated PEG	Photothermal	Modulation of electrical activity	Inhibition of neural activity in primary hippocampal neurons [16]
Au NSs	20 nm	532 nm	Functional groups that target voltage-gated sodium, TRPV1 and P2X3 ion channels	Photothermal	Modulation of electrical activity	Action potentials in dorsal root ganglion cells [28]

(*Continued*)

TABLE 13.1 (CONTINUED) Summary of NP Characteristics for Modulation of Neural Activity

NP	Size	Irradiation	Functionalization	Mechanism	Applications	Observed effects
Au NRs	48.6 × 13.8 nm	780 nm	Poly(4-styrenesulfonic acid)	Photothermal	Modulation of Ca^{2+} dynamics	Intracellular Ca^{2+} transients [13]
Au NRs	60 × 15 nm	780 nm	Cationic protein/lipid complex	Photothermal	Modulation of Ca^{2+} dynamics	Ca^{2+} influx by TRPV1 activation [12]
Au NRs	82.9 × 13.4 nm	982 nm	Streptavidin	Photothermal	Modulation of Ca^{2+} dynamics	Ca^{2+} transients in astrocytes [37]
Au NSs	96 nm	800 nm (off-resonance – fs pulses)	PEG+ antibodies against a hemagglutinin	Photothermal	Modulation of Ca^{2+} dynamics	Ca^{2+} transients in hippocampal neurons and in dendritic compartments [38]
Au NSs	30–40 nm	1040 nm (fs pulses)	con A-biotin and NHS-biotin with streptavidin and neutravidin	Photothermal	Modulation of electrical activity	Action potentials in cortical neurons *in vivo*, contractions in epitheliomuscular cells *in vivo* [39]
Electromagnetized Au NSs	35.6 nm	2 × 10–3 T/100 Hz (Magnetic field)	2 thiol ligands (mercapto(methoxypolyethylene glycol) and the heptapeptide, CYGRGDS) capable of magnetization + PEG + RGD	Electromagnetic	Cell lineage reprogramming	Alleviated symptoms in mouse Parkinson's disease models [40]
Carbon Nanotubes (CNTs)	–	–	In combination with Supramolecular hydrogels	Photoelectrical and mechanical	Nerve regeneration	Conductive tissue regeneration [41]
Mercury telluride (HgTe)	Film	532 nm	Stabilized with thioglycolic acid	Photoelectrical	Modulation of electrical activity	Action potential in NG108-15 cells [42]
Semiconducting NRs+CNTs	Film	405 nm	–	Photoelectrical	Modulation of electrical activity	Stimulation of a light-insensitive chick retina [43]

(*Continued*)

TABLE 13.1 (CONTINUED) Summary of NP Characteristics for Modulation of Neural Activity

NP	Size	Irradiation	Functionalization	Mechanism	Applications	Observed effects
CdS quantum dots	Film/NPs	–	–	Photoelectrical	Potential modulation of electrical activity	Film degradation [44], endocytosis and non-specific binding [45]
Superparamagnetic ferrite nanoparticles	6 nm	Radio frequency magnetic field (40 MHz, 8.4 G)	Functional groups to target TRPV1 channels	Magneto-thermal	Remote control of ion channels – modulation of electrical activity and Ca^{2+} dynamics	Ca^{2+} influx in HEK 293 cells and action potentials in hippocampal neurons [18]
Boron nitride nanotubes	200–600 × 50 nm	Ultrasound (20 W, 40 kHz, 5 s, 4 times a day for 9 days)	–	Piezoelectric	Peripheral nerve regeneration	Increased neurite outgrowth in PC12 [21]
Tetragonal barium titanate nanoparticles	479 nm	Ultrasound, 5 s, different intensities (0.1–0.8 W/cm^2)	–	Piezoelectric	Modulation of Ca^{2+} dynamics	Ca^{2+} transients in human neuroblastoma-derived cells [22]
Photo absorbers	6 μm	532 and 800 nm	–	Photothermal	Modulation of Ca^{2+} dynamics	Ca^{2+} transients in cortical tissue and slice cultures [17]

Note: Plasmon peaks have only been indicated when relevant to the study. In some cases, the mechanism may include more than one effect, or may not be fully understood.

damage to thermally sensitive tissues and limiting toxicity due to high concentrations of exogenous particles.

Indeed, all biological applications require a careful control over biocompatibility. It is well known that some of the most commonly used capping ligands for the synthesis of Au NPs are toxic to cells. A prominent example is the cationic surfactant CTAB, which is commonly used in the preparation of Au NRs [46]. CTAB is known to induce cytotoxicity both *in vitro* [47] and *in vivo* [48] and to interfere with the surface hydration of the particles [49]. Depositing additional surface coatings has been one of the main strategies to reduce the negative effects caused by residual chemicals used during particle synthesis [50, 51].

While Au NPs have attracted most attention to date, other materials such as graphene quantum dots can effectively convert light energy into heat, potentially acting as a photothermal agent [52]. Graphene quantum dots maintain the intrinsic layered structure of graphene, but with a smaller lateral size; they contain abundant carboxylic and hydroxyl groups, which are useful for surface functionalization and biomedical applications [53].

Li et al. investigated the phototriggerable release of biomolecules from a hydrogel-polymer composite using NIR light [54]. Their work demonstrated that polypyrole nanoparticles could act as photothermal transducers within a pNIPAM hydrogel, triggering the release of glutamate. *In vivo* experiments further showed that remote control of brain activity was possible by using both stimulating and inhibitory biomolecules. The use of NIR wavelengths is potentially significant for reduced invasiveness, compared to the use of UV and visible wavelengths to drive photochemical processes (see Section 13.2.4).

13.2.2 Photomechanical Effects

The photomechanical effects of nanoparticles are multifaceted and may vary depending on the type of nanoparticle, the type of target cell, the location of the nanoparticle relative to the target cell, and the type of optical stimulation [55, 56]. In neuromodulation, gold nanoparticles are the most prominent nanomaterial that has been used for photomechanical interactions. The main photomechanical effect of plasmonic nanoparticles is the formation of vapor bubbles upon high energy pulsed optical stimulation at the LSPR peak [57, 58].

Depending on the localization of the nanoparticles relative to the cell, the vapor bubbles can create transient pores in the cell membrane or organelles to modulate cellular behavior in a process known as optoporation (Figure 13.1a) [57, 58]. Xiong et al. [59] detailed two main mechanisms that are involved in vapor bubble formation (Figure 13.1b). In the first mechanism, photon absorption by the nanoparticle induces rapid heating of the nanoparticle. This, in turn, causes instantaneous evaporation of the water immediately surrounding the nanoparticle, resulting in the formation of a vapor bubble. Due to the insulating nature of the vapor bubble, the near-total mechanical conversion of the absorbed optical energy, and the rapidity of this phenomenon (<100 ps), no thermal heating of the environment occurs in this mechanism [59].

In the second mechanism, vapor bubbles can be created via plasma formation [59, 60]. Plasma is formed when the medium surrounding the nanoparticle undergoes multiphoton ionization due to near-field enhancement during LSPR [60]. The rise in temperature and

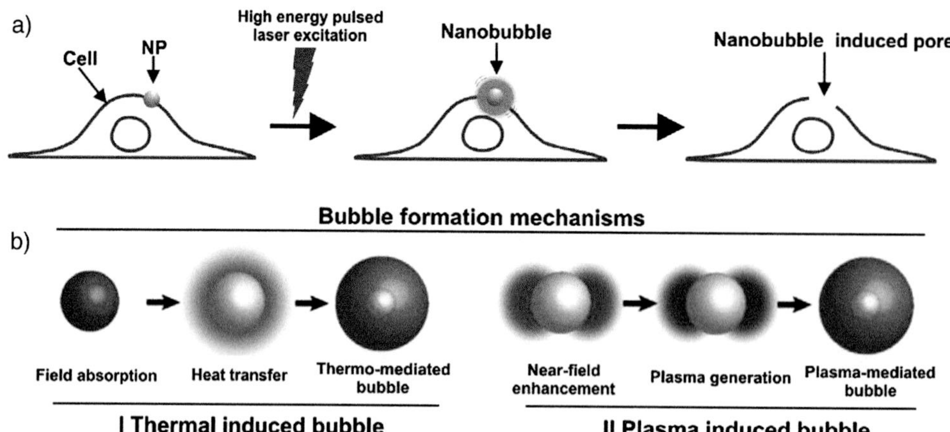

FIGURE 13.1 (a) Nanorod-mediated pore formation via vapor bubbles generated by photo-stimulation. (b) Schematic diagram of bubble formation via heating and via plasma formation. Reproduced with permission from [59], copyright 2016 the authors.

pressure on cooling of the plasma results in vapor bubble formation. As this mechanism is observed under longer laser pulses (>100 ps), thermal heating of the cellular environment occurs, and this may be damaging for the cell [59]. Other non-plasmonic nanomaterials, such as carbon black nanoparticles, have also shown to create vapor bubbles via laser-activated chemical reactions [61].

Irrespective of the mechanism involved, vapor bubbles can cause pore formation in the cell membrane by hydrodynamic stress upon expansion or by liquid jets and shockwaves upon collapsing under hydrostatic pressure [59]. In neuromodulation, this phenomenon has been used in several applications. Caprettini et al. [62] used optoporation to modulate drug intake by the neuronal N2A cell line. They fabricated 3D plasmonic nanostructures made of gold nanotubes, which were coupled to a microfluidics chip loaded with the molecule of interest for intracellular delivery. When N2A cells were grown on the nanostructures and stimulated with 8 ps near-infrared (NIR) pulses, the transient pore formation triggered spatio-temporal drug intake [62]. The advantage of this technique lies in the potential to deliver molecules that cannot naturally pass the cell membrane. Larger structures such as proteins, nanoparticles, and DNA can also be delivered, leading to the use of this technique as a virus-free approach to cell transfection [56, 60]. However, concerns arise around the time for pore closure which can take up to several minutes [62], thus exposing the cell to the extracellular environment. Also, the risk of cell damage is not negligible, which suggests the potential of this technique to mediate cell death in therapeutic applications [56].

The examples above indicate that the photomechanical effects of nanoparticles on neuronal cells can be significant. The application of such properties holds good potential for modulating drug delivery, targeted cell death, and cellular development in neuronal cell lines. The understanding of photomechanical effects also reinforces the importance of controlling this property of nanoparticles when they are used in other applications such as nanoparticle-enhanced NIR stimulation of neurons [6, 15]. Materials can also exhibit a range of shape and conformational changes upon photostimulation; these processes are considered further under photochemical effects in Section 13.2.4.

13.2.3 Photoelectric Effects

Photoconductive materials, such as silicon and conductive polymers, have the ability to generate electrical fields under optical stimulation and have therefore attracted attention as mediators for photoelectric neurostimulation [43]. Nanoparticles with photoelectric properties have also been evaluated for spatially selective neuronal stimulation. Nanoparticles with these properties can be semiconducting, metallic, or organic in nature [42, 63]. Thus, the mechanisms underlying the photoelectric properties of these nanoparticles will vary with the material type.

Semiconductors and metallic nanoparticles used in photoelectric neurostimulation are essentially quantum-confined nanoparticles and quantum dots (QDs) [5]. They are typically 2–6 nm in size and can absorb light in the ultraviolet to near-infrared range [6, 43]. In addition to their spatial resolution, QDs allow for tunability as their optical properties vary with their size, shape, and composition [43]. Examples of QDs used for photoelectric neurostimulation include cadmium telluride, mercury telluride, and cadmium selenide [5]. When these materials are stimulated at their excitation wavelength, free electrons and holes escape from the particles to generate dipole moments and electric fields [43]. If these localized fields are in close proximity to the neuronal cell membrane, they can activate voltage-gated ion channels and trigger an action potential [6].

There are two main ways that the neuron–QD interface has been established. Firstly, the QDs can be directly linked with the cell membrane via antibodies or by protein recognition (Figure 13.2a) [64]. In the former, primary antibodies are targeted to the cell membrane and their interaction with secondary antibodies carrying the QDs creates the neuron–QD interface [64]. In the latter, the QDs are protein-coated and cell membrane attachment is secured by linkage with membrane-bound receptors. The advantage of this method lies in the nanometer-scale proximity of the QD with the cell membrane and the high resolution of stimulation [64]. However, the technique is limited by attenuation of the electric field due the protein coating, non-specific binding, endocytosis of the QDs, and cytotoxicity [43].

The second method used to create the neuron–QD interface is by culturing the neurons on QD films made by drop-casting or layer-by-layer electrostatic deposition, and supported

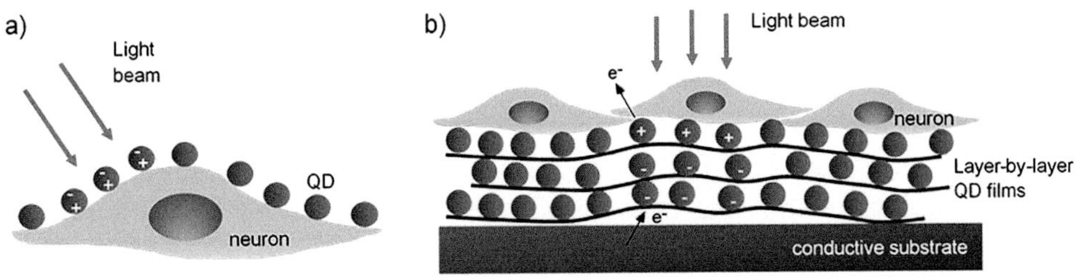

FIGURE 13.2 (a) Direct coupling of QDs to neuronal cells via antibody–antigen interactions or protein recognition, and b) photoelectric neuronal cell stimulation on QD films, as reported in [43].

by a conductive substrate (Figure 13.2b) [43]. The mechanism of neuronal stimulation in QD films can be more complex depending on the film structure and substrate used. For instance, in addition to the localized fields generated by optically excited electron-hole pairs, holes can also be generated by photochemical reactions between the photoelectrons and oxygen in the medium, thus creating a photocurrent augmented by electrons from the conductive substrate [42]. The main advantage of using QD films is the reduction in cytotoxicity, as endocytosis is considerably reduced. However, QD films can display low stimulation efficiency and reliability, presumably due to the increase in average distance between the QDs and the targeted ion channels [6].

More recently, some organic nanomaterials such as graphene have been shown to exhibit photoelectric properties. For example, Akhavan et al. [63] used the photoelectric properties of reduced graphene oxide nanomeshes and reduced graphene oxide to modulate neuronal cell differentiation. They demonstrated that under near-infrared stimulation, the excited photoelectrons from the graphene derivatives were injected into the neuronal cells and that cell differentiation could be enhanced or suppressed, depending on the energy level of the photoelectron [63]. The advantage of graphene-based nanomaterials is the higher biocompatibility compared to metallic QDs. On the other hand, carbon nanotubes can also generate a photocurrent when stimulated in the visible-near infrared range [65]. Thus, carbon nanotubes have been coupled to QDs with the aim of enhancing photocurrent generation [66].

These results indicate that the photoelectric effect of nanoparticles is an effective route for neuromodulation in terms of action potential generation and cell differentiation. However, challenges around the cytotoxicity of QDs still have to be overcome.

13.2.4 Photochemical Effects

Nanoparticle-based photochemical processes have been explored primarily for drug delivery in neurological research and neuromodulation. Nanoparticles that present photochemical effects typically contain photo-responsive pendant or functional groups that induce a chemical change within the structure. As summarized in Figure 13.3 [67], this is most commonly in the form of isomerization, dimerization, and bond cleavage reactions. UV and blue light are generally advantageous for driving photochemical processes, as these wavelengths can induce very rapid structural changes. However, they do not provide the same tissue transparency as NIR light. As shown in Figure 13.3a, NIR wavelengths can be used to drive drug release via the photothermal response of plasmonic nanoparticles (see Section 2.1), but these wavelengths can also generate photochemical responses through two photon events and up-conversion processes [67].

Photoreactive nanomaterials and structures have been investigated for promoting neural growth. In one study, Barillé et al. designed a substrate that was able to facilitate neuronal network outgrowth using phototriggerable bio-azopolymers [68]. The polymeric substrate underwent a conformational change that created nanostructures or grooves in the surface, which then provided a guide for PC12 cells to align and orient themselves. Moreover, photoswitching of the surface allowed for patterns to effectively be erased and redrawn in real time.

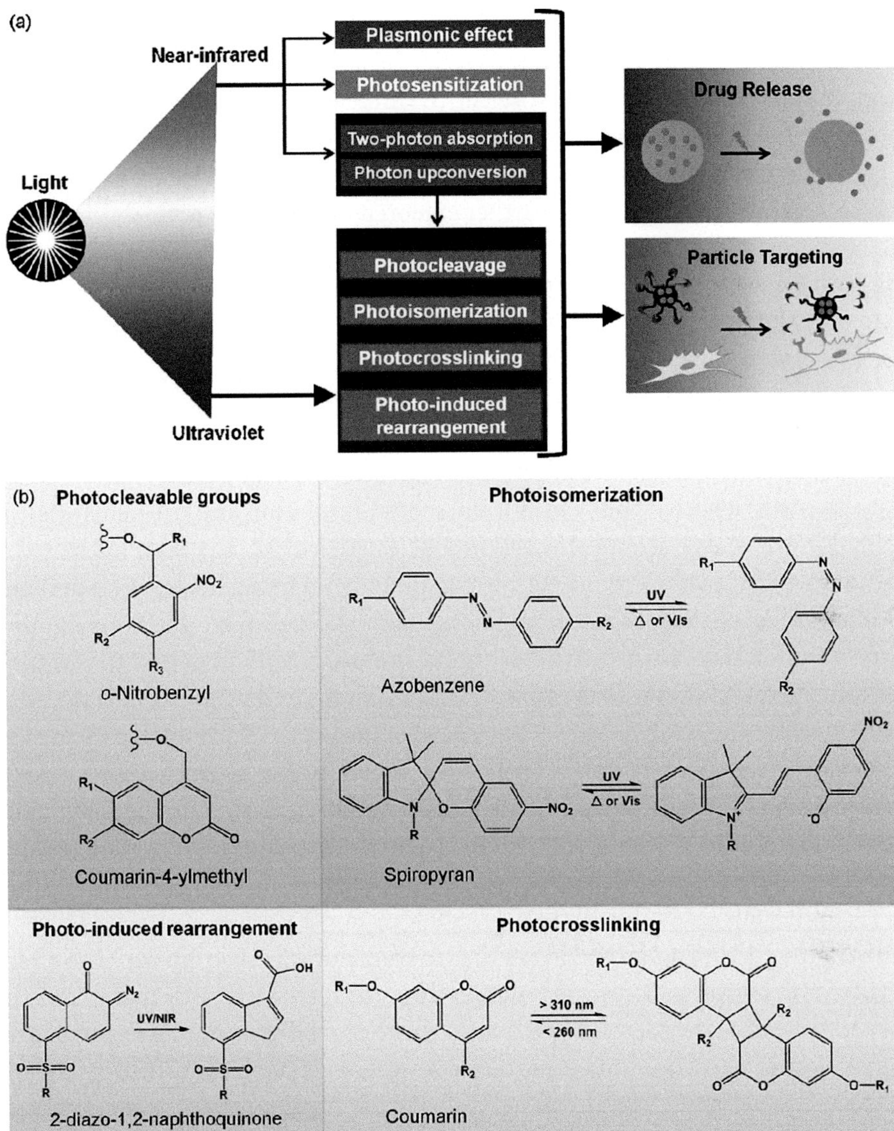

FIGURE 13.3 Mechanisms for phototriggered drug delivery. (a) Mechanisms of photo-responsiveness for nanoparticle targeting and drug release. (b) Selected chemical groups used for photochemical reactions, such as photocleavage, photoisomerization, photo-induced rearrangement, and photocrosslinking (reproduced with permission from [67], copyright 2015 Elsevier).

Photo-responsive nanomaterials and small molecules have also been applied for the photochemical control of neuromodulation. For example, bistable molecules have been used to create photoswitches with multiple configurations that can change upon exposure to light of different wavelengths [69–71]. This property has given rise to photoswitchable ligands that can control neurons by regulating K^+ channels and glutamate receptors [71, 72]. There is considerable potential to extend this approach by engineering new molecules for modulating a wider range of specific targets.

Photochemical control of the release of biomolecules has great potential in clinical and treatment applications due to the high degree of spatial and temporal precision. Several studies have explored the use of photosensitive ruthenium-based compounds as cages for biomolecules. Yang et al. showed that γ-aminobutyric acid, enclosed within a ruthenium-based cage, effectively reduced seizure activity when released *in vivo* by exposure to blue light [73]. An extension of this work further explored the effects of both concentration and light intensity on reduction of seizure activity for an implantable device [74]. A similar body of work used the same ruthenium compound (ruthenium-bipyridine-triphenylphosphine) to cage the seizure inducing compound 4-aminopyridine to study the treatment of epilepsy [75].

Caged glutamate is one of the most commonly used molecules for stimulation [76], but a number of neurotransmitters have been used [77]. The initial work relied on short wavelengths of about 380 nm to activate the cages. However, the high level of light-scattering in tissue at these wavelengths tends to limit the spatial resolution and the penetration depth is severely limited due to strong absorption [78]. Furthermore, the high intensity of light and high energy of photons required for uncaging can damage tissue. The development of two-photon responsive glutamate cages has allowed both finer spatial resolution and deeper penetration into tissue [79], as uncaging can occur with exposure to 800 nm light. Further refinement has reduced the optical radiation required by improving the efficiency of the two-photon uncaging process [78] or developing cages which release upon exposure to visible light [80, 81]. Caged molecules are also able to inhibit neural activity, through the release of GABA [78, 81]. However, this technique requires further development, as many of the compounds used also interact with the receptor before photoactivation [78].

13.3 PERIPHERAL NERVE REGENERATION

The primary function of a peripheral nerve is to transmit signals from the CNS to the rest of the body, or to convey sensory information from the rest of the body to the CNS. In case of injury or a health disorder, this pathway can be partially or totally disrupted, resulting in pain, loss of sensation, reduced muscular strength, poor coordination, atrophy, or complete paralysis. Even if peripheral nerves have the capacity to spontaneously regenerate following traumatic injuries, a clinical operation must be performed in case of a complete nerve transaction. Current clinical strategies include autografts, allografts, and nerve guides, yet the maximum regeneration distance is limited to 25 mm [3]. Researchers are currently focused on finding new methods and materials to improve this nerve regeneration distance. Even though the process of neural regeneration is well-known, nerve regeneration following injury remains a great challenge for neuroscientists and neurologists. The process involves outgrowth of neuronal branches (neurites) from the cell body. The neurites elongate, bifurcate, and connect to neighboring neurons to form an electrically functional network. Typically, one of the neurites differentiates into an axon, while the others either turn into dendrites or fail to become functional and retract [82].

It has been shown that the heat released by plasmon excitation of Au NRs can be used to stimulate neurite outgrowth in NG108-15 neuronal cells (Figure 13.4a). The greatest outgrowth was observed after irradiating the endocytosed particles with the highest laser

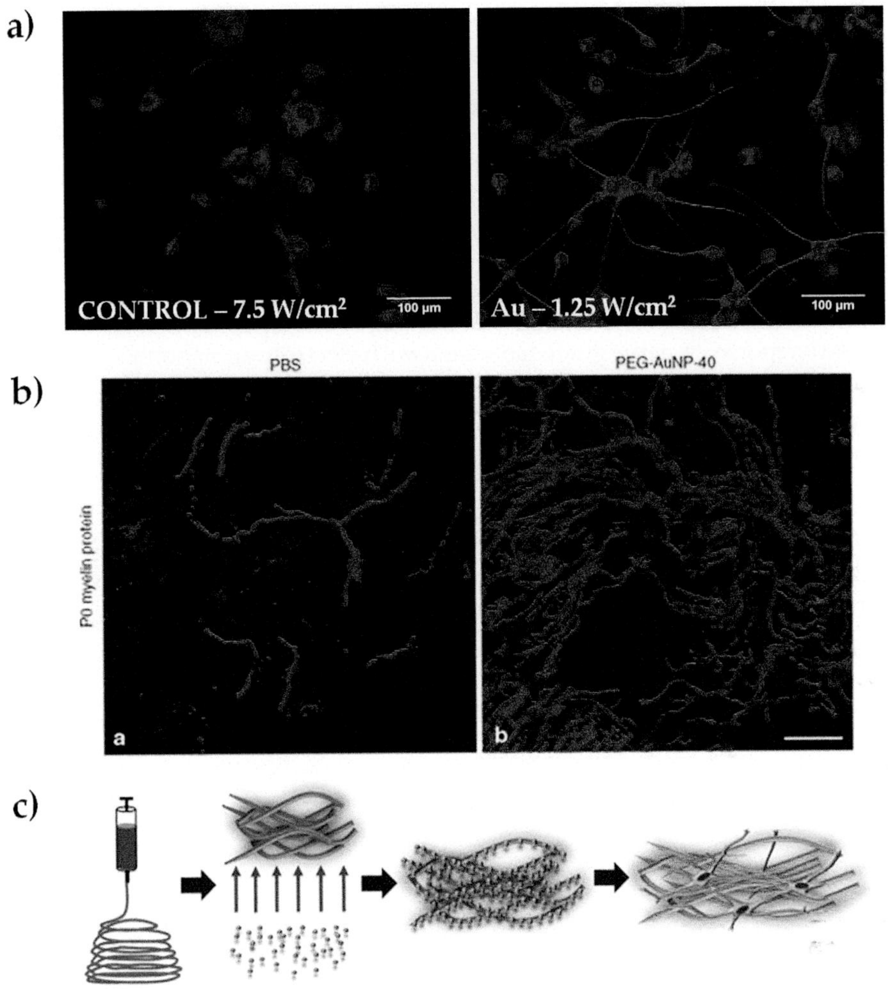

FIGURE 13.4 Representative results of Au NPs for peripheral nerve regeneration. (a) Examples of epifluorescence images of NG108-15 neuronal cells cultured alone or with Au NRs and exposed to different laser irradiances, as indicated in each panel. Cells were marked for β-III tubulin (in red) and DAPI (in blue, reproduced with permission from [10], copyright 2013 John Wiley and Sons). (b) Spontaneous remyelination by Schwann cells (myelin marker P0, in red) was enhanced in mice treated with polyethylene glycol-coated Au NPs (reproduced with permission from [11], copyright 2015 Elsevier). c) Schematic representation of electrospun nanofibers doped with 10 nm Au NPs (reproduced with permission from [34], copyright 2015 American Chemical Society).

dose (7.5 W/cm^2), obtaining an average increase in neurite length of almost 36% compared to the non-irradiated sample [10]. It was hypothesized that the mechanism underlying the outgrowth involves the activation of one or more transcription factors, supporting previous studies on iron oxide nanoparticles. Indeed, Kim and colleagues performed gene expression analysis in PC12 cells, observing changes in genes related to the cytoskeleton, signaling molecules, receptors for growth hormones, and ion channels [83]. These genes are known to be involved in neuronal differentiation [82]. Papastefanaki *et al.* used PEG-coated

Au NPs after mouse spinal cord injury, showing hind limb motor recovery, attenuation of microglial response, enhanced motor neuron protection, and increased remyelination 8 weeks after treatment (Figure 13.4b) [11]. In a different approach, Bhang and coworkers doped spherical Au NPs with manganese, which allows pH-triggered released of manganese ions after the endocytosis of the particles. They observed neurite outgrowth 24 hours after treatment, showing an increase of roughly 70% compared to control samples. They speculated that changes in intracellular signaling pathways were responsible for the outgrowth increase [33].

Au NPs have also been used for integration into nanocomposite nerve conduits. Recently Baranes et al. reported a nerve guide fabricated with electrospun nanofibers doped with 10 nm Au NPs (shown schematically in Figure 13.4c). The scaffolds encouraged a longer outgrowth of the neurites in primary neurons of the medicinal leech, preferring axonal elongation over the formation of complex networks [34]. Similarly, Das and coworkers reported on a nerve guide fabricated by adsorbing Au NPs onto silk fibers. This nano-hybrid material was successfully tested in a neurotmesis grade injury (complete axonal loss and conduction failure) of the sciatic nerve of Sprague-Dawley rats over a period of eighteen months. The nano-composites were found to promote adhesion and proliferation of Schwann cells *in vitro* and did not elicit any toxic or immunogenic responses *in vivo* [35]. Lin and colleagues tested chitosan-AuNP microgrooved nerve conduits both *in vitro* and *in vivo*. The results showed that the conduits pre-seeded with primary neuronal stem cells were able to support regeneration of the sciatic nerve better than the controls [36]. Taken together with the work of Barillé *et al.* showing neuronal network outgrowth on a phototriggered reconfigurable polymeric substrate [68] (see Section 13.2.4), these studies clearly show that neural regeneration is also influenced by the mechanical support of the guides. Nanoparticle-doped scaffolds open up new strategies to combine bio-materials and nanoparticles for providing physical and/or bioactive environments for neural regeneration. There is also potential to combine the electrical properties of Au NP and bio-materials to promote peripheral nerve elongation [84].

Although the interest in Au NPs for applications in nerve regeneration is expanding, *in vivo* studies are still limited by a lack of knowledge about the consequences of nanomaterials on intracellular pathways and inflammatory responses. It is known that a high concentration of metal nanoparticles in living organisms can cause cell oxidative stress and production of reactive oxygen species, leading to other serious cellular dysfunctions such as inflammation, cell membrane disruption, DNA damage, cancer, or apoptosis [86]. Söderstjerna et al. recorded a significantly higher number of apoptotic and oxidatively stressed cells after exposing Au NPs in a primary tissue model of the mouse retina [87]. In our laboratory we detected a significant oxidative stress increase after exposing NG108-15 neuronal cells to Au NRs for one hour [88]. This result confirmed a previously published report showing oxidative stress generated in the rat brain [89]. Au NPs have also been observed to cause a significant decrease in the levels of dopamine and serotonin *in vivo* [89]. Moreover, Au NPs have been imaged not only intracellularly, but also intranuclearly, raising questions of whether these nanomaterials can cause DNA damage and/or alter gene expression [87]. However, these effects can generally be minimized by reducing the concentration of NPs and using particles larger than about 15 nm [32].

13.4 MODULATION OF NERVE ELECTRICAL ACTIVITY

The use of light to modulate the electrical activity of neuronal cells, as shown schematically in Figure 13.5a, has attracted growing interest, due to the potential for less invasive neuronal interfaces, improved spatial resolution of stimulation, and avoiding electrical artifacts in associated neural recordings [90, 91]. The potential to use Au NPs as an exogenous light absorber in neural stimulation appears to have been first identified by [92], but we are not aware of any published demonstration by these workers, who have subsequently focused on the use of black photo-absorbers of ~6 μm diameter [17]. The initial suggestion was based on an analogy with infrared neural stimulation [93], where pulsed laser wavelengths in the range of approximately 1–6 μm have been used to stimulate action potentials in neurons. The primary mechanism in infrared neural stimulation appears to be the transient heating associated with absorption of light by water in the tissue [94, 95]. However, water absorption also limits the penetration depth of the infrared light to a few hundred microns [96], while cumulative heating effects tend to limit the stimulation site density and maximum repetition rates [97]. As discussed in Section 13.2.1, Au NRs allow highly localized photothermal heating through the absorption of wavelengths in the water transmission window from 600 to 1,200 nm.

Yong et al. [15] first confirmed that Au NRs can be used to stimulate cultured rat primary auditory neurons with near-infrared (780 nm) illumination. The laser-induced cell electrical activity was observed using whole cell patch clamp electrophysiology, as shown in Figure 13.5b. The open patch technique was used to show that action potentials were associated with transient temperature increases of about 6°C. The NRs were endocytosed by the neurons after 15–17 h incubation, as shown by dark field microspectroscopy. This work was soon followed by a demonstration that Au NRs could be used to elicit compound action potentials in the rat sciatic nerve *in vivo* (shown schematically in Figure 13.5c) [14]. NRs with peak absorption at 977 nm were introduced to the nerve bundle by microinjection. Subsequent TEM analysis of fixed cross-sectional slices showed Au NRs located near the surface of the axon plasma membrane. In contrast, Yoo et al. [16] found that

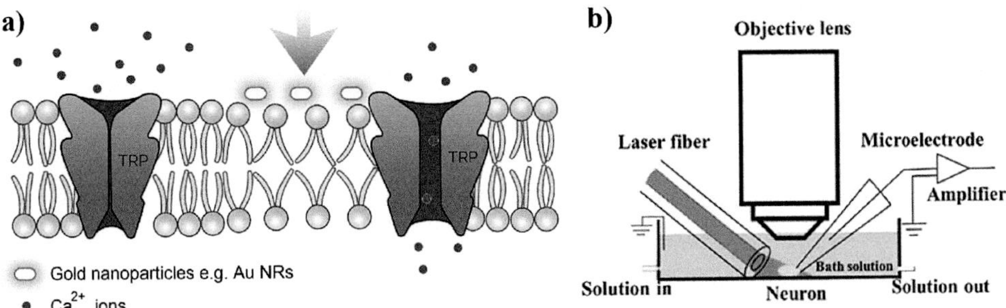

FIGURE 13.5 (a) Illustration of our current understanding of how nerve electrical activity is modulated by NIR excitation of Au NRs, based on changes in membrane capacitance and conduction of TRPV1 channels (TRP) due to plasmonic heating. b) Whole-cell patch clamp recording provides a convenient platform for studying optically stimulated neurons (reproduced with permission from [15], copyright 2014 John Wiley and Sons).

Au NRs inhibited neural activity in networks of primary cultured hippocampal neurons. This inhibitory effect was associated with longer laser exposures (1–30 minutes) and the NPs were coated with positively charged amine-terminated polyethylene glycol, which may have had an increased affinity for attachment to the cell membrane. The increased exposure time led to a sustained temperature rise of as much as 10°C at the plasma membrane. It is well known that a sustained increase in environmental temperature can have an inhibitory effect on neural activity [98] and similar effects have been observed in infrared neural stimulation [99].

Although the detailed mechanism is not yet understood, these effects appear to be broadly analogous to the processes observed in infrared neural stimulation. In particular, the local increase in temperature due to plasmonic heating produces a change in the electrical capacitance of the plasma membrane [15], in agreement with the observations of Shapiro et al. [94]. Recent work suggests that this increase in membrane capacitance is due to thermally driven changes in the membrane structural dimensions [100, 101]. In isolation, these changes in cell capacitance are unlikely to act as an excitatory stimulus, except in the most voltage sensitive cells [102, 103]. However, infrared-induced temperature changes have also been shown to modulate the responses of voltage- and temperature-sensitive (TRPV) ion channels [104, 105], as shown in Figure 13.5d. Modulation of Ca^{2+} dynamics in the soma may also be involved [13], again in analogy with effects observed for infrared neural stimulation [106]. Nakatsuji et al. [12] have subsequently confirmed that laser heating of Au NRs causes Ca^{2+} influx by TRPV1 activation. The surface chemistry of the NPs was modified with a cationic lipoprotein for non-cytotoxic targeting of the plasma membrane and the effect was also demonstrated in primary cultured dorsal root ganglion cells from wild type mice.

It has also been shown that relatively high levels of infrared laser exposure can lead to nanoporation of the cell membrane, with concomitant Ca^{2+} uptake and cellular swelling [109]. Once again, similar disruptions of the cell membrane have been observed in the presence of Au NPs, across a wide range from continuous wave to fs laser pulse lengths [110, 111]. However, it appears unlikely that this level of disruption (see Section 13.2.2) is generated by the relatively modest laser irradiances involved in NR-mediated neural modulation [14–16]. On the other hand, it is known that nanoscale heating with Au NSs can induce a gel-fluid phase transition in phospholipid giant unilamellar vesicles [112]. Subsequent studies have shown that membrane conductance can be controlled by plasmonic heating of single Au NPs over periods of several seconds, without a phase transition or nanopore formation. It was proposed that this effect is due to an increase in phospholipid mobility with increasing temperature and was observed in both artificial lipid bilayers formed on a planar patch clamp system and in HEK293 cells that lack temperature-sensitive ion channels [113]. Similar experiments with artificial membranes composed of asolectin have interpreted the transient current during initial heating in terms of capacitance changes, with a single NP producing a capacitive current of 0.75 pA under irradiance of 18 kW/cm^2 for ~1 ms [28]. Further work is needed to clarify the relative importance of these various contributions, their relationship to the ensuing biochemical pathways, and any long-term deleterious effects that may arise.

Recently it has been shown that less than 1 ms of NIR stimulation combined with Au NRs reliably produces strong Ca^{2+} transients in astrocytes [37]. While this interaction may help to facilitate minimally invasive studies of astrocyte function, it also points to the importance of targeting NPs to specific locations in the tissue. Eom et al. [37] targeted the astrocyte surface with biotinylated anti-thymocyte antigen-1 antibody and streptavidin-coated Au NRs. Carvalho-de-Souza et al. have demonstrated that Au NSs can be conjugated with functional groups that target voltage-gated sodium, TRPV1, and P2X3 ion channels, all of which are known to be expressed in the membrane of dorsal root ganglion neurons [28]. In each case, it was found that the Au NPs bound to the cultured neurons without impeding their excitatory capability and generated optically evoked action potentials at relatively low NP concentrations. In comparison, unconjugated Au NPs required higher concentrations to support optical stimulation and were readily washed out on solution exchange. Targeting membrane receptors is an important approach, as cells of different sensory specialization can express very different profiles of membrane receptors. However, to our knowledge, neuronal selectivity has yet to be demonstrated in mixed cultures or *in vivo*. Interestingly, it has been found that organically modified silica (ORMASIL) nanoparticles are preferentially taken up by neurons *in vivo* [114], but it is not yet clear what mechanism is involved and whether it can be extended to Au NPs.

13.5 OUTLOOK

Applications of nanotechnology in basic and clinical neuroscience are in the early stages of development, partly because of the complexities associated with neural cells and the CNS. Many of the biochemical, cellular, and genomic mechanisms of neural regeneration, modulation, and nanomaterial-tissue interactions are still not well understood. The challenges are numerous, but the impact that NPs can have on manipulating the physiology and the pathology of the nervous system and how we can intervene at a molecular level is significant. From a technical point of view, NPs need to be engineered to obtain a greater cellular specificity, multiple induced physiological functions (such as targeting of multiple cell receptors, ligands, or synapses), and minimal side-effects [115]. The current consensus [32] is that Au NPs larger than about 15 nm are less toxic than smaller particles, and that the primary cytotoxicity in Au NRs is associated with CTAB, the cationic capping ligand that is typically used in the preparation. Other materials, such as cadmium telluride, mercury telluride, and cadmium selenide quantum dots are intrinsically cytotoxic. The toxicity issues can be largely overcome by embedding the nanoparticles in biocompatible surfaces or uniformly coating them with biocompatible polymers or molecules [23, 88] and carefully targeting them to the required location. However, there remains a need to clarify the long-term effects of NPs on gene expression, activation of specific intracellular pathways, neurotransmitter release, and cellular inflammation.

In terms of electrical activity, careful studies are needed to investigate the influence of NPs on single ion channels and ionic currents. Carvalho-de-Souza et al. [28] have shown that Au NPs can be targeted to neuronal membrane receptors without hindering their excitatory functioning and resisting washout for periods of more than 30 minutes. However, the Au NP conjugates will have a limited lifetime due to the natural turnover of membrane proteins, and clearance and/or degradation of the NPs. Repeated treatment may lead to NP accumulation. It has been observed that intracellular Au NPs can increase neuronal excitability and aggravate seizure activation in hippocampal tissue, therefore suggesting that intracellular NPs might alter neuronal functions and cause hyper-excitability under pathological conditions [116].

Nanoparticles have attracted attention in a wide range of medical, therapeutic, and technological contexts [4, 23]. For example, Au NPs can be used as exogenous contrast agents for photoacoustic imaging [117]. Boulais *et al.* [118] review the interaction of short and ultrashort laser pulses with plasmonic NPs for the purpose of destroying, modifying, or manipulating molecular, subcellular, and cellular structures. In these applications, heating, low-density plasma generation, pressure wave release, and formation of vapor bubbles can be used to disrupt cells for drug delivery and cell transfection, typically by nanoporation. Unsurprisingly therefore, these shorter (sub-nanosecond) laser pulses are also more likely to generate a stress response in the target cells [119]. Considering the steadily growing knowledge and the impressive versatility demonstrated by nanoparticles in this wide range of applications, we confidently expect further progress and innovation in the field of neuromodulation.

REFERENCES

1. S. Mclaughlin, J. Podrebarac, M. Ruel, E.J. Suuronen, B. McNeill, E.I. Alarcon, Nano-engineered biomaterials for tissue regeneration: what has been achieved so far?, *Front. Mater.* 3(27) (2016).
2. A.L. Hodgkin, A.F. Huxley, Resting and action potentials in single nerve fibres, *J. Physiol.* 104(2) (1945) 176–195.
3. J.H.A. Bell, J.W. Haycock, Next generation nerve guides: materials, fabrication, growth factors, and cell delivery, *Tissue Eng. B* 18(2) (2012) 116–128.
4. G. Bodelón, C. Costas, J. Pérez-Juste, I. Pastoriza-Santos, L.M. Liz-Marzán, Gold nanoparticles for regulation of cell function and behavior, *Nano Today* 13 (2017) 40–60.
5. E. Colombo, P. Feyen, M.R. Antognazza, G. Lanzani, F. Benfenati, Nanoparticles: a challenging vehicle for neural stimulation, *Front. Neurosci.* 10 (2016).
6. Y. Wang, L. Guo, Nanomaterial-enabled neural stimulation, *Front. Neurosci.* 10(69) (2016).
7. J.M. Zuidema, R.J. Gilbert, D.J. Osterhout, Nanoparticle technologies in the spinal cord, *Cells Tissues Organs* 202(1–2) (2015) 102–115.
8. M. Shilo, M. Motiei, P. Hana, R. Popovtzer, Transport of nanoparticles through the blood-brain barrier for imaging and therapeutic applications, *Nanoscale* 6(4) (2014) 2146–2152.
9. C. Paviolo, P.R. Stoddart, Gold nanoparticles for modulating neuronal behavior, *Nanomaterials* 7(4) (2017) E92.
10. C. Paviolo, J.W. Haycock, J. Yong, A. Yu, P.R. Stoddart, S.L. McArthur, Laser exposure of gold nanorods can increase neuronal cell outgrowth, *Biotechnol. Bioeng.* 110(8) (2013a) 2277–2291.
11. F. Papastefanaki, I. Jakovcevski, N. Poulia, N. Djogo, F. Schulz, T. Martinovic, D. Ciric, G. Loers, T. Vossmeyer, H. Weller, M. Schachner, R. Matsas, Intraspinal delivery of polyethylene glycol-coated gold nanoparticles promotes functional recovery after spinal cord injury, *Mol. Ther.* 23(6) (2015) 993–1002.
12. H. Nakatsuji, T. Numata, N. Morone, S. Kaneko, Y. Mori, H. Imahori, T. Murakami, Thermosensitive ion channel activation in single neuronal cells by using surface-engineered plasmonic nanoparticles, *Angew. Chem.* 54(40) (2015) 11725–11729.
13. C. Paviolo, J.W. Haycock, P.J. Cadusch, S.L. McArthur, P.R. Stoddart, Laser exposure of gold nanorods can induce intracellular calcium transients, *J. Biophotonics* 7(10) (2014) 761–765.

14. K. Eom, J. Kim, J.M. Choi, T. Kang, J.W. Chang, K.M. Byun, S.B. Jun, S.J. Kim, Enhanced infrared neural stimulation using localized surface plasmon resonance of gold nanorods, *Small* 10(19) (2014) 3853–3857.
15. J. Yong, K. Needham, W.G.A. Brown, B.A. Nayagam, S.L. McArthur, A. Yu, P.R. Stoddart, Gold-nanorod-assisted near-infrared stimulation of primary auditory neurons, *Adv. Healthc. Mater.* 3(11) (2014) 1862–1868.
16. S. Yoo, S. Hong, Y. Choi, J. Park, Y. Nam, Photothermal inhibition of neural activity with near-infrared-sensitive nanotransducers, *ACS Nano* 8(8) (2014) 8040–8049.
17. N. Farah, A. Zoubi, S. Matar, L. Golan, A. Marom, C.R. Butson, I. Brosh, S. Shoham, Holographically patterned activation using photo-absorber induced neural-thermal stimulation, *J. Neural Eng.* 10(5) (2013) 056004.
18. H. Huang, S. Delikanli, H. Zeng, D.M. Ferkey, A. Pralle, Remote control of ion channels and neurons through magnetic-field heating of nanoparticles, *Nat. Nanotechnol.* 5(8) (2010) 602–606.
19. A. Tay, D. Di Carlo, Remote neural stimulation using magnetic nanoparticles, *Curr. Med. Chem.* 24(5) (2017) 537–548.
20. R. Guduru, P. Liang, J. Hong, A. Rodzinski, A. Hadjikhani, J. Horstmyer, E. Levister, S. Khizroev, Magnetoelectric 'spin' on stimulating the brain, *Nanomedicine* 10(13) (2015) 2051–2061.
21. G. Ciofani, S. Danti, D. D'Alessandro, L. Ricotti, S. Moscato, G. Bertoni, A. Falqui, S. Berrettini, M. Petrini, V. Mattoli, A. Menciassi, Enhancement of neurite outgrowth in neuronal-like cells following boron nitride nanotube-mediated stimulation, *ACS Nano* 4(10) (2010) 6267–6277.
22. A. Marino, S. Arai, Y.Y. Hou, E. Sinibaldi, M. Pellegrino, Y.T. Chang, B. Mazzolai, V. Mattoli, M. Suzuki, G. Ciofani, Piezoelectric nanoparticle-assisted wireless neuronal stimulation, *ACS Nano* 9(7) (2015) 7678–7689.
23. G. Chen, I. Roy, C. Yang, P.N. Prasad, Nanochemistry and nanomedicine for nanoparticle-based diagnostics and therapy, *Chem. Rev.* 116(5) (2016) 2826–2885.
24. D. Lovisolo, A. Gilardino, F.A. Ruffinatti, When neurons encounter nanoobjects: spotlight on calcium signalling, *Int. J. Environ. Res. Public Health* 11(9) (2014) 9621–9638.
25. R.A. Petros, J.M. DeSimone, Strategies in the design of nanoparticles for therapeutic applications, *Nat. Rev. Drug Discov.* 9(8) (2010) 615–627.
26. D. Jaque, L.M. Maestro, B. del Rosal, P. Haro-Gonzalez, A. Benayas, J.L. Plaza, E.M. Rodriguez J.G. Sole, Nanoparticles for photothermal therapies, *Nanoscale* 6(16) (2014) 9494–9530.
27. M. Das, K.H. Shim, S.S.A. An, D.K. Yi, Review on gold nanoparticles and their applications, *Toxicol. Environ. Health Sci.* 3(4) (2011) 193–205.
28. J.L. Carvalho-de-Souza, J.S. Treger, B. Dang, S.B.H. Kent, D.R. Pepperberg, F. Bezanilla, Photosensitivity of neurons enabled by cell-targeted gold nanoparticles, *Neuron* 86(1) (2015) 207–217.
29. V. Myroshnychenko, J. Rodriguez-Fernandez, I. Pastoriza-Santos, A.M. Funston, C. Novo, P. Mulvaney, L.M. Liz-Marzan, F.J.G. de Abajo, Modelling the optical response of gold nanoparticles, *Chem. Soc. Rev.* 37(9) (2008) 1792–1805.
30. S. Link, M.A. El-Sayed, Shape and size dependence of radiative, non-radiative and photothermal properties of gold nanocrystals, *Int. Rev. Phys. Chem.* 19(3) (2000) 409–453.
31. A.M. Funston, C. Novo, T.J. Davis, P. Mulvaney, Plasmon coupling of gold nanorods at short distances and in different geometries, *Nano Lett.* 9(4) (2009) 1651–1658.
32. H.Y. Chen, L. Shao, Q. Li, J. Wang, Gold nanorods and their plasmonic properties, *Chem. Soc. Rev.* 42(7) (2013) 2679–2724.
33. S.H. Bhang, J. Han, H.-K. Jang, M.-K. Noh, W.-G. La, M. Yi, W.-S. Kim, Y. Kim Kwon, T. Yu, B.-S. Kim, pH-triggered release of manganese from MnAu nanoparticles that enables cellular neuronal differentiation without cellular toxicity, *Biomaterials* 55 (2015) 33–43.

34. K. Baranes, M. Shevach, O. Shefi, T. Dvir, Gold nanoparticle-decorated scaffolds promote neuronal differentiation and maturation, *Nano Lett.* 16(5) (2016) 2916–2920.
35. S. Das, M. Sharma, D. Saharia, K.K. Sarma, M.G. Sarma, B.B. Borthakur, U. Bora, *In vivo* studies of silk based gold nano-composite conduits for functional peripheral nerve regeneration, *Biomaterials* 62 (2015) 66–75.
36. Y.-L. Lin, J.-C. Jen, S.-H. Hsu, I.-M. Chiu, Sciatic nerve repair by microgrooved nerve conduits made of chitosan-gold nanocomposites, *Surg. Neurol.* 70 (Supplement 1) (2008) S9–S18.
37. K. Eom, S. Hwang, S. Yun, K.M. Byun, S.B. Jun, S.J. Kim, Photothermal activation of astrocyte cells using localized surface plasmon resonance of gold nanorods, *J. Biophotonics* 10(4) (2017) 486–493.
38. F. Lavoie-Cardinal, C. Salesse, E. Bergeron, M. Meunier, P. De Koninck, Gold nanoparticle-assisted all optical localized stimulation and monitoring of Ca2+ signaling in neurons, *Sci. Rep.* 6 (2016).
39. W.D.A.M. de Boer, J.J. Hirtz, A. Capretti, T. Gregorkiewicz, M. Izquierdo-Serra, S.T. Han, C. Dupre, Y. Shymkiv, R. Yuste, Neuronal photoactivation through second-harmonic near-infrared absorption by gold nanoparticles, *Light Sci. Appl.* 7 (2018) 100.
40. J. Yoo, E. Lee, H.Y. Kim, D.H. Youn, J. Jung, H. Kim, Y.J. Chang, W. Lee, J. Shin, S. Baek, W. Jang, W. Jun, S. Kim, J.K. Hong, H.J. Park, C.J. Lengner, S.H. Moh, Y. Kwon, J. Kim, Electromagnetized gold nanoparticles mediate direct lineage reprogramming into induced dopamine neurons in vivo for Parkinson's disease therapy, *Nat. Nanotechnol.* 12(10) (2017) 1006–1014.
41. S. Marchesan, S. Bosi, A. Alshatwi, M. Prato, Carbon nanotubes for organ regeneration: an electrifying performance, *Nano Today* 11(4) (2016) 398–401.
42. T.C. Pappas, W.M.S. Wickramanyake, E. Jan, M. Motamedi, M. Brodwick, N.A. Kotov, Nanoscale engineering of a cellular interface with semiconductor nanoparticle films for photoelectric stimulation of neurons, *Nano Lett.* 7(2) (2007) 513–519.
43. L. Bareket-Keren, Y. Hanein, Novel interfaces for light directed neuronal stimulation: advances and challenges, *Int. J. Nanomedicine* 9 (Supplement 1) (2014) 65–83.
44. J.O. Winter, N. Gomez, B.A. Korgel, C.E. Schmidt, Quantum dots for electrical stimulation of neural cells, in: A.N. Cartwright, M. Osinski (Eds.) *Nanobiophotonics and Biomedical Applications II*, 2005, pp. 235–246.
45. N. Gomez, J.O. Winter, F. Shieh, A.E. Saunders, B.A. Korgel, C.E. Schmidt, Challenges in quantum dot-neuron active interfacing, *Talanta* 67(3) (2005) 462–471.
46. J. Pérez-Juste, I. Pastoriza-Santos, L.M. Liz-Marzán, P. Mulvaney, Gold nanorods: synthesis, characterization and applications, *Coord. Chem. Rev.* 249(17–18) (2005) 1870–1901.
47. E.E. Connor, J. Mwamuka, A. Gole, C.J. Murphy, M.D. Wyatt, Gold nanoparticles are taken up by human cells but do not cause acute cytotoxicity, *Small* 1(3) (2005) 325–327.
48. B. Isomaa, J. Reuter, B.M. Djupsund, The subacute and chronic toxicity of cetyltrimethylammonium bromide (CTAB), a cationic surfactant, in the rat, *Arch. Toxicol.* 35(2) (1976) 91–96.
49. I. Pastoriza-Santos, J. Perez-Juste, L.M. Liz-Marzan, Silica-coating and hydrophobation of CTAB-stabilized gold nanorods, *Chem. Mater.* 18(10) (2006) 2465–2467.
50. J.-J. Zhang, Y.-G. Liu, L.-P. Jiang, J.-J. Zhu, Synthesis, characterizations of silica-coated gold nanorods and its applications in electroanalysis of hemoglobin, *Electrochem. Commun.* 10(3) (2008) 355–358.
51. J.-J. Zhang, M.-M. Gu, T.-T. Zheng, J.-J. Zhu, Synthesis of gelatin-stabilized gold nanoparticles and assembly of carboxylic single-walled carbon nanotubes/Au composites for cytosensing and drug uptake, *Anal. Chem.* 81(16) (2009) 6641–6648.
52. F. Wo, R. Xu, Y. Shao, Z. Zhang, M. Chu, D. Shi, S. Liu, A multimodal system with synergistic effects of magneto-mechanical, photothermal, photodynamic and chemo therapies of cancer in graphene-quantum dot-coated hollow magnetic nanospheres, *Theranostics* 6(4) (2016) 485–500.

53. X. Yan, B.S. Li, L.S. Li, Colloidal graphene quantum dots with well-defined structures, *Acc. Chem. Res.* 46(10) (2013) 2254–2262.
54. W. Li, R. Luo, X. Lin, A.D. Jadhav, Z. Zhang, L. Yan, C.-Y. Chan, X. Chen, J. He, C.-H. Chen, P. Shi, Remote modulation of neural activities via near-infrared triggered release of biomolecules, *Biomaterials* 65 (2015) 76–85.
55. P. Desai, R.R. Patlolla, M. Singh, Interaction of nanoparticles and cell-penetrating peptides with skin for transdermal drug delivery, *Mol. Membr. Biol.* 27(7) (2010) 247–259.
56. S. Peeters, M. Kitz, S. Preisser, A. Wetterwald, B. Rothen-Rutishauser, G.N. Thalmann, C. Brandenberger, A. Bailey, M. Frenz, Mechanisms of nanoparticle-mediated photomechanical cell damage, *Biomed. Opt. Express* 3(3) (2012) 435–446.
57. B.S. Lalonde, E. Boulais, J.J. Lebrun, M. Meunier, Visible and near infrared resonance plasmonic enhanced nanosecond laser optoporation of cancer cells, *Biomed. Opt. Express* 4(4) (2013) 490–499.
58. E. Vanzha, T. Pylaev, A. Prilepskii, A. Golubev, B. Khlebtsov, V. Bogatyrev, N. Khlebtsov, Cell culture surfaces with immobilized gold nanostars: a new approach for laser-induced plasmonic cell optoporation, in: E.A. Genina, V.V. Tuchin (Eds.) Saratov Fall Meeting 2016 *Optical Technologies in Biophysics and Medicine XVIII*, Proc. SPIE, 10336 (2017) 103360L.
59. R. Xiong, S.K. Samal, J. Demeester, A.G. Skirtach, S.C. De Smedt, K. Braeckmans, Laser-assisted photoporation: fundamentals, technological advances and applications, *Adv. Phys. X* 1(4) (2016) 596–620.
60. Y.R. Davletshin, J.C. Kumaradas, The role of morphology and coupling of gold nanoparticles in optical breakdown during picosecond pulse exposures, *Beilstein J. Nanotechnol.* 7 (2016) 869–880.
61. P. Chakravarty, W. Qian, M.A. El-Sayed, M.R. Prausnitz, Delivery of molecules into cells using carbon nanoparticles activated by femtosecond laser pulses, *Nat. Nanotechnol.* 5(8) (2010) 607–611.
62. V. Caprettini, M. Dipalo, G.C. Messina, L. Lovato, F. Tantussi, F. De Angelis, 3D plasmonic nanostructures for in-vitro applications in neuroscience and cell biology, *IEEE* 16th International Conference on Nanotechnology, 2016, pp. 491–493.
63. O. Akhavan, E. Ghaderi, S.A. Shirazian, Near infrared laser stimulation of human neural stem cells into neurons on graphene nanomesh semiconductors, *Colloids Surf. B Biointerfaces* 126 (2015) 313–321.
64. J.O. Winter, T.Y. Liu, B.A. Korgel, C.E. Schmidt, Recognition molecule directed interfacing between semiconductor quantum dots and nerve cells, *Adv. Mater.* 13(22) (2001) 1673–1677.
65. M.A. El Khakani, V. Le Borgne, B. Aïssa, F. Rosei, C. Scilletta, E. Speiser, M. Scarselli, P. Castrucci, M. De Crescenzi, Photocurrent generation in random networks of multiwall-carbon-nanotubes grown by an "all-laser" process, *Appl. Phys. Lett.* 95(8) (2009) 083114–083114.
66. S. Banerjee, S.S. Wong, In situ quantum dot growth on multiwalled carbon nanotubes, *J. Am. Chem. Soc.* 125(34) (2003) 10342–10350.
67. A.Y. Rwei, W. Wang, D.S. Kohane, Photoresponsive nanoparticles for drug delivery, *Nano Today* 10(4) (2015) 451–467.
68. R. Barillé, R. Janik, S. Kucharski, J. Eyer, F. Letournel, Photo-responsive polymer with erasable and reconfigurable micro- and nano-patterns: an in vitro study for neuron guidance, *Colloids Surf. B Biointerfaces* 88(1) (2011) 63–71.
69. M. Banghart, K. Borges, E. Isacoff, D. Trauner, R.H. Kramer, Light-activated ion channels for remote control of neuronal firing, *Nat. Neurosci.* 7(12) (2004) 1381–1386.
70. A.A. Beharry, G.A. Woolley, Azobenzene photoswitches for biomolecules, *Chem. Soc. Rev.* 40(8) (2011) 4422.
71. R.H. Kramer, D.L. Fortin, D. Trauner, New photochemical tools for controlling neuronal activity, *Curr. Opin. Neurobiol.* 19(5) (2009) 544–552.

72. D.L. Fortin, M.R. Banghart, T.W. Dunn, K. Borges, D.A. Wagenaar, Q. Gaudry, M.H. Karakossian, Z.S. Otis, W.B. Kristan, D. Trauner, R.H. Kramer, Photochemical control of endogenous ion channels and cellular excitability, *Nat. Methods* 5(4) (2008) 331–338.
73. 73 X. Yang, D.L. Rode, D.S. Peterka, R. Yuste, S.M. Rothman, Optical control of focal epilepsy in vivo with caged γ-aminobutyric acid, *Ann. Neurol.* 71(1) (2012) 68–75.
74. D. Wang, Z. Yu, J. Yan, F. Xue, G. Ren, C. Jiang, W. Wang, Y. Piao, X. Yang, Photolysis of caged-GABA rapidly terminates seizures in vivo: concentration and light intensity dependence, *Front. Neurol.* 8 (2017) 215.
75. M. Zhao, L.M. McGarry, H. Ma, S. Harris, J. Berwick, R. Yuste, T.H. Schwartz, Optical triggered seizures using a caged 4-aminopyridine, *Front. Neurosci.* 9(25) (2015).
76. R. Wieboldt, K.R. Gee, L. Niu, D. Ramesh, B.K. Carpenter, G.P. Hess, Photolabile precursors of glutamate: synthesis, photochemical properties, and activation of glutamate receptors on a microsecond time scale, *Proc. Natl Acad. Sci. U S A* 91(19) (1994) 8752–8756.
77. G.C.R. Ellis-Davies, Caged compounds: photorelease technology for control of cellular chemistry and physiology, *Nat. Methods* 4(8) (2007) 619–628.
78. D. Warther, S. Gug, A. Specht, F. Bolze, J.F. Nicoud, A. Mourot, M. Goeldner, Two-photon uncaging: new prospects in neuroscience and cellular biology, *Bioorg. Med. Chem.* 18(22) (2010) 7753–7758.
79. M. Matsuzaki, G.C.R. Ellis-Davies, T. Nemoto, Y. Miyashita, M. Iino, H. Kasai, Dendritic spine geometry is critical for AMPA receptor expression in hippocampal CA1 pyramidal neurons, *Nat. Neurosci.* 4(11) (2001) 1086–1092.
80. E. Fino, R. Araya, D.S. Peterka, M. Salierno, R. Etchenique, R. Yuste, RuBi-glutamate: two-photon and visible-light photoactivation of neurons and dendritic spines, *Front. Neural Circuits* 3 (2009) 2.
81. E.M. Rial Verde, L. Zayat, R. Etchenique, R. Yuste, Photorelease of GABA with visible light using an inorganic caging group, *Front. Neural Circuits* 2 (2008) 2.
82. P. Polak, O. Shefi, Nanometric agents in the service of neuroscience: manipulation of neuronal growth and activity using nanoparticles, *Nanomed. Nanotech. Biol. Med.* 11(6) (2015) 1467–1479.
83. J.A. Kim, N. Lee, B.H. Kim, W.J. Rhee, S. Yoon, T. Hyeon, T.H. Park, Enhancement of neurite outgrowth in PC12 cells by iron oxide nanoparticles, *Biomaterials* 32(11) (2011) 2871–2877.
84. J.S. Park, K. Park, H.T. Moon, D.G. Woo, H.N. Yang, K.-H. Park, Electrical pulsed stimulation of surfaces homogeneously coated with gold nanoparticles to induce neurite outgrowth of PC12 cells, *Langmuir* 25(1) (2009) 451–457.
85. C. Paviolo, A.H.A. Clayton, S.L. McArthur, P.R. Stoddart, Temperature measurement in the microscopic regime: a comparison between fluorescence lifetime- and intensity-based methods, *J. Microsc.* 250(3) (2013b) 179–188.
86. M. Poljak-Blazi, M. Jaganjac, N. Zarkovic, Cell oxidative stress: risk of metal nanoparticles, in: K. Sattler (Ed.) *Handbook of Nanophysics: Nanomedicine and Nanorobotics*, CRC Press, 2010, pp. 1–17.
87. E. Söderstjerna, P. Bauer, T. Cedervall, H. Abdshill, F. Johansson, U.E. Johansson, Silver and gold nanoparticles exposure to in vitro cultured retina – studies on nanoparticle internalization, apoptosis, oxidative stress, glial- and microglial activity, *PLoS ONE* 9(8) (2014) e105359.
88. C. Paviolo, J.W. Haycock, P.R. Stoddart, S.L. McArthur, Effects of laser-exposed gold nanorods on biochemical pathways of neuronal cells, in: J. Friend, H.H. Tan (Eds.) *Micro/Nano Materials, Devices, and Systems, Proc. SPIE*, 8923 (2013c) 89231A.
89. N.J. Siddiqi, M.A.K. Abdelhalim, A.K. El-Ansary, A.S. Alhomida, W.Y. Ong, Identification of potential biomarkers of gold nanoparticle toxicity in rat brains, *J. Neuroinflammation* 9(1) (2012) 123.
90. C.-P. Richter, A.I. Matic, J.D. Wells, E.D. Jansen, J.T. Walsh, Neural stimulation with optical radiation, *Laser Photon. Rev.* 5(1) (2011) 68–80.

91. A.C. Thompson, P.R. Stoddart, E.D. Jansen, Optical stimulation of neurons, *Curr. Mol. Imaging* 3 (2014) 162–177.
92. S. Shoham, N. Farah, L. Golan, Method and system for optical stimulation of neurons, U.S. Patent No. 2010/0262212A1 (2010).
93. J. Wells, C. Kao, K. Mariappan, J. Albea, E.D. Jansen, P. Konrad, A. Mahadevan-Jansen, Optical stimulation of neural tissue in vivo, *Opt. Lett.* 30(5) (2005) 504–506.
94. M.G. Shapiro, K. Homma, S. Villarreal, C.P. Richter, F. Bezanilla, Infrared light excites cells by changing their electrical capacitance, *Nat. Comm.* 3 (2012), 736.
95. J. Wells, C. Kao, P. Konrad, T. Milner, J. Kim, A. Mahadevan-Jansen, E.D. Jansen, Biophysical mechanisms of transient optical stimulation of peripheral nerve, *Biophys. J.* 93(7) (2007) 2567–2580.
96. A.C. Thompson, S.A. Wade, W.G.A. Brown, P.R. Stoddart, Modeling of light absorption in tissue during infrared neural stimulation, *J. Biomed. Opt.* 17(7) (2012), 1–7.
97. A.C. Thompson, S.A. Wade, N.C. Pawsey, P.R. Stoddart, Infrared neural stimulation: influence of stimulation site spacing and repetition rates on heating, *IEEE Trans. Biomed. Eng.* 60(12) (2013) 3534–3541.
98. A.L. Hodgkin, B. Katz, The effect of temperature on the electrical activity of the giant axon of the squid, *J. Physiol.* 109(1–2) (1949) 240–249.
99. J.M. Cayce, R.M. Friedman, E.D. Jansen, A. Mahavaden-Jansen, A.W. Roe, Pulsed infrared light alters neural activity in rat somatosensory cortex in vivo, *NeuroImage* 57(1) (2011) 155–166.
100. M. Plaksin, E. Kimmel, S. Shoham, Correspondence: revisiting the theoretical cell membrane thermal capacitance response, *Nat. Comm.* 8(1) (2017a) 1431.
101. M. Plaksin, E. Kimmel, S. Shoham, Thermal transients excite neurons through universal intramembrane mechano-electrical effects, *Phys. Rev. X* 8(1) (2018b), 11043.
102. Q. Liu, M.J. Frerck, H.A. Holman, E.M. Jorgensen, R.D. Rabbitt, Exciting cell membranes with a blustering heat shock, *Biophys. J.* 106(8) (2014) 1570–1577.
103. E.J. Peterson, D.J. Tyler, Activation using infrared light in a mammalian axon model, *Annual International Conference of the IEEE Engineering in Medicine and Biology Society*, 2012, pp. 1896–1899.
104. E.S. Albert, J.M. Bec, G. Desmadryl, K. Chekroud, C. Travo, S. Gaboyard, F. Bardin, I. Marc, M. Dumas, G. Lenaers, C. Hamel, A. Muller, C. Chabbert, TRPV4 channels mediate the infrared laser-evoked response in sensory neurons, *J. Neurophys.* 107(12) (2012) 3227–3234.
105. X.Y. Li, J. Liu, S.S. Liang, K.W. Guan, L.J. An, X.F. Wu, S. Li, C.S. Sun, Temporal modulation of sodium current kinetics in neuron cells by near-infrared laser, *Cell Biochem. Biophys.* 67(3) (2013) 1409–1419.
106. G.M. Dittami, S.M. Rajguru, R.A. Lasher, R.W. Hitchcock, R.D. Rabbitt, Intracellular calcium transients evoked by pulsed infrared radiation in neonatal cardiomyocytes, *J. Physiol.* 589(6) (2011) 1295–1306.
107. C. Paviolo, P.R. Stoddart, Metallic nanoparticles for peripheral nerve regeneration: is it a feasible approach?, *Neural Regen. Res.* 10(7) (2015) 1065–1066.
108. H. Nakatsuji, T. Numata, N. Morone, S. Kaneko, Y. Mori, H. Imahori, T. Murakami, Thermosensitive ion channel activation in single neuronal cells by using surface-engineered plasmonic nanoparticles, *Angew Chem.* 54(40) (2015) 11725–11729.
109. H.T. Beier, G.P. Tolstykh, J.D. Musick, R.J. Thomas, B.L. Ibey, Plasma membrane nanoporation as a possible mechanism behind infrared excitation of cells, *J. Neural Eng.* 11(6) (2014), 066006.
110. J. Baumgart, L. Humbert, E. Boulais, R. Lachaine, J.J. Lebrun, M. Meunier, Off-resonance plasmonic enhanced femtosecond laser optoporation and transfection of cancer cells, *Biomaterials* 33(7) (2012) 2345–2350.
111. L. Tong, Y. Zhao, T.B. Huff, M.N. Hansen, A. Wei, J.X. Cheng, Gold nanorods mediate tumor cell death by compromising membrane integrity, *Adv. Mater.* 19(20) (2007) 3136–3141.

112. A.S. Urban, M. Fedoruk, M.R. Horton, J. Radler, F.D. Stefani, J. Feldmann, Controlled nanometric phase transitions of phospholipid membranes by plasmonic heating of single gold nanoparticles, *Nano Lett.* 9(8) (2009) 2903–2908.
113. P. Urban, S.R. Kirchner, C. Muhlbauer, T. Lohmuller, J. Feldmann, Reversible control of current across lipid membranes by local heating, *Sci. Rep.* 6 (2016), 22686.
114. F. Barandeh, P.L. Nguyen, R. Kumar, G.J. Iacobucci, M.L. Kuznicki, A. Kosterman, E.J. Bergey, P.N. Prasad, S. Gunawardena, Organically modified silica nanoparticles are biocompatible and can be targeted to neurons in vivo, *PLoS ONE* 7(1) (2012) e29424.
115. G.A. Silva, Neuroscience nanotechnology: progress, opportunities and challenges, *Nat. Rev. Neurosci.* 7(1) (2006) 65–74.
116. S. Jung, M. Bang, B.S. Kim, S. Lee, N.A. Kotov, B. Kim, D. Jeon, Intracellular gold nanoparticles increase neuronal excitability and aggravate seizure activity in the mouse brain, *PLoS ONE* 9(3) (2014) e91360.
117. W.W. Li, X.Y. Chen, Gold nanoparticles for photoacoustic imaging, *Nanomedicine* 10(2) (2015) 299–320.
118. E. Boulais, R. Lachaine, A. Hatef, M. Meunier, Plasmonics for pulsed-laser cell nanosurgery: fundamentals and applications, *J. Photochem. Photobiol. C* 17 (2013) 26–49.
119. S. Johannsmeier, P. Heeger, M. Terakawa, S. Kalies, A. Heisterkamp, T. Ripken, D. Heinemann, Gold nanoparticle-mediated laser stimulation induces a complex stress response in neuronal cells, *Sci. Rep.* 8 (2018) 6533.

CHAPTER 14

Optical Stimulation of Neural Circuits in Freely Moving Animals

Leore R. Heim and Eran Stark

CONTENTS

14.1 INTRODUCTION

Despite all that we have learned about the nervous system, from molecules to networks, we still cannot draw the schematics of, for example, curiosity. Nevertheless, bridging the gap between psychology and biology is a key goal for the neuroscience community (Human Brain Project, n.d.). In order to actually link a behavioral function to its underlying neural mechanisms, animal models must be free from anesthesia and constraints, thus allowing neural circuits to be probed during the behaviors that engage them. Hence, bridging the psychology-biology gap depends upon freely moving animals (Figure 14.1).

Merely observing which aspects of neural activity are correlated with a given behavior may provide invaluable information about the system, but to causally define it, the underlying circuit dynamics must be manipulated. Over the two centuries since Galvani established electricity as the "language" of the nervous system (Galvani, 1791), investigators have continuously perfected our capability to "converse" (i.e. record and manipulate) with neurons (Patil and Thakor, 2016). However, electrical stimulation, a canonical tool for manipulating neuronal activity, exhibits well-documented

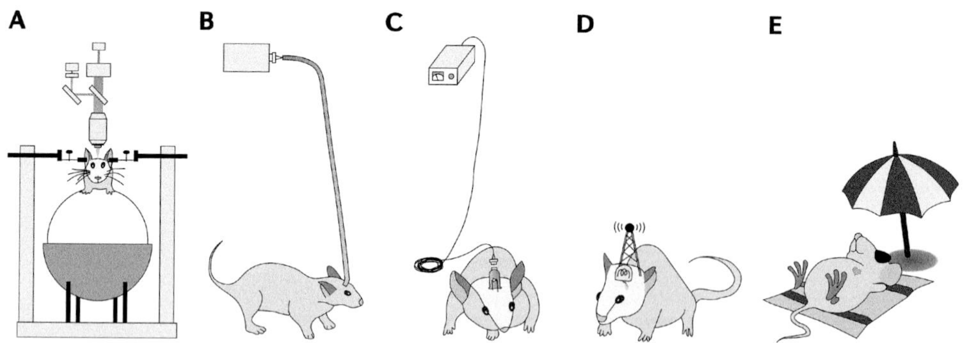

FIGURE 14.1 Concepts of optical stimulation systems. The delivery of light to the brains of behaving rodents can be divided into strategies based on the provided degrees of freedom. **(A)** Benchtop opto-stimulators (e.g. a two-photon microscope) can be used to record and/or manipulate neuronal activity in head-fixed mammals. (Redrawn with permission from Figure 14.1A in Dombeck et al., 2007.) **(B)** Animals large enough to receive implants can be tethered to optical fibers (Gradinaru et al., 2007). **(C)** By placing the light source/s on or in the head of the animal, rigid optical fibers can be replaced with flexible electrical wires, upscaling the number of independently controlled targets and providing many more degrees of freedom (Stark et al., 2012). **(D)** Head-mounted light sources can be powered without a tether, by a battery or wireless power transfer (Wentz et al., 2011). **(E)** Closing the loop between neural activity and optical stimulation on the head may enable pacemaker-like user-independent systems.

disadvantages (Tehovnik, 1996), including: (1) limited spatial resolution: we can either stimulate a few isolated neurons or an ill-defined cluster of cells, and (2) low specificity: it is almost impossible to control which cell type is excited. These detriments have been overcome by utilizing light as a stimulating agent (Figure 14.2).

In 2002, Zemelman and colleagues demonstrated that neurons can be genetically manipulated into becoming light-sensitive. Three years later, "optogenetics" took the neuroscience community by storm when the single-component, light-sensitive cation channel Channelrhodopsin-2 (ChR2) was expressed in neurons (Boyden et al., 2005). The ability to selectively activate and/or silence a genetically defined population of neurons with millisecond temporal precision and high throughput, in the absence of direct contact between stimulus and target, marked a new era in neural manipulation.

Since then, continuous efforts have been made to fulfill the promise of optogenetics. From the hardware perspective, the concept is simple: we must shed the right amount of light at the right place at the right time. In reality, the task has proven arduous: the ideal light delivery device should be able to utilize various tunable wavelengths to illuminate, with cellular spatial resolution and sub-millisecond temporal precision, multiple targets positioned deep within the brain while simultaneously recording multi-neuronal activity. And, all this must be done in freely moving animals. Given the technology available today, there are many tradeoffs between these requirements; some of them are even mutually exclusive. Consequently, devices for optical stimulation have typically been designed to answer a pre-defined set of research questions, and no single strategy can be used for all purposes.

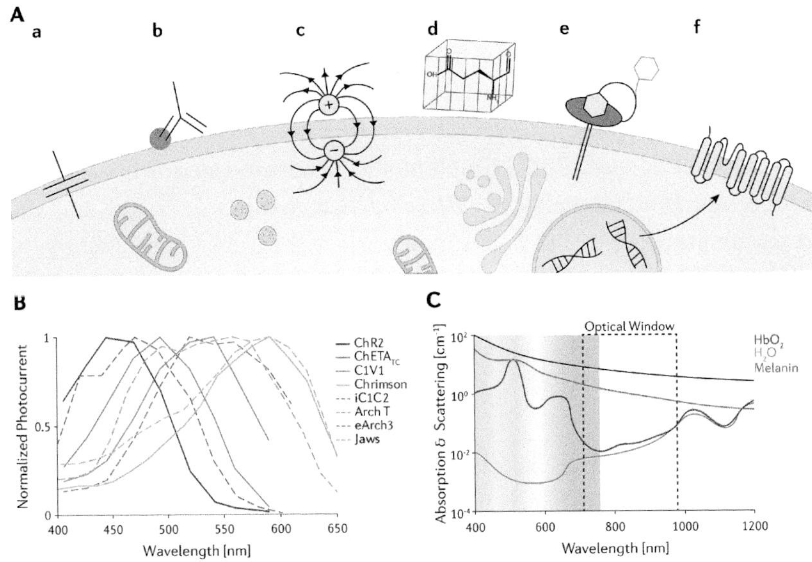

FIGURE 14.2 Optical control of neural activity. **(A)** Optical stimulation methods. **(a)** Infra-red (IR) radiation absorbed by water produces a temperature increase, which transiently alters membrane capacitance and generates depolarizing currents (Shapiro et al., 2012). **(b)** Photothermal effects can be augmented by gold nanoparticles conjugated to antibodies (Carvalho-de-Souza et al., 2015). **(c)** Optoelectric devices (e.g. quantum dots) act as photo-responsive electrodes that generate an electric field when exposed to light (Bareket-Keren and Hanein, 2014). **(d)** Bioactive compounds (e.g. glutamate) can be "caged" within photolabile protecting groups (Callaway and Katz, 1993). **(e)** Photoisomerizable molecules which undergo a reversible conformational change in response to the absorption of two different wavelengths can be tethered to a target protein (e.g. receptor), yielding a photosensitive actuator (Volgraf et al., 2006). **(f)** Microbial light-sensitive channels (opsins) can be encoded within the mammalian DNA (Boyden et al., 2005). **(B)** Action spectra of selected depolarizing (solid lines) and hyperpolarizing (dashed lines) opsins. Note coverage of the visible spectrum. Data adapted from Berndt et al., 2014; Chuong et al., 2014; Klapoetke et al., 2014; Mattis et al., 2012. **(C)** Absorption (colored lines) and scattering (black line) coefficients of biological tissue. Note that the ideal optical window is at near- and infra-red wavelengths (~650–950 nm, dashed box), which is distinct from the action spectra of microbial opsins. (Data adapted from Jacques, 2013.)

The goal of this chapter is to describe current approaches for optical stimulation of neural circuits in freely behaving animals. Specifically, we focus on opsin-based optogenetics and limit the exposition to miniaturized devices compatible with rodents, since these designs can then be adapted to larger animals. The content is organized according to the degrees of freedom for animal movement that the different techniques provide (Figure 14.1).

14.2 HEAD-FIXED

We begin our discussion on optical stimulation with optical recordings, since the two should be distinguished in the context of freely moving animals. Optical imaging of genetically encoded biosensors is complementary to electrical recordings in a similar way that optical stimulation in the context of optogenetics is complementary to electrical stimulation, i.e. it may provide better spatial resolution and specificity. However, one technical problem hinders imaging in freely moving animals, namely, the microscope. The detection of intrinsic signals and/or fluorescence requires optics that are difficult to miniaturize, and some optical recording methods (e.g. two-photon microscopy) also require high-energy, active photostimulation (Dombeck et al., 2007). Implantable, head-mounted "miniscopes" are under continuous development (Hamel et al., 2015; "UCLA Miniscope," n.d.), but such devices have yet to demonstrate optogenetic stimulation capabilities. Consequently, to preserve conventionally sized instrumentation, all-optical neurophysiology, i.e. using light to record and manipulate neural activity, has only been applied to freely moving worms (Leifer et al., 2011) or head-fixed mammals (Packer et al., 2015; Rickgauer et al., 2014; Figure 14.1A).

14.3 FIBER-TETHERED

The first freely moving rodents to be optogenetically manipulated were tethered to an optical fiber (Aravanis et al., 2007; Figure 14.1B, Figure 14.3, Figure 14.7Aa). With one end of the waveguide coupled to an external light source and the other inserted into the brain, this simple interface has become a standard approach for integrating optogenetics with behavioral studies.

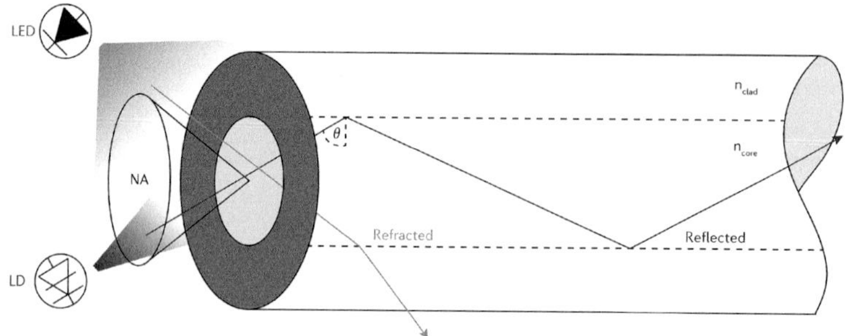

FIGURE 14.3 Optical waveguides. At the border between two media with different refractive indices (propagate light at different speeds), if the angle of the incident light ray is smaller than a particular critical angle (θ), light is transmitted through the second medium with a change in direction (refracted, blue); otherwise, it is reflected (red). The waveguide used most often in neuroscience research is the optical fiber (Keiser et al., 2014), composed of a core and cladding layer with different refractive indices $\left(n_{\text{core}} > n_{\text{clad}}\right)$. Light transmitted through the core is trapped by reflection at the core-cladding interface, allowing it to propagate along the fiber with minimal losses. Light is coupled into the fiber if it falls within its acceptance cone, described by the numerical aperture: $\text{NA} = \sqrt{n^2_{\text{core}} - n^2_{\text{clad}}}$. Efficient source-fiber coupling is greater when the source has a small area, when the fiber has a large input diameter and high NA, when the source and fiber are placed in close proximity, and when the source, medium, and fiber have as similar refractive indices as possible (Stark et al., 2012). Both light-emitting diodes (LEDs) and laser diodes (LDs) can be coupled to fibers, but since LDs produce ~Gaussian (i.e. coherent) beams and LEDs present ~Lambertian (i.e. wide angle) emission profiles, the coupling efficiency of LDs is at least an order of magnitude higher than of LEDs (Stark et al., 2012). Conversely, LDs may be larger, consume more current, and generate more heat than LEDs. See also Figure 14.6.

Although light can affect the nervous system without even touching the animal (Lima and Miesenböck, 2005), one of the greatest potentials of light as a stimulating agent – non-invasiveness – has limited practical applicability due to the optical properties of biological tissue. Even when the skull is removed and the furthest-reaching wavelengths (i.e. infra-red; Figure 14.2B) are utilized, surface illumination cannot penetrate deeper than ~500 µm while maintaining cellular spatial resolution, and is therefore typically accompanied by cortical excavation (Rickgauer et al., 2014). In contrast, waveguides not only release the animal from the bench, but also grant access to the entire depth of the nervous system while displacing no more than their own volume of tissue.

To better confine the distribution of exiting light and alleviate some of the pressure on neural tissue, waveguide dimensions can be greatly reduced by *de novo* fabrication (Wu et al., 2013), chemical etching in acidic solution (Royer et al., 2010), machining (Ozden et al., 2013), or heating and pulling ("tapering": Godwin et al., 1997, Figure 14.4A). Narrowing a single end of a fiber is advantageous also in terms of coupling efficiency: the larger input aperture can accept more light, while the other, smaller, output aperture can produce higher irradiance.

Fiber bundles, comprised of 1–100,000 fiber cores sharing a common cladding, have been used as brain-insertable microendoscopes, where each fiber core transmits one "pixel" of the final image (Flusberg et al., 2005). However, fiber bundles can also relay light in the other direction (Figure 14.5). For instance, multiple targets can be sequentially

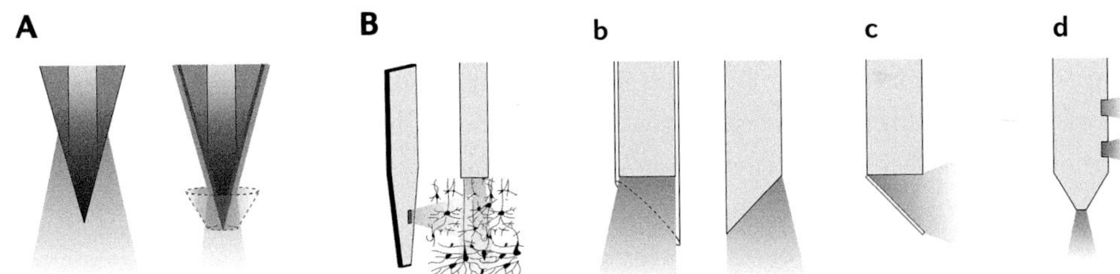

FIGURE 14.4 Waveguide manipulation. **(A)** During tapering of an optical fiber (left), core and cladding might be melted and mixed together, resulting in a homogeneous medium that can no longer serve as a waveguide. Consequently, a sizeable fraction of the light originally confined in the core is lost through the sidewalls of the taper (Godwin et al., 1997). This byproduct can be bypassed by coating the tapered fiber with reflective material (right), which can in turn serve as a recording electrode (Zhang et al., 2009). A problem with such coaxial optrodes is that the electronically monitored (within dashed line) and illuminated (blue shade) regions do not fully overlap. **(B)** Solutions to direct emitted light. **a.** µLEDs can provide lateral (laminar-specific) illumination (left), whereas waveguides preferentially illuminate a cone of tissue below the waveguide (right). **b.** Side-biased illumination can be produced by encasing a fiber in a beveled metal cannula (left; Tye et al., 2011) or by machining the fiber tip (right; Pashaie and Falk, 2013). **c.** Side-limited illumination can be achieved by placing a miniature mirror at the waveguide end (Zorzos et al., 2010). **d.** Multi-site illumination using a single waveguide is produced by etching multiple apertures and manipulating the coupling strategy (e.g. the power and angle of input light; Pisanello et al., 2014) or by wavelength division multiplexing (Segev et al., 2015).

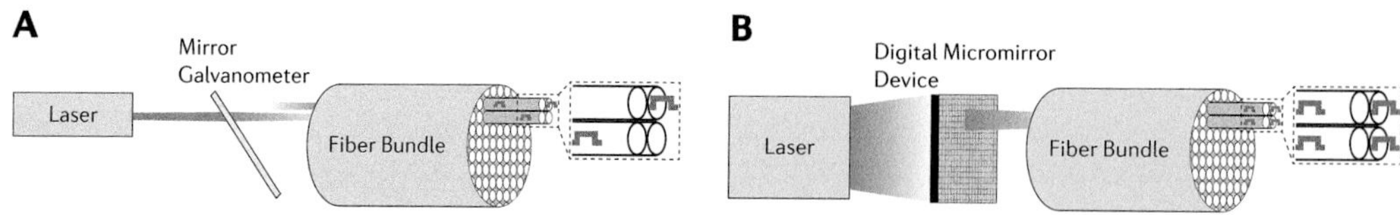

FIGURE 14.5 Patterned illumination using fiber bundles. **(A)** A mirror galvanometer, resonant scanner, or acousto-optic device can rapidly scan a single focused laser beam across multiple positions. **(B)** Parallel illumination techniques allow all targets to be illuminated simultaneously via spatial light modulators (e.g. a digital micromirror device). The temporal resolution of the latter approach is inherently higher, but available light power is divided between multiple targets, limiting the number of targets that can be illuminated at a given input power. All strategies based on fiber bundles require surface mounting, cortical excavation, and/or considerable volume of compressed/displaced neuronal tissue.

activated by rapidly positioning a galvanic mirror across different cores of the fiber bundle (Hayashi et al., 2012), or alternatively, they can be simultaneously activated by passing the laser beam through a digital micromirror device (DMD: Szabo et al., 2014; Zorzos et al., 2012). These two approaches for controlling multiple targets represent some of the possibilities for patterned illumination (Emiliani et al., 2015).

14.4 WIRE-TETHERED

Although optical fibers are commercially available and relatively straightforward to use, they are usually made of glass (silica) and thus stiff and intolerable to sharp bends. Consequentially, optical fibers tethered to a behaving animal may suffer from decreased transmission efficiency, irreversible physical damage (breakage), restriction of animal movement, or all three. While mechanical compliance can be improved using a rotary joint (optical "commutator"; Gradinaru et al., 2007), or fabricating flexible polymer fibers (Canales et al., 2015), such solutions are not scalable with respect to the number of independently illuminated targets.

Compared to optical waveguides, electrical wires provide many more degrees of freedom. Substituting the fiber with a wire enables using miniaturized light sources (light emitting diodes, LEDs, or laser diodes, LDs; Figure 14.6), moved from the bench onto the animal's head. Light can then reach the neuronal target through a cranial window (Huber et al., 2008) or, if deeper regions and/or higher resolutions are of interest, via optical fibers (Stark et al., 2012), or fabricated waveguides (Kampasi et al., 2016; Schwaerzle et al., 2013) confined to the implant. In this manner, instead of delivering light from a remote light source via rigid fibers, electricity can be delivered from a remote power source via highly flexible lightweight wires (e.g. multi-stranded Litz wires: Stark et al., 2012).

Moving the light source from the bench to the head is one conceptual leap; the natural next step is to insert the light source into the brain of the animal. In 2013, advances in micromachining techniques enabled the production of miniature light sources, i.e. μLEDs (Cao et al., 2013; Kim et al., 2013; McAlinden et al., 2013). When implanted within the brain, μLEDs present three unique properties. First, they bypass the delicate process of

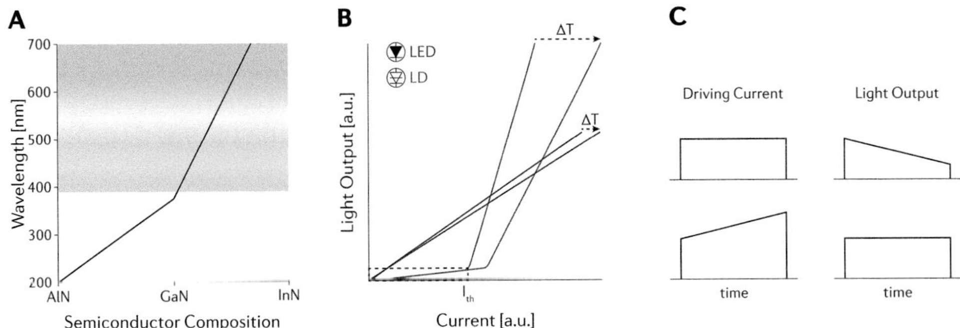

FIGURE 14.6 LDs and LEDs. **(A)** In LEDs, the emitted light wavelength depends on the energy gap between the p- and n-type semiconductors. Thus, a wide spectrum of colors from red (644 nm) down to UV (197 nm) can be achieved by varying the materials forming the diode. Semiconductors typically used include aluminum nitride (AlN), gallium nitride (GaN), indium nitride (InN) and their alloys. Values adapted from Schubert, 2006. **(B)** Current-to-light conversion (I-P transfer function) for LEDs (red) and LDs (blue) at two different temperatures. Although both LEDs and LDs exhibit a non-linear voltage-to-current relationship (V-I transfer function, not shown), these two diode types differ in the way they convert electric power to light. The I-P relationship of LEDs (red) is nearly linear (until the maximal light output), simplifying its control. In contrast, LDs (blue) initially behave as typical LEDs but then, upon crossing a threshold (I_{th}) for stimulated emission, rapidly increase their light output. Even though the light output at this "sub-threshold" domain (dashed box) is minuscule, it may affect low-light applications (e.g. optogenetics) and hence cannot be neglected. **(C)** I-P transfer functions, especially of LDs, are strongly dependent on device temperature, resulting in light power variance due to a poorly-controlled variable in a positive feedback loop: as temperature rises, more current must be applied to maintain light output constant, which further increases the heat. Temperature can be easily controlled on the bench, but is more challenging to stabilize in miniature implanted devices.

optical coupling by eliminating the need for a waveguide. Second, due to their size and lack of coupling losses, μLEDs have very low power requirements compared to LEDs or lasers, and efficient optical stimulation can be achieved with nano-watt scale light (Wu et al., 2015). Third, the illumination path emitted from on-probe μLEDs is perpendicular to the diode surface, providing a natural way to stimulate cells in a laminar-specific manner (Figure 14.4B). Two major concerns with implanted diodes are (1) tissue integrity, which strongly affects the performance of all implantable devices (Figure 14.7); and (2) thermal management, since all of the electrical energy that is not converted to light is released to the tissue as heat (typically over 95%; Wu et al., 2015). Nevertheless, experiments with mice implanted chronically with multi-site diode-probe arrays (Stark et al., 2015) and with μLED-probe arrays (Wu et al., 2015) have shown that the generated heat by itself is insufficient to modify neuronal activity.

14.5 WIRELESS

Flexible as they may be, electrical wires constrain animal movement, are at constant risk of damage, and limit the range of possible behaviors. The only way to circumvent these shortcomings is to remove the tether and provide power to head-mounted light sources by

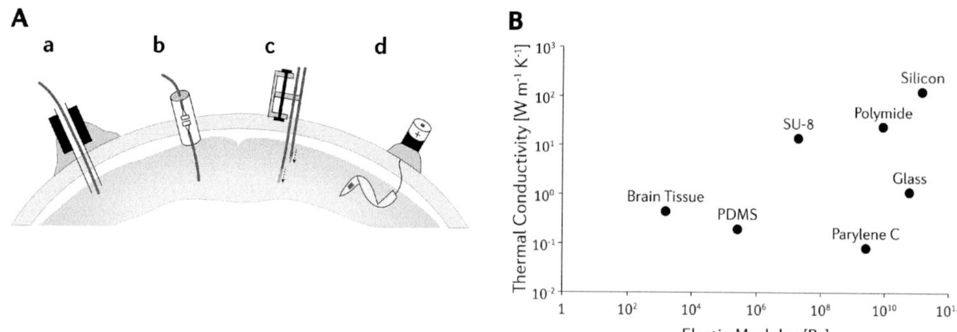

FIGURE 14.7 Implant-brain mechanical interaction. (A) Implantation strategies. **(a)** Removable fiber. A metal tube ("cannula") can be fixed to the skull, through which an optical fiber is inserted at the beginning of every experiment (Aravanis et al., 2007). **(b)** Permanently-implanted fiber. A mating sleeve ("ferrule") can be used to connect two fibers, one fixed inside the brain, the other coupled to an on-bench light source, thus reducing the tissue damage caused by repeated insertions (Sparta et al., 2012). **(c)** Movable fiber/s. A microdrive capable of independently maneuvering multiple devices (optical fibers, diode-tetrode, etc.) can be used to modify the depth of each device over extended durations (Stark et al., 2012). **(d)** Fiber-less devices. Instead of a stationary implant secured to the skull, an entire device can be implanted within the parenchyma, permitting movement with the brain rather than with the skull (Jeong et al., 2015). (B) Thermal conductivity and rigidity (elastic modulus) of the materials used for neural implants are major concerns when optimizing tissue integrity. (Data adapted from Hassler et al., 2011; Kaster et al., 2011; Scholten and Meng, 2015; Wu et al., 2015.)

a head-mounted battery (Iwai et al., 2011; Jeong et al., 2015) or via wireless power transfer (WPT; Figure 14.1D). WPT technologies can be divided into three strategies, according to the properties of the applied electromagnetic energy (Table 14.1). (1) At a certain distance from the emitting source ($r>2\lambda$), the "far-field", electromagnetic energy radiates as planar electric and magnetic waves perpendicular to each other and to the direction of propagation, and can be harvested using radio-frequency scavenging (Kim et al., 2013; Park et al., 2015). (2) At distances close to the emitting source ($r<<\lambda$), the "near-field", emitted energy is stored as either electric or magnetic fields, and can be harnessed via inductive coupling (Wentz et al., 2011). (3) At the transition zone between the near- and far-fields ($r\sim\lambda$), the "midfield", electromagnetic irradiance can be tailored to the dielectric properties of the animal, and thus provide power via inductive coupling to micro-implants (Ho et al., 2014; Montgomery et al., 2015).

TABLE 14.1 Wireless Power Transfer Techniques

	Near-field	**Midfield**	**Far-field**
EM source	Coils (LC circuits)	Resonant cavity (metal plates)	Antennas
Transmission frequency	<20 MHz	1.5 GHz	910 MHz
Dimensions	2 g, 1 cm^3	20 mg, 10 mm^3	~1 g, ~20 mm^3
Freedom of movement	20-cm diameter arena	21-cm diameter arena	~1 m
Reference	Wentz et al. (2011)	Montgomery et al. (2015)	Kim et al. (2013)

Despite these advances, currently available tether-free designs fail to provide the desired experimental freedom; WPT technologies suffer from low efficiencies and require confined environments (e.g. 21-cm diameter: Montgomery et al., 2015), while batteries add considerable weight to the animal's head and limit the duration of experiments. Moreover, when combining electrical readout with optical stimulation, removing the tether necessitates the recorded data to be stored on the animal's head and/or transferred telemetrically. A potentially promising third direction for wireless devices is based on biological energy harvesting (e.g. muscle movements; Dagdeviren et al., 2014).

14.6 ELECTRICAL READOUT

Manipulating neurons with light can make a fly jump (Lima and Miesenböck, 2005), a worm withdraw (Nagel et al., 2005), a mouse turn (Huber et al., 2008), and a monkey gaze sideways (Cavanaugh et al., 2012). Yet, understanding how this happens requires more than simply measuring the end result; neuronal activity must be registered. Electrical recordings enjoy a ground-truth status since they are inherent to neural communication, and indeed, many of the classical electrical recording techniques have been used alongside a separate optical stimulation device (e.g. tungsten electrode and optical fiber, Figure 14.9A; Han et al., 2009). "Optrodes", electrodes supplemented with optical stimulation capabilities, can be produced by gluing two instruments together (e.g. Gradinaru et al., 2007; Figure 14.8A) or by combining them in a single multifunctional apparatus (e.g. Wu et al., 2015; Figure 14.8B,C).

Combining electrical recordings with optical stimulation in an implantable device requires more than a compact mechanical design and an efficient optical interface. The proximity of the light sources and the recording electrodes gives rise to two distinct types of artifacts which resemble spontaneous neural activity (local field potentials and spikes) and hinders the interpretation of the recorded data. First, when light directly hits a metal electrode, a "light artifact" generated by the photovoltaic (Bequerel) effect may arise (Han et al., 2009; Figure 14.9A). Second, when current passes through a light source close to an un-buffered neuronal signal, electromagnetic interference (EMI) caused by capacitive coupling may produce artifacts (Stark et al., 2012; Figure 14.9B). Light artifacts are mitigated using low-power light (Stark et al., 2012), deflecting light away from the metal electrodes (Cardin et al., 2010), and/or using non-metal/specialized electrodes (Zorzos et al., 2011). EMI artifacts can be removed by offline adaptive filtering (Wu et al., 2015), or avoided completely by shielding each light source within a grounded Faraday cage (Stark et al., 2012), or by buffering the neuronal signal before the EMI source.

14.7 OUTLOOK

Once electrical recordings are combined with optical stimulation in a compact, noise-free device, the output of (or the activity in) a neuronal circuit can be harnessed to control the input it receives in a closed-loop manner (Grosenick et al., 2015; Stark et al., 2012). Such a closed-loop, "user-independent" system is especially appealing when applied to an un-tethered subject, enabling complete freedom from the experimental rig (Figure 14.1E). Since the late 1950s pacemakers have been implanted in the heart of patients (Kusumoto

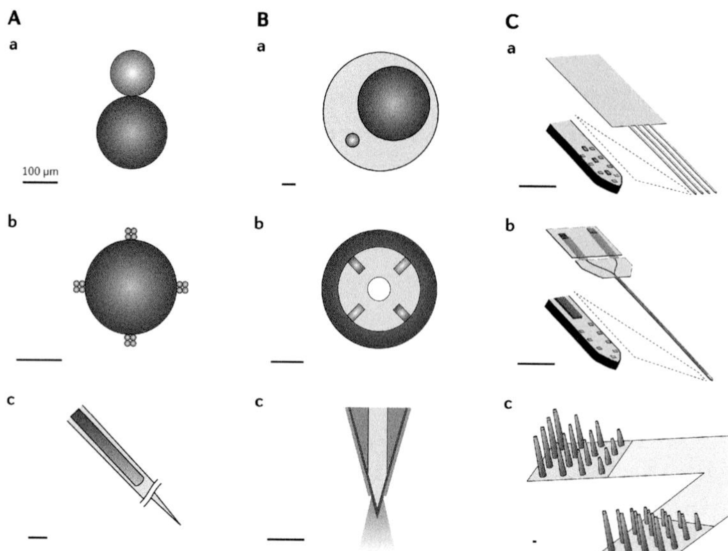

FIGURE 14.8 Optrode designs. Optical stimulation devices combined with electrical recording capabilities. Scale bars, 100 μm. **(A)** Commercially available optical fibers. **a.** Optrode constructed by gluing a 200 μm diameter optical fiber to a 125 μm diameter tungsten electrode (Gradinaru et al., 2007). **b.** "Optetrode", constructed by fixing four tetrode bundles to a single optical fiber (Anikeeva et al., 2012; English et al., 2012). **c.** "Optopatcher", constructed by inserting an optical fiber through a glass pipette (Katz et al., 2013). **(B)** Specialized optical fibers. **a.** Dual-core fiber, with a hollow core for extracellular recordings produced near a light-propagating optical core (LeChasseur et al., 2011). **b.** Multi-core, all-polymer flexible fiber, produced with integrating electrode material (Canales et al., 2015). **c.** Concentric electrode and fiber, produced by coating the tapered tip of an optical fiber with gold film (Zhang et al., 2009). **(C)** Fiber-less designs. **a.** Multi-shank (Michigan-like) array. Each monolithic shank is embedded with three neuron-sized μLEDs and eight Ti/Ir recording sites (Wu et al., 2015). **b.** Multi-color monolithic array, produced by coupling two LDs to a common waveguide via GRIN lens. This design is of special interest as it allows the same spot to be illuminated by different wavelengths, enabling, for instance, activating and silencing the same cell (Kampasi et al., 2016). **c.** Multi-electrode (Utah-like) array, containing multiple coaxial optrodes (Kwon et al., 2015).

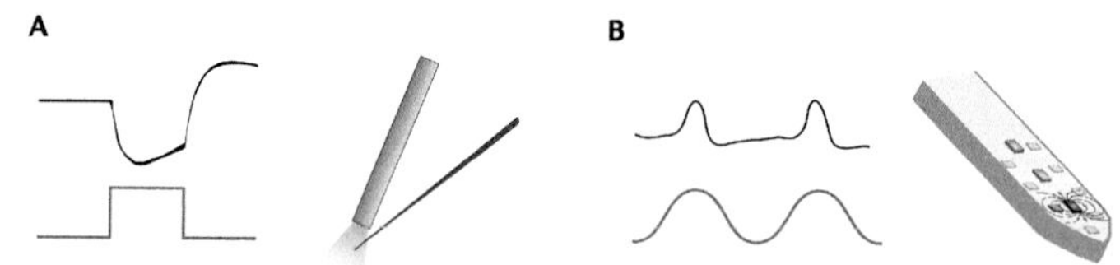

FIGURE 14.9 Opto-electrical artifacts. Combining optical manipulations with electrophysiological recordings can give rise to two distinct types of artifacts. **(A)** Photovoltaic artifacts. Direct illumination (blue line, applied voltage) of metal electrodes can induce large potentials on the recording electrodes (black line, recorded signal; Han et al., 2009). **(B)** Electromagnetic artifacts. When a diode (or any other load), is located close to an un-buffered neural potential recorded by an electrode, artifacts can be caused by electromagnetic interference (EMI). Any dielectric (including air) causes capacitive coupling of the light source and the electrode, and thus the resulting artifacts resemble the waveform of the current $\left(i = C \cdot dV / dt\right)$ passing through the diode.

and Goldschlager, 1996). In essence, pacemakers are wireless stimulators that can operate in a metronomical fashion or in response to an event sensed by the device itself (e.g. conduction failure). An equivalent opto-electrical implant, capable of sensing and blocking epileptic seizures (Krook-Magnuson et al., 2013), or supplementing/substituting a brain region permanently damaged following a stroke, may provide invaluable scientific insights and, in the long term, beneficial clinical applications.

REFERENCES

Anikeeva, P., Andalman, A.S., Witten, I., Warden, M., Goshen, I., Grosenick, L., Gunaydin, L.A., Frank, L.M., Deisseroth, K., 2012. Optetrode: a multichannel readout for optogenetic control in freely moving mice. *Nat. Neurosci.* 15, 163–170. doi:10.1038/nn.2992

Aravanis, A.M., Wang, L.-P., Zhang, F., Meltzer, L.A., Mogri, M.Z., Schneider, M.B., Deisseroth, K., 2007. An optical neural interface: in vivo control of rodent motor cortex with integrated fiberoptic and optogenetic technology. *J. Neural Eng.* 4, S143. doi:10.1088/1741-2560/4/3/S02

Bareket-Keren, L., Hanein, Y., 2014. Novel interfaces for light directed neuronal stimulation: advances and challenges. *Int. J. Nanomedicine* 9 Supplement 1, 65–83. doi:10.2147/IJN.S51193

Berndt, A., Lee, S.Y., Ramakrishnan, C., Deisseroth, K., 2014. Structure-guided transformation of channelrhodopsin into a light-activated chloride channel. *Science* 344, 420–424. doi:10.1126/science.1252367

Bovetti, S., Fellin, T., 2015. Optical dissection of brain circuits with patterned illumination through the phase modulation of light. *J. Neurosci. Methods* 241, 66–77. doi:10.1016/j.jneumeth.2014.12.002

Boyden, E.S., Zhang, F., Bamberg, E., Nagel, G., Deisseroth, K., 2005. Millisecond-timescale, genetically targeted optical control of neural activity. *Nat. Neurosci.* 8, 1263–1268. doi:10.1038/nn1525

Callaway, E.M., Katz, L.C., 1993. Photostimulation using caged glutamate reveals functional circuitry in living brain slices. *Proc. Natl. Acad. Sci. U. S. A.* 90, 7661–7665.

Canales, A., Jia, X., Froriep, U.P., Koppes, R.A., Tringides, C.M., Selvidge, J., Lu, C., Hou, C., Wei, L., Fink, Y., Anikeeva, P., 2015. Multifunctional fibers for simultaneous optical, electrical and chemical interrogation of neural circuits in vivo. *Nat. Biotechnol.* 33, 277–284. doi:10.1038/nbt.3093

Cao, H., Gu, L., Mohanty, S.K., Chiao, J.C., 2013. An integrated μLED optrode for optogenetic stimulation and electrical recording. *IEEE Trans. Biomed. Eng.* 60, 225–229. doi:10.1109/TBME.2012.2217395

Cardin, J.A., Carlén, M., Meletis, K., Knoblich, U., Zhang, F., Deisseroth, K., Tsai, L.-H., Moore, C.I., 2010. Targeted optogenetic stimulation and recording of neurons in vivo using cell-type-specific expression of channelrhodopsin-2. *Nat. Protoc.* 5, 247–254. doi:10.1038/nprot.2009.228

Carvalho-de-Souza, J.L., Treger, J.S., Dang, B., Kent, S.B.H., Pepperberg, D.R., Bezanilla, F., 2015. Photosensitivity of neurons enabled by cell-targeted gold nanoparticles. *Neuron* 86, 207–217. doi:10.1016/j.neuron.2015.02.033

Cavanaugh, J., Monosov, I.E., McAlonan, K., Berman, R., Smith, M.K., Cao, V., Wang, K.H., Boyden, E.S., Wurtz, R.H., 2012. Optogenetic inactivation modifies monkey visuomotor behavior. *Neuron* 76, 901–907. doi:10.1016/j.neuron.2012.10.016

Chuong, A.S., Miri, M.L., Busskamp, V., Matthews, G.A.C., Acker, L.C., Sørensen, A.T., Young, A., Klapoetke, N.C., Henninger, M.A., Kodandaramaiah, S.B., Ogawa, M., Ramanlal, S.B., Bandler, R.C., Allen, B.D., Forest, C.R., Chow, B.Y., Han, X., Lin, Y., Tye, K.M., Roska, B., Cardin, J.A., Boyden, E.S., 2014. Noninvasive optical inhibition with a red-shifted microbial rhodopsin. *Nat. Neurosci.* 17, 1123–1129. doi:10.1038/nn.3752

Dagdeviren, C., Yang, B.D., Su, Y., Tran, P.L., Joe, P., Anderson, E., Xia, J., Doraiswamy, V., Dehdashti, B., Feng, X., Lu, B., Poston, R., Khalpey, Z., Ghaffari, R., Huang, Y., Slepian, M.J., Rogers, J.A., 2014. Conformal piezoelectric energy harvesting and storage from motions of the heart, lung, and diaphragm. *Proc. Natl. Acad. Sci.* 111, 1927–1932. doi:10.1073/pnas.1317233111

Dombeck, D.A., Khabbaz, A.N., Collman, F., Adelman, T.L., Tank, D.W., 2007. Imaging large-scale neural activity with cellular resolution in awake, mobile mice. *Neuron* 56, 43–57. doi:10.1016/j.neuron.2007.08.003

Emiliani, V., Cohen, A.E., Deisseroth, K., Häusser, M., 2015. All-optical interrogation of neural circuits. *J. Neurosci.* 35, 13917–13926. doi:10.1523/JNEUROSCI.2916-15.2015

English, D.F., Ibanez-Sandoval, O., Stark, E., Tecuapetla, F., Buzsáki, G., Deisseroth, K., Tepper, J.M., Koos, T., 2012. GABAergic circuits mediate the reinforcement-related signals of striatal cholinergic interneurons. *Nat. Neurosci.* 15, 123–130. doi:10.1038/nn.2984

Flusberg, B.A., Cocker, E.D., Piyawattanametha, W., Jung, J.C., Cheung, E.L.M., Schnitzer, M.J., 2005. Fiber-optic fluorescence imaging. *Nat. Methods* 2, 941–950. doi:10.1038/nmeth820

Galvani, L., 1791. De viribus electricitatis in motu musculari, commentarius. *Bonon. Sci. Art. Inst. Acad.* 7, 364–415.

Godwin, D.W., Che, D., O'Malley, D.M., Zhou, Q., 1997. Photostimulation with caged neurotransmitters using fiber optic lightguides. *J. Neurosci. Methods* 73, 91–106. doi:10.1016/S0165-0270(96)02208-X

Gradinaru, V., Thompson, K.R., Zhang, F., Mogri, M., Kay, K., Schneider, M.B., Deisseroth, K., 2007. Targeting and readout strategies for fast optical neural control in vitro and in vivo. *J. Neurosci.* 27, 14231–14238. doi:10.1523/JNEUROSCI.3578-07.2007

Grosenick, L., Marshel, J.H., Deisseroth, K., 2015. Closed-loop and activity-guided optogenetic control. *Neuron* 86, 106–139. doi:10.1016/j.neuron.2015.03.034

Hamel, E.J.O., Grewe, B.F., Parker, J.G., Schnitzer, M.J., 2015. Cellular level brain imaging in behaving mammals: an engineering approach. *Neuron* 86, 140–159. doi:10.1016/j.neuron.2015.03.055

Han, X., Qian, X., Bernstein, J.G., Zhou, H., Franzesi, G.T., Stern, P., Bronson, R.T., Graybiel, A.M., Desimone, R., Boyden, E.S., 2009. Millisecond-timescale optical control of neural dynamics in the nonhuman primate brain. *Neuron* 62, 191–198. doi:10.1016/j.neuron.2009.03.011

Hassler, C., Boretius, T., Stieglitz, T., 2011. Polymers for neural implants. *J. Polym. Sci. Part B Polym. Phys.* 49, 18–33. doi:10.1002/polb.22169

Hayashi, Y., Tagawa, Y., Yawata, S., Nakanishi, S., Funabiki, K., 2012. Spatio-temporal control of neural activity in vivo using fluorescence microendoscopy. *Eur. J. Neurosci.* 36, 2722–2732. doi:10.1111/j.1460-9568.2012.08191.x

Ho, J.S., Yeh, A.J., Neofytou, E., Kim, S., Tanabe, Y., Patlolla, B., Beygui, R.E., Poon, A.S.Y., 2014. Wireless power transfer to deep-tissue microimplants. *Proc. Natl. Acad. Sci. U. S. A.* 111, 7974–7979. doi:10.1073/pnas.1403002111

Huber, D., Petreanu, L., Ghitani, N., Ranade, S., Hromádka, T., Mainen, Z., Svoboda, K., 2008. Sparse optical microstimulation in barrel cortex drives learned behaviour in freely moving mice. *Nature* 451, 61–64. doi:10.1038/nature06445

Human Brain Project, n.d. *The Human Brain Project* [WWW Document]. Available at https://www.humanbrainproject.eu/ (accessed 4.16.16).

Iwai, Y., Honda, S., Ozeki, H., Hashimoto, M., Hirase, H., 2011. A simple head-mountable LED device for chronic stimulation of optogenetic molecules in freely moving mice. *Neurosci. Res.* 70, 124–127. doi:10.1016/j.neures.2011.01.007

Jacques, S.L., 2013. Optical properties of biological tissues: a review. *Phys. Med. Biol.* 58, R37. doi:10.1088/0031-9155/58/11/R37

Jeong, J.-W., McCall, J.G., Shin, G., Zhang, Y., Al-Hasani, R., Kim, M., Li, S., Sim, J.Y., Jang, K.-I., Shi, Y., Hong, D.Y., Liu, Y., Schmitz, G.P., Xia, L., He, Z., Gamble, P., Ray, W.Z., Huang, Y., Bruchas, M.R., Rogers, J.A., 2015. Wireless optofluidic systems for programmable in vivo pharmacology and optogenetics. *Cell* 162, 662–674. doi:10.1016/j.cell.2015.06.058

Kampasi, K., Stark, E., Seymour, J., Na, K., Winful, H.G., Buzsáki, G., Wise, K.D., Yoon, E.,2016. Fiberless multicolor neural optoelectrode with micro-optic assembly of laser diodes, gradient-index lenses and dielectric waveguides, *Scientific Reports* 6(1), 30961. doi:10.1038/srep30961

Kaster, T., Sack, I., Samani, A., 2011. Measurement of the hyperelastic properties of ex vivo brain tissue slices. *J. Biomech.* 44, 1158–1163. doi:10.1016/j.jbiomech.2011.01.019

Katz, Y., Yizhar, O., Staiger, J., Lampl, I., 2013. Optopatcher—an electrode holder for simultaneous intracellular patch-clamp recording and optical manipulation. *J. Neurosci. Methods* 214, 113–117. doi:10.1016/j.jneumeth.2013.01.017

Keiser, G., Xiong, F., Cui, Y., Shum, P.P., 2014. Review of diverse optical fibers used in biomedical research and clinical practice. *J. Biomed. Opt.* 19, 080902–080902. doi:10.1117/1.JBO.19.8.080902

Kim, T., McCall, J.G., Jung, Y.H., Huang, X., Siuda, E.R., Li, Y., Song, J., Song, Y.M., Pao, H.A., Kim, R.-H., Lu, C., Lee, S.D., Song, I.-S., Shin, G., Al-Hasani, R., Kim, S., Tan, M.P., Huang, Y., Omenetto, F.G., Rogers, J.A., Bruchas, M.R., 2013. Injectable, cellular-scale optoelectronics with applications for wireless optogenetics. *Science* 340, 211–216. doi:10.1126/science.1232437

Klapoetke, N.C., Murata, Y., Kim, S.S., Pulver, S.R., Birdsey-Benson, A., Cho, Y.K., Morimoto, T.K., Chuong, A.S., Carpenter, E.J., Tian, Z., Wang, J., Xie, Y., Yan, Z., Zhang, Y., Chow, B.Y., Surek, B., Melkonian, M., Jayaraman, V., Constantine-Paton, M., Wong, G.K.-S., Boyden, E.S., 2014. Independent optical excitation of distinct neural populations. *Nat. Methods* 11, 338–346. doi:10.1038/nmeth.2836

Krook-Magnuson, E., Armstrong, C., Oijala, M., Soltesz, I., 2013. On-demand optogenetic control of spontaneous seizures in temporal lobe epilepsy. *Nat. Commun.* 4, 1376. doi:10.1038/ncomms2376

Kusumoto, F.M., Goldschlager, N., 1996. Cardiac pacing. *N. Engl. J. Med.* 334, 89–99. doi:10.1056/NEJM199601113340206

Kwon, K.Y., Lee, H.-M., Ghovanloo, M., Weber, A., Li, W., 2015. Design, fabrication, and packaging of an integrated, wirelessly-powered optrode array for optogenetics application. *Front. Syst. Neurosci.* 9, 69. doi:10.3389/fnsys.2015.00069

LeChasseur, Y., Dufour, S., Lavertu, G., Bories, C., Deschênes, M., Vallée, R., De Koninck, Y., 2011. A microprobe for parallel optical and electrical recordings from single neurons in vivo. *Nat. Methods* 8, 319–325. doi:10.1038/nmeth.1572

Leifer, A.M., Fang-Yen, C., Gershow, M., Alkema, M.J., Samuel, A.D.T., 2011. Optogenetic manipulation of neural activity in freely moving *Caenorhabditis elegans*. *Nat. Methods* 8, 147–152. doi:10.1038/nmeth.1554

Lima, S.Q., Miesenböck, G., 2005. Remote control of behavior through genetically targeted photostimulation of neurons. *Cell* 121, 141–152. doi:10.1016/j.cell.2005.02.004

Mattis, J., Tye, K.M., Ferenczi, E.A., Ramakrishnan, C., O'Shea, D.J., Prakash, R., Gunaydin, L.A., Hyun, M., Fenno, L.E., Gradinaru, V., Yizhar, O., Deisseroth, K., 2012. Principles for applying optogenetic tools derived from direct comparative analysis of microbial opsins. *Nat. Methods* 9, 159–172. doi:10.1038/nmeth.1808

McAlinden, N., Massoubre, D., Richardson, E., Gu, E., Sakata, S., Dawson, M.D., Mathieson, K., 2013. Thermal and optical characterization of micro-LED probes for in vivo optogenetic neural stimulation. *Opt. Lett.* 38, 992. doi:10.1364/OL.38.000992

Montgomery, K.L., Yeh, A.J., Ho, J.S., Tsao, V., Mohan Iyer, S., Grosenick, L., Ferenczi, E.A., Tanabe, Y., Deisseroth, K., Delp, S.L., Poon, A.S.Y., 2015. Wirelessly powered, fully internal optogenetics for brain, spinal and peripheral circuits in mice. *Nat. Methods* 12, 969–974. doi:10.1038/nmeth.3536

Nagel, G., Brauner, M., Liewald, J.F., Adeishvili, N., Bamberg, E., Gottschalk, A., 2005. Light activation of channelrhodopsin-2 in excitable cells of *Caenorhabditis elegans* triggers rapid behavioral responses. *Curr. Biol.* 15, 2279–2284. doi:10.1016/j.cub.2005.11.032

Ozden, I., Wang, J., Lu, Y., May, T., Lee, J., Goo, W., O'Shea, D.J., Kalanithi, P., Diester, I., Diagne, M., Deisseroth, K., Shenoy, K.V., Nurmikko, A.V., 2013. A coaxial optrode as multifunction write-read probe for optogenetic studies in non-human primates. *J. Neurosci. Methods* 219, 142–154. doi:10.1016/j.jneumeth.2013.06.011

Packer, A.M., Russell, L.E., Dalgleish, H.W.P., Häusser, M., 2015. Simultaneous all-optical manipulation and recording of neural circuit activity with cellular resolution in vivo. *Nat. Methods* 12, 140–146. doi:10.1038/nmeth.3217

Park, S.I., Shin, G., Banks, A., McCall, J.G., Siuda, E.R., Schmidt, M.J., Chung, H.U., Noh, K.N., Mun, J.G.-H., Rhodes, J., Bruchas, M.R., Rogers, J.A., 2015. Ultraminiaturized photovoltaic and radio frequency powered optoelectronic systems for wireless optogenetics. *J. Neural Eng.* 12, 56002. doi:10.1088/1741-2560/12/5/056002

Pashaie, R., Falk, R., 2013. Single optical fiber probe for fluorescence detection and optogenetic stimulation. *IEEE Trans. Biomed. Eng.* 60, 268–280. doi:10.1109/TBME.2012.2221713

Patil, A.C., Thakor, N.V., 2016. Implantable neurotechnologies: a review of micro- and nanoelectrodes for neural recording. *Med. Biol. Eng. Comput.* 54, 23–44. doi:10.1007/s11517-015-1430-4

Pisanello, F., Sileo, L., Oldenburg, I.A., Pisanello, M., Martiradonna, L., Assad, J.A., Sabatini, B.L., De Vittorio, M., 2014. Multipoint-emitting optical fibers for spatially addressable in vivo optogenetics. *Neuron* 82, 1245–1254. doi:10.1016/j.neuron.2014.04.041

Rickgauer, J.P., Deisseroth, K., Tank, D.W., 2014. Simultaneous cellular-resolution optical perturbation and imaging of place cell firing fields. *Nat. Neurosci.* 17, 1816–1824. doi:10.1038/nn.3866

Royer, S., Zemelman, B.V., Barbic, M., Losonczy, A., Buzsáki, G., Magee, J.C., 2010. Multi-array silicon probes with integrated optical fibers: light-assisted perturbation and recording of local neural circuits in the behaving animal. *Eur. J. Neurosci.* 31, 2279–2291. doi:10.1111/j.1460-9568.2010.07250.x

Scholten, K., Meng, E., 2015. Materials for microfabricated implantable devices: a review. Lab Chip 15, 4256–4272. doi:10.1039/C5LC00809C

Schubert, E.F., 2006. *Light-Emitting Diodes*. Cambridge University Press, Cambridge, UK.

Schwaerzle, M., Seidl, K., Schwarz, U.T., Paul, O., Ruther, P., 2013. Ultracompact optrode with integrated laser diode chips and SU-8 waveguides for optogenetic applications. Presented at the 2013 IEEE 26th International Conference on Micro Electro Mechanical Systems (MEMS), pp. 1029–1032. doi:10.1109/MEMSYS.2013.6474424

Segev, E., Fowler, T., Faraon, A., Roukes, M.L., 2015. Visible array waveguide gratings for applications of optical neural probes, in: Hirschberg, H., Madsen, S.J., Jansen, E.D., Luo, Q., MohantyS.K., Thakor, N.V. (Eds.), *Optical Techniques in Neurosurgery, Neurophotonics, and Optogenetics II*. p. 93052L. doi:10.1117/12.2078599

Shapiro, M.G., Homma, K., Villarreal, S., Richter, C.-P., Bezanilla, F., 2012. Infrared light excites cells by changing their electrical capacitance. *Nat. Commun.* 3, 736. doi:10.1038/ncomms1742

Sparta, D.R., Stamatakis, A.M., Phillips, J.L., Hovelsø, N., van Zessen, R., Stuber, G.D., 2012. Construction of implantable optical fibers for long-term optogenetic manipulation of neural circuits. *Nat. Protoc.* 7, 12–23. doi:10.1038/nprot.2011.413

Stark, E., Koos, T., Buzsáki, G., 2012. Diode probes for spatiotemporal optical control of multiple neurons in freely moving animals. *J. Neurophysiol.* 108, 349–363. doi:10.1152/jn.00153.2012

Stark, E., Roux, L., Eichler, R., Buzsáki, G., 2015. Local generation of multineuronal spike sequences in the hippocampal CA1 region. *Proc. Natl. Acad. Sci.* 112, 10521–10526. doi:10.1073/pnas.1508785112

Szabo, V., Ventalon, C., De Sars, V., Bradley, J., Emiliani, V., 2014. Spatially selective holographic photoactivation and functional fluorescence imaging in freely behaving mice with a fiberscope. *Neuron* 84, 1157–1169. doi:10.1016/j.neuron.2014.11.005

Tehovnik, E.J., 1996. Electrical stimulation of neural tissue to evoke behavioral responses. *J. Neurosci. Methods* 65, 1–17. doi:10.1016/0165-0270(95)00131-X

Tye, K.M., Prakash, R., Kim, S.-Y., Fenno, L.E., Grosenick, L., Zarabi, H., Thompson, K.R., Gradinaru, V., Ramakrishnan, C., Deisseroth, K., 2011. Amygdala circuitry mediating reversible and bidirectional control of anxiety. *Nature* 471, 358–362. doi:10.1038/nature09820

UCLA Miniscope [WWW Document], n.d. Available at http://miniscope.org/index.php?title=Main_Page (accessed 6.15.16).

Volgraf, M., Gorostiza, P., Numano, R., Kramer, R.H., Isacoff, E.Y., Trauner, D., 2006. Allosteric control of an ionotropic glutamate receptor with an optical switch. *Nat. Chem. Biol.* 2, 47–52. doi:10.1038/nchembio756

Wentz, C.T., Bernstein, J.G., Monahan, P., Guerra, A., Rodriguez, A., Boyden, E.S., 2011. A wirelessly powered and controlled device for optical neural control of freely-behaving animals. *J. Neural Eng.* 8, 46021. doi:10.1088/1741-2560/8/4/046021

Wu, F., Stark, E., Im, M., Cho, I.-J., Yoon, E.-S., Buzsáki, G., Wise, K.D., Yoon, E., 2013. An implantable neural probe with monolithically integrated dielectric waveguide and recording electrodes for optogenetics applications. *J. Neural Eng.* 10, 56012. doi:10.1088/1741-2560/10/5/056012

Wu, F., Stark, E., Ku, P.-C., Wise, K.D., Buzsáki, G., Yoon, E., 2015. Monolithically integrated μLEDs on silicon neural probes for high-resolution optogenetic studies in behaving animals. *Neuron* 88, 1136–1148. doi:10.1016/j.neuron.2015.10.032

Zemelman, B.V., Lee, G.A., Ng, M., Miesenböck, G., 2002. Selective Photostimulation of genetically ChARGed neurons. *Neuron* 33, 15–22. doi:10.1016/S0896-6273(01)00574-8

Zhang, J., Laiwalla, F., Kim, J.A., Urabe, H., Van Wagenen, R., Song, Y.-K., Connors, B.W., Zhang, F., Deisseroth, K., Nurmikko, A.V., 2009. Integrated device for optical stimulation and spatiotemporal electrical recording of neural activity in light-sensitized brain tissue. *J. Neural Eng.* 6, 55007. doi:10.1088/1741-2560/6/5/055007

Zorzos, A., Fonstad, C., Boyden, E., Franzesi, G.T., Dietrich, A., 2011. *Light-Proof Electrodes.* US20110087126 A1.

Zorzos, A.N., Boyden, E.S., Fonstad, C.G., 2010. A multi-waveguide implantable probe for light delivery to sets of distributed brain targets. *Opt. Lett.* 35, 4133–4135.

Zorzos, A.N., Scholvin, J., Boyden, E.S., Fonstad, C.G., 2012. 3-dimensional multiwaveguide probe array for light delivery to distributed brain circuits. *Opt. Lett.* 37, 4841–4843.

CHAPTER 15

Holographic Optical Neural Interfaces (HONIs)

Shani Rosen, Shir Paluch, and Shy Shoham

CONTENTS

15.1 INTRODUCTION

Optical neural stimulation has become a key neurophotonic technology, providing multiple benefits over electrical stimulation due to its non-contact nature, high spatiotemporal resolution, and potential for cell-type selective targeting. Early work on multi-site optical stimulation focused on sequentially accessing stimulation loci (Shoham et al., 2005; Nikolenko et al., 2007), an approach with relatively limited temporal resolution due to scan and dwell durations. To avoid scanning, the utilization of spatial light modulators (SLMs) for parallel cell activation was explored. First, amplitude SLMs like digital mirror devices were used (Knapczyk et al., 2005; Wang et al., 2007; Farah et al., 2007). Digital mirror-type amplitude modulating projectors shape the projected light pattern by directly switching each pixel on or off. These devices can generally produce high temporal and lateral resolution patterns but suffer from very high losses when the projected pattern is sparse (light hitting the "off" pixels is lost). In contrast, phase SLMs modulate the phase (wavefront) of the incoming beam such that the diffracted light power is divided solely between the "on" regions with minimal light loss in the modulation process. This enables efficient parallel scan-less optical "holographic" stimulation where arbitrary light patterns are created in the Fourier plane of a phase SLM simply by displaying the desired Computer-Generated Hologram (CGH) on the SLM. This benefit prompted several research teams independently to pursue the development of such holographic excitation optical neural interfaces (Lutz et al., 2008; Nikolenko et al., 2008; Golan et al., 2009).

Holographic illumination has multiple advantages over other light patterning techniques (Vaziri and Emiliani, 2012): it provides high efficiency modulation, enables generation of a variety of illumination patterns, and can provide a relatively large excitation FOV and a dynamic correction of the projected pattern. Another major advantage is the ability to produce three-dimensional light patterns while most other methods are restricted to two dimensions (Daria et al., 2009; Yang et al., 2011; Anselmi et al., 2011). The holographically generated light patterns can excite neurons using a number of different mechanisms, based on either a single- or multi-photon modes of excitation. A parallel line of research has focused on applying holographically generated light patterns towards functional microscopic imaging (Nikolenko et al., 2008; Daria et al., 2009; Dal Maschio et al., 2010).

15.1.1 Single-Photon and Two-Photon HONIs

The early realizations of holographic optical neural stimulation were applied towards non-optogenetic modes of excitation, including one photon (Lutz et al., 2008; Papagiakoumou et al., 2013) and two-photon (Nikolenko et al., 2008; Dal Maschio et al., 2010; Go et al., 2012) photolysis of caged compounds, where light absorption led to synaptic excitation, or direct visual excitation (Golan et al., 2009). However, because uncaging is typically a non-reversible process, optogenetic applications were soon pursued. The first demonstrations of single-photon HONIs for optogenetic stimulation included selectively activating multiple retinal ganglion cells as a path to restore vision lost following photoreceptor degeneration (Reutsky-Gefen et al., 2013) and single-photon activation of cultured neurons and HEK cells (Papagiakoumou, 2013). In another related development, photo-absorber induced neural-thermal stimulation (PAINTS) was demonstrated by illuminating holographically patterned light onto exogenous extracellular photo-absorbers and creating local thermal transients that stimulate the neurons (Farah et al., 2013).

Two-photon excitation (TPE) uses intense ultra-short laser pulses to generate a high photon flux around the focal plane, creating a localized excitation spot which improves axial sectioning and depth penetration inside the heavily scattering brain tissue. Optogenetic TPE has thus become the central mode of precisely stimulating neurons in brain tissue (Andrasfalvy et al., 2010; Papagiakoumou et al., 2010; Rickgauer and Tank, 2009), and its combination with digital holography is the main method used for simultaneously exciting multiple neurons, both *in vitro* (Packer et al., 2012; Paluch-Siegler et al., 2015; Chaigneau et al., 2016; Shemesh et al., 2017) and *in vivo* (Packer et al., 2015; Pégard et al., 2017). Holographic TPE's high efficiency generation of distributed light patterns and inherent ability to address 3D circuits make it uniquely suitable for this application. Nevertheless, Holographic TPE suffers from several critical challenges which need to be addressed in order to achieve more efficient stimulation. First, in order to increase the probability for excitation, high laser power and high-power efficiency are required while avoiding tissue heating. Second, TPE increases the speckle contrast (see Section 15.4.2), i.e. it increases the fluctuations in the signal since it is now proportional to $<I^2>$ rather than $<I>$. Third, perhaps the main challenge is the small excitation area of TPE (Shoham, 2010): in order to efficiently excite a neuron, the illuminated spot should be expanded or scanned to cover most of the neuron's surface area (Figures 15.1a, b). Diffraction limited spots maximize TPE but scanning them across the soma reduces the temporal resolution and thus limits applicability. Illumination of large spots can be achieved by several means: reducing the effective objective lens numerical aperture (which is inversely proportional to the lateral spot size and axial spot size squared) or creating shaped illumination patterns. Both techniques for expanding the spot size cause severe deterioration in axial resolution. Temporal Focusing (TF) is a possible solution for decoupling the lateral and axial dimensions by shaping the temporal profile of the pulse (see Section 15.2.4). This method enables projection of large illumination spots with improved axial resolution (Figure 15.3c) but constrains the excitation to a specific plane.

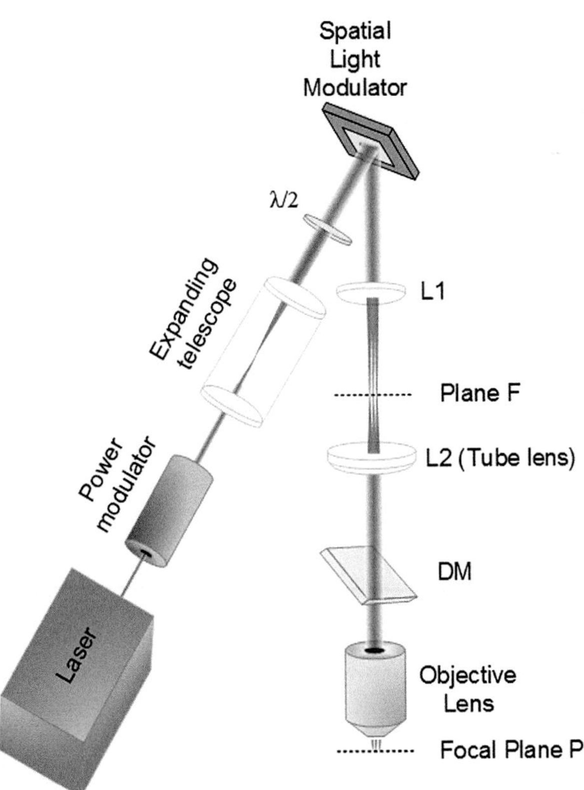

FIGURE 15.1 Scheme of a standard holographic system (see text for detailed description).

15.2 OPTICAL DESIGN AND CONSIDERATIONS

15.2.1 Basic Design

In a typical phase-only Fourier holographic system, the output pattern is the Fourier transform of the phase modulated wavefront reflected (or transmitted) by the SLM (Goodman, 2005):

$$E_{\text{pattern}}(x,y) = \mathcal{F}\left\{E_{\text{hologram}}(u,v)\right\} \tag{15.1}$$

Where $\mathcal{F}\{\cdot\}$is a Fourier transform, the coordinates u,v describe the lateral position in the hologram (SLM) plane, and the coordinates x,y describe the lateral position in the projected (focal) plane, also termed the "reconstruction plane".

In a standard holographic setup (Figure 15.1) a collimated laser beam first passes through a laser power modulator, and is then expanded by a telescope to fill the SLM. SLM phase modulation is generally polarization-dependent, and therefore a half-wave plate is usually placed before the SLM to control the polarization of the beam. The SLM's phase modulation physics dictate the specific optical arrangement that will be used to separate the incident input beam from the output beam (reflected by the SLM). In an oblique-incidence optical system, the angle between the input and output beam of the SLM should be as small as possible to minimize the system's distortion (angle of 10° or less) (Lesem et al., 1969). Following the SLM, an additional telescope (lenses L_1 & L_2) is used to match the size of the modulated beam to the back aperture of the objective lens, which serves as the Fourier lens (L_{obj}). This is achieved by choosing m, the telescope magnification to satisfy the following equation (subject to limiting apertures):

$$mL = 2 f_{obj} NA \tag{15.2}$$

where f_{obj} is the objective focal length and L is the lateral dimension of the SLM.

A 4-f configuration is common, and the optimal arrangement for this telescope since it maximizes light efficiency. At the focal plane of lens L1 (F), which is conjugated to the focal plane (P), an image of the desired pattern is obtained, and spatial filtering of undesired diffraction orders can be performed by placing a proper mask in this plane (Martín-Badosa et al., 2007; Spalding et al, 2008; Papagiakoumou et al., 2008; Golan et al., 2009). Importantly, the whole setup can be easily integrated into a standard microscope by placing a dichroic mirror (DM), or optionally a polarizing beam splitter before the objective lens.

15.2.2 Optical Elements' Characteristics

The fundamental properties of the optical elements comprising the system and primarily the SLM will ultimately determine its performance. Phase SLMs are optical elements that modulate light propagation in space and time by changing only the phase of the incident light and are thus more power-efficient than amplitude modulation devices. This is especially important for TPE methods that require very high peak power levels. The SLM may be any phase-modulating device based, among other options, on liquid-crystals (LCs), MEMS, or deformable mirrors. The liquid crystal on silicone (LCoS)-based SLM is the most mature and readily available technology, providing high pixel counts, almost no inter-pixel dead-space and simple interfaces, at a reasonable cost. Phase modulation is achieved in this device by applying locally defined voltages across the LC layer. The LC molecules realign in response to the voltage gradient and cause a polarization-dependent change in the optical properties of the LC layer.

The SLM's LC characteristics determine its properties: the LC layer's thickness affects the switching rate between different holograms (the time it takes for the LC molecules to realign is proportional to the square of the LC layer's thickness: Amako and Sonehara, 1991). Thus, the thinner the SLM the faster it operates. In addition, the SLM's thickness also limits its specified wavelength working range, as the maximal wavelength for which a full 2π phase modulation is achieved. Reflective SLMs have half the thickness of transmissive SLMs since light passes through the LC layer twice, so they can operate faster, and also usually provide higher resolutions and a larger fill factor. Due to lack of surface flatness, reflective SLMs usually aberrate the light beam more than transmissive SLMs, but this can be corrected in the hologram calculation process.

High diffraction efficiency is another critical factor to consider. Multi-level phase modulation is inherently more efficient than binary phase modulation (Goodman and Silvestri, 1970), for which the upper bound on diffraction efficiency is approximately 40%: another 40% of the light directed to a replicate image lying on the opposite side of the optical axis, and the remaining light is sent to higher diffraction orders (ghost orders). LC-based SLMs use either nematic or ferroelectric liquid crystals. Nematic LCs have multi-level phase modulation (typically 256 levels), but are relatively slow, while ferroelectric LCs on the other hand are binary (have only two phase levels, 0 or π), but provide a very fast switching rate of up to a few kFPS. For time-multiplexing or time averaging applications that require kHz refresh rates, a fast ferroelectric SLM should be considered and low diffraction efficiency can often be compensated by higher laser power. For two-photon systems and other applications that require high diffraction efficiency, nematic SLMs are more appropriate. Diffraction efficiency is also affected by the effectiveness of the anti-reflection coating of the front electrode of the SLM and by the SLM fill factor – the fraction of the SLM unit cell that is occupied by the actual modulating pixel. With LCoS, electronic circuitry can be moved below the LC layer, resulting in fill factors very close to 1.

Thus, in practical holographic setups, the maximum achievable diffraction efficiency is never 100% and a portion of the light is always directed to the zero order and to higher (ghost) diffraction orders. Typical SLMs contain square or rectangular pixels with a small dead-space between pixels. The area formed by the dead-spaces (in between modulating pixels) acts like a mirror, and the light incident on it will be reflected to the (so-called) zero order in the reconstruction plane (Figure 15.2). As mentioned in Section 15.2.1, these unwanted diffraction orders are typically blocked by placing a beam block or a slit in the conjugated reconstruction plane, F (see Figure 15.1). The beam block may be a small obstructing element that blocks only a small region around the zero order, resulting in an irregularly shaped excitation FOV containing a blind zone. For ferroelectric SLMs half of the FOV (including the zero order) should be blocked since light is divided equally between the 1st and –1st diffraction orders (Figure 15.2). As an alternative to beam blocking, hologram

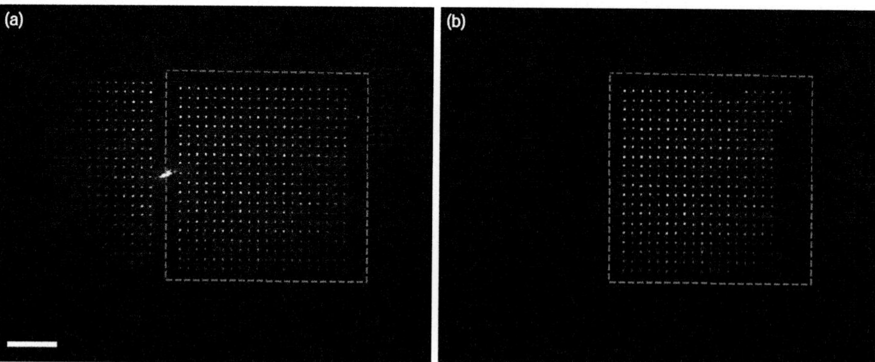

FIGURE 15.2 Limiting the FOV with a slit. **(a)** An unblocked output pattern of a 20 × 20 spot grid created by a binary SLM. The zero-order, replicate pattern and higher orders are visible. The dashed gray line indicates the desired excitation FOV. Scale bar is 200 µm. **(b)** Inserting a square slit at the intermediate image plane blocks the undesired orders. (From Golan et al., 2009.)

calculation can be modified to minimize the effect of the zero and ghost orders. In addition, the zero and ghost orders light may be suppressed by improving the projected hologram: by including a corrective beam to interfere with the zeroth order (Palima and Daria, 2007) or by filtering out the influence of physical cross-talk between pixels in the Fourier domain (manifested as high spatial frequencies). This requires a preliminary SLM characterization (Ronzitti et al., 2012). Another simple alternative to beam blocking, that is sometimes utilized, is to mathematically add a spherical phase shift to the computed hologram for axial shifting of the desired pattern away from the zeroth order (Zhang et al., 2009; Polin et al., 2005). This is applicable for 2D samples, but for 3D samples cells situated around the zeroth order may be excited. This approach can optionally also simplify zero-order blocking.

Another optical phenomenon that can significantly affect the quality of the projected hologram is optical aberrations. These can be caused due to several elements along the optical path (Martín-Badosa et al., 2007). The SLM, which is made flat at the pixel level is un-flat at the die level and causes an aberration which is similar to astigmatism. This can be compensated for by adding a correction phase to the hologram. Spherical aberrations in the focal plane are caused due to the lenses in the setup and can be minimized by properly choosing and placing the lenses. Lenses L1 and L2 can be simple plano-convex lenses; however, the choice of lenses for the beam-expanding telescope requires high-quality optics since the light hitting the SLM should be perfectly collimated.

15.2.3 System Characteristics

The desired characteristics of the projected holograms are generally considered in the first stages of system design and tailored to the specific application in mind. Two of the most prominent system characteristics are the resolution and size of the excitation field.

The system's resolution is determined by the diffraction-limited spot size, given by:

$$\Delta x \cong \frac{\lambda}{2NA}, \quad \Delta z \cong \frac{2n\lambda}{NA^2} \tag{15.3}$$

Where Δx and Δz define the lateral and axial resolution respectively, λ is the stimulation wavelength, n is refractive index, and NA is the **effective** objective lens numerical aperture (Anselmi et al., 2011; Knapczyk et al., 2005; Martín-Badosa et al., 2007). The effective NA is determined by the filling of the objective lens aperture by the illuminating laser beam. Underfilling will ensure minimal power loss; however, the objective NA will not be used to its maximum, resulting in an increased spot size. For photo-stimulation, light efficiency is typically more important than spot size, because the resolution of photo-stimulation is often affected by other factors such as tissue scattering, opsin saturation, etc. Therefore, underfilling with truncation ratios of 0.5–0.7 is often recommended. In this analysis it is assumed that the system is designed to generate diffraction limited spots. If extended circular patterns of size $W \gg \Delta x$ are used, the axial resolution scales linearly with W, while for diffraction limited beams it scales approximately with W^2 (as seen in Eq. (15.3)). The axial resolution of high-NA systems generating extended circular patterns is thus significantly better than that obtained with low-NA Gaussian beams when accessing the same stimulation area, both for 1PE and 2PE.

The stimulation field area, i.e. the Field Of View (FOV, also termed FOE, Field of Excitation: Ronzitti et al., 2017), is derived from the well-known formulation of the grating equation, which states that light diffracted from a grating will have maxima at angles $sin\theta = k\lambda / d$, where k is the diffraction order and d is the effective grating pitch size (i.e. its size in the BFP of the objective lens, after the telescope magnification m). In HONIs, generally only the first diffraction order ($k = 1$) is of interest, and the minimal effective grating pitch (which yields the largest diffraction angle) is given by $2md_{SLM}$, where d_{SLM} is the SLM pixel pitch size. This leads to the following definition of the FOV in terms of the maximal lateral size of the stimulation area:

$$x_{\max} = \frac{\lambda}{2md_{SLM}} f_{obj} \tag{15.4}$$

To access a three-dimensional FOV, a parabolic phase can be added to the SLM. A superposition of Fresnel lenses, each creating an individual beamlet with a predefined direction and curvature, results in an axial shift of the beamlet focus position out of the objective focal plane (Golan et al., 2009). The shortest Fresnel focal length that can be generated on the SLM is limited by its finite pixel pitch. This sets a maximal depth shift on the stimulation plane:

$$z_{\max} = \frac{\lambda f^2_{obj}}{N \cdot (md_{SLM})^2} \cong \frac{\lambda f_{obj}}{2NA \cdot md_{SLM}} \tag{15.5}$$

Where N is the effective number of pixels along the SLM side for a square SLM (typically 500–2,000). The diffraction efficiency should also be considered when defining the system's accessible volume. Due to the SLM's pixelization, the diffraction efficiency declines as the stimulation point is axially shifted from the focal plane, or laterally shifted from the optical axis (Golan et al., 2009; Yang et al., 2011). This decline may be compensated by applying an inverse amplitude correction during hologram design (at the expense of a lower average stimulation efficiency).

Thus, the effective number of accessible points in the lateral and axial FOV:

$$\left(\frac{x_{\max}}{\Delta x}\right)^2 \cong \left(\frac{L}{2d_{SLM}}\right)^2 = 0.25N^2 \ , \quad \left(\frac{z_{\max}}{\Delta z}\right) \cong \left(\frac{L}{8n \cdot d_{SLM}}\right) \cong \frac{N}{10} \tag{15.6}$$

Consequently, in a typical HONIs setup 10–100k points can be accessed inside the projection volume.

15.2.4 Two-Photon Holographic Systems: Scanning versus Temporal Focusing

In integrating HONIs into a multi-photon microscope, the small excitation spot inherent to these systems severely hinders excitation efficiency: to efficiently excite a neuron, the illuminated spot should be expanded or scanned to cover most of the neuron's surface area (Figure 15.3a).

One popular solution for multi-cell patterned illumination is to project small diffraction limited spots and to raster- or spiral-scan them over the cell soma. The scanning speed should be fast enough to cover the whole cell area according to the biological response time. The first demonstration of scanning the soma to induce two-photon activation in ChR2 expressing neurons was done with a single spot (Rickgauer and Tank, 2009) and was later combined with holographic patterning by simply adding an SLM into the two-photon scanning microscope path (Packer et al., 2012). However, scanning of diffraction limited spots across the soma reduces the temporal resolution and thus its general

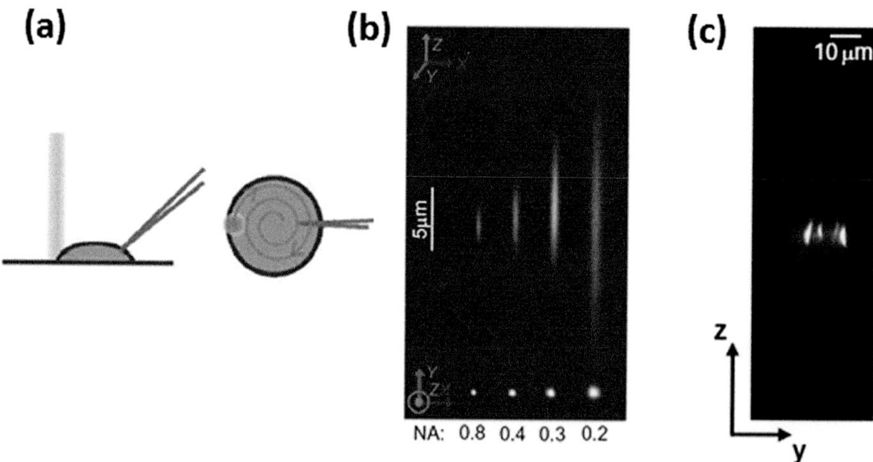

FIGURE 15.3 Two-photon modalities for optogenetic activation. **(a)** A schematic depiction of the geometry of whole-cell scanning 2P stimulation, shown in side- and top-views. **(b)** Measured 2P intensity profiles of the point spread function (PSF) at several effective NA values, along the optical axis z (top), and in the lateral (x–y) plane (bottom). **(c)** y–z section of the measured axial propagation of a 20µm temporally focused holographic spot. (From Oron et al., 2012.)

applicability. Circumventing this issue by spot expansion or holographic projection of multi-focal "patch" patterns across the soma generally compromises both the projected light intensity and (crucially) its axial resolution.

Temporal Focusing (TF) offers an alternative solution for efficient TPE, by extending the lateral dimensions of the illumination pattern without deteriorating its axial resolution. TF was first used to provide axial confinement in two-photon illumination of femtosecond pulses for wide-field two-photon microscopy applications (Oron et al., 2005; Zhu et al., 2005) and is increasingly being used for generation of axially confined light patterns (Papagiakoumou et al., 2008, 2009, 2010). To combine TF with holographic projection, a diffraction grating is introduced into the setup at plane F and the projected pattern created at this plane is imaged by a telescope (L_2 and L_{obj} in Figure 15.1) (Papagiakoumou et al., 2008). The grating disperses the rays corresponding to different spectral components of the laser beam. This results in stretching of the pulse in time since rays accumulate a relative phase offset while propagating. The rays are recombined at the focal plane and the pulse is compressed back, causing an especially localized two-photon absorption since it is inversely proportional to the pulse duration. In practice, TF leads to focusing in the temporal domain and hence allows projection of large two-dimensional illumination patterns or patches without deteriorating the axial resolution (Figure 15.1a).

The axial characteristics of the system change when TF is applied (Durst et al., 2006; Oron et al., 2012). For a temporally focused Gaussian beam, the Rayleigh range, z_R is:

$$z_R = \frac{2f_{obj}^{\ 2}}{k}\left(s^2 + \alpha^2\Omega^2\right) \approx \frac{\lambda}{NA^2_{\ obj}} \tag{15.7}$$

where k is the mean magnitude of the excitation wave vector, and s and $\alpha\Omega$ are the spot sizes at the back aperture of the objective in the direction orthogonal and parallel to the grating linear dispersion, respectively. This is in stark contrast to non-temporally focused two-photon point illumination in which the Rayleigh range, and therefore also the axial resolution, scale quadratically with the lateral size.

This relationship shows that the axial resolution is only dependent on the objective lens numerical aperture and not on the lateral spot size, hence TF decouples the axial and lateral resolution. Figure 15.4 shows the improvement of axial sectioning of holographically generated circular spots with different diameters and of a shaped spot due to the use of TF. When TF is applied, the measured FWHM of these patterns is constant and equals 5.5 ± 0.4 μm.

The experimentally measured FWHM is slightly larger than the theoretically expected FWHM for a Gaussian beam due to physical limitations of the optical system (aberrations, finite aperture of the objective, and dispersion of the pulse) and due to the rapid phase variations of the pattern generated on the grating. Smoother phase variations will cause fewer scattering angles of the different wavelengths and will result in a better depth confinement (Papagiakoumou et al., 2009).

TF also has the advantage of reducing the scattering effect of holographically generated patterns. This is due to the fact that each spectral component of the laser beam dispersed by

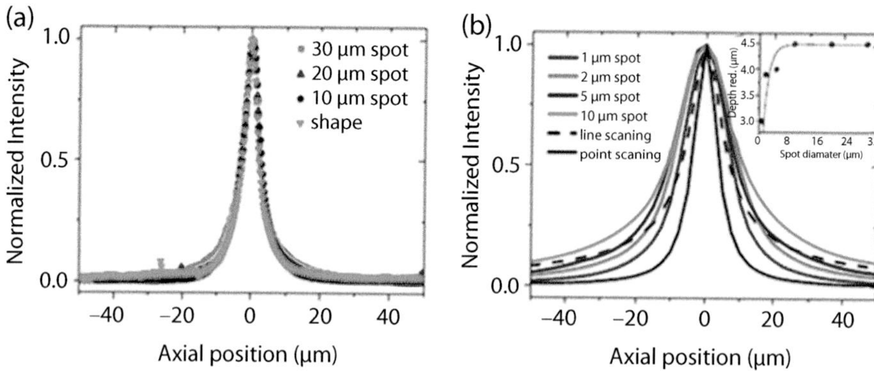

FIGURE 15.4 Axial confinement using TF (a) Measured axial profiles of the intensity distribution for the 30, 20, 10μm diameter spot and a dendritic shape with TF (b) Calculated axial intensity distribution compared to the two limiting cases of Gaussian circular illumination and line illumination with TF. (From Papagiakoumou et al., 2008.)

the grating travels a different optical path and acquires a slightly different speckle pattern causing overall smoothing of the intensity pattern. The use of temporal focusing is thus beneficial also for the preservation of the pattern shape, as demonstrated experimentally for up to 550 μm (Oron et al., 2012) (see Figure 15.5).

TF can also be combined with other phase modulation light patterning techniques such as generalized phase contrast (GPC) to produce axially confined shaped illumination patterns (Vaziri and Emiliani, 2012; Papagiakoumou, 2013; Papagiakoumou et al., 2010; Oron et al., 2012). GPC is an SLM based interferometric technique in which a binary (0 or π) phase modulated light interferes with the non-diffracted light after traveling through a phase contrast filter (PCF) that imposes phase retardation between the two beams. The interference between the waves creates the desired intensity pattern at the focal plane of the objective lens. GPC has several advantages over digital holography; since the diffracted beam interferes directly with the non-diffracted beam, no light is lost on the zero order (high power efficiency) and speckle-free patterns are formed. In GPC, simple binary phase

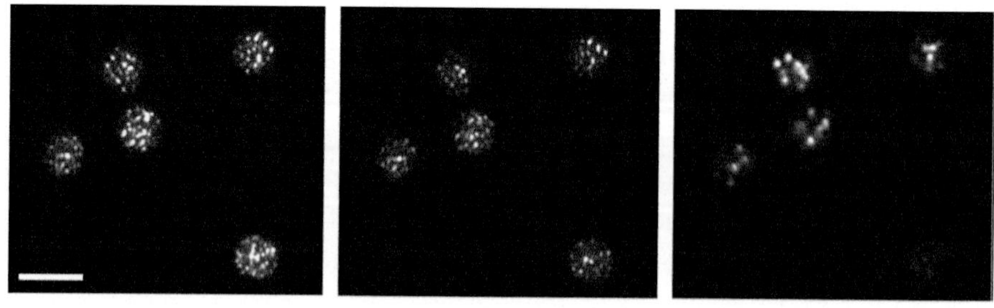

FIGURE 15.5 2P-excited fluorescence images on a thin fluorescent film of a configuration of multiple spots of 7-μm diameter without scattering (no slice; left) and after propagation through acute cortical brain slices of 550 μm with TF (middle) and without TF in the optical setup (right). (From Oron et al., 2012.)

masks can be projected on the SLM; this enhances the process speed: no need for iterative calculations and fast ferroelectric SLMs can be used. However, in order to maximize the interferometric contrast, a ratio of 1:4 is required between pixels modulated by π and the total number of pixels, thus patterns created by GPC are restricted to a small excitation FOV (approximately 100 μm²) compared to digital holography (several hundred μm²). In addition, three-dimensional patterns cannot be created via GPC.

15.3 HOLOGRAM CALCULATION AND DESIGN

15.3.1 Basic Algorithms and Considerations

In Fourier holography, any reconstructed pattern may be written as a sum of diffraction-limited spots. Here we assume that the total number of these points is $M \ll N^2$. The hologram generating such a pattern can be regarded as a superposition of M gratings multiplied by M Fresnel lenses, each diverting a fraction of the incident beam to create a beamlet with a defined direction in space. The Fourier lens then focuses each beamlet to a diffraction-limited spot on the sample (three-dimensional FOV), resulting in the following complex-amplitude hologram:

$$t(u,v) = \sum_{m=1}^{M} \alpha_m \underbrace{\exp\left(2\pi i \frac{x_m u + y_m v}{\lambda f_{obj}}\right)}_{\text{shift in } xy} \underbrace{\exp\left(-\pi i \cdot z_m \frac{u^2 + v^2}{\lambda f^2_{obj}}\right)}_{\text{shift in } z} \tag{15.8}$$

Where $[x_m, y_m, z_m]$ are the coordinates of the mth spot in the reconstruction volume and α_m is its relative amplitude (assuming that $z_m \ll f_{obj}$).

Although a simple Fourier transform relationship exists between the hologram and the desired pattern, calculation of CGHs is not a trivial task since in phase-only holographic systems only the phase of the incoming beam is modulated but not its amplitude. In HONIs, light patterns are usually sparsely scattered across the FOV and this property can be used to simplify the calculation. In addition, there are many degrees of freedom that can be used to optimize the result including the phase freedom of the intensity pattern and amplitude freedoms outside the FOV. Several performance parameters are typically optimized in CGH calculation: diffraction efficiency of the desired pattern, uniformity measures for measuring how evenly light is divided between the different spots, computational load, and number of iterations (Spalding et al., 2008; Dufresne et al., 2001; Polin et al., 2005). Finally, the SLM's physical limitations, for example, its finite pixel size and phase quantization, also have an impact on the resulting pattern, and when needed, the projected hologram can be adjusted to compensate for the system's aberrations or scattering effects. The quantization of phase levels reduces the diffraction efficiency and diverts light to higher, undesired orders (Goodman and Silvestri, 1970). This effect quickly becomes negligible as the number of quantization levels exceeds 16; however, for binary SLMs the maximum theoretical diffraction efficiency is 40.5%. Spatial pixilation results in a spatially varying sinc-shaped diffraction efficiency (Dammann, 1970; Tan et al., 2001), requiring correcting the spots amplitude α_m by an inverse correction factor.

Most algorithms start by calculating a basic hologram using the superposition method (S) according to Eq. (15.8). The amplitude assigned to the holographic beam (α_m) is simply the incoming laser beam amplitude (Dammann, 1970; Tan et al., 2001). The Random Superposition method (SR) achieves slightly better results by assigning random phases to each of the M spots (Lesem et al., 1969). The Gerchberg-Saxton (GS) algorithm (Gerchberg, 1972) starts by calculating the SR hologram but further iterates to improve the results. This is a computationally demanding algorithm since Fourier transforms need to be repeatedly calculated (Gallagher and Liu, 1973). For sparse intensity patterns, the SR and GS algorithms produce similar results; however, for patches of light or more complex patterns the GS performs better.

Among other algorithms are the Random Mask (RM) (Montes-Usategui et al., 2006), adaptive-additive (AA) (Dufresne et al., 2001), and direct search (DS) (Meister and Winfield, 2002); RM is a simple and fast algorithm in which each SLM pixel codes for a randomly chosen spot. The spot intensity is determined by the relative number of pixels that code for it. However, the light efficiency is inversely proportional to the number of pixels and is overall very low. AA is a generalization of the GS algorithm for improving the uniformity by optimizing a uniformity bias parameter ξ that is integrated into the phase argument (GS is a special case for which $\xi = 0$). The GS algorithm minimizes the error between the target intensity pattern and the actual intensity pattern by searching for the best hologram for this purpose and changing the intensity of all pixels in lexicographic or random order accordingly.

The weighted Gerchberg-Saxton (GSW) (Di Leonardo et al., 2007) algorithm is the most commonly used algorithm for the calculation since it produces the optimal efficiency and uniformity among the presented algorithms. GSW is a variation of the GS algorithm for correcting the uniformity of the pattern by assigning weights to each spot thus introducing M additional degrees of freedom. Table 15.1 summarizes the performance of the different algorithms and shows that although GSW demands higher computational load, it produces very high efficiency and uniformity.

TABLE 15.1 Summary of Performance of the Different Algorithms for Shaping of a Highly Symmetric 2D 10*10 Square Grid and a Less Symmetric 3D Structure of 18 Spots Located on the Sides of a Diamond-Lattice Unit Cell (in Brackets)

algorithm	e	u	σ(%)	K	scaling
RM	0.01 (0.07)	0.58 (0.79)	16 (13)	1 (1)	N
S	0.29 (0.60)	0.01(0.52)	257(40)	1 (1)	$N \times M$
SR	0.69 (0.72)	0.01 (0.57)	89 (28)	1 (1)	$N \times M$
GS	0.94 (0.92)	0.80 (0.75)	17 (14)	30 (30)	$K \times N \times M$
AA	0.93 (0.92)	0.79 (0.88)	9(6)	30 (30)	$K \times N \times M$
DS	0.68 (0.67)	1.00 (1.00)	0(0)	7.5×10^5 (1.7×10^5)	$K \times P \times M$
GSW	0.93 (0.93)	0.99 (0.99)	1 (1)	30 (30)	$K \times N \times M$

M is the number of spots, N is the number of pixels in the hologram, and P is the number of gray levels. (From Spalding et al., 2008.)

15.3.2 Holographic Speckle Mitigation Algorithms

Holographic speckle is encountered when larger, multi-spot patches or contiguous shapes are projected. In this case when adjacent spots have pseudo-random phases they interfere randomly. The resulting distribution of intensities obeys the well-known statistics of the speckle phenomenon (Goodman, 1976), justifying the term "speckle noise". The severity of speckle creation can be quantified by the speckle contrast, which is the ratio between the standard deviation to the mean of the intensity: $C = \frac{\sigma}{\langle I \rangle} = \frac{\sqrt{\langle I^2 \rangle - \langle I \rangle^2}}{\langle I \rangle}$

The non-uniformity of the desired intensity pattern can reach 20% in 1P excitation and 50% in 2P excitation (Papagiakoumou et al., 2008; Lutz et al., 2008).

Several time averaging solutions to the holographic speckle problem were proposed and were shown to provide a reduction of speckle contrast in one- and two-photon holographic projections. The speckle contrast can be reduced by a factor of $\sqrt{N}$ by time averaging over N intensity patterns with identical amplitude and different phase distributions. Mechanical averaging can be accomplished by adding an optical element such as a rotating diffuser (Papagiakoumou et al., 2008) that pseudo-randomly scrambles the resulting phases and reduces speckle contrast. Another option is to use fast switching SLMs and project different holograms rapidly to match the biological response time. One can either compute N different holograms (Amako and Sonehara, 1991; Goodman, 1976; Arrizón and Testorf, 1997) of the desired pattern or divide the spots comprising a contiguous shape into several sparse subframes, each containing a subset of the original shape, a method termed pixel separation (Takaki and Yokouchi, 2011; Makowski, 2013; Mori et al., 2014).

An alternative approach termed "shift averaging" requires the calculation of only a single hologram. In this approach a sequence of cyclic shifts of a single hologram are displayed and an intelligent choice of a specific shift sequence eliminates the speckle. Shift averaging has been shown to minimize speckle noise significantly by averaging over a small number of holograms both for one-photon (Golan and Shoham, 2009) and two-photon holographic projection systems (Matar et al., 2011).

However, all time averaging solutions require the projection of multiple holograms to reduce speckle noise, thereby sacrificing temporal resolution, which may be crucial for neural stimulation. Alternative approaches to speckle reduction are based on jointly controlling both the amplitude and phase of the projected patterns. Early attempts at speckle reduction were focused on applying phase-smoothing constraints in the projected plane (Wyrowski and Bryngdahl, 1989; Aagedal et al., 1996) but required over-sampling of the pattern, and postiterations with reduced phase freedom, thus increasing the computational load by orders of magnitude. A more recent approach (Yuan and Tao, 2014; Wu et al., 2015) addresses this challenge by incorporating phase information into the GS-type iteration process. Generally, adding a phase constraint in the projected plane comes at the expense of efficiency, and the importance of these factors has to be weighted in the algorithm design. In Yuan and Tao (2014) and Wu et al. (2015), a constant phase constraint is applied in the projected plane and a pattern-specific rate parameter is manually tuned to improve both accuracy and diffraction efficiency. In contrast, the algorithm presented in

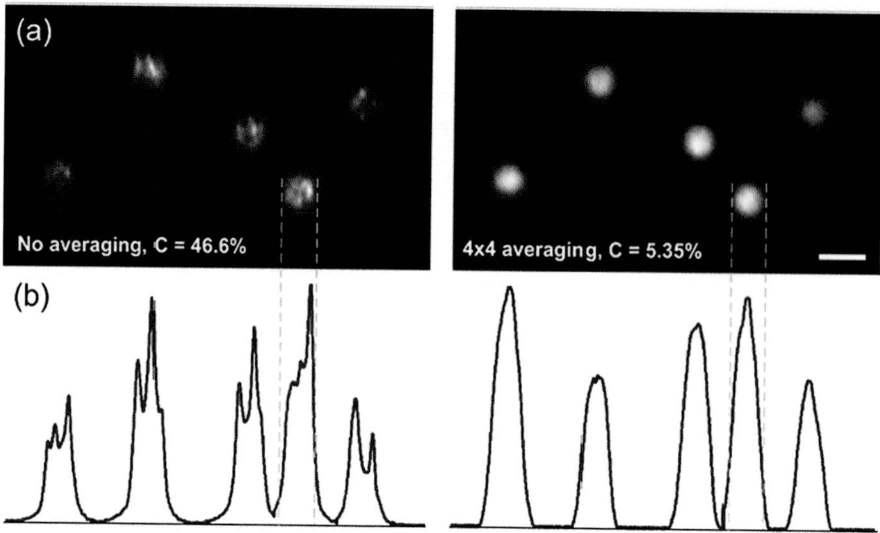

FIGURE 15.6 Demonstration of shift averaging in two-photon fluorescence microscopy. Using 4*4 shift averaging the speckle contrast was reduced almost tenfold. (From Matar et al., 2011.)

Aharoni and Shoham (2018) achieves high (>50%) diffraction efficiency one-dimensional speckle-free patterns with no need for pattern-specific parameter adjustment. In other implementations, the constant phase is replaced with a quadratic phase constraint, which is determined according to the geometric relations of optical elements in the system. This has the advantage of effectively spreading the light in the projected plane (without the need for speckle-generating phase randomization), allowing iterative (Pang et al., 2016) and non-iterative (Shimobaba and Ito, 2015; Pang et al., 2017) hologram calculation of speckle-free patterns, yet with lower diffraction efficiencies Figure 15.6.

15.4 RECENT DEVELOPMENTS, APPLICATIONS, AND OUTLOOK

The last few years have seen major advances in the emergence of techniques for precision bidirectional brain activity mapping studies, where holographic control systems are coupled with an imaging system to allow all optical probing of neural activity. These could be achieved by using new multi-wavelength probes for functional imaging (Akerboom et al., 2013) and stimulation (Yizhar et al., 2011; Zhang et al., 2008; Lin et al., 2013). Moreover, volumetric imaging and stimulation of three-dimensional neural ensembles typically requires the ability to decouple the imaging and stimulation planes. Among possible options are computational axial shift of the hologram or using remote focusing imaging for shifting or tilting the imaging plane (Botcherby et al., 2007; Anselmi et al., 2011). Related advances include the emergence of soma-targeted functional indicators and opsins that further enhance precision by minimizing off-target stimulation. Another important trend is the ability to use these precise neural ensemble stimulation patterns to probe behavioral readout in animal and human subjects.

As in other domains of optical imaging, we expect to see more and more developments towards miniaturizing of holographic setups, to create portable interfaces that can either be implanted or worn. The most critical factor that affects the holographic system size is the separation of the SLM incident and reflected beams. These can overlap by imposing a polarization change by a Polarizing Beam Splitter (PBS) that could direct the light to the desired path. Other adaptations for minimizing the total length of the telescopes while maintaining the same magnification include using small or negative curvature lenses for the telescopes or locating the SLM closer to lens L_1 (which results in lower efficiency due to lower practical numerical aperture) (Martín-Badosa et al., 2007; Golan et al., 2009). The system output can be coupled into a fiber and implanted inside the brain to achieve deep control of neural processes (Zhang et al., 2006). For controlling accessible cells like retinal ganglion cells, the system can be integrated into glasses that transmit the light directly through the eye. For fast calculation of the holograms, technologies for parallel calculations like Graphic Processing Units (GPUs) can be used to provide the ability for nearly live control of neural circuits.

REFERENCES

Aagedal, Harald, Michael Schmid, Thomas Beth, Stephan Teiwes, and Frank Wyrowski. 1996. "Theory of Speckles in Diffractive Optics and Its Application to Beam Shaping." *Journal of Modern Optics* 43 (7): 1409–21. doi:10.1080/09500349608232814.

Aharoni, Tal, and Shy Shoham. 2018. "Phase-Controlled, Speckle-Free Holographic Projection with Applications in Precision Optogenetics." *Neurophotonics* 5 (2). doi:10.1117/1.NPh.5.2.025004.

Akerboom, Jasper, Nicole Carreras Calderón, Lin Tian, Sebastian Wabnig, Matthias Prigge, Johan Tolö, Andrew Gordus, Michael B Orger, Kristen E Severi, and John J Macklin. 2013. "Genetically Encoded Calcium Indicators for Multi-Color Neural Activity Imaging and Combination with Optogenetics." *Frontiers in Molecular Neuroscience* 6: 2.

Amako, Jun, and Tomio Sonehara. 1991. "Kinoform Using an Electrically Controlled Birefringent Liquid-Crystal Spatial Light Modulator." *Applied Optics* 30 (32): 4622–28.

Andrasfalvy, Bertalan K, Boris V Zemelman, Jianyong Tang, and Alipasha Vaziri. 2010. "Two-Photon Single-Cell Optogenetic Control of Neuronal Activity by Sculpted Light." *Proceedings of the National Academy of Sciences* 107 (26): 11981–86.

Anselmi, Francesca, Cathie Ventalon, Aurélien Bègue, David Ogden, and Valentina Emiliani. 2011. "Three-Dimensional Imaging and Photostimulation by Remote-Focusing and Holographic Light Patterning." *Proceedings of the National Academy of Sciences* 108 (49): 19504–509.

Arrizón, V, and M Testorf. 1997. "Efficiency Limit of Spatially Quantized Fourier Array Illuminators." *Optics Letters* 22 (4): 197–99.

Botcherby, Edward J, Rimas Juskaitis, Martin J Booth, and Tony Wilson. 2007. "Aberration-Free Optical Refocusing in High Numerical Aperture Microscopy." *Optics Letters* 32 (14): 2007–2009.

Chaigneau, Emmanuelle, Emiliano Ronzitti, Marta A Gajowa, Gilberto J Soler-Llavina, Dimitrii Tanese, Anthony Y B Brureau, Eirini Papagiakoumou, Hongkui Zeng, and Valentina Emiliani. 2016. "Two-Photon Holographic Stimulation of ReaChR." *Frontiers in Cellular Neuroscience* 10: 234.

Dal Maschio, Marco, Francesco Difato, Riccardo Beltramo, Axel Blau, Fabio Benfenati, and Tommaso Fellin. 2010. "Simultaneous Two-Photon Imaging and Photo-Stimulation with Structured Light Illumination." *Optics Express* 18 (18): 18720–31.

Dammann, H. 1970. "Blazed Synthetic Phase-Only Holograms." *Optik* 31 (1): 95.

Daria, Vincent Ricardo, Christian Stricker, Richard Bowman, Stephen Redman, and Hans-A Bachor. 2009. "Arbitrary Multisite Two-Photon Excitation in Four Dimensions." *Applied Physics Letters* 95 (9): 93701.

Dufresne, Eric R, Gabriel C Spalding, Matthew T Dearing, Steven A Sheets, and David G Grier. 2001. "Computer-Generated Holographic Optical Tweezer Arrays." *Review of Scientific Instruments* 72 (3): 1810–16.

Durst, Michael E, Guanghao Zhu, and Chris Xu. 2006. "Simultaneous Spatial and Temporal Focusing for Axial Scanning." *Optics Express* 14 (25): 12243–54.

Farah, Nairouz, Inna Reutsky, and Shy Shoham. 2007. "Patterned Optical Activation of Retinal Ganglion Cells." In *2007* 29th Annual International Conference of the IEEE Engineering in Medicine and Biology Society, 6368–70, Lyon, France.

Farah, Nairouz, Alaa Zoubi, Suhail Matar, Lior Golan, Anat Marom, Christopher R Butson, Inbar Brosh, and Shy Shoham. 2013. "Holographically Patterned Activation Using Photo-Absorber Induced Neural-Thermal Stimulation." *Journal of Neural Engineering* 10 (5): 56004. doi:10.1088/1741-2560/10/5/056004.

Fukuchi, Norihiro, Y E Biqing, Yasunori Igasaki, Narihiro Yoshida, Yuji Kobayashi, and Tsutomu Hara. 2005. "Oblique-Incidence Characteristics of a Parallel-Aligned Nematic-Liquid-Crystal Spatial Light Modulator." *Optical Review* 12 (5): 372–77.

Gallagher, N C, and B Liu. 1973. "Method for Computing Kinoforms that Reduces Image Reconstruction Error." *Applied Optics* 12 (10): 2328–35.

Gerchberg, Ralph W. 1972. "A Practical Algorithm for the Determination of Phase from Image and Diffraction Plane Pictures." *Optik* 35: 237–46.

Go, Mary Ann, Christian Stricker, Stephen Redman, Hans-A Bachor, and Vincent Ricardo Daria. 2012. "Simultaneous Multi-Site Two-Photon Photostimulation in Three Dimensions." *Journal of Biophotonics* 5 (10): 745–53.

GolanLior, Inna Reutsky-Gefen, Nairouz Farah, and Shy Shoham. 2009. "Design and Characteristics of Holographic Neural Photo-Stimulation Systems." *Journal of Neural Engineering* 6: 66004. doi:10.1088/1741-2560/6/6/066004.

Golan, Lior, and Shy Shoham. 2009. "Speckle Elimination Using Shift-Averaging in High-Rate Holographic Projection." *Optics Express* 17 (3): 1330. doi:10.1364/OE.17.001330.

Goodman, Joseph W. 1976. "Some Fundamental Properties of Speckle." *JOSA* 66 (11): 1145–50.

Goodman, Joseph W. 2005. *Introduction to Fourier Optics*. Roberts and Company Publishers, Greenwood Village, CO.

Goodman, Joseph W, and A M Silvestri. 1970. "Some Effects of Fourier-Domain Phase Quantization." *IBM Journal of Research and Development* 14 (5): 478–84.

Knapczyk, M, A Krishnan, L Grave de Peralta, A Bernussi, and H Temkin. 2005. "High-Resolution Pulse Shaper Based on Arrays of Digital Micromirrors." *IEEE Photonics Technology Letters* 17 (10): 2200–202.

Leonardo, Roberto Di, Francesca Ianni, and Giancarlo Ruocco. 2007. "Computer Generation of Optimal Holograms for Optical Trap Arrays." *Optics Express* 15 (4): 1913–22.

Lesem, L B, P M Hirsch, and J A Jordan. 1969. "The Kinoform: A New Wavefront Reconstruction Device." *IBM Journal of Research and Development* 13 (2): 150–55.

Lin, John Y, Per Magne Knutsen, Arnaud Muller, David Kleinfeld, and Roger Y Tsien. 2013. "ReaChR: A Red-Shifted Variant of Channelrhodopsin Enables Deep Transcranial Optogenetic Excitation." *Nature Neuroscience* 16 (10): 1499.

Lutz, Christoph, Thomas S Otis, Vincent DeSars, Serge Charpak, David A DiGregorio, and Valentina Emiliani. 2008. "Holographic Photolysis of Caged Neurotransmitters." *Nature Methods* 5 (9): 821–27. doi:10.1038/nmeth.1241.

Makowski, Michal. 2013. "Minimized Speckle Noise in Lens-Less Holographic Projection by Pixel Separation." *Optics Express* 21 (24): 29205–16.

Martín-Badosa, E, M Montes-Usategui, A Carnicer, J Andilla, E Pleguezuelos, and I Juvells. 2007. "Design Strategies for Optimizing Holographic Optical Tweezers Set-Ups." *Journal of Optics A: Pure and Applied Optics* 9 (8). doi:10.1088/1464-4258/9/8/S22.

Matar, Suhail, Lior Golan, and Shy Shoham. 2011. "Reduction of Two-Photon Holographic Speckle Using Shift-Averaging." *Optics Express* 19 (27): 25891. doi:10.1364/OE.19.025891.

Meister, Martin, and Richard J Winfield. 2002. "Novel Approaches to Direct Search Algorithms for the Design of Diffractive Optical Elements." *Optics Communications* 203 (1–2): 39–49.

Montes-Usategui, Mario, Encarnación Pleguezuelos, Jordi Andilla, and Estela Martín-Badosa. 2006. "Fast Generation of Holographic Optical Tweezers by Random Mask Encoding of Fourier Components." *Optics Express* 14 (6): 2101–107.

Mori, Yutaka, Takahiko Fukuoka, and Takanori Nomura. 2014. "Speckle Reduction in Holographic Projection by Random Pixel Separation with Time Multiplexing." *Applied Optics* 53 (35): 8182–88.

Nikolenko, Volodymyr, Kira E Poskanzer, and Rafael Yuste. 2007. "Two-Photon Photostimulation and Imaging of Neural Circuits." *Nature Methods* 4 (11): 943.

Nikolenko, Volodymyr, Brendon O Watson, Roberto Araya, Alan Woodruff, Darcy S Peterka, and Rafael Yuste. 2008. "SLM Microscopy: Scanless Two-Photon Imaging and Photostimulation Using Spatial Light Modulators." *Frontiers in Neural Circuits* 2: 5.

Oron, Dan, Eirini Papagiakoumou, Francesca Anselmi, and Valentina Emiliani. 2012. "Two-Photon Optogenetics." *Progress in Brain Research*, 196: 119–43.

Oron, Dan, Eran Tal, and Yaron Silberberg. 2005. "Scanningless Depth-Resolved Microscopy." *Optics Express* 13 (5): 1468–76.

Packer, Adam M, Darcy S Peterka, Jan J Hirtz, Rohit Prakash, Karl Deisseroth, and Rafael Yuste. 2012. "Two-Photon Optogenetics of Dendritic Spines and Neural Circuits." *Nature Methods* 9 (12): 1202.

Packer, Adam M, Lloyd E Russell, Henry W P Dalgleish, and Michael Häusser. 2015. "Simultaneous All-Optical Manipulation and Recording of Neural Circuit Activity with Cellular Resolution in Vivo." *Nature Methods* 12 (2): 140.

Palima, Darwin, and Vincent Ricardo Daria. 2007. "Holographic Projection of Arbitrary Light Patterns with a Suppressed Zero-Order Beam." *Applied Optics* 46 (20): 4197–201.

Paluch-Siegler, Shir, Hod Dana, Tom Mayblum, Inbar Brosh, Inna Gefen, and Shy Shoham. 2015. "All-Optical Bidirectional Neural Interfacing Using Hybrid Multiphoton Holographic Optogenetic Stimulation." *Neurophotonics* 2 (3): 31208.

Pang, Hui, Jiazhou Wang, Axiu Cao, and Qiling Deng. 2016. "High-Accuracy Method for Holographic Image Projection with Suppressed Speckle Noise." *Optics Express* 24 (20): 22766. doi:10.1364/OE.24.022766.

Pang, Hui, Jiazhou Wang, Man Zhang, Axiu Cao, Lifang Shi, and Qiling Deng. 2017. "Non-Iterative Phase-Only Fourier Hologram Generation with High Image Quality." *Optics Express* 25 (13): 14323. doi:10.1364/OE.25.014323.

Papagiakoumou, Eirini. 2013. "Optical Developments for Optogenetics." *Biology of the Cell* 105 (10): 443–64.

Papagiakoumou, Eirini, Francesca Anselmi, Aurélien Bègue, Vincent De Sars, Jesper Glückstad, Ehud Y Isacoff, and Valentina Emiliani. 2010. "Scanless Two-Photon Excitation of Channelrhodopsin-2." *Nature Methods* 7 (10): 848.

Papagiakoumou, Eirini, Aurélien Bègue, Ben Leshem, Osip Schwartz, Brandon M Stell, Jonathan Bradley, Dan Oron, and Valentina Emiliani. 2013. "Functional Patterned Multiphoton Excitation Deep Inside Scattering Tissue." *Nature Photonics* 7 (4): 274.

Papagiakoumou, Eirini, Vincent De Sars, Valentina Emiliani, and Dan Oronb. 2009. "Temporal Focusing with Spatially Modulated Excitation." *Optics Express* 17 (7): 5391–5401.

Papagiakoumou, Eirini, Vincent De Sars, Dan Oron, and Valentina Emiliani. 2008. "Patterned Two-Photon Illumination by Spatiotemporal Shaping of Ultrashort Pulses." *Optics Express* 16 (26): 22039–47.

Pégard, Nicolas C, Alan R Mardinly, Ian Antón Oldenburg, Savitha Sridharan, Laura Waller, and Hillel Adesnik. 2017. "Three-Dimensional Scanless Holographic Optogenetics with Temporal Focusing (3D-SHOT)." *Nature Communications* 8 (1): 1228.

Polin, Marco, Kosta Ladavac, Sang-Hyuk Lee, Yael Roichman, and David G Grier. 2005. "Optimized Holographic Optical Traps." *Optics Express* 13 (15): 5831–45.

Reutsky-Gefen, Inna, Lior Golan, Nairouz Farah, Adi Schejter, Limor Tsur, Inbar Brosh, and Shy Shoham. 2013. "Holographic Optogenetic Stimulation of Patterned Neuronal Activity for Vision Restoration." *Nature Communications* 4. doi:10.1038/ncomms2500.

Rickgauer, John Peter, and David W Tank. 2009. "Two-Photon Excitation of Channelrhodopsin-2 at Saturation." *Proceedings of the National Academy of Sciences* 106 (35): 15025–30.

Ronzitti, Emiliano, Marc Guillon, Vincent de Sars, and Valentina Emiliani. 2012. "LCoS Nematic SLM Characterization and Modeling for Diffraction Efficiency Optimization, Zero and Ghost Orders Suppression." *Optics Express* 20 (16): 17843. doi:10.1364/OE.20.017843.

Ronzitti, Emiliano, Cathie Ventalon, Marco Canepari, Benoît C Forget, Eirini Papagiakoumou, and Valentina Emiliani. 2017. "Recent Advances in Patterned Photostimulation for Optogenetics." *Journal of Optics* 19 (11): 113001.

Shemesh, Or A, Dimitrii Tanese, Valeria Zampini, Changyang Linghu, Kiryl Piatkevich, Emiliano Ronzitti, Eirini Papagiakoumou, Edward S Boyden, and Valentina Emiliani. 2017. "Temporally Precise Single-Cell-Resolution Optogenetics." *Nature Neuroscience* 20 (12): 1796.

Shimobaba, Tomoyoshi, and Tomoyoshi Ito. 2015. "Random Phase-Free Computer-Generated Hologram." *Optics Express* 23 (7): 9549. doi:10.1364/OE.23.009549.

Shoham, Shy. 2010. "Optogenetics Meets Optical Wavefront Shaping." *Nature Methods* 7 (10): 798.

Shoham, Shy, Daniel H O'Connor, Dmitry V Sarkisov, and Samuel S H Wang. 2005. "Rapid Neurotransmitter Uncaging in Spatially Defined Patterns." *Nature Methods* 2 (11): 837.

Spalding, G C, J Courtial, and R D Leonardo. 2008. "Holographic Optical Tweezers," in: *Structured Light and Its Applications*, Andrews, D L (Ed.), Academic Press, London, UK.

Takaki, Yasuhiro, and Masahito Yokouchi. 2011. "Speckle-Free and Grayscale Hologram Reconstruction Using Time-Multiplexing Technique." *Optics Express* 19 (8): 7567–79.

Tan, Kim L, Stephen T Warr, Ilias G Manolis, Timothy D Wilkinson, Maura M Redmond, William A Crossland, Robert J Mears, and Brian Robertson. 2001. "Dynamic Holography for Optical Interconnections. II. Routing Holograms with Predictable Location and Intensity of Each Diffraction Order." *JOSA* A 18 (1): 205–15.

Vaziri, Alipasha, and Valentina Emiliani. 2012. "Reshaping the Optical Dimension in Optogenetics." *Current Opinion in Neurobiology* 22 (1): 128–37.

Wang, Sheng, Stephanie Szobota, Yuan Wang, Matthew Volgraf, Zhaowei Liu, Cheng Sun, Dirk Trauner, Ehud Y Isacoff, and Xiang Zhang. 2007. "All Optical Interface for Parallel, Remote, and Spatiotemporal Control of Neuronal Activity." *Nano Letters* 7 (12): 3859–63.

Wu, Liang, Shubo Cheng, and Shaohua Tao. 2015. "Complex Amplitudes Reconstructed in Multiple Output Planes with a Phase-Only Hologram." *Journal of Optics* 17 (12): 125603.

Wyrowski, Frank, and Olof Bryngdahl. 1989. "Speckle-Free Reconstruction in Digital Holography." *JOSA* A 6 (8): 1171–74.

Yang, Sunggu, Eirini Papagiakoumou, Marc Guillon, Vincent De Sars, Cha-Min Tang, and Valentina Emiliani. 2011. "Three-Dimensional Holographic Photostimulation of the Dendritic Arbor." *Journal of Neural Engineering* 8 (4): 46002.

Yizhar, Ofer, Lief E Fenno, Thomas J Davidson, Murtaza Mogri, and Karl Deisseroth. 2011. "Optogenetics in Neural Systems." *Neuron* 71 (1): 9–34. doi:10.1016/j.neuron.2011.06.004.

Yuan, Zhanzhong, and Shaohua Tao. 2014. "Generation of Phase-Gradient Optical Beams with an Iterative Algorithm." *Journal of Optics* 16 (10): 105701.

Zhang, Feng, Matthias Prigge, Florent Beyrière, Satoshi P Tsunoda, Joanna Mattis, Ofer Yizhar, Peter Hegemann, and Karl Deisseroth. 2008. "Red-Shifted Optogenetic Excitation: A Tool for Fast Neural Control Derived from Volvox Carteri." *Nature Neuroscience* 11 (6): 631.

Zhang, Feng, Li-Ping Wang, Edward S Boyden, and Karl Deisseroth. 2006. "Channelrhodopsin-2 and Optical Control of Excitable Cells." *Nature Methods* 3 (10): 785.

Zhang, Hao, Jinghui Xie, Juan Liu, and Yongtian Wang. 2009. "Elimination of a Zero-Order Beam Induced by a Pixelated Spatial Light Modulator for Holographic Projection." *Applied Optics* 48: 5834–41. doi:10.1364/AO.48.005834.

Zhu, Guanghao, James Van Howe, Michael Durst, Warren Zipfel, and Chris Xu. 2005. "Simultaneous Spatial and Temporal Focusing of Femtosecond Pulses." *Optics Express* 13 (6): 2153–59.

CHAPTER 16

Multi-Photon Nanosurgery

Anna Letizia Allegra Mascaro and Francesco Saverio Pavone

CONTENTS

16.1 INTRODUCTION

In this chapter we present basic features of multi-photon nanosurgery and show some examples that illustrate the advantages offered by this novel methodology. In the first part we will review selected literature on the application of pulsed-laser irradiation to biological samples, focusing on the nervous system. In the second part we will discuss hypotheses on the mechanism of nanosurgery. The third section will provide technical details on laser dissection.

16.2 NANODISSECTION IN THE NERVOUS SYSTEM OF LIVING ORGANISMS

Nonlinear optical interactions are intrinsically confined to the focal region of a focused laser beam. This leads to inherent 3D spatial resolution, but also to the possibility of localized targeted photophysics and photochemistry. In parallel to its application in imaging, multi-photon absorption was targeted to the manipulation of the biological sample under investigation. Laser-microdissection was used to carve out channels in hydrogels for directing neuronal growth (Sarig-Nadir et al., 2009). Importantly, high-energy pulsed-laser irradiation has been applied to the selective disruption of labeled cells, intracellular structures, and blood vessels. NIR radiation can induce the formation of reactive oxygen species (ROS) by nonlinear excitation of photosensitive molecules and subsequent photo-oxidation processes (Vogel and Venugopalan, 2003; Vogel et al., 2005). ROS-mediated lethal effects are used in photodynamic therapy of tumors and other diseases (Doiron and Gomer, 1984). Another two-photon induced photochemical effect is the highly localized photorelease of caged compounds with NIR laser radiation. The photorelease (or photocleavage, uncaging) of bioactive substances, as well as of photolabile ion-specific (e.g. Ca^{2+} and Mg^{2+}) chelators, have been reported (Denk, 1994; Callaway and Yuste, 2002; Soeller et al., 2003). Direct optical stimulation without the use of cages or dyes is also possible (Hirase et al., 2002), but may cause irreversible photodamage as a side-effect.

However, in certain scenarios irreversible photodamage might be the goal. In particular, multi-photon absorption can be exploited to selectively disrupt labelled cells and intracellular structures. Laser nanodissection has proven to be highly advantageous over the micromanipulators or needles used to dissect cells in the past, owing to its sub-micrometer precision and to the fact that no mechanical interaction with the sample is involved. Due to its capability of dissecting different structures at multiple scales, laser nanosurgery has been used to target and perturb intracellular structures, single cells, or a subpopulation of cells in an organism. The first published report of femtosecond laser subcellular nanosurgery was by Konig et al. (1999); the authors demonstrated the potential of the technique by ablating nano-scale regions of the genome within the nucleus of living cells. Following this seminal work, laser nanosurgery has been applied in living cells to investigate the biological function of subcellular compartments, like mitochondria or microtubules (Tolic-Norrelykke et al., 2004; Watanabe et al., 2004; Sacconi et al., 2005; Shen et al., 2005; Shimada et al., 2005; Kumar et al., 2006). A similar approach has been applied to living animals, where two-photon fluorescence (TPF) imaging and laser-induced lesions have been combined. For instance, femtosecond laser ablations allowed the characterization of the complex events associated with embryo development *in vivo* (Supatto et al., 2004; Kohli and Elezzabi, 2008).

In the nervous system, the opportunity provided by multi-photon nanosurgery to selectively injure single elements of neuronal networks is a valuable tool for dissecting the rules of brain remodeling. Ben-Yakar's group developed a laser axotomy lab-on-a-chip for *in vivo* nerve regeneration studies in living *C. elegans* (Guo et al., 2008). By taking advantage of multi-photon absorption to sever individual neurons, the authors demonstrated nerve regeneration after axotomy in the nematode (Figure 16.1a) (Yanik et al., 2004, Guo et al., 2008). Laser-mediated dissection was applied to define the role of specific neurons in behavior; as an example, a study by Chung et al. on the thermo-tactic behavior of nematodes identified the function of a single neuron by observing behavioral changes after the surgery (Chung et al., 2006). Thanks to this selectivity, targeted dissection of single neurons recently allowed to causally study the effects of single neurons on network coding (see for instance Barker and Baier, 2015; Yan et al., 2017).

One very important application of laser nanodissection is to study the mammalian central nervous system (CNS). One of the first *in vivo* applications to the neocortex was shown in a seminal study by Sacconi et al., where laser nanosurgery was performed on selected cortical neurons in living mice expressing fluorescent proteins (Sacconi et al., 2007). Different compartments of CNS neurons were disrupted: neural cell bodies, single neuronal branches, or even single spines (Sacconi et al., 2007; Allegra Mascaro et al., 2010).

Following this work, Allegra Mascaro et al. (2013) used two-photon imaging and laser axotomy to test the regenerative potential of laser-injured axons in the cerebellar cortex *in vivo* (Figure 16.1b). Neurons irradiated with a focused, controlled femtosecond energy dose were characterized with time lapse 3D two-photon imaging. The authors demonstrated growth associated protein-43 (GAP43) dependent plasticity of climbing fibers in adult mice after laser dissection of single axonal branches. Selective photo-ablation of cerebellar climbing fibers was used in mouse pups to study developmental synapse competition in the CNS (Carrillo et al., 2013). The authors demonstrated the role of different neuronal domains in regulating synapse elimination. In parallel, Canty and colleagues (Canty et al., 2013) tested the heterogeneity of response to injury of cortical axons (Figure 16.1c). Individually laser-lesioned axons from different neuronal classes in the adult cortical gray matter exhibited either a low or high level of regeneration after laser axotomy associated with the re-establishment of normal synaptic density. Axons of regenerated neurons showed dynamic changes several months after the initial damage (Canty et al., 2013).

Laser irradiation has been applied to the spinal cord to investigate the regenerative properties of a lesioned axon. By taking advantage of an *in vivo* preparation for TPF imaging in the mouse spinal cord (Davalos et al., 2008), Lorenzana and colleagues (Lorenzana et al., 2015) laser-axotomized a dorsal column sensory axon just before or after a major branch point. They found different axonal responses depending on the location of the axotomy within the branch, i.e. either near or distal to the branch point. This study illustrated that the axonal branch may inhibit retrograde degeneration and that a spared axonal branch stabilizes remaining axon architecture after axotomy to the other branch. In the developing peripheral nervous system, Turney and Lichtman showed that an axon that would have been eliminated was spared by removing with laser microsurgery another axon converging on the same synaptic site (Turney and Lichtman, 2012). The remaining axon not only survives, but rapidly grows to occupy the synaptic sites vacated by the removed axon.

Taken together, these results provide a framework for understanding synaptic rearrangements.

Furthermore, laser-induced lesions have been used to damage glial cells and the blood–brain barrier (BBB). A full characterization of the dynamics of microglial migration towards the lesion site allowed to confirm the role of these cells in the recovery process (Davalos et al., 2005, Nimmerjahn et al., 2005, Tsai et al., 2009). In addition, a recent study on the role of astrocytes on BBB permeability showed that blood vessels maintain their integrity even after removal of astrocytic endfoot by laser ablation, thus suggesting that they do not contribute to the immediate BBB barrier (Kubotera et al., 2019).

Nonlinear laser ablation provides a method to induce vascular lesions one vessel at a time and to study cerebral microvascular diseases. Laser dissection has been exploited to produce targeted photo-disruptions of blood vessels, defining a new model of highly confined stroke (Nishimura et al., 2006). Focal photothrombosis was used to occlude single penetrating arterioles to investigate the modification in flow of red blood cells in individual subsurface microvessels surrounding the occlusion. TPF microscopy, coupled to microvessel occlusion technique, enabled studying the impact of microvascular lesions at cellular level, e.g. by showing that penetrating arterioles are a bottleneck in the supply of blood to superficial layers of neocortex (Nishimura et al., 2007). Alternatively, femtosecond laser pulses focused on a blood vessel may lead to another form of vascular insult, namely

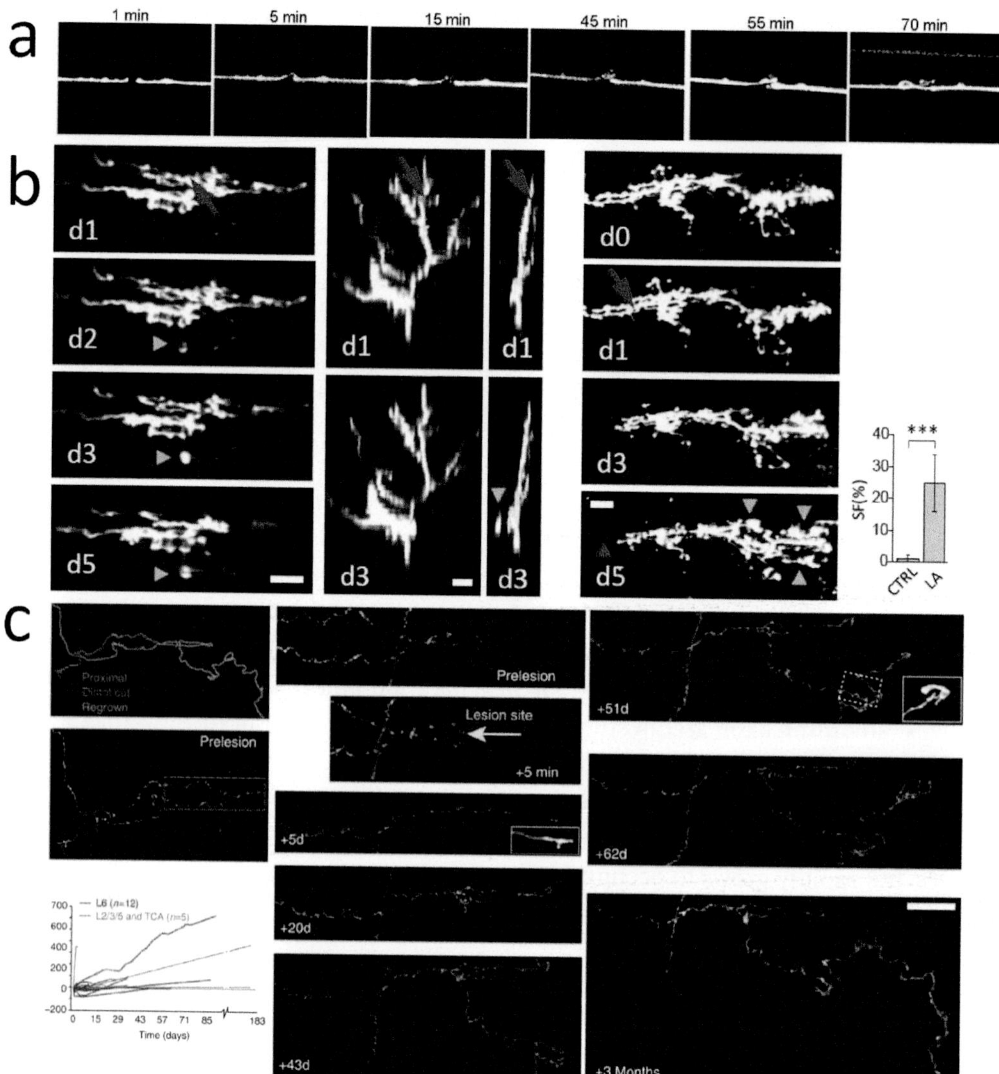

FIGURE 16.1 In vivo rewiring of axonal branches after laser nanosurgery. (a) Time-lapse imaging of C. elegans in a microfluidic chip showing axonal recovery of an injured neuron. The fluorescence images show the reconnection of the proximal to the distal end of an axon 70 minutes after laser-nanosurgery. Scale bar, 10 μm. (Reprinted with permission from Springer Nature Customer Service Centre GmbH: Springer, Nature Methods, Femtosecond laser nanoaxotomy lab-on-a-chip for in vivo nerve regeneration studies (Guo et al., 2008) (b) The panels show two time courses (from day 0 to day 5) of cerebellar axons before and after laser dissection. The first image (d0) was acquired one day before laser irradiation. The laser beam was focused on the axon where the red arrow points on d1. The red and green arrowheads highlight the degeneration of distal portion and the protrusion of new branches, respectively. Scale bar: 15 μm. The histogram compares the sprouting frequency (SF) in control and laser axotomized climbing fibers. (Reproduced with permission from Allegra Mascaro et al., 2013.) The lower panel shows an example of nanosurgery and regeneration of cortical pyramidal neurons. Following laser dissection (yellow arrow in the +5 min panel)

FIGURE 16.1 (CONTINUED)
and subsequent degeneration of the axonal ending (red line), few axons of pyramidal layer 6 (L6) neuron demonstrated sustained regrowth over a 3-month time period (green line). The red dotted line box indicates the area followed over time in the time-stamped panels. Insets at +5d and +51d show the growth cone. The graph on the lower left shows that 55% of L6 axons (red) demonstrated regrowth over several weeks compared with only ~20% of all other axons (blue). (Reproduced with permission from Canty et al., 2013.)

vessel rupture, which provides a model of hemorrhage (Nishimura et al., 2006). Time-lapse observation of the adjacent labeled neurons reveals a relatively high stability of the network in the case of confined damage, while loss of dendritic spines and degeneration of dendrites can be observed if the injury is more widespread (Zhang and Murphy, 2007).

This model of selective disruption of single blood vessels is highly relevant in clinical settings, since cortical micro-infarcts and micro-hemorrhages increase the risk of developing dementia (Cullen et al., 2005, Gold et al., 2007).

Finally, an interesting application of the method is for correlative light and electron microscopy experiments. Pulsed near-infrared lasers can be used to create a three-dimensional marking in a tissue to identify a cell of interest for correlative light and electron microscopy applications (Bishop et al., 2011; Allegra Mascaro et al., 2013; Canty et al., 2013; Karreman et al., 2014).

16.3 PHYSICAL MECHANISM OF NANOSURGERY

Crucial elements of nanosurgery are the laser and the method to target the structure of interest. Laser pulses can damage the cell or tissue at different levels. The extent of the damage strongly depends on the power, repetition rate, and duration of the pulses. Based upon the laser features, two regimes can be established for nanosurgery (Vogel and Venugopalan, 2003; Vogel et al., 2005). The high repetition rate (or "low-density plasma") regime uses long series of pulses from femtosecond oscillators with repetition rates of the order of 80 MHz and pulse energies well below the optical breakdown threshold (Konig et al., 1999; Oehring et al., 2000; Konig et al., 2001; Smith et al., 2001; Tirlapur and Konig, 2002; Zeira et al., 2003; Sacconi et al., 2005; Supatto et al., 2005; Sacconi et al., 2007). The low repetition rate regime uses amplified series of pulses at kHz repetition rates and pulse energies slightly above the optical breakdown threshold (Watanabe et al., 2004; Yanik et al., 2004; Heisterkamp et al., 2005; Shen et al., 2005).

The hypothetic mechanisms of these two regimes are explained below. The low repetition rate regime effects are related to thermoelastically induced formation followed by exponential growth of micrometer-sized plasma bubbles. An extensive review focused on this regime was published by Tsai et al. (2009). Otherwise, the dissection may be performed in a high repetition rate regime; in this case the dissection is mediated by quasi-free-electron-induced decomposition. Electrons are considered "quasi-free" if they have sufficient kinetic energy to be able to move without being captured by local potential energy barriers. As reported in a paper by Vogel and his collaborators (Vogel et al., 2005), the formation of quasi-free electrons depends on a combination of photoionization and impact ionization (Figure 16.2a).

The photoionization mechanism relies on a combination of multi-photon ionization (MPI) and tunneling; the interaction of a high-energy photon and a molecule results in the release of a quasi-free electron in the medium. Afterwards, the quasi-free electron can absorb photons in a non-resonant process, called "inverse Bremsstrahlung", whilst colliding with ions or atomic nuclei (Ready, 1971). After a sequence of inverse Bremsstrahlung absorption events, the kinetic energy of the electron is sufficiently large to produce another

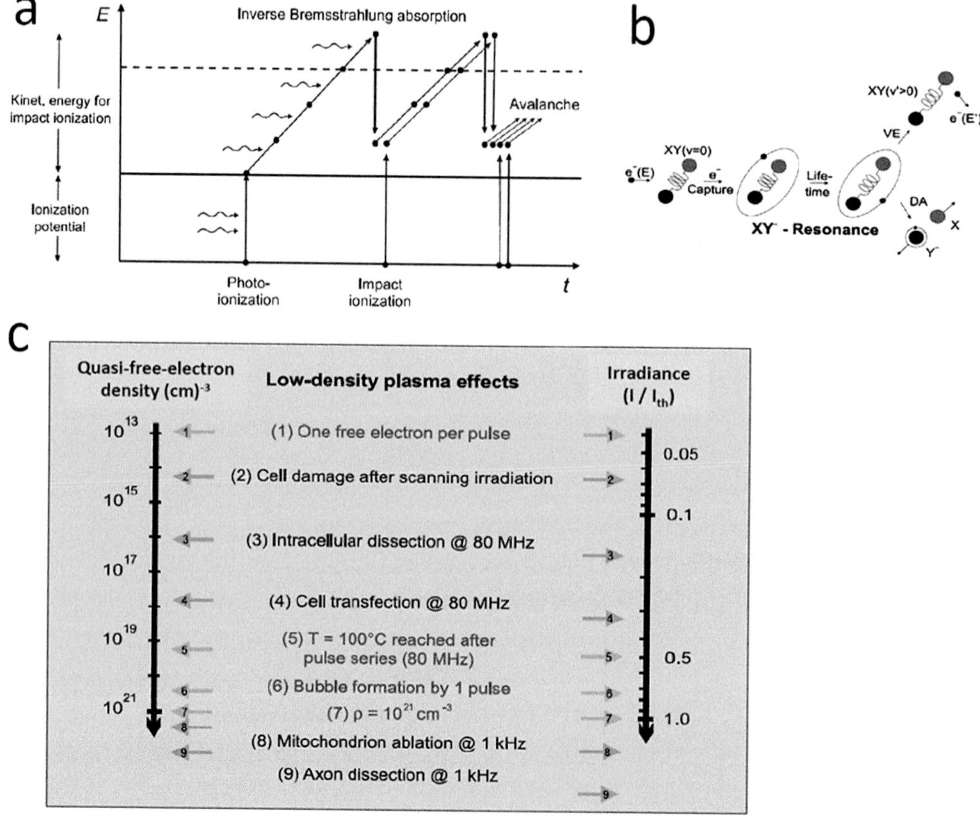

FIGURE 16.2 Mechanism of nanosurgery. (a) Interplay of photoionization, inverse Bremsstrahlung absorption, and impact ionization in the process of plasma formation. Recurring sequences of inverse Bremsstrahlung absorption events and impact ionization lead to an avalanche growth in the number of free electrons. (b) Mechanism of plasma-induced chemical decomposition. The image illustrates the proposed dynamics of the vibrational excitation and dissociative electron attachment in resonant electron–molecule model. (c) Overview of physical breakdown phenomena induced by femtosecond laser pulses, together with experimental damage, transfection, and dissection thresholds for cells. The different effects are depicted together with the corresponding values of quasi-free-electron density and irradiance. All data refer to plasma formation in water with femtosecond pulses of about 100 fs duration and 800 nm wavelength. (Reprinted by permission from Springer Nature Customer Service Centre GmbH: Springer, Applied Physics B: Lasers and Optics, Mechanisms of femtosecond laser nanosurgery of cells and tissues (Vogel et al., 2005).)

free electron through impact ionization (Arnold et al., 1992; Ridley, 1999; Kaiser et al., 2000). If the irradiance is high enough to overcome the losses of free electrons through diffusion out of the focal volume and through recombination, the recurring sequence of inverse Bremsstrahlung absorption events and impact ionization leads to an avalanche growth in the number of free electrons (so-called "avalanche ionization" or "cascade ionization", see Figure 16.2a).

Next, the plasma-mediated chemical decomposition of the fluorophore can be induced through two different mechanisms:

i) Capture of a quasi-free electron of the plasma in an antibonding molecular orbital causes the rupture of a chemical bond; this resonant interaction can initiate the fragmentation of the biological molecule (Boudaiffa et al., 2000; Hotop, 2001; Gohlke and Illenberger, 2002; Huels et al., 2003; Garrett et al., 2005). Capture can occur when the nuclear wave functions of the initial ground state and the final anion state overlap. For a molecule XY, this process corresponds to: $e^- + XY \rightarrow XY^{*-}$. Afterwards, the transient anion state decays either by electron detachment, leaving a vibrationally excited molecule (VE in Figure 16.2b), or by the dissociative process $XY^{*-} \rightarrow X^{\bullet} + Y^-$ (DA in Figure 16.2b).

ii) Fragmentation of the biomolecule derives from its interaction with reactive oxygen species (ROS, i.e. free radicals that contain oxygen atoms), originating from laser-induced ionization and dissociation of water molecules. The creation of ROS such as OH^* and H_2O_2 through various pathways following ionization and dissociation of water molecules has been investigated by Nikogosyan et al. (1983), and reviewed by Garret et al. (2005). Both oxygen species are known to cause cell damage (Konig et al., 1999). Heisterkamp et al. (2002) confirmed the dissociation of water molecules during femtosecond-laser-induced plasma formation by chemical analysis of the gas content of the bubbles.

Figure 16.2c summarizes different low-density plasma effects and physical breakdown phenomena, scaled by the corresponding values of quasi-free-electron density and irradiance. The reported examples correspond to specific studies: "cell damage" (point 2 in Figure 16.2c) refers to the irradiation of PtK2 cells with 800-nm pulses at 80-MHz repetition rate, leading to membrane dysfunction and, finally, to apoptosis-like cell death (Tirlapur et al., 2001); "chromosome dissection" (3) relates to intra-nuclear chromosome dissection (Konig et al. 1999); "cell transfection" (4) was performed inducing a transient membrane permeabilization through laser irradiation (Tirlapur and Konig, 2002); "mitochondrion ablation" (8) concerns the disruption of a single mitochondrion in a living cell using 1-kHz pulse trains (Shen et al., 2005); "axon dissection" (9) addresses experiments of axotomy performed in *C. elegans* worms (Yanik, et al., 2004).

As a general rule, the extent of the induced plasma depends on the pulse energy, pulse duration, and wavelength of the laser as well as the focal volume determined by the numerical aperture of the objective. Nanosurgery can be confined to a very small spot if performed with light of shorter wavelength, which can be focused into a smaller diffraction limited volume. Noteworthy, the energy threshold for plasma formation increases with the wavelength. In addition, more photons are needed to cross the photoionization threshold with lower frequency radiation (Vogel and Noack, 2001; Vogel et al., 2005). Other crucial factors to consider are the label of the target structure, which influences the ablation threshold significantly (Botvinick et al., 2004; Raabe et al., 2009), as well as the laser pulse duration that heavily impacts the efficiency of plasma formation in the focal volume. Shorter pulses allow nanosurgery at lower energy levels: while hundreds of μJ are required using nanosecond pulses (Vogel et al., 2005), three orders of magnitude less energy is needed with picosecond pulses (Aist et al., 1993), and only few nJ are sufficient with femtosecond pulses (Heisterkamp et al., 2005). Most lasers for two-photon microscopes deliver pulsed infrared light with a pulse width in the range of hundreds of femtoseconds. Using these lasers at power levels higher than usually employed for imaging allows performing nanosurgery on selected structures. High transmission values for objective lenses in the wavelength range employed in the experiment are very important, not only to reach sufficient energy levels in the focus, but also to prevent damage to the objective lens by the high energy deposition of the pulsed laser (Markaki et al., 2017). To induce a very precise ablation of the structure of interest at a sub-micron scale, most nanosurgery applications are based on plasma-induced ablation, because the damage it causes remains restricted to the diffraction limited focal volume (Vogel et al., 2005; Markaki et al., 2017). Pulsed lasers with high peak power in the femtosecond to a few nanosecond regimes can provide sufficient photon densities to generate plasma in a biological sample by light. Nonlinear absorption as used in two-photon microscopy confines the plasma formation and the resulting nanosurgery to the focal volume. To ablate larger volumes or complex structures either the laser spot can be targeted to the region of interest by a scanner or the sample can be moved by a motorized *xy*-stage.

16.4 OPTICAL NANOSURGERY PROCEDURE

The nanosurgery experiment starts with an *in vivo* imaging session of the exposed mouse brain under the optical window. Z-stacks are acquired to obtain 3D-reconstructions of the labeled sample. During the imaging session, the excitation wavelength (λ) is chosen to maximize the multi-photon cross-section of the labeled structures (in our experiment with EGFP-tagged neurons, $\lambda = 935$ nm). After selecting the *xyz* coordinates for the desired lesion site, laser dissection is performed by irradiating the selected point with a high energy dose of Ti:Sapphire laser at the same wavelength used for imaging.

In a typical experiment, the laser power is increased 5–10 times more than the power used for imaging and the shutter is opened for a period of the order of hundreds of milliseconds. The increase of energy dose delivered on the sample in the dissection with respect to imaging arises largely from an increase in exposure time, with a lower contribution from the increase in laser power itself. The pixel dwell time is $\approx$5 μs during imaging, whereas it is increased to hundreds of milliseconds to perform the nanosurgery. As a consequence, the number of pulses (and the associated energy dose) delivered per unit area is drastically increased (five orders of magnitude). After performing the nanosurgery, the laser power is decreased and a 3D-image of the sample is acquired to visualize the effects on the irradiated structure. It is worth mentioning that the experimental parameters to perform nanosurgery, i.e. laser power and exposure time, critically depend on the target. Among the many factors, depth, fluorescence level, and dimension of the target structure remarkably affect the energy dose threshold to perform the nanosurgery. In a study in 2011, laser cuts were performed under different irradiation conditions on the cortex of anesthetized rodents to characterize the damage produced by nanosurgery (Nguyen et al., 2011). The width of the laser cut was shown to decrease exponentially with depth and increase as the cube root of the laser energy. Cut width at fixed depth and laser energy was highly variable. The authors showed that the width of the laser cut (w) scaled as the cube root of the ballistic laser energy that reaches the focus unscattered, which is an exponential function of depth (d) beneath the cortical surface:

$$w = C\exp\left(\frac{-d}{3l_s}\right)$$

where C depends on hydrostatic and vapor pressures and on laser energy (Vogel et al., 1999) and l_s is the scattering length. In addition, they showed that the maximal achievable cut depth (d_{max}) increased logarithmically with laser energy:

$$d_{max} = l_s \ln\left(\frac{E}{E_{th}}\right)$$

where *Eth* is the threshold energy for optical breakdown in the brain. Although two-photon microscopy allows imaging of layer-5 pyramidal neurons (600–800 μm below the pia), nanosurgery is more limited in the achievable depth because of energy density loss along the beam path across the brain. The soma of layer-2/3 pyramidal neurons (200–450 μm below the pia) can be selectively disrupted by laser irradiation.

Finally, the anesthetic levels and the optical window preparation critically affect the stability of the sample and, thus, the outcome of the irradiation. Practically, the laser power is set close to the maximum available value (≈300 mW, measured after the objective lens) and an exposure time of 100 ms is tested to induce ablation. Then an imaging session is performed to observe the effects of the laser irradiation. If no morphological changes were induced, the laser surgery is repeated with an increased exposure time. This procedure is performed iteratively, until a detectable structural modification is observed. In a typical experiment the exposure time ranges from 100 to 1,000 ms.

16.5 CONCLUSIONS

Neuronal plasticity after damage has been intensively studied for more than a century in mature vertebrates. Thanks to recent advances in laser technology, injury paradigms achieved a really high resolution; the spatial localization of multi-photon absorption (Zipfel et al., 2003) can indeed be exploited to perform *in vivo* selective lesions on single cellular compartments (Sacconi et al., 2005) without causing any visible collateral damage to the surrounding structures (Sacconi et al., 2007). This micro-scale model of neural degeneration, combined with two-photon fluorescence microscopy, allows the study of the remodeling properties of neuronal and vascular networks after injury in optically accessible parts of the adult CNS *in vivo* (Helmchen and Denk, 2005; Svoboda and Yasuda, 2006). In comparison with other models of neural ablation, e.g. massive neural disruption through mechanical transection, nanosurgery is highly specific and can be finely tuned to a single neuron, leaving the surrounding structures unaltered.

Here we provided an extensive description of the nanosurgery technique, from recent applications in the nervous system of living organisms, to mechanistic insights, and, finally, to technical details for performing laser dissection.

BIBLIOGRAPHY

Aist, J. R., H. Liang and M. W. Berns (1993). "Astral and spindle forces in PtK2 cells during anaphase B: a laser microbeam study." *J Cell Sci* **104**(Pt 4): 1207–1216.

Allegra Mascaro, A. L., P. Cesare, L. Sacconi, G. Grasselli, G. Mandolesi, B. Maco, G. W. Knott, L. Huang, V. De Paola, P. Strata and F. S. Pavone (2013). "In vivo single branch axotomy induces GAP-43-dependent sprouting and synaptic remodeling in cerebellar cortex." *Proc Natl Acad Sci U S A* **110**(26): 10824–10829.

Allegra Mascaro, A. L., L. Sacconi and F. S. Pavone (2010). "Multi-photon nanosurgery in live brain." *Front Neuroenergetics* **2**: 21.

Arnold, D., E. Cartier and D. J. Dimaria (1992). "Acoustic-phonon runaway and impact ionization by hot-electrons in silicon dioxide." *Phys Rev B* **45**(3): 1477–1480.

Barker, A. J. and H. Baier (2015). "Sensorimotor decision making in the zebrafish tectum." *Curr Biol* **25**(21): 2804–2814.

Bishop, D., I. Nikic, M. Brinkoetter, S. Knecht, S. Potz, M. Kerschensteiner and T. Misgeld (2011). "Near-infrared branding efficiently correlates light and electron microscopy." *Nat Methods* **8**(7): 568–570.

Botvinick, E. L., V. Venugopalan, J. V. Shah, L. H. Liaw and M. W. Berns (2004). "Controlled ablation of microtubules using a picosecond laser." *Biophys J* **87**(6): 4203–4212.

Boudaiffa, B., P. Cloutier, D. Hunting, M. A. Huels and L. Sanche (2000). "Resonant formation of DNA strand breaks by low-energy (3 to 20 eV) electrons." *Science* **287**(5458): 1658–1660.

Callaway, E. M. and R. Yuste (2002). "Stimulating neurons with light." *Curr Opin Neurobiol* **12**(5): 587–592.

Canty, A. J., L. Huang, J. S. Jackson, G. E. Little, G. Knott, B. Maco and V. De Paola (2013). "In-vivo single neuron axotomy triggers axon regeneration to restore synaptic density in specific cortical circuits." *Nat Commun* **4**: 2038.

Carrillo, J., N. Nishiyama and H. Nishiyama (2013). "Dendritic translocation establishes the winner in cerebellar climbing fiber synapse elimination." *J Neurosci* **33**(18): 7641–7653.

Chung, S. H., D. A. Clark, C. V. Gabel, E. Mazur and A. D. Samuel (2006). "The role of the AFD neuron in *C. elegans* thermotaxis analyzed using femtosecond laser ablation." *BMC Neurosci* **7**: 30.

Cullen, K. M., Z. Kocsi and J. Stone (2005). "Pericapillary haem-rich deposits: evidence for microhaemorrhages in aging human cerebral cortex." *J Cereb Blood Flow Metab* **25**(12): 1656–1667.

Davalos, D., J. Grutzendler, G. Yang, J. V. Kim, Y. Zuo, S. Jung, D. R. Littman, M. L. Dustin and W. B. Gan (2005). "ATP mediates rapid microglial response to local brain injury in vivo." *Nat Neurosci* **8**(6): 752–758.

Davalos, D., J. K. Lee, W. B. Smith, B. Brinkman, M. H. Ellisman, B. Zheng and K. Akassoglou (2008). "Stable in vivo imaging of densely populated glia, axons and blood vessels in the mouse spinal cord using two-photon microscopy." *J Neurosci Methods* **169**(1): 1–7.

Denk, W. (1994). "Two-photon scanning photochemical microscopy: mapping ligand-gated ion channel distributions." *Proc Natl Acad Sci U S A* **91**(14): 6629–6633.

Doiron, D. R. and C. J. Gomer (1984). *Porphyrin Localisation and Treatment of Tumours*. AR Liss: New York.

Garrett, B. C., D. A. Dixon, D. M. Camaioni, D. M. Chipman, M. A. Johnson, C. D. Jonah, G. A. Kimmel, J. H. Miller, T. N. Rescigno, P. J. Rossky, S. S. Xantheas, S. D. Colson, A. H. Laufer, D. Ray, P. F. Barbara, D. M. Bartels, K. H. Becker, K. H. Bowen, Jr., S. E. Bradforth, I. Carmichael, J. V. Coe, L. R. Corrales, J. P. Cowin, M. Dupuis, K. B. Eisenthal, J. A. Franz, M. S. Gutowski, K. D. Jordan, B. D. Kay, J. A. Laverne, S. V. Lymar, T. E. Madey, C. W. McCurdy, D. Meisel, S. Mukamel, A. R. Nilsson, T. M. Orlando, N. G. Petrik, S. M. Pimblott, J. R. Rustad, G. K. Schenter, S. J. Singer, A. Tokmakoff, L. S. Wang, C. Wettig and T. S. Zwier (2005). "Role of water in electron-initiated processes and radical chemistry: issues and scientific advances." *Chem Rev* **105**(1): 355–390.

Gohlke, S. and E. Illenberger (2002). "Probing biomolecules: gas phase experiments and biological relevance." *Europhys News* **33**(6): 207–209.

Gold, G., P. Giannakopoulos, F. R. Herrmann, C. Bouras and E. Kovari (2007). "Identification of Alzheimer and vascular lesion thresholds for mixed dementia." *Brain* **130**(Pt 11): 2830–2836.

Guo, S. X., F. Bourgeois, T. Chokshi, N. J. Durr, M. A. Hilliard, N. Chronis and A. Ben-Yakar (2008). "Femtosecond laser nanoaxotomy lab-on-achip for in vivo nerve regeneration studies." *Nat Methods* **5**(6): 531–533.

Heisterkamp, A., I. Z. Maxwell, E. Mazur, J. M. Underwood, J. A. Nickerson, S. Kumar and D. E. Ingber (2005). "Pulse energy dependence of subcellular dissection by femtosecond laser pulses." *Opt Express* **13**(10): 3690–3696.

Heisterkamp, A., T. Ripken, T. Mamom, W. Drommer, H. Welling, W. Ertmer and H. Lubatschowski (2002). "Nonlinear side effects of fs pulses inside corneal tissue during photodisruption." *Appl Phys B* **74**(4–5): 419–425.

Helmchen, F. and W. Denk (2005). "Deep tissue two-photon microscopy." *Nat Methods* **2**(12): 932–940.

Hirase, H., V. Nikolenko, J. H. Goldberg and R. Yuste (2002). "Multiphoton stimulation of neurons." *J Neurobiol* **51**(3): 237–247.

Hotop, H. (2001). "Dynamics of low-energy electron collisions with molecules and clusters." In International Symposium on Gaseous Dielectrics IX. L. G. Christophorou and J. K. Olthoff. Ellicott City, MD, USA, Kluwer Academic/Plenum, New York.

Huels, M. A., B. Boudaïffa, P. Cloutier, D. Hunting and L. Sanche (2003). "Single, double, and multiple double strand breaks induced in DNA by 3–100 eV electrons." *J Am Chem Soc* **125**(15): 4467–4477.

Kaiser, A., B. Rethfeld, M. Vicanek and G. Simon (2000). "Microscopic processes in dielectrics under irradiation by subpicosecond laser pulses." *Phys Rev B* **61**(17): 11437–11450.

Karreman, M. A., L. Mercier, N. L. Schieber, T. Shibue, Y. Schwab and J. G. Goetz (2014). "Correlating intravital multi-photon microscopy to 3D electron microscopy of invading tumor cells using anatomical reference points." *PLoS One* **9**(12): e114448.

Kohli, V. and A. Y. Elezzabi (2008). "Laser surgery of zebrafish (Danio rerio) embryos using femtosecond laser pulses: optimal parameters for exogenous material delivery, and the laser's effect on short- and long-term development." *BMC Biotechnol* **8**: 7.

Konig, K., I. Riemann, P. Fischer and K. J. Halbhuber (1999). "Intracellular nanosurgery with near infrared femtosecond laser pulses." *Cell Mol Biol (Noisy-le-grand)* **45**(2): 195–201.

Konig, K., I. Riemann and W. Fritzsche (2001). "Nanodissection of human chromosomes with near-infrared femtosecond laser pulses." *Opt Lett* **26**(11): 819–821.

Kubotera, H., H. Ikeshima-Kataoka, Y. Hatashita, A. L. A. Mascaro, F. S. Pavone and T. Inoue (2019). "Astrocytic endfeet re-cover blood vessels after removal by laser ablation." *Sci Rep* **9**(1): 1263.

Kumar, S., I. Z. Maxwell, A. Heisterkamp, T. R. Polte, T. P. Lele, M. Salanga, E. Mazur and D. E. Ingber (2006). "Viscoelastic retraction of single living stress fibers and its impact on cell shape, cytoskeletal organization, and extracellular matrix mechanics." *Biophys J* **90**(10): 3762–3773.

Lorenzana, A. O., J. K. Lee, M. Mui, A. Chang and B. Zheng (2015). "A surviving intact branch stabilizes remaining axon architecture after injury as revealed by in vivo imaging in the mouse spinal cord." *Neuron* **86**(4): 947–954.

Markaki, Y., and H. Harz (2017). *Light Microscopy*. Springer.

Nguyen, J., J. Ferdman, M. Zhao, D. Huland, S. Saqqa, J. Ma, N. Nishimura, T. H. Schwartz and C. B. Schaffer (2011). "Sub-surface, micrometer-scale incisions produced in rodent cortex using tightly-focused femtosecond laser pulses." *Lasers Surg Med* **43**(5): 382–391.

Nikogosyan, D. N., A. A. Oraevsky and V. I. Rupasov (1983). "Two-photon ionization and dissociation of liquid water by powerful laser UV radiation." *Chem Phys* **77**(1): 131–143.

Nimmerjahn, A., F. Kirchhoff and F. Helmchen (2005). "Resting microglial cells are highly dynamic surveillants of brain parenchyma in vivo." *Science* **308**(5726): 1314–1318.

Nishimura, N., C. B. Schaffer, B. Friedman, P. D. Lyden and D. Kleinfeld (2007). "Penetrating arterioles are a bottleneck in the perfusion of neocortex." *Proc Nat Acad Sci* **104**(1): 365–370.

Nishimura, N., C. B. Schaffer, B. Friedman, P. S. Tsai, P. D. Lyden and D. Kleinfeld (2006). "Targeted insult to subsurface cortical blood vessels using ultrashort laser pulses: three models of stroke." *Nat Methods* **3**(2): 99–108.

Oehring, H., I. Riemann, P. Fischer, K. J. Halbhuber and K. Konig (2000). "Ultrastructure and reproduction behaviour of single CHO-K1 cells exposed to near infrared femtosecond laser pulses." *Scanning* **22**(4): 263–270.

Raabe, I., S. K. Vogel, J. Peychl and I. M. Tolic-Norrelykke (2009). "Intracellular nanosurgery and cell enucleation using a picosecond laser." *J Microsc* **234**(1): 1–8.

Ready, J. F. (1971). *Effects of High Power Laser Radiation*. Academic Press, Orlando, FL.

Ridley, B. K. (1999). *Quantum Processes in Semiconductors*. Oxford, Oxford University Press.

Sacconi, L., R. P. O'Connor, A. Jasaitis, A. Masi, M. Buffelli and F. S. Pavone (2007). "In vivo multiphoton nanosurgery on cortical neurons." *J Biomed Opt* **12**(5): 050502.

Sacconi, L., I. M. Tolic-Norrelykke, R. Antolini and F. S. Pavone (2005). "Combined intracellular three-dimensional imaging and selective nanosurgery by a nonlinear microscope." *J Biomed Opt* **10**(1): 14002.

Sarig-Nadir, O., N. Livnat, R. Zajdman, S. Shoham and D. Seliktar (2009). "Laser photoablation of guidance microchannels into hydrogels directs cell growth in three dimensions." *Biophys J* **96**(11): 4743–4752.

Shen, N., D. Datta, C. B. Schaffer, P. LeDuc, D. E. Ingber and E. Mazur (2005). "Ablation of cytoskeletal filaments and mitochondria in live cells using a femtosecond laser nanoscissor." *Mech Chem Biosyst* **2**(1): 17–25.

Shimada, T., W. Watanabe, S. Matsunaga, T. Higashi, H. Ishii, K. Fukui, K. Isobe and K. Itoh (2005). "Intracellular disruption of mitochondria in a living HeLa cell with a 76-MHz femtosecond laser oscillator." *Opt Express* **13**(24): 9869–9880.

Smith, N. I., K. Fujita, T. Kaneko, K. Katoh, O. Nakamura, S. Kawata and T. Takamatsu (2001). "Generation of calcium waves in living cells by pulsed-laser-induced photodisruption." *Appl Phys Lett* **79**(8): 1208–1210.

Soeller, C., M. D. Jacobs, K. T. Jones, G. C. Ellis-Davies, P. J. Donaldson and M. B. Cannell (2003). "Application of two-photon flash photolysis to reveal intercellular communication and intracellular Ca2+ movements." *J Biomed Opt* **8**(3): 418–427.

Supatto, W., E. Brouzes, E. Farge and E. Beaurepaire (2004). "In vivo micro-dissection and live embryo imaging by two-photon microscopy to study *Drosophila melanogaster* early developement." *Femtosec Laser Appl Biol* **5463**: 13–20.

Supatto, W., D. Debarre, B. Moulia, E. Brouzes, J. L. Martin, E. Farge and E. Beaurepaire (2005). "In vivo modulation of morphogenetic movements in Drosophila embryos with femtosecond laser pulses." *Proc Natl Acad Sci U S A* **102**(4): 1047–1052.

Svoboda, K. and R. Yasuda (2006). "Principles of two-photon excitation microscopy and its applications to neuroscience." *Neuron* **50**(6): 823–839.

Tirlapur, U. K. and K. Konig (2002). "Targeted transfection by femtosecond laser." *Nature* **418**(6895): 290–291.

Tirlapur, U. K., K. Konig, C. Peuckert, R. Krieg and K. J. Halbhuber (2001). "Femtosecond near-infrared laser pulses elicit generation of reactive oxygen species in mammalian cells leading to apoptosis-like death." *Exp Cell Res* **263**(1): 88–97.

Tolic-Norrelykke, I. M., L. Sacconi, G. Thon and F. S. Pavone (2004). "Positioning and elongation of the fission yeast spindle by microtubule-based pushing." *Curr Biol* **14**(13): 1181–1186.

Tsai, P. S., P. Blinder, B. J. Migliori, J. Neev, Y. Jin, J. A. Squier and D. Kleinfeld (2009). "Plasma-mediated ablation: an optical tool for submicrometer surgery on neuronal and vascular systems." *Curr Opin Biotechnol* **20**(1): 90–99.

Turney, S. G. and J. W. Lichtman (2012). "Reversing the outcome of synapse elimination at developing neuromuscular junctions in vivo: evidence for synaptic competition and its mechanism." *PLOS Biol* **10**(6): e1001352.

Vogel, A. and J. Noack (2001). "Numerical simulations of optical breakdown for cellular surgery at nanosecond to femtosecond time scales." In *BiOS 2001 The International Symposium on Biomedical Optics*. SPIE, San Jose, CA.

Vogel, A., J. Noack, G. Huttman and G. Paltauf (2005). "Mechanisms of femtosecond laser nanosurgery of cells and tissues." *Appl Phys B* **81**(8): 1015–1047.

Vogel, A., J. Noack, K. Nahen, D. Theisen, S. Busch, U. Parlitz, D. Hammer, G. Noojin, B. Rockwell and R. Birngruber (1999). "Energy balance of optical breakdown in water at nanosecond to femtosecond time scales." *Appl Phys B* **68**(2): 271–280.

Vogel, A. and V. Venugopalan (2003). "Mechanisms of pulsed laser ablation of biological tissues." *Chem Rev* **103**(2): 577–644.

Watanabe, W., N. Arakawa, S. Matsunaga, T. Higashi, K. Fukui, K. Isobe and K. Itoh (2004). "Femtosecond laser disruption of subcellular organelles in a living cell." *Opt Express* **12**(18): 4203–4213.

Yan, G., P. E. Vértes, E. K. Towlson, Y. L. Chew, D. S. Walker, W. R. Schafer and A.-L. Barabási (2017). "Network control principles predict neuron function in the *Caenorhabditis elegans* connectome." *Nature* **550**: 519.

Yanik, M. F., H. Cinar, H. N. Cinar, A. D. Chisholm, Y. Jin and A. Ben-Yakar (2004). "Neurosurgery: functional regeneration after laser axotomy." *Nature* **432**(7019): 822.

Zeira, E., A. Manevitch, A. Khatchatouriants, O. Pappo, E. Hyam, M. Darash-Yahana, E. Tavor, A. Honigman, A. Lewis and E. Galun (2003). "Femtosecond infrared laser-an efficient and safe in vivo gene delivery system for prolonged expression." *Mol Ther* **8**(2): 342–350.

Zhang, S. and T. H. Murphy (2007). "Imaging the impact of cortical microcirculation on synaptic structure and sensory-evoked hemodynamic responses in vivo." *PLoS Biol* **5**(5): e119.

Zipfel, W. R., R. M. Williams and W. W. Webb (2003). "Nonlinear magic: multiphoton microscopy in the biosciences." *Nat Biotechnol* **21**(11): 1369–1377.

III

Clinical and Human Neurophotonics

CHAPTER 17

High Resolution Diffuse Optical Tomography of the Human Brain

Muriah D. Wheelock and Adam T. Eggebrecht

CONTENTS

17.1 INTRODUCTION: HEMODYNAMIC IMAGING OF HUMAN BRAIN FUNCTION

Mapping spatially distributed brain activity has revolutionized our understanding of brain function (Corbetta and Shulman, 2002; Zhang and Raichle, 2010; Power et al., 2011; Yeo et al., 2011). Brain imaging via positron emission tomography (PET), functional magnetic resonance imaging (fMRI), and, more recently, diffuse optical tomography (DOT) have illuminated many aspects of the biological basis of human behavior. Brain systems that support all aspects of cognition – from sensing the visual world, to generating language, to interacting socially, to daydreaming or sleeping – are accessible to quantitative investigation because of these techniques (Petersen et al., 1988; Corbetta and Shulman, 2002; Raichle, 2010). Further, brain imaging has proven useful in clinical investigations of brain function. Specifically, several neurological disorders manifest as measurable alterations in distributed brain networks, including degenerative diseases such as Alzheimer's disease (Buckner et al., 2009), neurodevelopmental disorders such as autism spectrum disorder (ASD) (Kennedy and Courchesne, 2008; Eggebrecht et al., 2017; Marrus et al., 2017), or disorders due to an insult such as ischemic stroke (Carter et al., 2012; Baldassarre et al., 2016). While many clinical disorders are known to manifest in the brain, optimizing neuroimaging technologies as tools for understanding these disorders and tracking their progression presents significant challenges.

Traditional neuroimaging methods, though powerful, are limited in certain key situations: PET utilizes ionizing radiation that is forbidden as an experimental procedure in children, and fMRI involves exposure to strong magnetic fields and induced electric fields and so is contraindicated in patients with implanted electronic devices (e.g. pacemakers, deep brain stimulators, and cochlear implants). Optical neuroimaging provides a unique alternative human brain mapping technique for situations in which either fMRI or PET is contraindicated. Optical methods utilize functional near-infrared spectroscopy (fNIRS), a safe technique (employed in pulse oximeters) that leverages sensitivity to blood volume and oxygenation (Jobsis, 1977) to report blood-oxygen-level-dependent (BOLD) signals (Ogawa et al., 1992) via measurements of light absorption (Obrig and Villringer, 2003). Additionally, optical imaging with fNIRS provides an approach to mapping brain function in a setting more ecologically natural than MRI (Villringer et al., 1993; Maki et al., 1995; Bluestone et al., 2001; Boas et al., 2004a; Cui et al., 2011; Hassanpour et al., 2015), and may be better suited to studies of children due to the unconstraining fNIRS imaging environment. fNIRS may be especially suited to studying children with neurodevelopmental disorders such as ASD (Suda et al., 2011; Lloyd-Fox et al., 2013; Vanderwert and Nelson, 2014; Blasi et al., 2015; Lloyd-Fox et al., 2017). The portable nature of optical imaging technology opens the door to high-resolution bedside imaging of functional brain health (White et al., 2012; Singh et al., 2014; Chalia et al., 2016; Ferradal et al., 2016). For example, optical imaging technology is ideal in clinical settings in which the patient is too medically unstable to be moved to an MRI machine (e.g. on a ventilator, on vasopressors, or utilizing extracorporeal membranous oxygenation).

Both fMRI and optical imaging utilize changes in blood flow and oxygenation concentration to infer neuronal activity. When a part of the brain is active, the local firing of neurons triggers a complex neurovascular cascade (Raichle and Mintun, 2006; Hillman et al., 2011) that produces a dramatic increase in local blood flow and glucose use resulting in a large increase in oxygen availability, though the local use of oxygen is roughly constant (Fox and Raichle, 1986; Fox et al., 1988). This dynamic change in oxygen supply relative to oxygen demand in the capillaries is the source of the BOLD signal of both fMRI (Ogawa and Lee, 1990; Ogawa et al., 1990; Belliveau et al., 1991; Bandettini et al., 1992;

Kwong et al., 1992; Ogawa et al., 1992), as well as fNIRS. The increasing blood flow causes a local increase of absorbance of the tissue due to a higher concentration of hemoglobin. Multiple wavelengths are used in fNIRS measurements to spectroscopically un-mix the contribution of the two primary chromophores participating in these dynamics, oxygenated (HbO_2) and deoxygenated hemoglobin (HbR).*

Multiple overlapping fNRIS measurements can be tomographically reconstructed to produce three-dimensional maps of brain function. This technique is known as diffuse optical tomography (DOT), and the overlapping fNIRS measurements results in improved image quality as compared to sparse fNIRS systems. Recent developments in DOT design have dramatically increased the density of overlapping fNIRS measurements. Specifically, high-density DOT (HD-DOT) systems possess a first nearest neighbor source-detector distance of less than 15 mm (Zeff et al., 2007). These HD- DOT systems have been used to map brain function using both task-based (Joseph et al., 2006; Zeff et al., 2007; Koch et al., 2010; Liao et al., 2010; White and Culver, 2010b; Eggebrecht et al., 2012; Habermehl et al., 2012; Zhan et al., 2012b, 2012a; Eggebrecht et al., 2014; Ferradal et al., 2014; Wu et al., 2014; Hassanpour et al., 2015; Wu et al., 2015a, b; Ferradal et al., 2016; Hassanpour et al., 2017); and resting state functional connectivity techniques (White et al., 2009; White et al., 2012; Eggebrecht et al., 2014; Ferradal et al., 2016). Importantly, recent studies have methodically and quantitatively demonstrated HD-DOT recovers cortical activation and functional connectivity maps with a spatial resolution comparable to that of fMRI (Eggebrecht et al., 2014; Ferradal et al., 2016).

In this chapter, we will briefly describe the physical mechanisms underlying fNIRS measurements, as well as theoretical modelling of light propagation in tissue. We then focus on continuous-wave opto-electronic instrumentation utilized in HD-DOT systems. We then highlight some of the HD-DOT validation research that is crucial in establishing reproducible cortical activation maps. Finally, we turn our attention to the use of HD-DOT in clinically oriented applications.

17.2 OPTICAL IMAGING OF BRAIN FUNCTION: THEORETICAL BACKGROUND

17.2.1 Photon Diffusion through Biological Tissue

The potential of utilizing visible and near-infrared (NIR) photons to probe physiology deep (>1 cm) in living intact tissue became apparent in the late 1970s when Jöbsis observed a range of wavelengths in the electromagnetic spectrum (~600–1,000 nm) wherein photons penetrate multiple centimeters through biological tissue (Jobsis, 1977). This deeper penetration is possible because the primary chromophores in biological tissue (water, lipids, and hemoglobin) absorb photons relatively weakly within this "optical window" (Figure 17.1). However, though the photon absorption is low, the scattering of photons is high in biological tissue and can be well approximated as a diffusive process (Johnson, 1970;

* An additional chromophore, cytochrome-c-oxydase, which may provide a more direct measure of metabolic activity, may also be measurable with HD-DOT, though that topic is beyond the scope of this chapter (Wray et al., 1988; Wobst et al., 2001; Uludag et al., 2002; Bale et al., 2014; de Roever et al., 2017)

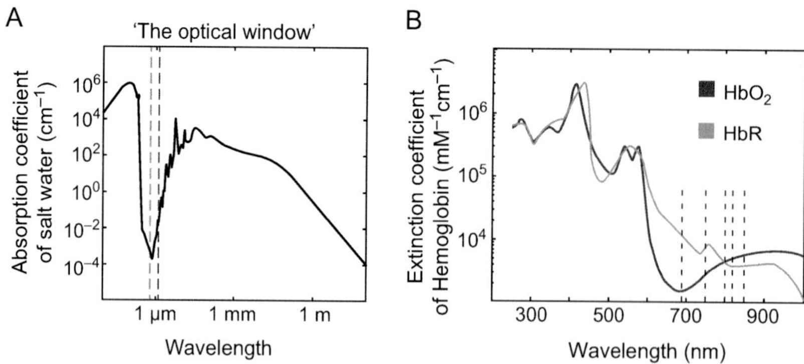

FIGURE 17.1 The optical window in biological tissue. **A** Within a narrow band of wavelengths, including the visible spectrum (~450–750 nm) and near-infrared range (blue line at 400 nm, red line at 900 nm), the absorption of photons is low in biological tissue. (Absorption spectrum data from Segelstein, 1981.) **B** Absorption spectra of oxygenated and deoxygenated hemoglobin (Wray et al., 1988). Green lines denote example popular wavelengths used in DOT (685, 750, 800, 830, 850 nm).

Groenhuis et al., 1983; Patterson et al., 1989; O'Leary et al., 1992; Boas et al., 1993). With a high-density array of sources and detectors, and an appropriate model for light propagation, it is possible to reconstruct relative changes in absorption and scattering properties within the tissue volume from a set of these measurements collected on the surface.

17.2.2 Forward Light Modeling

Given a set of source-detector pairs on the surface of the head (Figure 17.2A, B), we want to model how local variations in absorption within the head lead to changes in the measured light intensity (Figure 17.2C). The Boltzmann transport equation (BTE, equivalent here to the Radiative Transport Equation) is a conservation equation that can be utilized to describe the flow of photons through a scattering media. Let us define the energy radiance

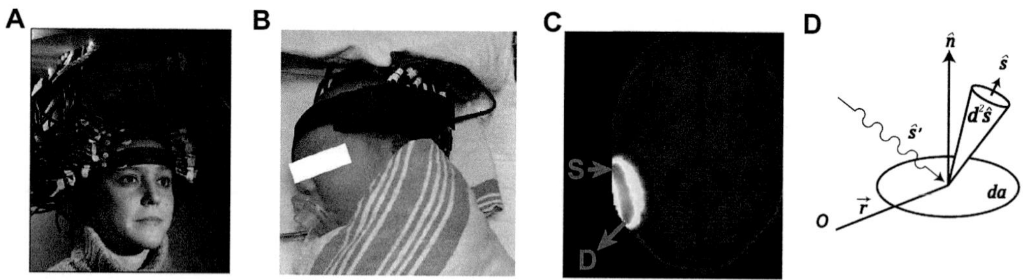

FIGURE 17.2 HD-DOT supported by high density arrays and proper modeling of photon diffusion through biological tissue. **A** An HD-DOT grid on a healthy adult participant. (Adapted from Eggebrecht et al., 2014.) **B** Bedside imaging in the neonatal unit with HD-DOT. (Adapted from Ferradal et al., 2016.) **C** Example sensitivity profile for a source (S)-detector (D) pair. Color map is logarithmic with a cutoff at 0.1% maximum sensitivity. **D** Schematic for forward light modeling. See text for details.

$\boldsymbol{I}(\vec{\boldsymbol{r}},t,\hat{\boldsymbol{s}})$ (in units of $\frac{W}{cm^2 \cdot sr}$), such that the differential energy dE flowing in a unit solid angle $d^2\hat{\boldsymbol{s}}$ through an elemental area da with associated normal $\hat{\boldsymbol{n}}$, at position $\vec{\boldsymbol{r}}$, in time dt is (Figure 17.2D):

$$dE = \boldsymbol{I}(\vec{\boldsymbol{r}},t,\hat{\boldsymbol{s}})\hat{\boldsymbol{s}}\cdot\hat{\boldsymbol{n}}\,da\,d^2\hat{\boldsymbol{s}}\,dt \tag{17.1}$$

Here, we are simplifying by considering energy at a specific wavelength of light as opposed to a range of wavelengths, and neglecting polarization, coherence, and non-linearities. (The radiance is proportional to the square of the electric field at position $\vec{\boldsymbol{r}}$, travelling in direction $\hat{\boldsymbol{s}}$ in time dt.) The BTE states that in each infinitesimal element of the volume of the media:

$$\begin{aligned}\frac{1}{v}\frac{\partial \boldsymbol{I}}{\partial t} &= \mu_s \int_{4\pi} f(\hat{\boldsymbol{s}},\hat{\boldsymbol{s}}')\boldsymbol{I}(\vec{\boldsymbol{r}},t,\hat{\boldsymbol{s}}')d^2\hat{\boldsymbol{s}}' + q(\vec{\boldsymbol{r}},t,\hat{\boldsymbol{s}}) \\ &\quad - \hat{\boldsymbol{s}}\cdot\nabla\boldsymbol{I}(\vec{\boldsymbol{r}},t,\hat{\boldsymbol{s}}) - (\mu_a+\mu_s)\boldsymbol{I}(\vec{\boldsymbol{r}},t,\hat{\boldsymbol{s}}),\end{aligned} \tag{17.2}$$

where v is the seed of light in the medium ($v = c/n$, where $c = 3\times10^8\,m/s$ is the speed of light in a vacuum and n is the index of refraction in the medium); μ_s is the scattering coefficient (in units of cm^{-1}); $f(\hat{\boldsymbol{s}},\hat{\boldsymbol{s}}')$ is the scattering phase function, which is essentially the probability density of a photon scattering from direction $\hat{\boldsymbol{s}}'$ into direction $\hat{\boldsymbol{s}}$; $q(\vec{\boldsymbol{r}},t,\hat{\boldsymbol{s}})$ is a source term (with units of $\frac{W}{cm^3 \cdot sr}$) representing power per volume emitted by sources at position $\vec{\boldsymbol{r}}$ in time dt in direction $\hat{\boldsymbol{s}}$; and μ_a is the absorption coefficient of the medium (in units of cm^{-1}). Conceptually, Equation (17.2) states that the change in radiance at time t in direction $\hat{\boldsymbol{s}}$ at position $\vec{\boldsymbol{r}}$ is due to four possible quantities: (i) gains in energy due to photons being scattered into direction $\hat{\boldsymbol{s}}$ and position $\vec{\boldsymbol{r}}$; (ii) gains in energy due to local sources of photons; (iii) changes in energy flow due to scattering; and (iv) losses in energy due to absorption and scattering. The absorption and scattering coefficients of the medium are wavelength dependent ($\mu_a(\lambda), \mu_s(\lambda)$, respectively), and correspond to the reciprocal of the typical distance traveled by a photon before it is either absorbed or scattered. These distances are distinct from (and much smaller than) the transport mean-free-path (also known as the random walk step), which represents the typical distance a collection of photons travels in a medium before their directions are randomized. The reciprocal of this random walk step is also wavelength dependent and is called the reduced scattering coefficient, $\mu_s'(\lambda)$. To simplify, we are treating the medium as if the index of refraction and the coefficients of absorption and scattering are constant.

If we make the assumption that that the radiance $\boldsymbol{I}(\vec{\boldsymbol{r}},t,\hat{\boldsymbol{s}})$ is isotropic, then Equation (17.2) can be simplified by expanding $\boldsymbol{I}(\vec{\boldsymbol{r}},t,\hat{\boldsymbol{s}})$ into spherical harmonics and truncating after the first term (Arridge and Schweiger, 1995; Schweiger et al., 1995; Durduran et al., 2010):

$$\boldsymbol{I}(\vec{\boldsymbol{r}},t,\hat{\boldsymbol{s}}) = \frac{1}{4\pi}\Phi(\vec{\boldsymbol{r}},t) + \frac{3}{4}\hat{\boldsymbol{s}}\cdot\boldsymbol{J}(\vec{\boldsymbol{r}},t), \tag{17.3}$$

where, $\Phi(\vec{r},t)$ is the fluence (in units of $\frac{W}{cm^2}$), defined as the total power per area radiating radially outward from the volume element at position $\vec{r}$ and time t:

$$\Phi(\vec{r},t) = \int_{4\pi} I(\vec{r},t,\hat{s})d^2\hat{s}, \tag{17.4}$$

and $J(\vec{r},t)$ is the photon current (in units of $\frac{W}{cm^2}$), essentially the vector sum of the radiance emerging from the volume element:

$$J(\vec{r},t) = \int_{4\pi} \hat{s}I(\vec{r},t,\hat{s})d^2\hat{s}. \tag{17.5}$$

Now, plugging Equation (17.3) into Equation (17.1) and integrating over all solid angles (using the assumption of isotropic scattering) obtains a scalar term:

$$\frac{1}{v}\frac{\partial\Phi(\vec{r},t)}{\partial t} + \nabla\cdot J(\vec{r},t) + \mu_a\Phi(\vec{r},t) = Q(\vec{r},t), \tag{17.6}$$

and a vector term,

$$\left(\frac{1}{v}\frac{\partial}{\partial t} + \mu_a + \mu_s - g\mu_s\right)J(\vec{r},t) = -\frac{1}{3}\nabla\Phi(\vec{r},t) + \int_{4\pi} q(\vec{r},t,\hat{s})\hat{s}d^2\hat{s}, \tag{17.7}$$

where $Q(\vec{r},t)$ is the total power per volume radiating radially outward from the volume element at position $\vec{r}$ and time t (in units of $\frac{W}{cm^3}$), and g is the ensemble average of the cosine of the scattering angle associated with a typical scattering event in the tissue:

$$g = \int_{4\pi} f(\hat{s},\hat{s}')\hat{s}\cdot\hat{s}'d^2\hat{s}' = \cos\theta, \tag{17.8}$$

where θ is the angle between the incident and outgoing scattering wave vectors. The anisotropy factor g reflects the probability that a photon is scattered in the forward direction and in soft mammalian tissue typically has a value around 0.9 (Cheong et al., 1990).

Our first key assumption (1) the scattering is isotropic, provides:

$$\int_{4\pi} q(\vec{r},t,\hat{s})\hat{s}d^2\hat{s} = 0. \tag{17.9}$$

If we now enforce a second key assumption (2) that variations in the photon current are slow relative to the time it takes the photons to travel a random walk step:

$$\frac{1}{v}\left|\frac{\partial J}{\partial t}\right| \ll (\mu_a + \mu_s - g\mu_s)|J|, \tag{17.10}$$

Equation (17.7) simplifies to Fick's law of diffusion:

$$J(\vec{r},t) = -\frac{v}{3(\mu_a + (1-g)\mu_s)} \nabla \Phi(\vec{r},t). \tag{17.11}$$

Now, defining the reduced scattering coefficient $\mu_s' = (1-g)\mu_s$, and substituting Equation (17.11) into Equation (17.6), we arrive at the diffusion approximation of the radiative transport equation for the photon fluence rate:

$$\frac{\partial \Phi(\vec{r},t)}{\partial t} - \nabla \cdot \left(D(\vec{r}) \nabla \Phi(\vec{r},t)\right) + v\mu_a(\vec{r})\Phi(\vec{r},t) = vQ(\vec{r},t), \tag{17.12}$$

where the diffusion coefficient (in units of $\frac{cm^2}{s}$) is defined as:

$$D(\vec{r}) = \frac{v(\vec{r})}{3\left(\mu_a(\vec{r}) + \mu_s'(\vec{r})\right)}, \tag{17.13}$$

where we are now explicitly noting that the index of refraction and coefficient of absorption and reduced coefficient of scattering may vary in the tissue. In practice, we often model the tissue using multiple tissue types (i.e. scalp/soft tissue, bone, gray matter, white matter, and cerebral spinal fluid), each with some set of estimated baseline optical properties. The validity of the diffusion approximation for photon propagation in biological tissue is appropriate as long as the above stated assumptions of isotropic scattering hold (Jacques and Pogue, 2008). As a rule of thumb, this is true when $\mu_s' > 10 \times \mu_a$, and when the depth of the location of interest in the tissue is much deeper than the transport mean-free-path, approximately $\frac{1}{\mu_s'}$. In the application of focus here, optical imaging of human brain function, these assumptions generally hold true at the depths of brain tissue.

Solutions to the diffusion equation for tissues with an arbitrary geometry and spatially varying optical properties (index of refraction, absorption, and scattering coefficients) can be calculated using a number of publicly available packages such as NIRFAST (Dehghani et al., 2009) or TOAST++ (Schweiger and Arridge, 2014) that utilize finite element modeling (FEM; Arridge et al., 1993; Paulsen and Jiang, 1995; Okada et al., 1996). Alternatively, Monte Carlo methods that do not rely on the assumptions of isotropic scattering or slow changes in photon currents can also be employed (Graaff et al., 1993; Hiraoka et al., 1993; Wang et al., 1995; Boas et al., 2002). Though Monte Carlo methods do provide more accurate solutions close to the surface of the tissue – where the assumptions required for the diffusion approximation break down – the diffusion approximation works well at the depth of the brain and is computationally far more efficient, allowing for solutions to be obtained for thousands of source-detector measurement pairs in a complex head geometry in just a few minutes (Wu et al., 2015b; Doulgerakis et al., 2017).

In solving the diffusion equation for photon propagation in biological tissue (Equation (17.12) above), we must distinguish between the three primary fNIRS measurement types: continuous wave (CW), frequency domain (FD), and time domain (TD) (Figure 17.3). In CW

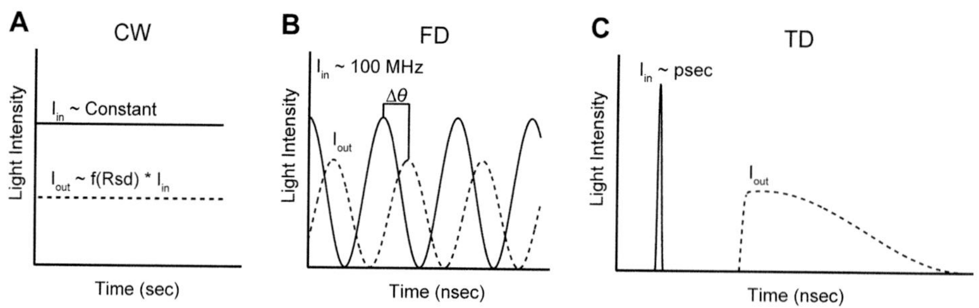

FIGURE 17.3 Three primary measurement strategies in diffuse optics. **A** Continuous Wave (CW), **B** Frequency Domain (FD), **C** Time Domain (TD). See text for details.

mode (the focus of this chapter), only the magnitude of the light intensity is measured at the detector (Figure 17.3A). In this case, because only one parameter is measured, relative changes in absorption in the modeled optical properties are all that can be accessed. In contrast, with FD systems that utilize source modulation in the hundreds of megahertz range, differences in both light intensity and phase relative to the source signal are measured at the detector (Figure 17.3B). This additional measurement provides access to relative (and potentially absolute) measures of scattering as well as absorption within the tissue. In the TD case, picosecond width pulses of light are administered to the tissue and fast electronics measure the distribution of photons that arrive at the detector. Both FD and TD strategies provide more rich detail about the underlying tissue dynamics. However, the current technology that supports these measurements is significantly more expensive and complex and a full discussion lies beyond the scope of this chapter (O'Leary et al., 1995; Culver et al., 2003; Liebert et al., 2003; Wolf et al., 2003; Yu et al., 2003; Liebert et al., 2004; Wabnitz et al., 2010; Torricelli et al., 2014; Wabnitz et al., 2014b, 2014a; Re et al., 2018). In the CW case, any variations in light intensity occur over a much longer timescale than the mean transit time of the light. Thus, the first term in Equation (17.12) is zero and the resulting time-independent diffusion equation can be solved using standard Green's functions methods (e.g. Arridge and Schweiger, 1995; Arridge, 1999; Ntziachristos et al., 1999). An additional simplification arises given that this fluence measurement contains relatively small variations in time (1–10%) due to small variations relative to a constant mean baseline value in the underlying hemodynamics. This approximation is not valid if the underlying absorption may be changing significantly, as in the case of active blood pooling or some other pathological processes within the tissue. However, for healthy head tissue, the approximation of a temporally constant spatial distribution of optical properties is sound. As such, we can model the (time independent) fluence at the boundary $\Phi(\vec{r})$ using perturbation methods with the Rytov approximation (Kak et al., 1988; O'Leary et al., 1995; Arridge, 1999):

$$\Phi(\vec{r}) = \Phi_0(\vec{r})e^{\Delta\Phi(\vec{r})}, \text{ where, } f_0(\vec{r}) \gg \Delta\Phi(\vec{r}), \text{ and, } \Delta\Phi = -\ln\left(\frac{\Phi}{\Phi_0}\right), \tag{17.14}$$

where $\Phi_0(\vec{r})$ is the baseline fluence measured at the surface arising from some baseline spatial distribution of optical properties $\mu_0(\vec{r})$ within the volume (Figure 17.4A).

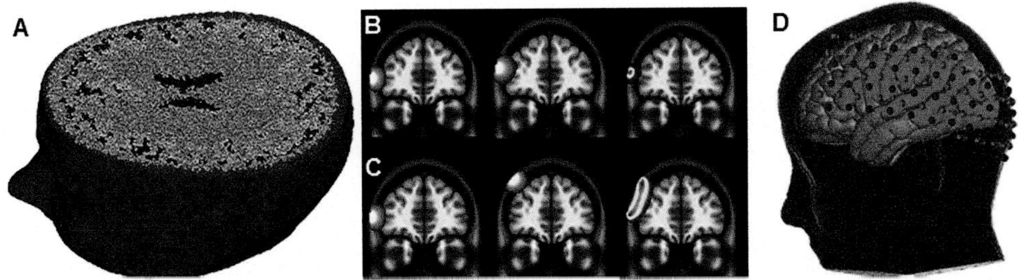

FIGURE 17.4 Finite Element Modeling for HD-DOT. **A** Segmented high-resolution mesh. **B** Green's functions for a source (left) and close detector (center) combine for a sensitivity profile that primarily samples superficial tissue. **C** Green's functions for a source (left) and distant detector (center) combine for a sensitivity profile that samples brain tissue. **D** Combining overlapping sensitivity profiles for 96 sources (red) and 92 detectors (blue) generates a smooth sensitivity at the level of the cortex (red coloring) within the volume of a subject's head.

The baseline fluence for a given source-detector measurement pair is typically estimated using the temporal mean of the time course of that measurement during an experiment (i.e., $\Phi_{0,sd} = \langle \Phi_{sd}(t) \rangle$) or during a time period immediately preceding inducing a perturbation with some task. Additionally, one can define effective Green's functions for the sources and detectors, respectively:

$$\tilde{G}_s(\vec{r}) = \int G(\boldsymbol{r}_s, \boldsymbol{r}) Q(\boldsymbol{r}_s) d\vec{r}_s, \tag{17.15}$$

$$\tilde{G}_d(\vec{r}) = \int G(\boldsymbol{r}, \boldsymbol{r}_d) Q(\boldsymbol{r}_d) d\vec{r}_d, \tag{17.16}$$

where, $\vec{r}_s$ and $\vec{r}_d$ are the source and detector positions, $G(\boldsymbol{r}_s, \boldsymbol{r})$ and $G(\boldsymbol{r}, \boldsymbol{r}_d)$ are the Green's functions for the sources and detectors (which we are assuming exist), and $Q(\boldsymbol{r}_s)$ and $Q(\boldsymbol{r}_d)$ are the source and detector spatial sensitivity distributions that we are assuming can take the same functional form – a fair assumption given the reciprocity theorem of electromagnetic radiation (Case and Zweifel, 1967), which directly gives rise to the adjoint formulation of the sensitivity relations (Arridge, 1999). (Reciprocity essentially states that transmitters and receivers of electromagnetic radiation can be modeled equivalently.) Thus, $\tilde{G}_s(\vec{r})$ and $\tilde{G}_d(\vec{r})$ provide a measure of some "effect" at voxel $\vec{r}$ in the tissue due to a source at position $\vec{r}_s$ or an "adjoint source" at position $\vec{r}_d$ (Figure 17.4B, C). Using the Rytov approximation with the Green's functions above and neglecting terms beyond the first order, one can arrive at the following solution:

$$\Delta\Phi(\vec{r}) = -\ln\left(\frac{\Phi}{\Phi_0}\right) = -\frac{v}{D}\int \frac{\tilde{G}_s(\vec{r})\tilde{G}_d(\vec{r})}{\tilde{G}_s(\vec{r}_d)} \Delta\mu_a(\boldsymbol{r}) d\vec{r}, \tag{17.17}$$

which states that ratiometric (i.e. differential) measurements of fluence at the boundary are related to spatial distribution of changes in internal absorption multiplied by the spatial sensitivity distributions for the source and detector and summed over all points in the tissue. The normalization term within the integral (also sometimes referred to as G_{sd}) is the Green's function of source evaluated at the position of the detector.

The next step is to discretize Equation (17.17) for some finite set of N_m source-detector pair measurements over a set of Nv voxels or nodes within a finite element mesh (Figure 17.4a). Using small-volume voxels (tetrahedral elements for the mesh) will facilitate more accurate solutions but will also add to the computational time required (Doulgerakis et al., 2017). While it is true that DOT is a relatively low-resolution imaging modality, it is important that the forward model be as accurate as possible such that image quality of the data is not reduced due to discretization errors in the model (Eggebrecht et al., 2014; Wu et al., 2015b; Doulgerakis et al., 2017). To achieve fMRI-comparable image quality, it is recommended to use meshes with mesh elements of a node distance of ~1–1.5 mm (around 600,000–1,000,000 nodes total in a head mesh). Equation (17.17) can be rewritten as:

$$\Delta\Phi(t) = -\ln\left(\frac{\Phi(t)}{\Phi_0}\right) = -\frac{vV}{D}\sum_{j}^{N_v}\frac{\tilde{G}_s(\boldsymbol{r}_j)\tilde{G}_d(\boldsymbol{r}_j)}{\tilde{G}_{sd}}\Delta\mu_a(\boldsymbol{r}_j,t), \tag{17.18}$$

where V is the volume of the discretization element. This matrix can be rewritten in a format where $\mathbf{y}(t) = -\ln\left(\frac{\Phi(t)}{\Phi_0}\right)$ is the vector of source-detector measurement pairs (each as a function of time), $\boldsymbol{A}$ is the sensitivity matrix (also called the Jacobian) that represents the full light model, and $\boldsymbol{x} = \Delta\mu_a(\boldsymbol{r}_j,t)$ is a vector representing the change in absorption in each voxel (also a function of time):

$$\boldsymbol{y}(t) = \boldsymbol{A}\boldsymbol{x}(t). \tag{17.19}$$

For simplicity, it is assumed here that the sensitivity is itself not a function of time. In practice, the measurements, the absorbance, and the sensitivity matrix are all a function of the wavelength of light emanating from the sources.

To calculate the light model $\boldsymbol{A}$, our lab uses NIRFAST (Dehghani et al., 2009) to model the Green's functions of Equation (17.17), which are primarily dependent upon three things: (1) the tissue boundary shape; (2) the internal distribution of baseline optical properties; and (3) the locations of the sources and detectors on the surface (Figure 17.4D). The tissue shape and optical property distributions are ideally generated from a subject-specific segmentation of the head (Gibson et al., 2003; Cooper et al., 2012; Eggebrecht et al., 2012), though atlas-based models can work quite well when subject-specific anatomy is not available (Custo et al., 2010; Brigadoi et al., 2014; Ferradal et al., 2014).

17.2.3 Image Reconstruction

As described above, the sensitivity matrix relates relative ratiometric changes in light-level measurements taken at the surface to relative changes in absorption within the volume. The sensitivity matrix can be directly inverted for image reconstruction using Tikhonov regularization along with spatially variant regularization to minimize the objective function:

$$\min\left\{\left\|\boldsymbol{y}_{\mathbf{meas}} - \boldsymbol{A}\boldsymbol{x}\right\|_2^2 + \lambda_1 \left\|\boldsymbol{L}\boldsymbol{x}\right\|_2^2\right\}. \tag{17.20}$$

The penalty term for image variance, $\lambda_1 \left\|\boldsymbol{L}\boldsymbol{x}\right\|_2^2$, incorporated the spatially variant regularization where,

$$\operatorname{diag}(\boldsymbol{L}) = \sqrt[2]{\left[\operatorname{diag}\left(\boldsymbol{A}^T\boldsymbol{A}\right) + \lambda_2\right]}. \tag{17.21}$$

A solution,

$$\boldsymbol{x} = \boldsymbol{A}^{\#}_{\lambda_1\lambda_2}\boldsymbol{y}_{\mathrm{meas}}, \tag{17.22}$$

can thus be obtained using a Moore–Penrose generalized inverse with,

$$\boldsymbol{A}^{\#}_{\lambda_1\lambda_2} = \boldsymbol{L}^{-1}\left(\tilde{\boldsymbol{A}}^T\tilde{\boldsymbol{A}} + \lambda_1\boldsymbol{I}\right)^{-1}\tilde{\boldsymbol{A}}^T\boldsymbol{y}_{\mathrm{meas}}, \tag{17.23}$$

where,

$$\tilde{\boldsymbol{A}} = \tilde{\boldsymbol{A}}\boldsymbol{L}^{-1}. \tag{17.24}$$

The optimal values of regulation parameters λ_1 and λ_2 depend upon the source-detector grid geometry and the underlying noise characteristics of the imaging system and model system. Optimal settings for these parameters are found through simulation and empirical studies to provide uniform imaging across the field of view as judged by evaluating point spread functions (in simulation) and, ideally, subject-matched comparisons to an alternate modality, such as functional MRI.

Relative changes in hemoglobin concentrations $\boldsymbol{C}$ can then be obtained from the absorption coefficients used in spectral decomposition,

$$\Delta\boldsymbol{C} = \boldsymbol{E}^{-1}\Delta\mu_a, \tag{17.25}$$

where, $\Delta\boldsymbol{\mu}_a$ is the measured differential change in absorption at 750 nm and 850 nm, equal to $\boldsymbol{x}$ listed above in Equation (17.4), $\boldsymbol{E}$ is a matrix containing the extinction coefficients of HbR and HbO_2, and $\Delta\boldsymbol{C} = [\Delta[HbO_2], \Delta[HbR]]$ is the matrix of concentration changes by time.

17.3 HIGH DENSITY DIFFUSE OPTICAL TOMOGRAPHY SYSTEM DESIGN

17.3.1 Challenges in Opto-Electric Designs

The image quality of DOT is fundamentally dependent upon utilizing multiple measurements that are sensitive to hemodynamics within overlapping regions of tissue. Focusing on CW HD-DOT design strategies, source-detector measurement pairs of various source-detector distances (R_{sd}) provide information about various depths, and, additionally, the overlapping distances at a given R_{sd} support improved lateral resolution (Toronov et al., 2000; Boas et al., 2004b; Zeff et al., 2007; White and Culver, 2010a). The HD grid provides

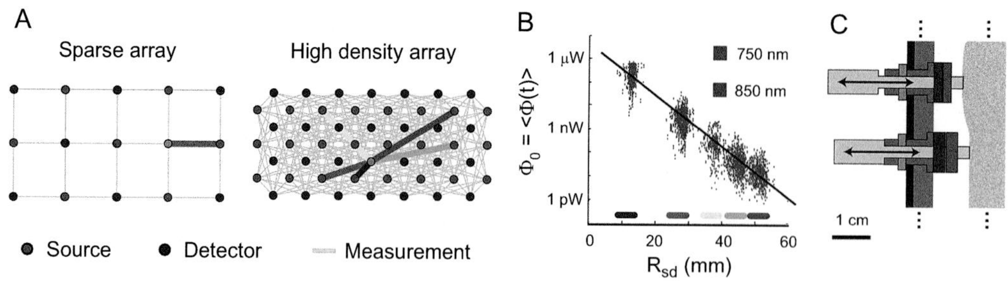

FIGURE 17.5 HD-DOT arrays provide **A** significantly more measurements within a given area and **B** require detectors to support a wide dynamic range to take advantage of various measurement distance. **C** Imaging cap design to support HD-DOT arrays use multiple strategies to ensure reliable and consistent coupling of optodes (adapted from Eggebrecht et al., 2014).

a given detector access to multiple sources at multiple distances (Figure 17.5A, B). This leads directly to significant challenges in managing crosstalk between detection channels and dynamic range of the detectors. For example, for the HD array shown in Figure 17.2A with a set of Rsd of 1.3, 3.0, 3.9, and 4.7 cm for the first four nearest neighbor separations, the detectors must maintain a linear response over at least six orders of magnitude of input optical power (Figure 17.5A, B). The crosstalk between detection channels must therefore also be less than 1×10^{-6}. To maintain these specifications, this system utilizes avalanche photodiodes coupled into 24-bit dedicated analogue to digital converters (Zeff et al., 2007; Eggebrecht et al., 2014). Other, commercially available HD-DOT systems, also use avalanche photodiodes for detection due to the large dynamic range required (Joseph et al., 2006; Chalia et al., 2016). To minimize crosstalk between source-detector measurement channels, time, frequency, and spatial encoding may be employed (Eggebrecht et al., 2014). The choice of wavelengths for the sources may be motivated by spectral width considerations, as light-emitting diodes emit photons over a relatively broad band around their characteristic center wavelength relative to laser diodes, which may impact the crosstalk of the spectroscopy (Uludag et al., 2002; Boas et al., 2004a; Eames and Dehghani, 2008; Eames et al., 2008).

17.3.2 Challenges in Optic-Scalp Coupling and Cap Design

Beyond challenges in optical and electrical components, reliable and consistent coupling of the optical elements to the scalp of the participant presents multiple significant challenges. Sources and detectors may be placed directly on the head (Zhao and Cooper, 2018) or coupled via optical fibers (Zeff et al., 2007; Liao et al., 2012). A general principle in ensuring reliable and comfortable imaging arrays is to provide a lightweight but rigid structure that maintains the optical fiber positions while minimizing torque on the fibers that can lead to coupling inconsistency over the course of an imaging session. A combination of foam and elastic pieces can help maintain a force perpendicular to the head surface to hold the optode directly coupled against the scalp while allowing for moderate translation normal to the head such that the imaging cap can conform to local variations in head shape (Figure 17.5C) (Eggebrecht et al., 2014). A crucial design detail, regardless of whether fibers are used or if the optical components are on the head, is to comb through the participant's hair to gain unimpeded access to the scalp. Hair (and product in hair) scatters light away from the optic-head system and lowers raw data quality. With this infrastructure in place, the cap maintaining the imaging array may be attached to the participant with hook-and-loop straps positioned to provide rigid yet comfortable stability on the head and to conform the curvature of the cap to a wide variety of head shapes and sizes. To ensure consistent placement of the imaging array on a given participant and across participants, measurements of the distance between specific fiducials on the imaging array and the head of the participant (e.g. the nasion, left and right tragus, and eyes) should be recorded.

17.3.3 Challenges in Data Quality Assurance

To ensure adequate coupling across the imaging array, a few simple metrics of data fidelity can help ensure a high-quality cap fit. First, the average light level for each source and detector can be displayed in a two-dimensional representation of the imaging array (Figure 17.6A). If the light level is low, or if there is significant spatial variance in mean light level (more than 2 orders of magnitude), then the associated optical element or fiber optic should be adjusted at the head to improve coupling. The adjustment typically involves improving either the combing through the hair and or ensuring the fiber/element is coupled to the scalp at a right angle. Second, as shown in Figure 17.5B, properly coupled elements will reflect a set of mean light levels that are logarithmically distributed as a function of distance, reflecting diffusion of photons through tissue. If the spread in light level at a given Rsd is more than 1–2 orders of magnitude, or if the slope of the fall-off is not approximately one order of magnitude in light level for every centimeter of additional Rsd, then the cap fit may not be optimal. Third, assuming the data is acquired at a frame rate of at least 4 Hz, the time course of individual source-detector pair measurements with a good signal-to-noise ratio will clearly exhibit characteristics consistent with the pulse (~1 Hz) frequency (Figure 17.6B, C). The relative magnitude of the pulse peak in a power spectral

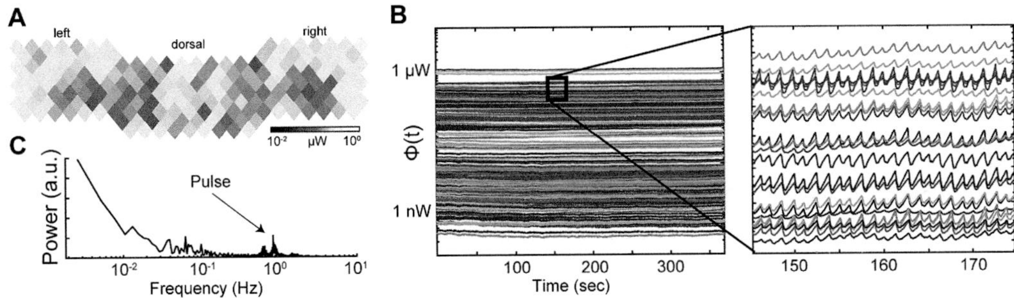

FIGURE 17.6 The many measurements provided by HD-DOT arrays leads to challenges in data quality assurance. **A** Average light level for each source and detector on the 96 source by 92 detector array of Eggebrecht et al., 2014. **B** Raw data time traces for a set of source-detector measurements highlight the clear pulsatility in the wave form. **C** Power spectral density of these measurements reveals a strong peak at the pulse frequency.

density plot is an excellent indicator of data quality: the more noise contamination, the lower the relative pulse peak power.

17.4 VALIDATION

17.4.1 Validation of HD-DOT with Traditional Retinotopy Paradigms

Method validation is an important process used to confirm the reliability, quality, and consistency of data analyses and results. For example, widespread adoption of fMRI methods exploded after high-fidelity retinotopic maps of visual cortex generated with fMRI both replicated and expanded on previous lesion-, electrophysiology-, and PET-based investigations of the topological organization of the visual cortex (Fox et al., 1986; Fox et al., 1987; Belliveau et al., 1992; Engel et al., 1994; Engel et al., 1997; Reppas et al., 1997; Wandell, 1999; Sereno et al., 2001; Wandell et al., 2007). Similarly, multiple HD-DOT studies also utilized mapping of the visual cortex to demonstrate advancements in image quality provided by the HD arrays in adults (Zhang et al., 2005; Zeff et al., 2007; White and Culver, 2010b; Eggebrecht et al., 2012) and even in infants (Liao et al., 2012). These retinotopy studies suggested that HD-DOT could be used to map the visual cortex, and replicated similar patterns of brain activity reported by PET and fMRI within the confines of the HD-DOT accessible field of view.

17.4.2 Validation of DOT with Functional Magnetic Resonance Imaging

While the previous HD-DOT studies reported activations consistent with expected patterns of brain activity in the literature, limited studies have provided a direct comparison across modalities. In recent neuroimaging research, fMRI has been held up as a "gold-standard" for generating non-invasive spatial brain activation and functional connectivity maps. In the following section, we briefly summarize some of the research articles using fMRI to validate DOT findings within the same human subject sample and same experimental paradigm. For more examples of DOT and HD-DOT studies using fMRI validation, see Table 17.1.

17.4.2.1 Concurrent DOT and fMRI Validation of Image Quality

In a seminal study in 2005, Zhang and colleagues acquired data using an MRI compatible DOT system. Concurrent DOT and fMRI were recorded in response to five blocks with alternating fixation followed by a flashing black and white checkerboard pattern. Results demonstrate bilateral visual cortex activation during concurrent fMRI and DOT (Figure 17.7) (Zhang et al., 2005). This extension of earlier retinotopy research into the realm of fMRI validation set a precedent for future DOT validation studies.

17.4.2.2 Non-Concurrent DOT and fMRI Validation of Image Quality

Due to the added complexity of designing MR compatible DOT equipment, the majority of HD-DOT validation studies have relied on non-concurrent fMRI. Habermehl and colleagues (2012) utilized vibrotactile stimulation to the thumb and fifth finger of the right hand. Motor strip activations corresponding to thumb and fifth finger were reconstructed using HD-DOT and fMRI within adult subjects. Within-subject data from this study

TABLE 17.1 Human Neuroimaging Studies Employing Diffuse Optical Tomography and fMRI Validation

Article	Year	s	d	NN1	Age	Brain Coverage
Chance et al.	1998	9	4	25 mm	A	Prefrontal cortex
Zhang et al.	2005	16	4	20 mm	A	Bilateral Visual Cortex
Joseph et al.	2006	32	32	25 mm	A	Motor Cortex
White et al.	2009	24	28	13 mm	A	Bilateral Visual Cortex
Habermehl et al.	2012	30	30	13 mm	A	Unilateral Parietal Cortex
Eggebrecht et al.	2012	24	28	13 mm	A	Bilateral Visual Cortex
Eggebrecht et al.	2014	92	96	13 mm	A	Visual, Parietal, Temporal, dlPFC
Ferradal et al.	2014	24	28	13 mm	A	Bilateral Visual Cortex
Ferradal et al.	2016	32	34	10 mm	I	Bilateral Temporal and Visual Cortex

S, sources; d, detectors; ch, channels; NN1, first nearest neighbor; N, number of subjects participating in study; dlPFC, dorsolateral prefrontal cortex. Age: A, Adult; I, Infant.

(Chance et al., 1998; Zhang et al., 2005; Joseph et al., 2006; White et al., 2009; Eggebrecht et al., 2012; Habermehl et al., 2012; Eggebrecht et al., 2014; Ferradal et al., 2014; Ferradal et al., 2016).

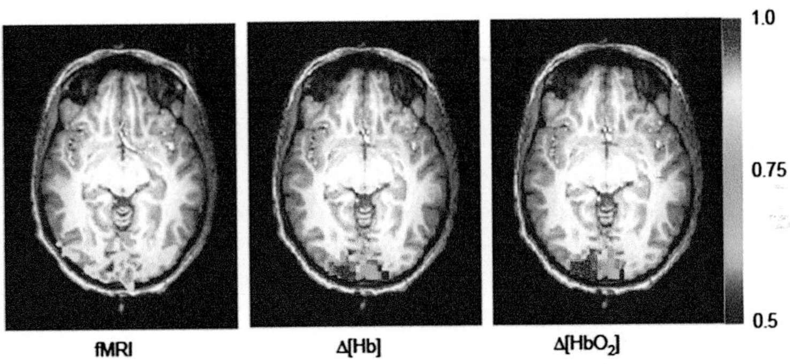

FIGURE 17.7 Validation of DOT with simultaneous fMRI. Visual cortex activations in response to flashing checkerboard stimuli are displayed for fMRI, deoxygenated hemoglobin [Hb] and oxygenated hemoglobin [HbO2] (Zhang et al., 2005).

suggests HD-DOT can be used to distinguish activation foci specific to different fingers, and that these activation foci were reconstructed with good spatial agreement to within-subject fMRI activations during the same task (Figure 17.8A) (Habermehl et al., 2012).

In the same year, a second research study utilized a retinotopy paradigm during separate HD-DOT and fMRI sessions in healthy adults. A rotating, flickering, checkerboard wedge was displayed on the screen in a phase-encoded paradigm. The location of activation during each quadrant of the checkerboard presentation was mapped to the visual cortex. Single-subject and group average data demonstrated that fMRI and HD-DOT retinotopic mapping boasted a high degree of correspondence in the visual cortex (Figure 17.8B) (Eggebrecht et al., 2012).

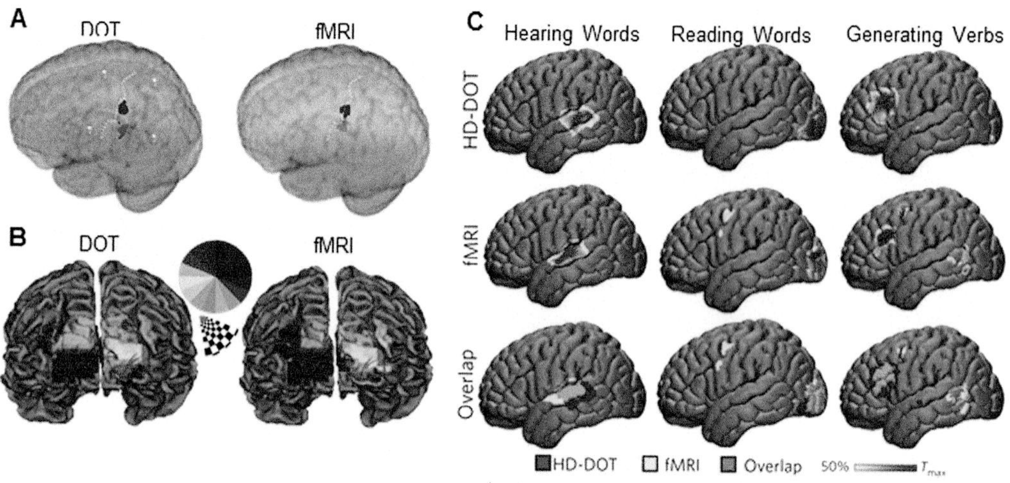

FIGURE 17.8 Task-based validation of DOT with non-concurrent fMRI. **A** Motor activation maps in a single subject undergoing vibrotactile stimulation. Pink represents the thumb and blue represents the pinky finger of the right hand (Habermehl et al., 2012). **B** Visual retinotopy activations in response to a rotating flashing checkerboard wedge. Location of the checkerboard wedge has been color-coded (pinwheel), and the corresponding activation on the visual cortex has been colored based on Eggebrecht et al., 2012. **C** Language-related activation maps in a group of five adults. Language paradigms included activations to hearing words played through a speaker, silent reading of words presented on a screen, and activations to covertly generating verbs in response to nouns printed to the screen. HD-DOT (top row) and fMRI (middle row) demonstrate a strong degree of overlap (bottom row) within primary auditory cortex, visual cortex, and dorsolateral prefrontal cortex (Eggebrecht et al., 2014).

Two years later, this validation work was extended into validation of a large field of view HD-DOT cap covering bilateral regions including aspects of the prefrontal cortex, temporal lobe, and visual cortex (Figures 17.2A, 17.4C) (Eggebrecht et al., 2014). This paper continued the strategy of replicating previously validated neuroimaging paradigms by employing hierarchical language paradigms established in a seminal PET study that first mapped out the spatial topology of single word processing in the brain (Petersen et al., 1988). During the task, several different experimental probes of language function were utilized. In the first task, participants listened to a pre-recorded list of single words played over a speaker. Next, participants silently read a series of simple words displayed one at a time in a block design on a screen. The third experimental run required participants to silently generate verbs in response to nouns presented on screen. Activation maps corresponding to each of these aspects of language were generated during a HD-DOT imaging session followed by an fMRI session on a separate day. Strong agreement between HD-DOT and fMRI was apparent with strong contrast-to-noise activations in the auditory cortex, visual cortex, and dorsolateral prefrontal cortex (Figure 17.8C). This study highlighted spatial correspondence throughout a spatially extended field of view that encompassed both sensory areas, known for exhibiting large signal to noise activations in response to a well-designed paradigm, as well as "higher-order" cognitive and association areas, known in the fMRI literature for exhibiting relatively smaller activation volumes and contrast levels.

While task-based studies were particularly useful for validating event-elicited brain activations between HD-DOT and fMRI, alternative neuroimaging experimental methods have also been used to validate HD-DOT against fMRI. An increasingly common method in the fMRI literature is the use of resting state-functional MRI (rs-fMRI), a technique that can be used to assess functional connectivity within the brain. These resting state methods are ideal for situations in which a participant may be unable to engage in a traditional task-based block-design or event-related neuroimaging paradigm, such as infants or those who are asleep or cognitively impaired. Functional connectivity can be inferred by assessing temporal correlations in low-frequency fluctuations of the BOLD signal (below 0.01 Hz) (Biswal et al., 1995; Lowe et al., 1998; Fox et al., 2005). Importantly, rs-fMRI data can be used to identify spatially distributed brain networks, comprising regions of the brain known to be activated by task, including primary cortical regions such as visual and motor cortex, as well as "higher order" cortical areas used for and attention and executive function (Fox and Raichle, 2007; Kahn et al., 2008; Vincent et al., 2008; Smith et al., 2009). The composition of these networks has been well-characterized using fMRI in adults and older pediatric populations (Buckner et al., 2009; Smith et al., 2009), and has also been increasingly established in infants (Doria et al., 2010; Smyser et al., 2010; Fransson et al., 2011; Gao et al., 2013; Smyser et al., 2013; Gao et al., 2015; Smyser et al., 2016b; Smyser et al., 2016a).

Validation of resting state functional connectivity methods for HD-DOT (fcDOT) were first demonstrated in healthy adults (White et al., 2009). This seminal paper showed that fcDOT maps were reproducible in participants across days and that bilateral maps of correlations within (but not between) visual and motor regions were replicated in fMRI in those same participants. More recently, subject-specific light modeling and an expanded field of view broadened the reach of fcDOT methods to map not just sensory or motor networks, but also spatially distributed "higher-order" cognitive networks including cortical aspects of the dorsal attention network, fronto-parietal control network, and the default mode network (Figure 17.9A) (Eggebrecht et al., 2014). Group level analyses demonstrated remarkable similarities in the topology of these brain networks between fcDOT and rs-fMRI. Cumulatively, these validation studies suggest DOT is capable of measuring task-based activations in healthy adults with comparable spatial specificity to that observed with fMRI.

17.5 DOT APPLICATIONS IN HUMAN CLINICAL POPULATIONS

We will finish this chapter with a brief overview highlighting studies that have applied DOT and HD-DOT technology to clinical applications that are beyond the reach of traditional neuroimaging methods such as fMRI (See Table 17.2). Infants in the neonatal intensive care unit (NICU) are a unique clinical population that presents significant challenges for monitoring brain health. Movement of the neonates to an fMRI imaging bay may be unsafe or impractical, especially for those who are profoundly infirm. Several studies have used DOT to assess brain activity and functional connectivity within preterm infants at a range of gestational ages recovering in the NICU. We also highlight several case reports of infants with brain injuries including hypoxic ischemic encephalopathy (HIE), intraventricular hemorrhage (IVH), stroke, and seizures.

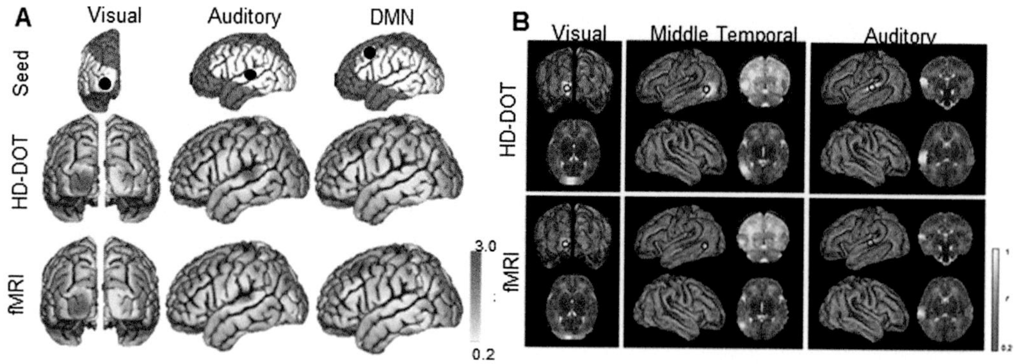

FIGURE 17.9 Resting state functional connectivity validation using high density DOT (HD-DOT) and fMRI. **A** Functional connectivity maps from generated from 11 adults using a visual, auditory, and default mode network (DMN) seed region (top row). HD-DOT (middle row), and fMRI (bottom row) functional connectivity maps bear striking similarity (Eggebrecht et al., 2014). **B** Functional connectivity maps generated from 14 term infants within the visual, middle temporal, and auditory cortex demonstrate similar spatial maps between functional connectivity DOT (top row) and functional connectivity fcMRI (bottom row) (Ferradal et al., 2016).

TABLE 17.2 Human Neuroimaging Studies Applying Diffuse Optical Tomography to Clinical Populations

Article	**Year**	**s**	**d**	**NN1**	**Age**	**Brain Coverage**
Chance et al.	1998	9	4	25 mm	EPT	Motor Cortex
Hintz et al.	2001	9	16	*	VPT	Motor cortex
Hebdon et al.	2002	32	32	*	HIE	Whole head (infant)
Hebden et al.	2004	32	32	*	EPT	Whole head (infant)
Gibson et al.	2006	32	32	*	VPT, IVH	Motor Cortex
Austin et al.	2006	32	32	*	IVH	Whole Brain (infant)
White et al.	2012	18	16	10 mm	PT, S	Bilateral Visual Cortex
Singh et al.	2014	16	16	20 mm	HIE	Whole Brain (infant)
Chalia et al.	2016	16	16	20 mm	HIE	Whole Brain (infant)

N, number of subjects participating in study; dlPFC, dorsolateral prefrontal cortex
Age: EPT, Extremely preterm infant; VPT, very preterm infant; HIE, infant with hypoxic ischemic encephalopathy; IVH, infant with intraventricular hemorrhage; PT, preterm; S, stroke.
*nearest source-detector distance not reported
(Chance et al., 1998; Hintz et al., 2001; Hebden et al., 2002; Hebden et al., 2004; Austin et al., 2006; Gibson et al., 2006; White et al., 2012; Singh et al., 2014; Chalia et al., 2016).

In a groundbreaking paper that reported a comparison of prefrontal fMRI and DOT activation in adults, the authors also reported DOT motor activations while stroking the left and right fingers of an infant born extremely preterm (27 weeks) (Chance et al., 1998). Three years later, Hintz and colleagues demonstrated motor cortex activity to passive arm movements in infants born moderately preterm (32–33 weeks gestational age) (Hintz et al., 2001). These early studies provided a proof that DOT could successfully be used within the NICU.

Neonatal brain injury, including the study of acute injury, such as HIE, is a particularly interesting application of DOT within the NICU. HIE occurs as a result of oxygen deprivation from fetal trauma either during gestation or during birth, and can result in long-term developmental complications including cerebral palsy, epilepsy, and sensory impairments. Hebdon and colleagues (2002) used a whole-head infant DOT cap to image a preterm infant with hypoxic ischemic encephalopathy (HIE) (Figure 17.10A). This infant experienced cerebral hemorrhage of the left ventricle, and images generated with whole-head DOT demonstrated an asymmetry consistent with the injury (Hebden et al., 2002). Further, Chalia and colleagues (2016) used HD-DOT to study hemodynamics associated with high-frequency "bursting" EEG activity, typically signifying pathological activity, in a group of term-born infants with HIE in the NICU. Infants presented with seizures in

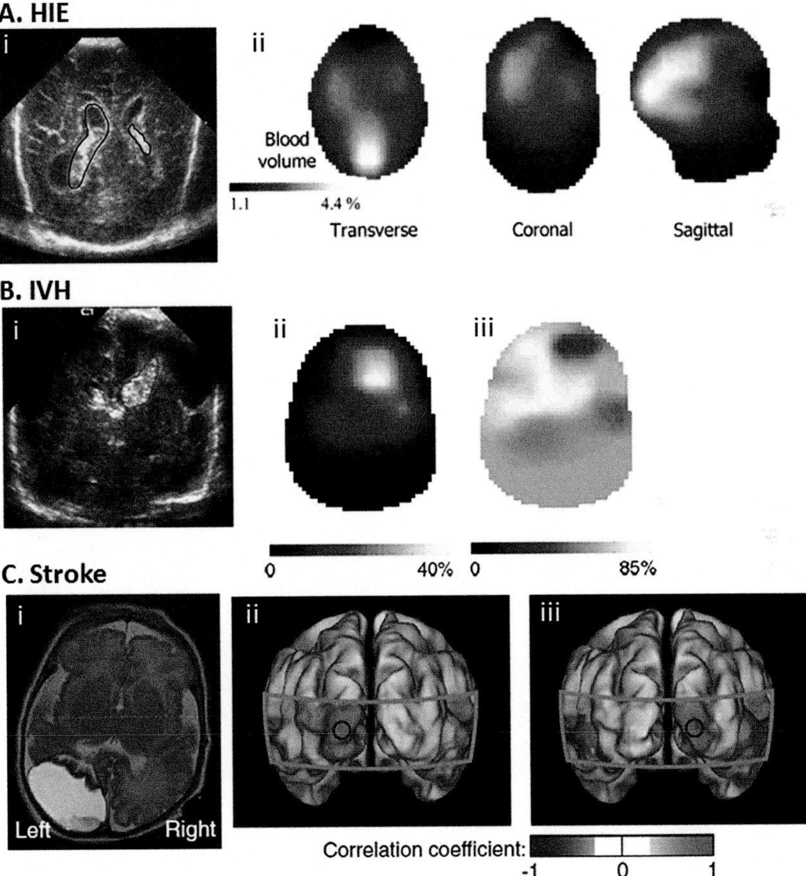

FIGURE 17.10 DOT studies of neonatal brain injury. **A** DOT whole brain reconstruction of preterm neonate with hypoxic ischemic encephalopathy (HIE) (Hebden et al., 2002). **B** DOT whole brain DOT reconstruction of preterm neonate with left sided intraventricular hemorrhage (IVH). (i) cranial ultrasound, (ii) regional blood volume, and (iii) regional oxygen saturation demonstrate disruptions in oxygen and hemoglobin in the region near the hemorrhagic parenchymal infarct (Austin et al., 2006). **C** Work from our group has assessed resting state functional connectivity using DOT (fcDOT). A seed placed in the left or right visual cortex will correlate with contralateral visual cortex in term-born neonates. Infants born preterm have weaker bilateral connectivity within the visual cortex. Bilateral visual network connectivity was completely absent in a preterm infant with a unilateral left occipital stroke (White et al., 2012).

the first 48 hours of life, and were scanned with combined EEG-DOT within 7 days of birth. Across infants, oxygenated hemoglobin initially declined during the EEG bursts, and peaked 10–12 seconds after the burst onset (Chalia et al., 2016). This study provides a powerful example of the clinical application of concurrent EEG and DOT methods (Figure 17.11).

Neonatal IVH has also been investigated using DOT. IVH is one of the leading forms of preterm brain injury, and typically occurs within the first 72 hours following birth. Among surviving infants, those with high-grade IVH will develop neurodevelopmental disability in greater than 50% of cases, with deficits across cognitive, language, and motor domains (Ancel et al., 2006; Adams-Chapman et al., 2008; McCrea and Ment, 2008). Austin and colleagues (2006) used DOT to scan 14 preterm infants in the NICU, several of whom were diagnosed with IVH. One such infant demonstrated increased regional blood volume and oxygen saturation using DOT (Figure 17.10B) corresponding to IVH in the left hemisphere, visible on ultrasound (Austin et al., 2006). These studies suggest imaging infants with DOT during this acute period of brain injury in the NICU may provide insights into neural disruptions leading to subsequent neurodevelopmental impairment later in life.

Resting state fcMRI and fcDOT methods provide powerful tools to investigate brain health via measurements of functional brain networks in newborn infants. A single high-quality dataset can be acquired in minutes, and data can be collected from subjects swaddled, resting quietly, sleeping, and under anesthesia or morphine without requirement of task performance or attention to stimulus. White and colleagues (2012) first collected

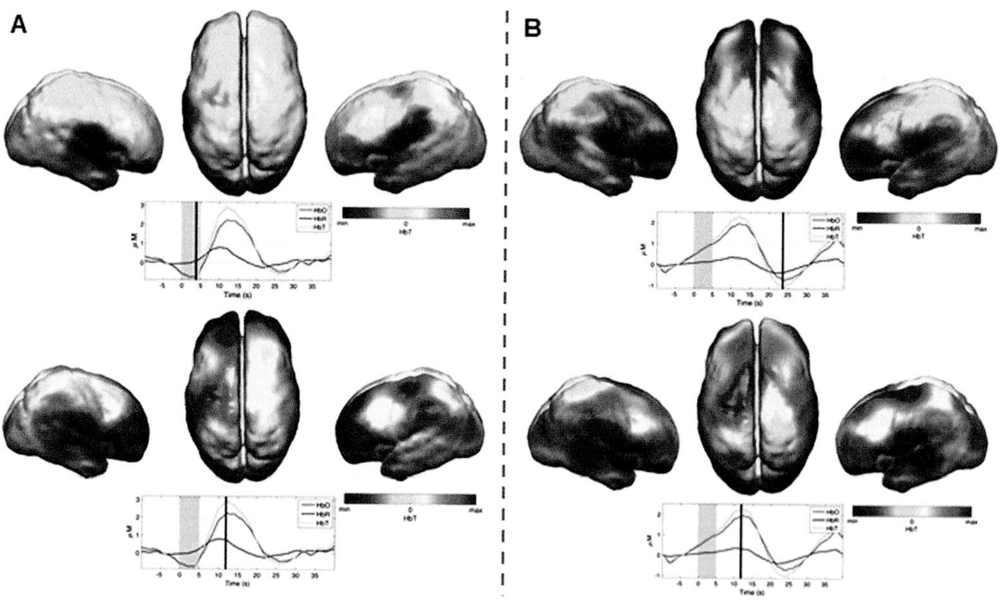

FIGURE 17.11 DOT reconstruction during seizures in term-born infants with hypoxic ischemic encephalopathy (HIE). **A** Whole brain DOT cortical reconstruction corresponding to a dip in total hemoglobin (HbT) just prior to (top row) and peak HbT (bottom row) following a seizure event in an infant with grade I HIE. **B** DOT corresponding to peak HbT (bottom row) and a subsequent dip in HbT (top row) following a seizure event in an infant with grade III HIE (Chalia et al., 2016).

fcDOT in term-born infants in the newborn nursery (White et al., 2012). They acquired fcDOT on three term-born infants and four preterm-born infants. One of these preterm infants demonstrated a large left occipital hemorrhage, apparent using structural MRI (Figure 17.10C). fcDOT revealed bilateral functional connectivity maps within visual cortex in the healthy term-born infants. This same pattern of bilateral visual cortex connectivity, although weaker, was also present in preterm-born subjects. However, bilateral visual cortex connectivity was absent in the preterm infant with left occipital hemorrhagic stroke (White et al., 2012). That same group expanded the field of view of their neonatal imaging cap and demonstrated similar patterns of bilateral visual, middle temporal, and auditory connectivity using HD-DOT and fMRI (Figure 17.9B). Cumulatively, these papers suggest DOT is capable of reliability and accurately measuring brain activity, and reinforces the utility of DOT in the study of infant populations in the NICU.

17.6 CONCLUSIONS

In this chapter, we have focused on the physical principles underlying optical neuroimaging in humans, and the challenges of design and implementation of high-density arrays. We have highlighted several papers that have demonstrated strong validation of the specificity and reliability of the technology. Finally, we summarized papers highlighting the unique potential for HD-DOT methods to profoundly impact clinical care. Advances in HD-DOT systems make this modality a viable alternative to fMRI, providing comparable spatial information about cerebral cortex activity and connectivity, with the added advantage of being portable (e.g. bedside data collection in populations that cannot be taken to a scanner). Significant challenges remain in technology development for these tools to be more widely utilized. Careful consideration is required in order to design HD-DOT devices that maintain good image quality over larger fields of view (high-density arrangements of sources and detectors that maintain a high dynamic range and low crosstalk) while also improving wearability (smaller and lighter weight imaging caps). Reliable and efficient anatomical coregistration methods are required to accurately estimate HD-DOT activations over the cortex. Finally, additional clinical and behavioral validation will be crucial for the establishment of HD-DOT as a widely adopted portable diagnostic device.

ACKNOWLEDGMENTS

This work was supported by the National Institutes of Health (NIH, grants: K01MH103594 and R21MH109775 (A.T.E.), and T32MH100019 (M.D.W.)).

REFERENCES

Adams-Chapman I, Hansen NI, Stoll BJ, Higgins R (2008) Neurodevelopmental outcome of extremely low birth weight infants with posthemorrhagic hydrocephalus requiring shunt insertion. *Pediatrics* 121:e1167–e1177.

Ancel PY, Livinec F, Larroque B, Marret S, Arnaud C, Pierrat V, Dehan M, N'Guyen S, Escande B, Burguet A, Thiriez G, Picaud JC, Andre M, Breart G, Kaminski M, Group ES (2006) Cerebral palsy among very preterm children in relation to gestational age and neonatal ultrasound abnormalities: the EPIPAGE cohort study. *Pediatrics* 117:828–835.

Arridge SR (1999) Optical tomography in medical imaging. *Inverse Problems* 15:R41–R93.

Arridge SR, Schweiger M (1995) Photon-measurement density functions. Part 2: finite-element-method calculations. *Applied Optics* 34:8026–8037.

Arridge SR, Schweiger M, Hiraoka M, Delpy DT (1993) A finite-element approach for modeling photon transport in tissue. *Medical Physics* 20:299–309.

Austin T, Gibson AP, Branco G, Yusof RM, Arridge SR, Meek JH, Wyatt JS, Delpy DT, Hebden JC (2006) Three dimensional optical imaging of blood volume and oxygenation in the neonatal brain. *Neuroimage* 31:1426–1433.

Baldassarre A, Ramsey LE, Siegel JS, Shulman GL, Corbetta M (2016) Brain connectivity and neurological disorders after stroke. *Current Opinion in Neurology* 29:706–713.

Bale G, Mitra S, Meek J, Robertson N, Tachtsidis I (2014) A new broadband near-infrared spectroscopy system for in-vivo measurements of cerebral cytochrome-c-oxidase changes in neonatal brain injury. *Biomedical Optics Express* 5:3450–3466.

Bandettini PA, Wong EC, Hinks RS, Tikofsky RS, Hyde JS (1992) Time course EPI of human brain function during task activation. *Magnetic Resonance in Medicine* 25:390–397.

Belliveau JW, Kennedy DN, Jr., McKinstry RC, Buchbinder BR, Weisskoff RM, Cohen MS, Vevea JM, Brady TJ, Rosen BR (1991) Functional mapping of the human visual cortex by magnetic resonance imaging. *Science* 254:716–719.

Belliveau JW, Kwong KK, Baker JR, C.E. S, Benson R, Goldberg IE, Cohen MS, Kennedy DN, T.J. B, Rosen BR (1992) MRI mapping of human visual cortex: retinotopic organization and frequency response of V1. In: Eleventh Annual Scientific Meeting of the Society of Magnetic Resonance in Medicine, p. 310. Berlin, Germany.

Biswal B, Yetkin FZ, Haughton VM, Hyde JS (1995) Functional connectivity in the motor cortex of resting human brain using echo-planar MRI. *Magnetic Resonance in Medicine* 34:537–541.

Blasi A, Lloyd-Fox S, Sethna V, Brammer MJ, Mercure E, Murray L, Williams SC, Simmons A, Murphy DG, Johnson MH (2015) Atypical processing of voice sounds in infants at risk for autism spectrum disorder. *Cortex* 71:122–133.

Bluestone A, Abdoulaev G, Schmitz C, Barbour R, Hielscher A (2001) Three-dimensional optical tomography of hemodynamics in the human head. *Optics Express* 9:272–286.

Boas D, Culver J, Stott J, Dunn A (2002) Three dimensional Monte Carlo code for photon migration through complex heterogeneous media including the adult human head. *Optics Express* 10:159–170.

Boas DA, Dale AM, Franceschini MA (2004a) Diffuse optical imaging of brain activation: approaches to optimizing image sensitivity, resolution, and accuracy. *Neuroimage* 23 Suppl 1:S275–S288.

Boas DA, Chen K, Grebert D, Franceschini MA (2004b) Improving the diffuse optical imaging spatial resolution of the cerebral hemodynamic response to brain activation in humans. *Optics Letters* 29:1506–1508.

Boas DA, O'Leary MA, Chance B, Yodh AG (1993) Scattering and wavelength transduction of diffuse photon density waves. *Physical Review. E, Statistical Physics, Plasmas, Fluids, and Related Interdisciplinary Topics* 47:R2999–R3002.

Brigadoi S, Aljabar P, Kuklisova-Murgasova M, Arridge SR, Cooper RJ (2014) A 4D neonatal head model for diffuse optical imaging of pre-term to term infants. *Neuroimage* 100:385–394.

Buckner RL, Sepulcre J, Talukdar T, Krienen FM, Liu H, Hedden T, Andrews-Hanna JR, Sperling RA, Johnson KA (2009) Cortical hubs revealed by intrinsic functional connectivity: mapping, assessment of stability, and relation to Alzheimer's disease. *The Journal of Neuroscience* 29:1860–1873.

Carter AR, Patel KR, Astafiev SV, Snyder AZ, Rengachary J, Strube MJ, Pope A, Shimony JS, Lang CE, Shulman GL, Corbetta M (2012) Upstream dysfunction of somatomotor functional connectivity after corticospinal damage in stroke. *Neurorehabilitation and Neural Repair* 26:7–19.

Case KM, Zweifel PF (1967) *Linear Transport Theory*. Reading, MA: Addison-Wesley Pub. Co.

Chalia M, Lee CW, Dempsey LA, Edwards AD, Singh H, Michell AW, Everdell NL, Hill RW, Hebden JC, Austin T, Cooper RJ (2016) Hemodynamic response to burst-suppressed and discontinuous electroencephalography activity in infants with hypoxic ischemic encephalopathy. *Neurophotonics* 3:031408.

Chance B, Anday E, Nioka S, Zhou S, Hong L, Worden K, Ovetsky MY, Pidikiti D, Thomals R (1998) A novel method for fast imaging of brain function, non-invasively, with light. *Optics Express* 2:411–423.

Cheong WF, Prahl SA, Welch AJ (1990) A review of the optical-properties of biological tissues. *IEEE Journal of Quantum Electronics* 26:2166–2185.

Cooper RJ, Caffini M, Dubb J, Fang Q, Custo A, Tsuzuki D, Fischl B, Wells W, 3rd, Dan I, Boas DA (2012) Validating atlas-guided DOT: a comparison of diffuse optical tomography informed by atlas and subject-specific anatomies. *Neuroimage* 62:1999–2006.

Corbetta M, Shulman GL (2002) Control of goal-directed and stimulus-driven attention in the brain. *Nature Reviews. Neuroscience* 3:201–215.

Cui X, Bray S, Bryant DM, Glover GH, Reiss AL (2011) A quantitative comparison of NIRS and fMRI across multiple cognitive tasks. *Neuroimage* 54:2808–2821.

Culver JP, Choe R, Holboke MJ, Zubkov L, Durduran T, Slemp A, Ntziachristos V, Chance B, Yodh AG (2003) Three-dimensional diffuse optical tomography in the parallel plane transmission geometry: evaluation of a hybrid frequency domain/continuous wave clinical system for breast imaging. *Medical Physics* 30:235–247.

Custo A, Boas DA, Tsuzuki D, Dan I, Mesquita R, Fischl B, Grimson WE, Wells W, 3rd (2010) Anatomical atlas-guided diffuse optical tomography of brain activation. *Neuroimage* 49:561–567.

de Roever I, Bale G, Cooper RJ, Tachtsidis I (2017) Functional NIRS measurement of cytochrome-C-oxidase demonstrates a more brain-specific marker of frontal lobe activation compared to the haemoglobins. *Advances in Experimental Medicine and Biology* 977:141–147.

Dehghani H, Eames ME, Yalavarthy PK, Davis SC, Srinivasan S, Carpenter CM, Pogue BW, Paulsen KD (2009) Near infrared optical tomography using NIRFAST: algorithm for numerical model and image reconstruction. *Communications in Numerical Methods in Engineering* 25:711–732.

Doria V, Beckmann CF, Arichi T, Merchant N, Groppo M, Turkheimer FE, Counsell SJ, Murgasova M, Aljabar P, Nunes RG, Larkman DJ, Rees G, Edwards AD (2010) Emergence of resting state networks in the preterm human brain. *Proceedings of the National Academy of Sciences of the United States of America* 107:20015–20020.

Doulgerakis M, Eggebrecht A, Wojtkiewicz S, Culver J, Dehghani H (2017) Toward real-time diffuse optical tomography: accelerating light propagation modeling employing parallel computing on GPU and CPU. *Journal of Biomedical Optics* 22:1–11.

Durduran T, Choe R, Baker WB, Yodh AG (2010) Diffuse optics for tissue monitoring and tomography. *Reports on Progress in Physics* 73: 076701.

Eames ME, Dehghani H (2008) Wavelength dependence of sensitivity in spectral diffuse optical imaging: effect of normalization on image reconstruction. *Optics Express* 16:17780–17791.

Eames ME, Wang J, Pogue BW, Dehghani H (2008) Wavelength band optimization in spectral near-infrared optical tomography improves accuracy while reducing data acquisition and computational burden. *Journal of Biomedical Optics* 13:054037.

Eggebrecht AT, Elison JT, Feczko E, Todorov A, Wolff JJ, Kandala S, Adams CM, Snyder AZ, Lewis JD, Estes AM, Zwaigenbaum L (2017) Joint attention and brain functional connectivity in infants and toddlers. *Cerebral Cortex* 27:1709–1720.

Eggebrecht AT, Ferradal SL, Robichaux-Viehoever A, Hassanpour MS, Dehghani H, Snyder AZ, Hershey T, Culver JP (2014) Mapping distributed brain function and networks with diffuse optical tomography. *Nature Photonics* 8:448–454.

Eggebrecht AT, White BR, Ferradal SL, Chen C, Zhan Y, Snyder AZ, Dehghani H, Culver JP (2012) A quantitative spatial comparison of high-density diffuse optical tomography and fMRI cortical mapping. *Neuroimage* 61:1120–1128.

Engel SA, Glover GH, Wandell BA (1997) Retinotopic organization in human visual cortex and the spatial precision of functional MRI. *Cerebral Cortex* 7:181–192.

Engel SA, Rumelhart DE, Wandell BA, Lee AT, Glover GH, Chichilnisky EJ, Shadlen MN (1994) fMRI of human visual cortex. *Nature* 369:525.

Ferradal SL, Eggebrecht AT, Hassanpour M, Snyder AZ, Culver JP (2014) Atlas-based head modeling and spatial normalization for high-density diffuse optical tomography: in vivo validation against fMRI. *Neuroimage* 85 Pt 1:117–126.

Ferradal SL, Liao SM, Eggebrecht AT, Shimony JS, Inder TE, Culver JP, Smyser CD (2016) Functional imaging of the developing brain at the bedside using diffuse optical tomography. *Cerebral Cortex* 26:1558–1568.

Fox MD, Raichle ME (2007) Spontaneous fluctuations in brain activity observed with functional magnetic resonance imaging. *Nature Reviews Neuroscience* 8:700–711.

Fox MD, Snyder AZ, Vincent JL, Corbetta M, Van Essen DC, Raichle ME (2005) The human brain is intrinsically organized into dynamic, anticorrelated functional networks. *Proceedings of the National Academy of Sciences of the United States of America* 102:9673–9678.

Fox PT, Miezin FM, Allman JM, Mintun MA, Van Essen DC, Raichle ME (1986) Retinotopic organization of human visual cortex using a PET analysis strategy that improves spatial resolution. *Society for Neuroscience Abstracts* 12:1181.

Fox PT, Miezin FM, Allman JM, Van Essen DC, Raichle ME (1987) Retinotopic organization of human visual cortex mapped with positron emission tomography. *Journal of Neuroscience* 7:913–922.

Fox PT, Raichle ME (1986) Focal physiological uncoupling of cerebral blood flow and oxidative metabolism during somatosensory stimulation in human subjects. *Proceedings of the National Academy of Sciences of the United States of America* 83:1140–1144.

Fox PT, Raichle ME, Mintun MA, Dence C (1988) Nonoxidative glucose consumption during focal physiologic neural activity. *Science* 241:462–464.

Fransson P, Aden U, Blennow M, Lagercrantz H (2011) The functional architecture of the infant brain as revealed by resting-state fMRI. *Cerebral Cortex* 21:145–154.

Gao W, Alcauter S, Smith JK, Gilmore JH, Lin W (2015) Development of human brain cortical network architecture during infancy. *Brain Structure and Function* 220:1173–1186.

Gao W, Gilmore JH, Shen D, Smith JK, Zhu H, Lin W (2013) The synchronization within and interaction between the default and dorsal attention networks in early infancy. *Cerebral Cortex* 23:594–603.

Gibson AP, Austin T, Everdell NL, Schweiger M, Arridge SR, Meek JH, Wyatt JS, Delpy DT, Hebden JC (2006) Three-dimensional whole-head optical tomography of passive motor evoked responses in the neonate. *Neuroimage* 30:521–528.

Gibson AP, Riley J, Schweiger M, Hebden JC, Arridge SR, Delpy DT (2003) A method for generating patient-specific finite element meshes for head modelling. *Physics in Medicine and Biology* 48:481–495.

Graaff R, Koelink MH, de Mul FF, Zijistra WG, Dassel AC, Aarnoudse JG (1993) Condensed Monte Carlo simulations for the description of light transport. *Applied Optics* 32:426–434.

Groenhuis RA, Ferwerda HA, Ten Bosch JJ (1983) Scattering and absorption of turbid materials determined from reflection measurements. 1: theory. *Applied Optics* 22:2456–2462.

Habermehl C, Holtze S, Steinbrink J, Koch SP, Obrig H, Mehnert J, Schmitz CH (2012) Somatosensory activation of two fingers can be discriminated with ultrahigh-density diffuse optical tomography. *Neuroimage* 59:3201–3211.

Hassanpour MS, Eggebrecht AT, Culver JP, Peelle JE (2015) Mapping cortical responses to speech using high-density diffuse optical tomography. *Neuroimage* 117:319–326.

Hassanpour MS, Eggebrecht AT, Peelle JE, Culver JP (2017) Mapping effective connectivity within cortical networks with diffuse optical tomography. *Neurophotonics* 4:041402.

Hebden JC, Gibson A, Austin T, Yusof RM, Everdell N, Delpy DT, Arridge SR, Meek JH, Wyatt JS (2004) Imaging changes in blood volume and oxygenation in the newborn infant brain using three-dimensional optical tomography. *Physics in Medicine and Biology* 49:1117–1130.

Hebden JC, Gibson A, Yusof RM, Everdell N, Hillman EM, Delpy DT, Arridge SR, Austin T, Meek JH, Wyatt JS (2002) Three-dimensional optical tomography of the premature infant brain. *Physics in Medicine and Biology* 47:4155–4166.

Hillman EM, Amoozegar CB, Wang T, McCaslin AF, Bouchard MB, Mansfield J, Levenson RM (2011) In vivo optical imaging and dynamic contrast methods for biomedical research. *Philosophical Transactions. Series A, Mathematical, Physical, and Engineering Sciences* 369:4620–4643.

Hintz SR, Benaron DA, Siegel AM, Zourabian A, Stevenson DK, Boas DA (2001) Bedside functional imaging of the premature infant brain during passive motor activation. *Journal of Perinatal Medicine* 29:335–343.

Hiraoka M, Firbank M, Essenpreis M, Cope M, Arridge SR, van der Zee P, Delpy DT (1993) A Monte Carlo investigation of optical pathlength in inhomogeneous tissue and its application to near-infrared spectroscopy. *Physics in Medicine and Biology* 38:1859–1876.

Jacques SL, Pogue BW (2008) Tutorial on diffuse light transport. *Journal of Biomedical Optics* 13:1–19.

Jobsis FF (1977) Noninvasive, infrared monitoring of cerebral and myocardial oxygen sufficiency and circulatory parameters. *Science* 198:1264–1267.

Johnson CC (1970) Optical diffusion in blood. *IEEE Transactions on Biomedical Engineering* 17:129–133.

Joseph DK, Huppert TJ, Franceschini MA, Boas DA (2006) Diffuse optical tomography system to image brain activation with improved spatial resolution and validation with functional magnetic resonance imaging. *Applied Optics* 45:8142–8151.

Kahn I, Andrews-Hanna JR, Vincent JL, Snyder AZ, Buckner RL (2008) Distinct cortical anatomy linked to subregions of the medial temporal lobe revealed by intrinsic functional connectivity. *Journal of Neurophysiology* 100:129–139.

Kak AC, Slaney M, IEEE Engineering in Medicine and Biology Society (1988) *Principles of Computerized Tomographic Imaging*. New York: IEEE Press.

Kennedy DP, Courchesne E (2008) The intrinsic functional organization of the brain is altered in autism. *Neuroimage* 39:1877–1885.

Koch SP, Habermehl C, Mehnert J, Schmitz CH, Holtze S, Villringer A, Steinbrink J, Obrig H (2010) High-resolution optical functional mapping of the human somatosensory cortex. *Frontiers in Neuroenergetics* 2:12.

Kwong KK, Belliveau JW, Chesler DA, Goldberg IE, Weisskoff RM, Poncelet BP, Kennedy DN, Hoppel BE, Cohen MS, Turner R (1992) Dynamic magnetic resonance imaging of human brain activity during primary sensory stimulation. *Proceedings of the National Academy of Sciences of the United States of America* 89:5675–5679.

Liao SM, Ferradal SL, White BR, Gregg N, Inder TE, Culver JP (2012) High-density diffuse optical tomography of term infant visual cortex in the nursery. *Journal of Biomedical Optics* 17:081414.

Liao SM, Gregg NM, White BR, Zeff BW, Bjerkaas KA, Inder TE, Culver JP (2010) Neonatal hemodynamic response to visual cortex activity: high-density near-infrared spectroscopy study. *Journal of Biomedical Optics* 15:026010.

Liebert A, Wabnitz H, Grosenick D, Macdonald R (2003) Fiber dispersion in time domain measurements compromising the accuracy of determination of optical properties of strongly scattering media. *Journal of Biomedical Optics* 8:512–516.

Liebert A, Wabnitz H, Steinbrink J, Obrig H, Moller M, Macdonald R, Villringer A, Rinneberg H (2004) Time-resolved multidistance near-infrared spectroscopy of the adult head: intracerebral and extracerebral absorption changes from moments of distribution of times of flight of photons. *Applied Optics* 43:3037–3047.

Lloyd-Fox S, Blasi A, Elwell CE, Charman T, Murphy D, Johnson MH (2013) Reduced neural sensitivity to social stimuli in infants at risk for autism. *Proceedings of the Royal Society B: Biological Sciences* 280:20123026.

Lloyd-Fox S, Blasi A, Pasco G, Gliga T, Jones EJH, Murphy DGM, Elwell CE, Charman T, Johnson MH, Team B (2017) Cortical responses before 6 months of life associate with later autism. *European Journal of Neuroscience* 47: 736–749.

Lowe MJ, Mock BJ, Sorenson JA (1998) Functional connectivity in single and multislice echoplanar imaging using resting-state fluctuations. *Neuroimage* 7:119–132.

Maki A, Yamashita Y, Ito Y, Watanabe E, Mayanagi Y, Koizumi H (1995) Spatial and temporal analysis of human motor activity using noninvasive NIR topography. *Medical Physics* 22:1997–2005.

Marrus N, Eggebrecht AT, Todorov A, Elison JT, Wolff JJ, Cole L, Gao W, Pandey J, Shen MD, Swanson MR, Emerson RW (2017) Walking, gross motor development, and brain functional connectivity in infants and toddlers. *Cerebral Cortex* :1–14.

McCrea HJ, Ment LR (2008) The diagnosis, management, and postnatal prevention of intraventricular hemorrhage in the preterm neonate. *Clinics in Perinatology* 35:777–792, vii.

Ntziachristos V, Chance B, Yodh A (1999) Differential diffuse optical tomography. *Optics Express* 5:230–242.

O'Leary MA, Boas DA, Chance B, Yodh AG (1992) Refraction of diffuse photon density waves. *Physical Review Letters* 69:2658–2661.

O'Leary MA, Boas DA, Chance B, Yodh AG (1995) Experimental images of heterogeneous turbid media by frequency-domain diffusing-photon tomography. *Optics Letters* 20:426–428.

Obrig H, Villringer A (2003) Beyond the visible--imaging the human brain with light. *Journal of Cerebral Blood Flow and Metabolism* 23:1–18.

Ogawa S, Lee TM (1990) Magnetic resonance imaging of blood vessels at high fields: in vivo and in vitro measurements and image simulation. *Magnetic Resonance in Medicine* 16:9–18.

Ogawa S, Lee TM, Nayak AS, Glynn P (1990) Oxygenation-sensitive contrast in magnetic resonance image of rodent brain at high magnetic fields. *Magnetic Resonance in Medicine* 14:68–78.

Ogawa S, Tank DW, Menon R, Ellermann JM, Kim SG, Merkle H, Ugurbil K (1992) Intrinsic signal changes accompanying sensory stimulation: functional brain mapping with magnetic resonance imaging. *Proceedings of the National Academy of Sciences of the United States of America* 89:5951–5955.

Okada E, Schweiger M, Arridge SR, Firbank M, Delpy DT (1996) Experimental validation of Monte Carlo and finite-element methods for the estimation of the optical path length in inhomogeneous tissue. *Applied Optics* 35:3362–3371.

Patterson MS, Chance B, Wilson BC (1989) Time resolved reflectance and transmittance for the non-invasive measurement of tissue optical properties. *Applied Optics* 28:2331–2336.

Paulsen KD, Jiang HB (1995) Spatially varying optical property reconstruction using a finite-element diffusion equation approximation. *Medical Physics* 22:691–701.

Petersen SE, Fox PT, Posner MI, Mintun M, Raichle ME (1988) Positron emission tomographic studies of the cortical anatomy of single-word processing. *Nature* 331:585–589.

Power JD, Cohen AL, Nelson SM, Wig GS, Barnes KA, Church JA, Vogel AC, Laumann TO, Miezin FM, Schlaggar BL, Petersen SE (2011) Functional network organization of the human brain. *Neuron* 72:665–678.

Raichle ME (2010) Two views of brain function. *Trends in Cognitive Sciences* 14:180–190.

Raichle ME, Mintun MA (2006) Brain work and brain imaging. *Annual Review of Neuroscience* 29:449–476.

Re R, Pirovano I, Contini D, Spinelli L, Torricelli A (2018) Time domain near infrared spectroscopy device for monitoring muscle oxidative metabolism: custom probe and in vivo applications. *Sensors* 18: e264.

Reppas JB, Niyogi S, Dale AM, Sereno MI, Tootell RBH (1997) Representation of motion boundaries in retinotopic human visual cortical areas. *Nature* 388:175–179.

Schweiger M, Arridge S (2014) The toast plus plus software suite for forward and inverse modeling in optical tomography. *Journal of Biomedical Optics* 19:1–16.

Schweiger M, Arridge SR, Hiraoka M, Delpy DT (1995) The finite element method for the propagation of light in scattering media: boundary and source conditions. *Medical Physics* 22:1779–1792.

Segelstein D (1981) The complex refractive index of water. Master Thesis, p. 167. Kansas City, MO: University of Missouri-Kansas City.

Sereno MI, Pitzalis S, Martinez A (2001) Mapping of contralateral space in retinotopic coordinates by a parietal cortical area in humans. *Science* 294:1350–1354.

Singh H, Cooper RJ, Wai Lee C, Dempsey L, Edwards A, Brigadoi S, Airantzis D, Everdell N, Michell A, Holder D, Hebden JC, Austin T (2014) Mapping cortical haemodynamics during neonatal seizures using diffuse optical tomography: a case study. *Neuroimage Clinical* 5:256–265.

Smith SM, Fox PT, Miller KL, Glahn DC, Fox PM, Mackay CE, Filippini N, Watkins KE, Toro R, Laird AR, Beckmann CF (2009) Correspondence of the brain's functional architecture during activation and rest. *PNAS* 106:13040–13045.

Smyser CD, Inder TE, Shimony JS, Hill JE, Degnan AJ, Snyder AZ, Neil JJ (2010) Longitudinal analysis of neural network development in preterm infants. *Cerebral Cortex* 20:2852–2862.

Smyser CD, Snyder AZ, Shimony JS, Blazey TM, Inder TE, Neil JJ (2013) Effects of white matter injury on resting state fMRI measures in prematurely born infants. *PloS one* 8:e68098.

Smyser CD, Snyder AZ, Shimony JS, Mitra A, Inder TE, Neil JJ (2016a) Resting-state network complexity and magnitude are reduced in prematurely born infants. *Cerebral Cortex* 26:322–333.

Smyser CD, Dosenbach NU, Smyser TA, Snyder AZ, Rogers CE, Inder TE, Schlaggar BL, Neil JJ (2016b) Prediction of brain maturity in infants using machine-learning algorithms. *Neuroimage* 136:1–9.

Suda M, Takei Y, Aoyama Y, Narita K, Sakurai N, Fukuda M, Mikuni M (2011) Autistic traits and brain activation during face-to-face conversations in typically developed adults. *PloS one* 6:e20021.

Toronov V, Franceschini MA, Filiaci M, Fantini S, Wolf M, Michalos A, Gratton E (2000) Near-infrared study of fluctuations in cerebral hemodynamics during rest and motor stimulation: temporal analysis and spatial mapping. *Medical Physics* 27:801–815.

Torricelli A, Contini D, Pifferi A, Caffini M, Re R, Zucchelli L, Spinelli L (2014) Time domain functional NIRS imaging for human brain mapping. *Neuroimage* 85 Pt 1:28–50.

Uludag K, Kohl M, Steinbrink J, Obrig H, Villringer A (2002) Cross talk in the Lambert-Beer calculation for near-infrared wavelengths estimated by Monte Carlo simulations. *Journal of Biomedical Optics* 7:51–59.

Vanderwert RE, Nelson CA (2014) The use of near-infrared spectroscopy in the study of typical and atypical development. *Neuroimage* 85 Pt 1:264–271.

Villringer A, Planck J, Hock C, Schleinkofer L, Dirnagl U (1993) Near infrared spectroscopy (NIRS): a new tool to study hemodynamic changes during activation of brain function in human adults. *Neuroscience Letters* 154:101–104.

Vincent JL, Kahn I, Snyder AZ, Raichle ME, Buckner RL (2008) Evidence for a frontoparietal control system revealed by intrinsic functional connectivity. *Journal of Neurophysiology* 100:3328–3342.

Wabnitz H, Moeller M, Liebert A, Obrig H, Steinbrink J, Macdonald R (2010) Time-resolved near-infrared spectroscopy and imaging of the adult human brain. *Advances in Experimental Medicine and Biology* 662:143–148.

Wabnitz H, Taubert DR, Mazurenka M, Steinkellner O, Jelzow A, Macdonald R, Milej D, Sawosz P, Kacprzak M, Liebert A, Cooper R (2014a) Performance assessment of time-domain optical brain imagers, part 1: basic instrumental performance protocol. *Journal of Biomedical Optics* 19:086010.

Wabnitz H, Jelzow A, Mazurenka M, Steinkellner O, Macdonald R, Milej D, Żołek N, Kacprzak M, Sawosz P, Maniewski R, Liebert A (2014b) Performance assessment of time-domain optical brain imagers, part 2: nEUROPt protocol. *Journal of Biomedical Optics* 19:086012.

Wandell BA (1999) Computational neuroimaging of human visual cortex. *Annual Review of Neuroscience* 22:145–173.

Wandell BA, Dumoulin SO, Brewer AA (2007) Visual field maps in human cortex. *Neuron* 56:366–383.

Wang L, Jacques SL, Zheng L (1995) MCML--Monte Carlo modeling of light transport in multi-layered tissues. *Computer Methods and Programs in Biomedicine* 47:131–146.

White BR, Culver JP (2010a) Quantitative evaluation of high-density diffuse optical tomography: in vivo resolution and mapping performance. *Journal of Biomedical Optics* 15:026006.

White BR, Culver JP (2010b) Phase-encoded retinotopy as an evaluation of diffuse optical neuro-imaging. *Neuroimage* 49:568–577.

White BR, Liao SM, Ferradal SL, Inder TE, Culver JP (2012) Bedside optical imaging of occipital resting-state functional connectivity in neonates. *Neuroimage* 59:2529–2538.

White BR, Snyder AZ, Cohen AL, Petersen SE, Raichle ME, Schlaggar BL, Culver JP (2009) Resting-state functional connectivity in the human brain revealed with diffuse optical tomography. *Neuroimage* 47:148–156.

Wobst P, Wenzel R, Kohl M, Obrig H, Villringer A (2001) Linear aspects of changes in deoxygenated hemoglobin concentration and cytochrome oxidase oxidation during brain activation. *Neuroimage* 13:520–530.

Wolf M, Franceschini MA, Paunescu LA, Toronov V, Michalos A, Wolf U, Gratton E, Fantini S (2003) Absolute frequency-domain pulse oximetry of the brain: methodology and measurements. *Advances in Experimental Medicine and Biology* 530:61–73.

Wray S, Cope M, Delpy DT, Wyatt JS, Reynolds EO (1988) Characterization of the near infrared absorption spectra of cytochrome aa3 and haemoglobin for the non-invasive monitoring of cerebral oxygenation. *Biochimica et Biophysica Acta* 933:184–192.

Wu X, Eggebrecht AT, Ferradal SL, Culver JP, Dehghani H (2014) Quantitative evaluation of atlas-based high-density diffuse optical tomography for imaging of the human visual cortex. *Biomedical Optics Express* 5:3882–3900.

Wu X, Eggebrecht AT, Ferradal SL, Culver JP, Dehghani H (2015a) Evaluation of rigid registration methods for whole head imaging in diffuse optical tomography. *Neurophotonics* 2:035002.

Wu X, Eggebrecht AT, Ferradal SL, Culver JP, Dehghani H (2015b) Fast and efficient image reconstruction for high density diffuse optical imaging of the human brain. *Biomedical Optics Express* 6:4567–4584.

Yeo BT, Krienen FM, Sepulcre J, Sabuncu MR, Lashkari D, Hollinshead M, Roffman JL, Smoller JW, Zollei L, Polimeni JR, Fischl B, Liu H, Buckner RL (2011) The organization of the human cerebral cortex estimated by intrinsic functional connectivity. *Journal of Neurophysiology* 106:1125–1165.

Yu G, Durduran T, Furuya D, Greenberg JH, Yodh AG (2003) Frequency-domain multiplexing system for in vivo diffuse light measurements of rapid cerebral hemodynamics. *Applied Optics* 42:2931–2939.

Zeff BW, White BR, Dehghani H, Schlaggar BL, Culver JP (2007) Retinotopic mapping of adult human visual cortex with high-density diffuse optical tomography. *Proceedings of the National Academy of Sciences of the United States of America* 104:12169–12174.

Zhan Y, Eggebrecht AT, Culver JP, Dehghani H (2012a) Singular value decomposition based regularization prior to spectral mixing improves crosstalk in dynamic imaging using spectral diffuse optical tomography. *Biomedical Optics Express* 3:2036–2049.

Zhan Y, Eggebrecht AT, Culver JP, Dehghani H (2012b) Image quality analysis of high-density diffuse optical tomography incorporating a subject-specific head model. *Frontiers in Neuroenergetics* 4:6.

Zhang D, Raichle ME (2010) Disease and the brain's dark energy. *Nature Reviews Neurology* 6:15–28.

Zhang X, Toronov VY, Webb AG (2005) Simultaneous integrated diffuse optical tomography and functional magnetic resonance imaging of the human brain. *Optics Express* 13:5513–5521.

Zhao H, Cooper RJ (2018) Review of recent progress toward a fiberless, whole-scalp diffuse optical tomography system. *Neurophotonics* 5:011012.

CHAPTER 18

Human Brain Imaging by Optical Coherence Tomography

Caroline Magnain, Jean C. Augustinack, David Boas, Bruce Fischl, Taner Akkin, Ender Konukoglu, and Hui Wang

CONTENTS

18.1 INTRODUCTION

Accurate neuroanatomical identification relies on understanding the composition and arrangement of the neurons and fibers that make up each brain structure. Numerous parcellations have been based on differences in cytoarchitecture, myeloarchitecture, chemoarchitecture, and pigmentoarchitecture to create architectonic brain maps (Campbell, 1905; Elliot Smith, 1907; Brodmann, 1909; Vogt and Vogt, 1919; von Economo and Koskinas, 1925; Flechsig, 1889; Vogt, 1910; Dietl et al., 1987; Zilles et al., 1988; Burkhalter and Bernardo, 1989; Campbell and Morrison, 1989; Jansen et al., 1989; Clarke, 1994; Hendry et al., 1994; Tootell and Taylor, 1995; Zilles et al., 1995; Braak and Braak, 1997; Braak, 1979; Ding et al., 2016). For this chapter, we will focus on the cyto- and myeloarchitecture of the human brain. The cytoarchitecture of the human brain is the study of the arrangement of cells within the tissue, while the myeloarchitecture is the study of the arrangement and density of myelin that surrounds the axons of the neurons in the cerebral cortex. Cytoarchitecture encompasses the nature of neurons themselves, their type, size, and morphology, as well as their organization into various cortical areas. In the cortex, the neurons are arranged in six layers, layer I being the closest to the pia mater and layer VI being adjacent to the white matter. The cerebral cortex of the human brain has been parcellated into cortical areas that exhibit variations – some slight and some considerable – in cytoarchitecture. The regions, frequently called Brodmann areas (BA), are named after Korbinian Brodmann, a German neurologist who published his work on brain parcellation based on cortical cytoarchitectonics in 1909. Each of these cortical regions also exhibits its own myeloarchitecture, as shown by Paul Emil Flechsig, a German neuroanatomist, pathologist, and psychiatrist (Flechsig 1889). The thickness and location of the layers in the gray matter vary between adjacent regions. The composition, such as the type, morphology, and density of neurons, as well as the myelin fiber density, vary across different regions. The most common stain to visualize the cortical architecture is a Nissl stain for neurons with a viable cytoplasm (Amunts et al., 1999; Augustinack et al., 2005) and to visualize myelinated fibers, a Gallyas stain (Pistorio et al., 2006). However, histology is a labor-intensive technique. It requires first sectioning the tissue with a vibratome or a cryomicrotome, then mounting the sections onto glass slides, which are dried, histologically stained, and coverslipped for preservation. This process is likely to introduce irremediable distortions at almost every step. Some of the distortions and artifacts from histology are demonstrated in Section 18.5.4 Advantages and Limitations. Full 3D reconstruction of the human brain from histological sections (stained for intact neurons) has been published by Amunts et al. (2013). Manual and automatic repairs on the damaged slices were performed as well as non-linear registration of the coronal slices to the MRI to reconstruct the entire volume. Even after this labor-intensive work, the BigBrain still has registration artifacts that are apparent in the other orientations (sagittal and axial). Registration between histology slices is difficult (Arsigny et al., 2005), even with the aid of intermediary images of the undistorted tissue, such as the blockface images or the MRI (Osechinskiy and Kruggel, 2010). Those distortions at the microscopic level can be avoided by imaging the blockface directly, prior to cutting (Odgaard et al., 1990). In contrast to histology, OCT probes the superficial part of a tissue block (a few hundred microns to a millimeter in depth depending on the tissue and the fixation), prior to any sectioning. Thus, OCT can be used to reconstruct brain samples of several cubic centimeters without distortion and with contrasts comparable to histology, as will be shown in the following sections, facilitating the reconstruction of a micron-resolution volume.

In this chapter, the principles and basis of Optical Coherence Tomography are first discussed and the various schemes, both in the Time Domain and the Fourier Domain, as well as Polarization-Sensitive OCT, are described (Section 18.2). Then, the mechanisms of OCT contrast in the post-mortem human brain are examined (Section 18.3). In order to acquire large volumetric data of brain samples, serial sectioning is integrated into the OCT system (Section 18.4). Next, OCT imaging of the human brain is shown (Section 18.5), first focusing on the human cortex, then on brain connectivity and finally on the imaging of individual neurons. The advantages and limitations of OCT for visualization of the human brain are reviewed. Finally, the future work necessary to achieve whole brain imaging at the micron level is considered and the potential applications are explored (Section 18.6).

18.2 OPTICAL COHERENCE TOMOGRAPHY

Tomographic techniques produce image slices of three-dimensional objects. Among those, optical coherence tomography (OCT) has been an active field during the past 25 years, leading to continuous breakthroughs in 3D diagnosis, real-time functional monitoring, high-resolution and fast pathological evaluation, and large-scale volumetric imaging. OCT collects ballistic and near-ballistic photons, which are back-scattered from the tissue. The intrinsic contrast and noninvasive measurement make it appealing for many clinical applications. Applications of OCT have shown great promise in ophthalmology, dermatology, cardiology, image-guided surgery, and neuroscience. Early OCT systems were based on low-temporal coherence interferometry (LCI) scanning in depth (Hess et al., 2002), which is performed in the time domain. Tomographic imaging of a cross-section was obtained by laterally stacking adjacent depth-scans. Later, the development of Fourier domain OCT removed the requirement for an axial scan and obtained full depth information in a single measurement. This technical advancement significantly improved the speed of OCT and enhanced system sensitivity for detecting weakly scattered signals.

18.2.1 Principle and Basic Properties of OCT

To achieve axial resolution along the optical axis of light propagation, OCT utilizes coherence gating strategies with a Michelson interferometer. The interferometer uses a low temporal coherence light source and splits the light into a reference path and a sample path. Light in the reference arm is reflected by mirror at a specified path length from the optical splitter. Light reflected back from the tissue returns to the interferometer and combines with the reference field at the optical splitter. A photodetector is located at the exit of the interferometer to collect the combined light. The signal collected by the photodetector is described as

$$\bar{I}_d(\tau) = I_s(t) + I_r(t) + G_{sr}(\tau) \tag{18.1}$$

where I_s and I_r represent the mean intensities returned from the sample and reference arms, respectively, and G_{sr} measures the amplitude of the interference fringes that represents the cross-correlation between the sample and reference arms as a function of time delay τ ($\tau = \Delta z/c$, where Δz is the optical path length difference between the reference and sample arms and c is the speed of light). The first two terms contribute to the DC signal at the detector. Assuming that the sample behaves as an ideal mirror that does not alter the sample beam, the third term becomes the real part of the temporal coherence function of the source, which forms a Fourier transform relation with the power spectrum density of the source based on the Wiener–Khinchin theorem. The shape and width of the source spectrum determines the temporal coherence length. Under the assumption of a Gaussian-shaped source spectrum with a bandwidth $\Delta\lambda$ (defined as full width at half maximum of the spectrum), the round-trip coherence length l_c is

$$l_c = \frac{2\ln 2}{\pi} \frac{\lambda_0^2}{\Delta\lambda} \tag{18.2}$$

Interfering fringes are visible when the optical path length difference between the sample and reference arms is within the coherence gate. Therefore, the coherence length determines the axial resolution of OCT. One important feature of OCT is that axial resolution is decoupled from transverse resolution. Therefore, ultra-high axial resolution (~1 μm) can be achieved without the necessity of highly focused beam.

The lateral resolution Δx of OCT is related to the numerical aperture of the objective focusing the light on the sample, and to the wavelength of light, and is given by

$$\Delta x = \frac{4\lambda_0}{\pi} \frac{f}{d} = 0.61 \frac{\lambda_0}{NA} \tag{18.3}$$

where d is the spot size on the objective lens, f is its focal length, and NA represents the numerical aperture of the microscope objective. For a OCT depth profile, the lateral resolution is maintained within a depth of focus, defined as twice the Rayleigh range z_R with a beam waist ω_0 (radius of the Gaussian beam defined at the $1/e^2$ peak intensity) at focus. The beam waist is considered to be 1/2 of the lateral resolution Δx.

$$z_R = \frac{\pi\omega_0^{\ 2}}{\lambda_0} \tag{18.4}$$

The assumption of an ideal reflector for a biological tissue is typically not realistic when measuring tissue. Instead, a back-scattering model is used in OCT, which assumes a fraction of power $R(z_s)$ is reflected from a specific layer located at the depth of z_s in the tissue. The interference signal can be expressed as a convolution,

$$I_s(\tau) = \sqrt{R(z_s)}\Delta G_{sr}(\tau) \tag{18.5}$$

System sensitivity, referring to the weakest sample reflectivity yielding a signal power equal to the noise of the system, is an important characteristic in OCT. The interferometric nature of OCT enables extraordinary sensitivity of 100 dB or higher. Therefore, OCT has the ability to image weakly scattered structures in a highly scattering (i.e. nontransparent) medium.

18.2.2 Time-Domain OCT

Time-domain (TD) OCT is characterized by two scans. An axial scan is conducted in the reference arm to localize scatters of the tissue in depth. As the optical path length in the reference arm cycles over a range, a depth profile called an "A-line" is acquired from the corresponding path lengths within the sample. The reference arm may consist of a mirror on an actuator, or a more complex rapid scanning delay line that is used for compensating dispersion mismatch between the reference and sample arms (Tearney, Bouma, and Fujimoto 1997). In order to create cross-sectional (B-line) or three-dimensional (3D) images of tissue, a second scan is performed laterally over the sample by galvanometer-based scanners. A schematic of a time-domain OCT is shown in Figure 18.1.

18.2.3 Full-Field OCT

Full-field (FF) OCT is a special type of time-domain OCT. Unlike conventional OCT, FF-OCT directly captures the en-face images (an entire plane of the tissue) on an imaging sensor such as a CCD or CMOS camera, without the need for lateral scanning of the sample beam. Volumetric imaging is generated by stacking consecutive optical sections, obtained by stepping through the depth of the sample matched with the axially scanned reference beam. The experimental setup is based on a Linnik interferometer (Dubois et al., 2002) with two identical microscope objectives in the sample and reference arms (Figure 18.2). Due to the broad bandwidth of the inherent light source, interference only occurs when the optical path lengths of the two arms are nearly equal. A piezoelectric stage actuator is generally used to oscillate the reference mirror to create a phase modulation of the coherence signal.

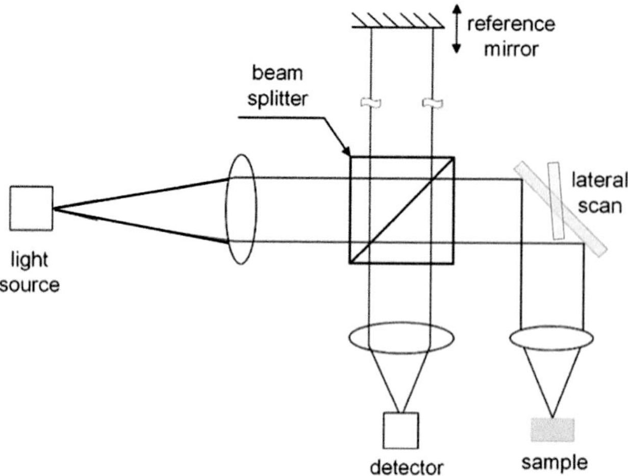

FIGURE 18.1 Basic schematic of a time-domain OCT setup.

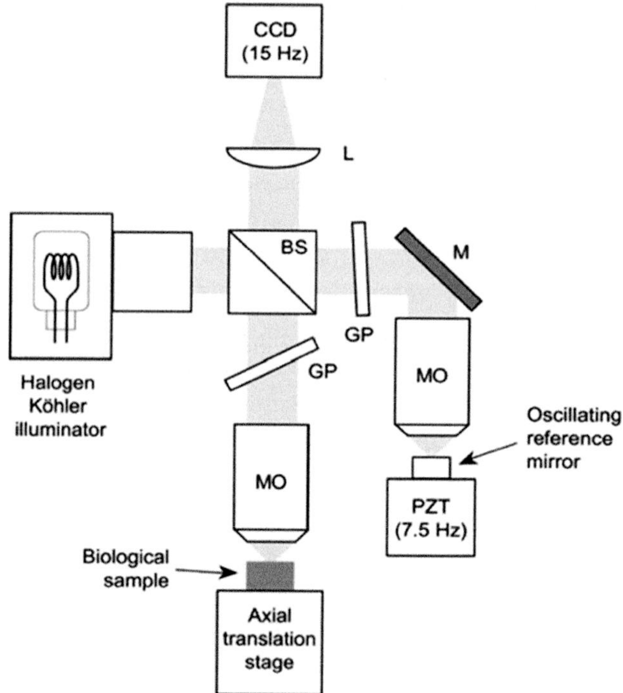

FIGURE 18.2 Schematic of a full-field OCT. The figure is reproduced from Figure 1 of Dubois et al., 2004, with permission.

FF OCT can achieve both ultra-high axial and transverse resolution that is comparable to traditional histology. An incoherent LED or white light is used as a low-coherence light source that enables an axial resolution down to 1 μm. The inclusion of a high numerical aperture microscope objective supports a submicron transverse resolution. Depth penetrations up to 200 μm have been achieved in fixed tissue (Dubois et al., 2004). A sensitivity of 80 dB is typically reported in FF OCT. Recent developments in FF OCT have achieved large fields of view comparable to histology, and have shown structural features of healthy and cancerous tissue in the breast and the brain (Assayag et al., 2013, 2014), with the goal of helping pathologists with intra-operative diagnoses.

18.2.4 Fourier-Domain OCT

The Wiener–Khinchin theorem forms the basis of Fourier-domain (FD) OCT. The interference of the electromagnetic fields from the sample and reference arms are spectrally resolved and the fringes of the spectral components are measured, containing the superposition of the light field reflected from all depths within the scattering sample that interfered with the reference field. A Fourier transform reconstructs the depth-wise scattering information from the spectral interference. In contrast to time-domain OCT, a stationary reference mirror is placed in the FD OCT, and the depth profile is created from a single spectral measurement. FD OCT is realized either by employing a spectrometer in the detection arm, called Spectral-Domain (SD) OCT (Fercher et al., 1995; Hausler and Lindner, 1998), or by sweeping a wavelength-tuning light source, called Swept-Source (SS) OCT (Chinn et al., 1997; Golubovic et al., 1997). Compared to TD OCT, the advantages of FD OCT manifest in faster acquisition speed and superior sensitivity (Choma et al., 2003; de Boer et al., 2003; Leitgeb et al., 2003). It has been shown that SD OCT and SS OCT share equivalent expressions for the enhanced signal-to-noise ratio, which usually results in a sensitivity advantage of 20–30 dB over TD OCT under the assumption of shot noise limited detection (Choma et al., 2003).

18.2.4.1 Spectral-Domain OCT

Spectral-domain (SD) OCT uses a broadband light source such as a superluminescent diode or an ultrafast laser. The interferometric signal is dispersed by a spectrometer and recorded by a line-scan camera such that the optical spectrum is obtained. Since the entire spectrum of the interference is acquired in a single measurement, SD OCT achieves significant detection efficiency. With a high-speed CCD or CMOS line-scan camera, the A-line rate can be up to two orders of magnitude faster than TD OCT (Potsaid et al., 2008; Grulkowski et al., 2009). 120 kHz and 150 kHz SD OCT have been reported for real-time ophthalmology applications (An et al., 2013; Zhang et al., 2015).

The light source power requirement for shot-noise limited detection is higher in SD OCT, as the reference power is dispersed onto the N total sensors of the line-scan camera. The practical design depends on the A-line rate requirement and the noise performance of the sensors. CCD cameras with low readout noise are critical in this type of system.

18.2.4.2 Swept-Source OCT

Swept-source (SS) OCT includes a frequency swept light source to create the broadband spectrum and a single or balanced photodetector with a high-speed A/D converter to record the interferometric signal. SS OCT records individual spectral interference components while the light source is swept in frequency domain. The key features of the SS OCT include rapid sweep repetition rates, single longitudinal mode operation for narrow instantaneous line-width and long coherence length, and low excess noise requirements for the swept source. Early swept source OCT systems used external cavity tunable laser technologies with galvanometers, resonant scanning mirrors, rotating polygons, gratings, dispersion prisms, and scanning filters (Yun et al., 2003, 2004; Choma et al., 2005; Huber et al., 2005; Goldberg et al., 2009; Chinn et al., 1997), which limit the sweep rate because of the long resonant cavity. Recently, the development of microelectromechanical systems (MEMS) has miniaturized the external cavity tunable lasers and increased the A-line rate to over a few hundred kHz (Liu and Zhang, 2007). Combining with a vertical-cavity surface emitting laser (VCSEL) technology, the coherence length of the MEMS-tunable lasers has been improved as well, which greatly increases the range of imaging depth with reduced sensitivity roll-off (Ireneusz Grulkowski et al., 2012). Another technology that has significantly improved the A-line rate of SSOCT is the Fourier-Domain Mode Locking (FDML) lasers, which use a long optical fiber delay and a tunable fiber Fabry–Perot filter to allow frequency swept propagation within the optical fiber delay and returning to the filter as it is tuned synchronously. With buffering or multiplexing, FDML lasers can achieve ultrahigh sweep rates of up to 5.2 MHz (Wolfgang Wieser et al., 2010).

18.2.5 Polarization-Sensitive OCT

18.2.5.1 Light Polarization

Light polarization describes the direction of oscillation of the electromagnetic wave. For an electromagnetic wave propagating along the z-direction, the field oscillates in the orthogonal plane, and its polarization state is described by the electric field components along the x and y directions. light with a frequency relationship $\Delta v \ll v_0$, the electric field of the wave can be written as

$$E(t) = E_x(t)\hat{x} + E_y(t)\hat{y}, \tag{18.6}$$

where $E_x(t) = E_{ox}\cos[2\pi v_o t]$, $E_y(t) = E_{oy}\cos[2\pi v_o t - \Delta\varphi]$, and $\Delta\varphi$ is the phase difference between the x and y components. The polarization state depends on the amplitude and phase relationships between $E_x(t)$ and $E_y(t)$. When the waves $E_x(t)$ and $E_y(t)$ are in-phase or out-of-phase with $\Delta\varphi = n\pi$ (n is an integer), light is linearly polarized at an angle depending on the ratio of E_{ox} and E_{oy}. When the amplitudes of the waves are equal ($E_{ox} = E_{oy}$) and the phase difference $\Delta\varphi$ equals $\pm n/2\pi$, light is circularly polarized. Otherwise, an elliptical polarization state is achieved.

18.2.5.2 Birefringence

Birefringence is an optical property of structures that introduce a change in the polarization state as polarized light propagates through the structure. It is commonly found in biological tissues such as muscle, tendon, and nerve, where the molecules form a regular alignment resulting in an anisotropic index of refraction that defines an optical axis (θ) of the birefringent sample. The propagating light waves oscillating parallel and perpendicular to the optic axis experience different indices of refraction (n_o – ordinary; n_e – extraordinary) and thus propagate at different speeds. Birefringence is defined as the difference between these two indices of refraction:

$$\Delta n = n_e - n_o. \tag{18.7}$$

Light passing through birefringent tissues experiences a delay between the orthogonal polarization states. The delay, known as retardance $\delta(z)$, is equivalent to the birefringence multiplied by the distance light travels

$$\delta(z) = \Delta n \Delta z, \tag{18.8}$$

and can be written in a phase representation ($\delta(z) = 360° \cdot \Delta n \cdot z/\lambda$) as the path length difference of one wavelength corresponds to phase difference of 360°.

18.2.5.3 PS-OCT Systems

Polarization-sensitive (PS) OCT (De Boer et al., 1997) utilizes polarization to gain sensitivity to additional tissue contrast mechanisms, including retardance and optic axis orientation. PS OCT systems can be divided into two categories, based on either Jones (Hitzenberger et al., 2001) or Mueller (de Boer et al., 1999) calculus.

The Jones formalism provides a convenient mathematical tool to describe light polarization and the influence of external factors (Jones, 1941). A 2 × 2 Jones matrix J characterizes the polarization property of any non-depolarizing optical component, which transforms the incident polarization state E to a transmitted state E′ given by E′ = JE. The Jones matrix for a birefringent material with an optic axis θ and a retardance δ is given by (Gil and Bernabeu, 1987),

$$J_b = \begin{bmatrix} e^{i\delta/2}\cos^2\theta + e^{-i\delta/2}\sin^2\theta & \left(e^{i\delta/2} - e^{-i\delta/2}\right)\cos\theta\sin\theta \\ \left(e^{i\delta/2} - e^{-i\delta/2}\right)\cos\theta\sin\theta & e^{i\delta/2}\sin^2\theta + e^{-i\delta/2}\cos^2\theta \end{bmatrix} \tag{18.9}$$

Jones calculus assumes that fully polarized light is not degraded in the tissue, therefore, the degree of polarization is maintaining at unity during light propagation. As the OCT interferometric signal typically arises from single back-scattered and not multiple scattered light, this assumption of maintaining the degree of polarization is valid in the majority of biological tissues. The Jones formalism has been commonly implemented in PS-OCT. The basic scheme is based on a low-coherence Michelson interferometer and a dual channel PS detection unit (Hee et al., 1992). de Boer et al. developed the first PS-OCT system for imaging biological tissue (De Boer et al., 1997). This was followed by developments described in Everett et al. (1998) and Schoenenberger et al. (1998) to obtain birefringence maps of the myocardium in porcine. The PS OCT technique was then advanced (Hitzenberger et al., 2001) to image both phase retardance and fast axis orientation in the myocardium.

The Mueller calculus along with the Stokes parameter provide a more comprehensive description of light polarization in turbid media, including the effect of depolarization. The Stokes parameter S is a vector $[I, Q, U, V]$ where I is the total intensity, and Q, U, and V represent the horizontal/vertical component, the ±45° linear component, and the left/right circular component, respectively. The Mueller matrix M is a 4 × 4 matrix which relates the Stokes vector of transmitted light S′ with that of the illuminating light S by S′ = MS (Huard, 1997). PS OCT systems capable of yielding a Mueller matrix have been presented in Yao and Wang (1999) and Yasuno et al. (2002). An approach to realize depth resolved Stokes parameters was presented in de Boer et al. (1999). The effects of multiple scattering and speckle noise on the polarization measurement have been evaluated using a Stokes parameter-based PS OCT, both resulting in a reduction in the degree of polarization (de Boer and Milner, 2002).

PS OCT techniques have been implemented in free space (Götzinger et al., 2005; Baumann et al., 2007) or using single-mode fiber optics (Cense et al., 2007). Recently, a polarization-maintaining fiber (PMF) based PS OCT technique has been introduced, which combines the advantages of fiber technology with the straightforward analyses of bulk setups. This PS OCT technique has been reported in time-domain (Al-Qaisi and Akkin, 2008), spectral-domain (Götzinger et al., 2009; Wang et al., 2010), and swept-source systems (Al-Qaisi and Akkin, 2010).

Here we describe a PMF-based spectral domain PS OCT at 1300 nm wavelength (black box of Figure 18.4A). The light source is a super-luminescence diode with a bandwidth 170 nm, yielding an axial (z-axis) resolution of 3.5 μm in tissue. Light from the source is linearly polarized and coupled into a PMF channel. A 2 × 2 PM coupler splits the incoming light into the reference and sample arms. In the reference arm, a quarter-wave plate (QWP) is oriented at 22.5° with respect to the incoming polarization state. In the sample arm, a QWP oriented at 45° ensures circular polarization on the sample. Birefringence alters the polarization into an elliptical state. Interferometric signals are split by a polarization splitter onto orthogonal polarization channels that are detected by two identical spectrometers. The line-scan cameras simultaneously acquire the spectra interference on the polarization channels with a maximum A-line rate of 150 kHz. A telescope and microscope objective are mounted in the sample arm, providing a tunable transverse resolution between 1.3 μm and 15 μm.

The depth profiles for the two polarization states are generated following a wavelength-to-k space remapping (Dorrer et al. 2000), dispersion compensation (Cense et al. 2004), and an inverse Fourier transform. The PMF based PS OCT system offers multiple contrasts in a single measurement, including reflectivity R(z), retardance $\delta(z)$, and optic axis orientation $\theta(z)$. The polarization contrasts parameters are computed based on the Jones formalism

$$R(z) \propto A_1(z)^2 + A_2(z)^2 \tag{18.10}$$

$$\delta(z) = \arctan\left(A_1(z) / A_2(z)\right) \tag{18.11}$$

$$\theta(z) = \left(\phi_1(z) - \phi_2(z)\right) / 2 \tag{18.12}$$

where A and ∅ denote the amplitude and phase of the complex depth profiles $A_{1,2} \exp\{i\varnothing_{1,2}(z)\}$ as a function of depth z, respectively, and the subscripts $_1$ and $_2$

correspond to the orthogonal polarization channels. It is noted that the optic axis orientation in Eq. (18.13) is a relative measurement bearing an offset with respect to the absolute axis orientation, due to an arbitrary phase delay between the orthogonal PMF channels. This offset is removed with post-processing (Wang et al., 2016), by placing a calibration retarder next to the sample (Wang et al., 2014), or by introducing paths for dynamic calibration (Liu et al., 2017).

18.3 OCT IMAGE CONTRAST MECHANISMS

Because the absorption in the near-infrared wavelength range is negligible, the attenuation of the incident light is primarily dictated by the scattering properties of the tissue constituents. In the case of the human brain, the main structures, neuronal cell bodies and fibers, cause the majority of the scatter when probed by OCT. The myelin sheath surrounding the fibers has a high refractive index with respect to the surrounding medium, resulting in high back-scattering and scattering of the light in the tissue, both in the white and the gray matter. In the gray matter, neurons are the main scatterers in the brain, with their size, shape, and density determining the scattering property of light.

18.3.1 Intensity Projections

To probe the optical properties of tissue, a 10× water immersion objective (Zeiss, N-Achroplan 10× W, NA 0.3) is used in the sample arm, giving a lateral resolution of 3.5 μm over a 1.5 mm × 1.5 mm field of view and a depth of focus of 100 μm. Each A-scan of the acquired volume shows the intensity profile with respect to the depth. Figure 18.3A shows examples of intensity profiles as a function of depth both in the white and gray matter of the primary visual cortex (BA17), where the profiles have been smoothed isotropically over 30 μm. The surface of the tissue is located at a depth of 0 μm on the plot. The white matter (WM) is made-up of myelinated fibers and shows the strongest attenuation of the light (purple curve). Since the layers of the cortex have different densities of neurons and fibers, this leads to different attenuation of the light (Figure 18.3A). The supragranular layers I

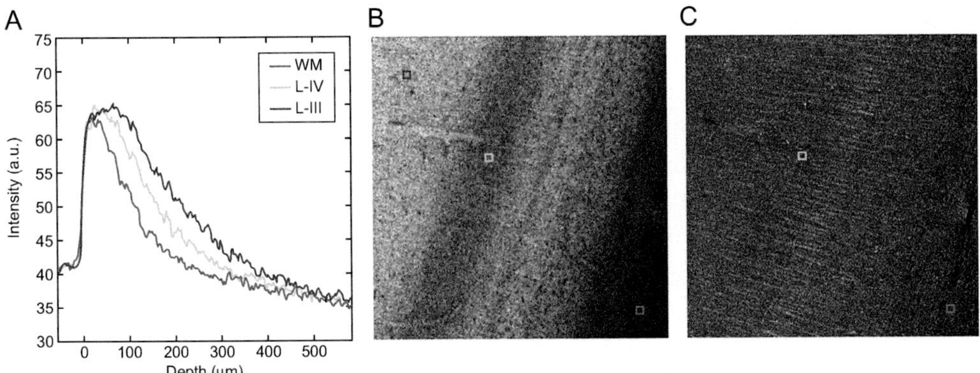

FIGURE 18.3 Optical coherence tomography acquisition in the primary visual cortex (BA17). A) Intensity profiles of layer III (orange), layer IV (blue), and white matter (purple), B) Average Intensity Projection and C) Maximum Intensity Projection. Field of view = 1.5 mm × 1.5 mm.

to III have the least myelinated fibers of the cortex. The attenuation in this part of the cortex is much lower than the white matter as shown by the orange curve of Figure 18.3A. In contrast, the infragranular layers IV to VI contain a plethora of fibers. In the primary visual cortex, layer IV has a special name, the stria of Gennari, named for the person who described it, Francesco Gennari (Gennari, 1782). Layer IV here is a highly myelinated band compared to other infragranular layers and all the supragranular layers. Therefore, the attenuation coefficient of the line of Gennari lies between that of white matter (purple line, Figure 18.3A) and layer III (orange line, Figure 18.3A). Also on Figure 18.3A, even though the depth of focus is approximately 100 μm with this optical objective, the penetration depth of light in the tissue and the separation between the different intensity profiles with respect to the brain regions can be resolved up to 400 μm. In order to visualize the difference in the attenuation, the average intensity projection (AIP) over the top 400 μm of the tissue is performed on each volume. Figure 18.3B shows the AIP of the primary visual cortex, from layer III (left) to the white matter (right). Due to the high attenuation, the white matter shows a dark contrast and thus the boundary between the white matter and the gray matter is easily identifiable. The wide dark band left of the white matter is the stria of Gennari, located in the layer IV and is well defined in the AIP image. The brightest layer observed in the OCT image is the layer III of the supragranular layers. Layers IV to VI, the infragranular layers, appear darker overall than the supragranular ones due to the higher density of myelinated fibers. The back-scattering of the tissue components can be partially assessed by performing the maximum intensity projection (MIP). The high refractive index of the myelin sheath surrounding the fibers allows for the imaging of fiber bundles, small and large. As an example, Figure 18.3C shows the MIP of the primary visual cortex, which highlights some of the fibers of the infragranular layer. However, the back-scattering signal is dependent on the fiber orientation: the more the fibers run in-plane, the higher the back-scattering intensity. The more the fibers go through-plane, the smaller the back-scattering. Moreover, if the neurons are larger than the lateral resolution and the speckle pattern size, the MIP images obtained with the 10× objective will be able to visualize individual neurons.

18.3.2 Polarization Contrasts

The polarization contrast results from birefringence that is commonly found in nerve fibers and associated sub-components. Oldenbourg et al. used a polarization microscope to characterize the peak retardance for a single microtubule and reported that the retardance linearly increased with the number of microtubules in the bundle (Oldenbourg et al., 1998). Katoh et al. reported that radial actin bundles of nerve growth cones could be observed on birefringence images (Katoh et al., 1999). The myelin sheath around axons produces strong birefringence that has been imaged in *ex vivo* mammalian and human brains with polarized light imaging (PLI) (Axer et al., 2011; Larsen et al., 2007) and PSOCT (Wang et al., 2011). PLI uses a rotating polarimeter to measure the retardance and the fiber axis orientations within brain slices (Axer et al., 2011). As the sample has to be physically sliced to permit optical transmission measurements, PLI suffers the same problems of tissue distortion as occurs with traditional histology for volumetric reconstruction and fiber tracking over long distances. A feasibility study using rotational-polarimeter based PS OCT to visualize fiber tracts was reported by (Nakaji et al., 2008) on rat brain slices. Subsequently, Wang et al. developed a polarization-maintaining fiber-based PS OCT to obtain the polarization contrasts of retardance and optic axis orientation in a single measurement and used it to systematically investigate the brain structures and fiber pathways in *ex vivo* brain (Wang et al., 2011).

The retardance contrast of PS OCT probes the fiber tracts in the brain with great specificity due to the high birefringence of the myelin sheath of the axons that make up these bundles. As Eq. (18.8) indicates, the cumulative retardance along the depth of light propagation is proportional to the level of birefringence. Figure 18.4 shows an en-face image of retardance for human primary visual cortex, which is created by taking the mean phase retardance value over a 200 μm depth range. The transverse resolution is 3.5 μm. The highest retardance is observed in white matter regions where myelinated fiber bundles are aligned and densely packed, whereas in regions dominated by neuronal cell bodies with only a few fibers, such as in Layer II–III, the retardance remains low as polarized light passes through the tissue. The upper layer IV, the line of Gennari, manifests a higher birefringence than the other cortical layers in the primary cortex, where a higher density of myelinated fibers is present. Layer VI fibers demonstrate a moderate retardance.

The optic axis orientation contrast of PSOCT represents the in-plane orientation, which is the projection of the 3D fiber orientation onto the plane perpendicular to the incident beam. One advantage of the optic axis orientation contrast is that it provides a direct measure of the fiber axis using an intrinsic optical property without the need for external labeling, which enables tracking of the trajectories using the orientation information. Figure 18.5 demonstrates an en-face optic axis orientation map of a human cerebellum section on a parasagittal view, with 3.5 μm transverse resolution. The optic axis orientation value is color-coded according to the color wheel shown at the lower right, and the intensity is

FIGURE 18.4 A retardance image of the primary visual cortex in the human brain. The color bar indicates retardance units of degrees. II–VI: layer; WM: white matter.

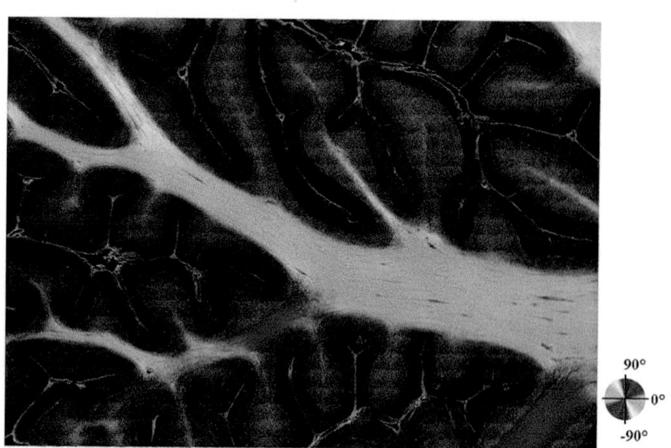

FIGURE 18.5 Optic axis orientation image of a human cerebellar section. The orientation is color-coded according to the color wheel, and the intensity is controlled by the retardance value.

proportional to the retardance. The en-face orientation is obtained by taking the histogram in depth with 3° binning and extracting the value with the highest occurrence, assuming the optic axis orientation does not change over the effective imaging depth of 200 μm. The in-plane orientation of fibers in the cerebellar lobule reveals the geometry of the bundles. In addition, the orientation of the sub-folial fissures that are only a few tens of microns in diameter are delineated as well. The transition of the colors indicates fiber turning, splitting, and merging. It is noted that the optic axis orientation only represents the axis of anisotropic structures, and is valid only when birefringence is above a threshold determined by the noise performance of the PS OCT system. Therefore, the background color in the gray matter regions where cell bodies are primarily seen does not imply an axis of directionality.

18.4 SERIAL SECTIONING OPTICAL COHERENCE TOMOGRAPHY

Serial blockface microscopy is a technique that includes a microscope operating in reflection mode and serial sectioning to achieve volumetric image reconstruction of tissue blocks. Unlike conventional microscopy that images a thin slice using transmitted light, blockface microscopy collects light reflected back from the surface of the tissue block that is mounted under the microscope and thin slices are cut between consecutive imaging sessions. First introduced in light microscopy (Odgaard et al., 1990), serial blockface imaging techniques have been used in confocal microscopy (Sands et al., 2005) and two-photon microscopy (Ragan et al., 2012; Tsai et al., 2003), as well as electron microscopy (Denk and Horstmann, 2004), with a particular emphasis on the study of large-scale synaptic connections and axonal networks in the brain. The most important advantage of the technique is the ability to automatically align stacks of images for volumetric reconstruction. In contrast to volumetric image reconstruction after slicing, which is extremely challenging to achieve due to slice-specific distortions that non-linear registration is not always able to fully compensate for, serial sectioning combined with imaging under the cut surface results in image slices that are intrinsically in register, making volumetric reconstruction straightforward.

The serial blockface imaging technique has been incorporated into OCT and PSOCT to establish large-scale volumetric imaging of the *ex vivo* brain and used to investigate neural pathways in 3D space (Wanget al., 2014). The experimental procedures include repeated scanning and slicing on a tissue block. The sample is mounted under the OCT. After acquiring one volumetric image, the sample is translated to a new position for the next acquisition. After the full surface is covered, the vibratome cuts off the top slice to allow deeper regions to be exposed. The first reports of a serial optical coherence scanning system include a home-built PS OCT with a 7 μm axial resolution and 15 μm transverse resolution and a commercial vibratome manually operated to realize volumetric imaging of *ex vivo* brain samples. Recently, an automatic serial-sectioning PS OCT (as-PS OCT) system has been developed to significantly improve the resolution and imaging speed. The as-PS OCT system is composed of a home-built spectral domain PS OCT with 3.5 μm axial resolution, motorized xyz stages to translate the sample, and a vibratome to section the tissue block. A local computer controls the acquisition, stage movement, and vibratome slicing for automatic imaging of tissue blocks. The acquisition software performs data processing in real-time and provides the options to save raw data and reconstructs volumetric data or en-face images. The local computer also communicates with a compute server to transfer the data for post-processing. Customized software has been developed to realize an optimized pipeline for data acquisition, tissue sectioning, and image reconstruction. Figure 18.6A shows the schematic of the as-OCT setup.

18.4.1 Automated Tiling and Serial Sectioning

as-OCT/as-PS OCT enables imaging of cubic centimeter blocks of tissue without the need for human intervention. As the surface area of the sample blocks are over one hundred times greater than that of the field-of-view of a single scan, multiple image tiles are stitched together to form a full section (Figure 18.6B, C). There is a 20–50% overlap between adjacent image tiles to facilitate registration and stitching and increase SNR. Motorized xyz stages are positioned under the sample arm to translate the tissue between image tiles. Upon completion of a full area scan, a vibratome (Ragan et al., 2012) is mounted adjacent to the OCT/PS OCT to cut off a superficial slice that is 50–200 μm thick. The acquisition and control software coordinates the positioning of the xyz stages and vibratome slicing between image tiles.

18.4.2 Data Transfer and Volumetric Reconstruction

A network pipeline is established between the local computer and a compute server at the beginning of image acquisition. The data is immediately transferred to the server upon completion of the acquisition of one OCT image tile. During the data transfer, the OCT acquisition begins acquiring the next imaging tile. Once transferred to the compute server, automatic scripts first stitch together the individual image tiles collected within the same imaging plane using rigid transformation. This registration is performed using a Fiji plug-in based on the Fourier shift theorem (Preibisch et al., 2009) and the stitching algorithm uses a linear blending of the overlap regions of the OCT image tiles. This creates a large

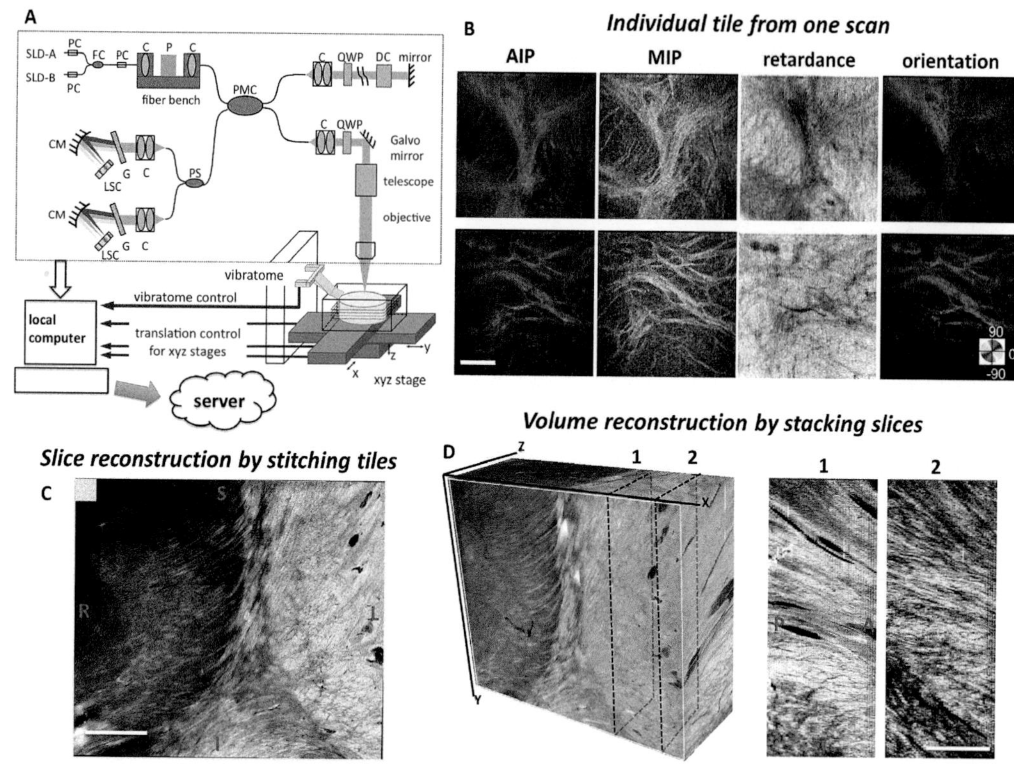

FIGURE 18.6 A) A schematic of an as-PSOCT system. The dashed block shows a home built spectral domain PSOCT (Wang et al., 2016). C, collimator; CM, concave mirror; DC, dispersion compensation block; FC, fiber coupler; G, grating; LSC, line scan camera; P, polarizer; PC, polarization controller; PMC, polarization-maintaining fiber coupler; PS, polarization splitter; QWP, quarter-wave plate; SLD, superluminescent diode. B) En-face images of two image tiles in the internal capsule of a human brain sample. Data is presented as average intensity projection (AIP), maximum intensity projection (MIP), retardance, and orientation. The orientation map was color-coded according to the color wheel, with the intensity weighted by AIP. Scale bar: 250 μm. C) Slice reconstruction in the thalamus and internal capsule by stitching the image tiles. The size of a single tile is indicated by the rectangle on the left-top corner. Scale bar: 2 mm. D) Volumetric reconstruction by stacking consecutive slices. The section shown in (C) is indicated by the yellow rectangle. Sections 1 and 2 show the fiber tracts of the internal capsule on the yz-plane, representing the cross sections of stacked slices. Scale bar: 2 mm. S: superior; I: inferior; L: left; R: right; A: anterior; P: posterior.

image slice. Successive image slices are then stacked to form a volume. Volumetric reconstruction can be carried out in one of two approaches, either by stacking the en-face images for every slice (Figure 18.6D) or by stitching the optical sections of consecutive imaging volumes (Figure 18.7). The first method provides an efficient way to manage the data for large-scale 3D reconstruction, while the voxel size in the z-direction is determined by the slice thickness, typically 50 μm. In contrast, the second method preserves the natural resolution of OCT/PS OCT to better than 5 μm, while the data size can be extremely large for further analysis and quantification. Figure 18.7A shows two cross-sections from

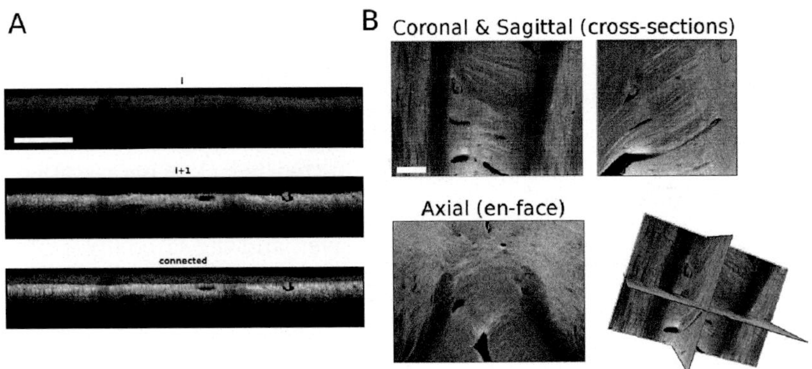

FIGURE 18.7 Volumetric reconstruction of part of an infant caudal medulla by stitching the optical sections of consecutive slices. A) The process of stitching cross-sectional images of adjacent slices. B) Orthogonal views of the volumetric reconstruction. Scale bar: 500 μm.

consecutive volumetric scans. As the effecting imaging depth is greater than the thickness of a slice, there is an overlapping region in the two sections and landmarks, such as fibers, vessels, and borders, are used to register consecutive volumes as shown in the bottom panel of Figure 18.7A. Figure 18.7B demonstrates a volumetric reconstruction of a portion of infant brainstem on orthogonal viewing planes with an isotropic voxel size of 8.7 μm (the data have been downsampled by a factor of 3). The fused image (cross-section of slices) exhibits a smooth transition between consecutive sections.

18.5 POST-MORTEM HUMAN BRAIN IMAGING

18.5.1 Human Cortex

18.5.1.1 Laminar Structure

Figure 18.8 shows all six layers of the entorhinal cortex (BA28) stained for Nissl substance (Figure 18.8A) and OCT on the same tissue: Figure 18.8B is the AIP image and Figure 18.8C is the MIP, both over a range of 400 μm of tissue. The Nissl stain, the gold standard, reveals the unique characteristics of the entorhinal cytoarchitecture. Layer I contains only a few neurons. The most notable layer in entorhinal cortex is layer II. The neurons of layer II are

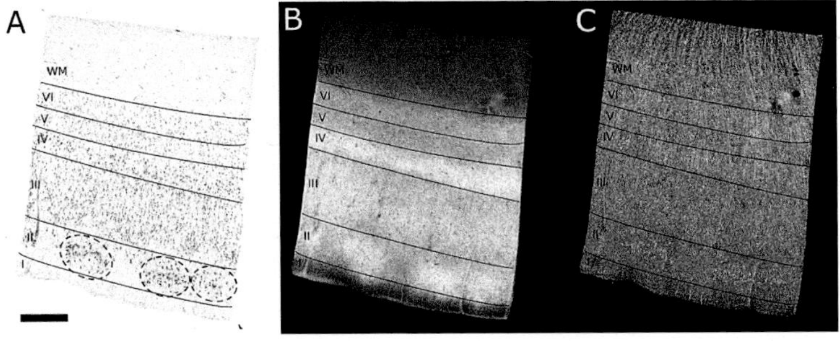

FIGURE 18.8 Laminar structure of entorhinal cortex imaged by A) Nissl staining, B) OCT AIP and C) OCT MIP. Scale bar = 500μm.

large pyramidal and stellate cells and are organized in groups called islands; Figure 18.8A illustrates three islands, shown in the dotted circles. Layer IV contains large pyramidal neurons as well; however, these cells are homogenously spaced. The laminar structure of the cytoarchitecture observed on the histologic slice is clearly visible on the AIP contrast obtained by OCT (Figure 18.8B). Layer I is dark, due to the lack of back-scatterers, such as neurons or fibers. Layer II can be easily identified; the neuron islands give a high intensity compared to the surrounding tissue of the same layer, and the same three islands can be distinguishable. Layer IV on the OCT image is a high, homogenous intensity band, which reflects the homogeneity of the neuron distribution of this layer, compared to layer II. In the OCT image, layers V and VI show a slight variation in intensity, with layer V being darker than layer VI. The boundary between the white and the gray matter is well-defined on the OCT image. Figure 18.8C shows the MIP contrast. The neurons from layers II and V are large enough to be visualized with the 10× objective. The neuronal arrangement with respect to the layers simulate with the histology: they are grouped on layer II and distributed in layer IV. Finally, some radial fibers can be observed in the gray matter, mainly in the infragranular layers (layers IV to VI). Fiber bundles can also be observed in the white matter.

The cortical architecture of large regions, encompassing multiple areas, can be revealed by OCT as well. Figure 18.9 shows an example of the temporal isocortex, the lateral side of the temporal lobe, which contains four Brodmann areas (BA20, 21, 22, and 36), with a surface area of 3 × 4.8 cm^2 (Magnain et al., 2014). The blockface image (Figure 18.9A) was taken after flat-facing of the tissue using a sliding cryo-microtome. In Figure 18.9B, the AIP over 400 μm of tissue of the whole lateral temporal lobe clearly distinguishes between the white matter and gray matter, which highly correlates with the photographic image of the blockface. However, the blockface image does not show the laminar architecture of the cortical ribbon. In contrast, the OCT AIP image shows different levels of intensity from the gray-white matter boundary (GWB) to the pial surface, revealing the laminar structure of the cortex. The cortical architecture image with OCT highly correlates with the histologic

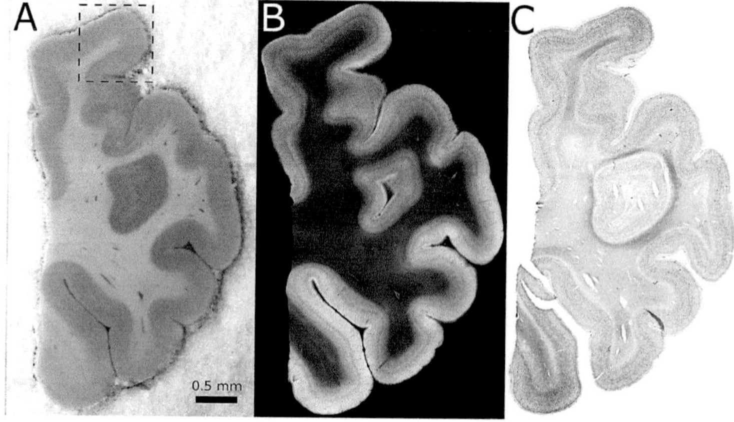

FIGURE 18.9 Temporal isocortex: A) photograph of the blockface, B) OCT AIP and C) Nissl stained slice. The dotted box represents part of BA36 and a close up is shown on Figure 18.10.

sections as shown in the Nissl stain (Figure 18.9C). Figure 18.10 is a close up of the top gyrus (BA36); the top images indicate AIP and the MIP from the OCT acquisition and the bottom row contains the Nissl and Gallyas stained sections. As shown in the previous figure, Figure 18.9, the AIP demonstrates a laminar structure comparable to the Nissl staining. However, the dark band in layer VI seen on the OCT corresponds to a slightly darker band in the Gallyas stain as well, both represented by the symbol* which contribute to the higher attenuation coefficient. This band (*) shows a bright contrast on the MIP image shown on Figure 18.10B due to the presence of myelinated fibers and their high back-scattering. Finally, the MIP reveals fanning fibers running from the white matter from the gray or vice versa. Gallyas stained fibers are demonstrated and visible on Figure 18.10D. The orientation of the fibers on both modalities are highly correlated.

The en-face retardance image of PS OCT quantifies the magnitude of the tissue birefringence. As birefringence specifically probes myelinated fiber tracts in the brain, the retardance image provides a positive correlation with the myelination content in the human brain. In addition to differentiating gray and white matter, the retardance also demonstrates contrast between cortical layers where distinctive myelinated fiber density is known. Figure 18.11 shows an en-face retardance map of the human primary somatosensory cortex. The tissue section with a total area of 2 × 1.5 cm is imaged by our home

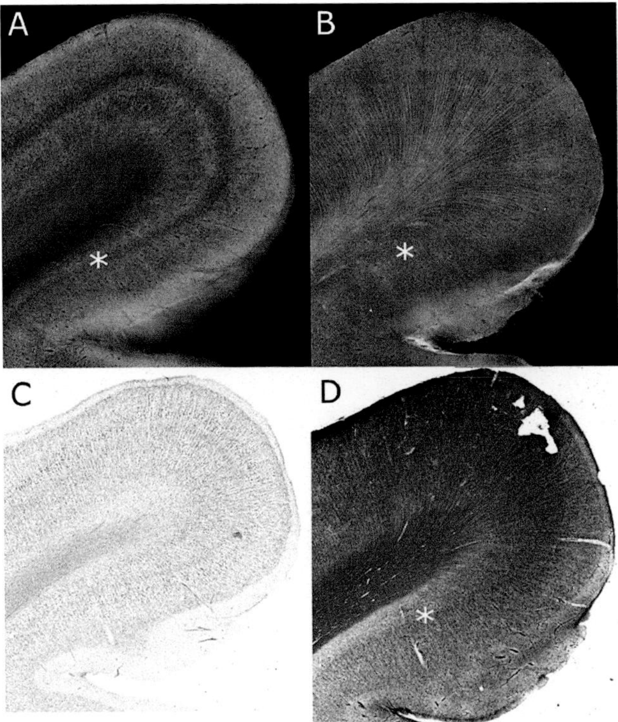

FIGURE 18.10 Upper gyrus (BA36) of the temporal isocortex presented in Figure 9 in the dotted box. Top row: A) AIP and B) MIP images acquired by OCT. Bottom row: C) Nissl and D) Gallyas stained sections

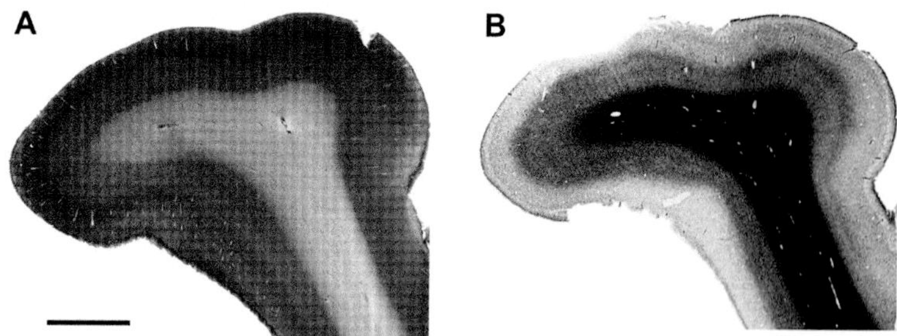

FIGURE 18.11 Retardance image of primary somatosensory cortex by A) PSOCT and validation by B) Gallyas stain. Scale bar = 4mm.

built as-PS OCT with a 10× water immersion objective. The retardance map demonstrates a clear differentiation of the supragranular (layers I–III) and infragranular (layers IV–VI) layers in the primary somatosensory cortex, validated by the Gallyas stain. The deeper layers (infragranular) exhibit a higher myelination which corresponds to a higher retardance than the superficial layers (supragranular).

18.5.1.2 Parcellating Cortex

Because OCT is capable of visualizing the cortical architecture of the human brain, it can be used to identify cortical boundaries based on layer intensity, cortical thickness and relative location.

Figure 18.12 shows the boundary between the entorhinal cortex (Brodmann area 28) and perirhinal cortex (Brodmann area 35). The boundary delineation is entirely based on the OCT contrast. In entorhinal cortex, the cell-dense areas occur in layers II and IV. This translates to bright bands in the AIP image in Figure 18.12A. The neurons of these layers (II and IV) are large enough to distinguish individually on the MIP as well on Figure 18.12B.

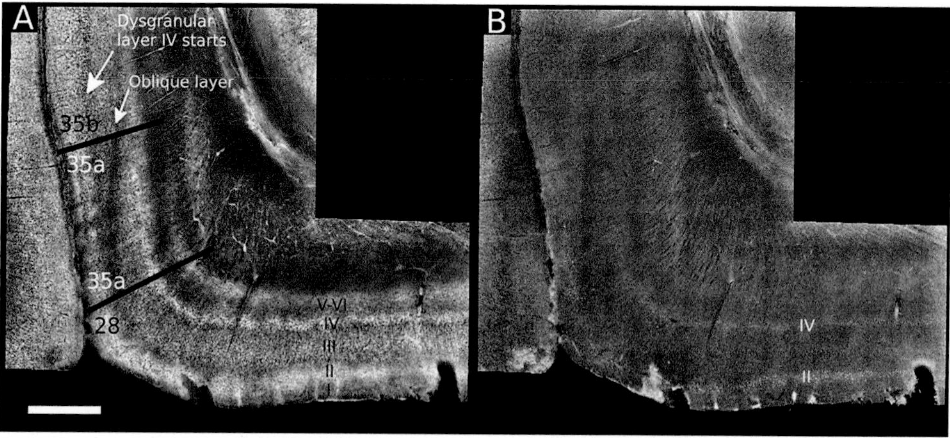

FIGURE 18.12 Optical coherence tomography reveals the laminar structure of the medial temporal lobe and delineates the entorhinal cortex (BA28) and the perirhinal cortex (BA35), as well as subdividing the latter into BA35a and 35b. Scale bar = 2 mm.

As shown previously in Figure 18.8, the contrast in layer II shows the islands created by the neuronal dense and non-dense areas. The boundary between BA28 and BA35 can be detected based on the arrangement of the neurons of layer II, with area 28 being rounder than area 35. In BA35, the neurons form columns instead of islands, and are illustrated on the AIP. Moreover, the subdivisions of perirhinal cortex into BA35a (closest to BA28) and BA35b can easily be seen on the AIP. Layer IV of BA28 continues into the transition area of perirhinal cortex, 35a. The oblique layer is another classic attribute of BA35a and appears as a dark layer, starting close to layer II and going deeper in the cortex along the length of the transition region. The subdivision 35b has brighter layers corresponding to the layers II and III and a dark dysgranular layer IV. The cortical boundaries between adjacent regions, with the exception of the primary cortices, are generally subtler than the boundary between entorhinal cortex and perirhinal cortex shown in the previous example. Other cortical boundaries will benefit from the 3D reconstruction of the cortex over several cubic centimeters.

18.5.2 Connectivity

18.5.2.1 Fiber Networks at the Microscopic Scale

The optic axis orientation offers a direct measurement of fiber orientation in the en-face image plane. Due to the high resolution of PS OCT, the orientation image allows clustering of fiber groups, which are not distinguishable in intensity and retardance images, and visualization of extensive fiber crossings at the microscopic scale. Figure 18.13 demonstrates the in-plane orientation map of 20 consecutive slices in the thalamus (left of the pink fibers on the first image) and internal capsule regions ($1.2 \times 1\ cm^2$). The serial sections are obtained by as-PS OCT with 50 μm slice thickness, and displayed at 200 μm intervals in depth. The transverse resolution of as-PS OCT is 3.5 μm. The orientation is color-coded as indicated by the color wheel, with an intensity weighted by the retardance.

The different colors on the orientation maps delineate small crossing fiber tracts in the internal capsule. These crossing patterns are barely distinguishable on retardance images

FIGURE 18.13 In-plane orientation maps in consecutive slices. The orientation is color-coded based on the color wheel, and the intensity is weighted by the retardance. The adjacent slices are 200μm apart. Th: thalamus; ic: internal capsule. Scale bar: 3 mm.

where all the fiber tracts are delineated regardless of orientation. The axis orientation of fibers in the thalamus region is visible as well, despite showing a lower retardance compared to the white matter. The fiber orientation maps with as-PS OCT offers an appealing tool to investigate fiber crossing at a resolution beyond other current fiber mapping technologies such as diffusion MRI, even *ex vivo* (best resolution of ≈150 μm). We note that the optic axis orientation does not provide an accurate estimation for through-plane fibers with a large inclined angle, where the SNR is low and the noise of the orientation measurement is high. Section 18.5.2.3 discusses the estimation of the 3D orientation where through-plane orientation is present.

18.5.2.2 Validation of Diffusion MRI with Serial Sectioning PS OCT

Diffusion MRI (dMRI) makes use of the diffusion of water molecules running preferentially along axonal axes in the brain, to characterize fiber organization and orientation. The technique provides a favorable solution to delineate white matter fiber bundles in the living human brain at resolutions of millimeter to sub-millimeter. Despite the broad applications in clinical studies such as Alzheimer's disease, schizophrenia, major depression, and presurgical planning (see reviews Fields, 2008; Thomason and Thompson, 2011), systematic validation of the dMRI techniques has been difficult. It is known that dMRI suffers from difficulty in resolving "fanning", crossing, and "kissing" fibers due to the limitation of spatial and angular resolution. However, the ability of dMRI to track fiber trajectories over long distance is an enormous advantage over traditional histology, in which fiber tracking has been extremely challenging. As a consequence, our understanding of where dMRI fails and how to improve the dMRI tracking with prior neuroanatomical pathway information is incomplete.

The advantages of direct accessibility to fiber orientation and the viability for volumetric reconstruction in serial sectioning PS OCT offer an appealing solution for systematic validation for dMRI. Wang et al. (2014) conducted a cross-validation of PS OCT and diffusion tensor imaging (DTI) in a $2 \times 1 \times 0.7$ cm^3 human medulla sample. The volumetric reconstruction by PS OCT has an in-plane resolution of 15 μm with a slice thickness of 150 μm, whereas the highest resolution DTI we were able to acquire had an isotropic resolution of 300 μm. Experimental procedures included scanning the same sample by DTI followed by serial sectioning PS OCT, and then voxel-wise co-registration was accomplished on the volumetric images using an affine registration (Reuter et al., 2010). The non-distortion feature of serial sectioning OCT facilitates an accurate and efficient registration with dMRI, resulting in a Pearson's correlation coefficient of 0.90. Figure 18.14 shows the co-registered PS OCT and DTI in-plane orientation images in one slice, which manifest excellent agreement in the majority of fiber bundles. The quantitative metric of fiber orientations permits a direct comparison between PS OCT and DTI. Statistical analysis of the voxel-wise orientation difference shows a mean of 0.1° with a standard deviation of 22.6° in the white matter region masked by fractional anisotropy and retardance.

Despite the positive correlations of DTI and PS OCT, complications remain in regions with intricate fiber patterns (Figure 18.14, right). The superior resolution of PS OCT reveals individual fiber traces, whereas pixelated color patches are observed on DTI orientation

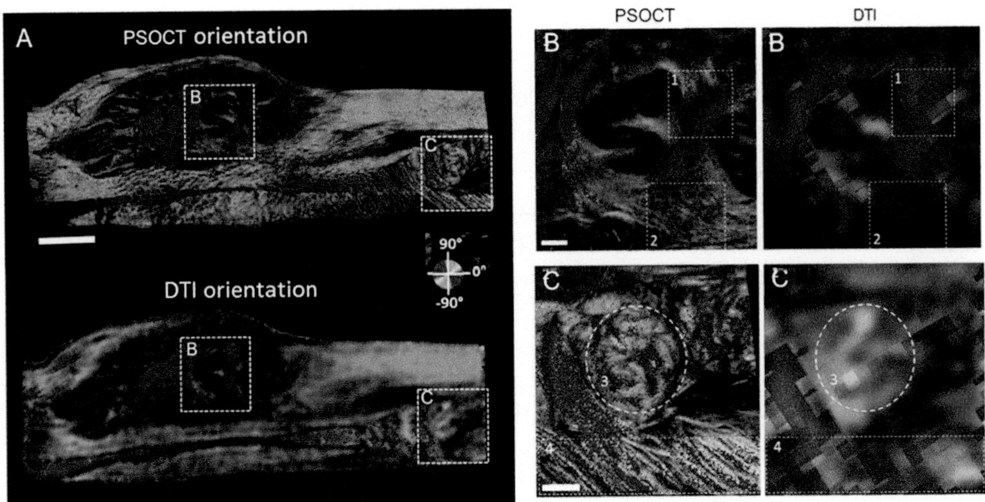

FIGURE 18.14 In-plane fiber orientation images of co-registered PSOCT and DTI on a human medulla sample. The DTI and PSOCT share the same orientation coding based on the color wheel. The intensity of the PSOCT orientation image is weighted by the retardance, while that of DTI is weighted by the fractional anisotropy. The right panel demonstrates two zoom-in regions where intricate fiber crossing patterns are revealed by PSOCT but lost with DTI. Scale bars: A: 2.5 mm; B and C: 500 μm.

image without interpretable structures (ROI-B, C and 3). Although the low-resolution estimation of DTI preserves the dominant orientation in regularly aligned fibers (ROI-4), it clearly loses the distinction of crossing features under complex geometries.

18.5.2.3 3D Orientation and Tractography

Comprehensive fiber tracking and connectivity investigations require quantification of 3D fiber orientation. This is not directly available in PS OCT, as the optic axis orientation measure only characterizes the orientation projected in the imaging plane. To overcome this problem, several techniques have been developed to obtain the through-plane orientation, including computational analysis on volumetric images and advancing optical models with multiple incident angles in PS OCT.

18.5.2.3.1 Computation-Based Method for Extracting the 3D Axis Computational algorithms have been widely used to quantify the orientation information on histological slices with myelin stains or neural tracers, and volumetric brains of small animals with clearing and labeling orientation (Budde and Frank, 2012; Choe et al., 2012; Schmitt et al., 2004; Budde et al., 2011; Leergaard et al., 2010). Those algorithms are exclusively based on digital signal processing of 2D or 3D images with features highlighting the fiber tracts, including gradient-based edge detection, Fourier transform, and structure tensor (Bigun, 1987). Structure tensor is a second-moment matrix that is derived from the gradient in an immediate neighborhood of a point and is capable of extracting the dominant directions of local features in the image. It has been used in histology and cleared brains to quantify fiber orientations in the brain (Ye et al., 2016; Budde and Frank, 2012). Wang et al. incorporated the structure tensor method in serial sectioning PS OCT and applied it to volumetric retardance images to obtain the 3D fiber orientation in the mouse brain (2015). Based on the orientation metric, tractography is conducted using a conventional dMRI tool to create fiber tracts at microscopic resolution. Here we briefly review the workflow for structure tensor analysis on volumetric PS OCT images.

The gradient (∇I_σ) of the image data (I) is computed using a convolution of the first-order Gaussian derivatives kernel with the 3D retardance images.

$$\nabla I_\sigma = \nabla K_\sigma * I$$

The structure tensor (J) is constructed by the outer product of the gradient vectors.

$$J = \nabla I_\sigma \nabla I_\sigma{}^T = \begin{bmatrix} I_{\sigma_x}^2 & I_{\sigma_{xy}} & I_{\sigma_{xz}} \\ I_{\sigma_{yx}} & I_{\sigma_y}^2 & I_{\sigma_{yz}} \\ I_{\sigma_{zx}} & I_{\sigma_{zy}} & I_{\sigma_z}^2 \end{bmatrix}$$

The components of the tensor matrix are smoothed through a convolution with a Gaussian kernel $K_{n,\rho}$, where n represents the number of neighboring points included, and ρ the standard deviation of the Gaussian kernel, which controls the weights of the neighboring points in smoothing.

$$J_{n,\rho} = K_{n,\rho} * J$$

Eigen-decomposition is performed on the tensor matrices, and the eigenvalues and eigenvectors are extracted. As the image gradient along fiber axis changes least, fiber orientation is represented by the eigenvector corresponding to the smallest eigenvalue.

Figure 18.15 demonstrates a 3D fiber orientation map and tractography of a subcortical region including the thalamus and the internal capsule. The sample having a volume of $1 \times 1.2 \times 0.45$ cm^3 is imaged by as-PS OCT with a 10× water immersion objective and a slice thickness of 50 μm. The 3D orientation is obtained by combining the optical measurement of in-plane orientation and computational through-plane angles, which is obtained by a structure tensor analysis as described above. The orientation map is shown with an isotropic resolution of 30 μm, which reveals the inter-mingled fibers running both in the plane and through the plane. Tractography is conducted using a Diffusion toolkit (Wang et al., 2007) developed for dMRI techniques, with an angular threshold of 45° and an intensity mask from the retardance value. Small seeds are selected in the internal capsule (C, D) and the thalamus (E) and the tracts passing through those ROIs are created. The trajectory of the small fiber groups reveals intricate pathways and extensive crossing in those regions. This optical tractography shows significant resolving power to investigate the neuroanatomical pathways that are beyond what is achievable with dMRI techniques; and therefore indicates the great potential to create a microscopic connectivity map in the human brain.

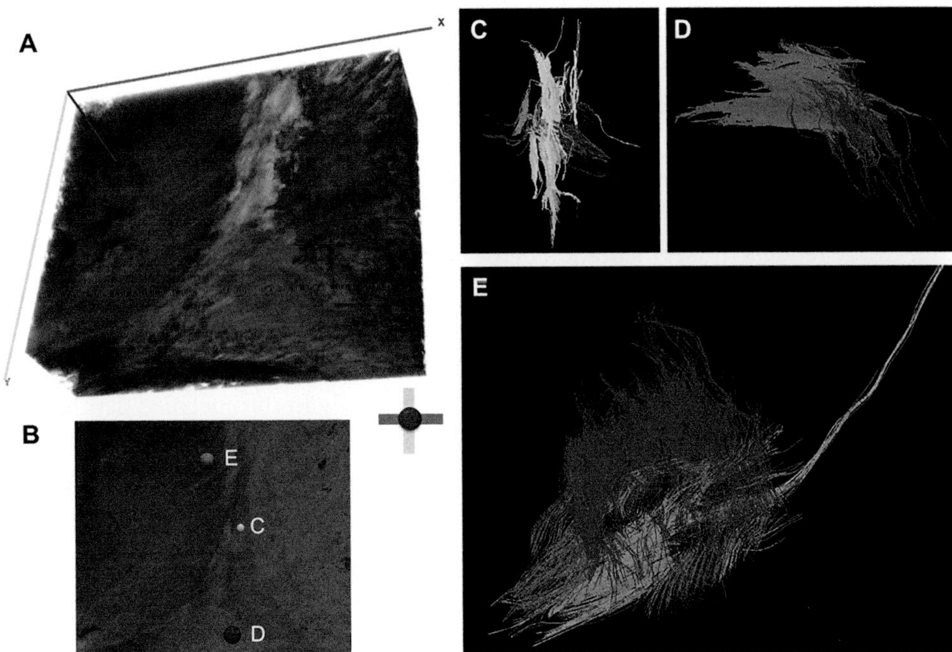

FIGURE 18.15 3D orientation and tractography with as-PSOCT. A) Volume rendering of a 3D orientation map in the thalamus and internal capsule region. The color-coding of the orientation value is indicated at the bottom-right corner. B) The retardance image of a slice showing the seeds where tracts are created and displayed in (C)–(E).

18.5.2.3.2 Measuring 3D Axis with PS OCT In parallel to computational based methods, PS OCT techniques have been advanced to measure the 3D orientation of birefringent structures. Multiple incident angles are incorporated to illuminate the sample and the 3D axis is extracted using Jones matrix PS OCT. Two models have been developed using the retardance and optic axis orientation information (Ugryumova et al., 2006; Ugryumova et al., 2009; Liu et al., 2016). The apparent retardance has a dependency on the through-plane angle α as:

$$\delta = 2\pi \frac{d \cdot \Delta n}{\lambda} \cos^2 \alpha \tag{18.14}$$

This relationship has been used in PLI with rotational polarimetry to extract 3D fiber orientation information. Ugryumova et al. (2006) adopted the method in PSOCT to measure the retardance with multiple illuminative angles, which has been further advanced with improved accuracy and generalization (Ugryumova et al., 2009). The through-plane orientation and tissue birefringence are extracted by fitting the retardance measurements. Another method has been developed based on the measurements of the optical axis orientation corresponding to different incident angles (Liu et al., 2016). The 3D axis is recovered by using a geometric transformation.

18.5.3 Imaging Neurons

To visualize individual neurons, a higher numerical aperture (NA) objective is required in order to gain higher lateral resolution (Magnain et al., 2015). For the following section, results are presented using a 40× water immersion objective with an NA of 0.8 (Olympus LUMPLANFL/IR 40W), providing a lateral resolution of 1.25 um over a 400 μm × 400 μm field of view. The speckle noise on the high-resolution optical coherence microscopy (OCM) volume image renders the segmentation of the neurons difficult. A frequency compounding technique is applied to the OCM data in order to reduce the speckle noise (Magnain et al., 2016). This method improves the contrast of the image while preserving the lateral resolution. The axial resolution, even though slightly reduced, remains high (~10 μm) because of the axial filtering provided by the high NA objective. Only in the specified depth of focus of ~10 μm are the neurons and fibers in focus. We thus extract an image from the volumetric OCM data by performing a MIP over 10 μm around the focus plane. The typical section thickness used for Nissl staining and the study of cytoarchitecture is 50 μm. To be able to compare Nissl stained slices and this OCM imaging of neurons, the top 50 μm of the tissue is imaged by OCM and then sectioned off using a vibratome. Five independent OCM acquisitions at different focus depths are obtained, separated by 10 μm each, starting at 5 μm under the surface as shown by the schematic in Figure 18.16. Figures 16A–E show the MIP images obtained at the five depths, from 5 μm under the surface to 45 μm deep,

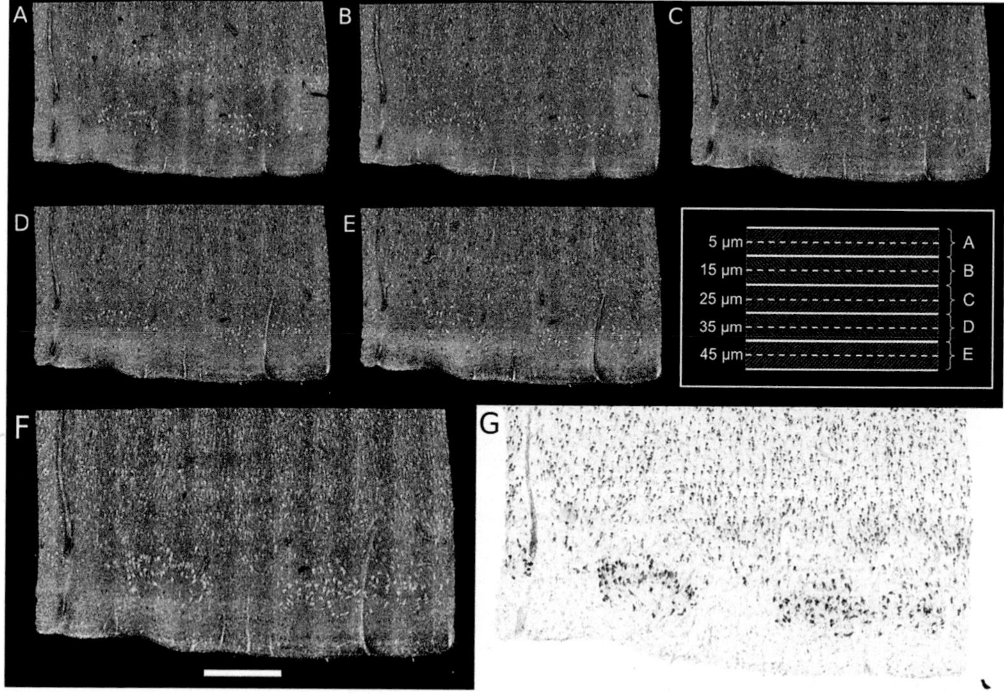

FIGURE 18.16 A–E) OCM images of the layers I to III of entorhinal cortex, acquired at five different depths as shown on the schematic. F) Combined OCM image covering the top 50 um of tissue and G) corresponding Nissl stain of the same tissue. Scale bar = 500 um.

in the entorhinal cortex, more specifically layers I to III. These five images are combined to visualize all the neurons in the 50 μm thick section as shown in Figure 18.16F. The 50 μm-thick section is then sliced, mounted, and processed for neurons using the Nissl stain. The digitized histological section is shown in Figure 18.16G.

As can be seen, the OCM and Nissl images are quite similar. First, the different vessels highlighted by background staining in the Nissl stained slice are shown on the combined OCM image, and even showing its 3D orientations through the different depth images. Second, the OCM images reveal the presence of neurons in both layers II and III and not in layer I. Third, as expected, the neuron sizes vary depending on their location. The neurons are larger in layer II and smaller in layer III, a finding confirmed by the Nissl stained slice image. Finally, the neuron islands in layer II can be identified, both in the histologic data and on the OCM image. In order for OCM to be competitive with standard histology such as Nissl staining, OCM need to be comparable with histology at the stereological level: the number, density, location, size, and shapes of neurons obtained by OCM need to match the data assessed by histology. The first step to validate OCM is to register both modalities at the neuron level. Due to the distortions introduced by histology protocols, a non-linear approach is needed. Landmarks on the OCM images and their correspondence on the Nissl stained slices are manually selected, with landmarks being neurons, vessels, or tissue borders. In Magnain et al. (2015), the accuracy of the non-linear registration has been assessed with respect to the number of landmarks. Approximately, 1,000 neurons in OCM have been selected, as well as the corresponding ones on the Nissl stained slice. The average registration error asymptotes to about 3 μm after 50 landmarks and is less than 4 μm after only 15 pairs of selected landmarks. Figure 18.17A shows the OCT image of one island from the entorhinal cortex presented on Figure 18.16, and Figure 18.17B shows the Nissl stained section after registration. To compare the neuronal content, we manually segmented neurons from each modality. Figure 18.17C shows both segmentations, red for OCM and green for Nissl, the yellow representing the overlapping neurons. As the segmentation overlap shows, an excellent agreement between the modalities is observed, and the neuronal information obtained by OCM is similar to the data recovered from Nissl staining in this region.

The type, shape, and size of neurons vary dramatically throughout the human brain, which influences the optical properties of the brain tissue. Figure 18.18 illustrates neurons from various location in the brain, both in the cortex and in subcortical regions. Each

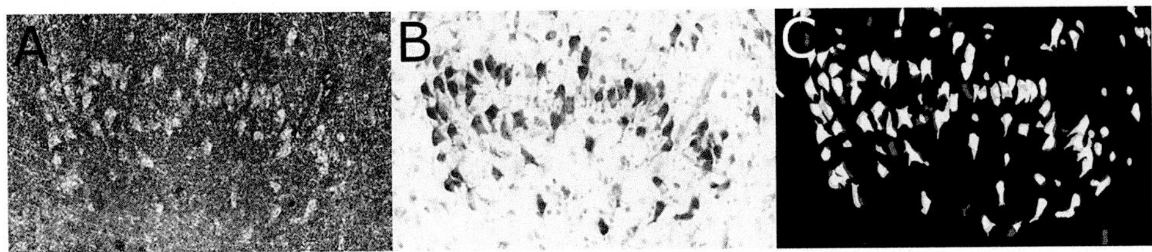

FIGURE 18.17 Layer II island of neurons in the entorhinal cortex observed by A) OCM and B) Nissl stain. C) The neurons have been manually segmented and registered: red for OCM, green for Nissl, and yellow for the overlap between the two modalities.

field of view is 500 μm × 500μm. Figures 18.18A and B shows pyramidal neurons from the temporal isocortex (BA21) and the pyramidal layers of the hippocampus in the medial temporal lobe, respectively. The size of the neurons is very different, from a few microns for BA21 to tens of microns for the hippocampus. Even though the diameter of the neurons covers a large range, the shape of the neurons, a triangular cell body with an apical dendrite, is plainly visible on both images. A subset of neurons found in two nuclei of the brainstem – the substantia nigra in the midbrain and the locus coeruleus in the pons – contain a pigment, called neuromelanin. This pigment gives the neurons a black hue that is visible to the naked eye on a gross sample, as well on a Nissl staining of these areas. The image in Figure 18.18C is the OCM acquisition of neurons located in the locus coeruleus. These neurons exhibit a high intensity contrast that is homogenous over the whole cell body, compared to the pyramidal cells in Figures 18.18A and B, where the center of the neurons is slightly darker than the borders. Finally, one of the largest neurons found in the brain are the Purkinje cells of the cerebellum, second only to the Betz cells found in primary motor cortex. Purkinje cells have a spherical cell body, and are packed in the narrow intermediate layer of the cerebellar cortex. These neurons have a single long axon in the granular layer, an extended dendritic tree in the molecular layer, which is organized in a 2D plane perpendicular to the long axis of the folium. Figure 18.18D shows the Purkinje cells visualized by OCM. The cell bodies are clearly distinguishable and a part of the dendritic tree can be visualized and resembles the well-known drawing made by Santiago Ramón y Cajal in 1899 (Ramón y Cajal, 1899).

18.5.4 Advantages and Limitations

One key advantage of the OCT technique is the imaging of the brain cyto- and myeloarchitecture at high resolution without the distortions or artifacts introduced by classic histological processing. The blockface image shown in Figure 18.9A represents the ground truth of brain geometry, similar to its natural state found in the skull (although distortions due to brain extraction and fixation are present). Because histology requires thin sections of tissue, ranging from 5 μm to 50 μm depending on the stain or dye used, handmounted on a glass slide, distortions are inherent to this technique, as discussed previously. Figure 18.9C demonstrate different types of distortions. First, during sectioning, the

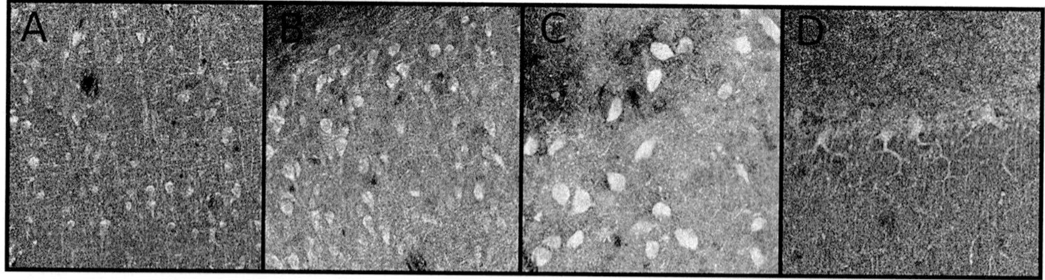

FIGURE 18.18 OCM images of various cell types – A) neurons from associative cortex BA21, B) neurons from the pyramidal layer of the hippocampus, C) neuromelanin-laden cells from the locus coeruleus of the pons, and D) Purkinje cells and part of their dendritic tree of the cerebellum. Each image has a field of view of 500 μm × 500 μm.

lower gyrus has been torn, which makes mounting in the correct anatomical position challenging. Overlapping of tissue can also be observed on the lower right, which prevents the visualization of the neurons in this region and makes registration of slices to each other or to MRI difficult. In contrast, the OCT image in Figure 18.9B does not exhibit any obvious distortion of the tissue geometry relative to the blockface image; because the acquisition is performed before the sectioning, the shape of the tissue is preserved.

The second advantage of OCT is the lack of tissue shrinkage; the effect of tissue shrinking during the drying phase of histology protocol can be appreciated around the vessels. The gap left by the vessels in the Nissl stained images is much larger than their size seen on the blockface. The cytoarchitecture is therefore distorted in a neighborhood around each vessel. No such distortions of the vessels are present in the OCT images; the tissue remains in water for the duration of the acquisition and never dehydrates.

Finally, the staining throughout the slice is not completely homogenous and background staining can be seen on the upper and lower gyri, as well as in the region surrounding the transitioning gyrus in the middle of the white matter, as shown in Figures 18.9 and 18.10. The uneven staining can render the study of the histologic slices challenging. For some stains, such as Gallyas, studying both the white matter and gray matter with the same protocol is challenging. For example, the Gallyas staining on Figure 18.11B has been optimized to visualize the myelin of the gray matter. As a result, the white matter is darkly stained and does not allow visualization of the fiber organization. The OCT contrast only depends on the tissue optical properties and therefore can be compared within the sample.

One major drawback of OCT and PS OCT for the characterization of the human brain is the limited number of contrasts that are available. OCT contrasts rely on the intrinsic optical properties of the tissue, namely, the attenuation coefficient and the birefringence. This allows us to probe the overall architecture (cyto- or myelo-) of the human cortex and the structures of subcortical regions. For the study of individual neurons, OCT can only rely on the shape and size of the cell bodies to determine the neuron types, which is sometimes incomplete, although the neuromelanin example above does show that OCT can be sensitive to some molecular markers of cell type. In contrast, histology and immunohistochemistry can make use of a wide variety of stains (dyes) and antibodies to differentiate neurons not only based on size and shape, but also on their neurotransmitter production.

18.6 FUTURE WORK

18.6.1 Quantifying Cellular Density and Myelin Content

Optical scattering arises from cell bodies and myelin content. If we can quantify spatial distributions in the optical scattering coefficient, we would then obtain spatial maps indicating the relative cell density and myelin content. But how do we disentangle myelin content from cell density? Fortunately, myelinated axons are birefringent and result in a retardance of the OCT signal that we can measure with PS OCT. Thus, it may be possible to develop models to fit the depth dependent optical attenuation and retardance to disentangle the relative cell density and myelin content.

As described by Izatt et al. (1994), the OCT signal attenuates with depth in the brain due to scattering, and there is additional attenuation above and below the focal depth of the measurement because of the confocal rejection of the scattered light. Fitting this signal model to the depth profile of the experimental data, it is possible to estimate local values for the scattering coefficient, achieving our first step in quantifying the relative cell density and myelin content. In addition, this fitting will also reveal information about the tissue back-scattering coefficient. The ratio of the back-scattering coefficient to the overall scattering coefficient reveals information about the anisotropy of the scattering direction, in that a larger/smaller ratio indicates more/less isotropic scattering. When the ratio of back-scattering to scattering becomes small, it indicates that the optical scattering is preferentially along the optic axis and typically results from larger scattering particles. Thus, it is possible to derive some information about the relative sizes of the scattering particles within the tissue.

Future work will integrate such model fitting of the tissue scattering and back-scattering coefficients with polarization measurements of tissue birefringence in order to quantify the relative spatial distributions of cell density and myelin content.

18.6.2 Three-Dimensional Tractography

PS OCT has a unique advantage of tracking fiber trajectories over long distance with a microscopic accuracy, due to the ability to directly assess fiber orientations and volumetric reconstruction in undistorted tissue blocks. The current approach of extracting 3D fiber orientation in the brain with PS OCT relies on digital processing, which inevitably has the drawback of losing resolution due to the requirement of spatial smoothing or integrating information from neighboring pixels, as well as an anisotropic point spread function and differential through-plane and in-plane sensitivities. Future developments of techniques for measuring 3D axis directly in the brain with improved PS OCT acquisition technologies that are scalable to large volumes is imperative. Based on the 3D orientation quantification and anatomical information provided by other contrasts of OCT/PS OCT, microscopic tractography methods can be developed to disentangle the fiber bundles with intricate crossing, and trace the fibers to their terminals. The creation of connectivity maps at high resolution has the potential to provide useful guidance for the development of more accurate *in vivo* tractography technologies such as in diffusion MRI to study the wiring system of the human brain in normal and diseased conditions.

18.6.3 Imaging Larger Volumes of Tissue

The imaging rig we utilized to obtain the results presented in this chapter was a 1300 nm spectral domain OCT system operating at a 47 kHz A-Scan line-rate, as detailed and utilized in Magnain et al. (2014); Srinivasan et al. (2010); Srinivasan et al. (2011); Lee et al. (2012); Magnain et al. (2016). This system has enabled us to produce stunning images, but is lacking several key advances critical for imaging of larger volumes of tissue. It presently takes us more than three days to image 1 cm^3 of tissue. At this rate, it would take over ten years to image the approximately 1200 cm^3 of the whole human brain. Two principal advances can be made to the optical acquisition system to reduce this to approximately ten weeks. First, the image acquisition rate can be increased 20× by utilizing commercially available swept source OCT systems that achieve A-Scan rates of 1 MHz (Wieser et al., 2014). In addition, the field of view (FOV) of each image can be increased and the pixel pitch reduced; For instance, increasing the FOV from 1.5 × 1.5 mm to 5 × 5 mm, and the pixel pitch from ~1.7 μm to 5 μm. The increased pixel pitch reduces the number of pixels by 10×, which combined with the faster acquisition rate results in a 200× faster volumetric image rate. This alone would reduce the projected greater than 3,600 days for imaging the whole human brain, down to 18 days. However, this projection ignores the dead time due to sample translation and sectioning. Increasing the FOV to 5 × 5 mm permits a 10× reduction in the number of sample translations, significantly reducing this dead time. For example, to scan a brain area typically spanning 13 × 11 cm, with no overlap between image tiles, would take 26 × 22 = 572 tiles and 21 minutes of data acquisition. Each tile will consist of 1,000 × 1,000 pixels and take 1.2 seconds to acquire, allowing for a 20% dead time on the fast axis sweep of the resonant scanner. Sample translation by 5 mm will take less than 1 second. Compare this with an estimated 5,720 tiles that would be acquired with a 1.5 mm FOV, which would add an additional approximately 80 minutes of sample translation time. Thus, the larger FOV allows us to speed up data acquisition by a factor of 4.

The volumetric imaging results presented in this chapter were obtained by repeatedly cutting 50 μm slices from the blockface after imaging the surface area. A vibratome can slice through fixed tissue with widths typically up to 2.5 cm, but that can extend up to 5 or 7 cm if one accepts a reduction in cutting thickness accuracy. To volumetrically image the entire blockface of the whole human brain requires cutting slices at least 11 cm wide. This cannot be done with a vibratome, but can be done with a microtome, as is used in histology to section coronally a whole human brain (Amunts et al., 2013; Ding et al., 2016), which requires that the tissue be hardened either by freezing or embedding in paraffin or PEG. The challenge is that these hardening methods increase optical scattering in the tissue. Freezing produces ice crystals which increases light scattering. Cryoprotectant is used to reduce the formation of large ice crystals and prevent tissue damage, but crystals nonetheless are formed, albeit smaller ones, and tissue becomes more scattering, which reduces OCT contrast for cell density and myelin content. Likewise, paraffin and PEG are white, which indicates that they also increase optical scattering and reduce OCT contrast to cell and myelin density. Different strategies need to be explored for how volumetric imaging of the whole human brain can be accomplished either by cutting large slices from the whole human brain possibly by freezing then warming the top 200 ums for OCT imaging, or blocking the whole human brain into smaller volumes which can be subsequently volumetrically imaged by cutting with a vibratome.

18.6.4 Applications

The capability of serial sectioning OCT/PS OCT to provide 3D reconstructions and permit microstructure exploration with better than 10 μm resolution advocates for the further development of automatic tools for segmentation, morphometric analysis, and statistical analysis of regional heterogeneity and cross-subject variations, which have only recently become available for macroscopic MRI with resolution of hundreds of micrometers to

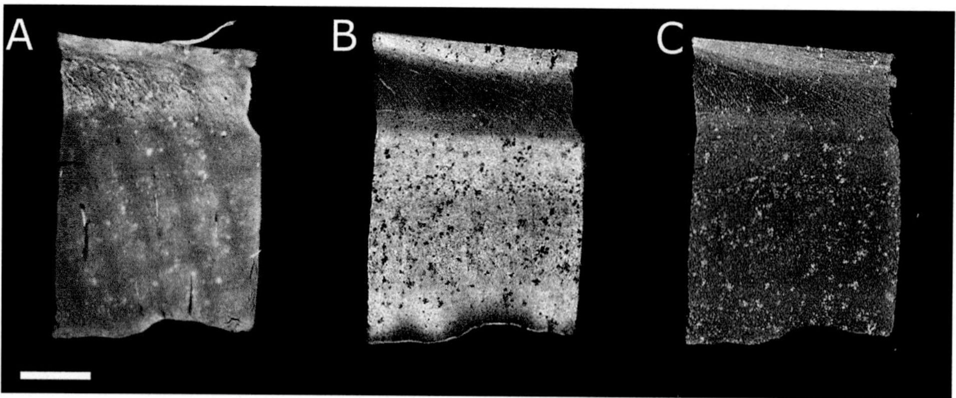

FIGURE 18.19 ß-amyloid plaques in the entorhinal cortex imaged by A) Thioflavin S and OCT: B) AIP and C) MIP. Scale bar = 1 mm.

1 mm (Bullmore and Sporns, 2009; Fischl, 2012; Smith et al., 2004). For example, as OCT allows distinguishing the laminar structure of cortical layers with clear contrasts, automatic algorithms can be developed to segment the cortical layers and quantify the layer thickness. Those morphological and quantitative characterizations of these layers will provide useful information to investigate the cortical boundaries among different brain regions.

The optical properties of scattering and birefringence originate from the underlying cellular and molecular content, such as size, cellular density, myelin content, and structural alignment. As a consequence, pathological processes may introduce alterations in these optical contrasts, which can be captured by OCT/PSOCT. Thus, important pathological processes such as neuron loss may be observed as a decrease in scattering or increase of light penetration, whereas deposit or aggregation of proteins may manifest as highly scattering spots. For example, ß-amyloid plaques, one of the pathological markers of Alzheimer's disease, exhibit a bright signal in the MIP and prevent the light from penetrating deeper, resulting in dark spots in the AIP (Bolmont et al., 2012). Figure 18.19 illustrates ß-amyloid plaques found in entorhinal cortex, stained by thioflavinS (Figure 18.19A) and imaged by OCT AIP and MIP in Figures 18.19B and 18.19C respectively. Another example is the degradation of myelin that may be correlated with a decrease in birefringence in multiple sclerosis and traumatic brain injury. The neuropathological progression also leads to a morphometric change in brain structures such as cortical thinning in specific layers (Du et al., 2007; Gómez-Isla et al., 1996), which could be quantitatively characterized by OCT with micrometer resolution.

REFERENCES

Al-Qaisi, Muhammad K., and Taner Akkin. 2008. "Polarization-Sensitive Optical Coherence Tomography Based on Polarization-Maintaining Fibers and Frequency Multiplexing." *Optics Express* 16 (17): 13032–41. doi:10.1364/OE.16.013032.

Al-Qaisi, Muhammad K., and Taner Akkin. 2010. "Swept-Source Polarization-Sensitive Optical Coherence Tomography Based on Polarization-Maintaining Fiber." *Optics Express* 18 (4): 3392–403.

Amunts, Katrin, Claude Lepage, Louis Borgeat, Hartmut Mohlberg, Timo Dickscheid, Marc-Étienne Rousseau, Sebastian Bludau, et al. 2013. "BigBrain: An Ultrahigh-Resolution 3D Human Brain Model." *Science (New York, N.Y.)* 340 (6139): 1472–75. doi:10.1126/science.1235381.

Amunts, Katrin, Axel Schleicher, Uli Bürgel, Hartmunt Mohlberg, Harry B. M. Uylings, and Karl Zilles. 1999. "Broca's Region Revisited: Cytoarchitecture and Intersubject Variability." *Journal of Comparative Neurology* 412 (2): 319–41.

An, Lin, Peng Li, Gongpu Lan, Doug Malchow, and Ruikang K. Wang. 2013. "High-Resolution 1050 Nm Spectral Domain Retinal Optical Coherence Tomography at 120 kHz A-Scan Rate with 6.1 Mm Imaging Depth." *Biomedical Optics Express* 4 (2): 245–59. doi:10.1364/BOE.4.000245.

Arsigny, Vincent, Xavier Pennec, and Nicholas Ayache. 2005. "Polyrigid and Polyaffine Transformations : A Novel Geometrical Tool to Deal with Non-Rigid Deformations – Application to the Registration of Histological Slices." *Medical Image Analysis* 9 (6): 507–23. doi:10.1016/j.media.2005.04.001.

Assayag, Osnath, Martine Antoine, Brigitte Sigal-Zafrani, Michael Riben, Fabrice Harms, Adriano Burcheri, Kate Grieve, Eugénie Dalimier, Bertrand Le Conte de Poly, and Claude Boccara. 2014. "Large Field, High Resolution Full-Field Optical Coherence Tomography: A Pre-Clinical Study of Human Breast Tissue and Cancer Assessment." *Technology in Cancer Research and Treatment* 13 (5): 455–68.

Assayag, Osnath, Kate Grieve, Bertrand Devaux, Fabrice Harms, Johan Pallud, Fabrice Chretien, Albert-Claude Boccara, and Pascale Varlet. 2013. "Imaging of Non-Tumorous and Tumorous Human Brain Tissues with Full-Field Optical Coherence Tomography." *NeuroImage: Clinical* 2: 549–57. doi:10.1016/j.nicl.2013.04.005.

Augustinack, Jean C., Andre J. W. van der Kouwe, Megan L. Blackwell, David H. Salat, Christopher J. Wiggins, Matthew P. Frosch, Graham C. Wiggins, et al. 2005. "Detection of Entorhinal Layer II Using 7Tesla Magnetic Resonance Imaging." *Annals of Neurology* 57 (4): 489–94. doi:10.1002/ana.20426.

Axer, Markus, Katrin Amunts, David Grässel, Christoph Palm, Jürgen Dammers, Hubertus Axer, Uwe Pietrzyk, and Karl Zilles. 2011. "A Novel Approach to the Human Connectome: Ultra-High Resolution Mapping of Fiber Tracts in the Brain." *NeuroImage* 54 (2): 1091–101. doi:10.1016/j.neuroimage.2010.08.075.

Baumann, Bernhard, Erich Götzinger, Michael Pircher, and Christoph K. Hitzenberger. 2007. "Single Camera Based Spectral Domain Polarization Sensitive Optical Coherence Tomography." *Optics Express* 15 (3): 1054–63.

Bigun, Josef. 1987. *Optimal Orientation Detection of Linear Symmetry.* Linköping University Electronic Press, Linköping, Sweden.

de Boer, Johannes F., Barry Cense, B. Hyle Park, Mark C. Pierce, Guillermo J. Tearney, and Brett E. Bouma. 2003. "Improved Signal-to-Noise Ratio in Spectral-Domain Compared with Time-Domain Optical Coherence Tomography." *Optics Letters* 28 (21): 2067–69.

de Boer, Johannes F., and Thomas E. Milner. 2002. "Review of Polarization Sensitive Optical Coherence Tomography and Stokes Vector Determination." *Journal of Biomedical Optics* 7 (3): 359–71. doi:10.1117/1.1483879.

de Boer, Johannes F., Thomas E. Milner, and J. Stuart Nelson. 1999. "Determination of the Depth-Resolved Stokes Parameters of Light Backscattered from Turbid Media by Use of Polarization-Sensitive Optical Coherence Tomography." *Optics Letters* 24 (5): 300–302. doi:10.1364/OL.24.000300.

de Boer, Johannes F., Thomas E. Milner, Martin J. C. van Gemert, and J. Stuart Nelson. 1997. "Two-Dimensional Birefringence Imaging in Biological Tissue by Polarization-Sensitive Optical Coherence Tomography." *Optics Letters* 22 (12): 934–936.

Bolmont, Tristan, Arno Bouwens, Christophe Pache, Mitko Dimitrov, Corinne Berclaz, Martin Villiger, Bettina M. Wegenast-braun, Theo Lasser, and Patrick C. Fraering. 2012. "Label-Free Imaging of Cerebral Beta-Amyloidosis with Extended-Focus Optical Coherence Microscopy." *Journal of Neuroscience* 32 (42): 14548–56. doi:10.1523/JNEUROSCI.0925-12.2012.

Braak, Heiko. 1979. "The Pigment Architecture of the Human Frontal Lobe." *Anatomy and Embryology* 157 (1): 35–68.

Braak, Heiko, and Eva Braak. 1997. "Diagnostic Criteria for Neuropathologic Assessment of Alzheimer's Disease." *Neurobiology of Aging* 18 (4) Supplement: S85–S88. doi:10.1016/S0197-4580(97)00062-6.

Brodmann, Korbinian. 1909. *Vergleichende Lokalisationslehre Der Grosshirnrinde in Ihren Prinzipien Dargestellt Auf Grund Des Zellenbaues*. Barth, Leipzig, Germany.

Budde, Matthew D., and Joseph A. Frank. 2012. "Examining Brain Microstructure Using Structure Tensor Analysis of Histological Sections." *NeuroImage* 63 (1): 1–10. doi:10.1016/j.neuroimage.2012.06.042.

Budde, Matthew D., Lindsay Janes, Eric Gold, Lisa Christine Turtzo, and Joseph A. Frank. 2011. "The Contribution of Gliosis to Diffusion Tensor Anisotropy and Tractography Following Traumatic Brain Injury: Validation in the Rat Using Fourier Analysis of Stained Tissue Sections." *Brain* 134 (8): 2248–60. doi:10.1093/brain/awr161.

Bullmore, Ed, and Olaf Sporns. 2009. "Complex Brain Networks: Graph Theoretical Analysis of Structural and Functional Systems." *Nature Reviews Neuroscience* 10 (3): 186–98. doi:10.1038/nrn2575.

Burkhalter, Andreas, and Kerry L. Bernardo. 1989. "Organization of Corticocortical Connections in Human Visual Cortex." *Proceedings of the National Academy of Sciences* 86 (3): 1071–75.

Campbell, A. W. 1905. *Histological Studies on the Localisation of Cerebral Function*. University Press, Cambridge, UK.

Campbell, Michael J., and John H. Morrison. 1989. "Monoclonal Antibody to Neurofilament Protein (SMI-32) Labels a Subpopulation of Pyramidal Neurons in the Human and Monkey Neocortex." *Journal of Comparative Neurology* 282 (2): 191–205.

Cense, Barry, Mircea Mujat, Teresa C. Chen, B. Hyle Park, and Johannes F. de Boer. 2007. "Polarization-Sensitive Spectral-Domain Optical Coherence Tomography Using a Single Line Scan Camera." *Optics Express* 15 (5): 2421–31.

Cense, Barry, Nader A. Nassif, Teresa C. Chen, Mark C. Pierce, Seok-hyun Yun, B. Hyle Park, Brett E. Bouma, Guillermo J. Tearney, and Johannes F. de Boer. 2004. "Ultrahigh-Resolution High-Speed Retinal Imaging Using Spectral-Domain Optical Coherence Tomography." *Optics Express* 12 (11): 2435–47.

Chinn, S. R., E. A. Swanson, and J. G. Fujimoto. 1997. "Optical Coherence Tomography Using a Frequency-Tunable Optical Source." *Optics Letters* 22 (5): 340–42. doi:10.1364/OL.22.000340.

Choe, A. S., I. Stepniewska, D. C. Colvin, Z. Ding, and A. W. Anderson. 2012. "Validation of Diffusion Tensor MRI in the Central Nervous System Using Light Microscopy: Quantitative Comparison of Fiber Properties." *NMR in Biomedicine* 25 (7): 900–908. doi:10.1002/nbm.1810.

Choma, Michael A., Kevin Hsu, and Joseph A. Izatt. 2005. "Swept Source Optical Coherence Tomography Using an All-Fiber 1300-Nm Ring Laser Source." *Journal of Biomedical Optics* 10 (4): 44009. doi:10.1117/1.1961474.

Choma, Michael A., M. Sarunic, C. Yang, and Joseph A. Izatt. 2003. "Sensitivity Advantage of Swept Source and Fourier Domain Optical Coherence Tomography." *Optics Express* 11 (18): 2183–89.

Clarke, Stephanie. 1994. "Modular Organization of Human Extrastriate Visual Cortex : Evidence from Cytochrome Oxidase Pattern in Normal and Macular Degeneration Cases." *European Journal of Neuroscience* 6 (5): 725–36.

Denk, Winfried, and Heinz Horstmann. 2004. "Serial Block-Face Scanning Electron Microscopy to Reconstruct Three-Dimensional Tissue Nanostructure." *PLoS Biology* 2 (11): e329.

Dietl, Monika M., Alphonse Probst, and Jose M. Palacios. 1987. "On the Distribution of Cholecystokinin Receptor Binding Sites in the Human Brain : An Autoradiographic Study." *Synapse* 1 (2): 169–83.

Ding, Song-lin, Joshua J. Royall, Susan M. Sunkin, Lydia Ng, Benjamin A. C. Facer, Phil Lesnar, Angie Guillozet-Bongaarts, et al. 2016. "Comprehensive Cellular-Resolution Atlas of the Adult Human Brain." *Journal of Comparative Neurology* 524 (16): 3127–481. doi:10.1002/cne.24080.

Dorrer, Christophe, Nadia Belabas, Jean-Pierre Likforman, and Manuel Joffre. 2000. "Spectral Resolution and Sampling Issues in Fourier-Transform Spectral Interferometry." *JOSA* B 17 (10): 1795–802.

Du, An-Tao, Norbert Schuff, Joel H. Kramer, Howard J. Rosen, Maria Luisa Gorno-Tempini, Katherine Rankin, Bruce L. Miller, and Michael W. Weiner. 2007. "Different Regional Patterns of Cortical Thinning in Alzheimer's Disease and Frontotemporal Dementia." *Brain: A Journal of Neurology* 130 (Pt 4): 1159–66. doi:10.1093/brain/awm016.

Dubois, Arnaud, Gael Moneron, Kate Grieve, and Albert-Claude Boccara. 2004. "Three-Dimensional Cellular-Level Imaging Using Full-Field Optical Coherence Tomography." *Physics in Medicine and Biology* 49: 1227–34.

Dubois, Arnaud, Laurent Vabre, Albert-Claude Boccara, and Emmanuel Beaurepaire. 2002. "High-Resolution Full-Field Optical Coherence Tomography with a Linnik Microscope." *Applied Optics* 41 (4): 805–12.

von Economo, C. F., and G. N. Koskinas. 1925. *Die Cytoarchitektonik Der Hirnrinde Des Erwachsenen Menschen*. J. Springer.

Elliot Smith, G. 1907. "A New Topographical Survey of the Human Cerebral Cortex, Being an Account of the Distribution of the Anatomically Distinct Cortical Areas and Their Relationship to the Cerebral Sulci." *Journal of Anatomy and Physiology* 41 (Pt 4): 237–54.

Everett, M. J., K. Schoenenberger, B. W. Colston, and L. B. Da Silva. 1998. "Birefringence Characterization of Biological Tissue by Use of Optical Coherence Tomography." *Optics Letters* 23 (3): 228–30.

Fercher, Adolph F., Ch K. Hitzenberger, G. Kamp, and Sy Y. El-Zaiat. 1995. "Measurement of Intraocular Distances by Backscattering Spectral Interferometry." *Optics Communications* 117 (1): 43–48.

Fields, R. Douglas. 2008. "White Matter in Learning, Cognition and Psychiatric Disorders." *Trends in Neurosciences* 31 (7): 361–70.

Fischl, Bruce R. 2012. "FreeSurfer." *NeuroImage* 62 (2): 774–81. doi:10.1016/j.neuroimage.2012.01.021.

Flechsig, P. 1889. "Neue Untersuchungen Über Die Marbildung in Den Menschlichen Grosshirnlappen." *Neurologisches Zentralblatt* 17: 977–96.

Gennari, F. 1782. *De Peculiari Structura Cerebri Nonnulisque Ejus Morbis*. Parma: Ex Regio Typographeo.

Gil, José J., and Eusebio Bernabeu. 1987. "Obtainment of the Polarizing and Retardation Parameters of a Non-Depolarizing Optical System from the Polar Decomposition of Its Mueller Matrix." *Optik* 76 (2): 67–71.

Goldberg, Brian D., S. M. Reza Motaghian Nezam, Priyanka Jillella, Brett E. Bouma, and Guillermo J. Tearney. 2009. "Miniature Swept Source for Point of Care Optical Frequency Domain Imaging." *Optics Express* 17 (5): 3619–29. doi:10.1364/OE.17.003619.

Golubovic, B., B. E. Bouma, G. J. Tearney, and J. G. Fujimoto. 1997. "Optical Frequency-Domain Reflectometry Using Rapid Wavelength Tuning of a Cr 4+:Forsterite Laser." *Optics Letters* 22 (22): 1704–706. doi:10.1364/OL.22.001704.

Gómez-Isla, T., Joseph L. Price, Daniel W. Mckeel, Jr, John C. Morris, John H. Growdon, and Bradley T. Hyman. 1996. "Profound Loss of Layer II Entorhinal Cortex Neurons Occurs in Very Mild Alzheimer's Disease." *The Journal of Neuroscience* 16 (14): 4491–500.

Götzinger, Erich, Bernhard Baumann, Michael Pircher, and Christoph K. Hitzenberger. 2009. "Polarization Maintaining Fiber Based Ultra-High Resolution Spectral Domain Polarization Sensitive Optical Coherence Tomography." *Optics Express* 17 (25): 22704–17. doi:10.1364/OE.17.022704.

Götzinger, Erich, Michael Pircher, and Christoph K. Hitzenberger. 2005. "High Speed Spectral Domain Polarization Sensitive Optical Coherence Tomography of the Human Retina." *Optics Express* 13 (25): 10217–29.

Grulkowski, I., M. Gora, Maciej Szkulmowski, Iwona Gorczynska, D. Szlag, S. Marcos, Andrzej Kowalczyk, and Maciej Wojtkowski. 2009. "Anterior Segment Imaging with Spectral OCT System Using a High-Speed CMOS Camera." *Optics Express* 17 (6): 4842–58.

Grulkowski, Ireneusz, Jonathan J. Liu, Benjamin Potsaid, Vijaysekhar Jayaraman, Chen D. Lu, James Jiang, Alex E. Cable, Jay S. Duker, and James G. Fujimoto. 2012. "Retinal, Anterior Segment and Full Eye Imaging Using Ultrahigh Speed Swept Source OCT with Vertical-Cavity Surface Emitting Lasers." *Biomedical Optics Express* 3 (11): 2733–51. doi:10.1364/BOE.3.002733.

Hausler, Gerd, and Michael Walter Lindner. 1998. "'Coherence Radar' and 'Spectral Radar'—New Tools for Dermatological Diagnosis." *Journal of Biomedical Optics* 3 (1): 21–31. doi:10.1117/1.429899.

Hee, Michael R., David Huang, Eric A. Swanson, and James G. Fujimoto. 1992. "Polarization-Sensitive Low-Coherence Reflectometer for Birefringence Characterization and Ranging." *Journal of the Optical Society of America B* 9 (6): 903–908. doi:10.1364/JOSAB.9.000903.

Hendry, S. H. C., M.-M. Huntsman, A. Vinuela, H. Mohler, A. L. de Blas, and E. G. Jones. 1994. "GABAA Receptor Subunit Immunoreactivity in Primate Visual Cortex: Distribution in Macaques and Humans and Regulation by Visual Input in Adulthood." *Journal of Neuroscience* 14 (4): 2383–401.

Hess, S. T., S. H. Huang, A. A. Heikal, and Watt W. Webb. 2002. "Biological and Chemical Applications of Fluorescence Correlation Spectroscopy: A Review." *Biochemistry* 41 (3): 697–705.

Hitzenberger, Christoph K., E. Goetzinger, M. Sticker, M. Pircher, and Adolf F. Fercher. 2001. "Measurement and Imaging of Birefringence and Optic Axis Orientation by Phase Resolved Polarization Sensitive Optical Coherence Tomography." *Optics Express* 9 (13): 780–90.

Huard, Serge. 1997. "Polarization of Light." In *Polarization of Light*, 348. Wiley-VCH.

Huber, Robert, Maciej Wojtkowski, K. Taira, James G. Fujimoto, and K. Hsu. 2005. "Amplified, Frequency Swept Lasers for Frequency Domain Reflectometry and OCT Imaging: Design and Scaling Principles." *Optics Express* 13 (9): 3513–28.

Izatt, Joseph A., Michael R. Hee, G. M. Owen, Eric A. Swanson, and James G. Fujimoto. 1994. "Optical Coherence Microscopy in Scattering Media." *Optics Letters* 19: 590–92.

Jansen, K. L. R., R. L. M. Faull, and M. Dragunow. 1989. "Excitatory Amino Acid Receptors in the Human Cerebral Cortex: A Quantitative Auroradiographic Study Comparing the Distribution of [3H]TCP, [3H]glycine, L-[3H]glutamate, [3H]AMPA and [3H]kainic Acid Binding Sites." *Neuroscience* 32 (3): 587–607.

Jones, R. Clark. 1941. "A New Calculus for the Treatment of Optical Systems. I. Description and Discussion of the Calculus." *Journal of the Optical Society of America* 31 (7): 488–93. doi:10.1364/JOSA.31.000488.

Katoh, Kaoru, Katherine Hammar, Peter J. S. Smith, and Rudolf Oldenbourg. 1999. "Arrangement of Radial Actin Bundles in the Growth Cone of Aplysia Bag Cell Neurons Shows the Immediate Past History of Filopodial Behavior." *Proceedings of the National Academy of Sciences* 96 (14): 7928–31. doi:10.1073/pnas.96.14.7928.

Larsen, Luiza, Lewis D. Griffin, David GRäßel, Otto W. Witte, and Hubertus Axer. 2007. "Polarized Light Imaging of White Matter Architecture." *Microscopy Research and Technique* 70 (10): 851–63. doi:10.1002/jemt.20488.

Lee, J., W. Wu, J. Y. Jiang, B. Zhu, and D. A. Boas. 2012. "Dynamic Light Scattering Optical Coherence Tomography." *Optics Express* 20: 22262–77.

Leergaard, Trygve B., Nathan S. White, Alex J. de Crespigny, Ingeborg Bolstad, Helen E. D'Arceuil, Jan G. Bjaalie, and Anders M. Dale. 2010. "Quantitative Histological Validation of Diffusion MRI Fiber Orientation Distributions in the Rat Brain." *PloS One* 5 (1): e8595. doi:10.1371/journal.pone.0008595.

Leitgeb, R., C. Hitzenberger, and Adolf Fercher. 2003. "Performance of Fourier Domain vs Time Domain Optical Coherence Tomography." *Optics Express* 11 (8): 889. doi:10.1364/OE.11.000889.

Liu, A. Q., and X. M. Zhang. 2007. "A Review of MEMS External-Cavity Tunable Lasers." *Journal of Micromechanics and Microengineering* 17 (1): R1. doi:10.1088/0960-1317/17/1/R01.

Liu, Chao J., Adam J. Black, Hui Wang, and Taner Akkin. 2016. "Quantifying Three-Dimensional Optic Axis Using Polarization-Sensitive Optical Coherence Tomography." *Journal of Biomedical Optics* 21 (7): 70501. doi:10.1117/1.JBO.21.7.070501.

Liu, Chao J., K. E. Williams, H. T. Orr, and T. Akkin. 2017. "Visualizing and Mapping the Cerebellum with Serial Optical Coherence Scanner." *Neurop* 4 (1): 11006.

Magnain, C., J. C. Augustinack, E. Konukoglu, M. P. Frosch, S. Sakadžić, A. Varjabedian, N. Garcia, V. J. Wedeen, D. A. Boas, and B. Fischl. 2015. "Optical Coherence Tomography Visualizes Neurons in Human Entorhinal Cortex." *Neurophotonics* 2 (1). doi:10.1117/1.NPh.2.1.015004.

Magnain, Caroline, Jean C. Augustinack, Martin Reuter, Christian Wachinger, Matthew P. Frosch, Timothy Ragan, Taner Akkin, Van Jay Wedeen, David A. Boas, and Bruce Fischl. 2014. "Blockface Histology with Optical Coherence Tomography: A Comparison with Nissl Staining." *NeuroImage* 84: 524–33. doi:10.1016/j.neuroimage.2013.08.072.

Magnain, Caroline, Hui Wang, Sava Sakadžić, Bruce R. Fischl, and David A. Boas. 2016. "En Face Speckle Reduction in Optical Coherence Microscopy by Frequency Compounding." *Optics Letters* 41 (9): 1925–28.

Nakaji, Haruo, Nobuo Kouyama, Yoshihiro Muragaki, Yoriko Kawakami, and Hiroshi Iseki. 2008. "Localization of Nerve Fiber Bundles by Polarization-Sensitive Optical Coherence Tomography." *Journal of Neuroscience Methods* 174 (1): 82–90. doi:10.1016/j.jneumeth.2008.07.004.

Odgaard, Anders, Kurt Andersen, Flemming Melsen, and Hans Jorgen G. Gundersen. 1990. "A Direct Method for Fast Three-Dimensional Serial Reconstruction." *Journal of Microscopy* 159 (3): 335–42.

Oldenbourg, R., E. D. Salmon, and P. T. Tran. 1998. "Birefringence of Single and Bundled Microtubules." *Biophysical Journal* 74 (1): 645–54. doi:10.1016/S0006-3495(98)77824-5.

Osechinskiy, Sergey, and Frithjof Kruggel. 2010. "Slice-to-Volume Nonrigid Registration of Histological Sections to MR Images of the Human Brain." *Anatomy Research International* 2011: 287860. doi:10.1155/2011/287860.

Pistorio, Ashley L., Stewart H. Hendry, and Xiaoqin Wang. 2006. "A Modified Technique for High-Resolution Staining of Myelin." *Journal of Neuroscience Methods* 153 (1): 135–46. doi:10.1016/j.jneumeth.2005.10.014.

Potsaid, Benjamin, Iwona Gorczynska, Vivek J. Srinivasan, Yueli Chen, James Y. Jiang, Alex E. Cable, and James G. Fujimoto. 2008. "Ultrahigh Speed Spectral/Fourier Domain OCT Ophthalmic Imaging at 70,000 to 312,500 Axial Scans per Second." *Optics Express* 16 (19): 15149–69.

Preibisch, Stephan, Stephan Saalfeld, and Pavel Tomancak. 2009. "Globally Optimal Stitching of Tiled 3D Microscopic Image Acquisitions." *Bioinformatics* 25 (11): 1463–65. doi:10.1093/bioinformatics/btp184.

Ragan, Timothy, Lolahon R. Kadiri, Kannan Umadevi Venkataraju, Karsten Bahlmann, Jason Sutin, Julian Taranda, Ignacio Arganda-Carreras, Yongsoo Kim, H. Sebastian Seung, and Pavel Osten. 2012. "Serial Two-Photon Tomography for Automated Ex Vivo Mouse Brain Imaging." *Nature Methods* 9 (3): 255–58. doi:10.1038/nmeth.1854.

Ramon y Cajal, Santiago. 1899. *Textura Del Sistema Nervioso Del Hombre Y de Los Vertebrados.* Madrid: Imprenta y Liberia de Nicolas Moya.

Reuter, Martin, H. Diana Rosas, and Bruce Fischl. 2010. "Highly Accurate Inverse Consistent Registration: A Robust Approach." *NeuroImage* 53 (4): 1181–96. doi:10.1016/j.neuroimage.2010.07.020.

Sands, Gregory B., Dane A. Gerneke, Darren A. Hooks, Colin R. Green, Bruce H. Smaill, and Ian J. Legrice. 2005. "Automated Imaging of Extended Tissue Volumes Using Confocal Microscopy." *Microscopy Research and Technique* 67 (5): 227–39.

Schmitt, O., M. Pakura, T. Aach, L. Hömke, M. Böhme, S. Bock, and S. Preusse. 2004. "Analysis of Nerve Fibers and Their Distribution in Histologic Sections of the Human Brain." *Microscopy Research and Technique* 63 (4): 220–43. doi:10.1002/jemt.20033.

Schoenenberger, Klaus, Bill W. Colston, Duncan J. Maitland, Luiz B. Da Silva, and Matthew J. Everett. 1998. "Mapping of Birefringence and Thermal Damage in Tissue by Use of Polarization-Sensitive Optical Coherence Tomography." *Applied Optics* 37 (25): 6026–36. doi:10.1364/AO.37.006026.

Smith, Stephen M., Mark Jenkinson, Mark W. Woolrich, Christian F. Beckmann, Timothy E. J. Behrens, Heidi Johansen-Berg, Peter R. Bannister, et al. 2004. "Advances in Functional and Structural MR Image Analysis and Implementation as FSL." *NeuroImage* 23 Suppl 1: S208–S219. doi:10.1016/j.neuroimage.2004.07.051.

Srinivasan, V. J., D. N. Atochin, H. Radhakrishnan, R. Jiang, J. Y. S. Uvinskaya W. Wu, S. Barry, et al. 2011. "Optical Coherence Tomography for the Quantitative Study of Cerebrovascular Physiology." *Journal of Cerebral Blood Flow and Metabolism* 31: 1339–45.

Srinivasan, Vivek J., Sava Sakadžić, Iwona Gorczynska, Svetlana Ruvinskaya, Weicheng Wu, James G. Fujimoto, and David A. Boas. 2010. "Quantitative Cerebral Blood Flow with Optical Coherence Tomography." *Optics Express* 18 (3): 2477–94.

Tearney, G. J., B. E. Bouma, and J. G. Fujimoto. 1997. "High-Speed Phase- and Group-Delay Scanning with a Grating-Based Phase Control Delay Line." *Optics Letters* 22 (23): 1811–13. doi:10.1364/OL.22.001811.

Thomason, Moriah E., and Paul M. Thompson. 2011. "Diffusion Imaging, White Matter, and Psychopathology." *Clinical Psychology* 7 (1): 63.

Tootell, Roger B. H., and J. B. Taylor. 1995. "Anatomical Evidence for MT and Additional Cortical Visual Areas in Humans." *Cerebral Cortex* 5 (1): 39–55.

Tsai, Philbert S., Beth Friedman, Agustin I. Ifarraguerri, Beverly D. Thompson, Varda Lev-Ram, Chris B. Schaffer, Qing Xiong, Roger Y. Tsien, Jeffrey A. Squier, and David Kleinfeld. 2003. "All-Optical Histology Using Ultrashort Laser Pulses." *Neuron* 39 (1): 27–41. doi:10.1016/S0896-6273(03)00370-2.

Ugryumova, Nadezhda, Sergei V. Gangnus, and Stephen J. Matcher. 2006. "Three-Dimensional Optic Axis Determination Using Variable-Incidence-Angle Polarization-Optical Coherence Tomography." *Optics Letters* 31 (15): 2305. doi:10.1364/OL.31.002305.

Ugryumova, Nadya, James Jacobs, Marco Bonesi, and Stephen J. Matcher. 2009. "Novel Optical Imaging Technique to Determine the 3-D Orientation of Collagen Fibers in Cartilage: Variable-Incidence Angle Polarization-Sensitive Optical Coherence Tomography." *Osteoarthritis and Cartilage* 17 (1): 33–42.

Vogt, C., and O. Vogt. 1919. *Allgemeine Ergebnisse Unserer Hirnforschung. Journal Für Psychologie Und Neurologie*. Barth, Leipzig, Germany.

Vogt, Oskar. 1910. "Die Myeloarchitektonische Felderung Des Menschlichen." *Journal Für Psychologie Und Neurologie* 15 (4/5): 221–32.

Wang, Hui, Taner Akkin, Caroline Magnain, Ruopeng Wang, Jay Dubb, William J. Kostis, Mohammad A. Yaseen, Avilash Cramer, Sava Sakadžić, and David A. Boas. 2016. "Polarization Sensitive Optical Coherence Microscopy for Brain Imaging." *Optics Letters* 41 (10): 2213–16.

Wang, Hui, Muhammad K. Al-Qaisi, and Taner Akkin. 2010. "Polarization-Maintaining Fiber Based Polarization-Sensitive Optical Coherence Tomography in Spectral Domain." *Optics Letters* 35 (2): 154–56.

Wang, Hui, Adam J. Black, Junfeng Zhu, Tyler W. Stigen, Muhammad K. Al-Qaisi, Theoden I. Netoff, Aviva Abosch, and Taner Akkin. 2011. "Reconstructing Micrometer-Scale Fiber Pathways in the Brain: Multi-Contrast Optical Coherence Tomography Based Tractography." *NeuroImage* 58 (4): 984–92. doi:10.1016/j.neuroimage.2011.07.005.

Wang, Hui, Christophe Lenglet, and Taner Akkin. 2015. "Structure Tensor Analysis of Serial Optical Coherence Scanner Images for Mapping Fiber Orientations and Tractography in the Brain." *Journal of Biomedical Optics* 20 (3): 036003. doi:10.1117/1.JBO.20.3.036003.

Wang, Hui, Junfeng Zhu, and Taner Akkin. 2014. "Serial Optical Coherence Scanner for Large-Scale Brain Imaging at Microscopic Resolution." *NeuroImage* 84: 1007–17. doi:10.1016/j.neuroimage.2013.09.063.

Wang, Hui, Junfeng Zhu, Martin Reuter, Louis N. Vinke, Anastasia Yendiki, David A. Boas, Bruce Fischl, and Taner Akkin. 2014. "Cross-Validation of Serial Optical Coherence Scanning and Diffusion Tensor Imaging: A Study on Neural Fiber Maps in Human Medulla Oblongata." *NeuroImage* 100: 395–404. doi:10.1016/j.neuroimage.2014.06.032.

Wang, Ruopeng, Thomas Benner, Alma Gregory Sorensen, and Van Jay Wedeen. 2007. "Diffusion Toolkit: A Software Package for Diffusion Imaging Data Processing and Tractography." In *Proceedings of the International Society for Magnetic Resonance in Medicine*, 15: 3720. Berlin, Germany.

Wieser, Wolfgang, Benjamin R. Biedermann, Thomas Klein, Christoph M. Eigenwillig, and Robert Huber. 2010. "Multi-Megahertz OCT: High Quality 3D Imaging at 20 Million A-Scans and 45 GVoxels per Second." *Optics Express* 18 (14): 14685–704. doi:10.1364/OE.18.014685.

Wieser, W., W. Draxinger, T. Klein, S. Karpf, T. Pfeiffer, and R. Huber. 2014. "High Definition Live 3D-OCT in Vivo: Design and Evaluation of a 4D OCT Engine with 1 GVoxel/s." *Biomedical Optics Express* 5: 2963–77.

Yao, Gang, and Lihong V. Wang. 1999. "Two-Dimensional Depth-Resolved Mueller Matrix Characterization of Biological Tissue by Optical Coherence Tomography." *Optics Letters* 24 (8): 537–39. doi:10.1364/OL.24.000537.

Yasuno, Y., S. Makita, Y. Sutoh, M. Itoh, and T. Yatagai. 2002. "Birefringence Imaging of Human Skin by Polarization-Sensitive Spectral Interferometric Optical Coherence Tomography." *Optics Letters* 27 (20): 1803–805. doi:10.1364/OL.27.001803.

Ye, Li, William E. Allen, Kimberly R. Thompson, Qiyuan Tian, Brian Hsueh, Charu Ramakrishnan, Ai-Chi Wang, et al. 2016. "Wiring and Molecular Features of Prefrontal Ensembles Representing Distinct Experiences." *Cell* 165 (7): 1776–88. doi:10.1016/j.cell.2016.05.010.

Yun, S. H., C. Boudoux, M. C. Pierce, J. F. de Boer, G. J. Tearney, and B. E. Bouma. 2004. "Extended-Cavity Semiconductor Wavelength-Swept Laser for Biomedical Imaging." *IEEE Photonics Technology Letters* 16 (1): 293–95. doi:10.1109/LPT.2003.820096.

Yun, S. H., C. Boudoux, G. J. Tearney, and B. E. Bouma. 2003. "High-Speed Wavelength-Swept Semiconductor Laser with a Polygon-Scanner-Based Wavelength Filter." *Optics Letters* 28 (20): 1981–83. doi:10.1364/OL.28.001981.

Zhang, Anqi, Qinqin Zhang, Yanping Huang, Zhiwei Zhong, and Ruikang K. Wang. 2015. "Multifunctional 1050 Nm Spectral Domain OCT System at 147 kHz for Posterior Eye Imaging." *Sovremennye Tekhnologii* v *Meditsine* 7 (1): 7–12. doi:10.17691/stm2015.7.1.01.

Zilles, Karl, Gottfried Schlaug, Massimo Matelli, Giuseppe Luppino, Axel Schleicher, Meishu Qu, Andreas Dabringhaus, Rudiger Seitz, and Per E. Roland. 1995. "Mapping of Human and Macaque Sensorimotor Areas by Integrating Architectonic, Transmitter Receptor, MRI and PET Data." *Journal of Anatomy* 187 (Pt 3): 515–37.

Zilles, K., A. Schleicher, M. Rath, and A. Bauer. 1988. "Quantitative Receptor Autoradiography in the Human Brain: Methodical Aspects." *Histochemistry* 90 (2): 129–37.

CHAPTER 19

Acousto-Optic Cerebral Monitoring

Michal Balberg and Revital Pery-Shechter

CONTENTS

19.1 INTRODUCTION

19.1.1 Clinical Need

Neurons require a constant and adequate supply of oxygen, glucose, and other metabolites in order to function properly. As the oxygen reserve in the brain is limited, and anaerobic metabolism of neurons is low, adequate perfusion of blood is crucial for the proper function and viability of these cells. A low blood flow rate may result in an ischemic deficit, whereas a high blood flow rate may increase the intracranial pressure (ICP) and disrupt the blood brain barrier (Tzeng and Ainslie, 2014). Consequently, in situations where the blood supply to the brain is altered, for example, when using a perfusion pump in a cardio-vascular surgery or when the patient suffers a traumatic brain injury or a stroke, it is important to continuously monitor the brain's perfusion.

The adequate perfusion of blood is regulated by the healthy brain in order to prevent situations where blood flow is higher or lower than needed for proper function (Czosnyka et al., 2009). However, when this autoregulation mechanism is compromised in the brain of patients suffering from traumatic brain injury, stroke, hypertension, or even under general anesthesia, it is crucial to continuously monitor the variations in cerebral blood flow (CBF) in order to manage and adjust cerebral perfusion. For example, recent clinical research using transcranial Doppler ultrasound (TCD) and optical assessment of cerebral autoregulation during cardio-vascular surgeries (Ono et al., 2012) has demonstrated that patients with impaired autoregulation are more likely than those with functional autoregulation to have perioperative stroke. The authors conclude that non-invasive monitoring of autoregulation may provide an accurate means to predict impaired autoregulation.

Clinical assessment of cerebral autoregulation (CA) has recently been demonstrated to be correlated with, predict, and even affect the outcome of patients suffering from severe brain illnesses (Czosnyka and Miller, 2014) including: sub-arachnoid hemorrhage (SAH) (Budohoski et al., 2012), patients undergoing cardio-vascular surgeries (Ono et al., 2014; Ono et al., 2013), and traumatic brain injury (Sorrentino et al., 2011). NIRS systems that measure cerebral oxygen saturation have been used in some cases (Brady et al., 2010; Joshi et al., 2012), as a surrogate for blood flow monitoring to assess autoregulation.

A clinically useful cerebral perfusion monitor (Dagal and Lam, 2011) should be simple to use, non-invasive, provide continuous, consistent, and reproducible measurement with a high temporal and spatial resolution, and be cost-effective.

Optical measurements, which provide non-invasive, simple to use readings of both spectral and temporal characteristics of the tissue's vasculature, with a high temporal resolution, can provide such necessary clinical information. However, it should be noted that non-invasive assessment of cerebral perfusion and oxygenation should differentiate between the contribution of cerebral and extra-cerebral perfusion to the acquired signals. This is particularly important when assessing autoregulatory function, as cerebral perfusion is regulated, while extra-cerebral perfusion is pressure passive (Kainerstorfer et al., 2015).

19.1.2 Optical Monitoring of Cerebral Perfusion and Oxygen Saturation

Near-infrared light is commonly used to monitor, and even image, hemodynamic, metabolic, and physiological changes within cerebral tissue (Fantini et al., 2016). Spatially resolved spectroscopy (NIRS) (or diffuse optical tomography (DOT)), that can be based on either constant wave (CW), time, or frequency varying sources, is able to track variations in the oxygen saturation of hemoglobin within cerebral tissue, but is also known to be affected by variations of oxygen saturation and blood flow through extra-cerebral, superficial tissue (Ohmae et al., 2006; Davie and Grocott, 2012). Currently, there are several clinically accepted CW NIRS systems,* based on spatially resolved spectroscopy, that are used

* For example: INVOS 5100C by Medtronic Inc., ForeSight Elite by Edwards Lifesciences Inc., SenSmart by Nonin Inc.

to monitor the brain of patients under general anesthesia, in particular, patients undergoing vascular and orthopedic surgeries, despite these limitations (Grocott and Davie, 2013).

Diffuse correlation spectroscopy, which measures CBF using temporal correlation of coherent light signals, can track changes in CBF, but is also affected by extra-cerebral perfusion (Mesquita et al., 2013). Initial attempts to overcome this contamination involve multi-source-detector elements, and measurements of the optical properties of the tissue prior to monitoring (Verdecchia et al., 2016)

Photoacoustic, or optoacoustic tomographic (PAT) imaging of cerebral vasculature was demonstrated to provide high-resolution measurements of hemodynamics and changes in local oxygen saturation in animal models (Yao and Wang, 2014). However, as high frequency ultrasound is highly attenuated by the skull, application of PAT to human subjects is still a challenge (Wang and Yao 2016).

Acousto-optic modulation of coherent, near-infrared light signals within scattering biological tissue enables better localization of the optical path of photons within the tissue, and thus reduces the effect of superficial contamination. Moreover, since coherent light is used for illumination, both spectral and temporal properties of the tissue can be investigated simultaneously, using the same hardware configuration. In this chapter, we describe the use of acousto-optic sensing in clinical applications involving measurements of changes in CBF and local oxygen saturation.

19.2 ACOUSTO-OPTICS IN BIOLOGICAL TISSUE

Light modulation by ultrasonic waves was introduced about 20 years ago (Mahan et al., 1998; Leutz and Maret, 1995) for imaging through turbid biological media. Wang (2001), Li and Wang (2002), and Sfez et al. extended the theory into practical *in vitro* demonstrations (Granot et al., 2001) (Lev and Sfez, 2002; Lev and Sfez, 2003; Granot et al., 2001) and preliminary *in vivo* applications (Lev et al., 2005; Lev and Sfez, 2003). Ramaz and his colleagues (Ramaz et al., 2004; Lesaffre et al., 2009) demonstrated how the signal-to-noise ratio (SNR) of ultrasound modulated light can be improved using photorefractive and other complex detection schemes.

During the last decade, several reviews of the technology and its benefits for imaging through turbid biological tissue were published (Elson et al., 2011; Wang, 2004; Wang and Yao, 2016). Most *in vitro* demonstrations rely on scanning a focused ultrasound beam while illuminating with a coherent light source, and detecting the modulated optical speckle pattern via a CCD camera or a single photodetector. These configurations did not provide the needed SNR that provides the spatio-temporal resolutions that are required for clinical applications.

19.2.1 Acousto-Optics in Turbid Media

Acousto-optic measurements are based on locally modulating light with a localized low power ultrasound beam (Mahan et al., 1998; Wang, 2001). There are several mechanisms through which light can be modulated using ultrasound waves (Elson et al., 2011): first, when an acoustic wave, i.e. a periodic pressure wave, is introduced into a turbid media it locally modulates the density and therefore the optical properties of the media, namely, the local absorption coefficient (μ_a), the reduced scattering coefficient (μs′), and the refractive index (n). Although this modulation does not depend on the temporal coherence of the light, it was shown to be usually too small for practical purposes for low coherence light sources (Wang, 2004)

Second, when coherent light is used to probe tissue, the different photons exiting the tissue travel through different trajectories within the turbid media, and accumulate a myriad of different optical phases. These photons interfere constructively and destructively at the plane of detector, creating time-varying random fluctuations, which are manifested in an interference pattern called a speckle pattern (Boas and Dunn, 2010). The acoustic wave, introduced into the illuminated media, is a moving pressure wave, which creates a periodic displacement of the scattering centers in the media and therefore changes the photonic optical path length and the associated optical phase, thus adding a modulation at the ultrasound frequency to the detected random speckle intensity. This is termed "tagging" of the light. Detecting such a modulation requires coherent light, with a coherence length much larger than the optical path in the medium.

The modulation created by the "tagging" of the light is temporally correlated with the pattern of the ultrasound wave as a function of the delayed time of propagation of the ultrasound wave. In this section, we explain the general concept of acousto-optic tagging, and in the following sections, we formulate the effect and its use for cerebral monitoring.

Figure 19.1 schematically shows the propagation of scattered photons in the presence of an ultrasound pulse and the resulting speckle image (Elson et al., 2011).

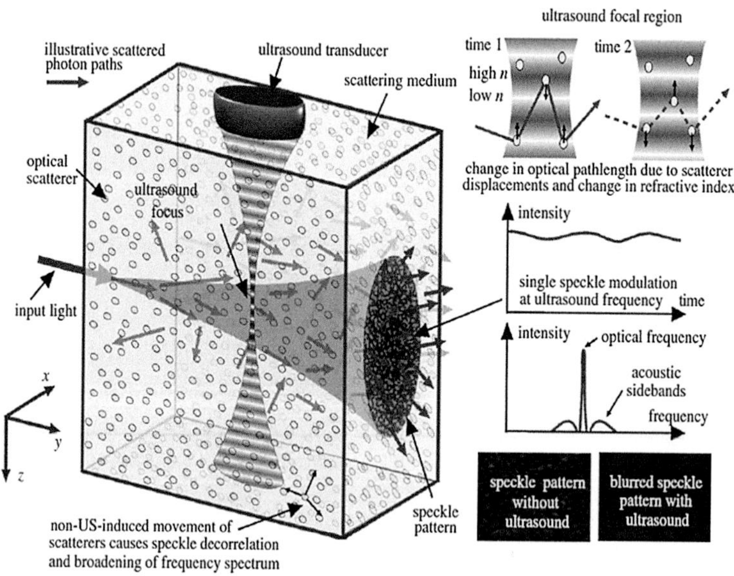

FIGURE 19.1 Schematic showing the propagation of representative highly scattered photons through a biological tissue in the presence of a focused ultrasound beam. The resulting speckle pattern at the output face is illustrated together with the modulation in intensity of a single speckle grain. (From Elson et al., 2011.)

Figure 19.2 shows schematics for an acousto-optic measurement setup. Figure 19.3a shows an example of a short train of ultrasound pulses introduced into a tissue phantom (Racheli et al., 2012). As shown in Figure 19.3b, the detected light (averaged over 10,000 pulses of ultrasound), using a single detector, carries, in addition to random fluctuations in time, which average to zero, a component corresponding to the ultrasound frequency. To extract the light intensity profile as a function of depth, a temporal cross-correlation between the transmitted ultrasound pulse and the speckle intensity fluctuations is performed, resulting in the solid black line shown in Figure 19.3c.

$$AO(\tau) = \left| \sum_{t} I(t)^{*} g(t-\tau) \right|^{2} \tag{19.1}$$

Equation (19.1) describes the acousto-optic signal (AO) as a function of the delay of the acoustic signal (τ), where I(t) is the detected light intensity as a function of time (t). If I(t) is not modulated in correlation with g(t), the AO amplitude will be constant, for each delay (τ). If we assume that only the light scattered by modulated particles, at a distance equal to the

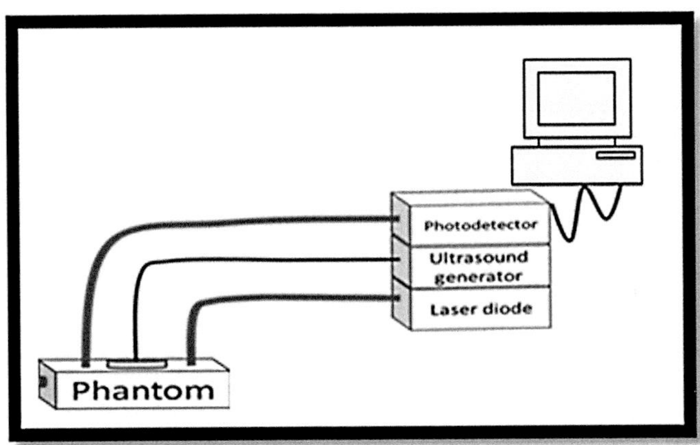

FIGURE 19.2 Illustration of an acousto-optic measurement setup in a reflection geometry.

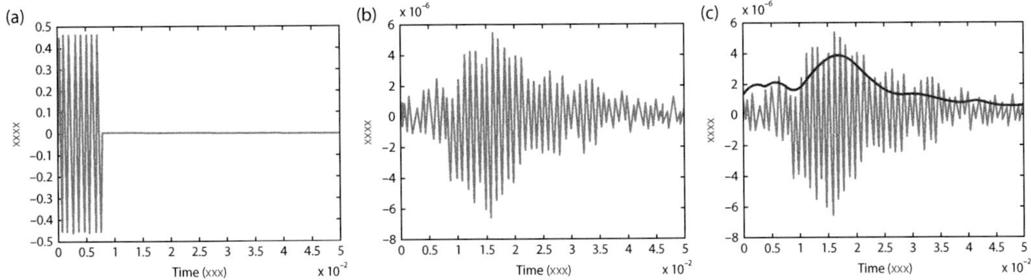

FIGURE 19.3 A demonstration of the AO signal composition (a) The generated ultrasound sequence. (b) The detected light. (c) The cross-correlation curve (black envelope line), calculated as the amplitude of the temporal cross-correlation between the ultrasound sequence in panel **a** and the detected light in panel **b**. (From Racheli et al., 2012.)

product of the speed of sound by τ (i.e. $z = v_s{}^*\tau$) from the ultrasound source, are in phase with the acoustic signal (or at a constant phase relative to it), only these photons will contribute to the AO signal.

Consequently, the amplitude of the AO line corresponds to the light intensity detected at a given point in time, which follows the propagation of the ultrasound pulse train in the phantom.

The induction of acousto-optically modulated light (coined "UTL") within cerebral tissue suffers from the high attenuation of the ultrasound wave that propagates through the skull bone. This results in a greatly reduced SNR. To increase the SNR, a low ultrasound frequency (lower than 1 MHz) should be used, as the attenuation increases with frequency (Pichardo et al., 2011). In addition, a coded series g(t) having the specific characterization of a narrow bandwidth autocorrelation function can be used (Racheli et al., 2012), in order to improve the SNR of the measured cross-correlation described above. Using such a temporally coded series for generating the ultrasound wave, the amplitude of the acousto-optic signal at delay τ is again defined by Eq. (19.1). The resulting cross-correlation is proportional to both the light distribution LD(t) and the acoustic pressure amplitude distribution PA(t). The pressure amplitude within the tissue depends on the amplitude of the generated acoustic signal g(t), the transfer function of the transducer, and the acoustic properties of the tissue. It does not depend on the wavelength of the light propagating through the tissue.

The modulated light intensity I(t) depends on the acoustic pressure amplitude (Leutz and Maret, 1995). Consequently, the amplitude of the measured AO signal as a function of time (or depth), depends on the acoustic pressure amplitude (PA(t)) and the local light distribution LD(t).

An example of an AO curve obtained with the coded series g(t) measured on a synthetic phantom model is presented in Figure 19.4. In this example, the carrier frequency f of the series was 1 MHz and the code bit length was 4 μsec. The code used was a Golay

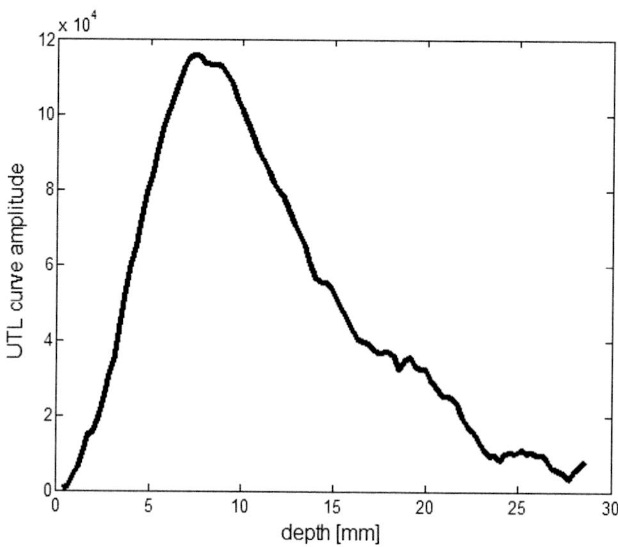

FIGURE 19.4 An example of an AO curve measured on a synthetic phantom model.

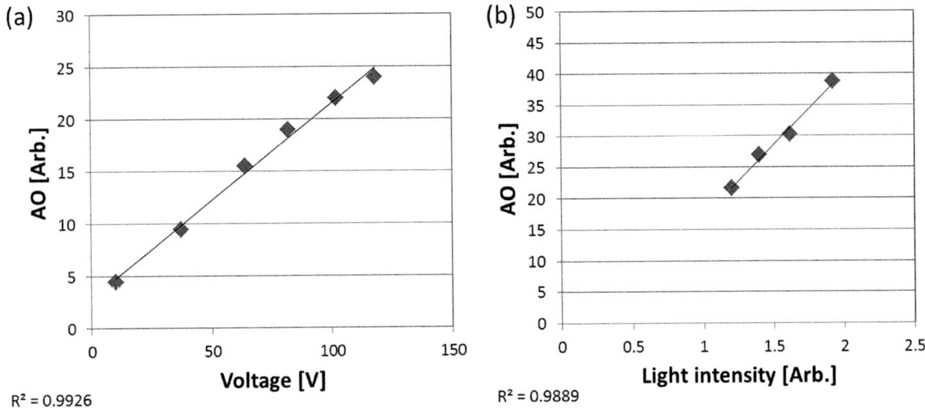

FIGURE 19.5 (a) Detected signal amplitude as a function of applied voltage (linear with acoustic pressure amplitude). (b) Detected signal amplitude as a function of light intensity.

series (Racheli et al., 2012). The total series length was 5 msec. The time units for the cross-correlation results are converted to depth using the known ultrasound speed of sound in the phantom. As discussed above, the detected signal is linearly proportional to the acoustic pressure amplitude, within a certain range (that depends on the acoustic properties of the tissue and the characteristics of the ultrasound transducer). Figure 19.5(a) shows a measurement of the amplitude of the AO signal (determined using Eq. (19.1) above), at a specific depth, as a function of the voltage applied to the transducer. The pressure amplitude generated by the transducer is linearly dependent on the voltage applied to it. The AO signal is linearly proportional to the input light intensity, and depends on the optical properties of the light source and the tissue. Figure 19.5(b) shows a measurement of detected acousto-optic signal (AO) as a function of the input light intensity.

19.2.2 Principles of Acousto-Optics for Physiologic Monitoring

19.2.2.1 Oxygen Saturation

As light propagates through biological tissues, it is scattered and absorbed. The molar extinction coefficients and the concentrations of both absorbing chromophores and scattering centers control the attenuation of light as it propagates a distance (d) in the tissue. The attenuation of light is determined by measuring the ratio between the incident light intensity (I_0) and the output light intensity (I) at a certain distance (L) from the light source. Optical density (OD) is a quantity defined by:

$$OD = -\log_{10}\left(\frac{I}{I_0}\right) \tag{19.2}$$

Under some assumptions (Zourabian et al., 2000) it can be shown that the optical density at wavelength λ (OD^{λ}) depends on the concentrations of the absorbing chromophores (C_i), on their molar extinction coefficient (ε_i), and on a wavelength-dependent path length parameter (B^{λ}) that results from scattering.

$$OD^{\lambda} = \sum_{i} \varepsilon_i C_i LB^{\lambda} + G \tag{19.3}$$

where the index "i" represents the ith chromophore, and G is a geometrical factor.

For blood in the infrared light wavelength range, there are mainly two chromophores that contribute significantly to OD, oxygenated hemoglobin (HbO), and deoxygenated hemoglobin (Hb). Using Eq. (19.3) OD, i.e. OD^{λ} is defined as:

$$OD^{\lambda} = \left\{ \varepsilon_{HBO}^{\lambda} \left[HBO \right] + \varepsilon_{HB}^{\lambda} \left[HB \right] \right\} LB^{\lambda} \tag{19.4}$$

Oxygen saturation is defined by the ratio of concentration of oxygenated hemoglobin [HBO] to the total concentration of hemoglobin [HBT] = [HBO] + [HB]. Therefore

$$S = \frac{[HBO]}{[HBO]+[HB]} = \frac{[HBO]}{[HBT]} = \frac{\Delta[HBO]}{\Delta[HBT]} \tag{19.5}$$

Thus, Eq. (19.4) can be written as: $OD^{\lambda} = \left\{ \varepsilon_{HBO}^{\lambda} S + \varepsilon_{HB}^{\lambda} \left(1-S\right) \right\} \left[HBT \right] LB^{\lambda}$

If R is the ratio between OD^{λ} measured at two different wavelengths, i.e. $R = OD^{\lambda 1}/OD^{\lambda 2}$, the oxygen saturation S, is given by:

$$S = \frac{\varepsilon_{HB}^{\lambda 2} R \left(B^{\lambda 2} / B^{\lambda 1} \right) - \varepsilon_{HB}^{\lambda 1}}{\left(\varepsilon_{HBO}^{\lambda 1} - \varepsilon_{HB}^{\lambda 1} \right) - R \left(B^{\lambda 2} / B^{\lambda 1} \right) \left(\varepsilon_{HBO}^{\lambda 2} - \varepsilon_{HB}^{\lambda 2} \right)} \tag{19.6}$$

The molar extinction coefficients $\varepsilon_{HBO}^{\lambda}$ and $\varepsilon_{HB}^{\lambda}$ are known for each wavelength λ, and B^{λ} can be determined empirically and is shown to depend on L and on the average scattering coefficient of blood. In practice, it is more common to measure the effective absorption coefficient $\mu^{i} = \sum_{j} \varepsilon_{j}^{i} \cdot C_{j}^{i}$ where i stands for a wavelength and j for the different chromophores affecting attenuation, where we assume that the scattering is similar for all wavelengths and can therefore be divided out in Eq. (19.6).

In order to extract the effective optical attenuation by the acousto-optic signal, it is assumed that the medium is diffusive, meaning that the scattering coefficient is much larger than the absorption coefficient, leading to a photon distribution dominated by scattering events (Ron et al., 2013).

As explained in Section 19.2.1, a volume within tissue is irradiated concurrently by a non-periodic coded ultrasound waveform (such as the Golay series mentioned above) and coherent light at wavelength λ_i. The light scattered from the medium is detected by a photodetector with a bandwidth double the central frequency of the ultrasound transducer.

A cross-correlation between the detected light and the transmitted ultrasound series is calculated for varying time delays τ, using Eq. (19.1). The amplitude of the cross-correlation for each time delay τ is denoted as $AO(\tau, \lambda)$.

Figure 19.6 shows the amplitude of the cross-correlation (i.e. $AO(z, \lambda)$) as a function of depth z and wavelength λ for different values of delay τ. The depth (distance from the

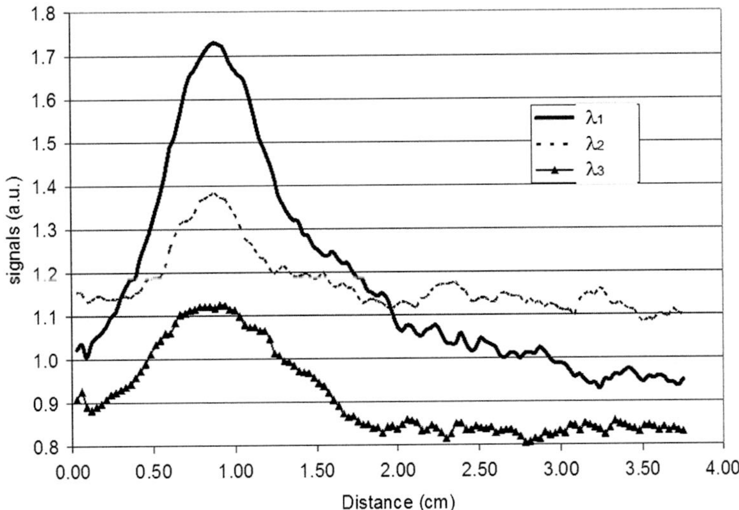

FIGURE 19.6 The amplitude of the cross-correlation C(z) for three different wavelengths, where $z = v_s*\tau$, where v_s is the speed of sound in the medium

transducer face) is calculated as the product of τ and the speed of sound in the medium (c_s). The graphs presented in Figure 19.6 show the cross-correlation amplitude obtained experimentally in a tissue-mimicking phantom at three different wavelengths $\lambda^1, \lambda^2, \lambda,^3$.

As can be seen in Figure 19.6, the amplitude of the calculated cross-correlation at each different wavelength λ^i, denoted as $AO(\tau,\lambda^i)$, is attenuated differently for each wavelength. The difference results from the variations in the optical parameters, such as absorption, scattering, and refractive index for the three wavelengths propagating in the medium.

Due to the acousto-optic effect, the correlation function $AO(\tau,\lambda^i)$ is proportional to the acoustic pressure PA(z), and to the light distribution $LD(\lambda^i)$ (derived as in Tuchin, 2007).

$$LD\left(z,\lambda^i\right) = K^* \prod_{\alpha=s,d} \left(1 + \frac{1}{\mu_e^i \sqrt{\left(\vec{r}-\vec{r}_\alpha\right)^2 + z^2}}\right) \frac{z}{\left(\vec{r}-\vec{r}_\alpha\right)^2 + z^2} \exp\left(-\mu_e^i \sqrt{\left(\vec{r}-\vec{r}_\alpha\right)^2 + z^2}\right) \quad (19.7)$$

where K^* is a constant, $\mu_e^i = \sqrt{3\mu_a^i\left(\mu_a^i + \mu_s^i\right)}$ is the effective decay rate of light in the medium, μ_a^i is the absorption coefficient, and μ_s^i is the scattering coefficient at wavelength λ^i; under the assumption of scattering dominated tissue $\mu_a^i << \mu_s^i$ and thus $\mu_e^i \cong \sqrt{3\mu_a^i\mu_s^i}$, $\vec{r}_\alpha$ is either the vector to the source ($\alpha = s$) or to the detector ($\alpha = d$) and z, the depth, is the distance in the direction parallel to the direction of propagation of the acoustic radiation into the medium.

For a large distance z from the transducer face, i.e. large compared to the medium mean free path and source-detector separation, the light distribution LD (z,λ^i) is simplified and can be written as proportional to $e^{-2\mu_e^i z}$, therefore the intensity of the light reaching the detector under this approximation can be written as $I\left(z,\lambda^i\right) = I_0^i e^{-2\mu_e^i z} + A_o$, where I_0^i is the initial light intensity upon entry into the medium and A_0 is an additive constant.

As explained above, the AO signal, calculated using Eq. (19.1), depends linearly on the pressure amplitude of the acoustic signal (PA($z = v_s{}^*\tau$), where v_s is the speed of sound within the media) and the detected light intensity reaching the detector (I(t)), the correlation is not zero only at the modulated locations, therefore I(t) can be replaced by $I(z,\lambda^i)$ given above.

Consequently,

$$AO\left(z,\lambda^i\right) \propto PA(z) I_0^i e^{-2\mu_e^i z} + C_o, \tag{19.8}$$

where C_0 is an additive constant.

For a known acoustic pressure amplitude distribution PA(z), the light distribution $LD(z,\lambda^i)$ can be extracted. In practice, however, the pressure profile is usually unknown, as it varies from medium to medium and is different for multilayered media having different acoustic impedances. This requires additional independent measurements of the cross-correlation to eliminate PA(z) and allow extraction of light distribution ratios (after eliminating the constant C_o). To achieve independent measurements, N different measurements (at least two), at N light wavelengths can be performed, assuming that the acoustic distribution is independent of the wavelength of light. When the measured $AO(z,\lambda^i)$ is divided by measured $AO(z,\lambda^j)$ for $i \neq j$, the resulting ratio of the light distributions at different wavelengths can be used for determining the oxygen saturation of blood in tissue.

In the case of a medium irradiated by three different wavelengths:

$$\frac{\tilde{I}^i}{\tilde{I}^j} = \frac{I_0^i}{I_0^j} e^{-2\Delta\mu_e^{ij} z} \tag{19.9}$$

where i,j = 1; 2; 3, ($i \neq j$) represents the three lasers, $\tilde{I}^i = \left(AO\left(z,\lambda^i\right) - C_o\right)$ is the amplitude of the signal at distance z, I_0^i, I_0^j are the input intensities of the ith and j^{th} wavelengths respectively, and $\Delta\mu_e^{ij} = \mu_e^i - \mu_e^j$.

The constant C_o is the added signal that is unrelated to the acousto-optic process and corresponds to the noise level of the system at the measured frequency bandwidth, it can be measured during periods without ultrasound modulation. Taking a logarithm of the equation above, $\Delta\mu_e^{ij}$ can be obtained

$$\Delta\mu_e^{ij} = -\frac{1}{2}\frac{\partial}{\partial z} \ln\left[\frac{I^i}{I^j}\right] \tag{19.10}$$

From Eq. (19.10) and Eqs (19.2) to (19.5) the tissue local saturation can be calculated.

19.2.2.2 *Blood Flow Measurements*

The movement of scattering particles causes phase shifts in the scattered coherent light and results in changes in the random speckle pattern, producing temporal fluctuations in the speckle pattern (Boas and Dunn, 2010). These movements can originate from several sources: one of the sources of this movement is Brownian motion due to temperature, which adds a random noise to the speckle pattern. A second source is the ultrasound wave itself, which adds a periodic temporal component to the speckle pattern at the ultrasound frequency. Movement of blood cells, i.e. blood flow, causes temporal fluctuations of the scattering centers that are not correlated with the generated ultrasound pattern. This adds additional frequency components to the speckle pattern that are related to the velocity of the moving blood cells via the Doppler Effect and cause a decrease in the temporal correlation between the trajectories of the photons that reach the detector, leading to a decrease in speckle contrast (Boas and Yodh, 1997). When analyzing the signal in the frequency domain, the spectrum obtained from the temporal fluctuations of the speckles is influenced by the entire photon path. At each scattering site along this path, the photons acquire a certain amount of phase change due to movement of the scattering centers. When blood cells are moving, the peak amplitude of the power spectrum of the light signal at the ultrasound frequency, $P(\Omega_{US})$, decreases and the second moment (or width) of the spectrum, $\left\langle (\Omega - \Omega_{US})^2 \right\rangle$ increases. The overall spectrum is a result of many such events that change the phase along the optical path. As shown in Figure 19.7 when blood flow (movement of blood cells) increases, the speckles' spectral component at the ultrasound frequency decreases, and the width of the spectral peak at the ultrasound frequency increases. A linear relationship between the amplitude of the AO signal and the flow rate has been determined and observed (Tsalach et al., 2016), for example, as shown in Figure 19.8. In the temporal domain, the AO curve resulting from cross-correlating the spatially averaged speckle pattern (over the area of a single detector) and the generated ultrasound is expected to decrease as flow increases, as shown in Figure 19.9.

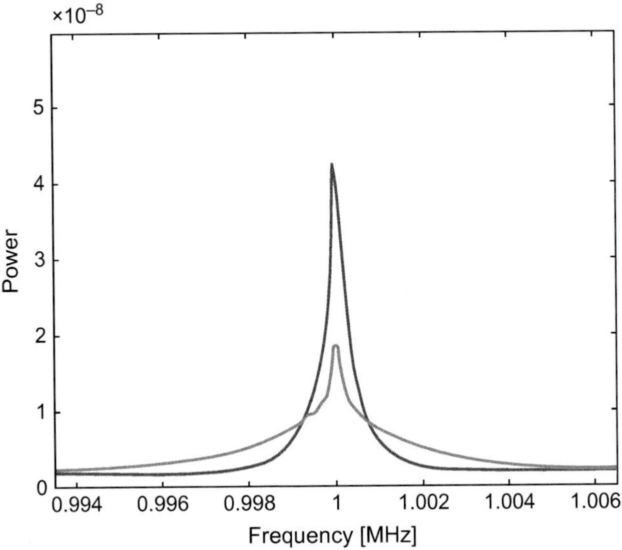

FIGURE 19.7 The power spectra obtained from acousto-optically modulated light with a 1MHz CW ultrasound wave (US) on a flow phantom model. The blue line is the power spectrum obtained for a low velocity and the green line is the power spectrum obtained from a high velocity experiment.

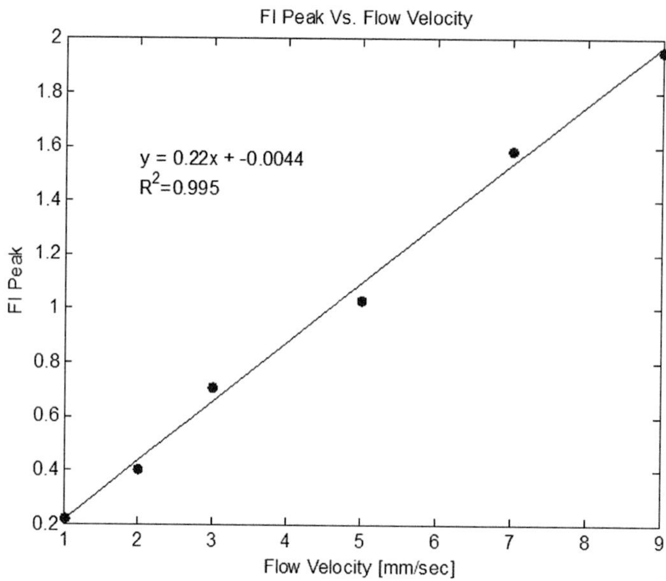

FIGURE 19.8 Flow index (FI) values, derived from the spectral broadening, obtained around the flow layer, for different magnitudes of flow velocity. A linear relation is apparent, with $R^2 = 0.995$. (From Tsalach et al., 2016.)

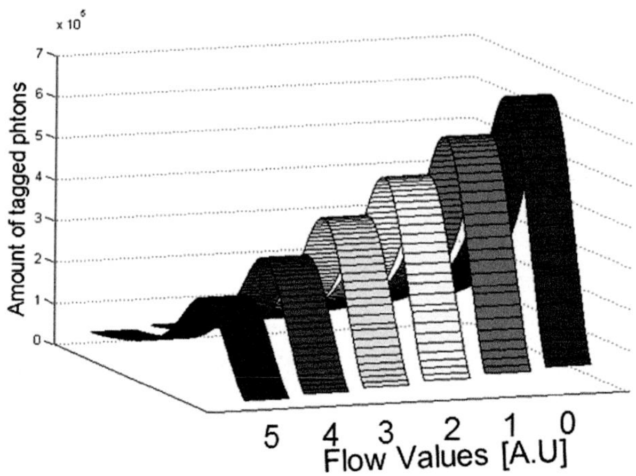

FIGURE 19.9 Illustrated AO curves as a function of flow magnitude. The amplitude decreases as flow increases.

Using either temporal or frequency analysis of the degree of local correlation over the random fluctuations, acousto-optics is able to measure changes in flow in the microcirculation in deeper and larger (about 0.1–0.5 cm³) tissue volumes compared to laser Doppler-based technologies (Ron et al., 2012; Racheli et al., 2012).

Monte Carlo simulations of acousto-optic modulated light (Wang, 2001; Sakadžić and Wang, 2002; Powell and Leung, 2012) enable demonstration of the modulation of the phase of the light by the acoustic field and its effect on the acquired signal. These papers examined

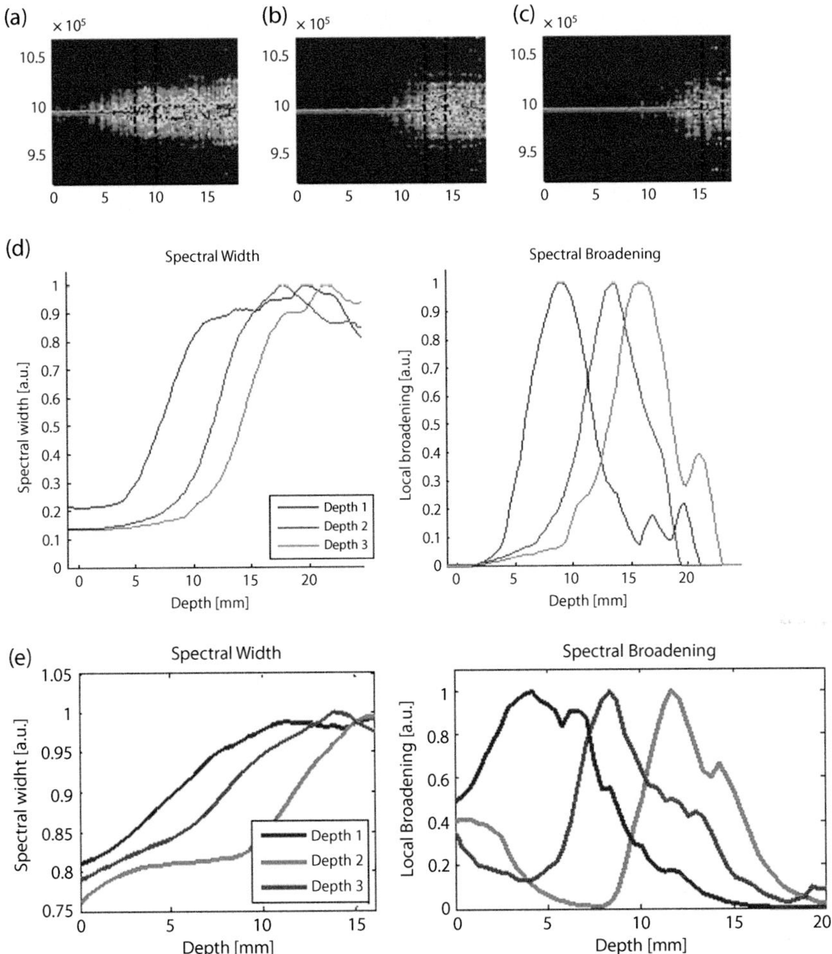

FIGURE 19.10 (A–C) Simulation results for three different depths of the 1 mm wide flow layers. Spectrograms (A), spectral width (B), and spectral broadening (C) curves exhibit apparent depth discrimination ability. (D–E) Experimental results obtained for flow in three channels at three different depths. Spectral width (D) and local broadening (E) both indicate a distinct depth discrimination. (From Tsalach et al., 2016.)

the amplitude of the modulation of the acquired signal by artificially separating the modulated and non-modulated signals, where the ultrasound beam was either homogenous or perpendicular to the direction of light. These simulations did not take into account coherent effects within the turbid media that are affected by movement. Tsalach et al. (2015) simulated a more realistic configuration where an ultrasound pulse propagated in a direction parallel to the direction of light illumination. The total signal that reaches a detector, in a reflection-based light detection, is simulated, while summing both the number of photons and the phases associated with their paths, generating a time-dependent heterodyne signal sensed by a single detector. The spectral width (defined by dividing the area under the curve of the power spectrum (PS) of the optical signal by the amplitude of the PS at the ultrasound frequency), and the differential spectral broadening (defined as the gradient of the spectral width along the direction of the ultrasound propagation) is shown to linearly correspond to the flow velocity within the channel (Tsalach et al., 2016). The simulation was compared to *in vitro* sensing of variable flow in millimeter-diameter flow channels at different depths. The work demonstrates the ability to selectively monitor changes in deep (>1 cm deep) channels, and the ability to separate the contribution from flow channels at different depths.

Figure 19.10 shows a comparison between simulation and *in vitro* acousto-optic measurements in channels at three different depths.

19.3 CLINICAL APPLICATIONS OF ACOUSTO-OPTIC CEREBRAL MONITORING

19.3.1 Acousto Optic Measurements of Cerebral Oxygen Saturation in Traumatic Brain Injured Patients

The use of non-invasive, acousto-optic monitoring of cerebral oxygen saturation was recently demonstrated in clinical settings (Rosenthal et al., 2014). The study was conducted using an acousto-optic monitor* that emits coherent, near-infrared light, at three wavelengths between 780–830 nm, through an optical fiber, and collects the back scattered light through an optical fiber to an avalanche photodiode. The acousto-optic modulation is conducted by an ultrasound transducer applied to the skin, with a central frequency around 1 MHz that is able to penetrate the skull with a relatively low attenuation. By analyzing the cross-correlation between the detected light signal and the generated, coded ultrasound wave, the system is able to measure a "virtual cross-section" of the light distribution (banana) between a source and a detecting optical fiber (Ron et al., 2013), as explained above. Analysis of the spatial decay of these signals enables extraction of the effective attenuation coefficient of the tissue, at selective depths, and from them, extraction of the absorption of different wavelengths by oxygenated and deoxygenated hemoglobin, and determination of the regional oxygen saturation of blood in the microvasculature of the brain, beneath the sensor. The correlation between the oxygen saturation values obtained by the non-invasive acousto-optic signals and invasive measurements by a catheter placed in the jugular vein bulb of 18 traumatic brain injured patients, was high ($r = 0.6$) and significant ($p<0.001$) for the non-invasive measurement obtained from the same hemisphere, where the invasive catheter was placed. This demonstrates the hemispheric localization of the acousto-optic measurement within cerebral tissue at least 1 cm deep below the skin and suggests the applicability of using the acousto-optic signal to measure cerebral oxygen saturation in TBI patients at risk of cerebral ischemia.

19.3.2 Acousto-Optic Cerebral Blood Flow Monitoring

Cerebral blood flow is not continuously monitored in routine clinical management of patients suffering from TBI, although it is recommended by the Brain Trauma Foundation's guidelines (Rossi and Castioni, 2008). This is mainly because there are currently no

* Cerox 3110 by Ornim Medical Ltd, Israel.

clinically validated protocols and devices for simple, easy to use, continuous monitoring of CBF available (Dagal and Lam, 2011).

A new acousto-optic CBF monitor* enables continuous monitoring of changes in microvascular blood flow in cerebral tissue. It was first validated on animal models, and its readings of relative changes in CBF (termed CFI) were shown to correlate significantly with invasive local readings of CBF index using laser Doppler (Ron et al., 2012) in controlled manipulations.

The acousto-optic monitor was further validated against quantitative regional-CBF readings of ^{133}Xe-SPECT of healthy volunteers. A good correlation between the acousto-optic CFI index measured underneath the sensor, and regional CBF measurement using ^{133}Xe-SPECT was demonstrated following administration of acetazolamide, a drug known to affect cerebral blood flow (Schytz et al., 2012). Previous measurements by the same clinical group, with CW NIRS with ICG as contrast agent, during similar manipulations, could not demonstrate a significant correlation between the NIRS signal and the SPECT measurement (Schytz et al., 2009).

Clinical measurements of acousto-optic CFI during induction of anesthesia and intubation of patients undergoing different surgeries (Schwarz et al., 2015) showed a similar hemodynamic response to the one previously reported while measuring with transcranial Doppler ultrasound (TCD) in large vessels (Kofke et al., 1994).

Further clinical validation of the acousto-optic CFI measurements is needed before the clinical community will accept this parameter as a valid clinical tool. Its non-invasive nature, ease of use, and simple interpretation render this modality a likely candidate for acceptance.

19.3.3 Acousto-Optic Cerebral Autoregulation Monitoring

Autoregulation monitors that are based on transcranial Doppler (TCD) ultrasound signals are available,† but are difficult to operate and are not widely used. A clinically useful monitor of autoregulation function should combine simple and direct measurement of CBF (and not a surrogate) with concurrent measurements of MAP, ICP, or both. The acousto-optic CFI measurement, combined with simultaneous measurement of MAP (either invasive or non-invasive) was recently recognized as a promising candidate (Brady et al. 2015).

Recently, an autoregulation acousto-optic monitor was presented‡ and validated in an animal experiment, where the blood pressure (MAP) was manipulated using different drugs. The acousto-optic CFI was measured concurrently with MAP, and a correlation index (ARI) between changes in CFI and MAP was calculated over the preceding 5 minutes for each measurement point. ARI, representing the maximal correlation during the preceding 5 minutes, was displayed in values between 0 and 100. An ARI close to zero (0) represented no correlation, and therefore active autoregulation, whereas an ARI close to 100 represents a maximal correlation, resulting from passive autoregulation. In order to

* C-Flow, by Ornim Medical, Ltd.
† Vittamed 505, by UAB Vittamed, Lithuania
‡ C-Flow-AR by Ornim Medical, Ltd.

reduce errors during null changes in MAP, ARI was calculated only during periods where a significant change in MAP was identified based on a Relevance Vector Machine (RVM) (Tipping, 2001) linear classifier based on the trend slope and its derivatives. This monitor awaits clinical validation.

In addition, post-analysis of variations in acousto-optic CFI relative to changes in MAP that occur during cardio-vascular surgeries demonstrated that CFI provides similar information about autoregulation phases and thresholds as blood flow velocity measurements by TCD (Murkin et al., 2015). However, while TCD is affected by operation of electro-cautery devices, the acousto-optic signals are not, which makes it possible to use the acousto-optic CFI signal during the surgical procedure itself.

A study of 64 patients undergoing cardio-pulmonary bypass (CPB) surgery (Hori et al., 2015) also demonstrated a significant correlation between the autoregulation indices calculated using TCD and CFI. Hori et. al. also reported a study on 110 patients that were monitored with CFI, and post-analysis of the autoregulation indices revealed that excursions below optimal blood pressure (defined as the lower limit of autoregulation) during perioperative periods are associated with cardiac surgery-associated acute kidney injury (Hori et al., 2016).

19.4 CONCLUSIONS AND FUTURE RESEARCH

Acousto-optic imaging was first introduced by Mahan and the team at GE about 20 years ago (Mahan et al., 1998) with a goal of developing a high-resolution optical mammography system. Since these initial efforts, acousto-optic imaging through biological tissue has been considered a less favorable modality compared to photoacoustics, primarily due to the low SNR associated with the modulated light reaching the detector. However, using coded ultrasound waves, and relatively low (less than 1 MHz) ultrasound frequencies, it was demonstrated that acousto-optic signals can reach cortical vasculature and obtain information regarding regional oxygen saturation, cerebral blood flow modulation, and autoregulation function in clinical and pre-clinical settings. Additional clinical validation is needed in order to establish this modality as a clinically viable, long-sought solution in the management of patients in critical care, under surgery, and even at the emergency room.

Functional brain imaging using NIRS (fNIRS) has been a long-standing goal for the neuroscience research community, as it allows simpler and more natural monitoring of hemodynamic changes associated with neuronal function than fMRI. Recently, fNIRS has been successfully used to communicate with locked-in patients (Chaudhary et al., 2017), in cases where electrical signals alone could not enable reliable communication. Integrating acousto-optic hemodynamic sensing into fNIRS systems can enable localized, simultaneous measurements of cerebral oxygen saturation and changes in CBF, extending the benefits of fNIRS from monitoring oxygen saturation only to blood flow and metabolic rates (i.e. $CMRO_2$). Such integration will require increasing the SNR of acousto-optic sensing to allow multi-location scanning at a rate of at least 10 Hz, compatible with current fNIRS systems. Such a system will allow implementation of existing (Fantini, 2014) hemodynamic models, which require simultaneous, high frequency measurements of both cerebral oxygenated and deoxygenated hemoglobin along with CBF. Such implementation may expand the use of NIRS based systems into clinical care in the coming years.

ACKNOWLEDGMENTS AND DISCLOSURE

MB and RS founded Ornim Medical Ltd, and hold shares in the company. MB acknowledges the support of the United States–Israel Binational Science Foundation (BSF grant 2015193).

REFERENCES

Boas, David A., and Yodh, Arjun G. 1997. "Spatially varying dynamical properties of turbid media probed with diffusing temporal light correlation." *JOSA A* 14 (1): 192–215.

Boas, David A., and Andrew K. Dunn. 2010. "Laser Speckle Contrast Imaging in Biomedical Optics." *Journal of Biomedical Optics* 15 (1): 11109. doi:10.1117/1.3285504.

Brady, Ken, Dean B. Andropoulos, Kathleen Kibler, and R. Blaine Easley. 2015. "A New Monitor of Pressure Autoregulation." *Anesthesia and Analgesia* 121 (5): 1121–23. doi:10.1213/ANE.0000000000000952.

Brady, K., B. Joshi, C. Zweifel, P. Smielewski, M. Czosnyka, R. B. Easley, and C. W. Hogue. 2010. "Real-Time Continuous Monitoring of Cerebral Blood Flow Autoregulation Using Near-Infrared Spectroscopy in Patients Undergoing Cardiopulmonary Bypass." *Stroke* 41 (9): 1951–56. doi:10.1161/STROKEAHA.109.575159.

Budohoski, Karol P., Marek Czosnyka, Peter Smielewski, Magdalena Kasprowicz, Adel Helmy, Diederik Bulters, John D. Pickard, and Peter J. Kirkpatrick. 2012. "Impairment of Cerebral Autoregulation Predicts Delayed Cerebral Ischemia After Subarachnoid Hemorrhage: A Prospective Observational Study." *Stroke* 43 (12): 3230–37. doi:10.1161/STROKEAHA.112.669788.

Chaudhary, Ujwal, Bin Xia, Stefano Silvoni, Leonardo G. Cohen, and Niels Birbaumer. 2017. "Brain–Computer Interface-Based Communication in the Completely Locked-In State." 1–25. doi:10.1371/journal.pbio.1002593.

Czosnyka, Marek, Ken Brady, Matthias Reinhard, Piotr Smielewski, and Luzius A. Steiner. 2009. "Monitoring of Cerebrovascular Autoregulation: Facts, Myths, and Missing Links." *Neurocritical Care* 10 (3): 373–86. doi:10.1007/s12028-008-9175-7.

Czosnyka, Marek, and Chad Miller. 2014. "Monitoring of Cerebral Autoregulation." *Neurocritical Care* 21 (2) Supplement 2: 95–102. doi:10.1007/s12028-014-0046-0.

Dagal, Armagan, and Arthur M. Lam. 2011. "Cerebral Blood Flow and the Injured Brain: How Should We Monitor and Manipulate It?" *Current Opinion in Anaesthesiology* 24 (2): 131–37. doi:10.1097/ACO.0b013e3283445898.

Davie, Sophie N., and Hilary P. Grocott. 2012. "Impact of Extracranial Contamination on Regional Cerebral Oxygen Saturation." *Anesthesiology* 116 (4): 834–40. doi:10.1097/ALN.0b013e31824c00d7.

Elson, D. S., R. Li, C. Dunsby, R. Eckersley, and M.-X. Tang. 2011. "Ultrasound-Mediated Optical Tomography: A Review of Current Methods." *Interface Focus* 1 (4): 632–48. doi:10.1098/rsfs.2011.0021.

Fantini, Sergio. 2014. "Dynamic Model for the Tissue Concentration and Oxygen Saturation of Hemoglobin in Relation to Blood Volume, Flow Velocity, and Oxygen Consumption: Implications for Functional Neuroimaging and Coherent Hemodynamics Spectroscopy (CHS)." *NeuroImage* 85: 202–21. doi:10.1016/j.neuroimage.2013.03.065.

Fantini, Sergio, Angelo Sassaroli, Kristen T. Tgavalekos, and Joshua Kornbluth. 2016. "Cerebral Blood Flow and Autoregulation: Current Measurement Techniques and Prospects for Noninvasive Optical Methods." *Neurophotonics* 3 (3): 31411. doi:10.1117/1.NPh.3.3.031411.

Granot, E., A. Lev, Z. Kotler, B. G. Sfez, and H. Taitelbaum. 2001. "Detection of Inhomogeneities with Ultrasound Tagging of Light." *Journal of the Optical Society of America a-Optics Image Science and Vision* 18 (8): 1962–67. doi:10.1364/JOSAA.18.001962.

Grocott, Hilary P., and Sophie N. Davie. 2013. "Future Uncertainties in the Development of Clinical Cerebral Oximetry." *Frontiers in Physiology* 4 (December): 1–4. doi:10.3389/fphys.2013.00360.

Hori, Daijiro, Charles Hogue, Hideo Adachi, Laura Max, Joel Price, Christopher Sciortino, Kenton Zehr, John Conte, Duke Cameron, and Kaushik Mandal. 2016. "Perioperative Optimal Blood Pressure as Determined by Ultrasound Tagged near Infrared Spectroscopy and Its Association with Postoperative Acute Kidney Injury in Cardiac Surgery Patients." *Interactive Cardiovascular and Thoracic Surgery* 22 (4): 445–51. doi:10.1093/icvts/ivv371.

Hori, Daijiro, Charles W. Hogue, Ashish Shah, Charles Brown, Karin J. Neufeld, John V. Conte, Joel Price, et al. 2015. "Cerebral Autoregulation Monitoring with Ultrasound-Tagged Near-Infrared Spectroscopy in Cardiac Surgery Patients." *Anesthesia and Analgesia* 121 (5): 1187–93. doi:10.1213/ANE.0000000000000930.

Joshi, Brijen, Masahiro Ono, Charles Brown, Kenneth Brady, R. Blaine Easley, Gayane Yenokyan, Rebecca F. Gottesman, and Charles W. Hogue. 2012. "Predicting the Limits of Cerebral Autoregulation During Cardiopulmonary Bypass." *Anesthesia and Analgesia* 114 (3): 503–10. doi:10.1213/ANE.0b013e31823d292a.

Kainerstorfer, Jana M., Angelo Sassaroli, Kristen T. Tgavalekos, and Sergio Fantini. 2015. "Cerebral Autoregulation in the Microvasculature Measured with Near-Infrared Spectroscopy." *Journal of Cerebral Blood Flow and Metabolism* 35 (6): 959–66. doi:10.1038/jcbfm.2015.5.

Kofke, W. A., M. L. Dong, M. Bloom, R. Policare, J. Janosky, and L. Sekhar. 1994. "Transcranial Doppler Ultrasonography with Induction of Anesthesia for Neurosurgery." *Journal of Neurosurgical Anesthesiology* 6 (2): 89–97.

Lesaffre, M., S. Farahi, M. Gross, P. Delaye, C. Boccara, and F. Ramaz. 2009. "Acousto-Optical Coherence Tomography Using Random Phase Jumps on Ultrasound and Light." *Optics Express* 17 (20): 18211–18. doi:10.1364/OE.17.018211.

Leutz, W., and G. Maret. 1995. "Ultrasonic Modulation of Multiply Scattered Light." *Physica B: Condensed Matter* 204 (1): 14–19. doi:10.1016/0921-4526(94)00238-Q.

Lev, A., E. Rubanov, B. Sfez, S. Shany, and A. J. Foldes. 2005. "Ultrasound-Modulated Light Tomography Assessment of Osteoporosis." *Optics Letters* 30 (13): 1692–94. doi:10.1364/OL.30.001692.

Lev, A., and B. G. Sfez. 2002. "Direct, Noninvasive Detection of Photon Density in Turbid Media." *Optics Letters* 27 (7): 473–75. doi:10.1364/OL.27.000473.

Lev, Aner, and Bruno Sfez. 2003. "In Vivo Demonstration of the Ultrasound-Modulated Light Technique." *Journal of the Optical Society of America. A, Optics, Image Science, and Vision* 20 (12): 2347–54. doi:10.1364/JOSAA.20.002347.

Lev, A., and B. G. Sfez. 2003. "Pulsed Ultrasound-Modulated Light Tomography." *Optics Letters* 28 (17): 1549–51. doi:10.1364/OL.28.001549.

Li, Hui, and Lihong V. Wang. 2002. "Autocorrelation of Scattered Laser Light for Ultrasound-Modulated Optical Tomography in Dense Turbid Media." *Applied Optics* 41 (22): 4739–42. doi:10.1364/AO.41.004739.

Mahan, G. D., W. E. Engler, J. J. Tiemann, and E. Uzgiris. 1998. "Ultrasonic Tagging of Light: Theory." *Proceedings of the National Academy of Sciences of the United States of America* 95 (November): 14015–19. doi:10.1073/pnas.95.24.14015.

Mesquita, Rickson C., Steven S. Schenkel, David L. Minkoff, Xiangping Lu, Christopher G. Favilla, Patrick M. Vora, David R. Busch, et al. 2013. "Influence of Probe Pressure on the Diffuse Correlation Spectroscopy Blood Flow Signal: Extra-Cerebral Contributions." *Biomedical Optics Express* 4 (7): 978–94. doi:10.1364/BOE.4.000978.

Murkin, John M., Moshe Kamar, Zmira Silman, Michal Balberg, and Sandra J. Adams. 2015. "Intraoperative Cerebral Autoregulation Assessment Using Ultrasound-Tagged Near-Infrared-Based Cerebral Blood Flow in Comparison to Transcranial Doppler Cerebral Flow Velocity: A Pilot Study." *Journal of Cardiothoracic and Vascular Anesthesia* 29 (5): 1187–93. doi:10.1053/j.jvca.2015.05.201.

Ohmae, Etsuko, Yasuomi Ouchi, Motoki Oda, Toshihiko Suzuki, Shuji Nobesawa, Toshihiko Kanno, Etsuji Yoshikawa, et al. 2006. "Cerebral Hemodynamics Evaluation by Near-Infrared Time-Resolved Spectroscopy: Correlation with Simultaneous Positron Emission Tomography Measurements." *NeuroImage* 29 (3): 697–705. doi:10.1016/j.neuroimage.2005.08.008.

Ono, M., G. J. Arnaoutakis, D. M. Fine, K. Brady, R. B. Easley, Y. Zheng, C. Brown, N. M. Katz, M. E. Grams, and C. W. Hogue. 2013. "Blood Pressure Excursions Below the Cerebral Autoregulation Threshold During Cardiac Surgery Are Associated with Acute Kidney Injury." *Critical Care Medicine* 41 (2): 464–71. doi:10.1097/CCM.0b013e31826ab3a1.

Ono, Masahiro, Kenneth Brady, R. Blaine Easley, Charles Brown, Michael Kraut, Rebecca F. Gottesman, and Charles W. Hogue. 2014. "Duration and Magnitude of Blood Pressure Below Cerebral Autoregulation Threshold During Cardiopulmonary Bypass Is Associated with Major Morbidity and Operative Mortality." *The Journal of Thoracic and Cardiovascular Surgery* 147 (1): 483–89. doi:10.1016/j.jtcvs.2013.07.069.

Ono, M., B. Joshi, K. Brady, R. B. Easley, Y. Zheng, C. Brown, W. Baumgartner, and C. W. Hogue. 2012. "Risks for Impaired Cerebral Autoregulation During Cardiopulmonary Bypass and Postoperative Stroke." *British Journal of Anaesthesia* 109 (3): 391–98. doi:10.1093/bja/aes148.

Pichardo, Samuel, Vivian W. Sin, and Kullervo Hynynen. 2011. "Multi-Frequency Characterization of the Speed of Sound and Attenuation Coefficient for Longitudinal Transmission of Freshly Excised Human Skulls." *Physics in Medicine and Biology* 56 (1): 219–50. doi:10.1088/0031-9155/56/1/014.

Powell, Samuel, and Terence S. Leung. 2012. "Highly Parallel Monte-Carlo Simulations of the Acousto-Optic Effect in Heterogeneous Turbid Media." *Journal of Biomedical Optics* 17 (4): 45002. doi:10.1117/1.JBO.17.4.045002.

Racheli, N., A. Ron, Y. Metzger, I. Breskin, G. Enden, M. Balberg, and R. Shechter. 2012. "Non-Invasive Blood Flow Measurements Using Ultrasound Modulated Diffused Light." *SPIE BiOS* 8223: 82232A–8. doi:10.1117/12.906342.

Ramaz, F., B. C. Forget, M. Atlan, A. C. Boccara, M. Gross, P. Delaye, and G. Roosen. 2004. "Photorefractive Detection of Tagged Photons in Ultrasound Modulated Optical Tomography of Thick Biological Tissues." *Optics Express* 12 (22): 5469–74. doi:10.1364/OPEX.12.005469.

Ron, A., N. Racheli, I. Breskin, Y. Metzger, Z. Silman, M. Kamar, A. Nini, R. Shechter, and M. Balberg. 2012. "Measuring Tissue Blood Flow Using Ultrasound Modulated Diffused Light." *SPIE* 8223: 82232J–7.

Ron, Avihai, Noam Racheli, Ilan Breskin, and Revital Shechter. 2013. "A Tissue Mimicking Phantom Model for Applications Combining Light and Ultrasound." In *Proc. SPIE 8583, Design and Performance Validation of Phantoms Used in Conjunction with Optical Measurement of Tissue V*, edited by Robert J. Nordstrom, 858307. SPIE. doi:10.1117/12.2003528.

Rosenthal, Guy, Alex Furmanov, Eyal Itshayek, Yigal Shoshan, and Vineeta Singh. 2014. "Assessment of a Noninvasive Cerebral Oxygenation Monitor in Patients with Severe Traumatic Brain Injury." *Journal of Neurosurgery* 120 (4): 901–907. doi:10.3171/2013.12.JNS131089.

Rossi, S., and C. A. Castioni. 2008. "Guidelines for the Management of Severe Traumatic Brain Injury: Still Needed?" *Minerva Anestesiologica* 74: 579–81. doi:10.1089/neu.2007.9997.

Sakadžić, Sava, and Lihong V. Wang. 2002. "Ultrasonic Modulation of Multiply Scattered Coherent Light: An Analytical Model for Anisotropically Scattering Media." *Physical Review E - Statistical, Nonlinear, and Soft Matter Physics* 66 (2): 14–19. doi:10.1103/PhysRevE.66.026603.

Schwarz, Marlon, Giovanni Rivera, Mary Hammond, Zmira Silman, Kirk Jackson, and W. Andrew Kofke. 2015. "Acousto-Optic Cerebral Blood Flow Monitoring During Induction of Anesthesia in Humans." *Neurocritical Care* 436–41. doi:10.1007/s12028-015-0201-2.

Schytz, Henrik W., Song Guo, Lars T. Jensen, Moshe Kamar, Asaph Nini, Daryl R. Gress, and Messoud Ashina. 2012. "A New Technology for Detecting Cerebral Blood Flow: A Comparative Study of Ultrasound Tagged NIRS and 133Xe-SPECT." *Neurocritical Care* 17 (1): 139–45. doi:10.1007/s12028-012-9720-2.

Schytz, H. W., T. Wienecke, L. T. Jensen, J. Selb, D. A. Boas, and M. Ashina. 2009. "Changes in Cerebral Blood Flow After Acetazolamide: An Experimental Study Comparing Near-Infrared Spectroscopy and SPECT." *European Journal of Neurology* 16 (4): 461–67. doi:10.1111/j.1468-1331.2008.02398.x.

Sorrentino, Enrico, Karol P. Budohoski, Magdalena Kasprowicz, Peter Smielewski, Basil Matta, John D. Pickard, and Marek Czosnyka. 2011. "Critical Thresholds for Transcranial Doppler Indices of Cerebral Autoregulation in Traumatic Brain Injury." *Neurocritical Care* 14 (2): 188–93. doi:10.1007/s12028-010-9492-5.

Tipping, Michael E. 2001. "Sparse Bayesian Learning and the Relevance Vector Machine." *Journal of Machine Learning Research* 1: 211–44. doi:10.1162/15324430152748236.

Tsalach, Adi, Zeev Schiffer, Eliahu Ratner, Ilan Breskin, Reuven Zeitak, Revital Shechter, and Michal Balberg. 2015. "Depth Selective Acousto-Optic Flow Measurement." *Biomedical Optics Express* 6 (12): 4871. doi:10.1364/BOE.6.004871.

Tsalach, Adi, Eliahu Ratner, Stas Lokshin, Zmira Silman, Ilan Breskin, Nahum Budin, and Moshe Kamar. 2016. "Cerebral autoregulation real-time monitoring." *PloS one* 11 (8): e0161907. doi: 10.1371/journal.pone.0161907.

Tsalach, A., Z. Schiffer, E. Ratner, I. Breskin, R. Zeitak, R. Shechter, and M. Balberg. 2016. "Depth Discrimination in Acousto-Optic Cerebral Blood Flow Measurement Simulation." *Progress in Biomedical Optics and Imaging - Proceedings of SPIE* 9708 (October). doi:10.1117/12.2211587.

Tuchin, Valery V. 2007. *Tissue Optics: Light Scattering Methods and Instruments for Medical Diagnosis*. SPIE Press: Bellingham, WA. doi:10.1117/3.684093.

Tzeng, Yu Chieh, and Philip N. Ainslie. 2014. "Blood Pressure Regulation IX: Cerebral Autoregulation Under Blood Pressure Challenges." *European Journal of Applied Physiology* 114 (3): 545–59. doi:10.1007/s00421-013-2667-y.

Verdecchia, Kyle, Mamadou Diop, Albert Lee, Laura B. Morrison, Ting-Yim Lee, and Keith St. Lawrence. 2016. "Assessment of a Multi-Layered Diffuse Correlation Spectroscopy Method for Monitoring Cerebral Blood Flow in Adults." *Biomedical Optics Express* 7 (9): 3659. doi:10.1364/BOE.7.003659.

Wang, Lihong V. 2001. "Mechanisms of Ultrasonic Modulation of Multiply Scattered Coherent Light: An Analytic Model." *Physical Review Letters* 87 (4): 43903. doi:10.1103/PhysRevLett.87.043903.

Wang, L. V. 2001. "Mechanisms of Ultrasonic Modulation of Multiply Scattered Coherent Light: A Monte Carlo Model." *Optics Letters* 26 (15): 1191–93. doi:10.1364/OL.26.001191.

Wang, Lihong V. 2004. "Ultrasound-Mediated Biophotonic Imaging: A Review of Acousto-Optical Tomography and Photo-Acoustic Tomography." *Disease Markers* 19 (2–3): 123–38. doi:10.1182/blood-2002-12-3791.

Wang, Lihong V., and Junjie Yao. 2016. "A Practical Guide to Photoacoustic Tomography in the Life Sciences." *Nature Methods* 13 (8): 627–38. doi:10.1038/nmeth.3925.

Yao, Junjie, and Lihong V. Wang. 2014. "Photoacoustic Brain Imaging: From Microscopic to Macroscopic Scales." *Neurophotonics* 1 (1): 11003. doi:10.1117/1.NPh.1.1.011003.

Zourabian, A., A. Siegel, B. Chance, N. Ramanujan, M. Rode, and D. A. Boas. 2000. "Trans-Abdominal Monitoring of Fetal Arterial Blood Oxygenation Using Pulse Oximetry." *Journal of Biomedical Optics* 5 (4): 391–405. doi:10.1117/1.1289359.

CHAPTER 20

Neurophotonic Vision Restoration

Adi Schejter Bar-Noam and Shy Shoham

CONTENTS

20.1 INTRODUCTION

By affecting the most prominent source of sensory information in humans, blindness inflicts a disability with profound personal and societal implications. Some of the most common causes of blindness are degenerative diseases of the outer retina, like Age-related Macular Degeneration (AMD) and Retinitis Pigmentosa (RP). Each year, upwards of 700,000 new patients are diagnosed with AMD, while a total of about 1.5 million individuals are affected by RP worldwide (Margalit et al., 2003).

Diseases of the outer retina result in the loss of photoreceptors, a population of nerve cells situated at the back of the retina that are specialized in detecting photons and converting them to chemical signals. Photoreceptors naturally convey signals to bipolar cells, amacrine cells, and horizontal cells, which in turn create contacts with the retinal ganglion cells (RGCs) that project to the thalamus. The relay of information between the retinal layers is not a simple signal cascade, but a signal transduction accompanied by complex neuronal processing (Aharoni, 2017). Despite the loss of photoreceptors in diseases of the outer retina, the adjacent inner retinal neurons and in particular the retinal ganglion cells and their optic nerve projections are largely maintained as functional. Artificial stimulation of these relatively well-preserved nerve cells, which forms a sort of an "information bypass", is one of the main approaches being pursued towards vision restoration. A *retinal neuroprosthesis* is a medical device aimed at providing an artificial sense of vision by translating visual scenes into appropriate spatio-temporal patterns of retinal neuronal activity. Current retinal prostheses largely rely on microelectrode array implants (Weiland and Humayun 2014); however, although early devices in this category are already being used to aid blind human subjects (recently gaining FDA and CE regulatory approval), it appears that their ultimate resolution may be severely limited by current spread. For example, recent clinical studies report best-case acuities of ~20/1200 for the 60 electrode epi-retinal (Humayun et al., 2012) and ~20/550 for 1,500 subretinal (Zrenner et al., 2011) systems. Other concerns with this general approach include long-term interface stability between electrode arrays and the fragile retina tissue and risks associated with extended surgery.

Photonic approaches are a natural choice in considering the challenges of scaling up to interfaces with thousands of channels in an attempt to approach highly functional vision. Photonic devices which rely on the introduction of retinal implants include photodiode arrays activated by pulsed infrared light (Mathieson et al., 2012), which are already nearing clinical application, or polymer-based photovoltaic implants (Ghezzi et al. 2013). Alternatively, approaches based on *direct light activation* for artificially controlling neural activity in a vision prosthesis (Figure 20.1) are expected to have some inherent biocompatibility advantages due to being implant-less; and since they do not rely on extracellular currents, they may potentially allow cellular-resolution stimulation. Neurophotonic interfacing could in principle allow rapid, massively parallel, light-efficient stimulation of retinal cells across macroscopic coverage areas, and could potentially be applied using one of multiple relevant strategies for *directly* exciting neurons with light (Callaway and Yuste, 2002). Photochemical methods which have been considered for this application

Essential elements:

Capturing device (video camera)

Processing unit:
- Image processing (retinal emulation)
- Design of projected pattern/image

Near-eye projection unit

Light-sensitive probe

FIGURE 20.1 The concept of a retinal prosthesis using direct light activation depicted in an illustration of the future optical-based prosthesis: camera video stream is processed and fed to a projection system, and excitation patterns are projected onto the photo-sensitized retina.

include uncaging of neurotransmitters such as glutamate (Shoham et al., 2005); "photopharmacology" where small light-sensitive molecules (photoswitches) render the membrane channels sensitive (Klapper et al., 2016); and the genetic expression of light-sensitive ion channels/pumps (Boyden et al., 2005; Yizhar et al., 2011). A photothermal method for stimulating neurons, known as infrared neural stimulation (INS) (Wells et al., 2005), eliminates the need for introducing exogenous components. This method relies on the absorption of light by water in order to generate heat transients. However, in order to minimize the intrinsic absorption of light in the water content of the eye prior to the retina, an alternative photothermal method has been proposed, which relies on the illumination of scattered photo-absorbers in the vicinity of retinal neurons (termed PAINTS (Farah et al., 2013)) using visible light. This photo-thermal approach is more suited for retinal stimulation, since the eye possesses the inherent property of relaying visible wavelengths to the retina.

The chapter will largely focus on the utilization of optogenetics as a means of achieving high spatiotemporal control of neuronal activity in the retina for vision restoration. Neurophotonic vision restoration and optogenetics are strongly coupled: this was the first medical application recognized and pursued in the context of optogenetics, and it has apparently motivated some of the earliest experiments examining the use of the algal opsin Channelrhodopsin-2 (ChR2) (Nagel et al., 2003) as an optogenetic probe. The toolbox of optogenetic probes available for excitation or inhibition of neuronal populations is already quite substantial and still rapidly growing, offering a large range of kinetic and spectral sensitivity characteristics (Yizhar et al., 2011). A number of studies by several research groups have explored the fundamental feasibility of using optogenetic probes in an optical retinal prosthesis, clearly suggesting that this technology may provide a viable path to vision restoration (Roska and Pepperberg 2014; Chuang et al., 2014; Barrett et al., 2014; Pan et al., 2015; Klapper et al., 2016).

20.2 OPTOGENETIC INTERFACES FOR VISION RESTORATION

Despite its many advantages, there are a number of major challenges for applying optogenetics for vision restoration which are actively being addressed by numerous studies in the field. The success of optogenetic strategies relies crucially on the specific selection of the optogenetic probe(s), method of transfection, and the choice of which cells to selectively target. These topics will be discussed in the following sections. In addition, optogenetic control for vision restoration relies on intense, patterned illumination, therefore requiring sophisticated illumination methods which must be carefully engineered.

20.2.1 Retinal Targets for Cell-Type Specific Stimulation

A central consideration for vision restoration is the retinal layer that is being targeted. Analogously to electrical prostheses, two main approaches are also employed with optogenetics: stimulation of the inner nuclear layer (Lagali et al., 2008; Doroudchi et al., 2011; van Wyk et al., 2015; Mace et al., 2015; Gaub et al., 2015; Cehajic-Kapetanovic et al., 2015) and stimulation of the ganglion cell layer (Bi et al., 2006; Nirenberg and Pandarinath, 2012; Reutsky-Gefen et al., 2013).

In patients suffering from RP the rods degenerate initially, followed by later cone degeneration which results in the presence of remnant cone bodies that are missing the light-sensitive outer segment. Since photoreceptors naturally hyperpolarize upon activation, one possible approach would be to express light-activated chloride pumps such as halorhodopsin (NpHR) in the surviving cone bodies (Figure 20.2a) (Busskamp et al. 2010). This solution has the advantage of exploiting the natural retinal circuitry; however, it is very limited due to the fact that the remnant cone bodies appear only in the fovea at later stages (Milam et al., 1998). Moreover, it has yet to be determined to what degree retinal remodeling occurs during retinal degeneration and how the re-activation of photoreceptors would delay such remodeling.

Alternatively, it is possible to transduce the bipolar cells which reportedly survive at much later stages of degenerative diseases of the outer retina. Studies in postmortem eyes showed that 78% of inner nuclear layer cells survived even in patients with severe RP (Santos et al., 1997), whereas in the late stage of the dry form of AMD (known as geographic atrophy (GA)), this layer remains relatively preserved (Kim et al., 2002). Despite the advantage of targeting cells responsible for the initial processing of the visualized image, some information may be lost; and as is the case for photoreceptors, retinal remodeling in late stages of

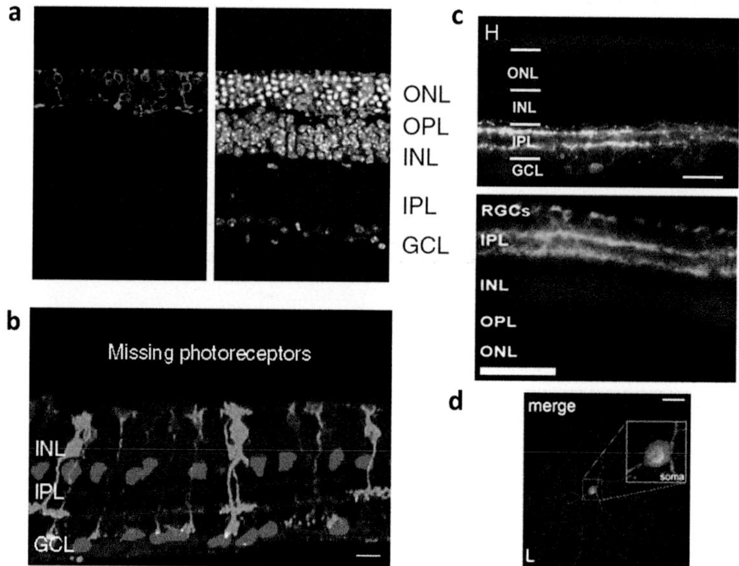

FIGURE 20.2 Optogenetic transduction of different layers in the retina. a) GFP-immunostained retinas from s-RD mice (left), and co-stained with DAPI (right) demonstrating expression of a light-sensitive chloride pump (mCAR-eNpHR-EYFP) in the photoreceptor layer. (From Busskamp et al., 2010.) b) Confocal image of electroporated rd1 mouse retina in which ChR2-EYFP was targeted to ON bipolar cells. (From Lagali et al., 2008.) c) Top – Fluorescence image of Chop2-GFP expression in RGC layer 12 months after injection of viral vectors, and bottom – frozen section of 8-week-old rd1/ChR2-eYFP transgenic mouse retina expressing ChR2 (green) in the RGC layer and DAPI (blue). (From Bi et al., 2006 and Reutsky-Gefen et al., 2013 respectively.) d) ChR2 (red) and eNpHR (green) differential transgene expression in RGC dendrites and soma, respectively. (From Greenberg et al., 2011.)

the diseases may prevent the use of bipolar cells as a target. Many studies have focused on stably transducing bipolar cells (Figure 20.2b) (Lagali et al., 2008; Doroudchi et al., 2011; van Wyk et al., 2015; Mace et al., 2015; Gaub et al., 2015) and have successfully measured responses to non-specific light flashes.

Finally, an additional attractive approach is to target the retinal ganglion cell layer (Bi et al., 2006; Zhang et al., 2009; Greenberg et al., 2011) (Figure 20.2c). These cells are also known to have a high survival rate in late-stage outer retinal degeneration diseases (30% in patients with severe RP (Santos et al., 1997) and 70% in GA-stage AMD patients (Kim et al., 2002)), and since they directly project to the thalamus, they provide an important alternative at advanced stages of retinal diseases when severe retinal remodeling hinders the ability to restore vision via stimulation of cells in the inner nuclear layer. The challenge of targeting this layer, however, is that it requires pre-processing of the projected image, as these cells naturally receive input following retinal processing. These cells are typically divided into ON, OFF, or ON–OFF cells according to the type of response to projected light onto their center-surround receptive field (Hubel and Wiesel, 1962, 1959, 1968). Moreover, there are over 20 different ganglion cell subtypes, each encoding distinct features of the visual scene (Huberman and Niell, 2011). Therefore, this raises the question whether it is necessary to individually target different cell types, or even subtypes, according to their natural encoding characteristics, or whether it would suffice to simply mimic natural responses in a majority of the RGCs.

A number of studies have demonstrated the potential of optogenetics as a powerful tool for engineering natural neuronal responses in retinal ganglion cells. For example, it is possible to transfect retinal neurons with light-sensitive channels or pumps that depolarize or hyperpolarize the cell in order to mimic ON and OFF RGC populations (Zhang et al., 2009). A separate study implemented sophisticated optogenetic tools to recreate antagonistic center-surround receptive fields by introducing ChR2 and NpHR into the soma and dendrites of RGCs (Greenberg et al., 2011) (Figure 20.2d).

However, is it necessary to transfect retinal ganglion cell subtypes, or cellular compartments for the visual cortex to interpret the projected image? An alternative view relies on targeting all RGC types with a single optogenetic probe together with the use of a sophisticated algorithm, which computes the processing carried out by the retina and enables to project light patterns that elicit activity in RGCs that mimic their activity in healthy retinas (see Section 20.3) (Farah et al., 2007; Reutsky-Gefen et al., 2013; Nirenberg and Pandarinath, 2012).

20.2.2 Optogenetic Probes for Vision Restoration

Currently, most studies rely on ChR2 as an optogenetic probe (Figure 20.3a–b). This seven-transmembrane-helix protein (Kato et al., 2012) is a proton channel that passively conducts cations following illumination of the light sensitive chromophore, all-trans-retinal, which is covalently linked to the protein. The most common method for transducing retinal cells with genes encoding for optogenetic proteins is by injecting Adeno-Associated Viruses (AAVs) into the subretinal or intravitreal space. These viruses are considered safe and can also be targeted to various cell types by using different serotypes (Surace and Auricchio, 2003; Rabinowitz et al., 2002).

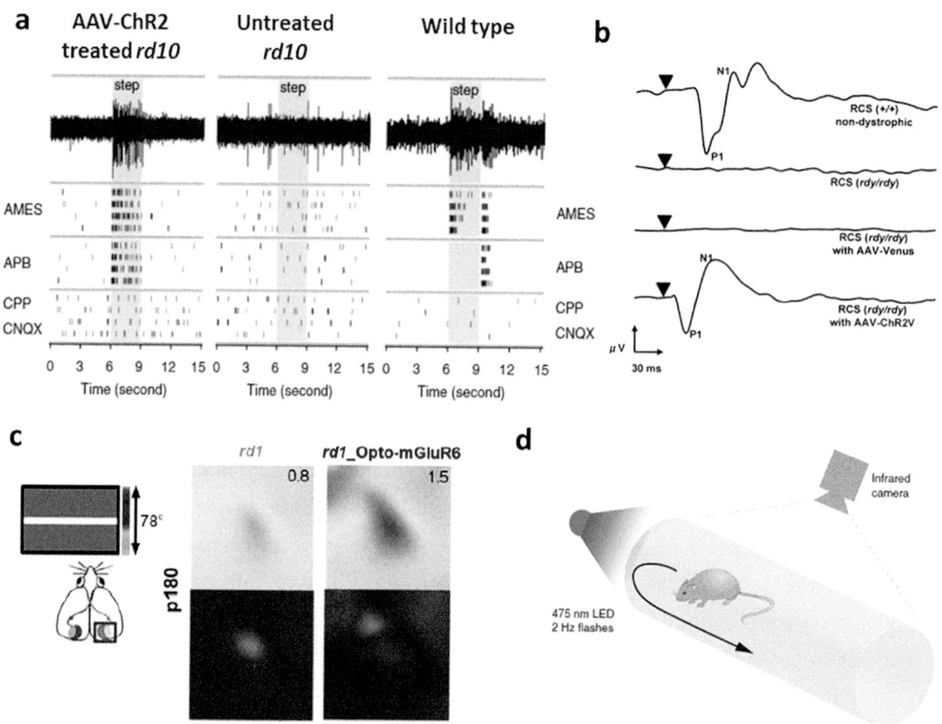

FIGURE 20.3 Demonstration of functional responses of optogenetic probes. a) An example of electrical recordings from isolated retinas. Blind retinas expressing ChR2 in the bipolar cells were able to mediate an optogenetically induced response to 3 second light flashes. In *in vivo* studies, optogenetic-induced responses were obtained using b) visual evoked potentials, c) intrinsic optical imaging, and d) behavioral assays, such as enabling the animal to navigate inside a maze towards a single light source. (Subfigures were adapted from Doroudchi et al., 2011 (a), Tomita et al., 2009 (b), van Wyk et al., 2015 (c) and Mace et al., 2015 (d).).

Although ChR2 is a common, robust light-sensitive channel used in many optogenetic studies, there is a strong motivation to seek alternative probes to be used in a retinal prosthesis. This is due to ChR2's requirement of intense light (1–10 mW mm-2) in the blue spectrum (peak ~460 nm) which can be harmful for biological tissues, and its tendency to be strongly desensitized over long periods of illumination.

One possibility is to create ChR variants through genetic mutation to obtain light-sensitive channels with improved sensitivities or reduced desensitization, such as ChIEF (Nirenberg and Pandarinath, 2012) and ChETA (Gunaydin et al., 2010). However, a major limitation in this domain is the difficulty in decoupling the tradeoff between the probe's sensitivity and its temporal resolution (Yizhar et al., 2011). One promising approach to resolving this issue is to increase the probe's Ca^{2+} permeability in order to promote the activation of voltage-sensitive Na^{+} channels, and thereby increase the sensitivity to light without effecting the channel's kinetics (Kleinlogel et al., 2011). For example, CatCh is ~70 times more sensitive to light than the WT ChR, but it has similar activation and deactivation time constants.

An additional motivation is to create a variant that is sensitive to wavelengths in the red or near-infrared spectrum. This is very beneficial in the human retina where the macula consists of a yellow pigment that absorbs excess blue and UV light, resulting in strong attenuation of the light introduced for stimulation of ChR2, requiring even higher (and potentially damaging) stimulation powers. This is exacerbated by the fact that the safety threshold for blue light is higher due to the photochemical damage that occurs at these wavelengths. Currently, there are a number of red-shifted probes which are activated by wavelengths as far as 640 nm, and the most popular are C1V1 (Yizhar et al., 2011; Lin et al., 2013), ReaChR (Lin et al., 2013) and Chrimson (Klapoetke et al., 2014). Sengputa et al. have demonstrated the ability to obtain responses in the mouse, macaque, and human retinas when expressing ReaChR in RGCs (Sengupta et al., 2016). Moreover, they showed the ability to restore responses to light in blind (*rd1*) mice by recording from the cortex and performing behavioral assays. An additional recent study tested the idea of creating a "white opsin" that is sensitive to a broad range of the visible spectrum with improved sensitivity (Batabyal et al., 2015). The "white opsin" is a single probe which combines three separate opsins: ChR2, C1V1, and ReaChR.

Recently, several research groups presented novel low-light optogenetic probes which were expressed in ON-bipolar cells (van Wyk et al., 2015; Gaub et al., 2015; Cehajic-Kapetanovic et al., 2015) (Figure 20.3c–d). All three probes relied on a light-sensing outer segment native to the retina and an intracellular G protein-coupled receptors (GPCR) protein which have inherent signal amplification properties. van Wyk et al. (2015) presented an interesting solution, Opto-mGluR6, which was shown to mediate responses in isolated blind retinas and in the visual cortex of intact mice (Figure 20.3c). Opto-mGluR6 is a chimera between two innate mouse proteins, consisting of the intracellular domains of the ON-bipolar cells receptor mGluR6, and the light-sensing domains of melanopsin, a photopigment naturally found in intrinsically photosensitive retinal ganglion cells (ipRGCs). The combined features of mGluR6 and elanopsin (peak sensitivity in mice is 467 nm) enable RGCs in Opto-mGluR6-expressing retinas to have a dynamic range that covers 2.5 log units which is equivalent to wild type retinas, and have light sensitives that are over three orders of magnitude in comparison with ChR2. However, a major disadvantage of this chimera is the slow kinetics of melanopsin, making Opto-mGluR6 difficult to use for navigation in dynamic surroundings. The following two studies relied on ectopic expression of rhodopsin, a native opsin GPCR which originated from the rat (Gaub et al., 2015) or human retina (Cehajic-Kapetanovic et al., 2015), and demonstrated optogenetic-induced activity in the isolated retina and visual cortex at light levels similar to van Wyk et al. (Figure 20.3d). One major drawback of this method is that rhodopsin requires 11-*cis* retinal for phototransduction, which is naturally obtained through chromophore recycling. Since the bipolar cells and RGCs are distant from the retinal pigment epithelium (RPE), the question remains whether the rhodopsin-based probe, which is prone to bleaching, will be able to function over long periods of time.

20.3 NEUROPHOTONIC VISION RESTORATION DEVICES: LIGHT PROJECTION AND COMPUTATION

An optical retinal prosthesis should be able to drive cellular activity of multiple neurons simultaneously with a high spatio-temporal resolution. Moreover, efficient optical stimulation of neurons in the retina requires the ability to project intense two-dimensional (or possibly three-dimensional) light patterns, with spectral and intensity levels which cannot be obtained from ambient light.

A number of systems have been utilized to stimulate neural populations, such as acousto-optic deflectors used for rapid scanning (Shoham et al., 2005), and digital micromirror devices (DMDs) for parallel illumination (Farah et al., 2007; Greenberg et al., 2011). However, the former method lacks the ability to excite a large number of cells simultaneously, and the latter method is highly inefficient in terms of power utilization.

Systems based on computer-generated holography (CGH), on the other hand, incorporate all the above requirements, and can be used to project rapidly changing 3D patterns for stimulation of neural populations (Golan et al., 2009a; Nikolenko et al., 2008; Nikolenko et al., 2010). Since phase-only modulation divides the power between the different spots at the focal point, theoretically no power is lost. By utilizing a liquid crystal, phase-only spatial light modulator (SLM), it is possible to control neuronal activity in neuronal cultures (Paluch-Siegler et al., 2015) and in retinal explants (Reutsky-Gefen et al., 2013) expressing optogenetic probes, at the resolution of a single cell (Figure 20.4).

As described previously (Section 20.2.1), by modeling the image processing carried out in the retina (Rodieck, 1965) (Figure 20.5a), it is possible to create an encoder that will

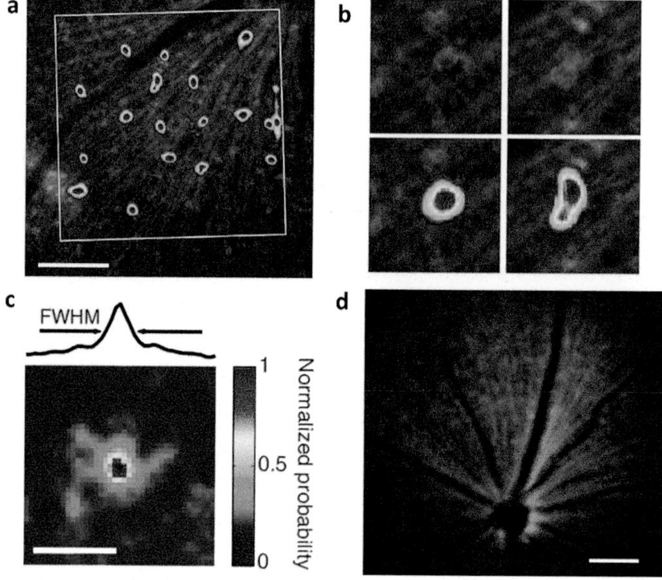

FIGURE 20.4 Single-cell-resolution control of ChR2 expressing RGCs. a) Superposed images of a retina expressing ChR2-eYFP in the RGCs flattened on an MEA (black dots) and representative stimulation fields. Scale bar = 200 μm. b) Blowup of two regions from (a). The stimulation fields (bottom panel) match the underlying visualized RGCs (top panel). c) Mean spatial distribution of the stimulation fields calculated for 202 units from 11 retinas. Scale bar = 50 μm. d) Fluorescent fundus image of a retina expressing ChR2-eYFP in the RGCs *in vivo* with an overlay of a holographic pattern projection (blue spots) on the retina. Scale bar = 250 μm. (Figure adapted from Reutsky-Gefen et al., 2013.)

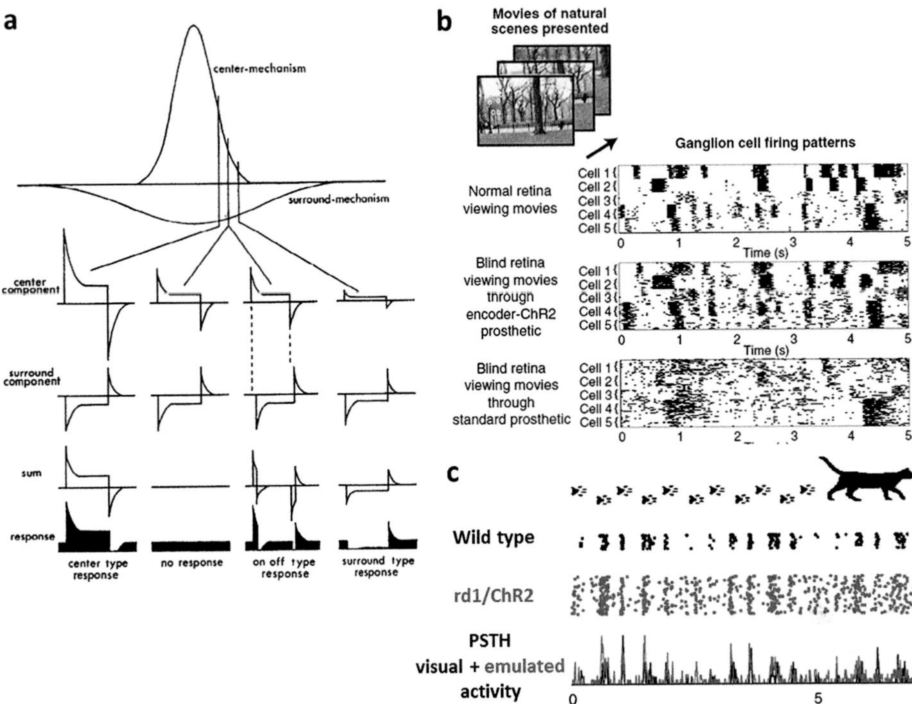

FIGURE 20.5 Visual information encoding. a) A model of the superposition carried out in RGCs when light falls on the center or surround of the receptive field, derived from Rodieck and Stone (Rodieck, 1965). b) A demonstration of a sophisticated encoding device that is capable of activating optogenetically transduced RGCs (ON and OFF) in isolated blind mouse retinas in the same manner as RGCs in wild type retinas responding to natural movie scenes. (Taken from Nirenberg and Pandarinath, 2012.) b) Using advanced optical projection techniques, such as computer generated holography, it is possible to induce activity in multiple RGCs expressing ChR2 in blind (*rd1*/ChR2) retinas at the level of a single cell and emulate RGC activity in wild type retinas. (Adapted from Reutsky-Gefen et al., 2013.)

generate a code (i.e. light pattern) for efficiently stimulating optogenetic-expressing retinal cells for conveying the correct information to the visual cortex. The ability to achieve high-resolution vision restoration using such an encoder largely relies on light-projection technologies capable of stimulating the retina with cellular resolutions. This guided two studies in the field of optogenetic vision restoration. Nirenberg and colleagues used an encoder to process images, before projecting patterned light onto a retina expressing solely ChR2 in both ON and OFF RGCs, using a mini-digital light projector (mini-DLP) (Nirenberg and Pandarinath, 2012) (Figure 20.5b). This group showed that combining their encoder with light projection at high spatio-temporal resolutions elicited responses from thousands of RGCs in blind retinas that closely match those of healthy retinas without the need for selective targeting of cell subtypes. In a separate study, normal activity of RGCs in isolated blinded retinas (identical to the former study) was emulated using a novel light projection technology for projecting light onto the retina (Reutsky et al., 2007; Reutsky-Gefen et al., 2013) (Figure 20.5c). This was carried out by first projecting holographic patterns onto sighted retinas, and then generating emulated patterns based on the RGCs' spatio-temporal activity that were projected onto pharmacologically blinded retinas.

To achieve a true optical retinal prosthesis, such projection technologies must be incorporated into head-mounted displays (HMDs), such as goggles (Figure 20.1). The general design of the goggles should include a camera for capturing images of the world, a display technology for projecting patterned light, and a processing unit for calculating the desired projected pattern, depending on the targeted retinal layer and optogenetic probe being used (Mathieson et al., 2012). When designing such goggles, many optical and safety aspects must be taken into consideration: the FOV size, the resolution of the pattern, and the required illumination power (Golan et al., 2009b; Goetz et al., 2013). Furthermore, the optics and inherent aberrations of the human eye have an effect on the resolution of the projected pattern that reaches the retina, which adds to the challenge of the design. In addition, the goggles will most likely include an eye-tracking device in order to track the position of the transfected cells in the retina and alter the direction of the light pattern accordingly.

20.4 CONCLUSIONS AND PERSPECTIVE

In summary, optogenetics provides many advantages over electronic solutions for vision restoration. To date, many studies demonstrated encouraging results obtained in isolated retinas and in downstream areas such as the visual cortex, in response to non-specific light flashes (Bi et al., 2006; Lagali et al., 2008) (Figure 20.3). However, further testing *in vivo* to complex stimuli is necessary to determine the extent to which the visual cortex is able to decipher the input from the optogenetically engineered retina.

20.4.1 Clinical Translation

Human trials sponsored by RetroSense Therapeutics and GenSight Biologics are currently underway. In early 2016, fifteen RP patients taking part in the RetroSense study were injected with a virus in order to express ChR2 in their RGCs in order to test safety. In 2018, GenSight began a Phase I/II study in which they intend to inject 18 RP patients with an adeno-associated virus (AAV) virus containing the red-shifted ChrimsonR gene in order to evaluate the safety of this gene expression and repeated light stimulation using DMD-based goggles. The objective of these clinical trials is to test safety of the method; however, we may gain insights into the actual potential of optogenetics as a means of restoring vision in humans.

There are still many challenges to overcome on the path towards clinical translation, among them safety issues of injecting viruses and foreign DNA to human subjects and their immunogenic responses. The virus used for transfecting human cells, adeno-associated virus (AAV), has already been used in multiple human trials and is considered safe (Simonelli et al., 2010), and has received FDA approval for gene therapy. However, it is yet to be established whether the optogenetic probe that is expressed in the human cell will be stable over time and whether it will cause damage to the cells expressing it. Moreover, the optogenetic probe, or combination of probes, determines the wavelengths required for artificially stimulating the retina. This has important consequences on the safety threshold, and may therefore have a major role in determining which probe will eventually be used for a human retinal prosthetic.

Furthermore, the method of delivery and targeting of specific retinal layers, especially in humans, has not been finalized. Unlike rodents in which most studies have been carried out, human retinas consist of an inner limiting membrane (ILM) which limits the penetration of AAV. It is most likely that this membrane will need to be removed, or a virus capable of crossing it will be required (Dalkara et al., 2009). In order to target optogenetic expression to the RGCs or bipolar cells, specific promotors must be used. There has been a great advancement in determining these promoters in animal models (Dalkara et al., 2013; Ivanova et al., 2010; Doroudchi et al., 2011); however, the efficacy of transfection needs to be tested in humans, as well as determining the confinement of transfection to the retina alone.

20.4.2 Future Neurophotonic HMDs

The very recent emergence of low-light vision restoration strategies, shown to enable an artificial visual stimulation in natural light conditions (van Wyk et al., 2015; Gaub et al., 2015; Cehajic-Kapetanovic et al., 2015) (Figure 20.3c–d), merits a re-examination of the need for sophisticated projection systems; nevertheless, there are arguably several good reasons why such a device will be required as part of a comprehensive vision restoration solution. First, the suggested methods for low-light vision restoration rely on genetic modification of bipolar cells. As the disease progresses, retinal remodeling may alter retinal processing, making it possible to achieve vision restoration only through RGC stimulation, and requiring the projection of intense, processed images onto the retina. Second, the opsins used in these methods (rhodopsin and melanopsin) are not sensitive to the entire visible spectrum (Kojima et al. 2011), meaning that not all features of the visual field will be viewed without some sort of spectral remapping. Third, results from the optomotor experiments of van Wyk et al. indicate that the use of low-level intensity is not sufficient for full vision restoration (van Wyk et al., 2015). This suggests that light amplification might still be required in many environments, and definitely at night-time. Therefore, light-projection systems mounted on goggles may be used as an accessory device, to enhance vision in patients treated with low-light optogenetic tools, by improving contrast and sensitivity to all visible wavelengths.

REFERENCES

Aharoni, T. 2017. Computational Platform for Portable Holographic Vision Restoration Neural Interfaces (Technion).

Barrett, J. M., R. Berlinguer-Palmini, and P. Degenaar. 2014. 'Optogenetic approaches to retinal prosthesis', *Vis Neurosci*, 31: 345–54.

Batabyal, S., G. Cervenka, D. Birch, Y. T. Kim, and S. Mohanty. 2015. 'Broadband activation by white-opsin lowers intensity threshold for cellular stimulation', *Sci Rep*, 5: 17857.

Bi, A., J. Cui, Y. P. Ma, E. Olshevskaya, M. Pu, A. M. Dizhoor, and Z. H. Pan. 2006. 'Ectopic expression of a microbial-type rhodopsin restores visual responses in mice with photoreceptor degeneration', *Neuron*, 50: 23–33.

Boyden, E. S., F. Zhang, E. Bamberg, G. Nagel, and K. Deisseroth. 2005. 'Millisecond-timescale, genetically targeted optical control of neural activity', *Nat Neurosci*, 8: 1263–8.

Busskamp, V., J. Duebel, D. Balya, M. Fradot, T. J. Viney, S. Siegert, A. C. Groner, E. Cabuy, V. Forster, M. Seeliger, M. Biel, P. Humphries, M. Paques, S. Mohand-Said, D. Trono, K. Deisseroth, J. A. Sahel, S. Picaud, and B. Roska. 2010. 'Genetic reactivation of cone photoreceptors restores visual responses in retinitis pigmentosa', *Science*, 329: 413–7.

Callaway, E. M., and R. Yuste. 2002. 'Stimulating neurons with light', *Curr Opin Neurobiol*, 12: 587–92.

Cehajic-Kapetanovic, J., C. Eleftheriou, A. E. Allen, N. Milosavljevic, A. Pienaar, R. Bedford, K. E. Davis, P. N. Bishop, and R. J. Lucas. 2015. 'Restoration of vision with ectopic expression of human rod opsin', *Curr Biol*, 25: 2111–22.

Chuang, A. T., C. E. Margo, and P. B. Greenberg. 2014. 'Retinal implants: a systematic review', *Br J Ophthalmol*, 98: 852–6.

Dalkara, D., L. C. Byrne, R. R. Klimczak, M. Visel, L. Yin, W. H. Merigan, J. G. Flannery, and D. V. Schaffer. 2013. 'In vivo-directed evolution of a new adeno-associated virus for therapeutic outer retinal gene delivery from the vitreous', *Sci Transl Med*, 5: 189ra76.

Dalkara, D., K. D. Kolstad, N. Caporale, M. Visel, R. R. Klimczak, D. V. Schaffer, and J. G. Flannery. 2009. 'Inner limiting membrane barriers to AAV-mediated retinal transduction from the vitreous', *Mol Ther*, 17: 2096–102.

Doroudchi, M. M., K. P. Greenberg, J. Liu, K. A. Silka, E. S. Boyden, J. A. Lockridge, A. C. Arman, R. Janani, S. E. Boye, S. L. Boye, G. M. Gordon, B. C. Matteo, A. P. Sampath, W. W. Hauswirth, and A. Horsager. 2011. 'Virally delivered channelrhodopsin-2 safely and effectively restores visual function in multiple mouse models of blindness', *Mol Ther*, 19: 1220–9.

Farah, N., I. Reutsky, and S. Shoham. 2007. 'Patterned optical activation of retinal ganglion cells', *Conf Proc IEEE Eng Med Biol Soc*, 2007: 6368–70.

Farah, N., A. Zoubi, S. Matar, L. Golan, A. Marom, C. R. Butson, I. Brosh, and S. Shoham. 2013. 'Holographically patterned activation using photo-absorber induced neural-thermal stimulation', *J Neural Eng*, 10: 056004.

Gaub, B. M., M. H. Berry, A. E. Holt, E. Y. Isacoff, and J. G. Flannery. 2015. 'Optogenetic vision restoration using rhodopsin for enhanced sensitivity', *Mol Ther*, 23: 1562–71.

Ghezzi, D., M. R. Antognazza, R. Maccarone, S. Bellani, E. Lanzarini, N. Martino, M. Mete, G. Pertile, S. Bisti, G. Lanzani, and F. Benfenati. 2013. 'A polymer optoelectronic interface restores light sensitivity in blind rat retinas', *Nat Photonics*, 7: 400–6.

Goetz, G. A., Y. Mandel, R. Manivanh, D. V. Palanker, and T. Cizmar. 2013. 'Holographic display system for restoration of sight to the blind', *J Neural Eng*, 10: 056021.

Golan, L., I. Reutsky, N. Farah, and S. Shoham. 2009. 'Design and characteristics of holographic neural photo-stimulation systems', *J Neural Eng*, 6: 066004.

Greenberg, K. P., A. Pham, and F. S. Werblin. 2011. 'Differential targeting of optical neuromodulators to ganglion cell soma and dendrites allows dynamic control of center-surround antagonism', *Neuron*, 69: 713–20.

Gunaydin, L. A., O. Yizhar, A. Berndt, V. S. Sohal, K. Deisseroth, and P. Hegemann. 2010. 'Ultrafast optogenetic control', *Nat Neurosci*, 13: 387–92.

Hubel, D. H., and T. N. Wiesel. 1959. 'Receptive fields of single neurones in the cat's striate cortex', *J Physiol*, 148: 574–91.

Hubel, D. H., and T. N. Wiesel. 1962. 'Receptive fields, binocular interaction and functional architecture in the cat's visual cortex', *J Physiol*, 160: 106–54.

Hubel, D. H., and T. N. Wiesel. 1968. 'Receptive fields and functional architecture of monkey striate cortex', *J Physiol*, 195: 215–43.

Huberman, A. D., and C. M. Niell. 2011. 'What can mice tell us about how vision works?', *Trends Neurosci*, 34: 464–73.

Humayun, M. S., J. D. Dorn, L. da Cruz, G. Dagnelie, J. A. Sahel, P. E. Stanga, A. V. Cideciyan, J. L. Duncan, D. Eliott, E. Filley, A. C. Ho, A. Santos, A. B. Safran, A. Arditi, L. V. Del Priore, R. J. Greenberg, and I. I. Study Group Argus. 2012. 'Interim results from the international trial of second sight's visual prosthesis', *Ophthalmology*, 119: 779–88.

Ivanova, E., G. S. Hwang, Z. H. Pan, and D. Troilo. 2010. 'Evaluation of AAV-mediated expression of Chop2-GFP in the marmoset retina', *Invest Ophthalmol Vis Sci*, 51: 5288–96.

Kato, H. E., F. Zhang, O. Yizhar, C. Ramakrishnan, T. Nishizawa, K. Hirata, J. Ito, Y. Aita, T. Tsukazaki, S. Hayashi, P. Hegemann, A. D. Maturana, R. Ishitani, K. Deisseroth, and O. Nureki. 2012. 'Crystal structure of the channelrhodopsin light-gated cation channel', *Nature*, 482: 369–74.

Kim, S. Y., S. Sadda, M. S. Humayun, E. de Juan, Jr., B. M. Melia, and W. R. Green. 2002. 'Morphometric analysis of the macula in eyes with geographic atrophy due to age-related macular degeneration', *Retina*, 22: 464–70.

Klapoetke, N. C., Y. Murata, S. S. Kim, S. R. Pulver, A. Birdsey-Benson, Y. K. Cho, T. K. Morimoto, A. S. Chuong, E. J. Carpenter, Z. Tian, J. Wang, Y. Xie, Z. Yan, Y. Zhang, B. Y. Chow, B. Surek, M. Melkonian, V. Jayaraman, M. Constantine-Paton, G. K. Wong, and E. S. Boyden. 2014. 'Independent optical excitation of distinct neural populations', *Nat Methods*, 11: 338–46.

Klapper, S. D., A. Swiersy, E. Bamberg, and V. Busskamp. 2016. 'Biophysical properties of optogenetic tools and their application for vision restoration approaches', *Front Syst Neurosci*, 10: 74.

Kleinlogel, S., K. Feldbauer, R. E. Dempski, H. Fotis, P. G. Wood, C. Bamann, and E. Bamberg. 2011. 'Ultra light-sensitive and fast neuronal activation with the Ca(2)+-permeable channelrhodopsin CatCh', *Nat Neurosci*, 14: 513–8.

Kojima, D., S. Mori, M. Torii, A. Wada, R. Morishita, and Y. Fukada. 2011. 'UV-sensitive photoreceptor protein OPN5 in humans and mice', *PLoS One*, 6: e26388.

Lagali, P. S., D. Balya, G. B. Awatramani, T. A. Munch, D. S. Kim, V. Busskamp, C. L. Cepko, and B. Roska. 2008. 'Light-activated channels targeted to ON bipolar cells restore visual function in retinal degeneration', *Nat Neurosci*, 11: 667–75.

Lin, J. Y., P. M. Knutsen, A. Muller, D. Kleinfeld, and R. Y. Tsien. 2013. 'ReaChR: a red-shifted variant of channelrhodopsin enables deep transcranial optogenetic excitation', *Nat Neurosci*, 16: 1499–508.

Mace, E., R. Caplette, O. Marre, A. Sengupta, A. Chaffiol, P. Barbe, M. Desrosiers, E. Bamberg, J. A. Sahel, S. Picaud, J. Duebel, and D. Dalkara. 2015. 'Targeting channelrhodopsin-2 to ON-bipolar cells with vitreally administered AAV restores ON and OFF visual responses in blind mice', *Mol Ther*, 23: 7–16.

Margalit, E., J. D. Weiland, E. De Juan, and M. Humayun. 2003. Chapter 7.5 *in Neuroprosthetics: Theory and Practice* (World Scientific Publishers: Hackensack, NJ).

Mathieson, K., J. Loudin, G. Goetz, P. Huie, L. Wang, T. I. Kamins, L. Galambos, R. Smith, J. S. Harris, A. Sher, and D. Palanker. 2012. 'Photovoltaic retinal prosthesis with high pixel density', *Nat Photonics*, 6: 391–97.

Milam, A. H., Z. Y. Li, and R. N. Fariss. 1998. 'Histopathology of the human retina in retinitis pigmentosa', *Prog Retin Eye Res*, 17: 175–205.

Nagel, G., T. Szellas, W. Huhn, S. Kateriya, N. Adeishvili, P. Berthold, D. Ollig, P. Hegemann, and E. Bamberg. 2003. 'Channelrhodopsin-2, a directly light-gated cation-selective membrane channel', *Proc Natl Acad Sci U S A*, 100: 13940–5.

Nikolenko, V., D. S. Peterka, and R. Yuste. 2010. 'A portable laser photostimulation and imaging microscope', *J Neural Eng*, 7: 045001.

Nikolenko, V., B. O. Watson, R. Araya, A. Woodruff, D. S. Peterka, and R. Yuste. 2008. 'SLM microscopy: scanless two-photon imaging and photostimulation with spatial light modulators', *Front Neural Circuits*, 2: 5.

Nirenberg, S., and C. Pandarinath. 2012. 'Retinal prosthetic strategy with the capacity to restore normal vision', *Proc Natl Acad Sci U S A*, 109: 15012–7.

Paluch-Siegler, S., T. Mayblum, H. Dana, I. Brosh, I. Gefen, and S. Shoham. 2015. 'All-optical bidirectional neural interfacing using hybrid multiphoton holographic optogenetic stimulation', *Neurophotonics*, 2.

Pan, Z. H., Q. Lu, A. Bi, A. M. Dizhoor, and G. W. Abrams. 2015. 'Optogenetic approaches to restoring vision', *Annu Rev Vis Sci*, 1: 185–210.

Rabinowitz, J. E., F. Rolling, C. Li, H. Conrath, W. Xiao, X. Xiao, and R. J. Samulski. 2002. 'Cross-packaging of a single adeno-associated virus (AAV) type 2 vector genome into multiple AAV serotypes enables transduction with broad specificity', *J Virol*, 76: 791–801.

Reutsky, I., D. Ben-Shimol, N. Farah, S. Levenberg, and S. Shoham. 2007. "Patterned optical activation of channelrhodopsin II expressing retinal ganglion cells". In *CNE '07*. 3rd International IEEE/EMBS Conference on Neural Engineering, *2007*, 50–52.

Reutsky-Gefen, I., L. Golan, N. Farah, A. Schejter, L. Tsur, I. Brosh, and S. Shoham. 2013. 'Holographic optogenetic stimulation of patterned neuronal activity for vision restoration', *Nat Commun*, 4: 1509.

Rodieck, R. W. 1965. 'Quantitative analysis of cat retinal ganglion cell response to visual stimuli', *Vision Res*, 5: 583–601.

Roska, B., and D. Pepperberg. 2014. 'Restoring vision to the blind: optogenetics', *Transl Vis Sci Technol*, 3: 4.

Santos, A., M. S. Humayun, E. de Juan, Jr., R. J. Greenburg, M. J. Marsh, I. B. Klock, and A. H. Milam. 1997. 'Preservation of the inner retina in retinitis pigmentosa. A morphometric analysis', *Arch Ophthalmol*, 115: 511–5.

Sengupta, A., A. Chaffiol, E. Mace, R. Caplette, M. Desrosiers, M. Lampic, V. Forster, O. Marre, J. Y. Lin, J. A. Sahel, S. Picaud, D. Dalkara, and J. Duebel. 2016. 'Red-shifted channelrhodopsin stimulation restores light responses in blind mice, macaque retina, and human retina', *EMBO Mol Med*, 8: 1248–64.

Shoham, S., D. H. O'Connor, D. V. Sarkisov, and S. S. Wang. 2005. 'Rapid neurotransmitter uncaging in spatially defined patterns', *Nat Methods*, 2: 837–43.

Simonelli, F., A. M. Maguire, F. Testa, E. A. Pierce, F. Mingozzi, J. L. Bennicelli, S. Rossi, K. Marshall, S. Banfi, E. M. Surace, J. Sun, T. M. Redmond, X. Zhu, K. S. Shindler, G. S. Ying, C. Ziviello, C. Acerra, J. F. Wright, J. W. McDonnell, K. A. High, J. Bennett, and A. Auricchio. 2010. 'Gene therapy for Leber's congenital amaurosis is safe and effective through 1.5 years after vector administration', *Mol Ther*, 18: 643–50.

Surace, E. M., and A. Auricchio. 2003. 'Adeno-associated viral vectors for retinal gene transfer', *Prog Retin Eye Res*, 22: 705–19.

Tomita, H., E. Sugano, Y. Fukazawa, H. Isago, Y. Sugiyama, T. Hiroi, T. Ishizuka, H. Mushiake, M. Kato, M. Hirabayashi, R. Shigemoto, H. Yawo, and M. Tamai. 2009. 'Visual properties of transgenic rats harboring the channelrhodopsin-2 gene regulated by the thy-1.2 promoter', *PLoS One*, 4: e7679.

van Wyk, M., J. Pielecka-Fortuna, S. Lowel, and S. Kleinlogel. 2015. 'Restoring the ON switch in blind retinas: opto-mGluR6, a next-generation, cell-tailored optogenetic tool', *PLoS Biol*, 13: e1002143.

Weiland, J. D., and M. Humayun. 2014. 'Retinal prosthesis', *IEEE Trans Biomed Eng*, 61: 1412–24.

Wells, J., C. Kao, E. D. Jansen, P. Konrad, and A. Mahadevan-Jansen. 2005. 'Application of infrared light for in vivo neural stimulation', *J Biomed Opt*, 10: 064003.

Yizhar, O., L. E. Fenno, T. J. Davidson, M. Mogri, and K. Deisseroth. 2011. 'Optogenetics in neural systems', *Neuron*, 71: 9–34.

Zhang, Y., E. Ivanova, A. Bi, and Z. H. Pan. 2009. 'Ectopic expression of multiple microbial rhodopsins restores ON and OFF light responses in retinas with photoreceptor degeneration', *J Neurosci*, 29: 9186–96.

Zrenner, E., K. U. Bartz-Schmidt, H. Benav, D. Besch, A. Bruckmann, V. P. Gabel, F. Gekeler, U. Greppmaier, A. Harscher, S. Kibbel, J. Koch, A. Kusnyerik, T. Peters, K. Stingl, H. Sachs, A. Stett, P. Szurman, B. Wilhelm, and R. Wilke. 2011. 'Subretinal electronic chips allow blind patients to read letters and combine them to words', *Proc Biol Sci*, 278: 1489–97.

CHAPTER 21

Optical Cochlear Implants

C.-P. Richter, Y. Xu, X. Tan, N. Xia, and N. Suematsu

CONTENTS

21.1 COCHLEAR IMPLANTS AND THEIR CHALLENGES

Cochlear implants (CIs) are considered one of the most successful neural prostheses. Today about 500,000 severe-to-profoundly deaf individuals have received a CI to restore some of their hearing. While some patients communicate over the phone in different languages, others receive little benefit from CIs. For all of the users, however, noisy listening environments and music perception constitute a challenge. It has been argued that performance could be improved by reducing the interaction between neighboring electrode contacts and subsequently creating more independent channels for stimulation. More selective stimulation with electric current can be achieved through multipolar stimulation. Multiple electrode contacts are used to narrow the current field and to achieve more spatially selective stimulation (Bierer et al., 2002; Mens et al., 2005; Snyder et al., 2004; Srinivasan et al., 2010). A different approach to increase the number of different pitch percepts is current steering (Koch et al., 2006; Nogueira et al., 2017; Wu et al., 2016). Neighboring electrodes are used simultaneously to "steer" the current to selected neuron populations, which sit between the two stimulating electrodes to introduce virtual channels (Berenstein et al., 2008; Choi et al., 2009; Koch et al., 2006; Landsberger et al., 2009; van den Honert et al., 2007). More recently, the use of photons has been suggested as a novel approach to evoke responses from small populations of neurons (Boyden et al., 2005; Hernandez et al., 2014; Izzo et al., 2006a; Wells et al., 2005). Depending on the tissue, photons can be delivered without direct contact between the optical source and the target structure. Light beams can be focused, allowing for selective irradiation of neural tissue and introducing a larger number of independent channels for stimulation. The underlying assumption for favoring optical energy to stimulate neuron populations over electrical current relates to the finding that optical radiation can be delivered more selectively to groups of target neurons (Moreno et al., 2011; Richter et al., 2011a). It is anticipated that with optical stimulation, neural prostheses with enhanced neural fidelity can be developed. In this chapter, we discuss some design parameters required for an optical cochlear implant.

21.2 NEURAL STIMULATION WITH PHOTONS

21.2.1 Infrared Neural Stimulation

Wells and coworkers published in 2005 that for selected radiation wavelengths in the range between 1,064 nm and 10,000 nm a pulsed laser could be used to stimulate nerves without damaging them (Wells et al., 2005). The wavelengths around 2,100 nm and in the range from 1,844–1,910 nm were especially interesting because lasers exist which emit light at those wavelengths. Inspired by the first results at Vanderbilt, we conducted similar experiments and irradiated the gerbil sciatic nerve with a Ho:YAG laser and could qualitatively validate Wells' results. In subsequent experiments we showed that stimulation of cranial nerves is possible by stimulating the facial nerve (Teudt et al., 2007), the auditory (Izzo et al., 2006a; Izzo et al., 2006b), and the vestibular system (Dittami et al., 2011; Rajguru et al., 2011). The underlying mechanism for neural stimulation with infrared light has not been fully determined. Today, everybody agrees that spatially and temporally confined heating results in the depolarization of the neuron. It is not fully clear how the heating results in an action potential. Experiments conducted to determine the mechanism for optical stimulation have shown that temperature-sensitive ion channels, the Transient Receptor Potential (TRV) vanilloid channels TRPV1 and TRPV4, are present in the cochlea and contribute to this (Albert et al., 2012; Balaban et al., 2003; Rhee et al., 2008; Suh et al., 2009; Takumida et al., 2005; Zheng et al., 2003), the heating changes in the membrane capacitance are associated with a depolarizing current (Plaksin et al., 2017; Shapiro et al., 2012), and the calcium homeostasis is affected (Dittami et al., 2011; Liu et al., 2014; Lumbreras et al., 2014; Rabbitt et al., 2016; Rajguru et al., 2011). It is not fully clear today how the spatially and temporally confined heating leads to an action potential.

21.2.2 Optogenetic and Thermogenetics

A different method to stimulate neurons with light has been presented: optogenetics. The method refers to the combination of optics and genetics to modulate the function of living cells. In 2005, Boyden and coworkers reported the possibility of expressing ion channels, which can be activated by light in neurons to control their neural activity (Boyden et al., 2005). The first ion channels or optogenetic tools were channelrhodopsin (ChR), a light-sensitive 7-transmembrane protein passing cations, and halorhodopsin, a light-sensitive 7-transmembrane protein pumping chloride ions. In addition to expressing ion channels that respond to radiation with wavelengths in the visible range, it is also possible to express ion channels that respond to temperature changes (Bernstein et al., 2012). In particular, two classes of candidates have been discussed: (1) the dominantly acting protein, Shibre^{ts1}, which inhibits endocytosis at temperatures above ~29 °C, and (2) the class of temperature sensitive TRP (Transient Receptor Potential) channels that can be activated by an increase in temperature (e.g. Caterina et al., 1997). Temperature sensitive channels have also been discussed in describing as a naturally occurring target during INS. TRP channels are expressed ubiquitously in many neurons and can be stimulated directly with infrared radiation (Albert et al., 2012; Suh et al., 2009; Yao et al., 2009). Moreover, it has been demonstrated that the Transient Receptor Potential Vanilloid 1 (TRPV1), TRPV2, and TRPV3 channels can be expressed in HEK293 cells Yao et al., 2009) and can be activated with temporally and spatially confined temperature changes created with near-infrared diode lasers (λ = 980 and 1460 nm).

Moser et al. (2013) studied the possibility of using an optogenetic approach for cochlear prostheses. To quantify the ability for stimulating mouse spiral ganglion neurons (SGNs) with blue light, the authors used transgenic mice, which express ChR2 in the somata, dendrites, and axons of SGNs. The blue light was delivered either via the round window or a cochleostomy. Irradiation of the SGNs resulted in a measurable optical auditory brainstem response (oABR) for either of the stimulation configuration (Hernandez et al., 2014).

To rescue mice from early onset genetic deafness, the authors crossed the *Thy1.2*-driven *ChR2* transgene with *Oto*$^{Pga/Pga}$ mice to express ChR2 in their SGNs. Likewise, the channel was expressed in SGNs using a viral vector. Efficient and cell-specific expression of the optogenetic tool was only observed for AAV2/6. Neural responses could be evoked by direct irradiation of the neurons up to 60 Hz, the current limit of their stimulation setup (Hernandez et al., 2014). The experiments showed that no acoustically evoked auditory brainstem responses (aABRs) could be evoked in the treated and untreated animals. However, in the treated animals, oABR could be evoked by irradiation of the cochlea with blue light (Hernandez et al., 2014).

Another group conducted short-term experiments in guinea pigs and rats to deliver ChR2 to the brainstem via a single injection of AVV. After a survival period of 2–3 weeks, the expression of ChR2 was examined. A broad pattern of expression of the channel was observed in the cochlear nucleus (Acker et al., 2011). Responses, which were evoked with a large diameter optical fiber at the cochlear nucleus, showed a broad pattern of excitation in the central nucleus of the inferior colliculus (ICC), similar to electric stimulation (Darrow et al., 2013a) or an acoustic click (Darrow et al., 2013b).

The potential of optogenetics for the auditory system lies in its known mechanism. The challenges lie in the speed of the channels. Moreover, for visible light, tissue will significantly scatter the photons and subsequently selective stimulation will require the light source in close proximity to the target structure.

21.2.3 Optoacoustics

Optoacoustics describe a technique in which light is used to generate sound waves in gas or other media (for an overview, see e.g. Tuchin, 2000). After the target tissue has absorbed the photons, their energy is converted into heat. The spatially confined heating leads to an expansion of the target volume and the formation of pressure waves. The pressure waves are particularly large if the pulse duration is shorter than the propagation time of the stress out of the irradiated zone. For pulse durations shorter than 1 μs the condition has been described as stress-confined (Niemz, 2004; Tuchin, 2000; Welch et al., 2012).

During optoacoustic stimulation in the cochlea, laser-induced pressure waves vibrate the basilar membrane and stimulate hair cells. The hair cells are required to transform the acoustic signal into action potentials on the auditory nerves. Since the acoustic signal is similar to an acoustic click, it is counterintuitive that the stimulation results in a selective stimulation of the cochlea. In one approach, Wenzel et al. (2009) used a Nd:YAG laser in the stress confined condition (λ = 532 nm, τp = 10 ns, f = 10 Hz, Q = 0–23 J/pulse) to stimulate the cochlea of guinea pigs. They were able to evoke ABRs in hearing animals. The response could not be evoked in deaf animals. In a recent development this group presented a method of how to encode a sequence of rapid laser pulses into different frequencies that can be represented in responses at different sites along the cochlea.

In a different approach, it has been reported that for pulse durations, which are orders of magnitudes longer than 1 μs, audible stress waves can be generated as well (Teudt et al., 2011). In the near-infrared, the responses correlated with the water absorption for the photons. Furthermore, the optically evoked responses in the cochlea correlated with the absorption coefficient of hemoglobin λ = 600–800 nm, (Schultz et al., 2012b). It has been suggested that light-induced pressure waves mechanically stimulated the cochlea (Schultz et al., 2012a; Schultz et al., 2012b; Teudt et al., 2011; Wenzel et al., 2009; Zhang et al., 2009).

The important message from the experiments is that all stimuli at radiation wavelength in the infrared will generate stress relaxation waves and will heat the target volume. Considering the technique for scientific applications or prosthetic means, issues of thermal damage and stress-related effects have to be addressed.

The intention is not to contrast photonic stimulation with the "gold standard", electrical stimulation, but to evaluate the pros and cons of the novel stimulation modality and its potential for neural prostheses.

21.2.4 Pros

21.2.4.1 Electrical Stimulation

- Mechanism is well understood.
- Power efficient.
- Gold standard for neural stimulation.

21.2.4.2 Infrared Neural Stimulation (INS)

- INS stimulation results from spatially confined heating of the neural tissue and does not require the introduction of an absorber for the photons.
- INS is spatially selective and promises more independent channels than electrical stimulation to simultaneously encode acoustic information at the neural tissue interface.

21.2.4.3 Optogenetics and Thermogenetics

- The mechanism for optogenetics is known; it is the activation of an ion channel through photons, which requires the introduction of an absorber for the photons.
- It is spatially selective if the distance between the light source and the target neurons is small.

21.2.4.4 Optoacoustics

- The absorption of the photons results in a mechanical event. An optical hearing aid can be envisioned.
- Various wavelengths can be used depending on the absorber.
- A single source emitter can be used to stimulate via the round window of the cochlea.

21.2.5 Challenges

21.2.5.1 Electrical Stimulation

- Spread of current in the tissue makes spatially confined stimulation difficult.

21.2.5.2 Infrared Neural Stimulation (INS)

- Heat is delivered to the target structure, which needs to be removed to prevent thermal damage during stimulation. Tissue heating limits the rate of stimulation.
- Energy requirements are large because the light sources are inefficient and the energy required for stimulation is about 100-times larger when compared to electrical stimulation.
- Optoacoustic events must be considered if such technology should be applied in patients with residual hearing.

21.2.5.3 Optogenetics and Thermogenetics

- Neurons must be manipulated genetically. This requires targeting of a selected population of neurons with a viral vector and the stable expression of temperature or light-sensitive ion channels.
- The efficiency of the optogenetic tool determines flux rates for stimulation and potential side-effects.
- The energy required for stimulation is about ten times larger than for electrical stimulation.

21.2.5.4 Optoacoustics

- Various wavelengths can be used, depending on the absorber.
- Each laser pulse results in a click-like response, which limits the ability to quickly encode the dynamic and complex frequency information in running speech or music.
- Sophisticated coding strategies need to be developed to convey frequency information to the cochlea.

21.3 THE OPTICAL COCHLEAR IMPLANT

The optical cochlear implant consists of three components, an external wearable speech processor and two implantable components: the receiver and spike generator and an array comprised of the optical sources and the contacts for electrical stimulation. The hybrid approach provides the possibility of decreasing the power requirements. In other words, an opto-electrical hybrid electrode is an ideal tool for cochlear implants. The acoustical signal is recorded by a single or by multiple microphones in the speech processor. It is then filtered using a pre-emphasis filter, which is typically a high pass filter with a corner frequency at 2 Hz. The filtered signal is then decomposed into many frequency bands. Several methods can be applied, such as calculating a gammatone-weighted spectrogram or gammatonegram, or filtering the acoustic signal with a gammatone filter, which can be described by an impulse response that is the product of a gamma distribution and sinusoidal tone.

The number of electrode contacts determines the number of selected frequency bands. The acoustical energy in each frequency band is then used to generate a series of electrical and optical pulses. Pulse tables that contain the timing and amplitude information are delivered across the skin to the implanted portion of the CI via a radio-frequency (RF) link. The RF link also serves to power the implanted portion of the optical implant. The receiver demodulates the RF signal and generates biphasic electrical current pulses that are delivered at corresponding sites along the cochlea for electrical stimulation. The receiver also generates current pulses that drive the optical sources. The current pulses are delivered via small platinum wires to the electrode contacts in the cochlea. The radiant energy is delivered with waveguides to sites of optical stimulation. The following several key aspects that determine the design and performance parameters of an opto-electrical hybrid implant are discussed.

21.4 SYSTEM REQUIREMENTS FOR AN OPTICAL COCHLEAR IMPLANT

21.4.1 The Light Delivery System

Optical methods to monitor and to modulate neural activity have been introduced into basic science and make their way into neuroprosthetics. The light delivery systems (LDSs) can be made of external light sources, implanted light sources, optical fibers, or waveguides. This chapter will not provide an exhaustive review on the light sources available today, but rather provide a discussion on the requirements and possibilities for CIs (Table 21.2).

21.4.2 Dimensions of an Optical Cochlear Implant Electrode

The dimensions of the cochlea limit the physical size of the light LDS, and the requirements for a human implant are similar to the dimension of contemporary cochlear implant electrodes. A detailed analysis based on micro-computed tomography (micro-CT) of human temporal bones using X-rays should be used as a guide for the design of an electrode (Avci et al., 2014; Erixon et al., 2009; Hatsushika et al., 1990). A circular electrode should fit into the scala tympani. Moreover, the electrode must be stiff enough that it can be inserted into the cochlea and it must be flexible enough that the insertion is not damaging cochlear structures, including surviving neurons in Rosenthal's canal. Since the electrode has a circular form, its diameter can be estimated from the largest circle, which can be housed in the scala tympani at different sites along the cochlea. For conservative measures, the smallest diameter which could be found in the literature was used. According to the measurements

TABLE 21.1 Width and Height Measures of Scala Tympani of a Human Cochlea

Human cochlear dimension				
angular distance [degree]	ST width [mm]	ST hight [mm]	ST-area [mm²]	Maximum circle diameter [mm]
0	3.01	1.62	2.00	0.97
100	1.63	1.08	1.57	0.91
200	1.59	0.93	1.24	0.71
300	1.53	0.96	1.13	0.76
400	1.41	0.92	0.95	0.73
500	1.30	0.83	0.85	0.63
600	1.23	0.74	0.79	0.56
700	1.19	0.61	0.74	0.48

The measures are from X-ray imaging and have been compiled from the literature (Avci et al., 2014; Erixon et al., 2009; Hatsushika et al., 1990).

the electrode at the cochlear base should be less than 0.97 mm and taper at its tip to 0.48 mm (Table 21.1).

21.4.2.1 Optical Fibers

The initial proof of concept experiments for INS were conducted using an open beam path at the free electron laser or with a single flat polished optical fiber, which was coupled to a solid-state laser and allowed delivering the photons to the target neurons. Specifically, in our experiments, the fiber was inserted with a 3D micromanipulator into the scala tympani of the basal cochlear turn (Izzo et al., 2006a). Subsequent chronic experiments were done in a cat animal model. The optical fibers were implanted into the cochlea, secured in the bulla and tunneled from the temporal bone of the cat to a cutaneous feed through between the scapulae (Matic et al., 2013). The animals carried a laser source in a backpack and stimulation over extended periods of time was possible (Figure 21.1).

The LDS made with optical fibers had limited survival time in cats. The two key failure points were at the anchor screwed and cemented to the bulla and at the cutaneous feed through. To advance from a single emitter to a multi-channel electrode, a three channel LDS was developed and fabricated by Lockheed Martin Aculight (Figure 21.2). The three channel LDS was too large to be inserted into the guinea pig cochlea but fit into the basal portion of a cat cochlea. It turned out that the three channel LDS was useful for acute experiments but it was too delicate for chronic implantation.

Despite the success with the optical fibers, they are not feasible as a chronic implantable LDS. A study using human cadaveric temporal bones showed that optical fibers fabricated from fused silica are stiff and unfit for deep insertion into the human cochlea (Balster et al., 2014). The results of their experiments demonstrated that fibers with a diameter larger than 50 μm break easily during insertion. They also showed that only optical fibers with a diameter of 25 μm and 50 μm could be inserted up to up to 20 mm (Balster et al., 2014).

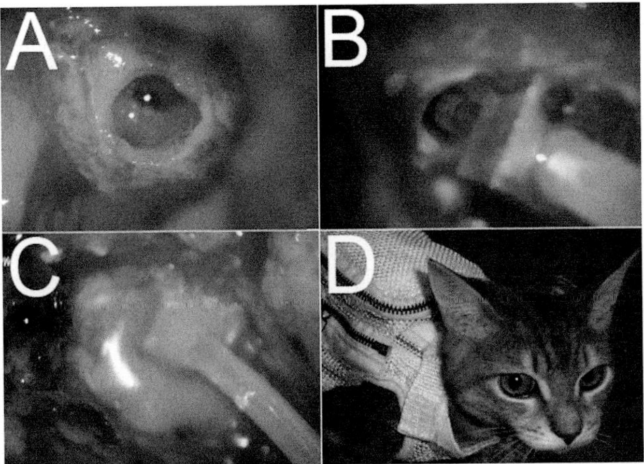

FIGURE 21.1 The image shows the implantation of an optical fiber light delivery system into a cat cochlea for chronic stimulation. **A** shows the cochleostomy created in the basal turn of a left cat cochlea. **B** shows the inserted optical fiber secured with a metal bracket to the bulla. In **C** the bulla is closed with dental acrylic. **D** shows the cat with a backpack carrying the stimulator.

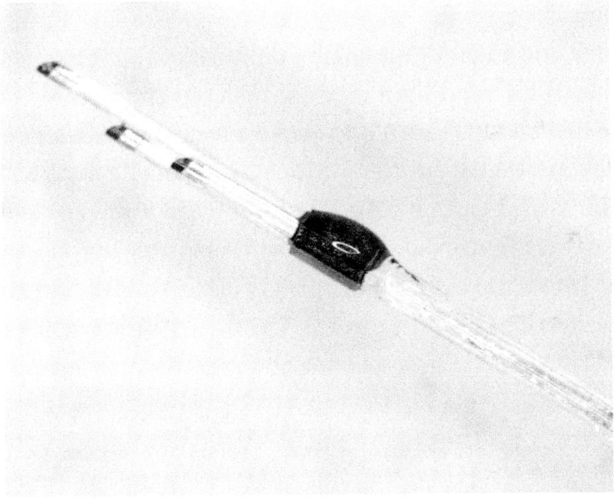

FIGURE 21.2 Three angle polished optical fibers, 100 μm in diameter were assembled to an LDS. The LDS was fabricated by Lockheed Martin Aculight.

Considering that infrared neural stimulation in the cochlea requires optical fibers with a diameter of 200 μm, using optical fibers 25 μm in diameter will be challenging. As a conclusion, silicone fibers are not a preferred solution for the LDS in the cochlea.

21.4.2.2 Light Sources

An alternative to silica fibers is the use of small light sources, which could be assembled to an array to be inserted into the cochlea. Possible light sources small enough to be inserted into the cochlea include vertical-cavity surface emitting lasers (VCSELs), VCSEL arrays, edge emitting laser diodes, and micro light emitting diodes (Figure 21.3, Table 21.2). The size of a single VCSEL emitting at 670–690 nm (red) is 250 μm × 250 μm × 200 μm (Figure 21.3). The maximum output power for those VCSELs is about 4 mW with an operation current of 15 mA. VCSELs are also available with a narrow emission spectrum at 1,860 nm. Their output power scales with the device's active diameter and is today up to 7.3 mW for continuous wave mode operation at room temperature. The power conversion efficiency is about 10% for VCSELs and decreases with increasing temperature. For example, a temperature increase to 40°C from room temperature results in an efficiency drop by 30%. The VCSELs (single aperture emitters) used to obtain the numbers above were 20–30 μm in diameter. Combining single aperture emitters to arrays on a single chip provides sources with a small footprint but higher output power. While 4 × 4 and 5 × 5 arrays were too limited in power to be used for INS, 5 × 7 arrays appeared promising. The array size is 250 μm × 450 μm × 200 μm, small enough for aligning them into an array that can be inserted into a cat cochlea. In pulsed operation mode the power measured in air in front of the sources was about 70 mW (pulse width (PW) = 10 μs, pulse repetition rate (RR) = 3,000 pulses per second (pps), and driving current amplitude of 300 mA). For our experiments the arrays were operated at PW = 100 μs and 200 Hz repetition rate, and 6 μJ/pulse energy. For the large arrays the power efficiency is decreased because of the self-heating of the arrays.

VCSEL technology has significantly advanced over the last years. However, VCSELs have limited efficiency and the radiant energy delivered to the tissue is too low for reliable optical stimulation with infrared light. For example, in the cochlea the radiant energy is less than 6 dB above the energy required to evoke a measurable response from the cochlea. Alternatives to VCSELs are edge emitters, which are available for 1,850 nm. The die is 300 μm wide, 100 μm thick, and can be 250, 350, or 450 μm long. For the testing, single and multiple-waveguide arrays were used. When operated in continuous wave mode the output power of the longest single waveguide (450 μm) is up to 40 mW at room temperature. The current requirement for the radiant power is 450 mA. While the output power increases

FIGURE 21.3 This figure shows different light sources. **A** and **E** show the image of an edge emitting laser diode. **B** shows the top and **F** the reverse side of an infrared VCSEL array (λ = 1860 nm). **C** and **D** show a blue μLED before and after being resized. **G** shows a VCSEL (λ = 680 nm). **H** shows the same VCSELs shown in **G** after wire bonding with a 25 μm diameter gold wire. Scale bar equals 1 mm.

TABLE 21.2 Comparison of Different Optical Sources regarding Their Power Efficiency and Radiation Wavelength

Small optical sources								
type	**λ (nm)**	**source size (μm × μm × μm)**	**output power (mW)**	**operation current (mA)**	**operating voltage (V)**	**power conversion efficiency (%)**	**operating temperature °C**	**operating mode**
μ-LED	465–475	1000 × 600 × 200	2.5	10	3.1	8.1	25	cw
VCSEL-single emitter	670–690	250 × 250 × 200	4	15	1.5	17.8	25	cw
VCSEL-single emitter	1860		7.3	90	1.5	5.4	25	cw
VCSEL-single emitter	1860		1.3	22	1.5	3.9	40	cw
VCSEL-5×7 emitter array	1860	250 × 450 × 200	70	300	1.5	15.6	25	pulsed (10 μs, 3000pps)
			60	200	1.5	20.0	25	pulsed (10 μs, 3000pps)
edge emitter	1875	20 × 450 × 100	70	600	1.5	7.8	25	

with the length of the die, the plug efficiency decreases. For the 450 μm long die the plug efficiency is ~12% at 100 mA (15 mW output power) and ~4% at 450 mA (40 mW output power). For the array, the output power of the longest single waveguide (450 μm) is up to 50 mW at room temperature in continuous wave mode operation. In pulsed mode operation, the output power could be increased by about a factor of 4.5. For the 250 μm die the maximum output power was increased to ~70 mW at 100 μs pulse width, 550 mA, and 1% duty cycle from 16 mW in continuous wave mode operation.

High-efficiency microscale Light Emitting Diodes (μLEDs) are an option for optogenetic approaches. Light sources were blue μLEDs (Pico LED, Rohm Semiconductor, Kyoto, Japan). The dimensions of the μLEDs (1,000 × 600 × 200 μm) are large when considering the size of the resulting multi-channel optrode, which has to be inserted into a cat or human cochlea. Therefore, the dimensions of the light source were minimized by trimming the carrier to 500 × 500 × 200 μm by either cutting them with a sharp blade or by grinding them down with a DREMEL rotary-tool. The chip has an InGaN structure with 33 mW of power dissipation. The peak forward current in constant wave mode is 10 mA, and is 50 mA in the pulsed mode. In pulsed mode the duty cycle is 1:10. The operating temperature is −100 to 85°C. The scanning angle is about 140 degrees. The radiation wavelength is 465–475 nm. The luminous intensity is about 25 mcd (millicandelas = millilumens/steradian). The radiant intensity or radiant flux I_e of the source can be estimated

$$I_e = \frac{I_V}{K(470\,\text{nm})} = \frac{25}{200}\frac{\text{mlm}}{\text{sr}}\frac{\text{W}}{\text{lm}} = 0.125\,\text{mW/sr}$$

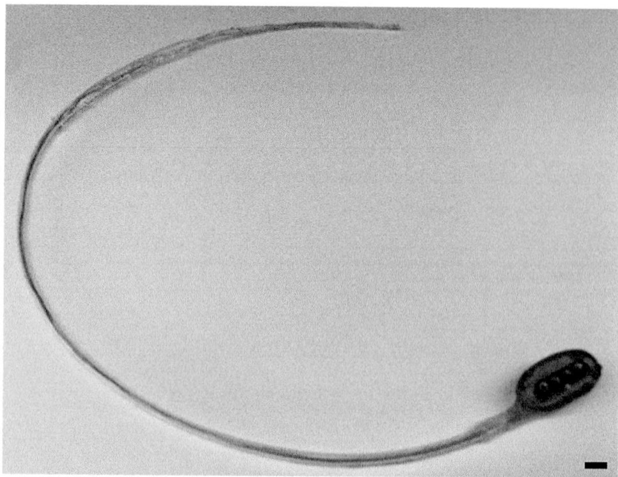

FIGURE 21.4 The image shows a fully assembled optrode, which has been implanted in cats. Scale bar is equivalent to 1 mm.

where I_v denotes the luminous intensity or visible flux (according to specs it is typically 25 mlm/sr) and $K(\lambda)$ is the luminous efficacy at a given wavelength (stated by the company to be 200 lm/W for the GaN wavers).

To build optrodes we have connected each of the light sources with the silver wire, 125 μm in diameter, as the backbone. The silver wire also acts as a heat sink and is connected to the cathode of the optical sources. The anode of each source is connected with a 25 μm platinum wire. After each of the wires has been connected with conductive epoxy to the light source the array is placed in a small mold, which has the dimension of the final electrode to be inserted in the cochlea (Figure 21.4). The mold is filled with silicone and is cured at 60°C overnight. The challenge for this solution is the power efficiency of the optical sources. In particular, the power efficiency for the VCSELs and edge emitting laser diodes is less than 10%. Taking into account the power required for stimulation, driving currents of up to 500 mA for each of the light sources would be necessary. Despite the sources being fully encapsulated in silicone this constitutes a significant risk. Moreover, the power requirements for the implant would significantly increase. At the current state of technology development small sources are not an option for INS. If a similar calculation is done for the LEDs and an optogenetic approach, the situation is more favorable; LEDs typically have wall-plug efficiency between 5 and 40% and the power required for stimulation with visible light is about ten times lower than INS. Currents up to 10 mA are expected for optogenetic stimulation for each of the optical sources. At the current state of the technology optical sources should be used wisely in an optogenetic approach.

24.4.2.3 Waveguides

In addition to optical fibers and small light sources, waveguides are available that can be used to deliver the radiant energy to the cochlea. Waveguides made of polyimide conduct infrared light well. Polyimides are also significantly more compliant when compared to optical fibers made of glass, yielding 20 to 40 times better compliance depending on the shape of the fiber. At the current state of technology waveguides appear to be the best option for an optical cochlear implant. To fabricate the LDS, between two and three waveguides are aligned such that the tips of the waveguides are located at the tip of the hybrid electrode. The waveguides will be completely encapsulated in silicone. After the silicone cured, platinum iridium bands will be distributed equally along the electrode. The bands are made out of platinum rings, which are cramped to hold the wire. After the bands are aligned, the electrode is placed in a small mold, which is subsequently filled with silicone. The optical electrical hybrid array will be cured overnight at 60°C. Before implantation each of the contacts will be tested for functionality. Although electrodes have been fabricated, the *in vivo* testing is still in progress.

21.4.2.4 Flexible Printed Circuit Board (FPCB)

A different solution used Flexible Printed Circuit Board (FPCB) technology. The single layer FPCB was designed as the light source carrier, which renders the fabrication process much easier. The substrate for the cochlear implant needs to be soft, flexible, and must have good biocompatibility. It has been demonstrated that polyimide polymers are suitable in stiffness and that they are biocompatible (Bae et al., 2012; Ejserholm et al., 2015; Mattioli-Belmonte et al., 2003; Starr et al., 2016; Sun et al., 2009) and were selected for the support base and the insulation cover layer. In our prototyping, copper was selected as the conductive material. Adhesive films provide the material to bond the copper foil to the base film.

To fabricate the multichannel optrode carrier, a 25 μm-thick copper foil was laminated on the upper surface of the polyimide substrate. Unwanted copper was etched from the copper layer, such that the resulting wire width was 80 μm. To isolate each channel, a 25 μm-thick polyimide film was laminated on the surface. Subsequently, the polyimide film applied for insulation was etched away on the top of the light source mounting areas and solder joints. Light source mounting areas and solder joints were further improved by electroplating a 25-μm-thick gold layer on the contact areas. Dictated by the number of current sources of our portable light stimulator, we only fabricated three channel optrodes. The number of contacts, however, can easily be expanded. Figure 21.5 shows the FPCB carrier

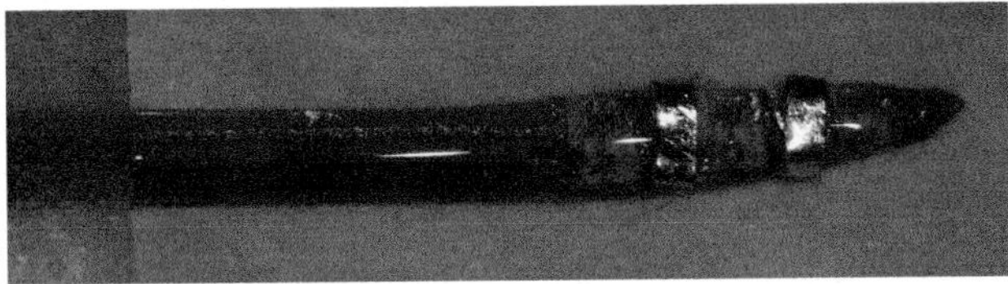

FIGURE 21.5 To reduce the power required for optical stimulation, a combination of electrical and optical stimulation has been proposed. Results by Austin Duke showed that energy required for stimulation can be reduced by hybrid stimulation. Initial results from co-stimulation in cat cochleae demonstrate that synergistic and antagonistic behavior can be seen. Experiments are required to determine the exact timing of the electrical and optical pulse for maximizing the reduction in power consumption for optical stimulation.

and a three-channel optrode (blue μLEDs). This carrier can also accommodate the red and infrared VCSELs, μLEDs, edge emitting laser diodes, or metal contacts for electrical stimulation.

21.4.2.5 Optoelectrical Hybrids

Pilot experiments have demonstrated that significant currents are required to drive the optical sources. Because of safety considerations, it is not desirable to drive such optical sources located in the cochlea. The possibility of decreasing the power requirements for stimulation is by opto-electrical hybrid stimulation. It has been shown by a group at Vanderbilt that the optical energy required for INS can be reduced by stimulation with electrical currents. To explore whether opto-electrical hybrid stimulation is beneficial for stimulating the auditory nerve, we have fabricated an electrode consisting of optical sources and electrical contacts. The electrode is shown in Figure 21.5. Electrical contacts formed from small platinum rings are placed in between the optical sources. The wires for the electrical contacts are at the backbone of the electrode so as not to interfere with the beam path for the light delivery. As an alternative to the FPC board, electrodes could also be fabricated by placing the optical sources on a wire backbone and placing electrical contacts in between optical ones.

21.4.3 Spot Size of the Light Beam

Experiments in the cochlea have demonstrated that optical fibers of 100 and 200 μm are large enough to deliver sufficient energy to the target structures for INS. For fibers with a smaller diameter it was difficult to evoke an auditory response. The challenges relate to the accurate placement of the optical fiber and the limited amount of radiant energy that could be coupled into the optical fiber.

For the 200 μm optical fibers the spot size at the target structure, the spiral ganglion neurons, was estimated from the spread of excitation recorded in the inferior colliculus in guinea pigs (Richter et al., 2011b). When spatial tuning curves (STCs) obtained from neural activity in the inferior colliculus evoked by INS were compared with similar curves obtained with acoustical stimulation, the average width of the optically evoked STCs was 357 ± 206 μm and that of pure tone STCs was 383 ± 131 μm (Richter et al., 2011b). The differences were statistically significant. In the guinea pig, the slope of the basilar membrane place-frequency map is about 3,200 μm/octave (Müller, 1996a). Considering that the spiral ganglion has only half the length of the basilar membrane, the corresponding distance for an octave would be 1600 μm along the spiral ganglion. In other words, the spot size at the spiral ganglion neurons obtained with infrared stimulation corresponds to a frequency range of about 0.22 octaves.

Similar numbers have been reported for the mouse cochlea using the optogenetic approach (Hernandez et al., 2014; Jeschke et al., 2015). In their experiments they used a 50 μm × 50 μm μLED as the light source and determined that the spatial spread of excitation is about 250 μm or about 0.33 octaves at about 200 μm from the light source (Hernandez et al., 2013; Hernandez et al., 2014).

21.4.4 Biocompatibility

Light emitting materials can be toxic and cannot be in direct contact with the perilymph in the scala tympani of the cochlea. Toxic compounds can diffuse into the fluid space, reach the spiral ganglion neurons, and damage spiral ganglion neurons. Therefore, it is important to encapsulate the optrode with a material that is inert to its environment, transparent for the light (infrared or visible light), and prevents toxic elements or compounds being released into the cochlea. The optrodes, which were designed to irradiate with infrared light were embedded with Silastic (MDX4-4210, Medical Grade Elastomere, base and curing agent (LOT 0006932899, Dow Corning Corp., USA)). After the silicone had cured, the electrodes as shown in Figures 21.4 and 21.5 were chronically implanted in cats. The remaining hearing after implantation was tested bi-weekly. No indication of cochlear damage through the placement of the electrodes was apparent.

21.4.5 Orientation of the Beam

The benefit of optical stimulation relates to the possibility of spatially selective stimulation of a population of neurons. This benefit comes with a price. The targeted neurons must be in the beam path. Therefore, the orientation of the light source or the LDS has to be precise for optimal stimulation. This is particularly difficult in a confined space such as the cochlea. Optical sources, optical fibers, or waveguides must be placed in the scala tympani along the cochlea such that the neurons in Rosenthal's canal are within the beam path. It has been shown that the deviation from the target should not be larger than 45° (Figure 21.6).

Each of the LDSs discussed above has its advantages and its challenges. In the following, the key points are summarized.

21.4.6 Pros

21.4.6.1 Optical Fibers

- Optical fibers have been used for pilot experiments.
- They can be easily mounted and pointed towards the target structures.
- They can be beveled for flexible adjustment of the orientation of the beam path.
- Lenses and other optics can easily be mounted.

21.4.6.2 Waveguides

- Waveguides are compliant.
- Bundles can be formed.
- Waveguides can transmit sufficient power.
- They have little bending losses.

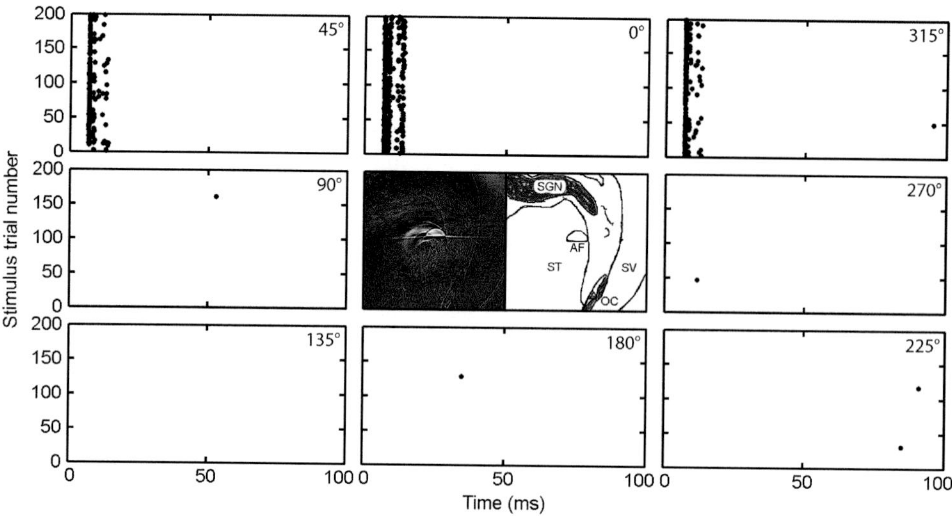

FIGURE 21.6 ICC single unit responses to different INS orientations in a normal hearing animal combined with imaging with synchrotron radiation. The single unit responded to acoustic stimulation as well, with the best frequency at 16 kHz. **Central panel**: A reconstructed slice perpendicular to the fiber axis and its sketch. The angle polished fiber (AF, the bright semi-circle in the center) was fixed with dental acrylic in the cochlea at the initial position (0°) for imaging with synchrotron radiation. Notice that the fiber was located between the spiral ganglion neurons (SGN) and the organ of Corti (OC) facing the SGNs. ST: Scala tympani; SV: Scala vestibuli. **Surrounding panels**: Single unit responses to different orientations of the INS changing in steps of 45°, as indicated in the upright corner of each panel, shown for 200 stimulus trials. This ICC unit showed in-phase responses at ~7ms following stimulus presentation. Only the action potentials detected in the first 100 ms following the stimulus were included in the raster plot.

21.4.6.3 Optical Sources

- Optical sources can easily be implanted.
- They are small in size.
- Initial devices have been fabricated for optogenetics and INS.
- Thin film electrodes have been built.
- The beam path can be focused on the target through lenses.

21.4.7 Challenges

21.4.7.1 Optical Fibers

- Optical fibers are stiff.
- Optical fibers have limited insertion depth into the cochlea.
- The fibers break easily.
- Insertion of the fibers bears the risk of damaging delicate cochlear structures.

- Flat polished fibers can only be placed in the cochlea base.
- Angle polished fibers have been tested in acute experiments.
- The size of the optical fiber should be larger than 100 μm for efficient stimulation.

21.4.7.2 Waveguides

- They are large in size.
- They have not been tested in the cochlea.

21.4.7.3 Optical Sources

- Optical sources have limited power to evoke auditory responses.
- Their wall-plug efficiency is poor and heat has to be dissipated in the cochlea.
- The large operation currents add risk.

21.4.8 The Power Requirements

The diameter of the optical fiber, the thickness of the modiolar bone, or its distance from the target structure determines the radiant energy required for optical stimulation. It is therefore not surprising that the reported threshold energies for stimulation vary significantly. Table 21.3 provides a comprehensive overview of data available. In the table we have focused on stimulation in the cochlea and have distinguished according to INS, optoacoustic, and optogenetic stimulation.

Experiments were conducted in mice, guinea pigs, and cats to determine the power required for INS. During the experiments a flat polished fiber was placed in front of Rosenthal's canal. Compound action potential amplitudes were measured for increasing radiant energies. After the experiments the fibers were fixed in place and the temporal bones were harvested and scanned with hard X-rays. The resulting microtomography reconstructions allowed for an exact measure between the tip of the optical fiber and the center of the spiral ganglion. This distance was used to correct for the energy losses caused by absorption between the optical fiber and the neurons. The results provided a good estimate of the energy and peak power required for stimulation.

Threshold levels were between 6 and 50 μJ/pulse (Figure 21.7). Pulse length was 100 μs and the pulse repetition rate was below 250 Hz. If translated into a human cochlea additional power might be required because the bone between the scala tympani and Rosenthal's canal is thicker in humans (0.35–1.15 mm) than in mice (4–13 μm, own measurements), guinea pigs (8–18 μm, own measurements), or cats (<40 μm, Shepherd et al., 2004).

21.4.8.1 Optogenetics

The energy required for optogenetic stimulation depends on several factors. The radiant energy is increased by absorption or scattering of the radiation. In the cochlea, light in the visible range is scattered and absorbed particularly by hemoglobin and by bone.

TABLE 21.3 Thresholds for Optical Stimulation in the Cochlea

Thresholds for optical stimulation in the cochlea																
publication	animal model	stimulation method	radiation wavelength (nm)	threshold	pulse length [ps]											
					5	10–20	30–35	50	100	200–250	300	400	600	800	1,000	2,000–5,000
					radiant exposure [mJ/cm²]											
Izzo et al. (2006) *Lasers in Surgery and Medicine* 38, 745–53	gerbil	INS	2120	50 µV CAP						18 ± 3						
Izzo et al. (2007) *IEEE TBE*, 54(6), 1108–14	gerbil	INS	1844–1873	50 µV CAP			5.3 ± 0.6		6.2 ± 1.5		21.7 ± 4.8		30.2 ± 8		58.4 ± 14.9	
Izzo et al. (2008) *Biophysical Journal*,94, 3159–66	gerbil	INS	1940	30 µV CAP	1.6 ± 0	2.1 ± 0.3	2.9 ± 0.6									
C-P Richter et al. (2008) *Hearing Research* 242, 42–51.	gerbil normal hearing	INS	1844–1873	50 µV CAP			6 ± 3		9 ± 2	14 ± 6		17 ± 1		20 ± 40		
Richter et al. (2008) *Hearing Research* 242, 42–51.	gerbil deaf	INS	1844–1873	50 µV CAP			12 ± 2		164 ± 3	272 ± 523		559 ± 104		719 ± 1262		
Rajguru et al. (2010) *Hearing Research* 269, 102–11.	cat normal hearing	INS	1860	20 µV CAP					2.4 (0.88)[a]	7.8–13.1						
Rajguru et al. (2010) *Hearing Research* 269, 102–11.	cat acute	INS	1860	20 µV CAP					53.2 (2.4)[a]	72.3–105.7						
Littlefield et al. (2010) *Laryngoscope* 120, 2071–82.	gerbil	INS	1844–1873	singel fiber >20%					31.8							
Richter et al. (2011) *J Neural Eng* 8, 056006 (11 pp).	guinea pig	INS	1860	ICC spatial tuning curve d′=2					15.3 ± 11.3							
Matic et al. (2013) *PLOS* ONE, 8(3), e58189.	cat	INS	1855	visible ABR					159							
Tan et al (2015) Scientific Reports 513273	guinea pig	INS	1862	visible CAP (3–10 µV)					189 ± 122							

(Continued)

TABLE 21.3 (CONTINUED) Thresholds for Optical Stimulation in the Cochlea

Thresholds for optical stimulation in the cochlea												
publication	**animal model**	**stimulation method**	**radiation wavelength (nm)**	**threshold**	**pulse length [ps]**							
					5 | 10–20 | 30–35 | 50 | 100 | 200–250 | 300 | 400 | 600 | 800 | 1,000 | 2,000–5,000							
					radiant exposure [mJ/cm²]							
Tan et al (2015) Scientific Reports 513273	guinea pig	INS	1862	ICC single				10.3 ± 4.9				
Guan et al. (2015) JBO Vol 20(8), 088004-1 to 088004-6.	guinea pig	INS	980	visible CAP			0.037					
Xie et al (2015) Biomedical Optics Express, Vol. 6, No. 6.	guinea pig	INS	1850	visible ABR			32.7 ± 6.7	42.4 ± 11.1				
Tian et al (2017) Laser Med Sci 32:357–362.	guinea pig	INS	980	0.3pV ABR wave 3		~5[b]	~5.5[b]	~6[b]	~8[b]	~12[b]		
Tian et al (2017) Laser Med Sci 32:357–362.	guinea pig	INS	810	0.3pVABR wave 3		~12[b]	~12.5[b]	~13[b]	~13.5[b]	~14[b]		
Wang et al (2017) Laser Med Sci, 32: 389–396.	guinea pig	INS	810	1 μV ABR wave 3	17.2 ± 3.6						316 ± 57.8	
Hernandez et al. (2014) The Journal of Clinical Investigation, Vol. 124, No. 3:1114–1129.	mouse	optogenetic	473	visible ABR								0.22 (2.2 ± 0.4 μJ/mm*²)

[a] low pulse repetition rates.
[b] approximated from figure.

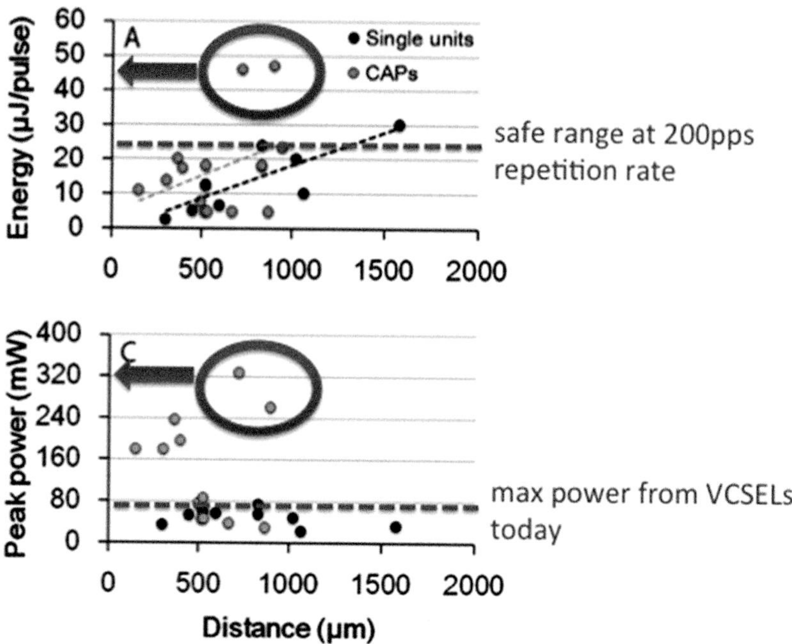

FIGURE 21.7 Panel A of this figure shows the radiant energy required to reach the threshold for stimulation with infrared light. Note, the distance given along the x-axis describes the physical distance between the tip of the optical fiber and the neurons and the y-axis provides the corresponding corrected radiant energy. Black circles show our data obtained from single unit recordings in the ICC, gray circles show data obtained from compound action potentials. For most of the cases the radiant energy was below 25 µJ per pulse. The same data as panel A are shown in panel B. that radiant energy is converted into peak power. Stimulation threshold is typically below 80 mW peak radiant power. The green broken line in panel A shows the largest energy that can be delivered per pulse and the repetition rate of 200 pulses per second. The green broken line in panel B shows the maximum energy that can be delivered edge emitting lasers today.

Thus, changes in, for example, the thickness of the modiolar wall can drastically affect spot size of the energy at the target structures. The second factor is the channel density in the cell membranes. The lower the expression of the ion channel, the more radiant energy is required for stimulation. In the mouse experiments, the energy is about 2 µJ over 1 mm^2 or 0.22 mJ/cm^2/pulse. The radiant exposure for optogenetics is about ten times smaller than the energy required for INS (see also Table 21.3). In contrast, electrical stimulation requires ten times less energy, 0.2 µJ (Zierhofer et al., 1995).

21.4.8.2 Optoacoustics

Optoacoustics describe a technique in which light generates sound waves in gas or another medium (for an overview, see, e.g. Tuchin, 2000). After the target tissue has absorbed the photons, their energy is converted into heat, leading to rapid heating of the optical zone. The resulting thermoelastic expansion causes stress, which will propagate as a stress wave out of the optical zone into the tissue. Pressure waves in particular are large if the pulse duration is shorter than the propagation time for stress wave out of the irradiated zone. For pulse durations shorter than 1 μs the condition has been described as stress-confined (Niemz, 2004; Tuchin, 2000; Welch et al., 2012). During optoacoustic stimulation in the cochlea, laser induced pressure waves vibrate the basilar membrane and stimulate hair cells. The hair cells are required to transform the acoustic signal into action potentials on the auditory nerve. It has been suggested that light induced pressure waves can be used to mechanically stimulate the cochlea (Schultz et al., 2012a; Schultz et al., 2012b; Teudt et al., 2011; Wenzel et al., 2009; Zhang et al., 2009).

Wenzel et al. (2009) used a Nd:YAG laser (λ = 532 nm, τp = 10 ns, f = 10 Hz, *the* Q = 0–23 J/pulse) to stimulate the cochlea of guinea pigs. They were able to evoke auditory brain stem responses in hearing animals. The response could not be evoked in deaf animals. For radiation wavelengths in the near-infrared, the responses correlated with the water absorption for the photons. Furthermore, a positive correlation for responses in the cochlea to optical stimulation occurred at λ = 600–800 nm, the absorption coefficient of hemoglobin (Schultz et al., 2012b).

It has also been reported for INS that for pulse durations, which are orders of magnitude longer than 1 μs, audible stress wave can be generated as well (Teudt et al., 2011; Xia et al., 2016). The peak equivalent value for a radiant exposure of 350 mJ/cm^2 in air was 62 dB (re 20 μPa) and was 31 mPa or 63.8 dB (re 20 μPa) in a swimming pool. In guinea pigs, the directly measured intracochlear pressure during laser stimulation (λ = 1,860 nm, PW = 100 μs, RR = 4 Hz, Q = 164 μJ/pulse) was between 96 and 106 dB (re 20 μPa). Radiant energy at the threshold for neural stimulation is typically ten times less, 17 μJ/pulse (Tan et al., 2015). The corresponding estimated pressure in the cochlea was 76–86 dB (re 20 μPa). With the assumption that the guinea pig middle ear transfer function has a gain of about 30 dB (Magnan et al., 1997; Magnan et al., 1999), the corresponding pressure in the ear canal can be estimated to be 46–56 dB (re 20 μPa).

21.4.8.3 The Dynamic Range

For INS, the radiant energy versus amplitude contours show a sigmoid increase. Once a response becomes visible it rapidly increases. Saturation typically occurs about 6 dB above the threshold energy. A similar energy versus response contour can be seen for the increase in the discharge rate of single auditory nerve fibers (ANFs) and for neural activity recorded from single units of the central nucleus of the inferior colliculus (ICC). For all, CAP recordings, ANFs, and neurons from the ICC, saturation can be achieved with optical stimulation. However, the dynamic range is about 6 dB, which is comparable to the increase in rate with electrical current and is clearly less than the dynamic range for acoustic stimulation, 20–40 dB SPL (re 20 μPa). The rate increases can be achieved over a wider range by applying a novel coding strategy which allows for a more gradual increase in the rate of action potentials with an increasing rate of the acoustic stimulus (see also below).

21.4.9 Safety of Cochlear INS

Irradiation with infrared light may result in heat accumulation in the exposed tissue leading to temperature damage of the tissue. For a neural interface the radiant energy per pulse should be as small as possible and the delivery rate of the pulses at a minimum low rate.

Two types of experiments were completed to determine the radiant energy and pulse repetition rate that allows for safe stimulation at 1,860 nm. The selection of the wavelength was dictated by the light sources available for stimulation. Initial studies were acute experiments in gerbils and cats. In gerbils, stimulation for 6 hours at 200 pps showed stable CAP amplitudes for the duration of the stimulation (Izzo et al., 2006a; Izzo et al., 2007). Similar results have been obtained in cats (Rajguru et al., 2010). While the laser parameters for the first two sets of the experiments were selected so that no damage occurs, another series of experiments was conducted to determine the threshold for damage. No functional or histological damage in the cochlea was observed for up to 5 hours of continuous irradiation if the selected laser parameters for INS were: $\lambda = 1.869$ nm; $\tau p = 100$; $f \leq 250$ Hz; $Q \leq 25$ μJ/pulse (Goyal et al., 2012). Functional loss was observed for radiant energies above 25 μJ/pulse. Corresponding cochlear histology from control animals and animals exposed to 98 or 127 μJ/pulse at 250 Hz pulse repetition rate did not show loss of spiral ganglion neurons, hair cells, or other soft tissue structures of the organ of Corti. Light microscopy did not reveal any structural changes in the soft tissue either. Note that the cochleae were harvested directly after the exposure, with no time for neural degeneration or structural changes to establish (Goyal et al., 2012).

A long-term study was conducted in cats during which they were chronically implanted with a 200 μm optical fiber for INS. Behavioral responses of the cats indicated that stimulation occurred and a perceptual event resulted (Matic et al., 2013). Six weeks of stimulation of the cochlea, 6 h per day, with a fixed set of laser parameters ($\lambda = 1{,}850$, $\tau p = 100$ μs, rate = 200 Hz, $Q = 12$ μJ/pulse) did not change the electrophysiological responses, either optically evoked or acoustically evoked. Spiral ganglion neuron counts and post implantation tissue growth, which is localized at the optical fiber, were similar in chronically stimulated and sham implanted cochleae (Matic et al., 2013).

A biocompatibility study in cats did not show changes in cochlear function after chronically implanting the electrode.

21.4.10 The Coding Strategy

A pre-emphasis filter was applied to the audio input. The filter was a highpass with a slope of 6 dB/octave and a corner frequency at 1,200 Hz. Next, a gammatone-like spectrogram with 64 frequency bands was calculated (Ellis, 2009). The selection of the gammatone-like spectrogram has been proposed previously to be used to mimic the frequency analysis performed by the ear (Patterson et al., 1992). Gammatone filters were conceived as a simple fit to experimental observations of the mammalian cochlea (de Boer, 1975; de Boer et al., 1978). For the calculation of the gammatone-like spectrogram, 128 sample points were used. The "frame-shift" or "hop size" for the calculation was 12 sample points. This corresponds to 272 μs. From the resulting gammatone-like spectrogram, a subset of frequency bands was selected. The number of frequency bands equaled the number of active contacts along the cochlear implant array. For example, for a Med-El device, the number was 12; for an Advanced Bionics Device, 16. Cochlear devices were not considered because a research interface to transmit the data was not available to us. The frequency selection for each frequency band also depended on the placement of the cochlear implant electrode and was determined by the location of an electrode contact along the cochlea. The distance along the cochlea was converted into a corresponding frequency by applying the corrected frequency place map for the spiral ganglion in humans (Stakhovskaya et al., 2007).

Although it has not been implemented in the initial version of the code, the gammatone-like spectrogram can be replaced by a bunch of single gammatone filters with the center frequency calculated according to the frequency place map of the human cochlea and the placement of the contacts of the CI electrode. Since the map can be distorted each patient should undergo a mapping procedure after activation of the CI to optimize the center frequency of the gamma tone filter according to the location where it should stimulate the cochlea. The bandwidth of the filters is adjusted to the critical bandwidth, which is in humans about 10% of the center frequency (e.g. Moore et al., 1987). The use of such a filter bench would decrease the computing power required.

21.4.10.1 Envelop Mapping

In previous research, an equation has been published that describes the relation between the sound level (p) at the outer ear canal and the corresponding rate of action potentials (R) that can be recorded from a single auditory nerve fiber (Köppl et al., 1999; Richter et al., 1995; Sachs et al., 1974; Winter et al., 1990; Yates, 1990; Yates et al., 1990; Yates et al., 2000). The equation includes cochlear nonlinearities and depends on five critical parameters as demonstrated in Figure 21.8: the spontaneous rate (a_0), the maximum rate (a_1), the threshold for stimulation (a_2), the level for nonlinear behavior (a_3), and a value describing the slope after the level for nonlinear behavior (a_4). The parameter a_0 shifts the curve towards larger values. The maximum rate a_1 limits the maximum rate to the number selected. The level for threshold has large effects on the mapping. Low threshold values result in a fast increase in the rate and quick saturation, whereas large threshold values slow the increase in rate, but limit the maximum achievable rate. Smaller effects are seen from the parameters a_3 and a_4. Default values are selected ($a_0 = 0$; $a_1 = 1$; $a_2 = 20$; $a_3 = 50$; $a_4 = 0.5$), which must be adjusted individually for each individual CI user. The selected values for $a_0 = 0$ and $a_1 = 1$ limit the rate R to the interval $0 \leq R \leq 1$. The input for the equation, the sound level p, is

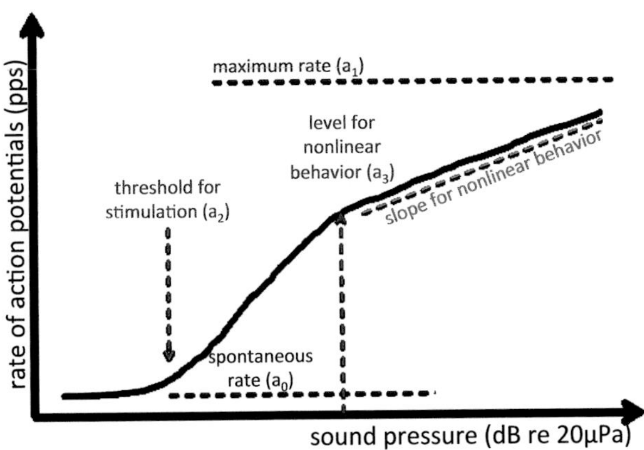

FIGURE 21.8 Rate-intensity functions have been explained by the nonlinearity in the basilar membrane input-output responses (Sachs et al., 1974; Yates, 1990; Yates et al., 1990). This function is used in our coding strategy to map sound pressure level to a normalized rate.

calculated from the acoustical power in each bin along a row of the gammatone-like spectrogram. In the present study the length of one bin corresponded to 272 μs (see also above).

21.4.10.2 TFS Mapping

In the normal hearing auditory system, it has been shown that phase is encoded on the auditory nerve by the timed occurrence of action potentials (phase-locking). After an action potential is generated, the probability for another action potential is at a maximum at integer numbers of cycles following the last action potential (Johnson, 1980; Kiang et al., 1965; Rose et al., 1967). Our code has been developed similarly. In its simplest version, at the occurrence of an electrical pulse, a multiplier called PhaseProb is set to 0 and increases linearly to 1 over the time of 1/(center frequency) of the selected frequency band in the spectrogram. At each integer, the number of cycle times PhaseProb is reset to 0. PowerProb is then modified by multiplying it by PhaseProb, which is a number of the interval $(0 \leq \text{PhaseProb} \leq 1) * w_{\text{phase}}$. The multiplier w_{phase} is a factor that can increase or decrease the phase effect. For the initial setting this weighing factor w_{phase} is selected to be 1.

21.4.11 Decision to Generate a Pulse and Pulse Table

The result of the PowerProb*PhaseProb is then compared with a random number (rn) of the interval $0 \leq rn \leq 1$, which is generated by MATLAB's random number generator. A pulse is generated if this random number is smaller than the number calculated from the normalized rate and the phase. A typical example for the spectrogram and the resulting electrical pulse pattern is shown for the following sentence "The boy did a handstand" (Figure 21.9). Each circle in the raster plot (Figure 21.9, lower panel) corresponds to the

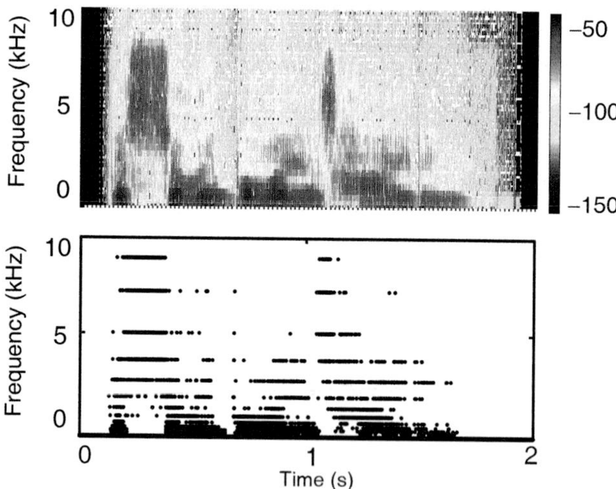

FIGURE 21.9 The top panel shows the spectrogram of the sentence "The silly boy is hiding". A pre-emphasis filter was applied. The bottom panel shows the pulse pattern for this sentence. Each circle shows the time of an electric biphasic pulse. Note: pulses can be delivered simultaneously at neighboring channels. Based on the slow pulse repetition rate and the stochastic behavior of the pulses, however, coincident events at neighboring electrodes are rare.

time an electrical pulse is delivered. The selection of the frequency band determines the rows of the spectrogram.

21.4.12 Outlook

Cochlear implants are considered one of the most successful neural prostheses. Directly stimulating the remaining auditory neurons in severe-to-profoundly deaf persons allows restoration of some of the hearing. Despite the success of the devices the performance of cochlear implants degrades in noisy listening environments and for music. Among many approaches to improve cochlear implants, it has been argued that increasing the number of independent stimulation sites along the cochlea may result in better performance of the patient. Spatially selective stimulation is possible with optical methods and makes them attractive for photonic stimulation of neurons. To translate the technology into a device, today's technical limitations of the light sources and other light delivery systems must be overcome to use a "light-alone" stimulation system. Rather, a combination of electrical and optical stimulation is attractive for incorporation in contemporary neural prostheses. Since little is known about the interaction between optical and electrical stimulation, research is required to explore this modality of stimulation. Furthermore, the improvement and development of optical sources in the near-infrared and infrared would facilitate the development of efficient light delivery systems small enough to be placed directly into the scala tympani along the cochlear spiral.

REFERENCES

Acker, L., Huang, B., Hancock, K.E., Hauswirth, W., Boyden, E.S., Brown, M.C., Lee, D.J. 2011. Channelrhodopsin-2 gene transfection of central auditory neurons: toward an optical prosthesis. *Abstr Assoc Res Otolaryngol* 34, 484.

Albert, E.S., Bec, J.M., Desmadryl, G., Chekroud, K., Travo, C., Gaboyard, S., Bardin, F., Marc, I., Dumas, M., Lenaers, G., Hamel, C., Muller, A., Chabbert, C. 2012. TRPV4 channels mediate the infrared laser-evoked response in sensory neurons. *J Neurophysiol* 107, 3227–34.

Avci, E., Nauwelaers, T., Lenarz, T., Hamacher, V., Kral, A. 2014. Variations in microanatomy of the human cochlea. *J Comp Neurol* 522, 3245–61.

Bae, S.H., Che, J.H., Seo, J.M., Jeong, J., Kim, E.T., Lee, S.W., Koo, K.I., Suaning, G.J., Lovell, N.H., Cho, D.I., Kim, S.J., Chung, H. 2012. In vitro biocompatibility of various polymer-based microelectrode arrays for retinal prosthesis. *Invest Ophthalmol Vis Sci* 53, 2653–7.

Balaban, C.D., Zhou, J., Li, H. 2003. Type 1 vanilloid receptor expression by mammalian inner ear ganglion cells. *Hear Res* 175, 165–170.

Balster, S., Wenzel, G.I., Warnecke, A., Steffens, M., Rettenmaier, A., Zhang, K., Lenarz, T., Reuter, G. 2014. Optical cochlear implant: evaluation of insertion forces of optical fibres in a cochlear model and of traumata in human temporal bones. *Biomed Tech (Berl)* 59, 19–28.

Berenstein, C.K., Mens, L.H., Mulder, J.J., Vanpoucke, F.J. 2008. Current steering and current focusing in cochlear implants: comparison of monopolar, tripolar, and virtual channel electrode configurations. *Ear Hear* 29, 250–60.

Bernstein, J.G., Garrity, P.A., Boyden, E.S. 2012. Optogenetics and thermogenetics: technologies for controlling the activity of targeted cells within intact neural circuits. *Curr Opin Neurobiol* 22, 61–71.

Bierer, J.A., Middlebrooks, J.C. 2002. Auditory cortical images of cochlear-implant stimuli: dependence on electrode configuration. *J Neurophysiol* 87, 478–92.

Boyden, E.S., Zhang, F., Bamberg, E., Nagel, G., Deisseroth, K. 2005. Millisecond-timescale, genetically targeted optical control of neural activity. *Nat Neurosci* 8, 1263–8.

Caterina, M.J., Schumacher, M.A., Tominaga, M., Rosen, T.A., Levine, J.D., Julius, D. 1997. The capsaicin receptor: a heat-activated ion channel in the pain pathway. *Nature* 389, 816–24.

Choi, C.T., Hsu, C.H. 2009. Conditions for generating virtual channels in cochlear prosthesis systems. *Ann Biomed Eng* 37, 614–24.

Darrow, K., Slama, M., Kempfle, J., Boyden, E.S., Polley, D., Brown, M.C., Lee, D.J. 2013a. A comparison of electrical and optical activation of midbrain and cortical pathways in mice expressing channelrhodopsin-2 in the cochlear nucleus. *Abstr Assoc Res Otolaryngol* 36, 265.

Darrow, K., Slama, M., Kempfle, J., Boyden, E.S., Polley, D., Brown, M.C., Lee, D.J. 2013b. Optogenetic control of central auditory neurons. *Abstr Assoc Res Otolaryngol* 36, 695.

de Boer, E. 1975. Synthetic whole-nerve action potentials for the cat. *J Acoust Soc Am* 58, 1030–45.

de Boer, E., de Jongh, H.R. 1978. On cochlear encoding: potentialities and limitations of the reverse-correlation technique. *J Acoust Soc Am* 63, 115–35.

Dittami, G.M., Rajguru, S.M., Lasher, R.A., Hitchcock, R.W., Rabbitt, R.D. 2011. Intracellular calcium transients evoked by pulsed infrared radiation in neonatal cardiomyocytes. *J Physiol* 589, 1295–306.

Ejserholm, F., Stegmayr, J., Bauer, P., Johansson, F., Wallman, L., Bengtsson, M., Oredsson, S. 2015. Biocompatibility of a polymer based on Off-Stoichiometry Thiol-Enes + Epoxy (OSTE+) for neural implants. *Biomater Res* 19, 19.

Ellis, D.P.W. 2009. Gammatone-like spectrograms. [Online] http://www.ee.columbia.edu/~dpwe/resources/matlab/gammatonegram/.

Erixon, E., Högstorp, H., Wadin, K., Rask-Andersen, H. 2009. Variational anatomy of the human cochlea: implications for cochlear implantation. *Otol Neurotol* 30, 14–22.

Goyal, V., Rajguru, S., Matic, A.I., Stock, S.R., Richter, C.P. 2012. Acute damage threshold for infrared neural stimulation of the cochlea: functional and histological evaluation. *Anat Rec (Hoboken)* 295, 1987–99.

Hatsushika, S., Shepherd, R.K., Tong, Y.C., Clark, G.M., Funasaka, S. 1990. Dimensions of the scala tympani in the human and cat with reference to cochlear implants. *Ann Otol Rhinol Laryngol* 99, 871–6.

Hernandez, V.H., Gehrt, A., Reuter, K., Jing, Z., Jeschke, M., Mendoza Schulz, A., Hoch, G., Bartels, M., Vogt, G., Garnham, C.W., Yawo, H., Fukazawa, Y., Augustine, G.J., Bamberg, E., Kugler, S., Salditt, T., de Hoz, L., Strenzke, N., Moser, T. 2014. Optogenetic stimulation of the auditory pathway. *J Clin Invest* 124, 1114–29.

Hernandez, V.H., Hoch, G., Reuter, K., Jing, Z., Bartels, M., Vogt, G., Garnham, C.W., Augustine, G.J., Kügler, S., Salditt, T., Strenzke, N., Moser, T. 2013. Optogenetic stimulation of the auditory nerve: towards an optical cochlear prosthetic. *Abstr Assoc Res Otolaryngol* 36, 694.

Izzo, A.D., Joseph T. Walsh, J., Jansen, E.D., Bendett, M., Webb, J., Ralph, H., Richter, C.-P. 2007. Optical parameter variability in laser nerve stimulation: a study of pulse duration, repetition rate, and wavelength. *IEEE Trans Biomed Eng* 54, 1108–14.

Izzo, A.D., Richter, C.-P., Jansen, E.D., Walsh, J.T. 2006a. Laser stimulation of the auditory nerve. *Laser Surg Med* 38, 745–53.

Izzo, A.D., Pathria, J., Suh, E., Walsh Jr., J.T., Whitlon, D.S., Jansen, E., Richter, C.-P. 2006b. Selectivity of optical stimulation in the auditory system. SPIE 6078, 60781P1–P7.

Jeschke, M., Moser, T. 2015. Considering optogenetic stimulation for cochlear implants. *Hear Res* 322, 224–34.

Johnson, D.H. 1980. The relationship between spike rate and synchrony in responses of auditory-nerve fibers to single tones. *J Acoust Soc Am* 68, 1115–22.

Kiang, N.Y.-S, Watanabe, T., Thomas, E.C. and Clark, L.F. 1965. Discharge patterns of single fibers in the cat's auditory nerve. *MIT Research Monograph No*, 35. MIT Press, Cambridge, MA.

Koch, D.B., Downing, M., Osberger, M.J., Litvak, I.M. 2007. Using current steering to increase spatial resolution in CII and HiRes 90K users. *Ear Hear.* 28, 38S–41S.

Köppl, C., Yates, G. 1999. Coding of sound pressure level in the barn owl's auditory nerve. *J Neurosci* 19, 9674–86.

Landsberger, D.M., Srinivasan, A.G. 2009. Virtual channel discrimination is improved by current focusing in cochlear implant recipients. *Hear Res* 254, 34–41.

Liu, Q., Frerck, M.J., Holman, H.A., Jorgensen, E.M., Rabbitt, R.D. 2014. Exciting cell membranes with a blustering heat shock. *Biophys J* 106, 1570–7.

Lumbreras, V., Bas, E., Gupta, C., Rajguru, S.M. 2014. Pulsed infrared radiation excites cultured neonatal spiral and vestibular ganglion neurons by modulating mitochondrial calcium cycling. *J Neurophysiol* 112, 1246–55.

Magnan, P., Avan, P., Dancer, A., Smurzynski, J., Probst, R. 1997. Reverse middle-ear transfer function in the guinea pig measured with cubic difference tones. *Hear Res* 107, 41–5.

Magnan, P., Dancer, A., Probst, R., Smurzynski, J., Avan, P. 1999. Intracochlear acoustic pressure measurements: transfer functions of the middle ear and cochlear mechanics. *Audiol Neurootol* 4, 123–8.

Matic, A.I., Robinson, A.M., Young, H.K., Badofsky, B., Rajguru, S.M., Stock, S., Richter, C.P. 2013. Behavioral and electrophysiological responses evoked by chronic infrared neural stimulation of the cochlea. *PLoS One* 8, e58189.

Mattioli-Belmonte, M., Giavaresi, G., Biagini, G., Virgili, L., Giacomini, M., Fini, M., Giantomassi, F., Natali, D., Torricelli, P., Giardino, R. 2003. Tailoring biomaterial compatibility: in vivo tissue response versus in vitro cell behavior. *Int J Artif Organs* 26, 1077–85.

Mens, L.H., Berenstein, C.K. 2005. Speech perception with mono- and quadrupolar electrode configurations: a crossover study. *Otol Neurotol* 26, 957–64.

Moore, B.C., Glasberg, B.R. 1987. Formulae describing frequency selectivity as a function of frequency and level, and their use in calculating excitation patterns. *Hear Res* 28, 209–25.

Moreno, L.E., Rajguru, S.M., Matic, A.I., Yerram, N., Robinson, A.M., Hwang, M., Stock, S., Richter, C.P. 2011. Infrared neural stimulation: beam path in the guinea pig cochlea. *Hear Res* 282, 289–302.

Moser, T., Hernandez, V.H., Hoch, G., Reuter, K., Jing, Z., Bartels, M., Vogt, G., Garnham, C.W., Augustine, G.J., Kügler, S., Salditt, T., Strenzke, N. 2013. Optogenetic stimulation of the auditory nerve. *Abstr Assoc Res Otolaryngol* 36, 268.

Müller, M. 1996a. *Frequenz- und Intensitätsanalyse im Innenohr der Säuger.* Habilitationsschrift Johann Wolfgang Goethe-Universität Frankfurt/Main.

Müller, M. 1996b. The cochlear place-frequency map of the adult and developing Mongolian gerbil. *Hear Res* 94, 148–56.

Niemz, M.H. 2004. *Laser Tissue Interactions: Fundamentals and Application.* 2nd ed. Springer, New York.

Nogueira, W., Litvak, L.M., Landsberger, D.M., Büchner, A. 2017. Loudness and pitch perception using dynamically compensated virtual channels. *Hear Res* 344, 223–34.

Patterson, R.D., Robinson, K., Holdsworth, J.W., McKeown, D., Zhang, C., Allerhand, M. 1992. Complex sounds and auditory images. In: Cazals, Y., Demany, L., Horner, K., (Eds.), *Auditory Physiology and Perception.* Pergamon, Oxford, pp. 429–46.

Plaksin, M., Kimmel, E., Shoham, S. 2017. Thermal transients excite neurons through universal intramembrane mechano-electrical effects. *bioRxiv* 111724.

Rabbitt, R.D., Brichta, A.M., Tabatabaee, H., Boutros, P.J., Ahn, J., Della Santina, C.C., Poppi, L.A., Lim, R. 2016. Heat pulse excitability of vestibular hair cells and afferent neurons. *J Neurophysiol* 116, 825–43.

Rajguru, S.M., Matic, A.I., Robinson, A.M., Fishman, A.J., Moreno, L.E., Bradley, A., Vujanovic, I., Breen, J., Wells, J.D., Bendett, M., Richter, C.P. 2010. Optical cochlear implants: evaluation of surgical approach and laser parameters in cats. *Hear Res* 269, 102–11.

Rajguru, S.M., Richter, C.P., Matic, A.I., Holstein, G.R., Highstein, S.M., Dittami, G.M., Rabbitt, R.D. 2011. Infrared photostimulation of the crista ampullaris. *J Physiol* 589, 1283–94.

Ren, T. 2002. Longitudinal pattern of basilar membrane vibration in the sensitive cochlea. *Proc Natl Acad Sci U S A* 99, 17101–6.

Rhee, A.Y., Li, G., Wells, J., Kao, Y.P.Y. 2008. Photostimulation of sensory neurons of the rat vagus nerve. SPIE 6854, 68540E1.

Richter, C.P., Heynert, S., Klinke, R. 1995. Rate-intensity-functions of pigeon auditory primary afferents. *Hear Res* 83, 19–25.

Richter, C.-P., Rajguru, S.M., Matic, A.I., Moreno, E.L., Fishman, A.J., Robinson, A.M., Suh, E., Walsh Jr., J.T. 2011a. Spread of cochlear excitation during stimulation with optical radiation: inferior colliculus measurements. *J Neural Eng* 8, 056006.

Richter, C.P., Rajguru, S.M., Matic, A.I., Moreno, E.L., Fishman, A.J., Robinson, A.M., Suh, E., Walsh, J.T. 2011b. Spread of cochlear excitation during stimulation with pulsed infrared radiation: inferior colliculus measurements. *J Neural Eng* 8, 056006.

Rose, J.E., Brugge, J.F., Anderson, D.J., Hind, J.E. 1967. Phase-locked response to low-frequency tones in single auditory nerve fibers of the squirrel monkey. *J Neurophysiol* 30, 769–93.

Sachs, M.B., Abbas, P.J. 1974. Rate versus level functions for auditory-nerve fibers in cats: tone-burst stimuli. *J Acoust Soc Am* 56, 1835–47.

Schultz, M., Baumhoff, P., Teudt, I.U., Maier, H., Kruger, A., Lenarz, T., Kral, A. 2012a. Pulsed wavelength-dependent laser stimulation of the inner ear. *Biomed Tech (Berl)* 57 Suppl 1.

Schultz, M., Baumhoff, P., Maier, H., Teudt, I.U., Kruger, A., Lenarz, T., Kral, A. 2012b. Nanosecond laser pulse stimulation of the inner ear-a wavelength study. *Biomed Opt Express* 3, 3332–45.

Shapiro, M.G., Homma, K., Villarreal, S., Richter, C.P., Bezanilla, F. 2012. Infrared light excites cells by changing their electrical capacitance. *Nat Commun* 3, 736.

Shepherd, R.K., Colreavy, M.P. 2004. Surface microstructure of the perilymphatic space: implications for cochlear implants and cell- or drug-based therapies. *Arch Otolaryngol Head Neck Surg* 130, 518–23.

Snyder, R.L., Bierer, J.A., Middlebrooks, J.C. 2004. Topographic spread of inferior colliculus activation in response to acoustic and intracochlear electric stimulation. *JARO* 5, 305–22.

Srinivasan, A.G., Landsberger, D.M., Shannon, R.V. 2010. Current focusing sharpens local peaks of excitation in cochlear implant stimulation. *Hear Res* 270, 89–100.

Stakhovskaya, O., Sridhar, D., Bonham, B.H., Leake, P.A. 2007. Frequency map for the human cochlear spiral ganglion: implications for cochlear implants. *JARO* 8, 220–33.

Starr, P., Agrawal, C.M., Bailey, S. 2016. Biocompatibility of common polyimides with human endothelial cells for a cardiovascular microsensor. *J Biomed Mater Res A* 104, 406–12.

Suh, E., Matic, A.I., Otting, M., Walsh Jr., J.T., Richter, C.-P. 2009. Optical stimulation in mice which lack the TRPV1 channel. *Proc of SPIE* 7180, 71800S, 1–5.

Sun, Y., Lacour, S.P., Brooks, R.A., Rushton, N., Fawcett, J., Cameron, R.E. 2009. Assessment of the biocompatibility of photosensitive polyimide for implantable medical device use. *J Biomed Mater Res A* 90, 648–55.

Takumida, M., Kubo, N., Ohtani, M., Suzuka, Y., Anniko, M. 2005. Transient receptor potential channels in the inner ear: presence of transient receptor potential channel subfamily 1 and 4 in the guinea pig inner ear. *Acta Otolaryngol* 125, 929–34.

Tan, X., Rajguru, S., Young, H., Xia, N., Stock, S.R., Xiao, X., Richter, C.P. 2015. Radiant energy required for infrared neural stimulation. *Sci Rep* 5, 13273.

Teudt, I.U., Maier, H., Richter, C.P., Kral, A. 2011. Acoustic events and "optophonic" cochlear responses induced by pulsed near-infrared laser. *IEEE Trans Biomed Eng* 58, 1648–55.

Teudt, I.U., Nevel, A.E., Izzo, A.D., Walsh, J.T., Jr., Richter, C.P. 2007. Optical stimulation of the facial nerve: a new monitoring technique? *The Laryngoscope* 117, 1641–7.

Tuchin, V. 2000. *Tissue Optics: Light Scattering Methods and Instruments for Medical Diagnosis.* SPIE Press, Bellingham, WA.

van den Honert, C., Kelsall, D.C. 2007. Focused intracochlear electric stimulation with phased array channels. *J Acoust Soc Am* 121, 3703–16.

Welch, A.J., van Gemert, M.J.C. 2012. *Optical-Thermal Response of Laser-Irradiated Tissue*. Second ed. Plenum Press, New York.

Wells, J., Kao, C., Jansen, E.D., Konrad, P., Mahadevan-Jansen, A. 2005. Application of infrared light for in vivo neural stimulation. *J Biomed Opt* 10, 064003.

Wenzel, G.I., Balster, S., Zhang, K., Lim, H.H., Reich, U., Massow, O., Lubatschowski, H., Ertmer, W., Lenarz, T., Reuter, G. 2009. Green laser light activates the inner ear. *J Biomed Opt* 14, 044007.

Winter, I.M., Robertson, D., Yates, G.K. 1990. Diversity of characteristic frequency rate-intensity functions in guinea pig auditory nerve fibres. *Hear Res* 45, 191–202.

Wu, C.C., Luo, X. 2016. Excitation patterns of standard and steered partial tripolar stimuli in cochlear implants. *J Assoc Res Otolaryngol* 17, 145–58.

Xia, N., Tan, X., Xu, Y., Hou, W., Mao, T., Richter, C.P. 2016. Pressure in the cochlea during infrared irradiation. *IEEE Trans Biomed Eng* 65, 1575–1584.

Yao, J., Liu, B., Qin, F. 2009. Rapid temperature jump by infrared diode laser irradiation for patch-clamp studies. *Biophys J* 96, 3611–9.

Yates, G.K. 1990. Basilar membrane nonlinearity and its influence on auditory nerve rate-intensity functions. *Hear Res* 50, 145–62.

Yates, G.K., Manley, G.A., Köppl, C. 2000. Rate-intensity functions in the emu auditory nerve. *J Acoust Soc Am* 107, 2143–54.

Yates, G.K., Winter, I.M., Robertson, D. 1990. Basilar membrane nonlinearity determines auditory nerve rate-intensity functions and cochlear dynamic range. *Hear Res* 45, 203–19.

Zhang, K.Y., Wenzel, G.I., Balster, S., Lim, H.H., Lubatschowski, H., Lenarz, T., Ertmer, W., Reuter, G. 2009. Optoacoustic induced vibrations within the inner ear. *Opt Express* 17, 23037–43.

Zheng, J., Dai, C., Steyger, P.S., Kim, Y., Vass, Z., Ren, T., Nuttall, A.L. 2003. Vanilloid receptors in hearing: altered cochlear sensitivity by vanilloids and expression of TRPV1 in the organ of corti. *J Neurophysiol* 90, 444–55.

Zierhofer, C.M., Hochmair-Desoyer, I.J., Hochmair, E.S. 1995. Electronic design of a cochlear implantant for multi-channel high rate pulsatle stimulation strategies. *IEEE Trans Rehab Eng* 3, 112–6.

CHAPTER 22

Label-Free Fluorescence Interrogation of Brain Tumors

Brad A. Hartl, Shamira Sridharan, and Laura Marcu

CONTENTS

22.1 INTRODUCTION

The National Cancer Institute projects that there will be 23,770 new cases of nervous system cancers, including brain cancers, in the United States in 2016 and 16,050 deaths due to the disease [1]. This disease has an extensive economic toll, estimated at $4.9 billion in treatment costs alone in 2014, and yet has one of the worst cancer-specific survival rates [1, 2]. Glioblastoma multiforme (GBM) is the most common primary brain tumor and also the most aggressive, with a five-year mean survival rate of only 3–5% [3, 4]. Without treatment, the average survival time for GBM patients is typically three months and with treatment, the median survival is 14.6 months [5, 6]. Gliomas arise due to an accumulation of genetic and chromosomal alterations in non-neuronal glial cells, which provide a variety of support functions for the neurons in the central nervous system. Gliomas are generally classified in two ways, by their cell type and tumor grade. Cell types can consist of either ependymoma, astrocytoma, oligodendroglioma, or a combination of these. A grading system is then used to further classify the tumor based on pathologic evaluation. The World Health Organization classification system is the most common, and consists of grades I through IV. Low grade gliomas (LGG) – grade I and grade II – are less aggressive and can often be treated with surgical resection alone or in combination with radio- and chemotherapy [7]. High grade gliomas (HGG) – grade III and grade IV – are considered very aggressive and are histologically highly heterogeneous, containing both stromal and neoplastic tissues [8]. Grade IV tumors, including GBM, are mitotically active, necrosis-prone neoplasms and are usually associated with swift disease progression and a fatal outcome.

Current standard of care for malignant primary brain tumors, particularly high grade gliomas, consists of maximal safe surgical resection of the tumor followed by radiation and/or chemotherapy [9]. The extent of resection is conventionally determined by comparing the volume of enhanced signal intensity in the pre-operative and post-operative T1-weighted contrast enhanced images (T2-weighted for non-enhancing tumors) obtained using magnetic resonance imaging (MRI) [10]. Gliomas invade the normal brain parenchyma in finger-like projections, which along with the anatomical location of the tumor itself can make complete tumor resection difficult to achieve without removing functionally significant normal brain tissue [9]. However, studies have shown that the extent of surgical removal of malignant tissue is the single most important factor determining long-term survival of the patient [11, 12]. Therefore, surgeons strive to find a balance between aggressive tissue resection and minimization of normal brain tissue loss in order to improve patient outcomes.

Various tools have been developed to aid neurosurgeons in this complex decision-making process. Image-guided stereotaxic surgical procedures that incorporate guidance from pre-operative MRI scans and ultrasound are commonly used in neurosurgery. However, non-uniform shifts in the brain tissue due to the opening of the skull, tissue resection, and cyst reduction limits the applicability of pre-operative imaging modalities for margin assessment. Techniques that incorporate pre-operative MRI and intra-operative ultrasound suffer from inaccuracies in registration of images across the two modalities, which has been reported to be up to 5 mm, in addition to those from the brain shift itself [13]. Intra-operative 3D ultrasound has the lowest overall inaccuracy, which is reported at 1.4 ± 0.45 mm [14]. However, in the brain, these seemingly small inaccuracies can lead to the loss of functional brain tissue. Therefore, while the MRI and ultrasound-based imaging modalities improve targeting of the lesions, there is still a need for a high precision locational tissue assessment of surgical margins. Intra-operative biopsies can provide a more definite pathological identification, but requires the tissue to be resected, and the whole process can take up to 20 minutes.

Optical imaging modalities providing pathological tissue assessment with a high degree of spatial precision, and working in conjunction with current image guided surgical modalities, can help bridge this gap in surgical tumor margin assessment in the brain. In this chapter, we review the potential applications of label-free optical interrogation techniques that exploit tissue autofluorescence for enhancing outcomes of brain tumor surgery.

22.2 ENDOGENOUS FLUORESCENCE

22.2.1 Steady State Fluorescence

22.2.1.1 Steady State Fluorescence Parameters

Steady state fluorescence measurements provide a relatively straightforward means to observe either the entire fluorescence spectra, or select wavelengths bands to facilitate ratiometric measurements and/or imaging. When fluorescence emissions are acquired, a variety of parameters can be extracted from the measurements, including ratios between different wavelengths (such as to calculate an approximation of redox ratio), peak shifts, peak broadening/narrowing, and intensity changes. One of the largest challenges in making steady state measurements of tissue is maintaining consistent excitation/collection geometries, as such fluctuations can cause changes to the acquired emission spectrum. These changes in excitation/collection will also make collecting absolute intensity measurements challenging unless the source-to-sample distances are accurately accounted for.

22.2.1.2 Steady State Instrumentation

The instrumentation necessary for steady state autofluorescence spectroscopy or imaging is relatively simple, the basic requirements are: an excitation source, a detector to measure the emission, and an optical setup to transmit and collect the light. For intra-operative applications, fiber optics and endoscopes can provide the most flexibility to access the targeted tissue; but for applications with more accessibility the transmission and collection of light can be accomplished with a camera and/or microscope lens. Excitation can be performed with lasers, LEDs, or lamps; the latter of which can provide the greatest spectral flexibility when paired with either filters or a monochromator to select wavelength(s). Correspondingly, the collected emission light can be selected using filters, a monochromator, or a liquid-crystal tunable filter, all of which can then be sampled and digitized using a photodiode, CCD, or photomultiplier tube. For steady state measurements, both excitation and emission wavelength bands can be readily tuned to provide information on the fluorophores of interest.

22.2.2 Time-Resolved Fluorescence

Fluorescence lifetime is an intrinsic property of a fluorophore and represents a non-destructive and robust means to provide a wider variety of information – relative to steady state measurements – about the molecule itself and its surrounding environment. Since it is not affected by the wavelength of excitation, one- or multi-photon excitation, duration of light exposure, or intensity of emission, fluorescence lifetime can be considered a state function [15]. Thus, it can provide information on a variety of local environmental factors including, but not limited to, temperature, pH, polarity, and the presence of fluorescence quenchers, as well as internal factors such as its molecular structure, all of which cannot be determined with steady state fluorescence measurements.

22.2.2.1 Fluorescence Lifetime Parameter

After a molecule has been energized to an excited state (S_1 or higher) and undergone internal conversion back to the lowest vibrational level of S_1, there are a variety of competing pathways it can take to return to its ground state, as outlined in Figure 22.1. Direct pathways include fluorescence and internal conversion, and indirect pathways (via intersystem crossing to a triplet state, T_1) include phosphorescence and internal conversion. If the rates of the nonradiative processes are combined as

$$k_{nr}^S = k_{ic}^S + k_{isc}, \tag{22.1}$$

then the fluorescence lifetime can be defined as

$$\tau_S = \frac{1}{k_r^S + k_{nr}^S}, \tag{22.2}$$

and the fluorescence quantum yield as

$$\Phi_F = \frac{k_r^S}{k_k^S + k_{nr}^S}. \tag{22.3}$$

Thus, if the rate at which a molecule undergoes internal conversion is reduced, the molecule's lifetime and quantum yield will increase. This provides the basis for the thermal dependence of lifetimes, where increases in lifetime and yield are generally observed as temperature decreases due to the reduced thermal agitations (including collisions with solvent molecules and intramolecular vibrations) [16].

For a mono-exponential fluorophore, the intensity can be represented as a single exponential function where I_0 is the intensity at $t = 0$ and the fluorescence lifetime, τ, is the length of time it takes for the intensity to decrease to $1/e$ of the original intensity

$$I(t) = I_0 e^{(-t/\tau)}. \tag{22.4}$$

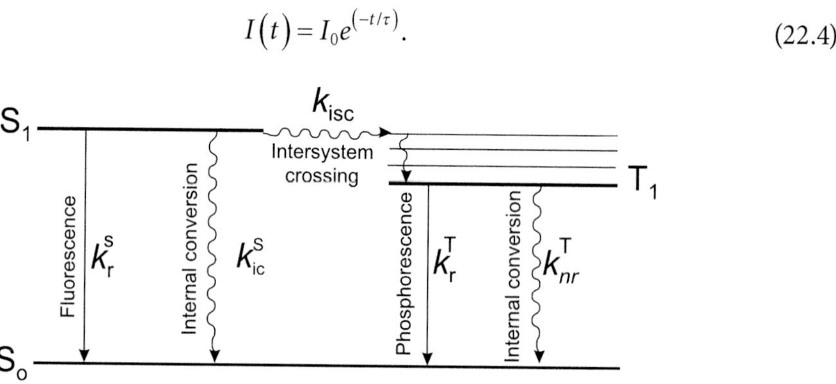

FIGURE 22.1 Jablonski diagram outlining the primary vibrational and electronic pathways – and their associated rate constants – an excited molecule can take to return back to the ground state.

For biological tissues composed of multiple fluorophores, the total intensity can be calculated as the sum of the M components

$$I_{\text{tot}}(t) = \sum_{i=1}^{M} \alpha_i e^{(-t/\tau_i)} \tag{22.5}$$

where i represents the ith fluorescence component. The pre-exponential factors α_i represent the individual amplitudes from each decay component at $t=0$, and $\Sigma\alpha_i$ is commonly normalized to unity. The average lifetime of the set of M components can be calculated as

$$\tau_{\text{avg}} = \frac{\sum_{i=1}^{M} \alpha_i \tau_i^2}{\sum_{i=1}^{M} \alpha_i \tau_i} \tag{22.6}$$

or as

$$\tau_{\text{avg}} = \sum_{i=1}^{M} f_i \tau_i \tag{22.7}$$

where

$$f_i = \frac{\alpha_i \tau_i}{\sum_{i=1}^{M} \alpha_i \tau_i} \tag{22.8}$$

represents the fractional contribution of each decay component to the steady state intensity. From such aggregates of multiple fluorophores, the average lifetime can be calculated as follows [17]

$$\tau_{\text{avg}} = \frac{\int_{t=0}^{\infty} t \cdot I(t)}{\int_{t=0}^{\infty} I(t)}. \tag{22.9}$$

22.2.2.2 Fluorescence Lifetime Instrumentation

To measure fluorescence lifetime a variety of instrumentation setups have been utilized, including time-correlated single-photon counting (TCSPC), pulse sampling, time-gated, and phase modulation (frequency domain) techniques [18]. The TCSPC method measures the timing of individual photons as they are detected, relative to the excitation pulse, whereas the other methods measure the intensity at the different time-points along the fluorescent decay. TCSPC commonly employs high-repetition-rate lasers with ultrashort pulse durations for excitation. Photon-counting multichannel photomultiplier tubes (PMT) or single-photon avalanche photodiodes are typically used as detectors, along with constant fraction discriminators and a time-to-amplitude converter to determine the photon-arrival time, with respect to the excitation laser pulse. The aggregate of arrival times is then converted into a histogram which represents the fluorescence intensity decay in time.

The pulse sampling method uses short pulses for excitation, after which the fluorescence response is detected with a photodetector, and electronics that have fast enough response times to resolve the fluorescence decay in time. In addition to the pulse width of the laser there are two primary instrumentation factors that affect fluorescence transients measured. The first factor is the response time of the electronic components such as the photodetector, amplifier, and digitizer. For the pulse sampling technique, multichannel plate photomultiplier tubes (MCP-PMT) are often preferred over traditional PMTs because they produce significantly less temporal broadening of the signal. The second factor is the intermodal dispersion that arises from the multimode fibers, which are commonly employed to temporally delay different wavelength bands as the autofluorescence signals are acquired [19]. The system response time is typically between 100 picoseconds and a few nanoseconds, so it is necessary to deconvolve the measured signal to obtain the true fluorescence decay. Due to the need for fast detector and digitizer electronics, point measurements are traditionally acquired and raster scanning is used to generate images, as can also be the case for TCSPC. Although each method has its own advantages and disadvantages when compared to other techniques, the pulse sampling method is perhaps the most conducive to clinical translation due to the fact that it can be performed in the ambient lighting of a surgical setting.

The time-gated technique works similarly, except that the sampling interval (also referred to as the gate) for individual excitation pulses is delayed in time to provide intensity information at select time-points along the decay curve. This method is commonly combined with imaging-based detectors such as gated optical imaging intensifiers with CCDs and intensified-CCDs. Phase modulation (frequency domain), uses a high-frequency (typically MHz) intensity modulated excitation source. The fluorescence signal is then delayed, e.g. phase shifted, relative to the excitation signal, thus providing an estimation for the lifetimes of the fluorophores being sampled.

22.3 SOURCES OF ENDOGENOUS FLUORESCENCE IN BRAIN TISSUE

It is well known that a wide variety of naturally occurring molecules inside the human body fluoresce [15, 20]. However, due to the unique composition and needs of the central nervous system, only a few of these fluorophores are present in the brain and most are located intracellularly. The fluorophores reviewed in this section are limited to those that can be appreciably excited using 300 nm to 400 nm wavelength excitation sources. A listing of studies that have measured these fluorophores in isolation is included at the end of the section (Table 22.1). Ultimately, when reviewing the lifetimes of the fluorophores in this table, it is important to keep in mind that most measurements were performed on different imaging setups and different techniques were used to calculate lifetime. Through this table we hope to convey the complexity and interconnectedness of the system of fluorophores that combine to generate the aggregated autofluorescence signals measured in brain tissue.

TABLE 22.1 Fluorescence Properties of the Endogenous Fluorophores Found in Central Nervous System Tissue

Fluorophore	Function in Brain	Ref.	λ_{ex}	$\lambda_{em\text{-}max\ free/bound}$	$\tau_{avg\ free/bound}$	$\tau_1\ (\alpha_1)$	$\tau_2\ (\alpha_2)$	$\tau_3\ (\alpha_3)$	$\tau_4\ (\alpha_4)$	ΔQE
NAD+ (free)	Cofactor coenzyme (oxidized) [76]	[22]	365	3.26*	0.6 (0.65) 4.0 (0.35)					
NADH (free)	Cofactor coenzyme (reduced) [76]	[22]	365		0.806*	0.55 (0.9)	1.6 (0.1)			
		[24]	370[2p]		0.444	0.350 (0.88*)	0.760 (0.12*)			
		[32]	370[2p]		0.428	0.356 (0.82)	0.754 (0.18)			
		[58]	355	460	0.63–0.70					
		[64]	375[2p]	465						
		[70]	350	470	2.92*	0.387 (0.73)	3.650 (0.27)			
		[77]	355	460	0.51*	0.400 (0.83*)	0.780 (0.17*)			
		[78]	370[2p]		0.460	0.395 (0.96*)	1.05 (0.04*)			
NADH (glyceraldehyde phosphate dehydrogenase)	Glycolysis [79]	[30]	340	465/465						0.6
NADH (lactate dehydrogenase)	Glycolysis [79]	[30]	340	465/440						3
		[31]	337		1.57*	0.55 (0.75)	2.3 (0.25)			
		[32]	370[2p]		0.428/1.08	0.338 (0.495*)	1.081 (0.472*)	2.53 (0.033*)		
		[34]	352							1.5
		[39]	360	460/435						
NADH (isocitrate dehydrogenase)	Krebs Cycle [79]	[33]	280		4.84					

(Continued)

TABLE 22.1 (CONTINUED) Fluorescence Properties of the Endogenous Fluorophores Found in Central Nervous System Tissue

Fluorophore	Function in Brain	Ref.	λ_{ex}	$\lambda_{em\text{-}max\ free/bound}$	$\tau_{avg\ free/bound}$	$\tau_1\ (\alpha_1)$	$\tau_2\ (\alpha_2)$	$\tau_3\ (\alpha_3)$	$\tau_4\ (\alpha_4)$	ΔQE
NADH (malate dehydrogenase)	Krebs Cycle [79]	[24]	370^{2p}		0.444/0.806	0.602 (0.85*)	1.33 (0.15*)			
		[27]	355	470/470	0.38/0.94					~2
		[31]	337		1.42*	0.8 (0.7)	2.0 (0.3)			
		[32]	370^{2p}		0.428/0.81	0.304 (0.307*)	0.767 (0.642*)	1.57 (0.051*)		
		[36]	385^{2p}							~4.8
		[40]	355^{2p}	468/446						
		[80]	385^{2p}		0.480/0.769					
NADH (complex I)	Electron Transport Chain [79]	[28]	340–360		5.7					10
NADH (alcohol dehydrogenase)	Detoxifying [79]	[25]	336		2.81	1.8 (0.76*)	4.2 (0.24*)			
		[33]	280		5.36					
		[37]	340							1.5–2
		[41]	337	469/451						
NADH (aldehyde dehydrogenase)	Detoxifying [79]	[38]	340							1.5–2
NADH (carboxyl-terminal binding protein (CtBP))	Transcriptional corepressor [79]	[29]	340	455/425						~5
$NADP^+$ (free)	Cofactor coenzyme (oxidized) [76]	[22]	365		3.92*	0.65 (0.5)	4.4 (0.50)			
NADPH (free)	Cofactor coenzyme (reduced) [76]	[22]	365		1.20*	0.53 (0.75)	1.8 (0.25)			
NADPH (glutamate dehydrogenase)	Detoxifying/ urea synthesis [79]	[51]	340							1.5–2

(Continued)

TABLE 22.1 (CONTINUED) Fluorescence Properties of the Endogenous Fluorophores Found in Central Nervous System Tissue

Fluorophore	Function in Brain	Ref.	λ_{ex}	$\lambda_{em\text{-}max\ free/bound}$	$\tau_{avg\ free/bound}$	$\tau_1\ (\alpha_1)$	$\tau_2\ (\alpha_2)$	$\tau_3\ (\alpha_3)$	$\tau_4\ (\alpha_4)$	ΔQE
NADPH (catalase)	Antioxidant [79]	[52]	340	460/440						~1.7
FAD (free)	Cofactor prosthetic group for oxidoreductases [81]	[54]	470	530						
		[55]	400	530						
		[58]	355	540	1.54					
		[59]	370		2.91					
		[60]	370		2.36					
		[70]	446	520	2.75*	0.330 (0.18)	2.810 (0.82)			
		[77]	440	525	2.8					
		[82]	450	540						
		[83]	460		2.50*	0.014 (0.81)	1.09 (0.07)	2.06 (0.11)	4.89 (0.02)	
		[84]	N/A	530	5.2/<1					
FAD (flavin reductase)	DNA synthesis (rxn also involves NADH) [79]	[56]	440		3.9/0.72*	0.028 (0.25)	0.20 (0.52)	0.46 (0.22)	3.3 (0.01)	
FAD (lipoamide dehydrogenase)	Mitochondrial enzyme [79]	[57]	438	522/504	2.47/3.81*	0.88 (0.35)	4.14 (0.65)			
		[85]	370		2.92*	0.8 (0.49)	3.4 (0.51)			
		[86]	458		2.0					2.4
FAD (D-amino acid oxidase)	Peroxisomal enzyme [79]	[87]	355		2.03*	0.04 (0.24)	0.13 (0.52)	2.3 (0.24)		0.26
FMN (free)	Cofactor prosthetic group for oxidoreductases [88]	[22]	436		5.0					
		[59]	370		4.27					
		[60]	370		4.50					
		[89]	436	520						
		[90]	470		0.43–4.68 depending on conc.					

(Continued)

TABLE 22.1 (CONTINUED) Fluorescence Properties of the Endogenous Fluorophores Found in Central Nervous System Tissue

Fluorophore	Function in Brain	Ref.	λ_{ex}	$\lambda_{em\text{-}max\ free/bound}$	$\tau_{avg\ free/bound}$	$\tau_1\ (\alpha_1)$	$\tau_2\ (\alpha_2)$	$\tau_3\ (\alpha_3)$	$\tau_4\ (\alpha_4)$	ΔQE
FMN (flavin reductase)	DNA synthesis (rxn also involves NADH) [79]	[56]	440		4.6/2.80*	0.008 (0.19)	0.05 (0.39)	0.25 (0.39)	4.5 (0.04)	
Vitamin B_2 – Riboflavin	Coenzyme, major component of FAD & FMN [91]	[59]	370	4.12						
		[64]	375^{2p}	530						
Vitamin B_6 – Pyridoxal phosphate (PLP)	Coenzyme, the only active form of Vitamin B6 [92, 93]	[61]	327	400						
		[62]	330	390						
		[63]	337		2–3					
Vitamin B_6 – Pyridoxine	[94]	[64]	375^{2p}	390						
Vitamin B_6 – Pyridoxamine	[95]	[65]	340	390						
Vitamin B_9 – folic acid	DNA synthesis/repair, cofactor [96]	[64]	375^{2p}	450						
		[66]	370	470						
Vitamin D_3 – cholecalciferol	Precursor to calcitriol [97]	[64]	375^{2p}	460						
Lipofuscin (lipopigments)	Lipid-containing byproduct of lysosomal digestion [98]	[67]	N/A	560						
		[68]	364	560						
		[69]	330–80	555						
		[70]	446		1.98*	0.390 (0.48)	2.240 (0.52)			
Aggrecan	Extracellular matrix [72]	**[58]**	355	420	3.5–5.5					
Chondroitin sulfate	Extracellular matrix [73]	**[58]**	355	425	2.8–4.4					
Cholesterol	Membrane, antioxidant [71]	**[58]**	355	400	3.0–5.7					
Serotonin	Neurotransmitter [99]	[75]	366	430						
Glutamate decarboxylase (GAD)	Neurotransmitter production (GABA) [100]	[74]	335	380						

Change in quantum yield, ΔQE, is defined as the ratio of the quantum yield of the fluorophore when bound to an enzyme divided by the quantum yield of the fluorophore when unbound. Asterisks (*) denote calculated values for τ_{avg} based on Equations 22.6 and 22.7, as well as α_i when f_i and/or τ_{avg} were not explicitly reported in the reference. For measurements using two-photon excitation (^{2p}), the equivalent single photon excitation wavelengths have been given. Bolded references indicate unpublished measurements.

22.3.1 NAD

Nicotinamide adenine dinucleotide (NAD) is one of the most significant endogenous fluorophores in the brain. It serves numerous functions in the cell, though its largest and most prominent role is as a coenzyme for redox reactions where it interacts with over 300 dehydrogenase enzymes [21]. It has two redox states, oxidized (NAD^+) and reduced (NADH). Although both forms have been shown to fluoresce with different lifetimes [22], much of the work performed has primarily been with NADH due to its much greater quantum yield. The ratio of NAD^+ to NADH in the cell can range between 1 and 700 [23]. When free in solution, NADH has two conformational isomers [24] that are attributed to the stretched and folded (when the fluorescent nicotinamide ring is closer to the adenine base) conformers [25, 26]. In the free form, it can also be partially quenched with collisions or adenine moiety stacking [17].

When bound to enzymes or proteins, NADH is typically held rigidly in the stretched conformation, reducing the nonradiative relaxation rate. As previously mentioned, this ultimately leads to an increase in both lifetime and quantum yield. A blue-shifting of the peak emission has also concurrently been observed. Although a limited number of studies have been performed to measure these parameters together using a single system [27–30], numerous studies – as reviewed in Table 22.1 – have demonstrated these effects independently for lifetime [24, 25, 31–33], quantum yield [28, 30, 34–38], and spectral shifts [39–41] using a wide variety of enzymes. In addition to changes in lifetime, quantum yield, and peak emission shifts, studies have also observed a red-shifting and twofold increase in absorbance when NADH was bound to malate dehydrogenase [36]. Most of the measured fluorescent signal from NADH is estimated to originate in the mitochondria, with negligible contribution from the cytoplasm [42]. Furthermore, more than 80% of fluorescence is assumed to comes from bound pools of NADH [43].

In addition to binding in binary complexes with an enzyme, NADH can also be a part of ternary complexes (enzyme and substrate). In these ternary complexes, NADH is often bound tighter and has been shown to also alter the observed lifetime [44, 45]. For alcohol dehydrogenase, the ternary complexes doubled the lifetime relative to the binary complexes [25]. Studies with lactate dehydrogenase have also demonstrated a concurrent increase quantum yield of ternary versus binary complexes [34]. Due to the same principle as when NADH is bound and constrained, an increase in lifetime and quantum yield has been observed when increasing the viscosity of the NADH solution [46].

Although a fluorescence lifetime temperature dependence for NADH has been observed [25], this effect is not significant when measuring tissue in an *in vivo* state. Furthermore, the lifetime and spectral shape of NADH is independent of pH for biological considerations (pH ranges 5–9) [47].

22.3.2 NADP

NAD's phosphorylated analogue – NADP, which is mostly used for anabolic cellular processes – has very similar fluorescent spectral and lifetime properties [22]. Similar to NADH, the reduced form – NADPH – is much more efficient and thus most work has been performed with it [22]. Studies have observed that the quantum yield of NADPH is roughly twice that of NADH in serum albumin [48]. In actual tissue, this is offset by the fact that the total amount of NADPH is estimated to be five [49] to ten [50] times less that of NADH. It is important to note that NADH and NADPH mostly bind to different collections of enzymes, each of which can affect their net fluorescence intensity and lifetime contribution to the overall measured signal. Although only a limited number of studies have been performed to investigate the effect of enzyme binding on the fluorescent properties of NADPH, results have indicated trends similar to that of NADH. When bound to enzymes, increases in quantum yield of 1.5–4.0 [45, 51, 52] and a 20 nm blue-shift in peak emission [52] have been observed.

22.3.3 Flavins

There are three main flavins present in brain tissue: riboflavin (vitamin B_2), flavin mononucleotide (FMN), and flavin adenine dinucleotide (FAD). Each of these flavins exists in one of three redox states: the oxidized or quinone state, the semiquinone (radical) state, and the reduced or hydroquinone state. Each of these electronic states has an effect on the optical properties of the molecules; as an example, pure riboflavin in solution appears yellow, blue, and transparent in the three redox states, respectively. All flavins, however, tend to produce green fluorescence emissions with peaks around 520–540 nm. Spectral fitting analysis of a suspension of cultured esophageal cells excited at 351 nm found that that the estimated relative spectral contributions of flavins amounted to between 3–10% of the total signal, depending upon cell type [53].

By far the most studied flavin is FAD, which is the primary counterpart to NAD for estimating the overall redox state of a cell. Although it is commonly stated, but not cited, that only the oxidized form of FAD is fluorescent [54], others have observed fluorescence using comparable excitation wavelengths in all three redox states including FAD_{ox}, FADH, $FADH_2$ (peak emissions of 530, 440, and 470, respectively) [55]. Similar to NADH, studies have found that free and bound FAD_{ox} have different lifetimes. Free FAD_{ox} in aqueous solution is partially quenched by its flexible adenine tail (which is not present in FMN or riboflavin); this effect, however, is negligible when the molecule is bound to a protein [56]. Regardless, most FAD_{ox} in the cell is found in the bound form. Limited studies have performed direct comparison between bound and unbound FAD_{ox}; however, Chorvat et al. observed only a minor increase in lifetime (3.0 versus 2.5 ns) when bound to lipoamide dehydrogenase [57]. Due to the differences in the setups used, it is difficult to make a more general comparison between the two forms. However, studies have demonstrated that FAD has a significantly higher lifetime than NADH in the free form [58]. Similarly, free FMN has been observed to have a lifetime of 4.3–5.0 ns [22, 56, 59, 60] and only 0.3 ns when bound to flavin reductase [56]. Riboflavin has not been studied in the bound form, but has a comparable unbound lifetime of 4.1 ns [59].

22.3.4 Other Fluorophores

Although most of the endogenous fluorescence in the brain comes from NAD(P)H and flavins, a variety of other fluorophores are also found in the brain. Another set of coenzymes found in nearly all cells of the human body are the many forms of vitamin B_6. The active form, pyridoxal phosphate (PLP), as well as pyridoxine (PL) and pyridoxamine (PM), all have a fluorescence emission peak around 390–400 nm [61–65]. Lifetime studies have only been performed on PLP, with an estimate of 2–3 ns [63]. Both vitamins B_9 (folic acid) and D_3 (cholecalciferol) are also fluorescent, with peak emissions around 460 nm [64, 66].

Another commonly cited fluorophore is lipofuscin (also termed lipopigments), which has a peak fluorescence emission similar to flavins of around 560 nm [67–69] and a lifetime of 1.35 ns [70]. Although lipopigments have been shown to fluoresce, their spectral contribution for cells measured in suspension has been estimated at only 2–4% [53].

Cholesterol in the brain (which accounts for approximately 25% of the body's total cholesterol) has many roles, including maintaining membrane function, acting as an antioxidant, and is a precursor to many molecules [71]. It has been shown to be fluorescent with peak emission around 400 nm and lifetimes of 3–6 ns [58].

Fluorescence has also been observed in two extracellular matrix (ECM) molecules, aggrecan, and chondroitin sulfate, each of which are found in the brain ECM [72, 73]. Both fluoresce at around 420 nm and have lifetime ranges of 3.5–5.5 ns for aggrecan and 2.8–4.4 ns for chondroitin sulfate [58].

The neurotransmitters glutamate decarboxylase and serotonin also fluoresce, with peak emissions at 380 and 430, respectively [74, 75]; however, their relative contribution to the overall fluorescence signal measured in the brain has not yet been quantitatively investigated.

22.4 STEADY STATE FLUORESCENCE APPLICATIONS

A limited number of *in vivo* experiments evaluating autofluorescence have been performed on rodent brains, though most are demonstrations of proof-of-concept for the instrumentation setup as opposed to studies with statistically powered sample sizes for delineating tissue types. Intensity-based measurements have been used to assess the effect of perturbing the NADH content via oral, intraperitoneal, and intravenous injections of NAD^+ and NADH [101]. Results showed small increases in intensity when either of the redox pairs were given to the animal. This work also highlights the important differences between *in vivo* and *ex vivo* measurements; when animals received a lethal dose of chloral hydrate, the fluorescence intensity from NADH increased significantly post-mortem, relative to the baseline while the animal was alive and under anesthesia. Intensity measurements have also been more recently used for assessing the redox ratio of orthotopic tumors in rodents [102]. The primary difference in the spectra acquired from the two tissue types was that the tumor tissue showed decreased intensity in the longer wavelengths beyond 550 nm, which translates to a more reductive, metabolically active state in the diseased tissue relative to healthy tissue. Studies have also been conducted with intracranially placed fibers in rats to assess sleep states; however, these experiments were conducted on a very limited number of animals [103]. A combined Raman spectroscopy and fluorescence lifetime probe was used to image the cortical surface of a single healthy rat; however, these measurements were performed post-mortem [104].

Bottiroli et al. completed the first clinical autofluorescence studies of the brain using a filtered mercury lamp (366 and 405 nm) as an excitation source, and a spectrograph for collection [105]. A follow -up study was conducted in which healthy white matter, grey matter, and neoplastic lesions were recorded in a larger set of 12 patients undergoing surgical resection [106]. The excitation and emission light were transmitted via bundles of optical fiber, but this technique required the operating room lights to be turned off during data acquisition. Relative to healthy tissues, the neoplastic tissue generally demonstrated a red-shift and spectral broadening relative to healthy tissues. White matter also showed similar trends relative to cortical grey matter.

Autofluorescence spectroscopy has also been combined with white light diffuse reflectance spectroscopy in the clinical setting for tumor margin detection [107–110]. These studies employed a nitrogen laser (337 nm) as an excitation source, and collected and transmitted light in bundles of fiber optics. The first study using this setup investigated the difference between normal brain and tumor tissue in 26 patients, providing 100% sensitivity and 76% specificity [107]. A concurrent study with 24 patients showed similar trends, with fluorescence peaks around 460 nm (Figure 22.2), and also that the tumor margins could be distinguished from healthy tissue with 94% sensitivity and 93% specificity using empirical discrimination algorithms applied post-operatively [108]. The technique, however, showed decreased performance (80% sensitivity and 89% specificity) for differentiating solid tumor from healthy tissue. A final follow-up study was also conducted using similar instrumentation, except for a liquid-crystal tunable filter and CCD that allowed for *in vivo* spectral imaging of brain tissue; the results, however, were limited to only a single patient [110].

More recent studies have utilized excitation at 320 and 410 nm to measure the spectral fluorescence intensity of recently excised diseased and healthy brain tissue samples [111].

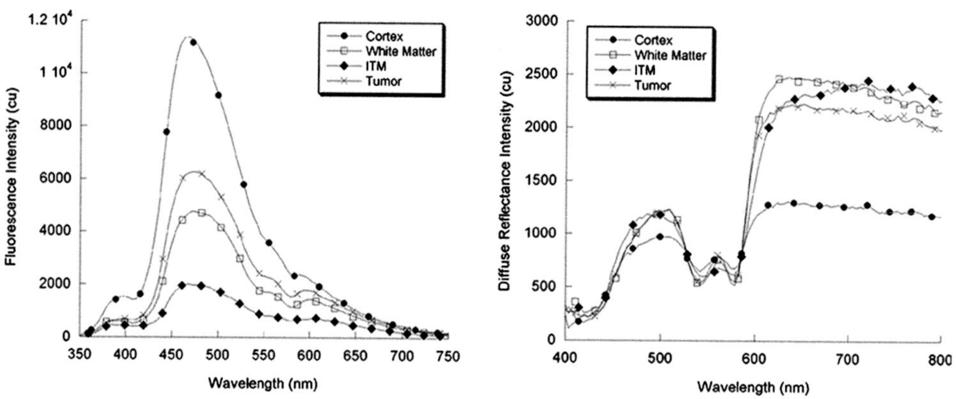

FIGURE 22.2 Fluorescence and diffuse reflectance spectra acquired from the surgical resection of a glioma tumor. For the fluorescence spectra acquired, the shapes of the spectra were generally consistent, with changes in intensity providing the largest differences between the tissue types. The diffuse reflectance spectra acquired were mostly consistent across the different tissue types, with all showing the characteristic dips corresponding to hemoglobin absorption between 500 and 600 nm. The cortex tissue, however, differed significantly in the red wavelengths with large depressions observed beyond 600 nm cu, calibrated units (calibrated with respect to a National Institute of Standards and Technology-calibrated tungsten lamp). (Adapted from Toms et al., 2005.)

Similar studies using the same instrumentation setup have found differences in fixed samples from a variety of brain tissue types [112]. These intensity measurements still suffer significantly from blood absorption on the surface of the samples, which manifests in the wide variability found in the healthy tissue measurements relative to previously published work.

22.5 TIME-RESOLVED FLUORESCENCE APPLICATIONS

22.5.1 Human Studies

To provide additional, more robust information content relative to steady state measurements, time-resolved fluorescence spectroscopy (TRFS) measurements can be acquired [18, 113]. Studies conducted thus far have primarily used the pulse sampling technique previously described, with a nitrogen laser (337 nm) as the excitation source and a monochromator and MCP-PMT for measuring fluorescence emission spectra. The first *ex vivo* studies to incorporate time-resolved measurements investigated normal white matter, normal cortex, glioblastoma multiforme, and meningioma tissues [114–116]. Although all tissue types were found to have peak intensities around 460 nm, normal tissue was found to have a lifetime of ~1.0 ns while the fluorescence lifetime of LGG and HGG tumors were longer at ~1.1 and ~1.3 ns, respectively. These findings suggested that NAD(P)H was primary source of fluorescence, with changes in the bound and unbound fractions corresponding to the contrasting lifetimes observed in the different tissue types.

A subsequent study using a similar instrumentation setup included patients undergoing surgical resection for variety of brain tumors including LGG (oligodendroglioma, oligodendrocytoma, diffuse astrocytoma) and HGG (anaplastic oligodendroglioma, anaplastic oligoastrocytoma, and glioblastoma multiforme) tumors. Specific sites that were interrogated were validated against pre-operative MRI scans, surgeon experience, and histopathological analysis when biopsies were concurrently taken. Tissues from 17 patients were measured both *in vivo* and *ex vivo*, for which significant differences in lifetime were observed, underscoring the importance of making *in vivo* measurements for future comparisons [117]. For the *in vivo* measurements made in this study (Figure 22.3), at the 460 nm peak – corresponding to NAD(P)H fluorescence – the normal white matter and normal cortex displayed a lifetime of ~1.0 ns. At 460 nm the LGG tissues were significantly lower at ~0.6 ns, while only minor changes were observed in the HGG relative to healthy tissue (Figure 22.3). Following this was a much larger 42 patient study conducted using the same instrumentation, and it was in good agreement with the trends from the previous 17 patient study (Figure 22.4) [118]. With the larger patient population, LGG was well discriminated from healthy tissue with 100% sensitivity and 98% specificity. However, due to the very heterogeneous nature of the HGG tumors, both in terms of histopathological classification and spatially throughout the tumor, only 47% sensitivity and 94% specificity were obtained.

A smaller intra-operative study on three patients with high grade gliomas was also completed using a slightly modified instrumentation setup [119]. While a nitrogen laser (337 nm) was still used as an excitation source, the means for collecting the fluorescence was replaced by a wide-field fluorescence lifetime imaging microscopy apparatus, including a

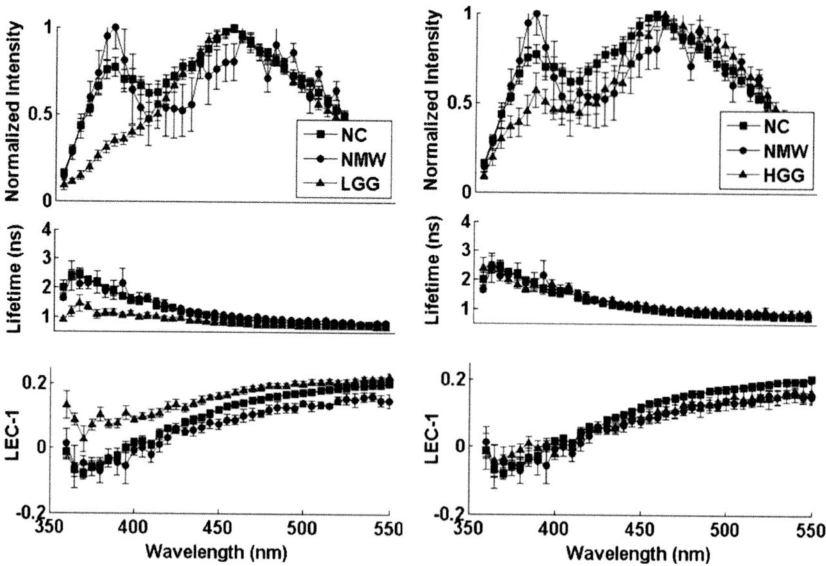

FIGURE 22.3 Comparison of fluorescence spectroscopic parameters between healthy brain tissues (normal cortex (NC, $n=16$) and normal white matter (NWM, $n=3$)) and diseased tissues (low grade glioma (LGG, $n=5$, left) and high grade glioma (HGG, $n=9$, right)). Fluorescence parameters include normalized intensity, average lifetime, and Laguerre expansion coefficient (LEC-1) across the emission wavelengths. Spectra were acquired *in vivo* during a surgical procedure. Values shown are mean ± standard error. (Adapted from Butte et al., 2010.)

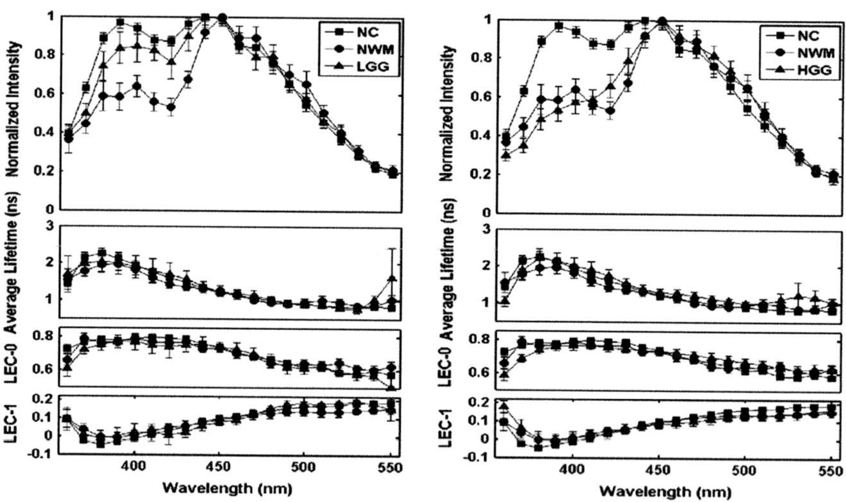

FIGURE 22.4 Comparison of fluorescence spectroscopic parameters between healthy brain tissues (normal cortex (NC, $n=35$) and normal white matter (NWM, $n=12$)) and diseased tissues (low grade glioma (LGG, $n=7$, left) and high grade glioma (HGG, $n=17$, right)). Fluorescence parameters include normalized intensity, average lifetime, and Laguerre expansion coefficients (LEC-0 and LEC-1) across the emission wavelengths. Spectra were acquired *in vivo* during a surgical procedure. Values shown are mean ± standard error. (Adapted from Butte et al., 2011.)

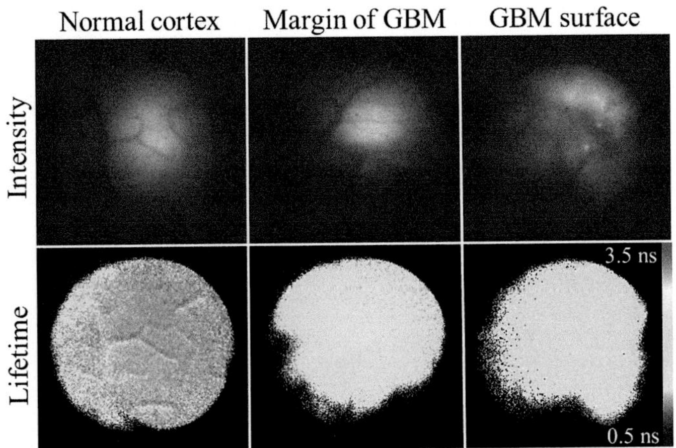

FIGURE 22.5 Representative fluorescence intensity (top) and average lifetime (bottom) images from the different tissue types. Left column: normal cortex; middle column: margin of a GBM; right column: primary GBM area. For each image the average lifetime pixel value was retrieved from the binning of four original pixels (2×2 square). Fluorescence measurements were acquired *in vivo* during a surgical procedure. (Adapted from Sun et al., 2010.)

gradient index lens attached to a 10,000 fiber optic imaging bundle and acquired on a fast-gated intensified CCD. Spectral selection was provided by bandpass filters centered at 390 and 450 nm, with bandwidths of 70 and 60 nm, respectively. A total of 13 locations were interrogated for each patient, including normal cortex, glioblastoma infiltrated cortex, and brain tumor margin (Figure 22.5). Although the results could not be directly compared with previous in-patient measurements, the overall trends agreed with the previous results, with GBM tissue having a longer lifetime of ~1.6 ns relative to the normal tissue at ~1.3 ns.

22.5.2 Radiation-Induced Brain Necrosis in a Rat Model

To assess whether fluorescence lifetime techniques could be used to delineate radiation-induced necrosis (an undesirable side-effect of radiation therapy) from normal brain tissue, a controlled study was performed using a rat model of the disease [120]. Animals were imaged with MRI to determine the location of the necrosis tissues prior to a live surgery for fluorescence lifetime assessment of the tissue. Optical fibers were stereotaxically inserted into the brain tissue during surgery to measure the fluorescence intensity and lifetime of both healthy and necrosis tissues using a high spectral resolution point-TRFS system as well as a real-time multispectral-TRFS acquisition system. At the end of surgery, the animals were perfused with formalin to fix the brain which was rescanned with MRI for data coregistration purposes. The corresponding coronal sections of the brain were stained with hematoxylin and eosin (H&E) for pathological validation. Measurements from both systems indicated significantly lower lifetimes in the necrosis tissue relative to healthy tissue (Figure 22.6). In addition, appreciable differences were observed in animals assessed just prior to the onset of pathological necrosis (Figure 22.6). The contrast observed in necrosis tissue, as well as the pre-necrosis tissue, was attributed

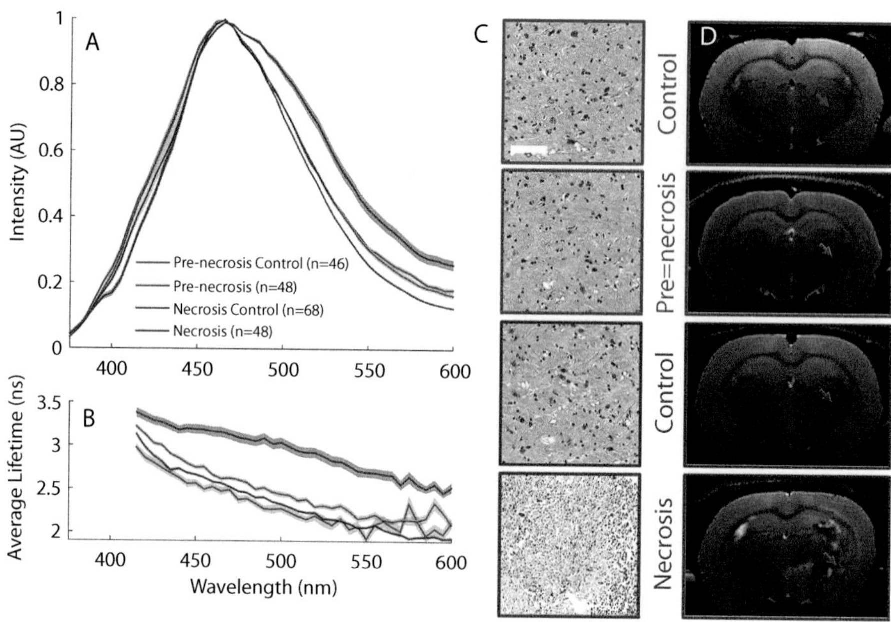

FIGURE 22.6 Averages of acquired spectra using the point-TRFS setup for pre-necrosis (orange) and necrosis (red) tissues, along with healthy tissues from their corresponding control groups (colored blue and purple, respectively). Normalized integrated intensity (A) and average lifetime (B); for both plots, solid lines indicate the mean, with the shaded region showing mean ± standard error of the mean. Representative coronal H&E stained histology at 10× magnification of the different tissue types is presented (C); the scale bar shown is 100 μm. Representative coronal MRI scans from each of the three tissue types are also shown; the location of the corresponding histology is indicated with an arrow (D).

to differences in the relative subpopulations of NAD(P)H and FAD as opposed to any additional fluorophores specifically present in the radiation-induced necrosis tissue. These results showed that radiation-induced brain necrosis tissue contains significantly different metabolic signatures that could be detectable in real-time with label-free fluorescence lifetime techniques.

22.6 CONCLUSION

Brain surgery presents one of the most challenging environments to remove tumors due to the critical importance of the adjacent healthy tissues. The trade-off between disease recurrence caused by insufficient tissue resection and functional losses due to excessive tissue resection makes this an area that could benefit significantly from intra-operative diagnostic technologies. The current gold standard of histology during surgery is both resource and time intensive. The label-free fluorescence methods reviewed in this chapter show strong potential to help the neurological surgeon navigate and assess different brain tissues in real-time – including normal cortex, normal white matter, radiation-induced necrosis, LGG, and HGG – during the pre-operative biopsy stage, as well as for intra-operative surgical navigation.

Despite the great deal of work already performed using this technique, further controlled studies in animals will help provide additional insight into the complicated endogenous sources of the fluorescence and reveal further how the signals measured correlate to the population dynamics of NAD(P)H, FAD, and the many other fluorophores present in brain tissue. Results thus far indicate that endogenous fluorescence lifetime techniques represent a promising method to delineate a wide variety of diseased tissue, in terms of the disease sub-type and its severity, while also identifying healthy tissues in real-time and ultimately have the potential to provide critically essential guidance for the neurological surgeon.

ACKNOWLEDGMENTS

This work has been supported in part by the NIH grant R21 CA178578, and the University of California, Davis Comprehensive Cancer Center's Brain Malignancies Innovation Group.

REFERENCES

1. NCI. SEER Stat *Fact Sheets: Brain and Other Nervous System Cancer.* 2016a; available from: http://seer.cancer.gov/statfacts/html/brain.html.
2. NCI. *A Snapshot of Brain and Central Nervous System Cancers.* 2016b; available from: http://www.cancer.gov/research/progress/snapshots/brain.
3. Gallego, O., Nonsurgical treatment of recurrent glioblastoma. *Current Oncology*, 2015. **22**(4): p. e273–81.
4. Stewart, B.W., C.P. Wild, and C. Wild, *World Cancer Report 2014.* 2014: World Health Organization.
5. Schapira, A.H.V., *Neurology and Clinical Neuroscience.* 2006: Elsevier Health Sciences.
6. Stupp, R., et al., Radiotherapy plus concomitant and adjuvant temozolomide for glioblastoma. *The New England Journal of Medicine*, 2005. **352**(10): p. 987–96.
7. Louis, D.N., et al., The 2007 WHO classification of tumours of the central nervous system. *Acta Neuropathologica*, 2007. **114**(2): p. 97–109.
8. Wen, P.Y. and S. Kesari, Malignant gliomas in adults. *The New England Journal of Medicine*, 2008. **359**(5): p. 492–507.
9. Robins, H.I., A.B. Lassman, and D. Khuntia, Therapeutic advances in malignant glioma: current status and future prospects. *Neuroimaging Clinics of North America*, 2009. **19**(4).
10. Lacroix, M., et al., A multivariate analysis of 416 patients with glioblastoma multiforme: prognosis, extent of resection, and survival. *Journal of Neurosurgery*, 2001. **95**(2): p. 190–8.
11. Stummer, W., et al., Extent of resection and survival on glioblastoma multiforme-identification of and adiustment for bias. *Neurosurgery*, 2008. **62**(3): p. 564–74.
12. Sanai, N. and M.S. Berger, Glioma extent of resection and its impact on patient outcome. *Neurosurgery*, 2008. **62**(4): p. 753–64.
13. Unsgård, G., Ultrasound-guided neurosurgery, in *Practical Handbook of Neurosurgery: From Leading Neurosurgeons*, M. Sindou, Editor. 2009, Springer Vienna, Vienna, pp. 907–26.
14. Langø, T., et al., Accuracy evaluation of a 3D ultrasound-based neuronavigation system, in *CARS 2002 Computer Assisted Radiology and Surgery: Proceedings of the 16th International Congress and Exhibition Paris, June 26–29, 2002*, H.U. Lemke, et al., Editors. 2002, Springer Berlin Heidelberg, Berlin, Heidelberg, pp. 63–8.
15. Berezin, M.Y. and S. Achilefu, Fluorescence lifetime measurements and biological imaging. *Chemical Reviews*, 2010. **110**(5): p. 2641–84.
16. Valeur, B. and M.N. Berberan-Santos, *Molecular Fluorescence: Principles and Applications.* 2013: Wiley.

17. Lakowicz, J.R., *Principles of Fluorescence Spectroscopy*. 2013: Springer US.
18. Marcu, L., P.M.W. French, and D.S. Elson, *Fluorescence Lifetime Spectroscopy and Imaging: Principles and Applications in Biomedical Diagnostics*. 2014: Taylor & Francis.
19. Yankelevich, D.R., et al., Design and evaluation of a device for fast multispectral time-resolved fluorescence spectroscopy and imaging. *Review of Scientific Instruments*, 2014. **85**(3).
20. Croce, A.C. and G. Bottiroli, Autofluorescence spectroscopy and imaging: a tool for biomedical research and diagnosis. *European Journal of Histochemistry*, 2014. **58**(4): p. 320–37.
21. Li, N. and G. Chen, Flow injection analysis of trace amounts of NADH with inhibited chemiluminescent detection. *Talanta*, 2002. **57**(5): p. 961–7.
22. Schneckenburger, H. and K. Konig, Fluorescence decay kinetics and imaging of NAD(P)H and flavins as metabolic indicators. *Optical Engineering*, 1992. **31**(7): p. 1447–51.
23. Lin, S.J. and L. Guarente, Nicotinamide adenine dinucleotide, a metabolic regulator of transcription, longevity and disease. *Current Opinion in Cell Biology*, 2003. **15**(2): p. 241–6.
24. Vishwasrao, H.D., et al., Conformational dependence of intracellular NADH on metabolic state revealed by associated fluorescence anisotropy. *Journal of Biological Chemistry*, 2005. **280**(26): p. 25119–26.
25. Gafni, A. and L. Brand, Fluorescence decay studies of reduced nicotinamide adenine-dinucleotide in solution and bound to liver alcohol-dehydrogenase. *Biochemistry*, 1976. **15**(15): p. 3165–71.
26. Couprie, M.E., et al., First use of the UV super-ACO free-electron laser - fluorescence decays and rotational-dynamics of the NADH coenzyme. *Review of Scientific Instruments*, 1994. **65**(5): p. 1485–95.
27. Lakowicz, J.R., et al., Fluorescence lifetime imaging of free and protein-bound NADH. *Proceedings of the National Academy of Sciences of the United States of America*, 1992. **89**(4): p. 1271–5.
28. Blinova, K., et al., Mitochondrial NADH fluorescence is enhanced by complex I binding. *Biochemistry*, 2008. **47**(36): p. 9636–45.
29. Fjeld, C.C., W.T. Birdsong, and R.H. Goodman, Differential binding of NAD(+) and NADH allows the transcriptional corepressor carboxyl-terminal binding protein to serve as a metabolic sensor. *Proceedings of the National Academy of Sciences of the United States of America*, 2003. **100**(16): p. 9202–7.
30. Verlick, S.F., Fluorescence spectra and polarization of glyceraldehyde-3-phosphate and lactic dehydrogenase coenzyme complexes. *Journal of Biological Chemistry*, 1958. **233**(6): p. 1455–67.
31. Rupley, J.A., et al., Comparison of lactate and malate dehydrogenases: fluorescence and thermodynamic properties. *Biochemical and Biophysical Research Communications*, 1980. **93**(3): p. 654–60.
32. Yu, Q.R. and A.A. Heikal, Two-photon autofluorescence dynamics imaging reveals sensitivity of intracellular NADH concentration and conformation to cell physiology at the single-cell level. *Journal of Photochemistry and Photobiology B-Biology*, 2009. **95**(1): p. 46–57.
33. Iweibo, I., Protein fluorescence and electronic energy transfer in the determination of molecular dimensions and rotational relaxation times of native and coenzyme-bound horse liver alcohol dehydrogenase. *Biochimica et Biophysica Acta*, 1976. **446**(1): p. 192–205.
34. Lee, C.Y., et al., Purification and biochemical studies of lactate dehydrogenase-X from mouse. *Molecular and Cellular Biochemistry*, 1977. **18**(1): p. 49–57.
35. Shore, J.D., et al., NADH binding to porcine mitochondrial malate-dehydrogenase. *Journal of Biological Chemistry*, 1979. **254**(18): p. 9059–62.
36. Tseng, C.H., et al., Using shaped ultrafast laser pulses to detect enzyme binding. *Optics Express*, 2011. **19**(24): p. 24638–46.
37. Iweibo, I. and H. Weiner, Coenzyme interaction with horse liver alcohol dehydrogenase. Evidence for allosteric coenzyme binding sites from thermodynamic equilibrium studies. *Journal of Biological Chemistry*, 1975. **250**(6): p. 1959–65.

38. Ambroziak, W., L.L. Kosley, and R. Pietruszko, Human aldehyde dehydrogenase - coenzyme binding-studies. *Biochemistry*, 1989. **28**(13): p. 5367–73.
39. McClendon, S., N. Zhadin, and R. Callender, The approach to the Michaelis complex in lactate dehydrogenase: the substrate binding pathway. *Biophysical Journal*, 2005. **89**(3): p. 2024–32.
40. Kasischke, K.A., et al., Two-photon NADH imaging exposes boundaries of oxygen diffusion in cortical vascular supply regions. *Journal of Cerebral Blood Flow and Metabolism*, 2011. **31**(1): p. 68–81.
41. Huber, R., et al., Protein binding of NADH on chemical preconditioning. *Journal of Neurochemistry*, 2000. **75**(1): p. 329–35.
42. Mayevsky, A. and G.G. Rogatsky, Mitochondrial function in vivo evaluated by NADH fluorescence: from animal models to human studies. *American Journal of Physiology-Cell Physiology*, 2007. **292**(2): p. C615–C640.
43. Blinova, K., et al., Distribution of mitochondrial NADH fluorescence lifetimes: steady-state kinetics of matrix NADH interactions. *Biochemistry*, 2005. **44**(7): p. 2585–94.
44. Piersma, S.R., et al., Optical spectroscopy of nicotinoprotein alcohol dehydrogenase from *Amycolatopsis methanolica*: a comparison with horse liver alcohol dehydrogenase and UDP-galactose epimerase. *Biochemistry*, 1998. **37**(9): p. 3068–77.
45. Brochon, J.C., et al., Pulse fluorimetry study of beef liver glutamate dehydrogenase reduced nicotinamide adenine dinucleotide phosphate complexes. *Biochemistry*, 1976. **15**(15): p. 3259–65.
46. Ghukasyan, V.V. and A.A. Heikal, *Natural Biomarkers for Cellular Metabolism: Biology, Techniques, and Applications*. 2014: CRC Press.
47. Ogikubo, S., et al., Intracellular pH sensing using autofluorescence lifetime microscopy. *Journal of Physical Chemistry B*, 2011. **115**(34): p. 10385–90.
48. Avidor, Y., et al., Fluorescence of pyridine nucleotides in mitochondria. *Journal of Biological Chemistry*, 1962. **237**(7): p. 2377–2383.
49. Klaidman, L.K., A.C. Leung, and J.D. Adams, Jr., High-performance liquid chromatography analysis of oxidized and reduced pyridine dinucleotides in specific brain regions. *Analytical Biochemistry*, 1995. **228**(2): p. 312–7.
50. Heikal, A.A., Intracellular coenzymes as natural biomarkers for metabolic activities and mitochondrial anomalies. *Biomarkers in Medicine*, 2010. **4**(2): p. 241–63.
51. Shafer, J.A., et al., Binding of reduced cofactor to glutamate dehydrogenase. *European Journal of Biochemistry*, 1972. **31**(1): p. 166–71.
52. Gaetani, G.F., et al., A novel NADPH:(bound) NADP(+) reductase and NADH:(bound) NADP(+) transhydrogenase function in bovine liver catalase. *Biochemical Journal*, 2005. **385**: p. 763–8.
53. Villette, S., et al., Ultraviolet-induced autofluorescence characterization of normal and tumoral esophageal epithelium cells with quantitation of NAD(P)H. *Photochemical and Photobiological Sciences*, 2006. **5**(5): p. 483–92.
54. Jahn, K., V. Buschmann, and C. Hille, Simultaneous fluorescence and phosphorescence lifetime imaging microscopy in living cells. *Scientific Reports*, 2015. **5**.
55. Zhao, R.K., et al., Ultrafast transient mid IR to visible spectroscopy of fully reduced flavins. *Physical Chemistry Chemical Physics*, 2011. **13**(39): p. 17642–8.
56. Yang, H., et al., Protein conformational dynamics probed by single-molecule electron transfer. *Science*, 2003. **302**(5643): p. 262–6.
57. Chorvat, D., Jr., and A. Chorvatova, Spectrally resolved time-correlated single photon counting: a novel approach for characterization of endogenous fluorescence in isolated cardiac myocytes. *European Biophysics Journal*, 2006. **36**(1): p. 73–83.
58. Fatakdawala, H., et al., Time-resolved fluorescence spectroscopic analysis of endogenous fluorophores. *Manuscript in preparation*.
59. Koziol, B., et al., Riboflavin as a source of autofluorescence in *Eisenia fetida* coelomocytes. *Photochemistry and Photobiology*, 2006. **82**(2): p. 570–3.

60. Wahl, P., et al., Time resolved fluorescence of flavin adenine dinucleotide. *FEBS Letters*, 1974. **44**(1): p. 67–70.
61. Honikel, K.O. and N.B. Madsen, Comparison of the absorbance spectra and fluorescence behavior of phosphorylase b with that of model pyridoxal phosphate derivatives in various solvents. *Journal of Biological Chemistry*, 1972. **247**(4): p. 1057–64.
62. Kempe, T.D. and G.R. Stark, Pyridoxal 5′-phosphate, a fluorescent probe in the active site of aspartate transcarbamylase. *Journal of Biological Chemistry*, 1975. **250**(17): p. 6861–9.
63. Greenaway, F.T. and J.W. Ledbetter, Fluorescence lifetime and polarization anisotropy studies of membrane surfaces with pyridoxal 5′-phosphate. *Biophysical Chemistry*, 1987. **28**(3): p. 265–71.
64. Zipfel, W.R., et al., Live tissue intrinsic emission microscopy using multiphoton-excited native fluorescence and second harmonic generation. *Proceedings of the National Academy of Sciences of the United States of America*, 2003. **100**(12): p. 7075–80.
65. Bueno, C. and M.V. Encinas, Photophysical and photochemical studies of pyridoxamine. *Helvetica Chimica Acta*, 2003. **86**(10): p. 3363–75.
66. Duggan, D.E., et al., A spectrophotofluorometric study of compounds of biological interest. *Archives of Biochemistry and Biophysics*, 1957. **68**(1): p. 1–14.
67. Wagnieres, G.A., W.M. Star, and B.C. Wilson, In vivo fluorescence spectroscopy and imaging for oncological applications. *Photochemistry and Photobiology*, 1998. **68**(5): p. 603–32.
68. Marmorstein, A.D., et al., Spectral profiling of autofluorescence associated with lipofuscin, Bruch's membrane, and sub-RPE deposits in normal and AMD eyes. *Investigative Ophthalmology and Visual Science*, 2002. **43**(7): p. 2435–41.
69. Mochizuki, Y., et al., The difference in autofluorescence features of lipofuscin between brain and adrenal. *Zoological Science*, 1995. **12**(3): p. 283–8.
70. Schweitzer, D., et al., Towards metabolic mapping of the human retina. *Microscopy Research and Technique*, 2007. **70**(5): p. 410–9.
71. Bjorkhem, I. and S. Meaney, Brain cholesterol: long secret life behind a barrier. *Arteriosclerosis Thrombosis and Vascular Biology*, 2004. **24**(5): p. 806–15.
72. Morawski, M., et al., Aggrecan: beyond cartilage and into the brain. *International Journal of Biochemistry and Cell Biology*, 2012. **44**(5): p. 690–3.
73. Kwok, J.C.F., P. Warren, and J.W. Fawcett, Chondroitin sulfate: a key molecule in the brain matrix. *International Journal of Biochemistry and Cell Biology*, 2012. **44**(4): p. 582–6.
74. Shukuya, R. and G.W. Schwert, Glutamic acid decarboxylase. II. The spectrum of the enzyme. *Journal of Biological Chemistry*, 1960. **235**: p. 1653–7.
75. Crespi, F., et al., Autofluorescence spectrofluorometry of central nervous system (CNS) neuromediators. *Lasers in Surgery and Medicine*, 2004. **34**(1): p. 39–47.
76. Ying, W.H., NAD(+)/ NADH and NADP(+)/NADPH in cellular functions and cell death: regulation and biological consequences. *Antioxidants and Redox Signaling*, 2008. **10**(2): p. 179–206.
77. De Beule, P.A., et al., A hyperspectral fluorescence lifetime probe for skin cancer diagnosis. *Review of Scientific Instruments*, 2007. **78**(12): p. 123101.
78. Yaseen, M.A., et al., In vivo imaging of cerebral energy metabolism with two-photon fluorescence lifetime microscopy of NADH. *Biomedical Optics Express*, 2013. **4**(2): p. 307–21.
79. Lehninger, A.L., D.L. Nelson, and M.M. Cox, *Lehninger Principles of Biochemistry*. 2005: W. H. Freeman.
80. Chia, T.H., et al., Multiphoton fluorescence lifetime imaging of intrinsic fluorescence in human and rat brain tissue reveals spatially distinct NADH binding. *Optics Express*, 2008. **16**(6): p. 4237–49.
81. Deluca, C. and N.O. Kaplan, Flavin adenine dinucleotide synthesis in animal tissues. *Biochimica Et Biophysica Acta*, 1958. **30**(1): p. 6–11.
82. Islam, M.S., et al., pH dependence of the fluorescence lifetime of FAD in solution and in cells. *International Journal of Molecular Sciences*, 2013. **14**(1): p. 1952–63.

83. Digris, A.V., et al., Thermal stability of a flavoprotein assessed from associative analysis of polarized time-resolved fluorescence spectroscopy. *European Biophysics Journal*, 1999. **28**(6): p. 526–31.
84. Konig, K. and I. Riemann, High-resolution multiphoton tomography of human skin with subcellular spatial resolution and picosecond time resolution. *Journal of Biomedical Optics*, 2003. **8**(3): p. 432–9.
85. Wahl, P., et al., A pulse fluorometry study of lipoamide dehydrogenase. Evidence for non-equivalent FAD centers. *European Journal of Biochemistry*, 1975. **50**(2): p. 413–8.
86. Bastiaens, P.I., et al., Conformational dynamics and intersubunit energy transfer in wild-type and mutant lipoamide dehydrogenase from *Azotobacter vinelandii*. A multidimensional time-resolved polarized fluorescence study. *Biophysical Journal*, 1992. **63**(3): p. 839–53.
87. Nakashima, N., et al., Picosecond fluorescence lifetime of the coenzyme of D-amino-acid oxidase. *Journal of Biological Chemistry*, 1980. **255**(11): p. 5261–3.
88. Spector, R., Riboflavin homeostasis in the central nervous-system. *Journal of Neurochemistry*, 1980. **35**(1): p. 202–9.
89. Asano, M. and H. Iwahashi, Caffeic acid inhibits the formation of 7-carboxyheptyl radicals from oleic acid under flavin mononucleotide photosensitization by scavenging singlet oxygen and quenching the excited state of flavin mononucleotide. *Molecules*, 2014. **19**(8): p. 12486–99.
90. Grajek, H., et al., Flavin mononucleotide fluorescence intensity decay in concentrated aqueous solutions. *Chemical Physics Letters*, 2007. **439**(1–3): p. 151–6.
91. Schaus, R. and J.E. Kirk, The riboflavin concentration of brain, heart, and skeletal-muscle in individuals of various ages. *Journals of Gerontology*, 1956. **11**(2): p. 147–50.
92. Ebadi, M. and J. Bifano, The synthesis of pyridoxal phosphate in rat brain regions. *The International Journal of Biochemistry*, 1978. **9**(8): p. 607–11.
93. Coursin, D.B., Vitamin B6 and brain function in animals and man. *Annals of the New York Academy of Sciences*, 1969. **166**(1): p. 7–15.
94. Tews, J.K. and R.A. Lovell, The effect of a nutritional pyridoxine deficiency on free amino acids and related substances in mouse brain. *Journal of Neurochemistry*, 1967. **14**(1): p. 1–7.
95. Loo, Y.H. and L. Badger, Spectrofluorimetric assay of vitamin B6 analogues in brain tissue. *Journal of Neurochemistry*, 1969. **16**(5): p. 801–4.
96. Reynolds, E.H., Benefits and risks of folic acid to the nervous system. *Journal of Neurology, Neurosurgery, and Psychiatry*, 2002. **72**(5): p. 567–71.
97. Kiraly, S.J., et al., Vitamin D as a neuroactive substance: review. *The Scientific World Journal*, 2006. **6**: p. 125–39.
98. Riga, D., et al., Brain lipopigment accumulation in normal and pathological aging. *Annals of the New York Academy of Sciences*, 2006. **1067**: p. 158–63.
99. Twarog, B.M., I.H. Page, and H. Bailey, Serotonin content of some mammalian tissues and urine and a method for its determination. *American Journal of Physiology*, 1953. **175**(1): p. 157–61.
100. Wu, J.Y., T. Matsuda, and E. Roberts, Purification and characterization of glutamate decarboxylase from mouse brain. *Journal of Biological Chemistry*, 1973. **248**(9): p. 3029–34.
101. Rex, A., M.P. Hentschke, and H. Fink, Bioavailability of reduced nicotinamide-adenine-dinucleotide (NADH) in the central nervous system of the anaesthetized rat measured by laser-induced fluorescence spectroscopy. *Pharmacology and Toxicology*, 2002. **90**(4): p. 220–5.
102. Liu, Q., et al., Compact point-detection fluorescence spectroscopy system for quantifying intrinsic fluorescence redox ratio in brain cancer diagnostics. *Journal of Biomedical Optics*, 2011. **16**(3).
103. Mottin, S., et al., Determination of NADH in the rat brain during sleep-wake states with an optic fibre sensor and time-resolved fluorescence procedures. *Neuroscience*, 1997. **79**(3): p. 683–93.
104. Dochow, S., et al., Combined fiber probe for fluorescence lifetime and Raman spectroscopy. *Analytical and Bioanalytical Chemistry*, 2015. **407**(27): p. 8291–301.

105. Bottiroli, G., et al., Brain tissue autofluorescence: an aid for intraoperative delineation of tumor resection margins. *Cancer Detection and Prevention*, 1998. **22**(4): p. 330–9.
106. Croce, A.C., et al., Diagnostic potential of autofluorescence for an assisted intraoperative delineation of glioblastoma resection margins. *Photochemistry and Photobiology*, 2003. **77**(3): p. 309–18.
107. Lin, W.C., et al., In vivo brain tumor demarcation using optical spectroscopy. *Photochemistry and Photobiology*, 2001. **73**(4): p. 396–402.
108. Toms, S.A., et al., Intraoperative optical spectroscopy identifies infiltrating glioma margins with high sensitivity. *Neurosurgery*, 2005. **57**(4 Suppl): p. 382–91.
109. Lin, W.C., et al., In vivo optical spectroscopy detects radiation damage in brain tissue. *Neurosurgery*, 2005. **57**(3): p. 518–25.
110. Gebhart, S.C., R.C. Thompson, and A. Mahadevan-Jansen, Liquid-crystal tunable filter spectral imaging for brain tumor demarcation. *Applied Optics*, 2007. **46**(10): p. 1896–1910.
111. Nazeer, S.S., et al., Fluorescence spectroscopy as a highly potential single-entity tool to identify chromophores and fluorophores: study on neoplastic human brain lesions. *Journal of Biomedical Optics*, 2013. **18**(6).
112. Nazeer, S.S., et al., Fluorescence spectroscopy to discriminate neoplastic human brain lesions: a study using the spectral intensity ratio and multivariate linear discriminant analysis. *Laser Physics*, 2014. **24**(2).
113. Marcu, L. and B.A. Hartl, Fluorescence lifetime spectroscopy and imaging in neurosurgery. *IEEE Journal of Selected Topics in Quantum Electronics*, 2012. **18**(4): p. 1465–77.
114. Marcu, L., et al., Fluorescence lifetime spectroscopy of glioblastoma multiforme. *Photochemistry and Photobiology*, 2004. **80**(1): p. 98–103.
115. Butte, P.V., et al., Diagnosis of meningioma by time-resolved fluorescence spectroscopy. *Journal of Biomedical Optics*, 2005. **10**(6): p. 064026.
116. Yong, W.H., et al., Distinction of brain tissue, low grade and high grade glioma with time-resolved fluorescence spectroscopy. *Frontiers in Bioscience*, 2006. **11**: p. 1255–63.
117. Butte, P.V., et al., Intraoperative delineation of primary brain tumors using time-resolved fluorescence spectroscopy. *Journal of Biomedical Optics*, 2010. **15**(2).
118. Butte, P.V., et al., Fluorescence lifetime spectroscopy for guided therapy of brain tumors. *NeuroImage*, 2011. **54** Supplement 1: p. S125–S135.
119. Sun, Y.H., et al., Fluorescence lifetime imaging microscopy for brain tumor image-guided surgery. *Journal of Biomedical Optics*, 2010. **15**(5).
120. Hartl, B.A., et al., Detection of radiation-induced brain necrosis in live rats using label-free time-resolved fluorescence spectroscopy, in *Proceedings of the SPIE BiOS*. 2017, International Society for Optics and Photonics, San Francisco, CA.

CHAPTER 23

Higher Harmonic Generation Imaging for Neuropathology

Nikolay Kuzmin, Sander Idema, Eleonora Aronica, Philip C. de Witt Hamer, Pieter Wesseling, and Marie Louise Groot

CONTENTS

23.1 INTRODUCTION

Currently there is a gap between pre-operative imaging techniques, such as magnetic resonance imaging (MRI), that provide a rough location and diagnosis of the disease (e.g. tumor, seizure focus), and the histopathology that provides an accurate diagnosis, but up to 24 hours after the surgery. Occasionally, intra-operative analysis using hematoxylin-and-eosin (H&E) stained sections of snap-frozen material or smear preparations is performed by the pathologist to help establish brain tumor boundaries, but this procedure only allows analysis of small, selected regions, can only be performed on tissue fragments that are already resected, and is rather time-consuming (frozen section diagnosis) or does not allow analysis of tumors in the histological context (smear preparations). Fluorescence imaging techniques are increasingly used during surgery [1, 2] but are associated with several drawbacks, such as heterogeneous delivery and nonspecific staining [3, 4]. In particular, low-grade gliomas and normal brain tissue have an intact blood–brain barrier and take up little circulating dye [5–7]. Alternative techniques are therefore required, that can diagnose tissue without fluorescent labels and with a speed that enables "live" feedback to the surgeon while he/she operates.

Multi-photon microscopies, (combinations of) second and third harmonic generation microscopy, two- and three-photon excited auto-fluorescence microscopy, and coherent Raman scattering microscopy (CARS/SRS), show great potential as clinical tool for the real time assessment of the pathological state of tissue during surgery: the relative speed of the imaging modalities approaches "real" time, and no preparation steps of the tissue are required [8–15]. In particular, third harmonic generation (THG) imaging [15–24] is an emerging label-free microscopy technique with strong potential. THG has been successfully applied to imaging of unstained samples such as insect embryos, plant seeds and intact mammalian tissue [17], epithelial tissues [25–28], zebra fish embryos [18], and zebrafish nervous systems [16]. Excellent agreement of THG images with standard histopathology has been demonstrated in case of skin cancer diagnosis [22, 29] and for *ex-vivo* human brain tumor tissue [15]. THG images could therefore enable the transition from the current practice, histopathological analysis of fixed tissue after surgery, to *in situ* optical biopsy, especially if THG signals are combined with other nonlinear signals such a two- or three-photon excited autofluorescence, CARS signals, or SHG signals to increase the information content of the images.

23.2 THEORETICAL BACKGROUND

THG and SHG are nonlinear optical processes that may occur in tissue depending on nonlinear susceptibility coefficients $\chi^{(3)}$ and $\chi^{(2)}$ of the tissue and upon satisfying phase-matching conditions [19, 30–37]. In the THG process, three incident photons are converted into one photon with triple energy and one-third of the wavelength (Figure 23.1A). SHG signals result in the conversion of an incident photon pair into one photon with twice the energy and half the wavelength. Two- and three-photon excited fluorescence signals (2PF, 3PF) may simultaneously be generated by intrinsic proteins (Figure 23.1B). As a result, a set of distinct (harmonic) and broadband (autofluorescence) spectral peaks is generated in the visible range, while illuminating the sample with near-IR light. The THG, SHG, and/or multi-photon autofluorescence signals can be detected separately by using narrow band interference filters.

The efficiency of THG depends mainly on the third-order susceptibility $\chi^{(3)}$ of the medium and the phase-matching conditions. The total generated THG intensity by a laser beam with intensity I_ω and angular frequency ω in a medium is given by [38]:

$$I_{\mathrm{THG}} = \left(\frac{3\omega}{2n_\omega c}\right)^2 \chi^{(3)} I_\omega^3 \int_{-21}^{z2} \frac{e^{i\Delta kz}}{(1+2iz/b)^2} dz \tag{23.1}$$

where n_ω is the refractive index of the medium for the incident beam, c is the speed of light,

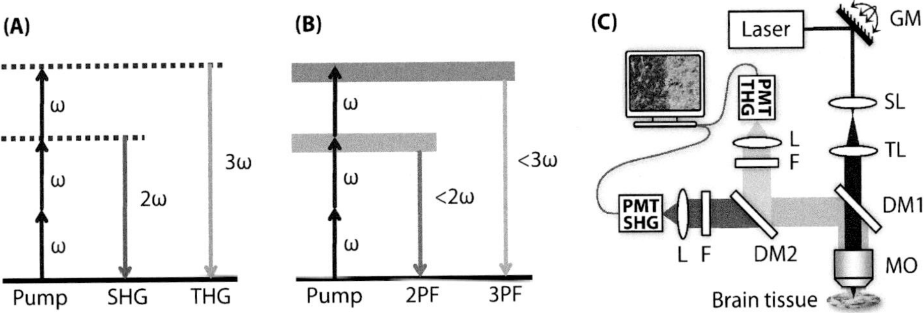

FIGURE 23.1 THG/SHG microscopy for brain tissue imaging. (A) Energy level diagram of the second (SHG) and third (THG) harmonic generation process. (B) Energy level diagram of the two- (2PF) and three-photon (3PF) excited autofluorescence process. (C) Multi-photon microscope setup: laser producing 200 fs pulses at 1,200 nm; GM – X–Y galvo-scanner mirrors; SL – scan lens; TL – tube lens; MO – microscope objective; DM1 – dichroic mirror reflecting back-scattered THG/SHG photons to the PMT detectors; DM2 – dichroic mirror splitting SHG and THG channels; F – narrow-band SHG and THG interference filters; L – focusing lenses; PMT – photomultiplier tube detectors.

$\Delta k = (n_{3\omega}\, 3\omega/c) - 3(n_{\omega}\omega/c)$ is the phase mismatch, z is the position along the beam axis, z_1 and z_2 are the boundaries of the medium, and b is the confocal parameter of the focused laser beam.

In a homogeneous medium with a positive phase mismatch Δk, the phase-matching integral in Eq. (23.1) goes to zero and no THG is produced, irrespective of the magnitude of I_{ω} and $\chi^{(3)}$ (note that practically all materials have a positive phase mismatch in the visible wavelength range). In contrast, partial phase matching can be achieved by introducing a small inhomogeneity at the focus, resulting in a finite THG signal. Detailed calculations show that the maximum value of the phase-matching integral is reached when a structure is half the size of the focal volume [38, 39].

As a result, the phase-matching conditions for THG are governed by the specific structure and composition of the material within the laser focus, and can be controlled through the focusing parameters of the laser. Brain tissue consists for a large part of axons and dendrites, with a diameter at the 1 μm scale. These structures contain a high concentration of lipids, which are known [17] to have a high $\chi^{(3)}$. By setting the focal volume of the incident laser beam to 2–3 times the size of a typical dendrite, we create a geometry where the phase-matching conditions enable efficient THG [39].

23.3 MATERIAL AND METHODS

23.3.1 Experimental Set-Up

The imaging setup in our lab (Figure 23.1C) consists of a commercial two-photon laser-scanning microscope (TriMScope I, LavisionBioTec GmbH) and a femtosecond laser source. The laser source is an optical parametric oscillator (Mira-OPO, APE) pumped at 810 nm by a Ti-sapphire oscillator (Coherent Chameleon Ultra II). The OPO generates 200 fs pulses at 1200 nm and repetition rate of 80 MHz. We selected this wavelength as it falls in the tissue transparency window, providing deeper penetration and reduced photodamage compared to the 700–1000 nm range, as well as harmonic signals generated in the visible wavelength range, facilitating their collection and detection with conventional objectives and detectors. The OPO beam is focused on the sample using a 25×× /1.10 (Nikon APO LWD) water-dipping objective (MO). The 1,200 nm beam has a focal spot size on the sample of d_{lateral} ~ 0.7 μm and d_{axial} ~ 4.1 μm, resulting in two- and three-photon resolution values oΔ $\Delta_{\text{2P,lateral}}$ ~ 0.5 μmΔ $\Delta_{\text{2P,axial}}$ ~ 2.9 μmΔ $\Delta_{\text{3P,lateral}}$ ~ 0.4 μmΔ $\Delta_{\text{3P,axial}}$ ~ 2.4 μm. Two high-sensitivity GaAsP photomultiplier tubes (PMT, Hamamatsu H7422-40) equipped with narrow-band filters at 400 nm and 600 nm are used to collect the THG and SHG signals, respectively, as a function of position of the focus in the sample. The signals are filtered from the 1,200 nm fundamental photons by a dichroic mirror (Chroma T800LPXRXT, DM1), split into SHG and THG channels by a dichroic mirror (Chroma T425LPXR, DM2), and passed through narrow-band interference filters (F) for SHG (Chroma D600/10X) and THG (Chroma Z400/10X) detection. For a practical optical biopsy, backward (epi-) detection is a fundamental requirement. Although THG is only produced in the forward direction, at 400 nm the scattering length is several times shorter than the absorption length. This ensures that a significant fraction of the generated THG photons is scattered back towards the sample surface, enabling efficient epi-detection [40] (Figure 23.1C). The laser beam is transversely scanned over the sample by a pair of galvo mirrors (GM). THG and SHG modalities are intrinsically confocal and therefore provide direct depth sectioning. A full 3D image of the tissue volume is obtained by scanning of the microscope objective with a stepper motor in vertical (*z*) direction. Transverse (*xy*) scanning of the sample with a motorized translation stage allows for larger scale imaging of the sample resulting in mosaics. Imaging data was acquired with the TriMScope I software ("ImSpector Pro"), images stacks were stored in 16-bit tiff-format and further processed and analyzed with "ImageJ" software (ver. 1.49 m, NIH, USA). All images were processed with logarithmic contrast enhancement.

23.3.2 Ethics Statement

All procedures on human surgical and autopsy tissue were performed with the approval of the Medical Ethical Committee of the VU University Medical Center and in accordance with Dutch license procedures, the AMC Research Code, and the declaration of Helsinki. All patients undergoing surgery gave a written informed consent for tissue biopsy collection and signed a declaration permitting the use of their biopsy specimens for scientific research. Post-mortem autopsies of structurally normal and Alzheimer's disease diagnosed brains were performed in the framework of the Netherlands Brain Bank with permission from the Ethical Committee of the Medical Faculty of the VU University Amsterdam. All experimental procedures with mouse brain tissues were carried out according to the animal welfare guidelines of the VU University Amsterdam, the Netherlands.

23.3.3 Tissue Processing

The surgical brain specimens used in this study include: 1) histologically normal temporal cortex and white matter tissue samples removed from patients undergoing surgery for intractable epilepsy; 2) specimens from patients with brain tumors (low- and high-grade glioma), including tumor margin areas (especially low-grade diffuse glioma cases), tumor core, and peritumoral areas; 3) hippocampal and cortical specimens obtained from patients undergoing surgery for intractable epilepsy (hippocampal sclerosis, HS; focal cortical dysplasia, FCD). After resection, the brain tissue samples were placed within 30 s in ice-cold artificial cerebrospinal fluid (ACSF) at 4° C containing (in mM): NaCl 125; KCl 3; NaH_2PO_4 1.25; $MgSO_4$ 2; $CaCl_2$ 2; $NaHCO_3$ 26; glucose 10; osmolarity 300 mosmol/kg [41], and transported to the laboratory, located within 200 m distance from the operating room. The transition time between resection of the tissue and the start of preparation of slices was less than 15 min. We prepared a 300–350 μm thick coronal slice of the freshly excised structurally normal tissue in ice-cold ACSF solution with a vibratome (Microm, HM 650 V, Thermo Fisher Scientific), placed it in a plastic Petri dish (diameter 50 mm), and covered it with a 0.17 mm thick glass cover slip to provide a flat sample surface during multi-photon imaging. Freshly excised tumor and epilepsy tissue samples were cut with a surgical scalpel in several parts to generate flat surfaces, rinsed with ACSF to remove blood contamination, embedded in 2 ×2×5 mm^3 agar blocks (40 mg/ml water solution of OXOID bacteriological agar, LP0011) and flattened with thin glass cover slips (0.17 mm thick, dia. 25 mm, Menzel-Gläser).

A piece of tissue of ~1 cm^3 was obtained from a brain autopsy performed in the framework of the Netherlands Brain Bank with permission from the Ethical Committee of the Medical Faculty of the VU University Amsterdam. Unfixed ~1 cm^3 brain autopsy tissue samples were submerged in original cerebrospinal fluid and were transported to the laboratory (delay 5 min). The blocks were rinsed with ACSF to remove blood contamination and 300–350 μm thick coronal slices were made with the vibratome. The slices were kept in the holding chamber with ACSF (0° C, no carbogen gas) and imaged in the same way as *ex vivo* human slices.

A brain of an adult C57/BL6 wild-type mouse, postnatal delay 60 days (P60), was rapidly removed after decapitation, dissected in ice-cold slicing solution (same as for human tissues) into 300–350 μm thick coronal slices, and kept in the holding chamber with carbogenated ACSF at 0° C before imaging.

23.3.4 Tissue Histology

After THG/SHG imaging brain tumor and epilepsy samples were fixed in 4% formaldehyde, embedded in paraffin, sliced in μ μm thick histological sections, and routinely stained with hematoxylin and eosin (H&E) and luxol fast blue (LFB) for microscopic examination. Low- and high-magnification images were obtained using a Leica DM4000B microscope, equipped with Leica digital camera (Leica DC500). Images were recorded and stored using IM50 imaging software and processed with ImageJ software (ver. 1.49 m, NIH, USA).

23.4 APPLICATIONS

23.4.1 Mouse Brain Morphology

In brain tissue of mice, the specific geometry and lipid content of brain tissue at the cellular level can be used to achieve partial phase-matching of THG [19]. Augmented by co-recording of SHG signals, THG was shown to visualize cells, nuclei, the inner and outer contours of axons, blood cells, vessels, resulting in the visualization of both gray and white matter as well as vascularization, up to a depth of 300 μm [19, 36]. A power of several tens of mW close to the sample surface was used, and increased up to 200 mW at higher depths to compensate for intensity losses due to light-scattering and focusing aberrations introduced by the tissue. Figure 23.2A shows a large-sized, × ×9 mm², THG/SHG image of a coronal section of an adult mouse brain. Brain areas such as cortex, hippocampus, and thalamus are clearly recognizable (indicated as C, D, E, respectively). The THG/SHG image is very similar to the histopathological image of a fixed and LFB stained mouse brain (Figure 23.2B). In this stain, myelin appears blue, cell somata as well as neuronal and astrocytic cell processes appear pink, and the nuclei of cells appear purple. In both images, myelinated axons give the brightest, most intense signals, but whereas in the histopathological images cell

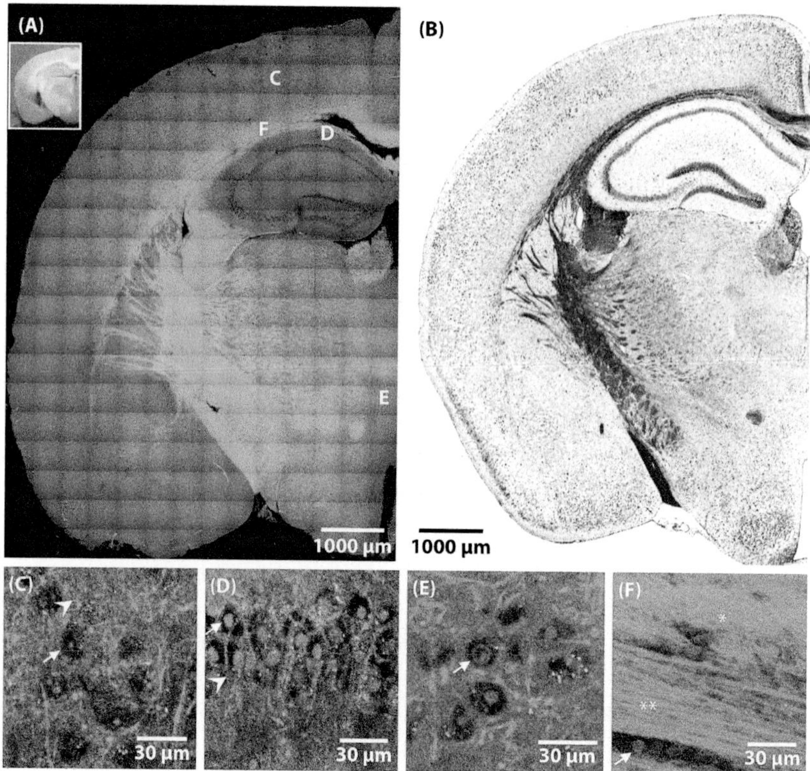

FIGURE 23.2 THG/SHG images of *ex vivo* unfixed and unstained coronal section of adult wild-type mouse (P60) brain. THG is shown in green, SHG in red. (A) 6.3××8.55 mm² mosaic image consisting of 1×19 tiles, each tile of 45×450 μm² is single 2 μm thick optical section taken at a depth of 20–30 μm. Full recording time was 11 min. The inset shows a photograph of the sample. (B) Myelin-stained (LFB) histology image of a similar brain region revealing cells (purple) and myelin (blue). (C–E) In the cortex (C), hippocampus (D), thalamus (E), the somata of brain cells (indicated by arrows) appear as dark holes in bright neuropil with nuclear/nucleolar outlines visualized by THG. The SHG modality reveals microtubules inside axons, marked by arrowheads. (F) Myelinated axon fibers in white matter, perpendicular (*) and parallel (**) to the image plane, and glial cells (arrow). Recording time of individual tiles (C–F) was 1 s.

nuclei are labeled and appear as dots, in the THG/SHG they appear as black holes in a bright background of neuropil. The THG/SHG images were recorded with full sub-cellular detail, which allows for the zooming in to areas of interest, to visualize individual brain cells and axon fibers (Figures 23.2 C–F). Cells, nuclei, nucleoli, and myelinated fibers are clearly visualized by THG, whereas in some cases ordered microtubule inside cell processes are revealed by SHG. The full THG/SHG image took 11 minutes to be recorded and required no preparation of the tissue sample, whereas a histopathological image may take up to 24 hrs to be generated.

23.4.2 Structurally Normal Human Brain

Similar to the mouse brain, in Kuzmin et al. [15] we showed that in the human brain THG reveals the neuropil (i.e. the dense network of neuronal and glial cell processes) with neuronal/glial cell somata, dispersed microvessels with intraluminal erythrocytes, and especially in the white matter many lipid-rich fibers representing myelinated axons. Some of the neuronal somata are filled with lipofuscin granules, that emit broad autofluorescence [42] signals that are detected in both the SHG and THG channel, and possibly also intrinsic THG signals and therefore appear yellow. Figure 23.3A shows a combined mosaic THG/SHG image of structurally normal human brain tissue sliced perpendicular to the brain surface with a transition from neocortex (top) to white matter (bottom). The transition between the neocortex (top) and lipid/myelin-rich white matter (bottom) is clearly visualized. The upper part of the image shows neurons and glial cells in the superficial part of the cortex as dark holes in a bright neuropil matrix with intracellular nuclear/nucleolar features outlined by THG (Figure 23.3B). Large pyramidal neurons appear filled with lipofuscin, except for their nuclei (shown with arrowheads in Figure 23.3C). In the white matter, glial cell somata are surrounded by a dense neuropil matrix consisting of axons. Blood vessels are visualized by both THG, sometimes with densely packed intraluminal erythrocytes, and SHG, highlighting the collagen in the blood vessel wall (SHG). A coincidence of THG and SHG signals reveals ball-shaped deposits in a blood vessel wall (indicated in yellow) – possibly atherosclerotic plaques, consisting of fat, cholesterol, and calcium and emitting strong THG and 2–3 photon excited autofluorescence photons, detected in the SHG channel.

Figure 23.4 shows a THG/SHG image of the hippocampus of a structurally healthy human brain. Here, we imaged post-mortem tissue with a significantly longer delay between excision and imaging than in Figure 23.3. After the THG/SHG imaging, the tissue was fixed, sliced, and stained with LFB for comparison with standard histopathology. As for the mouse tissue, the similarity of the THG/SHG images with the myelin stained images is excellent. In this case though, many neurons are visible by their dense lipofuscin deposits (indicated by the yellow signals), rather than as a black hole, or as a black contour with a nucleus/nucleolus inside. This allows for an even better visualization of structures within the hippocampus, which are characterized by alternating areas of high cell density and high myelin, or axon, density. In particular, the subiculum, the cornu ammonis, and the external limb of the dentate gyrus are easily identified. Upon magnification of each of these areas, a reasonably good correspondence with the histology is observed, though

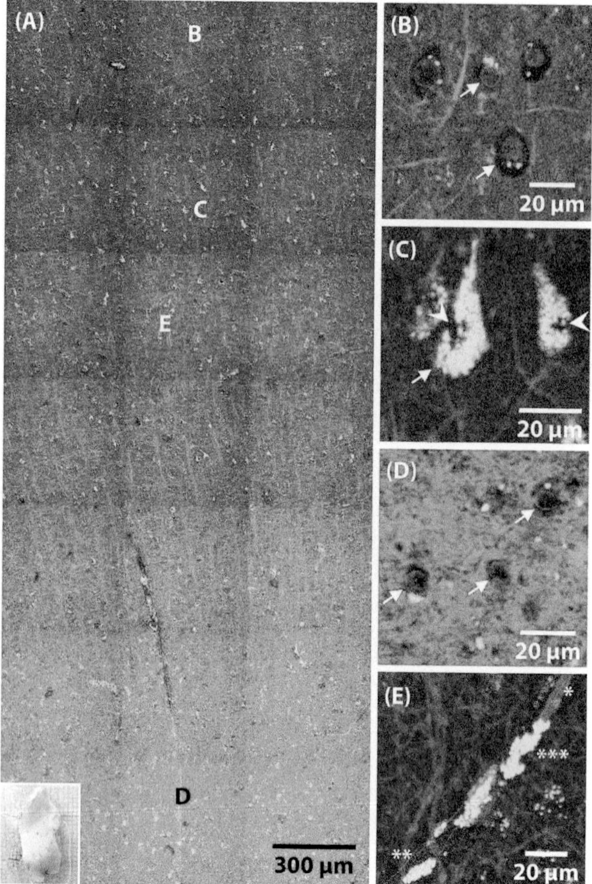

FIGURE 23.3 THG/SHG imaging (signal depicted in green and red, respectively) of structurally normal human brain tissue. The overlap of THG and SHG signals is shown in yellow and is generally produced by lipofuscin granules accumulated in neuronal somata as well as dispersed in the neuropil. (A) 1.3 × 3.15 mm^2 mosaic image consisting of ××7 tiles, of 45×450 μm^2 each, each tile is an optical section of 2 μm taken at depth of 20–30 μm. Recording time of the full image was 1.7 min. The inset shows a photograph of the sample. (B) Superficial layer of the neocortex with neuronal and glial cell somata appearing as black holes (arrows) in which nuclei are seen; (C) Large pyramidal neurons (arrow) filled with lipofuscin (arrowheads); (D) Glial somata (indicated by arrows) in white matter; (E) Blood vessel in gray matter: (*) densely-packed intraluminal erythrocytes (THG); (**) collagen in the blood vessel wall (SHG); (***) ball-shaped deposits in the blood vessel wall.

one-to-one comparison of the sections with THG images proved difficult because of the distortions of the tissue in the fixation process. However, parts of the sections cut from the volume analyzed with THG visualize a similar cell aspect and organization.

The images depicted in Figures 23.2, 23.3 and 23.4 suggest that much of the morphology of the brain can be revealed by THG/SHG imaging. However, can THG/SHG also serve to identify features typical for the pathology of brain diseases, such as epilepsy, cancer, or Alzheimer's disease? In the next sections we explore this issue by discussing images of diseased brain.

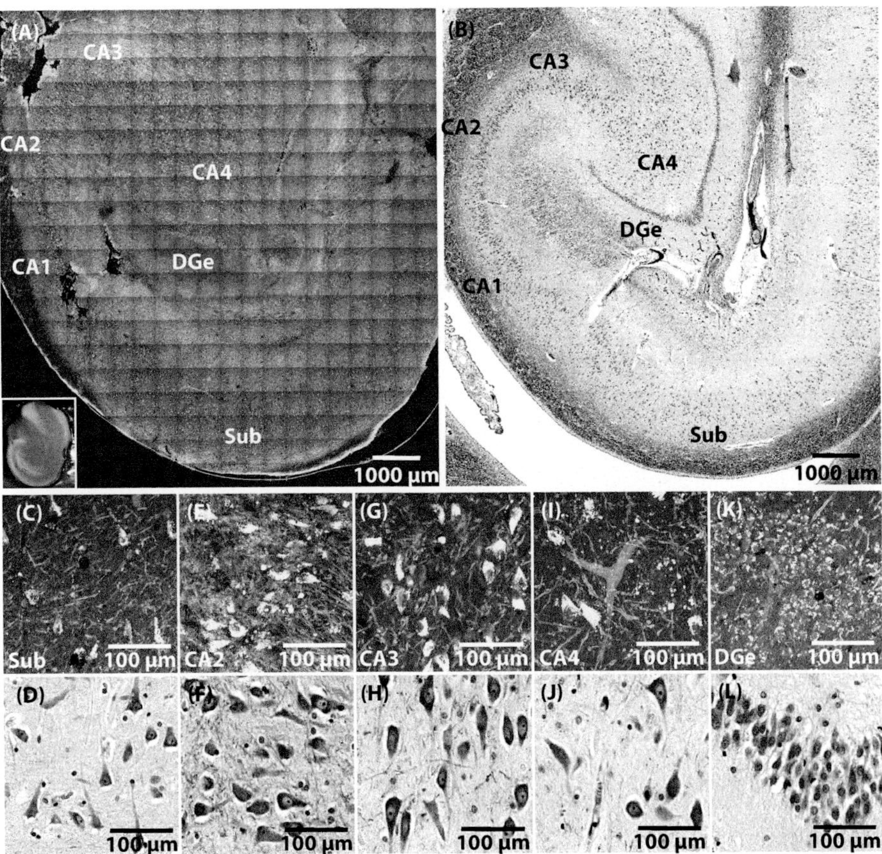

FIGURE 23.4 THG/SHG imaging of post-mortem unfixed and unstained structurally normal hippocampal tissue (A, C, E, G, I, K) and comparison to LFB histology (B, D, F, H, J, L). The overlap of THG (green) and SHG (red) signals is shown in yellow and is generally produced by lipofuscin granules accumulated in neuronal somata as well as in the neuropil. (A) ××9 mm^2 mosaic image consists of 2×20 tiles, each tile 45×450 μm^2, taken at a depth of 20–30 μm, full imaging time 16 min. (C, E, G, I, K) single 2 μm thick THG/SHG optical sections of different hippocampal cell layers taken at depth of 20–30 μm, imaging time 5 s.: Sub – subiculum; CA – *cornu ammonis*; DGe – external limb of the dentate gyrus; (D, F, H, J, L) corresponding LFB images. The inset of (A) shows a color photograph of the THG/SHG imaged hippocampus sample.

23.4.3 Epilepsy

Hippocampal specimens with hippocampal sclerosis (HS) are characterized by neuronal cell loss (with predominant damage of hippocampal sectors CA1, CA3, and CA4) and astrogliosis. The cortical specimens obtained from patients with epilepsy and focal cortical dysplasia (FCD) are characterized by distorted cortical lamination and the presence of balloon cells and dysmorphic neurons, as well as astrogliosis (FCD type IIb). A technique such as THG/SHG that visualizes the morphology of tissue may therefore be capable of visualizing these hallmarks. To our knowledge, no reports have been published of THG/SHG images of tissue specimens from HS or FCD cases. Here, we present two cases, one of HS and one of FCD type IIb.

Figure 23.5 shows THG/SHG images of unfixed and unstained sclerotic hippocampus tissue, and H&E stained histopathology images of the same tissue sample. In both images, severe loss of neurons in areas CA1, CA3, and CA4 and in the DGe layer are observed (compare for example with the healthy hippocampus displayed in Figure 23.4). Images with smaller FOV's show cellular details in the subiculum, *cornu ammonis*, and external limb of the dentate gyrus. The brain cells in the THG images, with a lot of lipofuscin deposits, show very similar morphology and organization to those in the corresponding H&E images. Also, other features, such as *corpora amylacea* near a blood vessel are visualized. Figure 23.6 shows an example of FCD IIb, characterized by loss of the classical

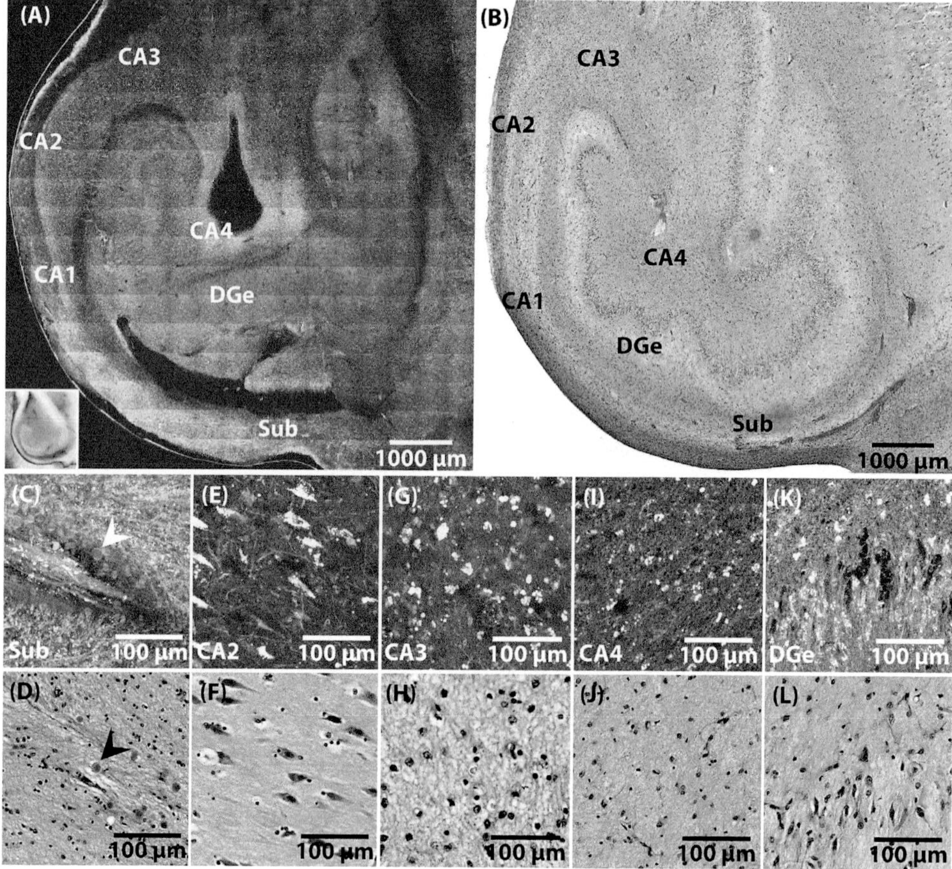

FIGURE 23.5 THG/SHG imaging of *ex vivo* unfixed and unstained sclerotic hippocampal tissue sample (A, C, E, G, I, K) with comparison to H&E histology (B, D, F, H, J, L) of the same sample. (A) 7.6×7.20 mm^2 mosaic image consists of 1×16 tiles, each tile 45×450 μm^2, taken at depth of 20–30 μm, full imaging time 11 min. The inset shows the photograph of the sample. (C, E, G, I, K) single 2 μm thick THG/SHG optical sections of different hippocampal cell layers taken at depth of 20–30 μm, imaging time 5 s.: Sub – subiculum; CA – *cornu ammonis*; DGe – external limb of the dentate gyrus; (D, F, H, J, L) corresponding H&E images. In the image of the subiculum (Sub), a somewhat larger blood vessel is seen surrounded by *corpora amylacea* indicated by white (C) resp. black (D) arrowheads.

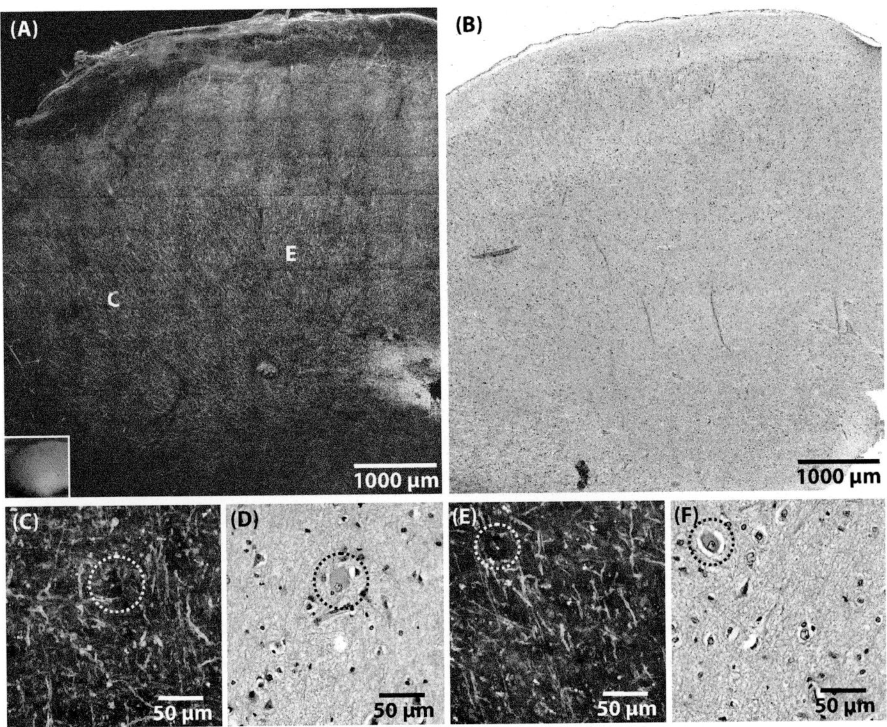

FIGURE 23.6 THG/SHG imaging (A, C, E) of *ex vivo* unfixed and unstained neocortical tissue sample (focal cortical dysplasia) of a patient diagnosed with left occipital dysplasia and comparison to H&E histology (B, D, F) of the same sample. (A) 5.4×5.85 mm^2 mosaic image consists of 1×13 tiles, each tile 45×450 μm^2, taken at depth of 20–30 μm, full imaging time 6.5 min. The inset shows the photograph of the sample. (C–F) THG/SHG and H&E images of the tissue areas with cortical dyslamination, gliosis and balloon cells (C, D) and dysplastic neurons (E, F) indicated by dashed circles. (C, E) singe 2 μm thick THG/SHG optical sections taken at depth of 20–30 μm, imaging time 5 s.

laminar organization with six layers, abnormal clustering, irregular localization of neuronal cells, and gliosis. Other histopathological findings of FCDs include the presence of cellular abnormalities, such as dysmorphic neurons (characterized by abnormal morphology and orientation) and balloon cells (immature cells with large size, in H&E staining characterized by abundant opalescent eosinophilic cytoplasm and one or more eccentric nuclei). In the smaller FOV images, a balloon cell identified in the H&E image appears also to be recognizable in the THG/SHG images (Figures 23.6C and D). A second abnormality is the presence of dysmorphic neurons with abnormal size and morphology of axons and dendrites, as well as increased accumulation of neurofilament proteins. An example of such a cell is given in Figure 23.6F. In the same tissue area, an abnormality in the THG/SHG image indicated by the dotted circle in Figure 23.6E may correspond to this same cell. We conclude that in these two examples typical hallmarks such as cortical dyslamination, gliosis, balloon cells, and, to a somewhat lesser extent in this example, dysmorphic neurons, are revealed by THG/SHG imaging and that these positive results warrant more thorough studies to further substantiate THG/SHG as a valid tool for fast epilepsy pathology.

23.4.4 Brain Tumors

Glial tumors (gliomas) account for almost 80% of the tumors originating from brain tissue. The vast majority of these tumors are so-called "diffuse gliomas" as they show very extensive ("diffuse") growth into the surrounding brain parenchyma. With surgical resection, irradiation, and/or chemotherapy it is impossible to eliminate all glioma cells without serious damage to the brain tissue. As a consequence, so far, patients with a diffuse glioma have a poor prognosis, a situation which strongly contributes to the fact that brain tumor patients experience more years of life lost than patients with any other type of cancer [43, 44].

Meanwhile it has also been demonstrated that the prognosis of patients with a diffuse glioma correlates with the extent of resection [45–47]. During brain surgery, however, it is extremely difficult for the neurosurgeon to determine the boundary of the tumor, i.e. whether a brain area contains tumor cells or not. If the neurosurgeon could have histopathological information on the tumor boundaries *during* brain surgery, then recognition of these tumor boundaries, and with that the surgical resection, could be significantly improved.

In Kuzmin et al. [15] THG/SHG imaging was shown provide label-free images of human tumor tissue of histopathological quality, in real time. In the THG tumor images increased cellularity, nuclear pleomorphism, and rarefaction of neuropil in fresh, unstained human brain tissue could be recognized. In Figure 23.7 we show two examples that further illustrate this: In Figure 23.7A a tumor has led to rarefication of the axon matrix (i.e. the intense green "wires"), and an increased number of (tumor) cell nuclei is visible in the resulting open spaces. Areas varying in cellular density can be recognized by an inverse change in THG intensity, because the tumor nuclei give a THG intensity lower than of the axon matrix. Figure 23.7B represents a focus of high-grade glioma in which the axon matrix has completely been replaced by tumor cells associated with vascular proliferation (the latter visible by the red SHG signals from the collagen in the vessel walls). In the high-grade

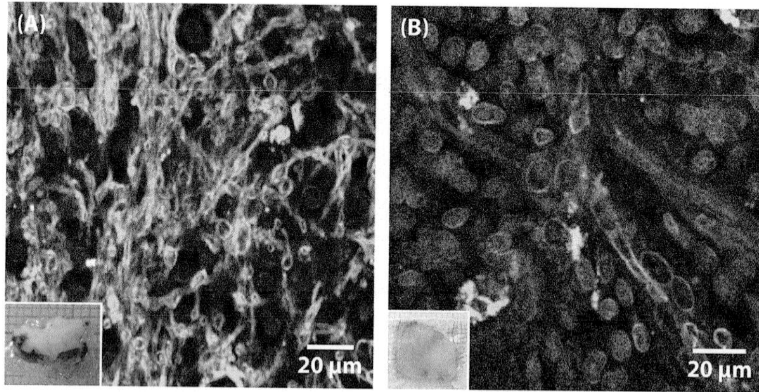

FIGURE 23.7 THG/SHG imaging of *ex-vivo* unfixed and unstained brain tumor samples. (A) Infiltrative low-grade glioma: brain tumor area with spatially varying cellular density. (B) High-grade glioma with high cellularity and vascular proliferation. Images are singe 2 μm thick THG/SHG optical sections taken at depth of 20–30 μm, imaging time 5 s. The insets show the photographs of the corresponding samples.

glioma focus, the tumor nuclei can be observed to have a more variable size and shape, consistent with the polymorphism of high-grade glioma cells. Also, the nuclei appear to generate more homogeneously intense THG signal over their whole volume than in the low-grade case, perhaps due to changes in chromatin organization.

23.4.5 Alzheimer's Disease

Alzheimer's disease (AD) is a debilitating disease that results in a progressive decline in cognitive function. Examinations on memory and cognitive abilities are standard tests for diagnosing AD. However, the definite diagnosis can only be reached by post-mortem histopathology. In case treatments become available, accurate diagnosis during the lifetime of the patient becomes important. AD brains are characterized by extracellular deposits of β-amyloid protein, intraneuronal accumulation of hyperphosphorylated tau protein, and the degeneration of dendrites and neurons. Here, multi-photon microscopy could play a role in either an (invasive) form of optical biopsy, or aid in a better understanding of the cell biology of Alzheimer's disease enabling the design of drug therapies or effective preventive measures, through studies in mouse models or acute brain slices. Webb and coworkers [42] have shown that senile plaques in transgenic Alzheimer's disease mouse models emit a broad autofluorescence, peaking at 525 nm, and generate SHG. This allowed for the label-free detection of senile plaques in acute brain slices of a thickness of 400 μm, as spherical objects with diameter ~30–70 μm. They suggested detection of the spectral emission signals, rather than generating a microscopic image, may be sufficient to make a diagnosis using GRIN lenses, or fiber bundles [42]. As an example of functional imaging, they characterized the polarity and the morphology of microtubule arrays, detected via SHG in the vicinity of senile plaques.

Here, in Figure 23.8 we show THG/SHG images of post-mortem human cortex visualizing brain cell somata with intracellular accumulations of lipofuscin granules, blood vessels, intraluminal erythrocytes, and an amyloid plaque. The age of the patient was 96 years, postmortem delay 5 h 40 m. The clinical diagnosis of AD was neuropathologically confirmed on formalin-fixed, paraffin-embedded tissues from different brain regions. In depth scans the plaques appear as spheres with outer diameter of 40–50 μm. SHG optical sections taken through a blood vessel show the blood vessel lumen, collagen fibers, and multiple ball-shaped deposits in the blood vessel wall (indicated by an arrowhead), consistent with amyloid angiopathy.

23.5 CONCLUSION

The excellent agreement between the THG/SHG and histopathological images for the neuroclinical cases discussed here and in the literature suggests that THG/SHG imaging could be a powerful diagnostic tool for fast characterization of the pathology of tissue. The advantages over classical histopathology are: 1) the tissue does not need time-consuming preparation such as fixation, slicing, and staining that can take up to 24 hrs in histopathology; 2) the images are 3D with typical depth of 100–20μ μm, therefore a whole volume of several cells layers deep can be analyzed at once without any distortions due to preparation; 3) the THG/SHG images appear to contain the same information as that of a myelin and H&E stain combined; 4) images can be recorded in real time, varying from 0.6 s for a 15×150 μm^2 surface

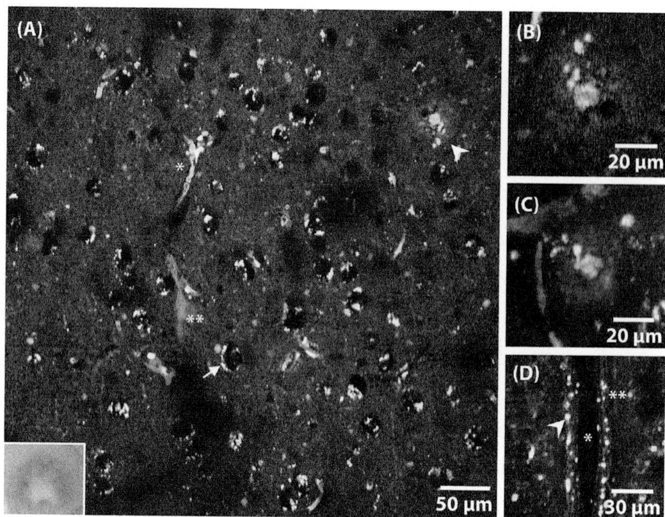

FIGURE 23.8 THG/SHG imaging of post-mortem unfixed and unstained Alzheimer's disease diagnosed cortical tissue. The overlap of THG (green) and SHG signals (red) is shown in yellow and is generally produced by lipofuscin granules accumulated in neuronal somata and neuropil. (A) Gray matter THG/SHG image containing brain cell somata (indicated by an arrow) with intracellular accumulations of lipofuscin granules, blood vessels (*), intraluminal erythrocytes (**), and an amyloid plaque (indicated by arrowhead). The inset shows the photograph of the sample. (B) Magnified image of the amyloid plaque indicated on image (A) by an arrowhead. (C) Amyloid plaque near to a blood vessel. The (B,C) images are singe 2 μm thick THG/SHG optical sections taken through the centers of the amyloid plaques to reveal both dense amyloid cores and peripheral halo, imaging times 1 s. (D) 2 μm thick THG/SHG optical sections taken through a blood vessel showing blood vessel lumen (*), collagen fibers (**), and multiple ball-shaped deposits in the blood vessel wall (indicated by an arrowhead).

to 10–15 min for 1×10 mm^2; 5) The technique can be miniaturized into a handheld tool [11, 48–53]. Now, more quantitative studies that compare THG/SHG with standard histopathology are required to substantiate the promise of THG/SHG reliability for diagnosis. If applied during surgery, either *ex vivo* or *in situ*, THG could help surgeons diagnose the tissue. Specifically, the technique could help spare healthy tissue, might help achieve a (more) complete resection of a lesion, or enable fast diagnosis and subsequent treatment. In addition, THG/SHG could be a valuable tool for studying cellular dynamics in acute brain slices, organoids, or mouse models and visualize the effects of drugs and disease processes over time.

REFERENCES

1. U. Pichlmeier, A. Bink, G. Schackert, W. Stummer, and A.G.S. Grp, 'Resection and survival in glioblastoma multiforme: An RTOG recursive partitioning analysis of ALA study patients.' *Neuro-Oncology*, 2008. 10(6): p. 1025–1034.
2. W. Stummer, J.C. Tonn, C. Goetz, W. Ullrich, H. Stepp, A. Bink, T. Pietsch, and U. Pichlmeier, '5-aminolevulinic acid-derived tumor fluorescence: The diagnostic accuracy of visible fluorescence qualities as corroborated by spectrometry and histology and postoperative imaging.' *Neurosurgery*, 2014. 74(3): p. 310–319.

3. Y.P. Li, R. Rey-Dios, D.W. Roberts, P.A. Valdes, and A.A. Cohen-Gadol, 'Intraoperative fluorescence-guided resection of high-grade gliomas: A comparison of the present techniques and evolution of future strategies.' *World Neurosurgery*, 2014. 82(1–2): p. 175–185.
4. T. Hollon, S.L. Hervey-Jumper, O. Sagher, and D.A. Orringer, 'Advances in the surgical management of low-grade glioma.' *Seminars in Radiation Oncology*, 2015. 25(3): p. 181–188.
5. A. Novotny, J. Xiang, W. Stummer, N.S. Teuscher, D.E. Smith, and R.F. Keep, 'Mechanisms of 5-aminolevulinic acid uptake at the choroid plexus.' *Journal of Neurochemistry*, 2000. 75(1): p. 321–328.
6. M. Hefti, H.M. Mehdorn, I. Albert, and L. Dorner, 'Fluorescence-guided surgery for malignant glioma: A review on aminolevulinic acid induced protoporphyrin IX photodynamic diagnostic in brain tumors.' *Current Medical Imaging Reviews*, 2010. 6(4): p. 254–258.
7. O. van Tellingen, B. Yetkin-Arik, M.C. de Gooijer, P. Wesseling, T. Wurdinger, and H.E. de Vries, 'Overcoming the blood-brain tumor barrier for effective glioblastoma treatment.' *Drug Resistance Updates*, 2015. 19: p. 1–12.
8. J.C. Jung and M.J. Schnitzer, 'Multiphoton endoscopy.' *Optics Letters*, 2003. 28(11): p. 902–904.
9. R.M. Williams, A. Flesken-Nikitin, L.H. Ellenson, D.C. Connolly, T.C. Hamilton, A.Y. Nikitin, and W.R. Zipfel, 'Strategies for high-resolution imaging of epithelial ovarian cancer by laparoscopic nonlinear microscopy.' *Translational Oncology*, 2010. 3(3): p. 181–194.
10. I. Pavlova, K.R. Hume, S.A. Yazinski, J. Flanders, T.L. Southard, R.S. Weiss, and W.W. Webb, 'Multiphoton microscopy and microspectroscopy for diagnostics of inflammatory and neoplastic lung.' *Journal of Biomedical Optics*, 2012. 17(3): p. 036014.
11. D.M. Huland, M. Jain, D.G. Ouzounov, B.D. Robinson, D.S. Harya, M.M. Shevchuk, P. Singhal, C. Xu, and A.K. Tewari, 'Multiphoton gradient index endoscopy for evaluation of diseased human prostatic tissue ex vivo.' *Journal of Biomedical Optics*, 2014. 19(11).
12. M. Ji, D.A. Orringer, C.W. Freudiger, S. Ramkissoon, X. Liu, D. Lau, A.J. Golby, I. Norton, M. Hayashi, N.Y. Agar, G.S. Young, C. Spino, S. Santagata, S. Camelo-Piragua, K.L. Ligon, O. Sagher, and X.S. Xie, 'Rapid, label-free detection of brain tumors with stimulated Raman scattering microscopy.' *Science Translational Medicine*, 2013. 5(201): p. 201ra119.
13. M.B. Ji, S. Lewis, S. Camelo-Piragua, S.H. Ramkissoon, M. Snuderl, S. Venneti, A. Fisher-Hubbard, M. Garrard, D. Fu, A.C. Wang, J.A. Heth, C.O. Maher, N. Sanai, T.D. Johnson, C.W. Freudiger, O. Sagher, X.S. Xie, and D.A. Orringer, 'Detection of human brain tumor infiltration with quantitative stimulated Raman scattering microscopy.' *Science Translational Medicine*, 2015. 7(309).
14. M. Jain, N. Narula, A. Aggarwal, B. Stiles, M.M. Shevchuk, J. Sterling, B. Salamoon, V. Chandel, W.W. Webb, N.K. Altorki, and S. Mukherjee, 'Multiphoton microscopy a potential "optical biopsy" tool for real-time evaluation of lung tumors without the need for exogenous contrast agents.' *Archives of Pathology and Laboratory Medicine*, 2014. 138(8): p. 1037–1047.
15. N.V. Kuzmin, P. Wesseling, P.C.d.W. Hamer, D.P. Noske, G.D. Galgano, H.D. Mansvelder, J.C. Baayen, and M.L. Groot, 'Third harmonic generation imaging for fast, label-free pathology of human brain tumors.' *Biomedical Optics Express*, 2016. 7(5): p. 1889–1904.
16. S.Y. Chen, C.S. Hsieh, S.W. Chu, C.Y. Lin, C.Y. Ko, Y.C. Chen, H.J. Tsai, C.H. Hu, and C.K. Sun, 'Noninvasive harmonics optical microscopy for long-term observation of embryonic nervous system development in vivo.' *Journal of Biomedical Optics*, 2006. 11(5).
17. D. Debarre, W. Supatto, A.M. Pena, A. Fabre, T. Tordjmann, L. Combettes, M.C. Schanne-Klein, and E. Beaurepaire, 'Imaging lipid bodies in cells and tissues using third-harmonic generation microscopy.' *Nature Methods*, 2006. 3(1): p. 47–53.
18. N. Olivier, M.A. Luengo-Oroz, L. Duloquin, E. Faure, T. Savy, I. Veilleux, X. Solinas, D. Debarre, P. Bourgine, A. Santos, N. Peyrieras, and E. Beaurepaire, 'Cell lineage reconstruction of early zebrafish embryos using label-free nonlinear microscopy.' *Science*, 2010. 329(5994): p. 967–971.

19. S. Witte, A. Negrean, J.C. Lodder, C.P.J. de Kock, G.T. Silva, H.D. Mansvelder, and M.L. Groot, 'Label-free live brain imaging and targeted patching with third-harmonic generation microscopy.' *Proceedings of the National Academy of Sciences of the United States of America*, 2011. 108(15): p. 5970–5975.
20. B. Weigelin, G.-J. Bakker, and P. Friedl, 'Intravital third harmonic generation microscopy of collective melanoma cell invasion.' *IntraVital*, 2012. 1(1): p. 32–43.
21. B. Weigelin, G.J. Bakker, and P. Friedl, 'Third harmonic generation microscopy of cells and tissue organization.' *Journal of Cell Science*, 2016.
22. S.Y. Chen, S.U. Chen, H.Y. Wu, W.J. Lee, Y.H. Liao, and C.K. Sun, 'In vivo virtual biopsy of human skin by using noninvasive higher harmonic generation microscopy.' *IEEE Journal of Selected Topics in Quantum Electronics*, 2010. 16(3): p. 478–492.
23. C.K. Tsai, T.D. Wang, J.W. Lin, R.B. Hsu, L.Z. Guo, S.T. Chen, and T.M. Liu, 'Virtual optical biopsy of human adipocytes with third harmonic generation microscopy.' *Biomedical Optics Express*, 2013. 4(1): p. 178–186.
24. M.R. Tsai, S.Y. Chen, D.B. Shieh, P.J. Lou, and C.K. Sun, 'In vivo optical virtual biopsy of human oral mucosa with harmonic generation microscopy.' *Biomedical Optics Express*, 2011. 2(8): p. 2317–2328.
25. F.S. Pavone and P.J. Campagnola, eds., *Second Harmonic Generation Imaging*. Series in Cellular and Clinical Imaging. Vol. 3. 2013, CRC Press: Boca Raton, FL. 476.
26. P.C. Wu, T.Y. Hsieh, Z.U. Tsai, and T.M. Liu, 'In vivo quantification of the structural changes of collagens in a melanoma microenvironment with second and third harmonic generation microscopy.' *Scietific Reports*, 2015. 5: p. 8879.
27. J. Adur, V.B. Pelegati, A.A. de Thomaz, M.O. Baratti, D.B. Almeida, L.A. Andrade, F. Bottcher-Luiz, H.F. Carvalho, and C.L. Cesar, 'Optical biomarkers of serous and mucinous human ovarian tumor assessed with nonlinear optics microscopies.' *PloS one*, 2012. 7(10): p. e47007.
28. M. Yildirim, N. Durr, and A. Ben-Yakar, 'Tripling the maximum imaging depth with third-harmonic generation microscopy.' *Journal of Biomedical Optics*, 2015. 20(9): p. 096013.
29. G.G. Lee, H.H. Lin, M.R. Tsai, S.Y. Chou, W.J. Lee, Y.H. Liao, C.K. Sun, and C.F. Chen, 'Automatic cell segmentation and nuclear-to-cytoplasmic ratio analysis for third harmonic generated microscopy medical images.' *IEEE Transactions on Biomedical Circuits and Systems*, 2013. 7(2): p. 158–168.
30. Y. Barad, H. Eisenberg, M. Horowitz, and Y. Silberberg, 'Nonlinear scanning laser microscopy by third harmonic generation.' *Applied Physics Letters*, 1997. 70(8): p. 922–924.
31. M. Muller, J. Squier, K.R. Wilson, and G.J. Brakenhoff, '3D microscopy of transparent objects using third-harmonic generation.' *Journal of Microscopy-Oxford*, 1998. 191: p. 266–274.
32. J.A. Squier, M. Muller, G.J. Brakenhoff, and K.R. Wilson, 'Third harmonic generation microscopy.' *Optics Express*, 1998. 3(9): p. 315–324.
33. J.X. Cheng and X.S. Xie, 'Green's function formulation for third-harmonic generation microscopy.' *Journal of the Optical Society of America B-Optical Physics*, 2002. 19(7): p. 1604–1610.
34. D. Debarre and E. Beaurepaire, 'Quantitative characterization of biological liquids for third-harmonic generation microscopy.' *Biophysical Journal*, 2007. 92(2): p. 603–612.
35. P. Mahou, N. Olivier, G. Labroille, L. Duloquin, J.M. Sintes, N. Peyrieras, R. Legouis, D. Debarre, and E. Beaurepaire, 'Combined third-harmonic generation and four-wave mixing microscopy of tissues and embryos.' *Biomedical Optics Express*, 2011. 2(10): p. 2837–2849.
36. H. Lim, D. Sharoukhov, I. Kassim, Y. Zhang, J.L. Salzer, and C.V. Melendez-Vasquez, 'Label-free imaging of Schwann cell myelination by third harmonic generation microscopy.' Proceedings of the National Academy of Sciences, 2014.
37. G.J. Tserevelakis, E.V. Megalou, G. Filippidis, B. Petanidou, C. Fotakis, and N. Tavernarakis, 'Label-free imaging of lipid depositions in *C. elegans* using third-harmonic generation microscopy.' *PloS one*, 2014. 9(1).

38. J.F. Ward and G.H.C. New, 'Optical third harmonic generation in gases by a focused laser beam.' *Physical Review*, 1969. 185(1): p. 57–72.
39. W. Supatto, D. Debarre, B. Moulia, E. Brouzes, J.L. Martin, E. Farge, and E. Beaurepaire, 'In vivo modulation of morphogenetic movements in Drosophila embryos with femtosecond laser pulses.' *Proceedings of the National Academy of Sciences of the United States of America*, 2005. 102(4): p. 1047–1052.
40. D. Debarre, N. Olivier, and E. Beaurepaire, 'Signal epidetection in third-harmonic generation microscopy of turbid media.' *Optics Express*, 2007. 15(14): p. 8913–8924.
41. G. Testa-Silva, M.B. Verhoog, N.A. Goriounova, A. Loebel, J. Hjorth, J.C. Baayen, C.P.J. de Kock, and H.D. Mansvelder, 'Human synapses show a wide temporal window for spike-timing-dependent plasticity.' *Frontiers in Synaptic Neuroscience*, 2010. 2: p. 12.
42. A.C. Kwan, K. Duff, G.K. Gouras, and W.W. Webb, 'Optical visualization of Alzheimer's pathology via multiphoton-excited intrinsic fluorescence and second harmonic generation.' *Optics Express*, 2009. 17(5): p. 3679–3689.
43. N.G. Burnet, S.J. Jefferies, R.J. Benson, D.P. Hunt, and F.P. Treasure, 'Years of life lost (YLL) from cancer is an important measure of population burden--and should be considered when allocating research funds.' *British Journal of Cancer*, 2005. 92(2): p. 241–5.
44. J.A. Schwartzbaum, J.L. Fisher, K.D. Aldape, and M. Wrensch, 'Epidemiology and molecular pathology of glioma.' *Nature Clinical Practice Neurology*, 2006. 2(9): p. 494–503.
45. J.S. Smith, E.F. Chang, K.R. Lamborn, S.M. Chang, M.D. Prados, S. Cha, T. Tihan, S. Vandenberg, M.W. McDermott, and M.S. Berger, 'Role of extent of resection in the long-term outcome of low-grade hemispheric gliomas.' *Journal of Clinical Oncology*, 2008. 26(8): p. 1338–45.
46. N. Sanai and M.S. Berger, 'Glioma extent of resection and its impact on patient outcome.' *Neurosurgery*, 2008. 62(4): p. 753–764.
47. I.Y. Eyupoglu, M. Buchfelder, and N.E. Savaskan, 'Surgical resection of malignant gliomas-role in optimizing patient outcome.' *Nature Reviews Neurology*, 2013. 9(3): p. 141–51.
48. M.J. Levene, D.A. Dombeck, K.A. Kasischke, R.P. Molloy, and W.W. Webb, 'In vivo multiphoton microscopy of deep brain tissue.' *Journal of Neurophysiology*, 2004. 91(4): p. 1908–1912.
49. T.A. Murray and M.J. Levene, 'Singlet gradient index lens for deep in vivo multiphoton microscopy.' *Journal of Biomedical Optics*, 2012. 17(2).
50. D.R. Rivera, C.M. Brown, D.G. Ouzounov, I. Pavlova, D. Kobat, W.W. Webb, and C. Xu, 'Compact and flexible raster scanning multiphoton endoscope capable of imaging unstained tissue.' *Proceedings of the National Academy of Sciences*, 2011. 108(43): p. 17598–17603.
51. S.-H. Chia, C.-H. Yu, C.-H. Lin, N.-C. Cheng, T.-M. Liu, M.-C. Chan, I.H. Chen, and C.-K. Sun, 'Miniaturized video-rate epi-third-harmonic-generation fiber-microscope.' *Optics Express*, 2010. 18(16): p. 17382–17391.
52. S.W. Chu, S.P. Tai, C.L. Ho, C.H. Lin, and C.K. Sun, 'High-resolution simultaneous three-photon fluorescence and third-harmonic-generation microscopy.' *Microscopy Research and Technique*, 2005. 197: p. 193–197.
53. C.M. Brown, D.R. Rivera, I. Pavlova, D.G. Ouzounov, W.O. Williams, S. Mohanan, W.W. Webb, and C. Xu, 'In vivo imaging of unstained tissues using a compact and flexible multiphoton microendoscope.' *Journal of Biomedical Optics*, 2012. 17(4).

Index